UNITEXT for Physics

UNITEXT for Physics series, formerly UNITEXT Collana di Fisica e Astronomia, publishes textbooks and monographs in Physics and Astronomy, mainly in English language, characterized of a didactic style and comprehensiveness. The books published in UNITEXT for Physics series are addressed to graduate and advanced graduate students, but also to scientists and researchers as important resources for their education, knowledge and teaching.

More information about this series at http://www.springer.com/series/13351

Kurt Lechner

Classical Electrodynamics

A Modern Perspective

Kurt Lechner
Department of Physics and Astronomy Galileo Galilei
University of Padua
Padua, Italy

and

Istituto Nazionale di Fisica Nucleare
Sezione di Padova
Padua, Italy

ISSN 2198-7882   ISSN 2198-7890   (electronic)
UNITEXT for Physics
ISBN 978-3-030-06301-6   ISBN 978-3-319-91809-9   (eBook)
https://doi.org/10.1007/978-3-319-91809-9

This Springer imprint is published by the registered company Springer Nature Switzerland AG
The registered company address is: Gewerbestrasse 11, 6330 Cham, Switzerland

Preface

The experimental evidence on the behavior of matter on subatomic scales collected so far leads to the conclusion that all microscopic physical phenomena can be explained by the assumption that matter is made up of *elementary particles*, structureless fundamental constituents of the Universe, which are subject to four types of fundamental interactions: gravitational, electromagnetic, weak, and strong. These interactions are, however, not transmitted via direct *contacts*, but they are rather *mediated* by a particular type of elementary particles, called *intermediate gauge bosons*. Whereas the gravitational interaction is one of the most *ancient* phenomena known in nature, the electromagnetic one has been studied and understood most thoroughly, having found a solid theoretical formulation known as *quantum electrodynamics*, during the first half of the last century. Most of the everyday physical phenomena – from the stability of matter to the plethora of phenomena related to the propagation of light – are, in fact, explained by this theory. The weak and strong interactions that, unlike the electromagnetic and gravitational ones, manifest themselves only at microscopic scales, met a similar firm theoretical foundation in the *standard model of elementary particles*, which includes quantum electrodynamics itself. In contrast, at present, gravity still appears to conflict with the laws of quantum physics, notwithstanding the recent progress accomplished within the promising framework of *superstring theory*.

Despite their common role of force mediators between the elementary constituents of nature, each fundamental interaction is characterized by unique features which imply peculiar physical phenomena: The strong interaction, mediated by gauge bosons called *gluons*, is the only one that gives rise to the phenomenon of *confinement*, which traps the quarks and the gluons themselves inside the nucleons, while the weak interaction is the only one to be mediated by *massive* gauge bosons, the particles $W^{\pm}$ and Z^0. On the other hand, the electromagnetic interaction is the only one to be mediated by particles, the *photons*, that, being electrically neutral, are not subject to a *mutual* electromagnetic interaction. Finally, the gravitational interaction is the only one that affects *all* elementary particles, including the intermediate gauge bosons themselves, and, in addition, it is mediated by particles of spin *two*, the *gravitons*, whereas the intermediate gauge bosons of the remaining interactions all carry spin *one*.

In the light of these important distinctions, it may appear somewhat surprising that all fundamental interactions are governed by a *common*, mathematically robust and elegant, theoretical framework, strongly constraining their general structure, a framework whose profound physical origin is still to be uncovered. Among the cornerstones of this unifying framework, let us quote the most significant ones: All fundamental interactions satisfy Einstein's *principle of relativity* and admit a manifestly covariant formulation that automatically implies the conservation of the total energy, momentum, and angular momentum of the Universe. Furthermore, each interaction is transmitted via exchange of one or more dedicated particles – the intermediate gauge bosons – that mathematically are represented by vector or tensor *fields*, whose dynamics is controlled by a fundamental symmetry, called *gauge invariance*. Noether's theorem then associates with each of these symmetries, and hence with each intermediate gauge boson, a *conserved quantity*. Last but not least, perhaps the most mysterious, but nonetheless, less fundamental cornerstone of the common theoretical framework is that the dynamics of all four fundamental interactions can be derived from a *variational principle*.

The core of this book on classical electrodynamics is a series of lectures on *electromagnetic fields*, held by the author in the years 2004–2011 for the master's degree in physics at the University of Padua. Its actual content covers, however, research topics that are far more advanced than what could be taught in a graduate course. A special concern of the book is to emphasize, on the one hand, those aspects that unite electrodynamics with the other fundamental interactions, including the cornerstones mentioned above, and, on the other, to highlight, where possible, their most significant differences. The major waiver implied by this perspective is an almost complete neglect of the important topic of *electromagnetic fields in matter*.

Classical electrodynamics is presented as a theory founded on a system of postulates: Einstein's principle of relativity, and the Maxwell and Lorentz equations. The whole broad phenomenology of the electromagnetic interactions follows, indeed, from these basic assumptions in a *stringent way*. Accordingly, special attention is paid to the issue of internal and physical *consistency* of classical electrodynamics, an issue that represents, indeed, the ultimate challenge of the postulates of any physical theory. In line with this purpose, from the outset, we highlight the traces left by the *twofold* ultraviolet divergences which *inevitably* accompany the dynamics of charged point-particles: the infinite self-force, and the infinite energy of the electromagnetic field. In fact, these divergences turn classical electrodynamics in an *internally inconsistent* theory, eventually. From a technical point of view, the resulting inconsistencies are codified by the so-called *radiation reaction*, a phenomenon of basic physical importance which *explicitly* violates time-reversal invariance. Correspondingly, in our presentation of the theory, this discrete symmetry will appear as a central theme, in its *metamorphosis* from an exact symmetry to a, first *spontaneously* and finally *explicitly*, violated symmetry. At first glance, the internal inconsistencies of classical electrodynamics are in apparent contradiction with the fact that, from an experimental point of view, the theory describes all classical electromagnetic phenomena with extreme precision. In fact, as long as one remains strictly within the axiomatic framework of classical electrodynamics, this contradiction allows only for a *pragmatic* solution, presented in Chaps. 15 and 16, which explains, nonetheless, the perfect agreement between theory and experiment. The ultimate solution of this paradox can, however, be found only within *quantum electrodynamics*.

For reasons of mathematical consistency, in order to conceive Maxwell's equations in a well-defined way, it is essential to formulate them in the *space of distributions*. In fact, as we shall deal mainly, although not exclusively, with *point-like* charged particles, our electric currents in general involve the Dirac δ-function. Therefore, it would make no sense to consider Maxwell's equations as partial differential equations for electromagnetic fields viewed as *functions*: They must rather be considered as differential equations for *distribution-valued* electromagnetic fields. Concerning the solutions of the resulting equations, for charged particles following strictly time-like trajectories, the usual *heuristic* application of the Green function method leads, indeed, to correct solutions for the fields. Conversely, for the light-like trajectories of *massless* charged particles, the Green function method *fails*, and in this case it is only the full-fledged distributional approach that allows for analytical, well-defined, solutions of Maxwell's equations.

As a rule, each theoretical subject presented in this book is illustrated with a series of physically relevant examples, worked out in detail, and the introduction of new mathematical tools is likewise motivated and accompanied by practical examples. The solution of the problems proposed at the end of each chapter may result in a better understanding of some of the topics covered in the text, although they do not affect the comprehension of the subsequent chapters.

Organization of the material. The book is divided into three parts. In Part I (Chaps. 1–4), we introduce the physical and mathematical foundations of the classical electrodynamics of a system of charged point-like particles. This initial part includes, in particular, the mathematical tools necessary for a precise formulation of the theory: the *theory of distributions*, which, as anticipated above, is essential for a correct handling of the singularities arising from the point-like nature of charged particles, and the *tensor calculus*, a standard tool of any relativistic theory. In Chap. 2, we introduce the *fundamental equations of electrodynamics*, i.e., the Maxwell and Lorentz equations, and analyze their basic structural properties and the conservation laws they imply. The first part ends with the presentation of the *variational method* in Chaps. 3 and 4. This method is introduced as an alternative approach for the formulation of a generic *relativistic field theory*, encoding its dynamics in a concise and elegant way, and as a basic prerequisite for the validity of *Noether's theorem*. The close and universal connection between this theorem and the variational method is then illustrated in detail for the electrodynamics of point-like charges in Chap. 4.

The main concern of Part II (Chaps. 5–14) is the derivation of the basic phenomenological predictions of classical electrodynamics, starting from a series of exact solutions of Maxwell's equations. This part includes a detailed study of the properties of the electromagnetic field in empty space, a systematic treatment of the electric and magnetic fields generated by a charged particle in arbitrary motion – the celebrated *Liénard-Wiechert fields* – and an extensive investigation of the *radiation* phenomenon, both in the non-relativistic and in the ultrarelativistic limits. We analyze in detail the angular and spectral distributions of the radiation emitted by a series of important physical systems, such as the high-energy accelerators, the collisions between charged particles, the linear antennas, the classical hydrogen atom, and the *Thomson scattering*. This part also includes some topics that rarely receive a systematic treatment in textbooks: the problem of the electromagnetic fields and potentials generated by a *massless* charged particle in uniform linear motion, a close comparison between the electromagnetic and gravitational radiations, the derivation of the multifaceted aspects of the synchrotron radiation, a detailed theoretical explanation of the *Čerenkov effect* and, finally, an extensive analysis of the radiation emitted by a prototypical *non-compact* charge distribution: the infinite conducting wire.

Part III (Chaps. 15–21) focuses both on more speculative and on more recent developments of theoretical high-energy physics. Chapter 15 addresses, with due care, the *radiation reaction* phenomenon, whose most problematic aspect is the ultraviolet divergent self-force arising in the Lorentz equation. The purpose of this chapter is twofold. In the first place, we emphasize the conceptual motivations that force us to replace the *divergent* Lorentz equation – a *dogma* of classical electrodynamics – with the *finite* Lorentz-Dirac equation which, in turn, violates explicitly *time-reversal* invariance. Then, we show how this violation introduces in classical electrodynamics an irreconcilable self-contradiction, presenting itself in different forms depending on the *pragmatic* point of view one takes: typically as a *violation of causality*. As already mentioned, this internal inconsistency can be definitively resolved only within quantum field theory. The subsequent chapter is devoted to the second *classical* ultraviolet problem of classical electrodynamics, namely, the infinite energy of the electromagnetic field of a charged point-particle. Surprisingly, this problem, whose persistency would undermine nothing less than the law of conservation of energy, was resolved definitively only about 40 years ago. In Chap. 16, we present the solution of this problem in a modern key, by relying, once more, on the theory of distributions, and clarify the inextricable link existing between the Lorentz-Dirac equation and the local conservation of energy and momentum. In Chap. 17, we address the delicate issue of the classical electrodynamics of *massless* charged particles, forced to move at the speed of light. In this regard, two main questions arise. The first is whether Maxwell's equations are also well posed in these extremal conditions, and the second regards the existence of *exact analytical solutions* of these equations, superseding the Liénard-Wiechert fields. The answer to the first question is affirmative. Only recently, also the considerably more complex second problem has found a constructive solution, which we report in a concise form in this chapter. The electromagnetic field produced by a massless charged particle exhibits singular contributions supported on strings and surfaces, which are, however, perfectly well behaved in the space of distributions. What makes the solution non-trivial is that, in this case, the Green function method fails. This method fails, in fact, for all trajectories that become light-like in the infinite *past*. Chapter 18 examines *massive* vector fields. The interest in this kind of fields relies on the fact that, in quantum theory, they describe *massive spin-one* particles, a sort of massive photons, as e.g., the vector bosons $W^{\pm}$ and Z^0 mediating the weak interactions. Although some characteristic features of these particles, as, for instance, their finite mean lifetime, are of genuine quantum mechanical nature, a classical analysis is, nevertheless, able to reveal the most significant differences occurring between the electromagnetic interaction and an interaction mediated by a massive vector field. Chapter 19 provides an introduction to the classical electro-dynamics of *extended* charged objects occupying a p-dimensional volume, the so-called *p-branes*. The most simple non-trivial p-brane is a *string*, corresponding to $p = 1$, while a point-particle corresponds to $p = 0$. The choice of this topic is partially motivated by the fact that p-branes constitute the elementary excitations of all *superstring theories*, which are candidates to reconcile *gravity* with *quantum mechanics*, and to unify the former with the other fundamental interactions. One of the basic purposes of this chapter is to show how the fundamental paradigms of the electrodynamics of charged particles, such as Lorentz invariance, gauge invariance, the Maxwell and Lorentz equations, and the main conservation laws, extend naturally to the electrodynamics of extended charged objects. In particular, in the mathematical language of *differential forms*, for which we provide a practical introduction at the beginning of the chapter, the generalization of Maxwell's equations from particles to p-branes appears straightforward. The two final chapters of the book deal with *magnetic monopoles* and, more generally, with *dyons*. In Chap. 20, we show that the framework of classical electrodynamics, although based on a rather rigid system of postulates, is perfectly compatible with the existence of this exotic type of particles in nature. Chapter 21 illustrates, instead, how the *quantum dynamics* of magnetic monopoles offers a solution to the *ancient* problem of the *quantization of the electric charge*, namely, the experimental observation that all electric charges present in nature are *integer* multiples of a fundamental charge.

Prerequisites. The reader should have a basic knowledge of the cornerstones of classical electromagnetism and relativistic kinematics, such as Maxwell's equations and the special Lorentz transformations. A minimal familiarity with Maxwell's equations in covariant form and, more in general, with four-dimensional tensors, is helpful, albeit not mandatory, as the physical origin and the basic ingredients of the tensor calculus are explained in detail in Chap. 1. Elementary notions of the theory of distributions, such as the concept of the Dirac δ-function, may be useful. However, the basic properties of the distributions used in the text are presented in a self-contained way in Chap. 2. Finally, a certain familiarity with the variational method for a Lagrangian system with a finite number of degrees of freedom is useful but, again, not mandatory.

Acknowledgements. The author would like to express his gratitude to his friend and colleague Prof. Pieralberto Marchetti for a long series of enlightening conversations on the foundations of classical electrodynamics. They became the actual breeding ground of this book.

Padua, Italy   Kurt Lechner
April 2018

Contents

Part I
Theoretical Foundations

Chapter 1
Foundations of Special Relativity

In the discovery of *special relativity* Electrodynamics, representing a relativistic
theory *par excellence*, played a key role. The *principle of special relativity*, stating
that all physical laws must have the same form in all inertial reference frames, was
indeed a cornerstone of Einstein's reanalysis of this theory, and – as our knowledge
of the microscopic world deepened – it gained increasing reliability as a fundamental paradigm of nature: all fundamental interactions respect, in fact, this basic
principle. The most simple and elegant way to implement it – actually the only one
of a *real* utility – is represented by the so called *manifest covariance paradigm*,
realized in the mathematical framework of the *tensor calculus*. This paradigm has
been applied successfully to all *basic* theories, i.e. those aiming to decode the elementary laws of nature, as the field theories describing the four fundamental interactions and, much more speculatively, *superstring theory*, while maintaining its full
effectiveness in classical as well as in quantum theory. Our exposition of *classical
electrodynamics* will thus be based a fortiori on this paradigm.

In the construction of a physical theory a basic concern should be the emphasis on
the a priori *assumptions* on which the theory is based, so that one can neatly distinguish the consequences of those assumptions, from the consequences of additional
hypotheses, added at later stages of the construction. For this reason, in Sect. 1.1
we first of all retrieve the logical path that conducted from the *postulates* of relativity to the manifest covariance paradigm, and to the tensor calculus. We present the
main ingredients of the tensor calculus with a certain degree of completeness, since
we will use it widely in the book. In the final part of this chapter we will analyze
in detail the structure of the *Poincaré group*, that is to say, the set of all coordinate transformations that connect a generic inertial reference frame to another such
frame. In a relativistic theory this symmetry group is, in fact, intimately tied to the
principal conservation laws of the underlying physical system – through *Noether's
theorem*. Being of fundamental importance for all physics, this relation will then be
investigated in depth in Chap. 3.

© Springer International Publishing AG, part of Springer Nature 2018   3
K. Lechner, *Classical Electrodynamics*, UNITEXT for Physics,
https://doi.org/10.1007/978-3-319-91809-9_1

1.1   Postulates of Relativity

Newtonian mechanics and *special relativity* are based on a few a priori assumptions which are *common* to the two, regarding in particular the intrinsic properties of space and time, while they differ substantially in the *relativity principles* on which each theory is based. The common assumptions regarding the space-time continuum are the homogeneity of time and the homogeneity and isotropy of the three-dimensional space. Further elements in common are that the laws of physics in both theories are formulated with respect to a special class of coordinate systems – the *inertial reference frames* – and that both implement the physical equivalence of all these coordinate systems through a *specific* relativity principle. The principle of *Galilean relativity* of Newtonian mechanics demands that the laws of *mechanics* maintain their form under the *Galilean transformations*

$$\mathbf{x}' = \mathbf{x} - \mathbf{v}t, \qquad t' = t,$$

where $\mathbf{v}$ is the velocity of one inertial reference frame with respect to another. *Einstein's relativity principle* demands, instead, that *all* laws of physics have the same form in all inertial reference frames, thereby not making any a priori assumption about the nature of the transformation laws of space and time. In particular, *special relativity* renounces to the absoluteness paradigm of space-intervals and time-intervals of Newtonian mechanics, replacing it – in a sense – with the request of *constancy of the speed of light*. Summarizing, the **postulates of relativistic physics** are:

(1) the space is isotropic and homogeneous, and the time is homogeneous;
(2) the speed of light is the same in all inertial reference frames;
(3) all laws of physics have the same form in all inertial reference frames.

To implement these postulates in practice – in particular Einstein's relativity postulate (3), which poses strong restrictions on the allowed physical laws – it is necessary to determine preliminarily the transformation laws of the space-time coordinates from one inertial frame to another. As a matter of fact, as we will show in Sect. 1.2, the form of these transformation laws is completely fixed by the postulates (1) and (2). Before proceeding any further, we specify the notations, and the conventions, adopted in the book.

Notations. We will denote the *contravariant* space-time coordinates of an *event* $\{t, \mathbf{x}\}$ by Greek indices μ, ν, ρ etc. assuming the values $0, 1, 2, 3$. More precisely, we introduce the space-time coordinate with an *upper* index

$$x^\mu = (x^0, x^1, x^2, x^3), \quad x^0 = ct.$$

The *spatial* coordinates $\mathbf{x}$ of the event will instead be denoted by Latin indices i, j, k etc. assuming the values $1, 2, 3$, that is to say, we set

$$x^i = (x^1, x^2, x^3).$$

In summary, we adopt the notation $x^\mu = (x^0, x^i) \leftrightarrow (x^0, \mathbf{x})$. If instead we write "$x$", with no index at all, in general we refer to the four-dimensional coordinate x^μ. A scalar field in four-dimensional space-time, for instance, will be denoted by the symbol $\Phi(x)$. While, in general, we shall indicate the magnitude of a generic spatial vector $\mathbf{V}$ by the symbol V, the magnitude $|\mathbf{x}|$ of the position vector $\mathbf{x}$ will frequently be denoted by r. The *Minkowski metric* and its inverse, denoted by $\eta_{\mu\nu}$ and $\eta^{\mu\nu}$, respectively, are diagonal 4×4 matrices defined through

$$\text{diag}(\eta_{\mu\nu}) = (1, -1, -1, -1) = \text{diag}(\eta^{\mu\nu}), \qquad \eta^{\mu\nu}\eta_{\nu\rho} = \delta^\mu_\rho. \tag{1.1}$$

The components of these symmetric matrices are therefore

$$\eta_{00} = \eta^{00} = 1, \qquad \eta_{ij} = \eta^{ij} = -\delta^j_i, \qquad \eta_{0i} = \eta^{0i} = 0. \tag{1.2}$$

According to the choice above the *signature* of our Minkowski metric is $1 - 3 = -2$. We adopt Einstein's *summation convention* over repeated "dummy" indices, suppressing the summation symbol Σ every time the summation is made over an index that appears *twice* in the same expression. With the expression $\eta^{\mu\nu}\eta_{\nu\rho}$ in (1.1), for example, we mean the sum

$$\eta^{\mu\nu}\eta_{\nu\rho} = \sum_{\nu=0}^{3} \eta^{\mu\nu}\eta_{\nu\rho}.$$

Similar conventions hold for expressions containing multiple sums. With the aid of the Minkowski metric we define the *covariant* space-time coordinates of an event, with a *lower* index, as

$$x_\mu = \eta_{\mu\nu}x^\nu = (x^0, -x^1, -x^2 - x^3).$$

We have thus $x_\mu = (x_0, x_i) = (x^0, -\mathbf{x})$, i.e. $x_0 = x^0$, and $x_i = -x^i$. From the definition of x_μ we derive the *inversion formula*

$$x^\mu = \eta^{\mu\nu}x_\nu.$$

It is usually said that the Minkowski metric *lowers* and *raises* the indices. In what follows, to simplify the notation, frequently we will set the speed of light c to 1.

1.2 Lorentz and Poincaré Transformations

As remarked above, unlike the postulates of Newtonian mechanics, the postulates of relativity do not specify a priori the nature of the coordinate transformations from one inertial reference frame to another: They rather *determine* uniquely the form of these transformations, that are going to become the *Poincaré transformations*. In this section we present the explicit derivation of these basic transformation laws from the postulates of relativity, illustrating thereby the extreme economy of the latter, together with the ineluctability of the former.

1.2.1 Linearity

First of all, we prove that postulate (1) implies that the transformations from an inertial reference frame to another are necessarily *linear* in the coordinates. Let us consider an inertial reference frame K with coordinates x^μ. The coordinates x'^μ of another inertial reference frame K' are then related to the coordinates of K through an invertible map $f^\mu : \mathbb{R}^4 \to \mathbb{R}^4$, such that

$$x'^\mu(x) = f^\mu(x). \tag{1.3}$$

We then consider two generic events whose coordinates in K are x^μ and y^μ, respectively. In K' the coordinates of these events are then $x'^\mu = f^\mu(x)$ and $y'^\mu = f^\mu(y)$. According to postulate (1) there are no privileged positions and instants, and consequently a change of the origin of space and time in K, i.e. a translation $x^\mu \to x^\mu + b^\mu$, $y^\mu \to y^\mu + b^\mu$, where b^μ is an arbitrary constant vector, cannot modify the spatial and temporal "distances" $x'^\mu - y'^\mu$ between the two events in K'. More precisely, we must have

$$x'^\mu - y'^\mu = f^\mu(x) - f^\mu(y) = f^\mu(x+b) - f^\mu(y+b), \tag{1.4}$$

for every vector b^μ. Assuming that f^μ is a differentiable map, taking the derivative of (1.4) with respect to the coordinate x^ν we find

$$\frac{\partial f^\mu(x)}{\partial x^\nu} = \frac{\partial f^\mu(x+b)}{\partial x^\nu},$$

for every b^μ. This means that the partial derivatives of the functions $f^\mu(x)$ must be independent of x:

$$\frac{\partial f^\mu(x)}{\partial x^\nu} = \text{constant} = \Lambda^\mu{}_\nu.$$

Integrating these relations, from (1.3) we deduce that the coordinates of K' are related to those of K through a generic inhomogeneous *linear* transformation

$$x'^\mu \doteq \Lambda^\mu{}_\nu x^\nu + a^\mu. \tag{1.5}$$

The four parameters a^μ correspond to arbitrary *translations* of space and time, which are indeed possible physical operations that can connect an inertial reference frame to another. Conversely, if $\Lambda^\mu{}_\nu$ is an arbitrary matrix, the relations (1.5) in general do *not* correspond to a transformation between two inertial reference frames. Choosing, for example, $a^\mu = 0$ and $\Lambda^\mu{}_\nu = k\,\delta^\mu{}_\nu$ with k a constant, one obtains the *scale transformation* $x'^\mu = kx^\mu$ and, as we will see below, if two inertial reference frames are related by a transformation of this kind, at most one of the two can be *inertial*. Before moving to the determination of the allowed matrices $\Lambda^\mu{}_\nu$, we derive the transformation law of the *covariant* coordinates x_μ. Multiplying (1.5) with $\eta_{\rho\mu}$ we obtain

$$x'_\rho = \eta_{\rho\mu}x'^\mu = \eta_{\rho\mu}\Lambda^\mu{}_\nu x^\nu + \eta_{\rho\mu}a^\mu = \eta_{\rho\mu}\Lambda^\mu{}_\nu \eta^{\nu\sigma}x_\sigma + \eta_{\rho\mu}a^\mu,$$

or, equivalently,

$$x'_\rho = \widetilde{\Lambda}_\rho{}^\sigma x_\sigma + a_\rho, \quad \text{where} \quad \widetilde{\Lambda}_\rho{}^\sigma = \eta_{\rho\mu}\Lambda^\mu{}_\nu \eta^{\nu\sigma}, \quad a_\rho = \eta_{\rho\mu}a^\mu. \quad (1.6)$$

1.2.2 Invariance of the Space-Time Interval

In the determination of the matrices $\Lambda^\mu{}_\nu$ corresponding to *actual* transformations between two inertial reference frames, the postulate (2), that distinguishes definitively Newtonian physics from relativistic physics, plays a crucial role, as it allows to prove a fundamental theorem. Given an inertial reference frame K, and two events whose space-time coordinates in K are x^μ and y^μ, we denote their *distance vector* in K by $dx^\mu = y^\mu - x^\mu$. We then define the *space-time interval*, or more simply the *interval*, between the two events in K as the positive, negative, or vanishing quantity (see (1.2))

$$ds^2 = dx^\mu dx^\nu \eta_{\mu\nu} = dx^0 dx^0 \eta_{00} + \eta_{ij}dx^i dx^j = dt^2 - |d\mathbf{x}|^2.$$

The postulates (1) and (2) then allow to prove the following theorem.

Theorem I (invariance of the interval). The interval between two events is independent of the inertial reference frame, that is to say

$$ds'^2 = ds^2 \qquad (1.7)$$

for all inertial reference frames K and K'.

Proof Consider two generic events, whose coordinates and distance vector in K are x^μ, y^μ, and $dx^\mu = y^\mu - x^\mu$, respectively. According to (1.5) the distance vector dx'^μ of the same events in K' is then related to dx^μ by the relation

$$dx'^\mu = y'^\mu - x'^\mu = \Lambda^\mu{}_\nu y^\nu + a^\mu - (\Lambda^\mu{}_\nu x^\nu + a^\mu) = \Lambda^\mu{}_\nu dx^\nu.$$

The interval ds'^2 between the same two events in K' can then be written as

$$ds'^2 = dx'^\mu dx'^\nu \eta_{\mu\nu} = \Lambda^\mu{}_\alpha dx^\alpha \Lambda^\nu{}_\beta dx^\beta \eta_{\mu\nu} = G_{\alpha\beta} dx^\alpha dx^\beta, \qquad (1.8)$$

where we introduced the symmetric matrix

$$G_{\alpha\beta} = \Lambda^\mu{}_\alpha \Lambda^\nu{}_\beta \eta_{\mu\nu},$$

which is *independent* of the chosen events. We now recall that two generic events with distance vector $dx^\mu = (dt, d\mathbf{x})$ can be connected by a light ray, if and only if one has $v = |d\mathbf{x}/dt| = 1$, that is if and only if $ds^2 = 0$. Since – according to postulate (2) – the speed of light is the same in all inertial reference frames, it follows that

$$ds^2 = 0 \quad \Leftrightarrow \quad dt = \pm|d\mathbf{x}| \quad \Leftrightarrow \quad dt' = \pm|d\mathbf{x}'| \quad \Leftrightarrow \quad ds'^2 = 0.$$

We conclude, therefore, that the quadratic form ds'^2 in (1.8), considered as a polynomial of second order in dt, possesses the zeros $dt = \pm|d\mathbf{x}|$. This implies that ds'^2 entails the decomposition

$$ds'^2 = G_{00}(dt - |d\mathbf{x}|)(dt + |d\mathbf{x}|) = G_{00} ds^2, \qquad (1.9)$$

where the coefficient G_{00} can depend only on the *relative velocity* $\mathbf{v}$ of K' with respect to K. Moreover, thanks to postulate (1) – that requires rotation invariance – G_{00} can depend on the vector $\mathbf{v}$ only through its magnitude $|\mathbf{v}|$. The decomposition (1.9) reduces therefore to

$$ds'^2 = G_{00}(|\mathbf{v}|)\, ds^2. \qquad (1.10)$$

If we invert now the roles played by K and K', in the relation (1.10) we must perform the replacements $\mathbf{v} \to -\mathbf{v}$, $s \to s'$, $s' \to s$, obtaining

$$ds^2 = G_{00}(|\mathbf{v}|)\, ds'^2.$$

Combining this relation with (1.10) we deduce that $G_{00}(|\mathbf{v}|) = \pm 1$, and, since $G_{00}(0) = 1$, for continuity we must have $G_{00}(|\mathbf{v}|) = 1$. This proves the equality (1.7) and hence the theorem. $\qquad\qquad\square$

The theorem of the invariance of the interval imposes on the matrix $\Lambda^\mu{}_\nu$ strong restrictions. From the relation (1.8) it follows, in fact, that for arbitrary distance vectors dx^μ we must have

$$ds'^2 = dx^\alpha dx^\beta (\Lambda^\mu{}_\alpha \Lambda^\nu{}_\beta \eta_{\mu\nu}) = ds^2 = dx^\alpha dx^\beta \eta_{\alpha\beta}.$$

But this is possible only if $\Lambda^\mu{}_\nu$ satisfies the constraints

$$\Lambda^\mu{}_\alpha \Lambda^\nu{}_\beta \eta_{\mu\nu} = \eta_{\alpha\beta}. \qquad (1.11)$$

Lorentz group. The 4×4 matrices $\Lambda^\mu{}_\nu \equiv \Lambda$, appearing in the transformation law (1.5) between two inertial reference frames, are therefore not arbitrary: They must satisfy the supplementary conditions (1.11), which in matrix notation read

$$\Lambda^T \eta \Lambda = \eta \quad \leftrightarrow \quad \Lambda^\mu{}_\alpha \Lambda^\nu{}_\beta \eta_{\mu\nu} = \eta_{\alpha\beta}. \tag{1.12}$$

Multiplying these relations from the left with $\Lambda\eta$ and from the right with $\Lambda^{-1}\eta$, they can be recast in the equivalent form

$$\Lambda \eta \Lambda^T = \eta \quad \leftrightarrow \quad \Lambda^\alpha{}_\mu \Lambda^\beta{}_\nu \eta^{\mu\nu} = \eta^{\alpha\beta}. \tag{1.13}$$

The set of all these matrices forms a *group*, called the *Lorentz group*, which is usually denoted by the symbol

$$O(1,3) = \{\text{real} 4 \times 4 \text{ matrices } \Lambda / \Lambda^T \eta \Lambda = \eta\}, \tag{1.14}$$

see Problem 1.9. The symbol "O" indicates commonly that the considered matrices are (pseudo)-orthogonal, and the entries $(1,3)$ refer to the fact that the (pseudo)-metric η has the diagonal elements $(1,-1,-1,-1)$. Taking the determinant of both members of the constraint (1.12) we derive the condition

$$(\det \Lambda)(-1)(\det \Lambda) = -1 \quad \Leftrightarrow \quad (\det \Lambda)^2 = 1.$$

Therefore, the determinant of a matrix belonging to the Lorentz group can take only the values

$$\det \Lambda = \pm 1. \tag{1.15}$$

It follows in particular that every element of $O(1,3)$ admits an inverse, which is of course a fundamental property of every *group*.

Poincaré group. What we have shown so far is that the postulates of relativity imply that two generic inertial reference frames are related by a nonhomogeneous *linear* transformation of the type (1.5)

$$x' = \Lambda x + a, \tag{1.16}$$

where Λ is an element of the Lorentz group, and a is an arbitrary constant vector. The set of these transformations forms again a group $\mathcal{P}$, that is called the *Poincaré group*. An element of this group is represented by a pair (Λ, a), more precisely

$$\mathcal{P} = \{(\Lambda, a) / \Lambda \in O(1,3), a \in \mathbb{R}^4\}. \tag{1.17}$$

The composition law of two elements of $\mathcal{P}$ can be derived by iterating the transformation (1.16), namely

$$(\Lambda_1, a_1) \circ (\Lambda_2, a_2) = (\Lambda_1 \Lambda_2, a_1 + \Lambda_1 a_2).$$

The Lorentz group is isomorphic to the subgroup of $\mathcal{P}$ formed by the elements $(\Lambda, 0)$, and the elements of $\mathcal{P}$ of the form $(\mathbf{1}, a)$ form the subgroup of the *translations*. The coordinate transformations (1.16) are called *Poincaré transformations*, and the transformations corresponding to $a = 0$ are called *Lorentz transformations*. Strictly speaking, what we have shown so far is that a transformation that connects two inertial reference frames is necessarily a Poincaré transformation. We ought still to convince ourselves that the contrary is also true, i.e. that every Poincaré transformation *actually* realizes a transition from one inertial reference frame to another. In reality this problem affects only the Lorentz transformations, as the translations have a self-evident physical meaning. We will address this issue in Sect. 1.4.

1.3 Physical Laws and Manifest Covariance

Once we have determined the transformation laws of the space-time coordinates from one inertial reference frame to another, we can proceed to the implementation of postulate (3), namely the development of a systematic strategy to formulate physical laws that respect automatically Einstein's *relativity principle*. As a first step in this direction we must identify the general *modus* according to which physical quantities change in the transition from one inertial reference frame to another. We will address this problem by taking a cue from Newtonian mechanics – formulated in a Cartesian coordinate system – in which the role of Lorentz transformations is played by spatial *rotations*.

Spatial rotations and three-dimensional tensors. The rotations of the axes of a Cartesian coordinate system are represented by 3×3 orthogonal matrices $R^i{}_j$ with unit determinant, forming the group

$$SO(3) = \{\text{real } 3 \times 3 \text{ matrices } R \,/\, R^T R = \mathbf{1}, \ \det R = 1\}.$$

The symbol "S" indicates commonly that the determinant of the matrices is equal to one. Under a rotation the coordinates $\mathbf{x}$ transform according to the law

$$x'^i = R^i{}_j \, x^j, \tag{1.18}$$

whereas the time remains invariant, $t' = t$. Let us now consider Newton's second law $\mathbf{F} = m\mathbf{a}$ in a Cartesian coordinate system K, and a rotation R that connects K to another Cartesian coordinate system K'. Since we have $\mathbf{a} = d^2\mathbf{x}/dt^2$, the acceleration, as well as the force, both being *vectors*, transform in the same way as $\mathbf{x}$. Multiplying then both sides of Newton's second law with the rotation matrix R, we obtain

$$F^i = ma^i \quad \Rightarrow \quad R^j{}_i F^i = m R^j{}_i a^i \quad \leftrightarrow \quad F'^j = ma'^j, \tag{1.19}$$

so that in K' it has the same form as in K. The real reason – of essentially geometrical origin – of this property is that Newton's second law equates a *vector* to another *vector*, i.e. a geometrical object that under rotations transforms in a well-defined way, namely as in (1.18). Due to the apparent simplicity of the argument, one usually says that Newton's second law is *manifestly covariant*, or also *manifestly invariant*, under rotations, in the sense that it has *evidently* the same form in all Cartesian coordinate systems. In the same way, all *fundamental* equations of Newtonian mechanics turn out to be manifestly covariant under rotations. We add two further examples. The first is *Euler's second law*

$$\frac{d\mathbf{L}}{dt} = \mathbf{r} \times \mathbf{F} \quad \leftrightarrow \quad \frac{dL^i}{dt} = \varepsilon^{ijk} r^j F^k, \tag{1.20}$$

where $\mathbf{L} = \mathbf{r} \times m\mathbf{v}$ is the angular momentum of the particle, and ε^{ijk} denotes the three-dimensional *Levi-Civita tensor*, defined by

$$\varepsilon^{ijk} = \begin{cases} 1, & \text{if } ijk \text{ is an even permutation of 1, 2, 3,} \\ -1, & \text{if } ijk \text{ is an odd permutation of 1, 2, 3,} \\ 0, & \text{if at least two indices are equal.} \end{cases} \tag{1.21}$$

The second example is the *angular momentum* formula for a rigid body

$$\mathcal{L}^i = I^{ij} \omega^j, \tag{1.22}$$

where ω is the angular velocity vector and I^{ij} is the rank-two *inertia tensor*

$$I^{ij} = \sum_n m_n \left(r_n^2 \delta^{ij} - r_n^i r_n^j \right). \tag{1.23}$$

Under a rotation the components of this tensor transform according to the law

$$I'^{ij} = R^i{}_m R^j{}_n I^{mn}, \tag{1.24}$$

whereas the vectors $\mathbf{r}$, $\mathbf{v}$, ω and $\mathbf{L}$ transform in the same way as $\mathbf{x}$ in (1.18), see Problem 1.8. From (1.24) we see that the inertia tensor transforms as if it were the product of two vectors, a characteristic that qualifies it as a *rank-two tensor*. Proceeding in a way analogous to (1.19), and resorting to property (1.24) and to the fact that $R \in SO(3)$, it is easy to show that equations (1.20) and (1.22) take in K' the form

$$\frac{dL'^i}{dt} = \varepsilon^{ijk} r'^j F'^k, \tag{1.25}$$

$$\mathcal{L}'^i = I'^{ij} \omega'^j, \tag{1.26}$$

respectively, see Problem 1.8. These equations are, hence, manifestly covariant.

Poincaré transformations and four-dimensional tensors. From the analysis above
we see that the physical quantities of Newtonian mechanics are grouped into three-
dimensional vectors and tensors, which transform *linearly* under rotations, and that
in their transformation laws each index is associated with a matrix R, see (1.18)
and (1.24). As we will see in Sect. 1.4, the rotations constitute actually a subgroup
of the Lorentz group, and consequently it is natural to assume that also in a relativistic theory the physical quantities are grouped into *multiplets*, which transform
linearly under the Lorentz group. In the language of *group theory* this circumstance
is expressed by saying that each multiplet hosts a – reducible or irreducible – *representation* of the Lorentz group. Then, from a basic result of group representation theory, it follows that these multiplets constitute necessarily *four-dimensional
tensors of rank* (m, n), where m and n are non-negative integers. By definition a
four-dimensional tensor $T^m{}_n$ of rank (m, n) is an object carrying m contravariant
(upper) indices and n covariant (lower) indices,

$$T^m{}_n \leftrightarrow T^{\mu_1 \cdots \mu_m}{}_{\nu_1 \cdots \nu_n}, \tag{1.27}$$

subject to a peculiar transformation law under the action of the Poincaré group
(1.17), that will be specified in a moment. Tensors of rank (0,0) are called *scalars*,
tensors of rank (1,0) and (0,1) are called *contravariant* (four-)vectors and *covariant*
(four-)vectors, respectively, and tensors of rank $(2, 0)$, $(0, 2)$ and $(1, 1)$ are called
double tensors. More generally, we will have to consider *tensor fields* $T^{\mu_1 \cdots \mu_m}{}_{\nu_1 \cdots \nu_n}$
(x) of rank (m, n), which with respect to *tensors* carry also a dependence on the
space-time coordinate x. By definition, a tensor field of rank (m, n) transforms
under a Poincaré transformation $x' = \Lambda x + a$ according to the law

$$T'^{\mu_1 \cdots \mu_m}{}_{\nu_1 \cdots \nu_n}(x') = \Lambda^{\mu_1}{}_{\alpha_1} \cdots \Lambda^{\mu_m}{}_{\alpha_m} \widetilde{\Lambda}_{\nu_1}{}^{\beta_1} \cdots \widetilde{\Lambda}_{\nu_n}{}^{\beta_n} T^{\alpha_1 \cdots \alpha_m}{}_{\beta_1 \cdots \beta_n}(x), \tag{1.28}$$

which represents a natural generalization of the transformation laws (1.18) and
(1.24). The matrix $\widetilde{\Lambda}$ appearing in (1.28) has been defined in (1.6),

$$\widetilde{\Lambda} = \eta \, \Lambda \, \eta. \tag{1.29}$$

Notice in particular that under a translation $x' = x + a$, a tensor field remains unal-
tered. By definition, the transformation law of a *tensor* (1.27) is obtained from (1.28)
dropping the dependence on the space-time coordinates. In the following for simplicity we will use the term *tensor* for *tensor fields* as well as for *tensors*, since it
will be clear from the context to which kind of object we are referring to.

Once we accept that the physical observables of a relativistic theory should be
grouped into four-dimensional tensors, the implementation of postulate (3) – Einstein's relativity principle – can be realized in analogy to the implementation of rotation invariance in Newtonian mechanics: In the same way as the laws of the latter,
equating three-dimensional vectors to three-dimensional vectors, maintain automatically their form under spatial rotations, the laws of relativistic physics will have
automatically the same form in all inertial reference frames, if they are written in

the four-dimensional tensor language, i.e. if they equate four-dimensional tensors to four-dimensional tensors of the same rank. More precisely, if $S^m{}_n$ and $T^m{}_n$ are two such tensors, schematically we have the implication

$$S^m{}_n(x) = T^m{}_n(x) \text{ in } K \;\Rightarrow\; S'^m{}_n(x') = T'^m{}_n(x') \text{ in } K'. \tag{1.30}$$

In fact, thanks to the transformation law (1.28), the equation in K' follows from the one in K, upon multiplying the latter by an appropriate series of matrices Λ and $\widetilde{\Lambda}$. In analogy to what happens in classical mechanics, a physical law written in the tensorial notation (1.30) proves thus to be *manifestly covariant* under Poincaré transformations, and correspondingly it satisfies *evidently* Einstein's relativity principle. This *manifest covariance paradigm* represents the most direct and efficient way to implement the third postulate in any relativistic theory. As a matter of fact this paradigm is *equivalent* to the third postulate, to the extent that there are no known physical laws having the same form in all inertial reference frames, but which cannot be cast in a manifestly covariant form. Given the widespread use we make of tensors in this book, in the next section we introduce the basic elements of the *tensor calculus*.

1.3.1   Tensor Calculus

In this section we introduce the basic operations that can be performed in the space of tensors, namely operations that applied to tensors give rise again to tensors. Obviously, due to the very definition of the Lorentz-group, in the derivations of these operations the constraint (1.12) will play a crucial role. Multiplying it from the right with η, and recalling the definition (1.29), it can be cast also in the equivalent form

$$\Lambda^T \widetilde{\Lambda} = \mathbf{1} = \widetilde{\Lambda}\,\Lambda^T \quad\leftrightarrow\quad \Lambda^\alpha{}_\mu \widetilde{\Lambda}_\alpha{}^\nu = \delta_\mu^\nu = \widetilde{\Lambda}_\mu{}^\alpha \Lambda^\nu{}_\alpha. \tag{1.31}$$

Raising and lowering of indices. A tensor of rank (m, n) can be transformed into a tensor of rank $(m \pm k, n \mp k)$, raising or lowering k indices with the aid of the Minkowski metric. The new tensor is usually denoted by the same symbol as the original one. Starting, for instance, from a tensor $T^{\mu\nu}{}_\rho$ of rank $(2, 1)$, and lowering two indices ($k = 2$), the resulting tensor is of rank $(0, 3)$ and we set

$$T_{\alpha\beta\rho} = \eta_{\alpha\mu}\eta_{\beta\nu}T^{\mu\nu}{}_\rho.$$

A tensor of rank (m, n) is hence for all purposes equivalent to a tensor of rank $(m \pm k, n \mp k)$; for this reason one defines sometimes the rank of a tensor as the integer $m + n$. To illustrate the consistency of the lowering and raising operations we show that, if $T^\mu{}_\nu$ is a tensor of rank $(1, 1)$, the object defined by $T_{\mu\nu} = \eta_{\mu\alpha}T^\alpha{}_\nu$ is indeed a tensor of rank $(0, 2)$, as it entails the correct transformation law (1.28):

$$T'_{\mu\nu} = \eta_{\mu\alpha}T'^{\alpha}{}_{\nu} = \eta_{\mu\alpha}\Lambda^{\alpha}{}_{\beta}\widetilde{\Lambda}_{\nu}{}^{\rho}T^{\beta}{}_{\rho} = \eta_{\mu\alpha}\Lambda^{\alpha}{}_{\beta}\widetilde{\Lambda}_{\nu}{}^{\rho}\eta^{\beta\gamma}T_{\gamma\rho}$$
$$= (\eta_{\mu\alpha}\Lambda^{\alpha}{}_{\beta}\eta^{\beta\gamma})\,\widetilde{\Lambda}_{\nu}{}^{\rho}T_{\gamma\rho} = \widetilde{\Lambda}_{\mu}{}^{\gamma}\widetilde{\Lambda}_{\nu}{}^{\rho}T_{\gamma\rho},$$

where in the last step we used the definition (1.29).

Products of tensors. The product between a tensor $T^{m}{}_{n}$ of rank (m, n) and a tensor $S^{k}{}_{l}$ of rank (k, l) is a tensor of rank $(m + k, n + l)$. This property follows directly from (1.28).

Scalar product and contraction of indices. Given a contravariant vector T^{μ} and a covariant vector U_{ν}, *contracting* their indices one can form the *scalar product* $T^{\mu}U_{\mu}$, which is a *scalar* under Lorentz transformations. Thanks to the constraints (1.31) we find, in fact,

$$T'^{\mu}U'_{\mu} = \Lambda^{\mu}{}_{\nu}T^{\nu}\widetilde{\Lambda}_{\mu}{}^{\rho}U_{\rho} = (\Lambda^{\mu}{}_{\nu}\widetilde{\Lambda}_{\mu}{}^{\rho})T^{\nu}U_{\rho} = \delta^{\rho}_{\nu}T^{\nu}U_{\rho} = T^{\nu}U_{\nu}.$$

We denote the *square* of a (covariant or contravariant) four-vector V by $V^2 = V^{\mu}V_{\mu}$. We say that a vector V is *time-like*, *light-like*, or *space-like*, if $V^2 > 0$, $V^2 = 0$, or $V^2 < 0$, respectively. More generally, starting from a tensor of rank (m, n), contracting k covariant indices with k contravariant indices one obtains a tensor of rank $(m - k, n - k)$. Starting, for example, from a tensor $T^{\mu\nu}{}_{\rho}$ of rank $(2, 1)$, by contracting the second index with the third one, we gain the contravariant vector

$$W^{\mu} = T^{\mu\nu}{}_{\nu}. \tag{1.32}$$

Using (1.31) one verifies in fact easily that $W'^{\mu} = \Lambda^{\mu}{}_{\nu}W^{\nu}$.

Derivative of a tensor field. The derivative of a tensor field of rank (m, n) is a tensor field of rank $(m, n + 1)$. Denoting the partial derivative with the shorthand symbol

$$\partial_{\mu} = \frac{\partial}{\partial x^{\mu}},$$

we will write the derivative of a tensor field $T^{m}{}_{n}$ as

$$\partial_{\mu}T^{\mu_1\cdots\mu_m}{}_{\nu_1\cdots\nu_n}(x).$$

To prove that this object is a tensors of rank $(m, n + 1)$, we must show that the operator ∂_{μ} corresponds to a *covariant* vector, that is to say, that it transforms according to the law

$$\partial'_{\mu} = \widetilde{\Lambda}_{\mu}{}^{\nu}\partial_{\nu}. \tag{1.33}$$

Using the relations (1.5) and (1.31) one finds indeed

$$\partial_{\nu} = \frac{\partial x'^{\alpha}}{\partial x^{\nu}}\,\partial'_{\alpha} = \Lambda^{\alpha}{}_{\nu}\,\partial'_{\alpha} \quad \Rightarrow \quad \widetilde{\Lambda}_{\mu}{}^{\nu}\partial_{\nu} = \widetilde{\Lambda}_{\mu}{}^{\nu}\Lambda^{\alpha}{}_{\nu}\,\partial'_{\alpha} = \delta^{\alpha}_{\mu}\partial'_{\alpha} = \partial'_{\mu}.$$

Symmetries. A double tensor $S^{\mu\nu}$ is called *symmetric* if $S^{\mu\nu} = S^{\nu\mu}$, and a double tensor $A^{\mu\nu}$ is called *antisymmetric* if $A^{\mu\nu} = -A^{\nu\mu}$; these properties are preserved under Lorentz transformations. The double contraction of the product between a symmetric and an antisymmetric tensor vanishes:

$$A^{\mu\nu} S_{\mu\nu} = 0. \tag{1.34}$$

One has, in fact,

$$\Phi = A^{\mu\nu} S_{\mu\nu} = -A^{\nu\mu} S_{\mu\nu} = -A^{\nu\mu} S_{\nu\mu} = -\Phi \quad \Rightarrow \quad \Phi = 0.$$

Given a generic double tensor $T^{\mu\nu}$, one defines its *symmetric* and *antisymmetric* parts, again double tensors, as

$$T^{(\mu\nu)} = \frac{1}{2}\left(T^{\mu\nu} + T^{\nu\mu}\right), \qquad T^{[\mu\nu]} = \frac{1}{2}\left(T^{\mu\nu} - T^{\nu\mu}\right), \tag{1.35}$$

respectively, where the first is a symmetric tensor, and the second an antisymmetric one. Correspondingly, one has the general decomposition

$$T^{\mu\nu} = T^{(\mu\nu)} + T^{[\mu\nu]}.$$

The double contractions between a generic tensor $T^{\mu\nu}$, and a symmetric one $S^{\mu\nu}$ and an antisymmetric one $A^{\mu\nu}$, respectively, fulfil the identities

$$T^{\mu\nu} S_{\mu\nu} = T^{\nu\mu} S_{\mu\nu} = T^{(\mu\nu)} S_{\mu\nu}, \quad T^{\mu\nu} A_{\mu\nu} = -T^{\nu\mu} A_{\mu\nu} = T^{[\mu\nu]} A_{\mu\nu}, \tag{1.36}$$

whose proofs are left as exercises.

Completely antisymmetric tensors. A tensor of rank $(n, 0)$ $A^{\mu_1 \cdots \mu_n}$ is said completely (anti)symmetric, if it is (anti)symmetric under the interchange of an arbitrary pair of its indices, properties that are preserved under Lorentz transformations. The double contraction between a completely symmetric (antisymmetric) tensor of rank $(n, 0)$ and an antisymmetric (symmetric) tensor of rank $(0, 2)$ vanishes; these properties generalize (1.34). The *completely antisymmetric part* of a tensor $T^{\mu_1 \cdots \mu_n}$ of rank $(n, 0)$ is defined as the tensor of the same rank

$$T^{[\mu_1 \cdots \mu_n]} = \frac{1}{n!}\left(T^{\mu_1 \mu_2 \cdots \mu_n} - T^{\mu_2 \mu_1 \cdots \mu_n} + \cdots\right),$$

where within parentheses appear $n!$ terms corresponding to the $n!$ permutations of the indices $\mu_1 \cdots \mu_n$, each term being multiplied by the sign $(-)^p$, where p is the *order* of the permutation. By construction $T^{[\mu_1 \cdots \mu_n]}$ is a completely antisymmetric tensor, and it vanishes if $T^{\mu_1 \cdots \mu_n}$ is symmetric even in a single pair of indices. Finally, the total contraction between a completely antisymmetric tensor $A^{\mu_1 \cdots \mu_n}$ and a generic tensor $T^{\mu_1 \cdots \mu_n}$ fulfils the identity

$$T^{\mu_1\cdots\mu_n}A_{\mu_1\cdots\mu_n} = T^{[\mu_1\cdots\mu_n]}A_{\mu_1\cdots\mu_n}, \tag{1.37}$$

which generalizes the second formula in (1.36). Specular properties hold for the *completely symmetric part* of a tensor of rank $(n, 0)$, denoted by

$$T^{(\mu_1\cdots\mu_n)} = \frac{1}{n!}\left(T^{\mu_1\mu_2\cdots\mu_n} + T^{\mu_2\mu_1\cdots\mu_n} + \cdots\right).$$

Invariant tensors. A constant tensor $T^m{}_n$ is called an *invariant* tensor, if for every $\Lambda \in O(1, 3)$ one has

$$T'^m{}_n = T^m{}_n.$$

The Lorentz group $O(1, 3)$ admits the *fundamental* invariant tensors

$$\eta^{\alpha\beta}, \qquad \eta_{\alpha\beta}, \qquad \varepsilon^{\alpha\beta\gamma\delta},$$

where $\varepsilon^{\alpha\beta\gamma\delta}$ denotes the four-dimensional *Levi-Civita tensor* – a completely anti-symmetric tensor – defined by

$$\varepsilon^{\alpha\beta\gamma\delta} = \begin{cases} 1, & \text{if } \alpha\beta\gamma\delta \text{ is an even permutation of } 0, 1, 2, 3, \\ -1, & \text{if } \alpha\beta\gamma\delta \text{ is an odd permutation of } 0, 1, 2, 3, \\ 0, & \text{if at least two indices are equal.} \end{cases} \tag{1.38}$$

The invariance of the Minkowski metric, and its inverse, follows directly from the constraints (1.12) and (1.13). For $\eta^{\mu\nu}$, for instance, one obtains

$$\eta'^{\alpha\beta} = \Lambda^\alpha{}_\mu \Lambda^\beta{}_\nu \eta^{\mu\nu} = \eta^{\alpha\beta}.$$

The invariance of the Levi-Civita tensor follows, on the other hand, from the *determinant identity*

$$\Lambda^\alpha{}_\mu \Lambda^\beta{}_\nu \Lambda^\gamma{}_\rho \Lambda^\delta{}_\sigma \, \varepsilon^{\mu\nu\rho\sigma} = (\det \Lambda) \, \varepsilon^{\alpha\beta\gamma\delta}, \tag{1.39}$$

holding for an arbitrary 4×4 matrix Λ. Actually, according to (1.15), in general one has $\det \Lambda = \pm 1$, so that as a matter of fact the Levi-Civita tensor is invariant solely under Lorentz transformations for which $\det \Lambda = 1$; we will come back to this point in Sect. 1.4.3. The Levi-Civita tensor obeys furthermore the algebraic identities (see Problem 1.3)

$$\varepsilon^{\mu\nu\rho\sigma}\varepsilon_{\alpha\beta\gamma\delta} = -4! \, \delta^\mu_{[\alpha}\delta^\nu_\beta\delta^\rho_\gamma\delta^\sigma_{\delta]}, \quad \varepsilon^{\mu\nu\rho\sigma}\varepsilon_{\alpha\beta\gamma\sigma} = -3! \, \delta^\mu_{[\alpha}\delta^\nu_\beta\delta^\rho_{\gamma]}, \tag{1.40}$$

$$\varepsilon^{\mu\nu\rho\sigma}\varepsilon_{\alpha\beta\rho\sigma} = -2!2! \, \delta^\mu_{[\alpha}\delta^\nu_{\beta]}, \quad \varepsilon^{\mu\nu\rho\sigma}\varepsilon_{\alpha\nu\rho\sigma} = -3! \, \delta^\mu_\alpha, \quad \varepsilon^{\mu\nu\rho\sigma}\varepsilon_{\mu\nu\rho\sigma} = -4! \tag{1.41}$$

The form of a *generic* invariant tensor is strongly constrained by the following theorem, that can be proven by relying on arguments from *group theory*.

Theorem II. A generic tensor $T^m{}_n$ which is invariant under Lorentz transformations, is necessarily an algebraic combination of the fundamental invariant tensors $\eta^{\alpha\beta}$, $\eta_{\alpha\beta}$ and $\varepsilon^{\alpha\beta\gamma\delta}$.

We illustrate the theorem with some examples.

(a) There exist no invariant tensors with *odd* total rank $m + n$. In fact, since the rank of the Minkowski metric as well as of the Levi-Civita tensor is *even*, an arbitrary algebraic combination of them is an even-rank tensor. In particular there are no invariant vectors or invariant rank-three tensors.

(b) An invariant double tensor $T^{\mu\nu}$ is necessarily of the form $T^{\mu\nu} = b\,\eta^{\mu\nu}$, with b a constant. In fact, $\eta^{\mu\nu}$ is the unique non-vanishing algebraic combination of rank $(2,0)$, that can be formed with $\eta^{\alpha\beta}$, $\eta_{\alpha\beta}$ and $\varepsilon^{\alpha\beta\gamma\delta}$. In the same way an invariant double tensor $T^\mu{}_\nu$ is necessarily of the form $T^\mu{}_\nu = b\,\delta^\mu_\nu$; notice that the *Kronecker symbol* can written as $\delta^\mu_\nu = \eta^{\mu\alpha}\eta_{\alpha\nu}$.

(c) The most general form of an invariant tensor $T^{\alpha\beta\gamma\delta}$ of rank $(4,0)$ is

$$T^{\alpha\beta\gamma\delta} = a_1\,\varepsilon^{\alpha\beta\gamma\delta} + a_2\,\eta^{\alpha\beta}\eta^{\gamma\delta} + a_3\,\eta^{\alpha\gamma}\eta^{\beta\delta} + a_4\,\eta^{\alpha\delta}\eta^{\beta\gamma},$$

where a_1, a_2, a_3 and a_4 are constants. If moreover it is known, for example, that $T^{\alpha\beta\gamma\delta}$ is antisymmetric in α and β, it follows in addition that $a_2 = 0$ and $a_3 = -a_4$. If, on the other hand, it is known that $T^{\alpha\beta\gamma\delta}$ is symmetric in α and β, then we must have $a_1 = 0$ and $a_3 = a_4$.

1.4   Structure of the Lorentz Group

In this section we analyze the structure of the Lorentz group $O(1,3)$, formed by all matrices Λ satisfying the constraint (1.12). On one hand, our aim is to find an explicit parameterization for a generic matrix Λ subject to this constraint, and, on the other, we want to identify the physical operations – connecting two inertial reference frames – associated with these matrices, thereby answering the question left open in Sect. 1.2.2. As we shall see, for this purpose it will be particularly useful to perform a detailed analysis of the Lorentz transformations close to the identity.

1.4.1   The Proper Lorentz Group

We begin the analysis of the Lorentz group with the observation that the constraint (1.12) implies the conditions

$$|\det \Lambda| = 1, \qquad |\Lambda^0{}_0| \geq 1.$$

The first condition has been obtained previously, see (1.15), and the second can be derived by setting in (1.12) $\alpha = \beta = 0$:

$$1 = (\Lambda^0{}_0)^2 - \Lambda^i{}_0\Lambda^i{}_0 \quad \Rightarrow \quad (\Lambda^0{}_0)^2 = 1 + |\mathbf{L}|^2, \quad \text{where } L^i \equiv \Lambda^i{}_0. \quad (1.42)$$

It follows that $|\Lambda^0{}_0| \geq 1$. If a matrix $\Lambda^\mu{}_\nu$ belongs to $O(1,3)$ we have, therefore, $\Lambda^0{}_0 \geq 1$ or $\Lambda^0{}_0 \leq -1$, and $\det \Lambda = 1$ or $\det \Lambda = -1$. Consequently, the Lorentz group (1.14) breaks up into four disjoint subsets,

$$O(1,3) = SO(1,3)_c \cup \Sigma_1 \cup \Sigma_2 \cup \Sigma_3, \quad (1.43)$$

where

$$SO(1,3)_c = \{\Lambda \in O(1,3)/ \det \Lambda = 1, \Lambda^0{}_0 \geq 1\}, \quad (1.44)$$
$$\Sigma_1 = \{\Lambda \in O(1,3)/ \det \Lambda = -1, \Lambda^0{}_0 \geq 1\}, \quad (1.45)$$
$$\Sigma_2 = \{\Lambda \in O(1,3)/ \det \Lambda = -1, \Lambda^0{}_0 \leq -1\}, \quad (1.46)$$
$$\Sigma_3 = \{\Lambda \in O(1,3)/ \det \Lambda = 1, \Lambda^0{}_0 \leq -1\}. \quad (1.47)$$

Out of these subsets only $SO(1,3)_c$ forms a *subgroup* of $O(1,3)$, which is called the *proper Lorentz group*. As mentioned earlier, the symbol "S" refers to the fact that the determinant of the matrices is equal to 1, and the subscript "c" signals that the proper Lorentz group appears to be continuously *connected* to the identity matrix – contrary to $O(1,3)$. In Sect. 1.4.3 we will see that each subset Σ_i $(i = 1, 2, 3)$ can be retrieved multiplying each element of $SO(1,3)_c$ with a fixed element Λ_i of Σ_i. For this reason, in what follows we shall analyze in detail the *proper* Lorentz group, postponing the analysis of the sets Σ_i to Sect. 1.4.3. We know already two classes of important elements of $SO(1,3)_c$. The first one is represented by the spatial *rotations*, corresponding to matrices Λ with elements

$$\Lambda^0{}_0 = 1, \quad \Lambda^i{}_j = R^i{}_j, \quad \Lambda^0{}_i = 0 = \Lambda^i{}_0,$$

where the rotation matrices R satisfy the relations $R^T R = \mathbf{1}$ and $\det R = 1$, i.e. $R \in SO(3)$. It is indeed immediately checked that the matrices Λ defined in this way satisfy the constraint (1.12). In this respect we recall that a generic rotation depends on *three* arbitrary parameters, which can be identified, for example, with the three *Euler angles*. A second class of important elements of $SO(1,3)_c$ is represented by the *special Lorentz transformations*, also called *boosts*, corresponding to a uniform linear motion of an inertial reference frame with respect to another. If the motion takes place along the x axis with velocity v, the space-time coordinates of the two frames are related by the known relations

$$t' = \gamma(t - vx), \quad x' = \gamma(x - vt), \quad y' = y, \quad z' = z, \quad (1.48)$$

where $\gamma = 1/\sqrt{1 - v^2}$, corresponding to the matrix

$$\Lambda = \begin{pmatrix} \gamma & -v\gamma & 0 & 0 \\ -v\gamma & \gamma & 0 & 0 \\ 0 & 0 & 1 & 0 \\ 0 & 0 & 0 & 1 \end{pmatrix}. \tag{1.49}$$

Again it is easy to verify that the constraint $\Lambda^T \eta \Lambda = \eta$ is satisfied. In general we can perform a special Lorentz transformation with velocity $\mathbf{v}$ in an arbitrary direction, and consequently the corresponding matrix Λ depends on *three* independent parameters, namely the components (v^x, v^y, v^z) of the vector $\mathbf{v}$.

The operations introduced so far – spatial rotations and special Lorentz transformations – involve altogether 6 parameters, and they are obviously continuously connected to the identity. We expect therefore that the 16 elements of a generic matrix $\Lambda \in SO(1,3)_c$ can be expressed in terms of 6 independent parameters. In other words, the *Lie group* $SO(1,3)_c$ ought to have dimension 6. To verify the correctness of this prediction we rewrite the constraint (1.12) in the form

$$H \equiv \Lambda^T \eta \Lambda - \eta = 0, \tag{1.50}$$

which amounts to a system of 16 equations in the 16 unknowns $\Lambda^\mu{}_\nu$. However, by construction H is a 4×4 *symmetric* matrix, so that only 10 out of those 16 equations are linearly independent. The general solution $\Lambda^\mu{}_\nu$ of the system (1.50) can therefore be expressed in terms of just $16 - 10 = 6$ independent parameters.

1.4.2 Finite and Infinitesimal Proper Lorentz Transformations

To determine a possible choice of these six parameters we begin by considering a generic Lorentz transformation close to the identity

$$\Lambda^\mu{}_\nu = \delta^\mu{}_\nu + \Omega^\mu{}_\nu, \qquad |\Omega^\mu{}_\nu| \ll 1, \ \forall \mu, \forall \nu.$$

Enforcing the constraint (1.50), that is equivalent to the constraint (1.12), we obtain the relation

$$(\delta^\alpha{}_\mu + \Omega^\alpha{}_\mu)\, \eta_{\alpha\beta} \left(\delta^\beta{}_\nu + \Omega^\beta{}_\nu\right) - \eta_{\mu\nu} = 0.$$

Keeping only the terms linear in $\Omega^\mu{}_\nu$, we deduce that this matrix must satisfy, in turn, the constraint

$$\eta_{\nu\alpha}\Omega^\alpha{}_\mu + \eta_{\mu\beta}\Omega^\beta{}_\nu = 0. \tag{1.51}$$

Introducing the matrix

$$\omega_{\mu\nu} = \eta_{\mu\beta}\Omega^\beta{}_\nu, \quad \text{which amounts to set } \Omega^\mu{}_\nu = \eta^{\mu\alpha}\omega_{\alpha\nu}, \tag{1.52}$$

the condition (1.51) translates into

$$\omega_{\mu\nu} = -\omega_{\nu\mu}. \tag{1.53}$$

The matrix $\omega_{\mu\nu}$ must thus be antisymmetric, and as such it has only six independent elements. Therefore, in agreement with the conclusion of the preceding section we infer that a generic Lorentz transformation close to the identity, that is to say, an *infinitesimal* Lorentz transformation, depends on six arbitrary parameters as it can be written as

$$\Lambda^{\mu}{}_{\nu} = \delta^{\mu}{}_{\nu} + \eta^{\mu\alpha}\omega_{\alpha\nu}. \tag{1.54}$$

We are now in a position to furnish also the explicit expression of a generic *finite* element Λ of $SO(1,3)_c$, thanks to the following theorem.

Theorem III. A generic element $\Lambda \in SO(1,3)_c$ can be expressed as the exponential

$$\Lambda = e^{\Omega}, \tag{1.55}$$

where the matrix Ω satisfies the constraint (1.51) or, equivalently, the matrix defined as $\omega = \eta\Omega$ is antisymmetric.

Proof In what follows we limit ourselves to prove that, if Ω satisfies (1.51), the matrices of the form (1.55) belong to the proper Lorentz group. For this purpose we must first of all prove that these matrices belong to the Lorentz group, i.e. that they satisfy the constraint $\Lambda^T \eta \Lambda = \eta$. In view of this it is convenient to rewrite (1.51) in matrix notation as

$$\eta\Omega = -\Omega^T\eta \quad \leftrightarrow \quad \Omega^T = -\eta\Omega\eta,$$

and to use then the identity (remember that $\eta^2 = 1$)

$$e^{-\eta\Omega\eta} = \sum_{N=0}^{\infty} \frac{(-)^N}{N!}(\eta\,\Omega\,\eta)^N = \sum_{N=0}^{\infty} \frac{(-)^N}{N!}\eta\,\Omega^N\eta$$

$$= \eta\left(\sum_{N=0}^{\infty} \frac{(-)^N}{N!}\Omega^N\right)\eta = \eta\,e^{-\Omega}\,\eta.$$

It then follows that

$$\Lambda^T\eta\Lambda = e^{\Omega^T}\eta\,e^{\Omega} = e^{-\eta\Omega\eta}\eta\,e^{\Omega} = \left(\eta\,e^{-\Omega}\,\eta\right)\eta\,e^{\Omega} = \eta,$$

meaning that $\Lambda \in O(1,3)$. Given that the exponential of a matrix is a *continuous* function of its elements, the set of matrices $\{\Lambda = e^{\Omega}\}$ is continuously connected to the identity matrix, and – since these matrices belong to the Lorentz group – they are elements of $SO(1,3)_c$. $\square$

We conclude this section with an analysis of the physical meaning of the six parameters $\omega_{\mu\nu}$. Each of these parameters should in fact be related to one of the six operations introduced in Sect. 1.4.1, connecting one inertial reference frame with another. Consider then a generic infinitesimal Lorentz transformation (1.54), from an inertial reference frame K to an inertial reference frame K'. Thanks to the antisymmetry constraint (1.53) we can then set in full generality

$$\omega_{00} = 0, \tag{1.56}$$

$$\omega_{i0} = V^i = -\omega_{0i}, \tag{1.57}$$

$$\omega_{ij} = \varphi\, \varepsilon^{ijk} u^k, \quad |\mathbf{u}| = 1. \tag{1.58}$$

To analyze the meaning of the six quantities $\mathbf{V}$, $\mathbf{u}$ and φ, we write out the infinitesimal coordinate transformations induced by the matrix (1.54)

$$x'^{\mu} = \Lambda^{\mu}{}_{\nu} x^{\nu} = x^{\mu} + \eta^{\mu\alpha} \omega_{\alpha\nu} x^{\nu},$$

which amount to

$$t' = t + \eta^{00} \omega_{0i}\, x^i = t - \mathbf{V} \cdot \mathbf{x}, \tag{1.59}$$

$$x'^i = x^i + \eta^{ij}(\omega_{j\,0} t + \omega_{j\,k} x^k) = x^i - V^i t + \varphi\,(\mathbf{u} \times \mathbf{x})^i. \tag{1.60}$$

For $\mathbf{V} = 0$ these transformations reduce to an infinitesimal spatial rotation of an angle φ around the direction $\mathbf{u}$, while for $\varphi = 0$ they reduce to an infinitesimal special Lorentz transformation[1] with relative velocity $\mathbf{V}$, see (1.48). The axes of K' appear thus rotated with respect to the ones of K by an infinitesimal angle φ around $\mathbf{u}$, and furthermore K' is moving with respect to K with a constant infinitesimal velocity $\mathbf{V}$ – in agreement with the general analysis of Sect. 1.4.1.

Finite special Lorentz transformations. Finally, we want to show in which way the *finite* special proper Lorentz transformation (1.49) can be retrieved from the general formula (1.55). Since (1.49) is a special Lorentz transformation along the x axis, in the general parameterization (1.56)–(1.58) we must set $\varphi = 0$ and $\mathbf{V} = (V(v), 0, 0)$, where $\mathbf{v} = (v, 0, 0)$ is the *finite* velocity of K' with respect to K. Moreover, since $V(v)$ represents the infinitesimal velocity, it is tied to v by the relation $V(v) = v + O(v^2)$. The non-vanishing elements of $\omega_{\mu\nu}$ are thus

$$\omega_{10} = V(v) = -\omega_{01},$$

so that from (1.52) we obtain the non-vanishing elements of Ω

$$\Omega^0{}_1 = -V(v) = \Omega^1{}_0. \tag{1.61}$$

---

[1] In equations (1.59) and (1.60) the factors $1/\sqrt{1 - V^2}$ are absent, since these transformation laws are valid only at first order in $\omega_{\mu\nu}$, i.e. in $\mathbf{V}$ and φ.

The explicit evaluation of e^{Ω} can be accomplished most conveniently through a Taylor-series expansion of the exponential, see Problem 1.7, and the result is

$$
e^{\Omega} = \begin{pmatrix} \cosh V(v) & -\sinh V(v) & 0 & 0 \\ -\sinh V(v) & \cosh V(v) & 0 & 0 \\ 0 & 0 & 1 & 0 \\ 0 & 0 & 0 & 1 \end{pmatrix}. \tag{1.62}
$$

This matrix matches actually with the expected matrix (1.49) if we set $\tanh V(v) = v/c$, i.e.

$$
V(v) = \operatorname{arctan}\left(\frac{v}{c}\right),
$$

where we restored the speed of light. From the particular form of the matrix (1.62) we recognize that a special Lorentz transformation with velocity v along the x axis, can be interpreted as a *hyperbolic rotation* of an "angle" $\operatorname{arctan}(v/c)$ in the (ct, x)-plane.

1.4.3   Parity, Time Reversal, and Pseudotensors

It remains to supply a physical interpretation for the three subsets Σ_i of the Lorentz group, introduced in (1.45)–(1.47). As anticipated in Sect. 1.4.1, these subsets can be retrieved from the proper Lorentz group, multiplying all elements of $SO(1,3)_c$ with a fixed matrix $\Lambda_i \in \Sigma_i$. The simplest choice for these fixed matrices is (see Problem 1.10)

$$
\Lambda_1 = \mathcal{P}, \qquad \Lambda_2 = \mathcal{T}, \qquad \Lambda_3 = -\mathbf{1}, \tag{1.63}
$$

where $\mathcal{P}$ is the matrix associated with the *parity* operation, with elements

$$
\mathcal{P}^0{}_0 = 1, \quad \mathcal{P}^i{}_j = -\delta^i{}_j, \quad \mathcal{P}^0{}_i = 0 = \mathcal{P}^i{}_0, \tag{1.64}
$$

and $\mathcal{T}$ is the matrix associated with the *time reversal* operation, with elements

$$
\mathcal{T}^0{}_0 = -1, \quad \mathcal{T}^i{}_j = \delta^i{}_j, \quad \mathcal{T}^0{}_i = 0 = \mathcal{T}^i{}_0. \tag{1.65}
$$

Thanks to the group relation $\Lambda_3 = \mathcal{PT}$, it is sufficient to analyze the meaning of the operations associated with parity and with time reversal. Usually one refers to the particular elements $\mathcal{P}$, $\mathcal{T}$ and $\mathcal{PT}$ of the Lorentz group as *discrete symmetries*.

Parity. The parity transformation $x'^{\mu} = \mathcal{P}^{\mu}{}_{\nu} x^{\nu}$ reflects all three[2] Cartesian axes, while leaving time invariant:

---

[2] A transformation reflecting *two* axes, say the x and y axes, corresponds to a *rotation* of $180°$ around the z axis, and belongs hence to $SO(1,3)_c$. On the contrary, the reflection of a *single* axis represents an operation that belongs to Σ_1, and that can be thought of as composed by $\mathcal{P}$ and by a rotation of $180°$ around the same axis, an operation belonging to $SO(1,3)_c$.

$$t' = t, \qquad \mathbf{x}' = -\mathbf{x}.$$

Under parity the Minkowski metric $\eta^{\mu\nu}$ remains invariant, i.e.

$$\mathcal{P}^\alpha{}_\mu \mathcal{P}^\beta{}_\nu \eta^{\mu\nu} = \eta^{\alpha\beta},$$

simply because $\mathcal{P} \in O(1,3)$, see (1.13). On the other hand, in virtue of the determinant identity (1.39) and of the relation $\det \mathcal{P} = -1$, under parity the Levi-Civita tensor changes its *sign*:

$$\mathcal{P}^\alpha{}_\mu \mathcal{P}^\beta{}_\nu \mathcal{P}^\gamma{}_\rho \mathcal{P}^\delta{}_\sigma \, \varepsilon^{\mu\nu\rho\sigma} = -\varepsilon^{\alpha\beta\gamma\delta}. \tag{1.66}$$

Accordingly, the Levi-Civita tensor constitutes an invariant *pseudotensor*. In general, we call an object $T^m{}_n$ a *pseudotensor under parity*, if under $SO(1,3)_c$ it transforms as in (1.28), while under parity it transforms as in (1.28) modulo a *minus sign*, namely

$$T'^{\mu_1 \cdots \mu_m}{}_{\nu_1 \cdots \nu_n}(\mathcal{P}x) = -\mathcal{P}^{\mu_1}{}_{\alpha_1} \cdots \mathcal{P}^{\mu_m}{}_{\alpha_m} \widetilde{\mathcal{P}}_{\nu_1}{}^{\beta_1} \cdots \widetilde{\mathcal{P}}_{\nu_n}{}^{\beta_n} T^{\alpha_1 \cdots \alpha_m}{}_{\beta_1 \cdots \beta_n}(x). \tag{1.67}$$

In this case we have of course $\widetilde{\mathcal{P}} = \eta \mathcal{P} \eta = \mathcal{P}$, see (1.29). It is of crucial importance that the *modified* transformation law (1.67) respects the group property: the composition of two such transformations amounts, indeed, to the identity map, in agreement with the corresponding composition law of $O(1,3)$, namely $\mathcal{P}\mathcal{P} = \mathbb{1}$. Notice, in particular, that (1.67) implies that the *product* of two pseudotensors is again a *tensor*. Due to the modified transformation law (1.67), and to the identity (1.66), the invariance of $\varepsilon^{\alpha\beta\gamma\delta}$ as a pseudotensor under parity is expressed simply by the equality

$$\varepsilon'^{\alpha\beta\gamma\delta} = \varepsilon^{\alpha\beta\gamma\delta}.$$

Starting from the Levi-Civita pseudotensor we can build additional pseudotensors. If, for example, $T^{\alpha\beta}$ is a double tensor, then $\varepsilon^{\alpha\beta\gamma\delta} T_{\gamma\delta}$ is a double pseudotensor, and $\varepsilon^{\alpha\beta\gamma\delta} T_{\alpha\beta} T_{\gamma\delta}$ is a pseudoscalar. In fact, since under parity $T^{\alpha\beta}$ transforms according to $T'^{\alpha\beta} = \mathcal{P}^\alpha{}_\mu \mathcal{P}^\beta{}_\nu T^{\mu\nu}$, we have the transformation laws

$$\begin{aligned}
(\varepsilon^{\alpha\beta\gamma\delta} T_{\gamma\delta})' &= \varepsilon^{\alpha\beta\gamma\delta} T'_{\gamma\delta} = -\mathcal{P}^\alpha{}_\rho \mathcal{P}^\beta{}_\sigma (\varepsilon^{\rho\sigma\gamma\delta} T_{\gamma\delta}), \\
(\varepsilon^{\alpha\beta\gamma\delta} T_{\alpha\beta} T_{\gamma\delta})' &= \varepsilon^{\alpha\beta\gamma\delta} T'_{\alpha\beta} T'_{\gamma\delta} = -\varepsilon^{\alpha\beta\gamma\delta} T_{\alpha\beta} T_{\gamma\delta},
\end{aligned} \tag{1.68}$$

where for simplicity we omitted the dependence on the space-time coordinates.

As previously established, physical laws that equate tensors to tensors are invariant under the complete Lorentz group $O(1,3)$. We can now generalize this manifest covariance paradigm, by asserting that also physical laws that equate pseudotensors to pseudotensors are invariant under $O(1,3)$, while laws that equate a tensor to a pseudotensor violate parity, being invariant only under the proper Lorentz group $SO(1,3)_c$. This raises naturally the question of whether the laws of physics are, or should be, invariant under the complete Lorentz group, or just under its proper subgroup. According to our current knowledge concerning the microscopic proper-ties of nature, the answer to this question is that the laws that govern the electro-magnetic, gravitational, and strong interactions respect indeed the complete Lorentz group, while – as Chien-Shiung Wu discovered in 1957 analyzing the characteris-tics of the *beta decay* [1] – the laws that govern the *weak interactions violate parity invariance*.

Time reversal. The other independent discrete symmetry of the Lorentz group is the time reversal transformation $x'^\mu = T^\mu{}_\nu x^\nu$, which reflects the time axis, while leaving the space coordinates invariant:

$$t' = -t, \qquad \mathbf{x}' = \mathbf{x}.$$

This operation entails features similar to the ones of parity illustrated above. In particular, under time reversal the Minkowski metric remains invariant,

$$T^\alpha{}_\mu T^\beta{}_\nu \eta^{\mu\nu} = \eta^{\alpha\beta},$$

while, given that $\det T = -1$, the Levi-Civita tensor changes its *sign*:

$$T^\alpha{}_\mu T^\beta{}_\nu T^\gamma{}_\rho T^\delta{}_\sigma \, \varepsilon^{\mu\nu\rho\sigma} = -\varepsilon^{\alpha\beta\gamma\delta}.$$

Consequently, $\varepsilon^{\alpha\beta\gamma\delta}$ is an invariant *pseudotensor under time reversal*. By definition, an object $T^m{}_n$ is a *pseudotensor under time reversal*, if under $SO(1,3)_c$ it trans-forms as indicated in (1.28), whereas under time reversal it transforms as indicated in (1.67) with $\mathcal{P}$ replaced by $\mathcal{T}$. As in the case of parity, physical laws equating a tensor to a *pseudotensor under time reversal* would violate time-reversal invariance. From the experiments conducted in 1964 by J. Cronin and V. Fitch on the decay modes of the neutral K *mesons* [2], we know now that this discrete symmetry is in reality *violated by the weak interactions*,[3] whereas, according to our present-day knowledge of the microscopic laws of nature, it is preserved by the other three fundamental interactions. Without going into the details of these experiments, we observe that the violation in nature of time-reversal invariance has important physi-cal consequences, one of the most striking being that this violation is indispensable in order to explain the *asymmetry* between matter and antimatter in our Universe. To prevent a possible confusion we anticipate that the *spontaneous* violation of time-reversal invariance that we shall encounter in electrodynamics in Sect. 6.2.3 does by no means regard the fundamental *equations* of electrodynamics – which are invari-ant – but rather their *solutions*.

---

[3]The experiments conducted by Cronin and Fitch revealed, actually, that the weak interactions break the $\mathcal{CP}$-symmetry, where the symbol "$\mathcal{C}$" stands for *charge conjugation*. But since, according to the $\mathcal{PCT}$ *theorem* [3], in a Lorentz-invariant world the *product* of the three discrete transforma-tions $\mathcal{P}, \mathcal{C}$ and $\mathcal{T}$ is always a preserved symmetry, a violation of $\mathcal{CP}$ implies necessarily a violation of $\mathcal{T}$.

1.5 Problems

1.1 Using the techniques introduced in Sect. 1.4, write a generic matrix R belonging to the rotation group

$$SO(3) = \{\text{real } 3 \times 3 \text{ matrices } R \,/\, R^T R = 1,\ \det R = 1\},$$

in terms of three independent parameters.

1.2 Prove that the object W^μ defined in (1.32) is a contravariant vector.

1.3 Show that the Levi-Civita tensor satisfies the algebraic identities between invariant tensors (1.40) and (1.41).
Hint: Verify first the last identity in (1.41), and then apply Theorem II of Sect. 1.3.1.

1.4 Prove the relations (1.36).

1.5 Verify that the matrix Λ given in (1.49) satisfies the constraint (1.12).

1.6 Given a generic tensor $T^{\mu\nu\rho}$ of rank $(3,0)$, prove the double implication

$$T^{[\mu\nu\rho]} = 0 \quad \Leftrightarrow \quad \varepsilon_{\mu\nu\rho\sigma} T^{\mu\nu\rho} = 0.$$

1.7 Consider the matrix $\Omega^\mu{}_\nu$ corresponding to the elements (1.61),

$$\Omega = \begin{pmatrix} 0 & -V(v) & 0 & 0 \\ -V(v) & 0 & 0 & 0 \\ 0 & 0 & 0 & 0 \\ 0 & 0 & 0 & 0 \end{pmatrix}.$$

Show that the exponential e^Ω equals the matrix given in (1.62).
Hint: Expand the exponential in a Taylor series, and notice that the matrix

$$M = \begin{pmatrix} 0 & 1 \\ 1 & 0 \end{pmatrix}$$

satisfies for every positive integer n the algebraic identities

$$M^{2n} = \begin{pmatrix} 1 & 0 \\ 0 & 1 \end{pmatrix}, \qquad M^{2n+1} = M.$$

1.8 Consider a generic rotation matrix $R \in SO(3)$, see Problem 1.1.

(a) Verify that under a rotation the *inertia tensor* (1.23) transforms as indicated in (1.24).
Hint: The relation $R^T R = 1$ is equivalent to $R^i{}_k R^j{}_k = \delta^{ij}$.

(b) Prove that under a rotation the angular momentum (1.22) of a rigid body transforms according to $\mathcal{L}'^i = R^i{}_j \mathcal{L}^j$.
(c) Prove that under a rotation the angular momentum of a particle $L^i = m\varepsilon^{ijk} r^j v^k$
transforms according to $L'^i = R^i{}_j L^j$.
Hint: Resort to the *determinant identity*

$$\varepsilon^{jnl} R^i{}_j R^m{}_n R^k{}_l = (\det R)\, \varepsilon^{imk}, \tag{1.69}$$

holding for an arbitrary 3×3 matrix R.
(d) Show that under a rotation *Euler's second law* (1.20) goes over to equation
(1.25).

1.9 Show that the set of matrices $O(1,3)$ defined in (1.14) forms a *group* – the
Lorentz group – proving in particular that

(a) if $\Lambda_1 \in O(1,3)$ and $\Lambda_2 \in O(1,3)$, also $\Lambda_1 \Lambda_2 \in O(1,3)$;
(b) if $\Lambda \in O(1,3)$, also $\Lambda^{-1} \in O(1,3)$.

1.10 Consider the three disjoint subsets Σ_i of the Lorentz group given in (1.45)–
(1.47). Show that an element $\Lambda \in \Sigma_i$ can be written in a unique way as $\Lambda = \Lambda_i \Lambda_0$,
where the matrices Λ_i are given in (1.63), and Λ_0 is an element of $SO(1,3)_c$. Proceed as indicated below.

(a) Notice that if two matrices B_1 and B_2 satisfy $\det B_r = 1$ $(r = 1, 2)$, it follows
that $\det(B_1 B_2) = 1$. Analogous properties hold if $\det B_r = \pm 1$.
(b) Prove that if two matrices B_1 and B_2 belong to $O(1,3)$, and satisfy the inequality $(B_r)^0{}_0 \geq 1$ $(r = 1, 2)$, then also the matrix $C = B_1 B_2$ satisfies the inequality $C^0{}_0 \geq 1$. Analogous properties hold if one of the two matrices, or both, satisfy instead the inequality $(B_r)^0{}_0 \leq -1$.
Hint: Use (1.42) and notice that, thanks to the decomposition (1.43), it suffices
to show that $C^0{}_0 \geq 0$.
(c) In order to prove the main assertion show that, if $\Lambda \in \Sigma_i$, then the matrix $\Lambda_0 =
\Lambda_i^{-1}\Lambda = \Lambda_i \Lambda$ belongs to $SO(1,3)_c$.

References

1. C.S. Wu, E. Ambler, R.W. Hayward, D.D. Hoppes, R.P. Hudson, Experimental test of parity
   conservation in beta decay. Phys. Rev. **105**, 1413 (1957)
2. J.H. Christenson, J.W. Cronin, V.L. Fitch, R. Turlay, Evidence for the 2π decay of the K_2^0
   meson. Phys. Rev. Lett. **13**, 138 (1964)
3. R.F. Streater, A.S. Wightman, *PCT, Spin and Statistics, and All That* (Princeton University
   Press, Princeton, 2000)

Chapter 2
The Fundamental Equations
of Electrodynamics

In this chapter we present the *fundamental equations* that govern the dynamics of
a system of charged particles interacting with the electromagnetic field, namely
the *Maxwell and Lorentz equations*, clarifying their role and analyzing their general characteristics. For the reasons explained in the previous chapter we will write
these equations in a *manifestly covariant* form, as well as in the standard three-
dimensional notation. We will in particular highlight their *distributional nature* and
derive the main conservation laws they imply. A considerable part of the book will
then be devoted to a detailed analysis of the solutions and physical consequences
of these equations. We begin the chapter with a description of the kinematics of a
relativistic particle.

2.1 Kinematics of a Relativistic Particle

Causal world lines. In Newtonian mechanics the motion of a particle is described by
the time-dependent position vector $\mathbf{y}(t) = (x^1(t), x^2(t), x^3(t))$, describing a three-
dimensional curve.[1] In a relativistic theory – according to the manifest covariance
paradigm – it is mandatory to introduce a four-dimensional position vector describing a four-dimensional curve, called the *world line* of the particle. It is represented
by the four functions of a generic real parameter λ

$$y^\mu(\lambda) = (y^0(\lambda), \mathbf{y}(\lambda)).$$

---

[1] In textbooks the time-dependent position vector of a particle is usually denoted by the symbol
$\mathbf{x}(t)$. We prefer the notations $\mathbf{y}(t)$ and $y^\mu(\lambda)$ – in place of $\mathbf{x}(t)$ and $x^\mu(\lambda)$ – in order to avoid
confusion with the symbol $x^\mu = (t, \mathbf{x})$ denoting a generic space-time point, as for instance the
argument x of the electromagnetic field $F^{\mu\nu}(x)$.

© Springer International Publishing AG, part of Springer Nature 2018
K. Lechner, *Classical Electrodynamics*, UNITEXT for Physics,
https://doi.org/10.1007/978-3-319-91809-9_2

We shall assume that these functions are sufficiently regular, to be more precise, of class C^2. A world line $y^\mu(\lambda)$ represents a *physically acceptable* motion, if its *tangent vector*

$$U^\mu(\lambda) = \frac{dy^\mu(\lambda)}{d\lambda}$$

satisfies for every λ the conditions:

(1) $U^2 \geq 0$,
(2) $U^0 > 0$.

A world line satisfying condition (1) is called *causal* and, as we will see below in (2.1), for such a trajectory the speed of the particle is always less than or equal to the speed of light. Condition (2) ensures instead that $y^0(\lambda)$ – the time coordinate – is a monotonically *increasing* function of λ, a property whose meaning will be clarified shortly. A world line obeying condition (2) is said to be *future directed*, whereas if this condition is replaced with $U^0 < 0$ it is said to be *past directed*. The *light cone* of $\mathbb{R}^4$ is defined as the set of all four-vectors U^μ satisfying the equation $U^2 = 0$. From a geometric point of view, condition (1) then selects for each λ the *interior* of the light cone, while the addition of condition (2) singles out the forward half of this cone, i.e. the *future light cone*. From now on we shall assume that the world line swept out by an arbitrary relativistic particle is *causal* and *future directed*, that is to say, that its tangent vector U^μ belongs for every λ to the interior of the future light cone.

Since $y^0(\lambda)$ is a monotonically increasing function, it can be inverted so as to determine λ as a function of time:

$$y^0(\lambda) = t \quad \Rightarrow \quad \lambda(t).$$

The position vector $\mathbf{y}(t)$ can then be obtained by replacing in the spatial world line $\mathbf{y}(\lambda)$ the parameter λ with $\lambda(t)$. We will use the simplified notation

$$\mathbf{y}(\lambda(t)) = \mathbf{y}(t).$$

In the following we will denote the three-dimensional spatial velocity and acceleration, as usual, by

$$\mathbf{v} = \frac{d\mathbf{y}}{dt}, \qquad \mathbf{a} = \frac{d\mathbf{v}}{dt}.$$

Writing out the causality condition (1) we obtain finally the inequality

$$U^2 = \frac{dy^\mu}{d\lambda}\frac{dy_\mu}{d\lambda} = \left(\frac{dt}{d\lambda}\right)^2 \frac{dy^\mu}{dt}\frac{dy_\mu}{dt} = \left(\frac{dt}{d\lambda}\right)^2 (1 - v^2) \geq 0, \tag{2.1}$$

stating that the speed of a particle can never exceed the speed of light.

Reparameterization invariance. If compared to Newtonian mechanics, it could seem that the relativistic world line introduces in the dynamics of the particle a *fourth* degree of freedom: the function $y^0(\lambda)$. This degree of freedom is, however, *spurious*, i.e. not physically observable, in that it reflects merely the arbitrariness in the choice of the parameter λ. Two world lines $y_1^\mu(\lambda)$ and $y_2^\mu(\lambda)$ are, in fact, physically equivalent if they can be connected by means of a redefinition of the parameter, namely if there exists a function $f : \mathbb{R} \to \mathbb{R}$, invertible and of class C^2 together with its inverse, such that

$$y_1^\mu(f(\lambda)) = y_2^\mu(\lambda). \tag{2.2}$$

In this case one says that the two world lines are connected by a *reparameterization*. The physical equivalence of these world lines follows from the fact that, thanks to the four relations (2.2), the trajectories associated with them are the same:

$$\mathbf{y}_1(t) = \mathbf{y}_2(t).$$

For this reason, the dynamics of a relativistic particle can be described in terms of a world line, in place of the three-dimensional trajectory, provided that the equations of motion are *reparameterization invariant*. Notice that the trajectory $\mathbf{y}(t)$ itself – an observable quantity – is reparameterization invariant, while the *functions* $\mathbf{y}(\lambda)$ and $y^0(\lambda)$ are not.

If all physical laws are reparameterization invariant, we can choose the parameter we prefer. A choice with a direct physical interpretation, to which we will frequently resort, is the $\mu = 0$ component of the world line, i.e. the time $\lambda = y^0(\lambda) = t$. In this case the world line is parameterized as

$$y^\mu(t) = (t, \mathbf{y}(t)). \tag{2.3}$$

Another convenient choice is the so-called *proper time* s, which has the advantage of being invariant under Lorentz transformations as well as under reparameterizations. Formally it is given by the symbolic expression

$$ds = \sqrt{dy^\mu\, dy_\mu}, \tag{2.4}$$

which represents a shorthand notation for the Lorentz *scalar*

$$s(\lambda) = \int_0^\lambda \sqrt{\frac{dy^\mu}{d\lambda'} \frac{dy_\mu}{d\lambda'}}\, d\lambda'. \tag{2.5}$$

The Lorentz invariance of s is in fact manifest, and its reparameterization invariance follows from the fact that in (2.5) the factors $d\lambda'$ formally cancel out. Notice also that, owing to the causality condition (1), the radicand in (2.5) is never negative. The proper time permits, in particular, to introduce the Lorentz and reparameterization invariant derivative

$$\frac{d}{ds} = \frac{1}{\sqrt{\frac{dy^\mu}{d\lambda}\frac{dy_\mu}{d\lambda}}} \frac{d}{d\lambda}. \tag{2.6}$$

Owing to the reparameterization invariance of s, in the expressions (2.4)–(2.6) we may use as parameter the time coordinate, obtaining thereby

$$ds = \sqrt{1 - v^2}\, dt, \quad s(t) = \int_0^t \sqrt{1 - v^2(t')}\, dt', \quad \frac{d}{ds} = \frac{1}{\sqrt{1 - v^2(t)}} \frac{d}{dt}. \tag{2.7}$$

Four-velocity and four-acceleration. With the aid of the invariant derivative (2.6) we can now build the *four-velocity*, *four-acceleration* and *four-momentum* of a particle with mass m, defined by

$$u^\mu = \frac{dy^\mu}{ds}, \qquad w^\mu = \frac{du^\mu}{ds}, \qquad p^\mu = mu^\mu, \tag{2.8}$$

respectively. By construction, all three objects are *four-vectors* – quotients between four-vectors and scalars – and they obey the identities

$$u^\mu u_\mu = 1, \qquad u^\mu w_\mu = 0, \qquad p^2 = p^\mu p_\mu = m^2. \tag{2.9}$$

The first identity follows from (2.4), the second is obtained differentiating the first with respect to s, and the third is equivalent to the first. Considering as parameter the time coordinate, and using the relations (2.3) and (2.7), the time and space components of the four-velocity in (2.8) turn out to be given by

$$u^\mu = \left(\frac{1}{\sqrt{1 - v^2}}, \frac{\mathbf{v}}{\sqrt{1 - v^2}} \right). \tag{2.10}$$

Correspondingly, the components of the four-momentum, i.e. the *relativistic momentum* $\mathbf{p}$ and the *relativistic energy* ε, assume the known expressions

$$\mathbf{p} = m\mathbf{u} = \frac{m\mathbf{v}}{\sqrt{1 - v^2}}, \qquad \varepsilon = p^0 = mu^0 = \frac{m}{\sqrt{1 - v^2}}. \tag{2.11}$$

Rest frame of a particle. It often occurs that a physical phenomenon can be most easily understood by analyzing it in the so-called instantaneous *rest frame*. For every *fixed* instant t there exists, in fact, an inertial reference frame K in which at that instant the particle has vanishing velocity, $\mathbf{v} = 0$. Indeed, at that instant, in K the above kinematical quantities assume the particularly simple form

$$u^\mu = (1, 0, 0, 0), \qquad p^\mu = (m, 0, 0, 0), \qquad w^\mu = (0, \mathbf{a}),$$

which may facilitate considerably the analysis.

2.2 Electrodynamics of Point-Particles

We introduce now the physical system which is the main object of investigation of this book, i.e. a set of N *charged particles interacting with the electromagnetic field*. The independent kinematical variables describing this system are the $4N$ functions $y_r^\mu(\lambda_r)$ $(r = 1, \ldots, N)$, parameterizing the N world lines γ_r swept out by the particles, and the antisymmetric *Maxwell tensor* $F^{\mu\nu}(x)$ representing the electromagnetic field:

$$F^{\mu\nu} = -F^{\nu\mu}.$$

This tensor is tied to the electric and magnetic fields $\mathbf{E}$ and $\mathbf{B}$ through the relations

$$F^{00} = 0, \tag{2.12}$$

$$F^{i0} = -F^{0i} = E^i, \tag{2.13}$$

$$F^{ij} = -\varepsilon^{ijk}B^k \quad \leftrightarrow \quad B^i = -\frac{1}{2}\varepsilon^{ijk}F^{jk}. \tag{2.14}$$

With the components of the tensor $F^{\mu\nu}$ we can build the two algebraically independent relativistic invariants

$$\varepsilon^{\mu\nu\rho\sigma}F_{\mu\nu}F_{\rho\sigma} = -8\,\mathbf{E}\cdot\mathbf{B}, \qquad F^{\mu\nu}F_{\mu\nu} = 2(B^2 - E^2), \tag{2.15}$$

the proof of their explicit expressions in terms of $\mathbf{E}$ and $\mathbf{B}$ being left as an exercise. Denoting the mass of the rth particle by m_r, for each particle we can introduce the kinematical quantities defined in Sect. 2.1: the proper time s_r, the four-velocity u_r^μ, the four-acceleration w_r^μ, the four-momentum $p_r^\mu = m_r u_r^\mu$, the position vector $\mathbf{y}_r(t)$, the velocity $\mathbf{v}_r$, the spatial acceleration $\mathbf{a}_r$, and the relativistic energy and momentum ε_r and $\mathbf{p}_r$.

Four-current. Denoting the electric charge of the rth particle with e_r we define the electric *four-current* of the system as

$$j^\mu(x) = \sum_r e_r \int_{\gamma_r} \delta^4(x - y_r)\, dy_r^\mu = \sum_r e_r \int \frac{dy_r^\mu}{d\lambda_r}\, \delta^4(x - y_r(\lambda_r))\, d\lambda_r, \tag{2.16}$$

where the symbol $\delta^4(\cdot)$ denotes Dirac's four-dimensional δ-function, actually a *distribution* (see Sect. 2.3.1). The general properties of the expression (2.16) will be analyzed in detail in Sect. 2.3.2. Here we limit ourselves to anticipating that j^μ is a *four-vector* field, that it is *reparameterization invariant*, and that it is identically *conserved*, i.e. it satisfies the *continuity equation*

$$\partial_\mu j^\mu = 0. \tag{2.17}$$

In Sect. 2.3.2 we will moreover show that the time and space components of the four-vector (2.16) can be written as

$$j^0(t, \mathbf{x}) = \sum_r e_r \, \delta^3(\mathbf{x} - \mathbf{y}_r(t)), \tag{2.18}$$

$$\mathbf{j}(t, \mathbf{x}) = \sum_r e_r \mathbf{v}_r(t) \, \delta^3(\mathbf{x} - \mathbf{y}_r(t)). \tag{2.19}$$

From the known formal properties of the three-dimensional δ-function, more precisely $\delta^3(\mathbf{x}) = 0$ if $\mathbf{x} \neq 0$, and $\int_{\mathbb{R}^3} \delta^3(\mathbf{x}) \, d^3x = 1$, one infers that j^0 is the *charge density* ρ of the particle system, and that $\mathbf{j}$ is the familiar three-dimensional *current density*.

We stress from the beginning that the current (2.16) cannot be regarded as a vector *field* in the strict sense, because its components, involving δ-functions, are not *functions*, but rather elements of $\mathcal{S}'(\mathbb{R}^4)$, namely *distributions*: j^μ is, actually, a *distribution-valued vector field*. The precise meaning of this statement, together with its physical consequences, will be clarified in Sect. 2.3, where we introduce the basics of *distribution theory* and investigate, hence, the distributional nature of Maxwell's equations.

2.2.1 The Fundamental Equations

We introduce now the *fundamental equations of electrodynamics* in the manifestly covariant form

$$\frac{dp_r^\mu}{ds_r} = e_r F^{\mu\nu}(y_r) u_{r\nu}, \qquad (r = 1, \dots, N), \tag{2.20}$$

$$\varepsilon^{\mu\nu\rho\sigma} \partial_\nu F_{\rho\sigma} = 0, \tag{2.21}$$

$$\partial_\mu F^{\mu\nu} = j^\nu. \tag{2.22}$$

We shall call these equations *Lorentz equations, Bianchi identity,* and *Maxwell equation,* respectively.[2] The role of these equations is to determine uniquely the fields $F^{\mu\nu}(x)$ and the world lines $y_r^\mu(\lambda_r)$, once certain initial conditions have been fixed: According to the *Newtonian determinism* the system of differential equations (2.20)–(2.22) should, in fact, give rise to a well-posed *Cauchy probl4em*. The Cauchy problem concerning the coordinates y_r^μ will be specified in Sect. 2.2.3, whereas the one associated with the Maxwell tensor $F^{\mu\nu}$ will be formulated later on in Sect. 5.1.3. We call the above system of equations *fundamental,* in that, in principle, any of its solutions should predict a physically observable phenomenon and,

---

[2] With the term *Maxwell equations* one usually refers to the system of equations (2.21) and (2.22) or, equivalently, to the system (2.24) and (2.25). Because of the *distinct* roles played, actually, by the Eqs. (2.21) and (2.22), in this chapter we prefer the different terminology introduced above.

conversely, any observed phenomenon occurring in nature should satisfy it. Unfortunately, as we shall see, these expectations are damped by two basic aspects inherent
to classical electrodynamics: the *spontaneous violation* of time-reversal invariance,
see Chap. 6, and the *ultraviolet divergences* inextricably tied to radiation reaction,
see Chap. 15.

Before proceeding any further, we rewrite the fundamental equations in the traditional three-dimensional notation:

$$\frac{d\mathbf{p}_r}{dt} = e_r(\mathbf{E} + \mathbf{v}_r \times \mathbf{B}), \qquad \frac{d\varepsilon_r}{dt} = e_r\mathbf{v}_r \cdot \mathbf{E}, \tag{2.23}$$

$$\frac{\partial \mathbf{B}}{\partial t} + \boldsymbol{\nabla} \times \mathbf{E} = 0, \qquad\qquad \boldsymbol{\nabla} \cdot \mathbf{B} = 0, \tag{2.24}$$

$$-\frac{\partial \mathbf{E}}{\partial t} + \boldsymbol{\nabla} \times \mathbf{B} = \mathbf{j}, \qquad\qquad \boldsymbol{\nabla} \cdot \mathbf{E} = \rho. \tag{2.25}$$

Sometimes we will refer to the second set of equations in (2.23) as *work-energy*
theorems, as they are, in fact, relativistic generalizations of the classical equation

$$\frac{d}{dt}\left(\frac{1}{2}mv^2\right) = \mathbf{v} \cdot \mathbf{F}.$$

Below we verify explicitly that Eqs. (2.23)–(2.25) are equivalent to the system
(2.20)–(2.22). In doing so we resort repeatedly to the defining relations (2.12)–
(2.14), as well as to formulas (2.7) and (2.8).

Lorentz equation. Setting in (2.20) $\mu = i$, and omitting the index r, we obtain

$$\frac{dp^i}{ds} = \frac{1}{\sqrt{1-v^2}}\frac{dp^i}{dt} = eF^{i\nu}u_\nu = e\left(F^{i0}u_0 + F^{ij}u_j\right) = \frac{e\left(E^i + \varepsilon^{ijk}B^k v^j\right)}{\sqrt{1-v^2}},$$

which is equivalent to the first equation of (2.23). Setting in (2.20) $\mu = 0$ we obtain
instead

$$\frac{dp^0}{ds} = \frac{1}{\sqrt{1-v^2}}\frac{d\varepsilon}{dt} = eF^{0\nu}u_\nu = eF^{0i}u_i = \frac{eE^i v^i}{\sqrt{1-v^2}},$$

which corresponds to the second equation in (2.23).

Bianchi identity. Setting in (2.21) $\mu = i$ we obtain

$$\varepsilon^{i\nu\rho\sigma}\partial_\nu F_{\rho\sigma} = \varepsilon^{i0jk}\partial_0 F_{jk} + \varepsilon^{ij0k}\partial_j F_{0k} + \varepsilon^{ijk0}\partial_j F_{k0}$$
$$= -\varepsilon^{ijk}\partial_0 F^{jk} + 2\,\varepsilon^{ijk}\partial_j F^{k0} = 2(\partial_0 B^i + \varepsilon^{ijk}\partial_j E^k) = 0,$$

which is the first equation of (2.24). Setting in (2.21) $\mu = 0$ we obtain instead

$$\varepsilon^{0\nu\rho\sigma}\partial_\nu F_{\rho\sigma} = \varepsilon^{0ijk}\partial_i F_{jk} = \varepsilon^{ijk}\partial_i F_{jk} = -2\partial_i B^i = 0,$$

which is the second equation in (2.24).

Maxwell equation. Choosing in (2.22) $\mu = i$ we obtain

$$j^i = \partial_\mu F^{\mu i} = \partial_0 F^{0i} + \partial_j F^{ji} = -\partial_0 E^i + \varepsilon^{ijk} \partial_j B^k,$$

which is the first equation of (2.25). Finally, setting in (2.22) $\mu = 0$ we obtain

$$\partial_\mu F^{\mu 0} = \partial_i F^{i0} = \partial_i E^i = j^0 = \rho,$$

which is the second equation of (2.25).

2.2.2   Parity and Time Reversal

The system of equations (2.20)–(2.22) is manifestly invariant under the *proper* Lorentz group. In order to establish its invariance under the complete Lorentz group $O(1, 3)$ it remains to verify its invariance under the discrete symmetries, i.e. under parity $\mathcal{P}$ and under time reversal $\mathcal{T}$ (see Sect. 1.4.3 and in particular the definitions (1.64) and (1.65)). For this purpose we must establish beforehand the transformation laws under these symmetries of all fundamental variables.

The coordinates x^μ and the world lines $y^\mu(\lambda)$ are obviously vectors under $\mathcal{P}$ as well as under $\mathcal{T}$. The proper time

$$ds = \sqrt{1 - v^2}\, dt$$

is instead a scalar under $\mathcal{P}$ and a pseudoscalar under $\mathcal{T}$, i.e. it changes its sign if one sends t in $-t$. Accordingly, the four-velocity $u^\mu = dy^\mu/ds$ is a vector under $\mathcal{P}$ and a pseudovector under $\mathcal{T}$, i.e. we have the transformation laws

$$\mathcal{P}:\ u'^\mu = \mathcal{P}^\mu{}_\nu u^\nu, \qquad \mathcal{T}:\ u'^\mu = -\mathcal{T}^\mu{}_\nu u^\nu. \tag{2.26}$$

The four-acceleration $w^\mu = d^2 y^\mu/ds^2$ and the derivative of the four-momentum $dp^\mu/ds = m w^\mu$ involve two derivatives and are, therefore, vectors under $\mathcal{P}$ as well as under $\mathcal{T}$. The four-current j^μ in (2.16), on the same footing as u^μ, constitutes a vector under $\mathcal{P}$ and a pseudovector under $\mathcal{T}$:

$$\mathcal{P}:\ j'^\mu = \mathcal{P}^\mu{}_\nu j^\nu, \qquad \mathcal{T}:\ j'^\mu = -\mathcal{T}^\mu{}_\nu j^\nu, \tag{2.27}$$

where for simplicity we did not write out the coordinate dependence. To verify the transformation laws (2.27) it is convenient to rewrite them in three-dimensional notation,

$$\mathcal{P}: \begin{cases} j'^0 = j^0, \\ \mathbf{j}' = -\mathbf{j}, \end{cases} \qquad \mathcal{T}: \begin{cases} j'^0 = j^0, \\ \mathbf{j}' = -\mathbf{j}, \end{cases} \tag{2.28}$$

because in this form they follow directly from the expressions (2.18) and (2.19). The transformation laws of $F^{\mu\nu}$ must now be determined so that the fundamental equations (2.20)–(2.22) maintain their form, if possible, under $\mathcal{P}$ as well as under $\mathcal{T}$. In this respect, the Bianchi identity (2.21) gives us no cue at all since – being a linear *homogeneous* equation – it is invariant under the complete group $O(1,3)$, whatever transformation law under $\mathcal{P}$ and $\mathcal{T}$ one chooses for $F^{\mu\nu}$. Conversely, given that the operator ∂_μ is a vector under $\mathcal{P}$ as well as under $\mathcal{T}$, while j^μ is a vector under $\mathcal{P}$ and a pseudovector under $\mathcal{T}$, the invariance of the Maxwell equation (2.22) forces us to consider $F^{\mu\nu}$ as a *tensor under $\mathcal{P}$* and as a *pseudotensor under $\mathcal{T}$*. In this way both members of this equation transform, in fact, in the same way: as vectors under $\mathcal{P}$ and as pseudovectors under $\mathcal{T}$. Finally, these non-trivial assignments are such that also the Lorentz equations (2.20) are invariant under the complete Lorentz group: both their members are, in fact, vectors under $\mathcal{P}$ as well as under $\mathcal{T}$. Their right-hand sides are, in particular, products of two objects that are tensors under $\mathcal{P}$ and pseudotensors under $\mathcal{T}$. In conclusion, *the fundamental equations of electrodynamics are invariant under the complete Lorentz group*, provided that we assign to the Maxwell tensor the transformation laws

$$\mathcal{P}: \ F'^{\alpha\beta} = \mathcal{P}^\alpha{}_\mu \mathcal{P}^\beta{}_\nu F^{\mu\nu}, \qquad \mathcal{T}: \ F'^{\alpha\beta} = -\mathcal{T}^\alpha{}_\mu \mathcal{T}^\beta{}_\nu F^{\mu\nu}. \tag{2.29}$$

According to the identifications (2.12)–(2.14), from these transformation laws we deduce the corresponding ones for the electric and magnetic fields (see Problem 2.15)

$$\mathcal{P}: \begin{cases} \mathbf{E}' = -\mathbf{E}, \\ \mathbf{B}' = \mathbf{B}, \end{cases} \qquad \mathcal{T}: \begin{cases} \mathbf{E}' = \mathbf{E}, \\ \mathbf{B}' = -\mathbf{B}. \end{cases} \tag{2.30}$$

Taking the obvious transformation laws of $\mathbf{v}_r$, $\mathbf{p}_r$ and ε_r into account, it is immediately seen that under the transformations (2.28) and (2.30) Eqs. (2.23)–(2.25) are in fact invariant, as they must be by construction.

The invariance under time reversal of the fundamental equations of electrodynamics entails a far-reaching consequence: if the configuration

$$\Sigma = \{\mathbf{y}_r(t), \ \mathbf{E}(t, \mathbf{x}), \ \mathbf{B}(t, \mathbf{x})\} \tag{2.31}$$

satisfies Eqs. (2.20)–(2.22), then these equations are also satisfied by the configuration

$$\Sigma^* = \{\mathbf{y}_r(-t), \ \mathbf{E}(-t, \mathbf{x}), \ -\mathbf{B}(-t, \mathbf{x})\}. \tag{2.32}$$

This peculiar property of electrodynamics – shared, as mentioned in Sect. 1.4.3, with the gravitational and strong interactions, but not with the weak ones – bears important and, under certain aspects, unexpected physical consequences which we shall analyze in detail in Sects. 5.4.2 and 6.2.3. The general characteristics of the fundamental equations (2.20)–(2.22) will be investigated in the next three sections.

2.2.3 *Lorentz Equation*

In order to simplify the notation we omit the index r and rewrite the Lorentz equation (2.20) in the equivalent form

$$H^\mu \equiv \frac{dp^\mu}{ds} - eF^{\mu\nu}(y)u_\nu = 0, \tag{2.33}$$

where the electromagnetic field is evaluated along the world line $y^\mu = y^\mu(\lambda)$ of the particle. Once the field $F^{\mu\nu}(x)$ is known, (2.33) constitutes a system of four ordinary second-order differential equations in the four unknown functions $y^\mu(\lambda)$, which in turn are equivalent to the four equations appearing in (2.23). On the other hand, since the variable λ appears only through the proper time s, these equations are manifestly invariant under reparameterizations. In accordance with our kinematical analysis of Sect. 2.1 they determine hence the functions $y^\mu(\lambda)$ only modulo a reparameterization.

Cauchy problem. We now address the Cauchy problem associated with equations (2.33), also called *initial value problem*. This equation represents the equation of motion of a particle and, in line with the classical Newtonian determinism, it should determine the position vector $\mathbf{y}(t)$ at each time t in a consistent and unique way, once the initial conditions $\mathbf{y}(0)$ and $\mathbf{v}(0) = \dot{\mathbf{y}}(0)$ have been assigned. To verify the correctness of this hypothesis we take advantage from reparameterization invariance which allows us to choose as parameter the time: $\lambda = t$. In this way the world line takes the form

$$y^\mu(t) = (t, \mathbf{y}(t)).$$

Within this choice the unknown functions are the three components of the position vector $\mathbf{y}(t)$, which must, nevertheless, satisfy the four equations (2.33). On the other side, only three out of these four equations are functionally independent, as the four-vector H^μ defined in (2.33) satisfies the identity

$$u_\mu H^\mu = u_\mu \left(\frac{dp^\mu}{ds} - eF^{\mu\nu}u_\nu \right) = 0 \quad \leftrightarrow \quad u^0 H^0 - \mathbf{u} \cdot \mathbf{H} = 0. \tag{2.34}$$

In fact, the scalar $u_\mu u_\nu F^{\mu\nu}$ vanishes for symmetry reasons, see (1.34), while the kinematical relations (2.9) imply the identity $u_\mu(dp^\mu/ds) = mw^\mu u_\mu = 0$. The quantity H^0 is thus a simple function of $\mathbf{H}$

$$H^0 = \frac{\mathbf{u} \cdot \mathbf{H}}{u^0} = \mathbf{v} \cdot \mathbf{H}.$$

The equation $H^0 = 0$, i.e. the work-energy theorem, see (2.23),

$$\frac{d\varepsilon}{dt} = e\,\mathbf{v} \cdot \mathbf{E}, \tag{2.35}$$

is therefore automatically satisfied once one imposes the equation $\mathbf{H} = 0$, i.e. the spatial Lorentz equation

$$\frac{d\mathbf{p}}{dt} = e\left(\mathbf{E} + \mathbf{v} \times \mathbf{B}\right). \tag{2.36}$$

This equation plays, for all purposes, the role of *Newton's second law*, and the work-energy theorem (2.35) is, thus, an automatic consequence of this law, in exactly the same way as it happens in non-relativistic mechanics. The implication $\mathbf{H} = 0 \Rightarrow H^0 = 0$ can also be verified directly, by differentiating the identity $p^\mu p_\mu = m^2$, rewritten as $\varepsilon^2 = |\mathbf{p}|^2 + m^2$, with respect to the time, and by using the kinematical relations (2.11)

$$\varepsilon \frac{d\varepsilon}{dt} = \mathbf{p} \cdot \frac{d\mathbf{p}}{dt} \quad \Rightarrow \quad \frac{d\varepsilon}{dt} = \frac{\mathbf{p}}{\varepsilon} \cdot \frac{d\mathbf{p}}{dt} = \mathbf{v} \cdot \frac{d\mathbf{p}}{dt}.$$

Using in the last term Eq. (2.36) we retrieve (2.35).

We end this section underlying that the precise meaning of the three independent equations (2.36) is

$$\frac{d}{dt}\left(\frac{m\mathbf{v}(t)}{\sqrt{1 - v(t)^2}}\right) = e\big(\mathbf{E}(t, \mathbf{y}(t)) + \mathbf{v}(t) \times \mathbf{B}(t, \mathbf{y}(t))\big). \tag{2.37}$$

Notice in particular that the electric and magnetic fields at the right-hand side are evaluated at the particle's position $\mathbf{y}(t)$ at time t. In order to compare the system of differential equations (2.37) with the dynamics of the *Newtonian mechanics*, we remark that it can be recast in the *normal form* $\ddot{\mathbf{y}} = \mathbf{f}(\mathbf{y}, \dot{\mathbf{y}}, t)$, namely (see Problem 2.10)

$$m\mathbf{a} = e\sqrt{1 - v^2}\left(\mathbf{E} - (\mathbf{v} \cdot \mathbf{E})\mathbf{v} + \mathbf{v} \times \mathbf{B}\right), \tag{2.38}$$

which traces closely the structure of Newton's equation $\mathbf{F} = m\mathbf{a}$. In the absence of *singularities*, if the fields $\mathbf{E}(x)$ and $\mathbf{B}(x)$ are known, and once the initial conditions $\mathbf{y}(0)$ and $\mathbf{v}(0)$ have been chosen, this system would admit a *unique* solution $\mathbf{y}(t)$, as appropriate for a well-defined Cauchy problem. Finally, once $\mathbf{y}(t)$ is known, relations (2.7) allow to determine $s(t)$ and to reconstruct thus the four-dimensional world line $y^\mu(s)$. Unfortunately, as we will see in Chap. 15, the right-hand side of equation (2.38) *will* develop infinities – arising from the singularities of $\mathbf{E}(x)$ and $\mathbf{B}(x)$ precisely at the particle's positions – that shall turn the above Cauchy problem in an ill-defined one.

2.2.4 Bianchi Identity

In three-dimensional notation the Bianchi identity (2.21) corresponds to the four equations (2.24). In manifestly covariant notation it can be written in the three equivalent ways

$$\varepsilon^{\mu\nu\rho\sigma}\partial_\nu F_{\rho\sigma} = 0, \tag{2.39}$$

$$\partial_{[\mu}F_{\nu\rho]} = 0, \tag{2.40}$$

$$\partial_\mu F_{\nu\rho} + \partial_\nu F_{\rho\mu} + \partial_\rho F_{\mu\nu} = 0, \tag{2.41}$$

the proofs being left as exercises, see Problem 2.2. In the terminology commonly used in electrodynamics these equations represent *half* of the Maxwell equations, more precisely, the half constraining the form of the electromagnetic field, while bearing no link with the charge distribution. The term *identity* is related to the fact that Eq. (2.39) admits a canonical *basis* of solutions, which satisfy it *identically*. These solutions are constructed by introducing an arbitrary vector field $A_\mu(x)$, called *vector potential*, *four-potential*, or also *gauge field*, and by setting

$$F_{\mu\nu} = \partial_\mu A_\nu - \partial_\nu A_\mu. \tag{2.42}$$

In fact, substituting this expression in (2.39) one finds

$$\varepsilon^{\mu\nu\rho\sigma}\partial_\nu F_{\rho\sigma} = \varepsilon^{\mu\nu\rho\sigma}(\partial_\nu\partial_\rho A_\sigma - \partial_\nu\partial_\sigma A_\rho) = 0, \tag{2.43}$$

the conclusion stemming from the fact that in both terms one contracts a pair of symmetric indices – those of the derivatives – with a pair of antisymmetric ones – those of the Levi-Civita tensor. If the Maxwell tensor is expressed in terms of a vector potential as in (2.43), it is said that $F_{\mu\nu}$ is the *field strength* of A_μ. Resorting to the methods of *differential geometry* one can, however, prove a much stronger result: for every antisymmetric tensor field $F_{\mu\nu}$ satisfying Eq. (2.39), there exists a vector field A_μ such that $F_{\mu\nu}$ can be written as in (2.42). The corresponding theorem goes under the name of *Poincaré lemma*, see Sect. 19.1.1, which holds provided that the considered space-time is "topologically trivial", an important example of such space-times being just $\mathbb{R}^4$. We will come back to the mathematical aspects of this important lemma in Chap. 19, where we shall present its rigorous formulation in the more appropriate framework of *differential forms*.

Gauge transformations. The – perhaps surprising – conclusion of the above analysis is that (2.42) represents the *general* solution of the Bianchi identity. There remains, however, an important issue that we must face: different vector potentials can give rise to the same field strength, i.e. to the same Maxwell tensor. Given an arbitrary scalar field $\Lambda(x)$, called *gauge function*, one can, in fact, introduce a new vector potential A'_μ by setting

$$A'_\mu = A_\mu + \partial_\mu\Lambda, \tag{2.44}$$

and, thanks to the commutativity of the partial derivatives, the new Maxwell tensor $F'_{\mu\nu}$ equals the original one

$$F'_{\mu\nu} = \partial_\mu A'_\nu - \partial_\nu A'_\mu = F_{\mu\nu} + \partial_\mu\partial_\nu\Lambda - \partial_\nu\partial_\mu\Lambda = F_{\mu\nu}. \tag{2.45}$$

The transformations (2.44) are called *gauge transformations*, and the relation (2.45) states thus that the Maxwell tensor is gauge invariant. In conclusion, the Bianchi identity admits a general solution in terms of a vector potential, but the latter is determined only modulo gauge transformations. Schematically we have therefore

$$\varepsilon^{\mu\nu\rho\sigma}\partial_\nu F_{\rho\sigma} = 0 \quad\Leftrightarrow\quad F_{\mu\nu} = \partial_\mu A_\nu - \partial_\nu A_\mu, \quad \text{with } A_\mu \approx A_\mu + \partial_\mu\Lambda. \tag{2.46}$$

In most cases our strategy to address the system of equations (2.20)–(2.22) will start with the solution of the Bianchi identity in terms of a vector potential A^μ, and then proceed with the substitution of the expression (2.42) in the Maxwell equation (2.22). In this way we are left with the solution of the *coupled* equations (2.20) and (2.22), in the unknown functions $y_r^\mu(\lambda_r)$ and $A_\mu(x)$, respectively.

Three-dimensional notation. Setting as usual $A^\mu = (A^0, \mathbf{A})$, in three-dimensional notation Eqs. (2.42) correspond to the known relations

$$\mathbf{E} = -\boldsymbol{\nabla} A^0 - \frac{\partial \mathbf{A}}{\partial t}, \qquad \mathbf{B} = \boldsymbol{\nabla} \times \mathbf{A}. \tag{2.47}$$

We have, in fact,

$$E^i = F^{i0} = \partial^i A^0 - \partial^0 A^i = -\partial_i A^0 - \partial_0 A^i,$$
$$B^i = -\varepsilon^{ijk}F^{jk} = -\varepsilon^{ijk}(\partial^j A^k - \partial^k A^j) = \varepsilon^{ijk}(\partial_j A^k - \partial_k A^j) = (\boldsymbol{\nabla} \times \mathbf{A})^i.$$

From the first relation in (2.47) we infer that the field $A^0(t, \mathbf{x})$ represents a dynamical generalization of the electrostatic potential. Similarly, in three-dimensional notation the gauge transformation (2.44) amounts to

$$A'^0 = A^0 + \frac{\partial\Lambda}{\partial t}, \qquad \mathbf{A}' = \mathbf{A} - \boldsymbol{\nabla}\Lambda.$$

Parity and time reversal. In Sect. 2.2.2 we saw that the Maxwell tensor is a tensor under parity and a *pseudotensor* under time reversal, see the transformation laws (2.29). In the light of the field-strength relation (2.42) – given that the operator ∂_μ is a vector under $\mathcal{P}$ as well as under $\mathcal{T}$ – it follows that A^μ is a vector under $\mathcal{P}$ and a *pseudovector* under $\mathcal{T}$. Hence, under these discrete symmetries the vector potential transforms according to

$$\mathcal{P}: \ A'^\mu = \mathcal{P}^\mu{}_\nu A^\nu, \qquad \mathcal{T}: \ A'^\mu = -\mathcal{T}^\mu{}_\nu A^\nu. \tag{2.48}$$

In virtue of formulas (1.64) and (1.65) the fields A^0 and $\mathbf{A}$ are thus subject to the transformation laws

$$\mathcal{P}: \begin{cases} A'^0 = A^0, \\ \mathbf{A}' = -\mathbf{A}, \end{cases} \qquad \mathcal{T}: \begin{cases} A'^0 = A^0, \\ \mathbf{A}' = -\mathbf{A}. \end{cases} \tag{2.49}$$

2.2.5 Maxwell Equation

The Maxwell equation

$$\partial_\mu F^{\mu\nu} = j^\nu, \tag{2.50}$$

linking $F^{\mu\nu}$ to the four-current j^μ, represents the *actual equation of motion* of the electromagnetic field: It specifies the modality in which a generic charge distribution generates an electromagnetic field. This equation is, in particular, consistent with the continuity equation $\partial_\mu j^\mu = 0$. In fact, taking the four-divergence of both members of Eq. (2.50), i.e. contracting it with ∂_ν, both members vanish identically: the left-hand side vanishes thanks to the continuity equation, while, in turn, the right-hand side vanishes thanks to the antisymmetry of the Maxwell tensor. As noted in Sect. 2.2.4, once we have resolved the Bianchi identity according to (2.42), the Maxwell equation becomes an equation for the vector potential. It would, hence, correspond to four partial differential equations in the four unknown functions A_μ. However, this counting is only partially meaningful. In the first place, the components of the vector potential are not all *physical*, as they are subject to the gauge transformations (2.44): Different vector potentials may, therefore, correspond to the same electric and magnetic fields, but only the latter, being experimentally observable, own physical reality. It is therefore not correct to consider the four components of the vector potential as independent physical variables. In the second place, the four components of the Maxwell equation are not functionally independent. To show it we introduce the four-vector

$$G^\nu = \partial_\mu F^{\mu\nu} - j^\nu$$

and rewrite Eq. (2.50) in the form $G^\nu = 0$. From the identities recalled above it then follows that G^ν satisfies identically the constraint

$$\partial_\nu G^\nu = 0 \quad \leftrightarrow \quad \partial_0 G^0 = -\mathbf{\nabla} \cdot \mathbf{G}. \tag{2.51}$$

This means that the time component of the Maxwell equation is linked to its spatial components. However, the link (2.51) is not of the algebraic type – it does not involve directly the equations of motion, but their derivatives – and therefore it is not straightforward to identify a set of functionally independent differential equations. The formulation of the *Cauchy problem* for the electromagnetic field appears, thus, more involved than the one associated with the Lorentz equation, and so we postpone it to Sect. 5.1.3 where we will have at our disposal the appropriate mathematical tools.

Degrees of freedom of the electromagnetic field. Qualitatively, the concept of *degrees of freedom* of a physical system refers to the *independent* variables needed to *thoroughly* describe its dynamics. In particular it is required that, once one has picked certain initial data of these variables, for example at time $t = 0$, the equations of motion of the system determine their values at any time t. It is therefore clear that there exists a deep connection between the degrees of freedom of a system and the associated Cauchy problem. We shall give a precise definition of the concept of *degree of freedom* in a general field theory in Sect. 5.1. Here we perform a *preliminary* analysis of the degrees of freedom of the electromagnetic field, without making reference to the vector potential. For this purpose we rewrite Maxwell's equations in the three-dimensional notation

$$-\frac{\partial \mathbf{E}}{\partial t} + \nabla \times \mathbf{B} = \mathbf{j}, \tag{2.52}$$

$$\frac{\partial \mathbf{B}}{\partial t} + \nabla \times \mathbf{E} = 0, \tag{2.53}$$

$$\nabla \cdot \mathbf{E} = \rho, \tag{2.54}$$

$$\nabla \cdot \mathbf{B} = 0. \tag{2.55}$$

The vectorial equations (2.52) and (2.53) correspond to six equations in the six unknown functions $\mathbf{E}(t, \mathbf{x})$ and $\mathbf{B}(t, \mathbf{x})$, involving the *time* derivatives of $\mathbf{E}$ and $\mathbf{B}$. Correspondingly, they represent *dynamical* equations which admit a unique solution once the initial values $\mathbf{E}(0, \mathbf{x})$ and $\mathbf{B}(0, \mathbf{x})$ are known for all $\mathbf{x}$. Conversely, the scalar equations (2.54) and (2.55) – which do not involve time derivatives – represent *constraints*, rather than dynamical equations. In particular the initial data $\mathbf{E}(0, \mathbf{x})$ and $\mathbf{B}(0, \mathbf{x})$ cannot be chosen arbitrarily since these equations, evaluated at $t = 0$, generate between these data the constraints

$$\nabla \cdot \mathbf{E}(0, \mathbf{x}) = \rho(0, \mathbf{x}), \tag{2.56}$$

$$\nabla \cdot \mathbf{B}(0, \mathbf{x}) = 0. \tag{2.57}$$

Therefore, at the instant $t = 0$ we can choose arbitrarily only $6 - 2 = 4$ components of the electromagnetic field, for at the same instant the remaining 2 components are determined in terms of the former by the relations (2.56) and (2.57). Finally, it can be shown that, if Eqs. (2.52) and (2.53) are satisfied for all t, and Eqs. (2.54) and (2.55) for $t = 0$, then the latter are automatically satisfied for all t, see Problem 2.11. In summary, the electromagnetic field is expected not to bear six, but only *four first-order degrees of freedom*, a prediction that we will confirm in Sect. 5.1.3.

On the solutions of the fundamental equations. The system (2.20)–(2.22) constitutes a system of "strongly" coupled non-linear differential equations which – except for very rare cases – cannot be solved analytically: The electromagnetic fields determine the particles' trajectories according to Eqs. (2.20), and the fields, in turn, are determined by the particles' trajectories according to Eqs. (2.21) and (2.22). Nevertheless, in many physical situations the problem is reduced, in practice, to one of the following two circumstances, in which the equations *de facto* decouple from each other.

(1) An *external* electromagnetic field $F^{\mu\nu}$ satisfies in a certain region of space equations (2.21) and (2.22) with $j^\mu = 0$. Examples are the constant electric field between the plates of a capacitor, the magnetic field in the vicinity of the poles of an electromagnet, the electric and magnetic fields associated with an electromagnetic wave, and the electromagnetic field in a particle accelerator. There is a single charged particle subjected to this field, and one is asked to determine the motion of the particle. This problem reduces to the solution of the single equation (2.20) in the four unknowns $y^\mu(\lambda)$ or, equivalently, to the solution of equation (2.37) in the three unknowns $\mathbf{y}(t)$.

(2) The world line of a charged particle, or of more charged particles, is known, and one is asked to determine the electromagnetic field created by this system of moving charges. This issue then regards solely Eqs. (2.21) and (2.22), which in turn can be solved in full generality in terms of the *Liénard-Wiechert fields*, see Chap. 7.

In both these situations, however, we must keep in mind that the real dynamics of the system is governed by the complete *system* of equations (2.20)–(2.22), and that the procedures described in (1) and (2) represent mere *idealizations* of the actual physical situations, whose validity must be examined on a case-by-case basis. In particular, in the idealization (1) we have neglected the field that the accelerated particle exerts on itself, and in the idealization (2) we have neglected the fields that the charges exert one on another, modifying their pre-established motion.

In the abstract, the appropriate strategy to solve the system of the fundamental equations of electrodynamics – and that *in principle* we will adopt in this book – is as follows. Having solved the Bianchi identity in terms of a vector potential A^μ, and choosing in the first instance generic world lines y_r^μ for all particles, we determine the exact solution of the Maxwell equation for A^μ, and hence for $F^{\mu\nu}$, which then become highly non-local functions of the y_r^μ. Subsequently, we substitute the field $F^{\mu\nu}$ so determined in the Lorentz equations (2.20), which then become *closed*, but rather complicated, equations in the unknowns y_r^μ. Once the latter have been resolved, we substitute the resulting y_r^μ in $F^{\mu\nu}$, thus obtaining the electromagnetic field as a function of x.

Self-interaction and ultraviolet divergences. As mentioned above, this program rarely can be accomplished analytically because of the *technical* difficulties involved in the solution of the differential equations (2.20)–(2.22). In addition to these difficulties, during the implementation of this program we will, however, meet also a *conceptual* problem with far more dramatic consequences: We shall in fact see that, because of the *self-interaction* of the charged particles, the quantities $F^{\mu\nu}(y_r)$ appearing in the Lorentz equations (2.20) are ultraviolet *divergent*. Ultimately these equations are therefore meaningless, and consequently the internal consistency of classical electrodynamics – as a theory describing the dynamics of charges and fields in a deterministic way – is hopelessly compromised. Most part of Chaps. 15 and 16 will be devoted to this fundamental issue of electrodynamics. For the appearance of ultraviolet divergences – and their cure – in *non-relativistic* electrodynamics see Problem 2.8. On the other hand, this consistency problem does not affect, at least not directly, the generation and propagation of electromagnetic fields, basic physical phenomena that will be covered in the central chapters of the book.

Active and passive charges. For the moment we have tacitly assumed that the electric charges of the particles constitute an arbitrary set of constants $\{e_r\}$. This hypothesis is supported by the analysis carried out so far on the system of equations (2.20)–(2.22), which appears consistent for any values of the charges. Actually, in these equations the electric charges appear in two different places: at the right-hand side of the Lorentz equation, and in the current of the Maxwell equation. On the other hand, there is no a priori reason that the charges appearing in these two positions be identical. In fact, if we introduce in the definition of the current (2.16) a set of *active* charges $\{e_r\}$, and at the right-hand side of the Lorentz equations an – a priori different – set of *passive* charges $\{e_r^*\}$, the fundamental equations would maintain all their good properties discussed so far. It remains, therefore, to understand what has led us to identify from the outset the passive charges with the active ones. A key indication in this regard emerges from a *non-relativistic* analysis of the fundamental equations, where all particles are assumed to have speeds $v_r \ll 1$.

Let us consider, in this limit, two particles with charges (e_1, e_1^*) and (e_2, e_2^*), respectively, and denote the position vector joining particle 1 with particle 2 by $\mathbf{r}$. The Maxwell equation (2.54), in which the charge density (2.18) contains the active charges e_r, then yields for the quasi-static electric field $\mathbf{E}_2$ ($\mathbf{E}_1$) created by particle 1 (2) at the position of particle 2 (1) the expressions

$$\mathbf{E}_2 = \frac{e_1}{4\pi} \frac{\mathbf{r}}{r^3}, \qquad \mathbf{E}_1 = -\frac{e_2}{4\pi} \frac{\mathbf{r}}{r^3}.$$

In addition, in the non-relativistic limit the magnetic fields are negligible: $\mathbf{B}_1 = 0 = \mathbf{B}_2$. The Lorentz equations (2.23) where, conversely, the passive charges appear, then become

$$\frac{d\mathbf{p}_1}{dt} = e_1^* \mathbf{E}_1 = -\frac{e_1^* e_2}{4\pi} \frac{\mathbf{r}}{r^3} = \mathbf{F}_{21}, \qquad \frac{d\mathbf{p}_2}{dt} = e_2^* \mathbf{E}_2 = \frac{e_2^* e_1}{4\pi} \frac{\mathbf{r}}{r^3} = \mathbf{F}_{12}, \quad (2.58)$$

where $\mathbf{F}_{12}$ ($\mathbf{F}_{12}$) is the force exerted by particle 1 (2) on particle 2 (1). The law of *action and reaction* $\mathbf{F}_{12} = -\mathbf{F}_{21}$, i.e. *Newton's third law* – a basic postulate of Newtonian mechanics – is therefore fulfilled only if

$$\frac{e_1}{e_1^*} = \frac{e_2}{e_2^*}.$$

By repeating the reasoning for an arbitrary pair of particles we see that the ratio e_r/e_r^* must be independent of the particle, i.e. a universal constant

$$e_r^* = Ce_r. \tag{2.59}$$

Performing a similar analysis – again in the non-relativistic regime – to impose also the conservation of the total energy, one finds that C must be equal to unity, see Problem 2.8. Eventually, we obtain thus $e_r^* = e_r$. We see, therefore, that at the non-relativistic level the identification of active and passive charges is intimately related to the law of action and reaction. Moreover, in non-relativistic physics this law is also equivalent to the conservation of the total momentum. From Eqs. (2.58), which are valid for an isolated system, we deduce, in fact, that

$$\frac{d}{dt}(\mathbf{p}_1 + \mathbf{p}_2) = \mathbf{F}_{21} + \mathbf{F}_{12} = 0.$$

On the other hand, in a *relativistic* theory the momentum combines with energy to form a four-vector, the *four-momentum*. We expect therefore that at the relativistic level the identification between the charges $\{e_r\}$ and $\{e_r^*\}$ will be imposed by the conservation of the total four-momentum, a prediction that will be confirmed in Sect. 2.4.3.

2.3   Theory of Distributions and Electromagnetic Fields

As pointed out in Sect. 2.2, the components of the four-current j^μ in (2.16) are not *functions*, but rather *distributions*, whose *support* are the world lines of the particles. From the form of the Maxwell equation (2.22) it then follows that the components of $F^{\mu\nu}$ cannot be differentiable functions along the world lines, because in that case the components of the four-vector $\partial_\mu F^{\mu\nu}$ would be *functions*, and hence they could not match the components of j^ν. So we draw the following conclusions: (i) the tensor $F^{\mu\nu}$ is necessarily *singular* along the world lines[3] and – as we shall see – in general it diverges as $1/r^2$ where r is the distance from a particle; (ii) the Maxwell equation (2.22) cannot be considered as a differential equation in the *space of functions*. Conversely, in the following sections we will see that the Maxwell equation is perfectly well defined if regarded as a differential equation in the *space of distributions*, that is to say, in $\mathcal{S}'(\mathbb{R}^4)$. In this new perspective, the components of $F^{\mu\nu}$ must thus be considered as elements of $\mathcal{S}'(\mathbb{R}^4)$ and, correspondingly, the derivatives appearing in Eq. (2.22) must be regarded as *derivatives in the sense of distributions*. For consistency, also the Bianchi identity (2.21) must then be regarded as a differential equation in $\mathcal{S}'(\mathbb{R}^4)$. Notice that this reinterpretation of Maxwell's equations as distributional equations is possible, because they are *linear* in $F^{\mu\nu}$. An analogous interpretation for Einstein's equations which, conversely, are *non-linear*, would actually fail.

Once we have given a precise mathematical meaning to the Maxwell and Bianchi equations, we may ask whether, at this point, the Lorentz equations for the particles

---

[3]In a sense, the only purpose of the Maxwell equation (2.22) is to *qualify* the singularities of $F^{\mu\nu}$ along the world lines of the particles, given that in their complement $\mathcal{C}$ the current is zero. In fact, in $\mathcal{C}$ the electromagnetic field satisfies the trivial equation $\partial_\mu F^{\mu\nu} = 0$.

are also well defined. As anticipated in Sect. 2.2.5, the answer to this question is unfortunately negative since the quantity $F^{\mu\nu}(y_r)$ appearing in Eqs. (2.20) represents the distribution $F^{\mu\nu}$ evaluated at the point of a trajectory, and in general the value of a distribution at a given point is not a well-defined quantity. In the case at hand, in the vicinity of a trajectory $F^{\mu\nu}$ diverges as $1/r^2$, and $F^{\mu\nu}(y_r)$ is therefore an antisymmetric matrix whose elements are infinite numbers: The Lorentz equations remain, hence, ill-defined.

2.3.1   Elements of the Theory of Distributions

In this section we recall some of the *operational* elements of the theory of *tempered distributions* – henceforth called simply *distributions* – in a space of arbitrary dimension D. We will provide the main concepts and results, mostly without proofs, referring for a more exhaustive presentation of the theory of distributions to a textbook of *functional analysis*, as, for instance, [1, 2].

Distributions and test functions. The Schwartz space $\mathcal{S} = \mathcal{S}(\mathbb{R}^D)$ of *test functions* of rapid decrease is defined as the vector space of the functions $\varphi : \mathbb{R}^D \to \mathbb{C}$ of class C^∞, which at infinity decay together with all their derivatives more rapidly than the inverse of any power of the coordinates. In other words, $\varphi \in \mathcal{S}$ if and only if all its *seminorms*, defined as

$$\|\varphi\|_{\mathcal{P},\mathcal{Q}} = \sup_{x \in \mathbb{R}^D} |\mathcal{P}(x)\mathcal{Q}(\partial)\varphi(x)|, \tag{2.60}$$

are finite. Above, $\mathcal{P}$ is a generic monomial of the coordinates x^μ, and $\mathcal{Q}$ is a generic monomial of the partial derivatives ∂_μ.[4] The vector space $\mathcal{S}$ is then endowed with the *topology induced by the seminorms* (2.60). For further details on this space, and on its topology, we refer the reader to [1]. The *space of tempered distributions* $\mathcal{S}' = \mathcal{S}'(\mathbb{R}^D)$ – the topological dual space of $\mathcal{S}$ – is defined as the set of all *linear and continuous* functionals F on $\mathcal{S}$

$$F : \mathcal{S} \to \mathbb{C},$$

$$\varphi \to F(\varphi).$$

A generic distribution $F \in \mathcal{S}'$ is completely determined by the complex values $F(\varphi)$ it assumes when applied to a generic test function $\varphi \in \mathcal{S}$. Below we recall a theorem that is of great practical utility to establish whether a given linear functional represents a distribution.

Theorem I. A linear functional F on $\mathcal{S}$ is continuous, that is to say it belongs to $\mathcal{S}'$, if and only if it can be bounded by a *finite* sum of seminorms of φ, i.e. if there exists a finite set of φ-independent positive constants $C_{\mathcal{P},\mathcal{Q}}$ such that

---

[4]In this section the indices μ, ν etc. assume the values $1, \ldots, D$, or equivalently $0, 1, \ldots, D-1$.

$$|F(\varphi)| \leq \sum_{\mathcal{P},\mathcal{Q}} C_{\mathcal{P},\mathcal{Q}} \, \|\varphi\|_{\mathcal{P},\mathcal{Q}}, \qquad \forall \varphi \in \mathcal{S}. \tag{2.61}$$

An important class of distributions is given by the so-called *regular distributions*, which are distributions that are represented by a function f. We say that a distribution F is represented by the function $f : \mathbb{R}^{\mathbb{D}} \to \mathbb{C}$, if

$$F(\varphi) = \int f(x) \, \varphi(x) \, d^D x, \qquad \forall \varphi \in \mathcal{S}. \tag{2.62}$$

Using Theorem I it is easy to prove that all *bounded* functions, and all functions with *integrable* singularities which diverge at infinity at most as some power of x, are examples of regular distributions, see Problem 2.4. In particular every test function φ represents a regular distribution, so that we may write the formal inclusion $\mathcal{S} \subset \mathcal{S}'$.

Symbolic notation. In general we are not allowed to multiply, or divide, distributions, nor is the value of a distribution F at a point x a well-defined quantity. Nevertheless, certain properties of distributions are more transparent if one resorts to the so-called *symbolic notation*, which consists in the introduction of a *formal* function $F(x)$ such that

$$F(\varphi) = \int F(x) \, \varphi(x) \, d^D x.$$

This way of writing mimics the rigorous expression (2.62), which is valid for regular distributions, and it usually leads to considerable simplifications of the mathematical manipulations. Henceforth, in this book we will mainly resort to the symbolic notation. The operations we present below refer to distributions F applied to test functions $\varphi \in \mathcal{S}$. Nevertheless, in many cases these operations will maintain their validity even if the distributions are applied to functions that are less regular than those of $\mathcal{S}$.

Sequences of distributions. If we endow the space $\mathcal{S}'$ with the so-called *weak topology*, see [1], it makes sense to analyze the convergence properties of sequences of distributions. By the very definition of this topology, a sequence of distributions $F_n \in \mathcal{S}'$ converges to a distribution F,

$$\mathcal{S}' - \lim_{n \to \infty} F_n = F, \tag{2.63}$$

if and only if for any $\varphi \in \mathcal{S}$ one has the ordinary limits in $\mathbb{C}$

$$\lim_{n \to \infty} F_n(\varphi) = F(\varphi). \tag{2.64}$$

The symbol "$\mathcal{S}' - \lim$" appearing in (2.63) will, henceforth, denote the distributional limit in (the weak topology of) $\mathcal{S}'$.

Derivative of a distribution. Every element $F \in \mathcal{S}'$ admits partial derivatives with respect to any coordinate x^{μ}, and the partial derivatives $\partial_{\mu} F$ still belong to $\mathcal{S}'$. They are defined by

$$(\partial_\mu F)(\varphi) = -F(\partial_\mu \varphi), \qquad \forall \varphi \in \mathcal{S}. \tag{2.65}$$

This definition entails the fundamental property that distributional derivatives *always commute*:

$$\partial_\mu \partial_\nu F = \partial_\nu \partial_\mu F. \tag{2.66}$$

We have, in fact,

$$(\partial_\mu \partial_\nu F)(\varphi) = -(\partial_\nu F)(\partial_\mu \varphi) = F(\partial_\nu \partial_\mu \varphi) = F(\partial_\mu \partial_\nu \varphi) = (\partial_\nu \partial_\mu F)(\varphi),$$

where we used the commutativity of partial derivatives if applied to functions of $\mathcal{S}$. Since the above equality holds for all φ, relation (2.66) follows. Once the space $\mathcal{S}'$ has been equipped with the weak topology, one can moreover prove that the derivatives are *continuous* operations in $\mathcal{S}'$. A direct consequence of this property is the following very useful theorem.

Theorem II. If a sequence F_n of distributions converges to F, i.e.

$$\mathcal{S}' - \lim_{n \to \infty} F_n = F,$$

then also the sequence of its derivatives $\partial_\mu F_n$ is convergent, and furthermore its limit equals $\partial_\mu F$:

$$\mathcal{S}' - \lim_{n \to \infty} \partial_\mu F_n = \partial_\mu F. \tag{2.67}$$

This property is of fundamental importance in that it allows to *interchange limits with derivatives*.

In practice, the explicit evaluation of the derivative of a distribution F can be simplified if in a subset B of $\mathbb{R}^D$ it is represented by a function $f : \mathbb{R}^D \to \mathbb{C}$ of class C^∞. In this case, for $x \in B$ the derivative of F can be computed simply in the sense of functions, so that the evaluation of the derivative of F is reduced essentially to the determination of $\partial_\mu F$ in the subset $\mathbb{R}^D \setminus B$, which is the *locus* where F is singular. This strategy will prove to be particularly efficient since the singularities of the distributions with which we will have to deal in practice, always constitute sets of *measure zero*.

The spaces O_M. A function $f : \mathbb{R}^D \to \mathbb{C}$ is said to be *polynomially bounded*, if there exists a positive polynomial $\mathcal{P}(x)$, such that $|f(x)| \leq \mathcal{P}(x)$ for all $x \in \mathbb{R}^D$. We define the spaces $O_M(\mathbb{R}^D) = O_M$ as the set of all functions $f : \mathbb{R}^D \to \mathbb{C}$ of class C^∞, which are polynomially bounded together with all their derivatives. In particular we have the inclusions $\mathcal{S} \subset O_M \subset \mathcal{S}'$.

Leibniz's rule. If $\varphi \in \mathcal{S}$ and $F \in \mathcal{S}'$, the product φF belongs to $\mathcal{S}'$. On the other hand, if $f \in O_M$ and $F \in \mathcal{S}'$, it can be shown that the product fF still belongs to $\mathcal{S}'$ and obeys Leibniz's rule

$$\partial_\mu(fF) = (\partial_\mu f)F + f\partial_\mu F. \tag{2.68}$$

A fortiori this rule holds when f is replaced with a test function φ.

Convolution. The convolution $F * \varphi$ between a distribution F and a test function φ is a *regular* distribution, which in symbolic notation can be written as

$$(F * \varphi)(x) = \int F(y)\,\varphi(x-y)\,d^D y. \tag{2.69}$$

The function $f(x) = (F * \varphi)(x)$ that represents it can be shown to belong to O_M, and its derivatives obey the rules

$$\partial_\mu(F * \varphi) = \partial_\mu F * \varphi = F * \partial_\mu \varphi. \tag{2.70}$$

If also $F \in \mathcal{S}$, one has moreover $F * \varphi = \varphi * F$.

Support of a distribution. Roughly speaking – in symbolic language – the support of a distribution F, denoted by $\mathrm{supp}(F)$, is the set of all points where $F(x)$ is different from zero. More formally, we define firstly the support of a test function $\varphi \in \mathcal{S}$ as the set $\mathrm{supp}(\varphi) = \{x \in \mathbb{R}^D / \varphi(x) \neq 0\}$. Given a distribution F and an open set $\Omega \subset \mathbb{R}^D$, we say that F vanishes on Ω if $F(\varphi) = 0$ whenever $\mathrm{supp}(\varphi) \subset \Omega$. The set $\mathrm{supp}(F)$ is defined as the complement of the largest open set Ω on which F vanishes – always a closed set. Notice that if we look at φ as a *distribution*, its support is the closure $\overline{\mathrm{supp}(\varphi)}$.

One-dimensional δ-function. Dirac's one-dimensional δ-function with support in a point $a \in \mathbb{R}$ is the element δ_a of $\mathcal{S}'(\mathbb{R})$ defined by $\delta_a(\varphi) = \varphi(a)$, $\forall\,\varphi \in \mathcal{S}(\mathbb{R})$. With this distribution one associates the symbolic function $\delta(x-a)$ such that

$$\delta_a(\varphi) = \int \delta(x-a)\,\varphi(x)\,dx = \varphi(a). \tag{2.71}$$

We will use the relation (2.71) every time we have to perform an integral over a δ-function. Hence, the rule is the following: eliminate the δ-function as well as the integral sign, and evaluate the function $\varphi(x)$ at the point which is the support of the δ-function. Below, resorting mainly to the symbolic notation, we list some of the most important properties of this distribution, which can easily be proven by applying both members to a test function. We begin with the expression of the nth derivative of the δ-function, for which the definition (2.65) gives

$$\left(\frac{d^n\delta_a}{dx^n}\right)(\varphi) = \int \left(\frac{d^n}{dx^n}\,\delta(x-a)\right)\varphi(x)\,dx = (-)^n\frac{d^n\varphi}{dx^n}(a).$$

For every $f \in O_M$ we then have

$$f(x)\,\delta(x-a) = f(a)\,\delta(x-a). \tag{2.72}$$

This relation implies some simple identities, as for instance

$$x\,\delta(x) = 0, \qquad x^2\frac{d}{dx}\,\delta(x) = 0, \qquad x\frac{d}{dx}\,\delta(x) = -\delta(x),$$

which can be verified by applying both members to a test function. In the same way one proves the relation

$$\frac{dH(x)}{dx} = \delta(x), \tag{2.73}$$

in which $H(x)$ denotes the discontinuous *Heaviside function*

$$H(x) = \begin{cases} 1, & \text{for } x \geq 0, \\ 0, & \text{for } x < 0. \end{cases} \tag{2.74}$$

Another discontinuous function that we will use is the *characteristic function* of an interval $[a, b]$, defined by

$$\chi_{[a,b]}(x) = \begin{cases} 1, & \text{for } a \leq x \leq b, \\ 0, & \text{for } x < a \text{ or } x > b, \end{cases} \tag{2.75}$$

which can also be expressed as $\chi_{[a,b]}(x) = H(x - a) - H(x - b)$. Given a real function $f : \mathbb{R} \to \mathbb{R}$, under certain conditions it is possible to confer a well-defined meaning to the formal expression $\delta(f(x))$. Namely, if f is differentiable and has a *finite* number of zeros $\{x_n\}$ such that the first derivatives $f'(x_n)$ are all different from zero, one sets

$$\delta(f(x)) = \sum_n \frac{\delta(x - x_n)}{|f'(x_n)|}. \tag{2.76}$$

The origin of this definition becomes self-evident if one applies both members to a test function and performs in the integral of the first member the change of variables $x \to y = f(x)$. A function that often appears is $f(x) = x^2 - a^2$, with $a \neq 0$, for which (2.76) yields

$$\delta(x^2 - a^2) = \frac{1}{2|a|}\left(\delta(x - a) + \delta(x + a)\right). \tag{2.77}$$

If, instead, one considers the function $f(x) = c(x - a)$, with $c \neq 0$, (2.76) gives

$$\delta(c(x - a)) = \frac{1}{|c|}\,\delta(x - a).$$

Finally, from the definition of the convolution (2.69) follows the identity, valid for all $\varphi \in \mathcal{S}$,

$$\delta * \varphi = \varphi. \tag{2.78}$$

D-dimensional δ-function. The concept of a distribution with a point-like support generalizes naturally to a space of arbitrary dimension. Given a vector $a^\mu \in \mathbb{R}^D$, the distribution δ_a supported in $x^\mu = a^\mu$ is an element of $\mathcal{S}'(\mathbb{R}^D)$ defined by $\delta_a(\varphi) = \varphi(a), \forall \varphi \in \mathcal{S}(\mathbb{R}^D)$. It is represented by the symbolic function

$$\delta^D(x - a) = \delta(x^0 - a^0)\, \delta(x^1 - a^1) \cdots \delta(x^{D-1} - a^{D-1}). \tag{2.79}$$

In four dimensions we will also use the shorthand notation

$$\delta^4(x - a) = \delta(x^0 - a^0)\, \delta^3(\mathbf{x} - \mathbf{a}). \tag{2.80}$$

In symbolic notation we have

$$\delta_a(\varphi) = \int \delta^D(x - a)\, \varphi(x)\, d^D x = \varphi(a). \tag{2.81}$$

For the derivatives of the δ-function we obtain

$$(\partial_\mu \delta_a)(\varphi) = -\delta_a(\partial_\mu \varphi) = -\partial_\mu \varphi(a),$$

an equality that in symbolic notation becomes

$$\int \partial_\mu \delta^D(x - a)\, \varphi(x)\, d^D x = -\partial_\mu \varphi(a).$$

For every $f \in O_M$ we have moreover

$$f(x)\, \delta^D(x - a) = f(a)\, \delta^D(x - a).$$

This relation, in conjunction with Leibniz's rule (2.68), implies the identities

$$x^\mu \delta^D(x) = 0, \qquad x^\mu x^\nu \partial_\rho \delta^D(x) = 0, \qquad x^\mu \partial_\nu \delta^D(x) = -\delta^\mu_\nu \delta^D(x).$$

If $C^\mu{}_\nu$ is a real invertible $D \times D$ matrix, one has also the symbolic relation

$$\delta^D(C(x - a)) = \frac{\delta^D(x - a)}{|\det C|}. \tag{2.82}$$

A further generalization of (2.76) and (2.82) is represented by the relation

$$\delta^D(f(x)) = \frac{\delta^D(x - a)}{|\det \partial_\mu f^\nu(a)|}, \tag{2.83}$$

where $f : \mathbb{R}^D \to \mathbb{R}^D$ is a differentiable function with a single zero in $x^\mu = a^\mu$, such that the matrix $\partial_\mu f^\nu(a)$ has a non-vanishing determinant.

Distributions with point-like support. We complete our list of properties of the δ-function enouncing a theorem that constrains heavily the form of distributions which are *different from zero* only in a finite set of points, that is whose *support* consists of a finite set of *points*.

Theorem III. A distribution $F \in \mathcal{S}'(\mathbb{R}^D)$ whose support is the point $x^\mu = a^\mu$ is necessarily a *finite* linear combination of the distribution $\delta^D(x-a)$ and of its derivatives, that is to say

$$F = c\delta^D(x-a) + c^\mu \partial_\mu \delta^D(x-a) + \cdots + c^{\mu_1 \cdots \mu_n} \partial_{\mu_1} \ldots \partial_{\mu_n} \delta^D(x-a), \quad (2.84)$$

where $c^{\mu_1 \cdots \mu_n}$ are constant coefficients.

If the support of a distribution consists of N points, it is given by a sum of N expressions like (2.84). Theorem III will prove very useful for the solution of algebraic equations for distributions.

Principal value. The function of a real variable $f(x) = 1/x$ does not define a distribution, due to the non-integrable singularity at $x = 0$. Nevertheless, in $\mathcal{S}'(\mathbb{R})$ we can define a distribution that represents a *regularization* of this function, called *principal value* of $1/x$, that is denoted by the symbol $\mathcal{P}\frac{1}{x}$. It is defined by

$$\left(\mathcal{P}\frac{1}{x}\right)(\varphi) = \int_0^\infty \frac{\varphi(x) - \varphi(-x)}{x}\, dx. \quad (2.85)$$

It is easy to see that for a test function $\varphi(x)$ which vanishes at $x = 0$ the integral (2.85) reduces to the expected expression

$$\left(\mathcal{P}\frac{1}{x}\right)(\varphi) = \int_{-\infty}^\infty \frac{\varphi(x)}{x}\, dx.$$

Further on we will also encounter a distribution of $\mathcal{S}'(\mathbb{R})$ that we call *composite principal value*, which is defined for any non-vanishing real number a by

$$\mathcal{P}\frac{1}{x^2 - a^2} = \frac{1}{2a}\left(\mathcal{P}\frac{1}{x-a} - \mathcal{P}\frac{1}{x+a}\right), \quad (2.86)$$

where the principal values of the second member appear translated by $\pm a$. Notice that, formally, the right-hand side of (2.86) is obtained from its left-hand side through a partial fraction decomposition.

Fourier transform. The *Fourier transform* constitutes a bijection from $\mathcal{S}$ to itself and extends naturally to a bijection from $\mathcal{S}'$ to itself. We denote the Fourier transform of a generic element $\varphi \in \mathcal{S}$ by $\widehat{\varphi}$ and, similarly, the one of a generic element $F \in \mathcal{S}'$ by $\widehat{F}$. In a D-dimensional space-time with Minkowskian signature

$(1, -1, \ldots, -1)$ we define the Fourier transform of a test function $\varphi(x)$ by

$$\widehat{\varphi}(k) = \frac{1}{(2\pi)^{D/2}} \int e^{-ik \cdot x} \, \varphi(x) \, d^D x, \tag{2.87}$$

where we have introduced the D *dual* variables k^μ, and we have set[5]

$$k \cdot x = k^0 x^0 - k^1 x^1 - \cdots - k^{D-1} x^{D-1}.$$

The definition (2.87) implies the *inversion formula*

$$\varphi(x) = \frac{1}{(2\pi)^{D/2}} \int e^{ik \cdot x} \, \widehat{\varphi}(k) \, d^D k,$$

which defines the *inverse Fourier transform*. As one sees, the inverse transform of a function can be simply computed by performing its transform and then changing the sign of the argument. From these relations it follows furthermore that the *square* of the Fourier transform amounts to the reflection of all coordinates

$$\widehat{\widehat{\varphi}}(x) = \varphi(-x). \tag{2.88}$$

Once we have introduced the Fourier transform of a test function, the Fourier transform $\widehat{F}$ of a distribution F is defined unambiguously by the relation

$$\widehat{F}(\varphi) = F(\widehat{\varphi}), \quad \forall \varphi \in \mathcal{S}. \tag{2.89}$$

From this definition one can deduce that the *symbols* $F(x)$ and $\widehat{F}(k)$ are tied by the *symbolic* relations

$$\widehat{F}(k) = \frac{1}{(2\pi)^2} \int e^{-ik \cdot x} F(x) \, d^4 x, \qquad F(x) = \frac{1}{(2\pi)^2} \int e^{ik \cdot x} \widehat{F}(k) \, d^4 k. \tag{2.90}$$

However, it is important to keep in mind that the Fourier transform of F and of its inverse can be determined by evaluating the integrals (2.90) only if the distributions F and $\widehat{F}$ are *sufficiently regular*. In the general case these integrals are, in fact, ill-defined as Lebesgue integrals, and to determine $\widehat{F}$ one must resort to the definition (2.89).

In the following we report the Fourier transforms of some important distributions. The Fourier transform of the derivative of a distribution and of a distribution multiplied by a coordinate are given by

---

[5]With respect to the spatial variables $(x^1, \cdots, x^{D-1})$ the function $\widehat{\varphi}(k)$ in (2.87) corresponds, actually, to the *inverse* Fourier transform. In a Minkowski space-time this choice has the advantage to preserve the manifest Lorentz invariance. For example, if $\varphi(x)$ is a scalar field, and k^μ is considered as a four-vector, then also $\widehat{\varphi}(k)$ is a scalar field, see Sect. 5.2.

$$\widehat{\partial_\mu F}(k) = ik_\mu \widehat{F}(k), \qquad \widehat{x^\mu F}(k) = i\frac{\partial}{\partial k_\mu}\widehat{F}(k), \tag{2.91}$$

respectively. The generalization of (2.91) to the Fourier transform of a generic ordered polynomial $P(\partial, x)$ applied to F reads

$$[\widehat{P(\partial, x)F}](k) = P\left(ik, i\frac{\partial}{\partial k}\right)\widehat{F}(k). \tag{2.92}$$

We recall also the transforms of the δ-function $\delta^D(x)$, of its derivatives, and of the constant distribution

$$\widehat{\delta^D}(k) = \frac{1}{(2\pi)^{D/2}}, \quad \widehat{\partial_\mu \delta^D}(k) = \frac{ik_\mu}{(2\pi)^{D/2}}, \quad \widehat{(1)}(k) = (2\pi)^{D/2}\delta^D(k). \tag{2.93}$$

In the one-dimensional distribution space $\mathcal{S}'(\mathbb{R})$ we sometimes also need the transforms of the *sign function*

$$\varepsilon(x) = H(x) - H(-x), \tag{2.94}$$

and of the principal value distribution (2.85)

$$\widehat{\varepsilon}(k) = -i\sqrt{\frac{2}{\pi}}\,\mathcal{P}\frac{1}{k}, \qquad \left(\widehat{\mathcal{P}\frac{1}{x}}\right)(k) = -i\sqrt{\frac{\pi}{2}}\,\varepsilon(k). \tag{2.95}$$

Notice that these formulas respect the general identity (2.88). Furthermore, there are simple rules to calculate the Fourier transform of a translated distribution $F_a(x) = F(x + a)$,

$$\widehat{F_a}(k) = e^{ik\cdot a}\widehat{F}(k), \tag{2.96}$$

and of a distribution with inverted argument $F_-(x) = F(-x)$,

$$\widehat{F_-}(k) = \widehat{F}(-k). \tag{2.97}$$

Finally, we remark that the third relation in (2.93) amounts to the symbolic representation of the δ-function

$$\int e^{-ik\cdot x}\,d^D x = (2\pi)^D\,\delta^D(k), \tag{2.98}$$

which is of common use in several branches of theoretical physics. The integral at the left-hand side is, of course, divergent, but one can give a rigorous meaning to the *formal* relation (2.98) by noting the distributional limit, see Problem 2.14,

$$S' - \lim_{L \to \infty} \int_{|x^\mu| < L} e^{-ik \cdot x} \, d^D x = (2\pi)^D \delta^D(k). \tag{2.99}$$

Fourier transform of the convolution. There exists a simple expression for the Fourier transform of the convolution $F * \varphi$ between an element F of $\mathcal{S}'$ and an element φ of $\mathcal{S}$. In a D-dimensional space – in symbolic notation – it is given by

$$\widehat{F * \varphi}(k) = (2\pi)^{D/2} \widehat{F}(k) \, \widehat{\varphi}(k). \tag{2.100}$$

Since a distribution multiplied by a function $f \in O_M$ remains a distribution, we cite a last useful theorem – actually a corollary of the *Paley-Wiener theorem*, see, for instance, Ref. [1] – regarding precisely those functions.

Theorem IV. The Fourier transform $\widehat{F}$ of a distribution F with *compact support* is an element of O_M.

The spaces L^p. In addition to the spaces O_M we will sometimes also encounter the *Banach spaces*

$$L^p(\mathbb{R}^D) = \left\{ f : \mathbb{R}^D \to \mathbb{C} \Big/ \int |f(x)|^p \, d^D x < \infty \right\}, \tag{2.101}$$

where $1 \le p < \infty$ is a real number. With respect to the functions of O_M the functions of L^p are less regular, in the sense that they do not need to be continuous, or even differentiable, but, since their absolute value is integrable if elevated to a positive power p, they behave much better than the former for large $|\mathbf{x}|$. Like the spaces O_M, the spaces L^p entail the inclusions $\mathcal{S} \subset L^p \subset \mathcal{S}'$ for all p.

2.3.2 Maxwell's Equations in the Space of Distributions

Once established that the equations of the electromagnetic field (2.21) and (2.22) can be formulated correctly only in the space of distributions, it is advisable to reexamine them briefly in this new framework.

Conservation and covariance of the four-current. As a preliminary step we analyze the properties of the four-current of a system of point-like charges (2.16)

$$j^\mu(x) = \sum_r e_r \int \frac{dy_r^\mu}{d\lambda_r} \, \delta^4(x - y_r(\lambda_r)) \, d\lambda_r, \tag{2.102}$$

that is now considered as a *distribution-valued vector field*. With this we mean that each of the four components of j^μ is a distribution in $\mathcal{S}'(\mathbb{R}^4)$. If we apply the symbolic expression (2.102) to a test function $\varphi(x)$ we obtain, in fact, the four linear functionals

$$j^\mu(\varphi) = \int j^\mu(x)\,\varphi(x)\,d^4x = \sum_r e_r \int \frac{dy_r^\mu}{d\lambda_r}\,\varphi(y_r(\lambda_r))\,d\lambda_r, \qquad (2.103)$$

where we interchanged the integrations with respect to x and λ_r, and used the symbolic relation (2.81). The proof that for each of the four values of μ the functional (2.103) defines, actually, a distribution in $\mathcal{S}'(\mathbb{R}^4)$ is left as an exercise, see Problem 2.13. First of all, we show that the so-defined current is a four-vector. For this purpose we consider the test functions as scalar fields, i.e. we require that under a Poincaré transformation $x' = \Lambda x + a$ from K to K' it transforms as $\varphi'(x') = \varphi(x)$. We must then prove that the functionals (2.103) obey the transformation law

$$j'^\mu(\varphi') = \Lambda^\mu{}_\nu j^\nu(\varphi). \qquad (2.104)$$

Since in the reference frame K' these functionals are given by

$$j'^\mu(\varphi') = \sum_r e_r \int \frac{dy_r'^\mu}{d\lambda_r}\,\varphi'(y_r'(\lambda_r))\,d\lambda_r,$$

the relations (2.104) then follow from the standard transformation laws

$$y_r'^\mu = \Lambda^\mu{}_\nu y_r^\nu + a^\mu, \qquad dy_r'^\mu = \Lambda^\mu{}_\nu dy_r^\nu, \qquad \varphi'(y_r'(\lambda_r)) = \varphi(y_r(\lambda_r)).$$

Next we prove that the current (2.103) is conserved in the *distributional sense*, namely that $(\partial_\mu j^\mu)(\varphi) = 0$ for all test functions. Recalling the definition of the distributional derivative (2.65), and using the explicit expression (2.103) with the replacement $\varphi \to \partial_\mu \varphi$, we obtain

$$
\begin{aligned}
(\partial_\mu j^\mu)(\varphi) = -j^\mu(\partial_\mu \varphi) &= -\sum_r e_r \int \frac{dy_r^\mu}{d\lambda_r}\,\partial_\mu\varphi(y_r(\lambda_r))\,d\lambda_r \\
&= -\sum_r e_r \int \frac{d\varphi(y_r(\lambda_r))}{d\lambda_r}\,d\lambda_r \qquad (2.105)\\
&= -\sum_r e_r\big(\varphi(y_r(\infty)) - \varphi(y_r(-\infty))\big) = 0.
\end{aligned}
$$

Both terms in the last line vanish because for $\lambda_r \to \pm\infty$ we have that $y_r^0(\lambda_r) \to \pm\infty$, and the test functions vanish at infinity in all directions, in particular along the time-line.

Finally, we illustrate the use of the δ-distribution as a symbolic function, deriving the expressions (2.18) and (2.19) starting from the symbolic representation (2.102). Since the current is *reparameterization invariant* we can write out the rth integral in (2.102) choosing as integration variable the time coordinate of the rth particle, i.e. choosing $\lambda_r = y_r^0$. Using the factorization (2.80) we obtain thus

$$j^\mu(t, \mathbf{x}) = \sum_r e_r \int_{\gamma_r} \frac{dy_r^\mu(y_r^0)}{dy_r^0} \, \delta^4\big(x - y_r(y_r^0)\big) \, dy_r^0 \tag{2.106}$$

$$= \sum_r e_r \int_{\gamma_r} \frac{dy_r^\mu(y_r^0)}{dy_r^0} \, \delta\big(t - y_r^0\big) \, \delta^3\big(\mathbf{x} - \mathbf{y}_r(y_r^0)\big) \, dy_r^0. \tag{2.107}$$

We use now (2.71) to perform the integral of $\delta(t - y_r^0)$ over dy_r^0, considering the rest of the integrand as a "test function" which correspondingly must be evaluated at $y_r^0 = t$. Accordingly, we obtain

$$j^\mu(t, \mathbf{x}) = \sum_r e_r \frac{dy_r^\mu(t)}{dt} \, \delta^3(\mathbf{x} - \mathbf{y}_r(t)), \tag{2.108}$$

where the identification $y_r^0(t) = t$ is understood. Writing out the time and space components of (2.108) we obtain finally the expressions anticipated in (2.18) and (2.19)

$$j^0(t, \mathbf{x}) = \sum_r e_r \, \delta^3(\mathbf{x} - \mathbf{y}_r(t)), \tag{2.109}$$

$$\mathbf{j}(t, \mathbf{x}) = \sum_r e_r \mathbf{v}_r(t) \, \delta^3(\mathbf{x} - \mathbf{y}_r(t)). \tag{2.110}$$

Bianchi identity. We move now on to the reanalysis of the Bianchi identity and the Maxwell equation. According to the conclusions drawn in the introduction to Sect. 2.3, we must consider the Maxwell tensor $F^{\mu\nu}$ as a distribution-valued tensor field, and the derivatives $\partial_\nu F_{\rho\sigma}$ appearing in Eqs. (2.21) and (2.22) as derivatives in the sense of distributions. As for the Bianchi identity, in the first place we have to face the problem of whether the expression $F_{\mu\nu} = \partial_\mu A_\nu - \partial_\nu A_\mu$ is still a solution of equation (2.21), i.e. if the formal computation done in (2.43) is still valid. If we want to give this question a well-defined meaning we must, first of all, qualify also A^μ as a *distribution-valued* vector field. The correctness of the computation performed in (2.43) – independently of the presence or absence of singularities in A^μ, provided they are of the *distributional type* – is then guaranteed by the commutativity of the distributional derivatives, see (2.66),

$$\partial_\nu \partial_\rho A_\sigma = \partial_\rho \partial_\nu A_\sigma.$$

The expression $F_{\mu\nu} = \partial_\mu A_\nu - \partial_\nu A_\mu$ satisfies, therefore, the Bianchi identity also in the *sense of distributions*. In Chap. 19 we will actually see that also the converse is true, namely that *every* distributional solution of the Bianchi identity in $\mathcal{S}'(\mathbb{R}^4)$ can be written in the form (2.42), with A^μ a certain distribution-valued vector field. Moreover, one can prove that if two such vector fields A'_μ and A_μ give rise to the same electromagnetic field $F_{\mu\nu}$, there exists always a scalar distribution Λ such that $A'_\mu = A_\mu + \partial_\mu \Lambda$. In this way we confirm that the general solution of the Bianchi identity in the *space of distributions* is still represented by formulas (2.46), notwithstanding their original derivation in the *space of functions*.

Maxwell equation. Thanks to the fact that, by construction, the current (2.102) is a distribution-valued field and that, by hypothesis, so is also $F^{\mu\nu}$, the Maxwell equation (2.22) is a well-defined partial differential equation in the space of distributions. Since in $\mathcal{S}'$ we also have $\partial_\mu j^\mu = 0$, the consistency of this equation requires again the constraint $\partial_\mu \partial_\nu F^{\mu\nu} = 0$. This time this constraint is satisfied identically in $\mathcal{S}'$ in virtue of the commutativity of the distributional derivatives, independently of the *distributional* singularities that may be present in $F^{\mu\nu}$. Eventually, once one has solved the Bianchi identity in terms of a vector potential A^μ according to (2.46), the Maxwell equation reduces to a distributional partial differential equation for A^μ.

2.3.3 The Electromagnetic Field of a Static Particle

The need to consider the equations that govern the dynamics of the electromagnetic field in the space of distributions emerges very clearly from the simple example of a static particle. For this reason we reanalyze now this *ancient* case in some detail. The world line associated with a static particle sitting at the origin, parameterized with time, has the form

$$y^0(t) = t, \qquad \mathbf{y}(t) = 0,$$

so that $\mathbf{v}(t) = 0$. According to formulas (2.109) and (2.110) such a particle then entails the charge and current densities

$$\rho(t, \mathbf{x}) = e\delta^3(\mathbf{x}), \qquad \mathbf{j}(t, \mathbf{x}) = 0.$$

In this case the magnetic field is zero, $\mathbf{B} = 0$, and the electric field is static. In the three-dimensional notation (2.52)–(2.55), the Maxwell equation and the Bianchi identity then reduce to the *fundamental equations of electrostatics*

$$\nabla \cdot \mathbf{E} = e\,\delta^3(\mathbf{x}), \qquad \nabla \times \mathbf{E} = 0, \tag{2.111}$$

respectively, where we recall that the latter is equivalent to $\partial_i E^j - \partial_j E^i = 0$. As is well known, the solution of this system of equations ought to be the *Coulomb field*

$$\mathbf{E}(t, \mathbf{x}) = \frac{e\mathbf{x}}{4\pi r^3}, \qquad r = |\mathbf{x}|, \tag{2.112}$$

a statement that in the following we will reexamine critically in the distributional framework. As a preliminary step, we determine the derivatives of $\mathbf{E}$ in the sense of *functions*. Since $\partial_i r = x^i/r$, for $\mathbf{x} \neq 0$ we obtain

$$\partial_i E^j = \frac{e}{4\pi r^3} \left(\delta^{ij} - 3\frac{x^i x^j}{r^2} \right). \tag{2.113}$$

The Bianchi identity in (2.111) seems therefore to be satisfied, since the derivatives (2.113) are symmetric in i and j. Conversely, the Maxwell equation seems violated since (2.113) would imply a divergence-less electric field: $\partial_i E^i = 0$. Our mistake consists apparently in having considered the fields E^j, as well as the derivatives ∂_i, in the sense of *functions*.

The equations of electrostatics in the space of distributions. We analyze now the solutions of the system (2.111) in the *three*-dimensional distribution space $\mathcal{S}' = \mathcal{S}'(\mathbb{R}^3)$, which is appropriate for *static* charged systems. First of all we must ask whether the components of the electric field $\mathbf{E}$ in (2.112) belong to $\mathcal{S}'$. The answer is affirmative because the components E^j have an *integrable* singularity in $\mathbf{x} = 0$, and at infinity they are bounded by a constant, see Problem 2.4. As a consequence of E^j being distributions, the derivatives $\partial_i E^j$ are well-defined in $\mathcal{S}'$, but they must be computed in the sense of distributions, i.e. relying on the very definition (2.65). Choosing a test function $\varphi(\mathbf{x})$ we find

$$
\begin{aligned}
(\partial_i E^j)(\varphi) &= -E^j(\partial_i \varphi) = -\frac{e}{4\pi}\int \frac{x^j}{r^3}\partial_i\varphi\, d^3x = -\frac{e}{4\pi}\lim_{\varepsilon\to 0}\int_{r>\varepsilon} \frac{x^j}{r^3}\partial_i\varphi\, d^3x \\
&= -\frac{e}{4\pi}\lim_{\varepsilon\to 0}\int_{r>\varepsilon}\left(\partial_i\left(\frac{x^j}{r^3}\varphi\right) - \partial_i\left(\frac{x^j}{r^3}\right)\varphi\right)d^3x \\
&= \frac{e}{4\pi}\lim_{\varepsilon\to 0}\left(\int_{r>\varepsilon}\frac{1}{r^3}\left(\delta^{ij} - 3\frac{x^i x^j}{r^2}\right)\varphi\, d^3x + \int_{r=\varepsilon} n^i n^j\, \varphi\, d\Omega\right) \\
&= \frac{e}{4\pi}\int \frac{1}{r^3}\left(\delta^{ij} - 3\frac{x^i x^j}{r^2}\right)\varphi\, d^3x + \frac{e}{3}\delta^{ij}\varphi(0,0,0).
\end{aligned}
\tag{2.114}
$$

We explain now the various steps. For fixed indices i and j the integrand in the first line is an element of the space of *absolutely integrable* functions $L^1(\mathbb{R}^3)$, see (2.101). According to a known result of *Mathematical Analysis*, the integrals of these functions over $\mathbb{R}^D$ – without the modulus – can be calculated by introducing an arbitrary sequence of integration domains converging to $\mathbb{R}^D$. Above we have chosen the sequence $V_\varepsilon = \mathbb{R}^3 \backslash S_\varepsilon$, where S_ε is the ball of radius ε centered at the origin. Since in V_ε the integrand is of class C^∞, in this domain we can use the standard differential calculus. Correspondingly, in the second line we applied Leibniz's rule to compute the derivatives of the first term and to the second term we applied Gauss's theorem. The boundary of V_ε consists of the sphere at infinity – which gives a vanishing contribution to the flux since φ vanishes more rapidly than the inverse of any power of r – and of the sphere of radius ε centered at the origin. To evaluate the integral over this sphere we went over to polar coordinates $\mathbf{x} \leftrightarrow (r, \phi, \vartheta)$, with solid angle $d\Omega = \sin\vartheta\, d\vartheta\, d\phi$, and we introduced the radial outgoing unit vector $n^i = x^i/r = x^i/\varepsilon$. The surface element then becomes $d\Sigma^i = n^i \varepsilon^2 d\Omega$. Finally, we used the *invariant integral* over angles, see Problem 2.6,

$$\int n^i n^j \, d\Omega = \frac{4\pi}{3} \delta^{ij}.$$

The limit of the first integral in the third line gives rise trivially to the corresponding integral in the fourth line, but the latter must be understood as a *conditionally convergent* integral. This property is defined as follows.

Definition. A multiple integral is called *conditionally convergent*, if it converges once the integrals over the single variables are performed in a specific *order*, or if it converges once a well-defined sequence of integration domains approaching the actual integration region has been chosen.

Regarding the case at hand, in the first integral of the fourth line of (2.114) we must first integrate over the polar angles ϕ and ϑ, and afterwards over the radius r, otherwise the integral would not even converge, see Problem 2.12. Rewriting the relation (2.114), valid for any φ, in symbolic notation, for the distributional derivatives of the electric field eventually we obtain the expression

$$\partial_i E^j = \frac{e}{4\pi r^3} \left(\delta^{ij} - 3\frac{x^i x^j}{r^2} \right) + \frac{e}{3} \delta^{ij} \delta^3(\mathbf{x}). \qquad (2.115)$$

By construction the distributional derivative of a distribution is again a distribution, and so the second member of (2.115) is surely a distribution. However, the first two terms on the right-hand side both diverge as $1/r^3$ for $r \to 0$, and hence they are non-integrable functions in every neighborhood of $r = 0$: Taken separately, they are, therefore, *not* distributions. It is only the particular combination appearing in (2.115), with relative coefficient -3, to be an element of $\mathcal{S}'$, see Problem 2.12. Comparing Eqs. (2.115) and (2.113), we see that the naive calculation reproduces the correct result in the region $\mathbf{x} \neq 0$, whereas it fails to reveal the presence of the δ-function supported in $\mathbf{x} = 0$, where the electric field is, in fact, singular. The expression (2.115) satisfies now both equations of the system (2.111) – the term involving the δ-function is symmetric in i and j and therefore does not affect the Bianchi identity – the first one being in particular equivalent to the identity in $\mathcal{S}'$

$$\nabla \cdot \frac{\mathbf{x}}{r^3} = 4\pi \delta^3(\mathbf{x}). \qquad (2.116)$$

Finally, we may reread our results in terms of the electrostatic potential A^0. The general solution of the Bianchi identity $\nabla \times \mathbf{E} = 0$ has in fact the form

$$\mathbf{E} = -\nabla A^0, \qquad (2.117)$$

where A^0 is the scalar potential. Using the same procedure that we employed above to prove (2.115), it is easy to verify that the relation (2.117) is satisfied in the distributional sense if one chooses for the potential the known expression

$$A^0 = \frac{e}{4\pi r},$$ (2.118)

still an element of $\mathcal{S}'$. In this way the Maxwell equation $\mathbf{\nabla} \cdot \mathbf{E} = e\,\delta^3(\mathbf{x})$ goes over to the Poisson equation $-\nabla^2 A^0 = e\,\delta^3(\mathbf{x})$, which in light of (2.118) is equivalent to the important distributional identity

$$\nabla^2 \frac{1}{r} = -4\pi\delta^3(\mathbf{x}),$$ (2.119)

of frequent use in many branches of theoretical physics.

Inconsistency of the Lorentz equation. Once we have solved Maxwell's equations we face the solution of the Lorentz equation, whose independent components are given in (2.37). Since in this case we have $\mathbf{v}(t) = \mathbf{y}(t) = 0$, the first member of this equation is identically zero, whereas its second member reduces to $e\,\mathbf{E}(t,0,0,0)$ – a formal expression that, due to the singular behavior of the electric field (2.112) in $\mathbf{x} = 0$, is *divergent*. Hence, we experience first-hand a phenomenon that we anticipated several times and that we will investigate in depth in Chap. 15: the force exerted by the electromagnetic field produced by a particle on the particle itself, the so-called *self-force*, is of infinite strength, and so the Lorentz equation, as it stands, is meaningless. Actually, in the case at hand of a static particle, there exists a well-known *pragmatic* solution to this problem: in agreement with experience, one sets the – diverging – quantity $\mathbf{E}(t,0,0,0)$ simply to zero, because in reality one observes that an isolated charged particle is not subjected to any acceleration and, hence, to any force. We will see, however, that for a *dynamical* particle this simple prescription would violate the basic paradigm of total four-momentum conservation and so, in the general case, it cannot be adopted.[6]

Ultraviolet divergent energy of the electromagnetic field. We conclude the analysis of the static particle with a comment regarding energy conservation, anticipating the expression of the *energy density* of the electromagnetic field (2.133)

$$w_{\mathrm{em}} = \frac{1}{2}\left(E^2 + B^2\right).$$

Substituting for $\mathbf{E}$ the Coulomb field (2.112), and setting $\mathbf{B} = 0$, for the total energy of the electromagnetic field of a static particle we would therefore obtain the divergent "constant"

$$\varepsilon_{\mathrm{em}} = \int w_{\mathrm{em}}\,d^3x = \frac{1}{2}\left(\frac{e}{4\pi}\right)^2 \int \frac{d^3x}{r^4},$$

where the divergence is caused – once more – by the singular behavior of the electric field at the particle's position $\mathbf{x} = 0$, and so it represents an *ultraviolet* divergence. On the other hand, in this case the energy of the particle $\varepsilon_{\mathrm{p}} = m/\sqrt{1 - v^2} = m$

---

[6]Setting, in the general case, $\mathbf{E}(t,0,0,0)$ equal to zero would, in particular, eliminate *tout court* the fundamental energy-loss phenomenon of accelerated particles.

is constant and finite. Consequently, if one wants the total energy $\varepsilon_{em} + \varepsilon_p$ to be conserved and finite, the energy of the electromagnetic field should be constant and *finite*. In Chap. 16 we will show that in the case of a static particle the only value of ε_{em} compatible with Lorentz invariance is, indeed, $\varepsilon_{em} = 0$. However, we will also see that in the case of a particle in arbitrary motion this simple prescription would fail, for it would violate, once more, the conservation of the total four-momentum as well as Lorentz invariance.

Both the problem of the infinite energy of the electromagnetic field and that of the infinite self-force arise from the singular behavior of the electric field at the position of the particle, and it is not difficult to realize that the origin of these ultraviolet divergences, that is to say, divergences arising from the laws that govern the physics at small distances, is precisely the point-like nature of the charged particles. While the second problem is, ultimately, still unsolved, see Chap. 15, the former has met an ultimate solution, although only recently [3], in the framework of the *theory of distributions*. We shall present it in a physically more transparent form in Chap. 16.

Singularities of the electric field and of the charge density. We illustrate the link between the singularities of the electrostatic field and the singularities of the charge distribution, listing the following known examples.

(1) Point-like particle:

$$\mathbf{E} = \frac{e}{4\pi} \frac{\mathbf{x}}{|\mathbf{x}|^3}, \qquad \rho = e\,\delta(x)\delta(y)\delta(z).$$

(2) Linear charge distribution:

$$\mathbf{E} = \frac{\lambda}{2\pi} \frac{(x,y,0)}{x^2 + y^2}, \qquad \rho = \lambda\,\delta(x)\delta(y).$$

(3) Planar charge distribution:

$$\mathbf{E} = \frac{\sigma}{2}\left(\varepsilon(x), 0, 0\right), \qquad \rho = \sigma\,\delta(x).$$

Recall that $\varepsilon(\cdot)$ denotes the *sign* function. One sees that the more regular a charge distribution is *locally*, i.e. at finite distances, the more regular is the electric field it produces in the vicinity of the charges. Conversely, a charge distribution that extends over regions becoming larger and larger, reaching infinity, produces fields with a weaker and weaker decrease at infinity. Notice that in all three cases the equations of electrostatics $\nabla \cdot \mathbf{E} = \rho$ and $\nabla \times \mathbf{E} = 0$ are satisfied in the *distributional* sense, the proofs being left as exercises.

2.4 Conservation Laws

Among the various laws of nature a special role is reserved in physics to the so-called *conservation laws*. These laws assert that during the time evolution of a system certain observable quantities, called *constants of motion*, do not vary. Examples are the energy, the momentum and the angular momentum of an isolated system. There is a very close connection between the *conservation laws* and the *continuous symmetries* of a physical system, which is implemented in a universal manner by *Noether's theorem*. The conceptual importance of this theorem, which in addition to establishing the existence of constants of motion provides also their explicit form, resides in its general validity: it is valid in any theory whose equations of motion stem from a *variational principle*, a valuable prototype of such a theory being precisely electrodynamics. In this section, we will identify the main constants of motion of electrodynamics relying on *heuristic* methods, i.e. without resorting to this theorem, using rather basic notions of electromagnetism. This alternative route is feasible, because the fundamental equations of electrodynamics are relatively simple. In Chap. 4 we will then verify that the constants of motion derived in this way are in agreement with those provided by Noether's theorem.

2.4.1 Conservation and Invariance of the Electric Charge

As prototype of a *local* conservation law, i.e. a conservation law based on a continuity equation relative to a certain four-current j^μ, we consider the conservation of the electric charge. If the charged matter consists of point-like particles the current is given by (2.16); if conversely the charge is spread out *smoothly*, as in macroscopic systems, the current will have a generic form. For what follows, the particular form of the current $j^\mu(x)$ will be irrelevant in that we will make reference only to the following assumptions:

(1) j^μ is a vector field;
(2) j^μ satisfies the continuity equation $\partial_\mu j^\mu = 0$;
(3) $\lim_{|\mathbf{x}| \to \infty} \left(|\mathbf{x}|^3 j^\mu(t, \mathbf{x}) \right) = 0$.

The meaning of the condition (3) is that for a fixed t the current decreases at spatial infinity more rapidly than $1/|\mathbf{x}|^3$, a property certainly possessed by the expressions (2.109) and (2.110). Assuming that j^μ satisfies the conditions (1)–(3), we now prove that there exists a *conserved* as well as *Lorentz invariant* total charge Q. The construction of Q follows a standard procedure which consists in integrating the continuity equation over a generic spatial volume V

$$\int_V \partial_0 j^0 \, d^3x = -\int_V \boldsymbol{\nabla} \cdot \mathbf{j} \, d^3x. \tag{2.120}$$

Applying to the second member Gauss's theorem, and defining the charge contained in a volume V as

$$Q_V = \int_V j^0 \, d^3x,$$

we obtain the *local* conservation law

$$\frac{dQ_V}{dt} = -\int_{\partial V} \mathbf{j} \cdot d\mathbf{\Sigma}. \tag{2.121}$$

This equation states that if the charge contained in a volume V varies, its variation equals exactly minus the amount of charge flowing across the boundary of V. This is, by its very definition, the precise meaning of *local* conservation of a physical quantity. In Eq. (2.121) we may now enlarge the volume V until it covers the whole space $\mathbb{R}^3$. For this purpose we choose for V a ball of radius r and then take of both members of (2.121) the limit for $r \to \infty$. Thanks to property (3) for $\mu = 0$, the limit

$$\lim_{r \to \infty} Q_V = \int j^0 \, d^3x = Q$$

gives rise to a convergent integral, defining the *total* charge Q. On the other hand, the flux $\int_{\partial V} \mathbf{j} \cdot d\mathbf{\Sigma}$ goes to zero as $r \to \infty$. In fact, going over to polar coordinates we have $d\mathbf{\Sigma} = \mathbf{n} \, r^2 d\Omega$ where, as in (2.114), $\mathbf{n}$ is the versor normal to the sphere and $d\Omega$ the solid angle, so that

$$\lim_{r \to \infty} \int_{\partial V} \mathbf{j} \cdot d\mathbf{\Sigma} = \int \mathbf{n} \cdot \lim_{r \to \infty} \left(r^2 \mathbf{j}\right) d\Omega = 0. \tag{2.122}$$

The integral of the second member vanishes because, thanks to property 3) for $\mu = i$, the limit between parentheses is zero. In this limit, Eq. (2.121) thus implies that the total charge is also *conserved*

$$\frac{dQ}{dt} = 0. \tag{2.123}$$

Relativistic invariance of the total charge. The fact that the total charge Q is a scalar under Lorentz transformations – a property certainly not owned by the charge Q_V in a finite volume V – is less obvious. To show it we evaluate in a certain reference frame K the total (time-independent) charge at the instant $t = 0$, rewriting it in the form

$$Q = \int j^0(0, \mathbf{x}) \, d^3x = \int j^0(x) \, \delta(t) \, d^4x$$

$$= \int j^0(x) \, \partial_0 H(t) \, d^4x = \int j^\mu(x) \, \partial_\mu H(t) \, d^4x. \tag{2.124}$$

In (2.124) we introduced the Heaviside function $H(t)$ and made use of the distributional identity (2.73). We consider now the total charge Q' in another reference frame K', linked to K by means of a *proper* Lorentz transformation $x'^\mu = \Lambda^\mu{}_\nu x^\nu$. We then have in particular $\Lambda^0{}_0 \geq 1$. In the same way as above we find now

$$Q' = \int j'^\mu(x')\, \partial'_\mu H(t')\, d^4x'.$$

Making use of the Lorentz transformations

$$j'^\mu(x') = \Lambda^\mu_\nu j^\nu(x),$$
$$\partial'_\mu = \widetilde{\Lambda}_\mu{}^\nu \partial_\nu,$$
$$d^4x' = |\det \Lambda|\, d^4x = d^4x,$$

we then obtain

$$Q' = \int j^\mu(x)\, \partial_\mu H(t')\, d^4x,$$

where the time $t' = t'(t, \mathbf{x})$ is given by

$$t' = \Lambda^0{}_0\, t + \Lambda^0{}_i\, x^i. \tag{2.125}$$

We can now evaluate the difference

$$Q' - Q = \int j^\mu(x)\, \partial_\mu \big(H(t') - H(t)\big) d^4x = \int \partial_\mu \big(j^\mu(x)\, (H(t') - H(t))\big) d^4x, \tag{2.126}$$

where to get the last expression we used the continuity equation of j^μ. Next we break up the four-divergence in its spatial and temporal parts, applying Gauss's theorem to the first one and the *fundamental theorem of calculus* for the time coordinate to the second one. Assuming that we can interchange the orders of integration we obtain

$$Q' - Q = \int dt \int_{\Gamma_\infty} (H(t') - H(t))\, \mathbf{j}(\mathbf{x}) \cdot d\mathbf{\Sigma} + \int d^3x \big[j^0(x)(H(t') - H(t))\big]_{t=-\infty}^{t=+\infty}. \tag{2.127}$$

In the first integral Γ_∞ is a spherical surface placed at spatial infinity, where $\mathbf{j}$ decreases more rapidly than $1/|\mathbf{x}|^3$; this integral is therefore zero, see (2.122). In the second integral we must evaluate the difference $H(t') - H(t)$ in the limit for $t \to \pm\infty$ for a fixed $\mathbf{x}$. Thanks to the inequality $\Lambda^0{}_0 \geq 1$, from Eq. (2.125) we deduce that if $t \to +\infty$, then also $t' \to +\infty$, so that in this limit both Heaviside functions tend to 1. Conversely, if $t \to -\infty$, then also $t' \to -\infty$, and in this case both Heaviside functions tend to zero. In conclusion, also the second integral in equation (2.127) vanishes. We thus obtain the equality

$$Q' = Q,$$

that we wanted to prove.

2.4.2 *Energy-Momentum Tensor and Four-Momentum*

In Sect. 2.4.3 we will show how the conservation of energy and momentum – cornerstone of any fundamental physical theory – occurs in electrodynamics as a consequence of equations (2.20)–(2.22). In this section, before considering this particular case, we set the problem of the realization of these conservation laws in a *generic relativistic theory*. In a relativistic theory the energy constitutes the fourth component of a four-vector, namely of the four-momentum. Since in such a theory a Lorentz transformation mixes energy and momentum, it is natural to expect that the conservation of the first cannot take place without the simultaneous conservation of the second. So we are looking for *four* constants of motion, grouped in a four-momentum P^ν whose time component $P^0 = \varepsilon$ represents the total energy of the system. In line with the conservation paradigm of the electric charge exposed in Sect. 2.4.1, we envisage *local* conservation laws also for the four-momentum. We require, therefore, each of the four components of P^ν to be associated with a conserved four-current $j^{\mu(\nu)}(x)$. These four currents altogether form a double tensor – called *energy-momentum tensor* – which commonly is denoted by the symbol

$$T^{\mu\nu} = j^{\mu(\nu)}. \tag{2.128}$$

As for the electric charge, we postulate that in a relativistic theory the conservation of the four-momentum occurs as consequence of the existence of an energy-momentum tensor $T^{\mu\nu}(x)$, possessing the following properties:

(1) $T^{\mu\nu}$ is a tensor field;
(2) $T^{\mu\nu}$ satisfies the continuity equation $\partial_\mu T^{\mu\nu} = 0$;
(3) $\lim_{|\mathbf{x}|\to\infty} \left(|\mathbf{x}|^3 T^{\mu\nu}(t,\mathbf{x}) \right) = 0$.

Condition (1) ensures, in particular, the Lorentz invariance of the continuity equation. We now derive the existence of conserved quantities along the same lines of Sect. 2.4.1. Integrating the continuity equation over a finite volume V we obtain

$$\frac{d}{dt} \int_V T^{0\nu} d^3x = - \int_V \partial_i T^{i\nu} d^3x. \tag{2.129}$$

On the basis of the identification (2.128) the components $T^{0\nu}$ correspond to the four-momentum *density*, so that the four-momentum contained in a volume V is given by

$$P_V^\nu = \int_V T^{0\nu} d^3x. \tag{2.130}$$

Then, from relation (2.129) we infer that the quantities $T^{i\nu}$ have to be interpreted as four-momentum *current densities*. Applying to the second member of (2.129) Gauss's theorem we recover in fact the flux equations

$$\frac{dP_V^\nu}{dt} = -\int_{\partial V} T^{i\nu} d\Sigma^i. \tag{2.131}$$

Writing the $\nu = 0$ components of equations (2.130) and (2.131), which regard the energy $\varepsilon_V = P_V^0$ contained in a volume V, we obtain

$$\varepsilon_V = \int_V T^{00} \, d^3x, \qquad \frac{d\varepsilon_V}{dt} = -\int_{\partial V} T^{i0} \, d\Sigma^i.$$

T^{00} represents thus the *energy density*, whereas the three-dimensional vector T^{i0} represents the *energy flux* – physical quantities that in the following will play a fundamental role. Analogous interpretations hold for the components $T^{\mu j}$, referring to the (relativistic) momentum. The three-dimensional vector T^{0j} represents the *momentum density*, and the three-dimensional tensor T^{ij}, also called *Maxwell's stress tensor*, represents the *momentum flux*. If, finally, in the local conservation equation (2.131) we extend the volume to the whole space, by virtue of property 3) we find that the *total* four-momentum P^ν is *finite* as well as *conserved*,

$$P^\nu = \int T^{0\nu} d^3x, \qquad \frac{dP^\nu}{dt} = 0. \tag{2.132}$$

Covariance of the total four-momentum. As in the case of the electric charge, in general the four-momentum P_V^ν contained in a finite volume V is time-dependent and it is not a tensor under Lorentz transformations. Conversely, the total four-momentum P^ν is a *four-vector*. The proof is based on the properties (1)–(3) and parallels the proof of the Lorentz invariance of the electric charge of Sect. 2.4.1. Performing in (2.132) the same operations that led to (2.124) one obtains straightforwardly

$$P^\nu = \int T^{\mu\nu}(x) \, \partial_\mu H(t) \, d^4x, \qquad P'^\nu = \int T'^{\mu\nu}(x') \, \partial'_\mu H(t') \, d^4x'.$$

Since by property (1) the energy-momentum tensors in the two reference frames are linked by the relation

$$T'^{\mu\nu}(x') = \Lambda^\mu{}_\alpha \Lambda^\nu{}_\beta T^{\alpha\beta}(x),$$

it follows that

$$P'^\nu = \Lambda^\nu{}_\beta \int T^{\mu\beta}(x) \, \partial_\mu H(t') \, d^4x.$$

We evaluate now the difference

$$P'^\nu - \Lambda^\nu{}_\beta P^\beta = \Lambda^\nu{}_\beta \int T^{\mu\beta}(x)\, \partial_\mu(H(t') - H(t))\, d^4x$$

$$= \Lambda^\nu{}_\beta \int \partial_\mu \left(T^{\mu\beta}(x)\, (H(t') - H(t)) \right) d^4x,$$

where in the last step we used the continuity equation. The integral we obtained is of the same form of the (vanishing) integral appearing in Eq. (2.126) and so it vanishes, too. We have therefore

$$P'^\nu = \Lambda^\nu{}_\beta P^\beta,$$

that is to say, P^ν is a four-vector.

2.4.3 Energy-Momentum Tensor in Electrodynamics

In this section we provide a constructive proof of the existence in the electrodynamics of point-like particles of an energy-momentum tensor $T^{\mu\nu}$ with the properties (1)–(3) postulated in Sect. 2.4.2. We first derive *heuristically* the form of the energy density T^{00}, and then we resort to Lorentz invariance to reconstruct the whole tensor. We begin recalling the known expression of the energy density of the electromagnetic field

$$T_{em}^{00} = \frac{1}{2}\left(E^2 + B^2\right). \tag{2.133}$$

Obviously the conserved total energy cannot simply be given by the integral of T_{em}^{00}, because the electromagnetic field exchanges energy with the charged particles. To quantify this exchange we calculate the time derivative of T_{em}^{00}, using the Maxwell equations in the form (2.52)–(2.55):

$$\frac{\partial T_{em}^{00}}{\partial t} = \mathbf{E} \cdot \frac{\partial \mathbf{E}}{\partial t} + \mathbf{B} \cdot \frac{\partial \mathbf{B}}{\partial t} = \mathbf{E} \cdot (\nabla \times \mathbf{B} - \mathbf{j}) - \mathbf{B} \cdot \nabla \times \mathbf{E}$$

$$= -\mathbf{j} \cdot \mathbf{E} - \nabla \cdot (\mathbf{E} \times \mathbf{B}).$$

We integrate now this equation over the whole space. Applying to the last term Gauss's theorem and assuming that $\mathbf{E}$ and $\mathbf{B}$ decrease sufficiently fast at spatial infinity, we see that it gives a vanishing contribution. Recalling the form of the current (2.110) and of the work-energy theorems (2.23), we obtain so

$$\frac{d}{dt} \int T_{em}^{00}\, d^3x = -\int \mathbf{j} \cdot \mathbf{E}\, d^3x = -\sum_r e_r \mathbf{v}_r \cdot \int \mathbf{E}(t, \mathbf{x})\, \delta^3(\mathbf{x} - \mathbf{y}_r(t))\, d^3x$$

$$= -\sum_r e_r \mathbf{v}_r \cdot \mathbf{E}(t, \mathbf{y}_r(t)) = -\frac{d}{dt}\left(\sum_r \varepsilon_r\right),$$

where $\varepsilon_r = m_r/\sqrt{1 - v_r^2}$ is the relativistic energy of the rth particle. This relation tells us that the conserved total energy is of the form

$$\varepsilon = \int T_{\mathrm{em}}^{00}\, d^3x + \sum_r \varepsilon_r = \int \left(\frac{1}{2}\left(E^2 + B^2\right) + \sum_r \varepsilon_r\, \delta^3(\mathbf{x} - \mathbf{y}_r(t)) \right) d^3x,$$

which in turn suggests to define the total energy density as

$$T^{00} = \frac{1}{2}\left(E^2 + B^2\right) + \sum_r \varepsilon_r\, \delta^3(\mathbf{x} - \mathbf{y}_r(t)).$$

It is then natural to assume that the total energy-momentum tensor is the sum of two contributions

$$T^{\mu\nu} = T_{\mathrm{em}}^{\mu\nu} + T_{\mathrm{p}}^{\mu\nu}, \tag{2.134}$$

the first due to the electromagnetic field and the second due to the particles, subject to the conditions

$$T_{\mathrm{em}}^{00} = \frac{1}{2}\left(E^2 + B^2\right), \qquad T_{\mathrm{p}}^{00} = \sum_r \varepsilon_r\, \delta^3(\mathbf{x} - \mathbf{y}_r(t)). \tag{2.135}$$

By exploiting this information we now determine the two contributions separately, relying on the fact that under Lorentz transformations both must behave as *tensors*. We start with the determination of $T_{\mathrm{em}}^{\mu\nu}$. The component T_{em}^{00} is bilinear in $\mathbf{E}$ and $\mathbf{B}$ and, since these fields constitute the components of the tensor $F^{\mu\nu}$, under Lorentz transformations they mix up linearly. But, under Lorentz transformations, also the components of $T_{\mathrm{em}}^{\mu\nu}$ transform linearly, and so the tensor $T_{\mathrm{em}}^{\mu\nu}$ must be *bilinear* in $F^{\mu\nu}$. Lorentz invariance imposes, therefore, the general structure[7]

$$T_{\mathrm{em}}^{\mu\nu} = a F^{\mu}{}_{\alpha} F^{\alpha\nu} + b\eta^{\mu\nu} F^{\alpha\beta} F_{\alpha\beta} + c F^{\mu\nu} F^{\alpha}{}_{\alpha}, \tag{2.136}$$

where a, b and c are constants. Thanks to the antisymmetry of $F^{\alpha\beta}$ the last term is identically zero, in that $F^{\alpha}{}_{\alpha} = F^{\alpha\beta}\eta_{\alpha\beta} = 0$. To determine a and b we compute T_{em}^{00} from the *ansatz* (2.136) using (2.15), and then compare the result with (2.135)

$$\begin{aligned} T_{\mathrm{em}}^{00} &= a F^{0}{}_{\alpha} F^{\alpha 0} + 2b\eta^{00}(B^2 - E^2) \\ &= a F^{0}{}_{i} F^{i0} + 2b(B^2 - E^2) = (a - 2b)E^2 + 2b B^2. \end{aligned}$$

---

[7] A priori $T_{\mathrm{em}}^{\mu\nu}$ could also include terms involving the Levi-Civita tensor, as for instance

$$\widetilde{T}_{\mathrm{em}}^{\mu\nu} = \eta^{\mu\nu} \varepsilon^{\alpha\beta\gamma\delta} F_{\alpha\beta} F_{\gamma\delta} = -8\eta^{\mu\nu}\mathbf{E}\cdot\mathbf{B}.$$

However, since $T_{\mathrm{em}}^{\mu\nu}$ must be a tensor, and $\widetilde{T}_{\mathrm{em}}^{\mu\nu}$ is a *pseudotensor*, see Sect. 1.4.3, such terms would violate the parity and time-reversal invariances of electrodynamics.

The comparison yields the values $a = 1$ and $b = 1/4$, and consequently the *energy-momentum tensor of the electromagnetic field* is given by

$$T_{\text{em}}^{\mu\nu} = F^\mu{}_\alpha F^{\alpha\nu} + \frac{1}{4}\eta^{\mu\nu}F^{\alpha\beta}F_{\alpha\beta}. \tag{2.137}$$

To determine $T_{\text{p}}^{\mu\nu}$ we rewrite its component T_{p}^{00} (2.135) in the form

$$T_{\text{p}}^{00} = \sum_r \int \varepsilon_r\, \delta^4(x - y_r)\, dy_r^0 = \sum_r m_r \int u_r^0 u_r^0\, \delta^4(x - y_r)\, ds_r,$$

where we used the relations $\varepsilon_r = m_r u_r^0$ and $dy_r^0 = u_r^0\, ds_r$. The last integral suggests to define the *energy-momentum tensor of the particles* as

$$T_{\text{p}}^{\mu\nu} = \sum_r m_r \int u_r^\mu u_r^\nu\, \delta^4(x - y_r)\, ds_r, \tag{2.138}$$

an expression that is manifestly covariant under Lorentz transformations and reproduces the correct component T_{p}^{00}. In (2.138) it is understood that the line integrals extend over the whole world line of each particle, as in the definition (2.16) of the four-current. Performing in (2.138) the integral over the δ-function $\delta(t - y_r^0)$, as in Eqs. (2.106)–(2.108), we can rewrite $T_{\text{p}}^{\mu\nu}$ in a non-covariant form which will be useful in the following

$$T_{\text{p}}^{\mu\nu} = \sum_r \frac{p_r^\mu p_r^\nu}{\varepsilon_r}\, \delta^3(\mathbf{x} - \mathbf{y}_r(t)). \tag{2.139}$$

Notice, finally, that the tensors (2.137) and (2.138) are *symmetric* in μ and ν, and so is the total energy-momentum tensor:

$$T^{\mu\nu} = T^{\nu\mu}.$$

This property, which at the moment seems rather *incidental*, will nevertheless play a crucial role in the following.

Continuity equation. The above expressions of $T_{\text{p}}^{\mu\nu}$ and $T_{\text{em}}^{\mu\nu}$ have been deduced relying essentially on heuristic arguments. Their ultimate legitimation derives from the fact that the total energy-momentum tensor $T^{\mu\nu}$ (2.134) is *conserved* – i.e. satisfies the continuity equation – provided that:

(a) the tensor $F^{\mu\nu}$ satisfies the Bianchi identity and the Maxwell equation;
(b) the particles' coordinates y_r^μ satisfy the Lorentz equations.

To prove, it we calculate separately the four-divergences of the two tensors, beginning with the electromagnetic one[8]

$$
\begin{aligned}
\partial_\mu T_{\text{em}}^{\mu\nu} &= \partial_\mu F^{\mu\alpha} F_\alpha{}^\nu + F^{\mu\alpha} \partial_\mu F_\alpha{}^\nu + \frac{1}{2} F_{\alpha\beta} \, \partial^\nu F^{\alpha\beta} \\
&= -F^{\nu\alpha} j_\alpha + \frac{1}{2} F_{\alpha\beta} \left(\partial^\alpha F^{\beta\nu} - \partial^\beta F^{\alpha\nu} \right) + \frac{1}{2} F_{\alpha\beta} \partial^\nu F^{\alpha\beta} \\
&= -F^{\nu\alpha} j_\alpha + \frac{1}{2} F_{\alpha\beta} \left(\partial^\alpha F^{\beta\nu} + \partial^\beta F^{\nu\alpha} + \partial^\nu F^{\alpha\beta} \right) \\
&= -\sum_r e_r \int F^{\nu\alpha}(y_r) \, u_{r\alpha} \, \delta^4(x - y_r) \, ds_r.
\end{aligned}
\tag{2.140}
$$

In the second line we used the Maxwell equation (2.22), in the third the Bianchi identity in the form (2.41), and in the last the definition of the four-current (2.16), choosing for the parameter λ_r the proper time s_r. To compute the four-divergence of the particles' energy-momentum tensor (2.138) we proceed as for the computation of the four-divergence of the electric current, see (2.105). This time, however, to avoid too cumbersome expressions we resort to the more simple *symbolic* notation. To verify the correctness of our formal manipulations it is sufficient to apply the intermediate results to a test function $\varphi \in \mathcal{S}(\mathbb{R}^4)$. Recalling that $m_r u_r^\nu = p_r^\nu$, from (2.138) we obtain

$$
\begin{aligned}
\partial_\mu T_{\text{p}}^{\mu\nu} &= \sum_r \int p_r^\nu u_r^\mu \partial_\mu \delta^4(x - y_r) \, ds_r = -\sum_r \int p_r^\nu \frac{d}{ds_r} \delta^4(x - y_r) \, ds_r \\
&= \sum_r \int \frac{dp_r^\nu}{ds_r} \delta^4(x - y_r) \, ds_r - \sum_r p_r^\nu \delta^4(x - y_r) \Big|_{s_r = -\infty}^{s_r = +\infty} \\
&= \sum_r \int \frac{dp_r^\nu}{ds_r} \delta^4(x - y_r) \, ds_r.
\end{aligned}
\tag{2.141}
$$

The second term in the second line is zero because for a fixed x the δ-function $\delta(x^0 - y_r^0(s_r))$ goes to zero as $s_r \to \pm\infty$, because for *physical* world lines $\lim_{s \to \pm\infty} y^0(s) = \pm\infty$, see Sect. 7.1. Summing up (2.141) and (2.140), and using the Lorentz equations (2.20), we finally obtain

$$
\partial_\mu T^{\mu\nu} = \sum_r \int \left(\frac{dp_r^\nu}{ds_r} - e_r F^{\nu\alpha}(y_r) u_{r\alpha} \right) \delta^4(x - y_r) \, ds_r = 0,
\tag{2.142}
$$

i.e. the continuity equation for $T^{\mu\nu}$.

---

[8]In (2.140) we made the implicit assumption that the components of the tensor $F^{\mu\nu}(x)$ are sufficiently regular functions, so that the derivatives of $T_{\text{em}}^{\mu\nu}$ could be computed by applying Leibniz's rule. Actually we will see that – for point-like particles – the tensors $F^{\mu\nu}(x)$ solving Maxwell's equations are so singular that the components of $T_{\text{em}}^{\mu\nu}$ are not even *distributions*, so that eventually it makes no sense at all to compute $\partial_\mu T_{\text{em}}^{\mu\nu}$. For the moment we will ignore this fundamental problem and continue with our *formal* analysis, postponing its solution to Chap. 16.

Formulas (2.134), (2.137) and (2.138) identify hence an energy-momentum tensor endowed with the properties (1) and (2) postulated in Sect. 2.4.2. Furthermore, under very general hypotheses one can show that $T^{\mu\nu}$ satisfies also property (3) regarding its asymptotic behavior at large distances. The term $T_{\mathrm{p}}^{\mu\nu}$ satisfies this property trivially, since for fixed t it is of compact spatial support. Concerning $T_{\mathrm{em}}^{\mu\nu}$, we anticipate that also this tensor satisfies property (3), provided that in the limit $t \to -\infty$ the accelerations of the charged particles vanish sufficiently fast. In Chap. 6 we will, in fact, show that in this case the electromagnetic field has the asymptotic behavior of a *Coulomb field*

$$F^{\mu\nu} \sim \frac{1}{|\mathbf{x}|^2}, \quad \text{for } |\mathbf{x}| \to \infty.$$

Consequently, the tensor $T_{\mathrm{em}}^{\mu\nu}$, being quadratic in $F^{\mu\nu}$, for large $|\mathbf{x}|$ decreases as

$$T_{\mathrm{em}}^{\mu\nu} \sim \frac{1}{|\mathbf{x}|^4}. \tag{2.143}$$

It satisfies, therefore, the envisaged limit $\lim_{|\mathbf{x}|\to\infty} \left(|\mathbf{x}|^3 \, T_{\mathrm{em}}^{\mu\nu} \right) = 0$.

Active and passive charges. Before proceeding any further, we remark that in the proof of the continuity equation (2.142) above, we tacitly assumed the identification between active and passive charges discussed at the end of Sect. 2.2.5, namely we have set $e_r = e_r^*$. We recall that $\{e_r\}$ are the charges appearing a priori in the four-current, whereas $\{e_r^*\}$ are the charges appearing at the second member of the Lorentz equations (2.20). Maintaining the expressions (2.137) and (2.138), and retracing the proof above without resorting to this identification, in place of (2.142) we would now arrive at the equation

$$\partial_\mu T^{\mu\nu} = \sum_r \int \left(\frac{dp_r^\nu}{ds_r} - e_r F^{\nu\alpha}(y_r) u_{r\alpha} \right) \delta^4(x - y_r) \, ds_r$$

$$= \sum_r (e_r^* - e_r) \int F^{\nu\alpha}(y_r) \, \delta^4(x - y_r) \, dy_{r\alpha}.$$

The continuity equation would therefore be violated, unless we impose $e_r^* = e_r$. We conclude therefore that – at the relativistic level – the identification between active and passive charges is implied by the *local conservation of the four-momentum*, as anticipated in Sect. 2.2.5.

Physical meaning of the components of $T^{\mu\nu}$. We analyze now the meaning of the components of $T^{\mu\nu}$ from a three-dimensional point of view, beginning again with the electromagnetic contribution. Inserting the components (2.12)–(2.14) of $F^{\mu\nu}$ in (2.137) we find the expressions

$$T_{\text{em}}^{00} = \frac{1}{2} \left(E^2 + B^2 \right),$$

(2.144)

$$T_{\text{em}}^{i0} = T_{\text{em}}^{0i} = (\mathbf{E} \times \mathbf{B})^i,$$

(2.145)

$$T_{\text{em}}^{ij} = \frac{1}{2} \left(E^2 + B^2 \right) \delta^{ij} - E^i E^j - B^i B^j.$$

(2.146)

We retrieved obviously the energy density T_{em}^{00} which was our starting point. In the $0i$ components we recognize the *Poynting vector* S^i, which notoriously represents the *energy flux* of the electromagnetic field,

$$T_{\text{em}}^{i0} = S^i, \qquad \mathbf{S} = \mathbf{E} \times \mathbf{B}.$$

(2.147)

In addition, we see that the Poynting vector equals also the *momentum density* T_{em}^{0i}. The space-space components T_{em}^{ij} form instead a three-dimensional symmetric tensor – *Maxwell's stress tensor* – representing the *momentum flux*. Finally, we observe that the energy-momentum tensor (2.137) of the electromagnetic field is *traceless*:

$$T_{\text{em}}^{\mu\nu} \eta_{\mu\nu} = 0.$$

(2.148)

As for the particles' energy-momentum tensor $T_{\text{p}}^{\mu\nu}$, from (2.139) we see that the four-momentum density has the expected form ($p_r^0 = \varepsilon_r$)

$$T_{\text{p}}^{0\mu} = \sum_r p_r^\mu \, \delta^3(\mathbf{x} - \mathbf{y}_r(t)).$$

(2.149)

In fact, the four-momentum of the particles which at a time t find themselves in a certain volume V is given by

$$\int_V T_{\text{p}}^{0\mu} \, d^3x = \sum_r p_r^\mu \int_V \delta^3(\mathbf{x} - \mathbf{y}_r(t)) \, d^3x = \sum_{r \in V} p_r^\mu,$$

where the sum extends over all particles contained at time t in V. Finally, from the formulas above we derive that the total conserved four-momentum (2.132) of the system is the sum of two terms:

$$P^\mu = P_{\text{em}}^\mu + P_{\text{p}}^\mu, \qquad P_{\text{em}}^\mu = \int T_{\text{em}}^{0\mu} \, d^3x, \qquad P_{\text{p}}^\mu = \sum_{r=1}^{N} p_r^\mu.$$

(2.150)

We conclude this section returning to the four-momentum balance equation relative to a volume V. According to (2.131), the four-momentum that leaves in the unit time interval the volume V is given

$$\frac{dP_V^\mu}{dt} = -\int_{\partial V} T^{i\mu} d\Sigma^i.$$

The integral of the second member is a surface integral and receives – a priori – contributions from $T_{\text{em}}^{i\mu}$ as well as from $T_{\text{p}}^{i\mu}$. Nevertheless, since at the considered instant the particles find themselves inside or outside V, the term $T_{\text{p}}^{i\mu}$ does not contribute, and we remain with

$$\frac{dP_V^\mu}{dt} = -\int_{\partial V} T_{\text{em}}^{i\mu} \, d\Sigma^i. \tag{2.151}$$

In other words, the variation of the total four-momentum contained in V – sum of the four-momenta of the particles and of the electromagnetic field – is determined solely by the flux of the latter. In particular, for $\mu = 0$ Eq. (2.151) yields for the energy *emitted* per unit time from the volume V the expression

$$-\frac{d\varepsilon_V}{dt} = \int_{\partial V} T_{\text{em}}^{i0} \, d\Sigma^i = \int_{\partial V} \mathbf{S} \cdot d\mathbf{\Sigma}. \tag{2.152}$$

This important relation is the key equation for the energy-balance analysis of all radiation phenomena, see Chaps. 7 and 8.

2.4.4 Four-Dimensional Angular Momentum

In this section we analyze the issue of the conservation of the *four-dimensional* angular momentum in a relativistic theory. We will provide a *constructive* proof of the existence of a four-dimensional conserved angular momentum in a *generic* relativistic theory, if only it is endowed with a *conserved* and *symmetric* energy-momentum tensor $T^{\mu\nu}$. We will then apply this general construction to the electrodynamics of point-like particles.

In an isolated system of non-relativistic particles, in addition to the energy and the momentum, also the angular momentum

$$\mathbf{L} = \sum_r \mathbf{y}_r \times \mathbf{p}_r, \tag{2.153}$$

where $\mathbf{p}_r = m_r \mathbf{v}_r$, is a conserved quantity. In a relativistic theory the three quantities (2.153), appropriately generalized, should become components of a four-dimensional tensor. To overcome the obstacle related to the fact that the exterior product of two vectors does not admit a four-dimensional extension, we recall that in three dimensions each vector is equivalent to a double antisymmetric tensor, rewriting hence (2.153) as

$$L^{ij} \equiv \varepsilon^{ijk} L^k = \sum_r \left(y_r^i p_r^j - y_r^j p_r^i \right). \tag{2.154}$$

This expression admits now a natural extension in terms of the four-dimensional antisymmetric tensor

$$L_{\mathrm{p}}^{\alpha\beta} = \sum_r \left(y_r^\alpha p_r^\beta - y_r^\beta p_r^\alpha \right), \qquad L_{\mathrm{p}}^{\alpha\beta} = -L_{\mathrm{p}}^{\beta\alpha}, \tag{2.155}$$

where $p_r^\alpha = m_r u_r^\alpha$ is now the four-momentum of the rth particle. It is easy to check that for a system of free particles, for which $dp_r^\alpha/dt = 0$, the six quantities (2.155) are indeed conserved. Using the identities

$$\frac{dy_r^\alpha}{dt} = \frac{u_r^\alpha}{u_r^0} = \frac{p_r^\alpha}{\varepsilon_r},$$

we find in fact

$$\frac{dL_{\mathrm{p}}^{\alpha\beta}}{dt} = \sum_r \left(\frac{dy_r^\alpha}{dt} p_r^\beta - \frac{dy_r^\beta}{dt} p_r^\alpha \right) = \sum_r \frac{1}{\varepsilon_r} \left(p_r^\alpha p_r^\beta - p_r^\beta p_r^\alpha \right) = 0.$$

General construction. The example just examined leads us to conclude that the four-dimensional angular momentum of a relativistic system must be represented by an *antisymmetric* tensor $L^{\alpha\beta}$, which encodes *six* conserved quantities. As for the electric charge and for the four-momentum, also for the angular momentum we insist on a local conservation law, by postulating the existence of an angular momentum *density current* $M^{\mu\alpha\beta}(x)$, or simply angular momentum *density*, such that

(1) $M^{\mu\alpha\beta}$ is a three-index tensor field;
(2) $M^{\mu\alpha\beta} = -M^{\mu\beta\alpha}$;
(3) $\partial_\mu M^{\mu\alpha\beta} = 0$.

We do not impose any condition on the asymptotic behavior of $M^{\mu\alpha\beta}$ at large distances, since it will be determined by the corresponding behavior of $T^{\mu\nu}$. The requirements (1) and (2) are, in fact, automatically satisfied, if we assume that $M^{\mu\alpha\beta}$ has the *standard form*

$$M^{\mu\alpha\beta} = x^\alpha T^{\mu\beta} - x^\beta T^{\mu\alpha}. \tag{2.156}$$

This choice is motivated in particular by the fact that, as we will see shortly, for a system of free particles (2.156) reduces in fact to (2.155). To verify the property (3) we evaluate the divergence of $M^{\mu\alpha\beta}$ by assuming that the energy-momentum tensor satisfies the continuity equation $\partial_\mu T^{\mu\nu} = 0$:

$$\partial_\mu M^{\mu\alpha\beta} = \delta_\mu^\alpha T^{\mu\beta} + x^\alpha \partial_\mu T^{\mu\beta} - \delta_\mu^\beta T^{\mu\alpha} - x^\beta \partial_\mu T^{\mu\alpha} = T^{\alpha\beta} - T^{\beta\alpha}. \tag{2.157}$$

We conclude therefore that, if the energy-momentum tensor is also *symmetric*,

$$T^{\alpha\beta} = T^{\beta\alpha},$$

then $M^{\mu\alpha\beta}$ satisfies indeed the continuity equation[9]

$$\partial_\mu M^{\mu\alpha\beta} = 0. \tag{2.158}$$

Proceeding in the usual way, and assuming that the fields involved decay sufficiently fast at spatial infinity,[10] from (2.158) we deduce the existence of the six conserved quantities

$$L^{\alpha\beta} = \int M^{0\alpha\beta} d^3x = \int \left(x^\alpha T^{0\beta} - x^\beta T^{0\alpha} \right) d^3x, \qquad L^{\alpha\beta} = -L^{\beta\alpha}. \tag{2.159}$$

Finally, performing the usual steps, one proves that under Lorentz transformations $L^{\alpha\beta}$ behaves as a rank-two *tensor*. To conclude we show that, as anticipated above, for a system of free particles (2.159) turns into (2.155). In this case the energy-momentum tensor reduces to the contribution $T_p^{\mu\nu}$ (2.138), whose components $T_p^{0\nu}$ have been evaluated in (2.149). Inserting the latter in the general expression (2.159) we find

$$L^{\alpha\beta} = \int \left(x^\alpha \sum_r p_r^\beta \, \delta^3(\mathbf{x} - \mathbf{y}_r(t)) - x^\beta \sum_r p_r^\alpha \, \delta^3(\mathbf{x} - \mathbf{y}_r(t)) \right) d^3x$$

$$= \sum_r \left(\left(\int x^\alpha \delta^3(\mathbf{x} - \mathbf{y}_r(t)) \, d^3x \right) p_r^\beta - \left(\int x^\beta \delta^3(\mathbf{x} - \mathbf{y}_r(t)) \, d^3x \right) p_r^\alpha \right)$$

$$= \sum_r \left(y_r^\alpha p_r^\beta - y_r^\beta p_r^\alpha \right), \square$$

Translation invariance. Actually, the field $M^{\mu\alpha\beta}$ behaves as a *tensor* only under Lorentz transformations, whereas under a translation $x^\mu \to x'^\mu = x^\mu + a^\mu$ it transforms *anomalously*

$$M'^{\mu\alpha\beta}(x') = x'^\alpha T'^{\mu\beta}(x') - x'^\beta T'^{\mu\alpha}(x') = M^{\mu\alpha\beta}(x) + a^\alpha T^{\mu\beta}(x) - a^\beta T^{\mu\alpha}(x).$$

We recall that under translations a tensor should, instead, remain invariant. Similarly the total angular momentum (2.159) transforms according to (see (2.132))

$$L'^{\alpha\beta} = L^{\alpha\beta} + a^\alpha P^\beta - a^\beta P^\alpha. \tag{2.160}$$

These *anomalies* are easily cured if we realize that in the definition of the angular momentum density (2.156) implicitly we considered as *pole Q* the origin, with coor-

---

[9]In Sect. 3.4 we will show that in any relativistic theory which is based on a *variational principle* there exists a *conserved* as well as *symmetric* energy-momentum tensor.

[10]Since in electrodynamics $T^{\mu\nu}$ at large distances decreases as $1/r^4$, see (2.143), the tensor $M^{\mu\alpha\beta}$ in (2.156) decreases only as $1/r^3$. Nevertheless, it can be seen that in electrodynamics – as in all other fundamental theories – the integrals (2.159) converge anyway, thanks to the peculiar behavior of the constituent fields at large distances.

dinates $x_Q^\mu = (0,0,0,0)$. For a generic pole the definition (2.156) must be replaced by the expression

$$M_Q^{\mu\alpha\beta} = (x^\alpha - x_Q^\alpha)T^{\mu\beta} - (x^\beta - x_Q^\beta)T^{\mu\alpha},$$

which still satisfies the properties (1)–(3) and is, in addition, translation invariant.

Physical meaning of the components of $L^{\alpha\beta}$. We investigate now the explicit form of the six constants of motion $L^{\alpha\beta}$. We analyze separately the three components L^{ij}, better the vector $L^i = \frac{1}{2}\varepsilon^{ijk}L^{jk}$ representing the *spatial* angular momentum, and the three new constants of motion L^{0i}, which are sometimes called *boost generators*, because of their relation to the special Lorentz transformations, see Sect. 3.3.2. Beginning with the spatial angular momentum, and specializing to the case of electrodynamics, from Eqs. (2.159), (2.147) and (2.149) we obtain the expression

$$L^i = \frac{1}{2}\varepsilon^{ijk}L^{jk} = \varepsilon^{ijk}\int x^j\, T^{0k}d^3x = \varepsilon^{ijk}\int x^j\left(S^k + \sum_r p_r^k\,\delta^3(\mathbf{x}-\mathbf{y}_r)\right)d^3x,$$

which in vector notation becomes

$$\mathbf{L} = \int (\mathbf{x} \times \mathbf{S})\,d^3x + \sum_r \mathbf{y}_r \times \mathbf{p}_r = \mathbf{L}_{\text{em}} + \mathbf{L}_{\text{p}}. \qquad (2.161)$$

The total spatial angular momentum is therefore composed of a contribution $\mathbf{L}_{\text{p}}$ which depends only on the particles, and of a contribution $\mathbf{L}_{\text{em}}$ which depends only on the electromagnetic field. The vector $\mathbf{L}_{\text{p}}$ reduces to the non-relativistic expression (2.153) if one neglects the factors $1/\sqrt{1-v_r^2}$. Also the vector $\mathbf{L}_{\text{em}}$ has a form that is not unexpected, since the Poynting vector $\mathbf{S} = \mathbf{E} \times \mathbf{B}$ equals not only the energy flux, but also the *momentum density* of the electromagnetic field.

Boost generators and center of mass motion. We now analyze the meaning of the boost components $L^{0i} \equiv K^i$, in a *generic* relativistic theory. For these constants of motion from (2.159) we recover the expressions

$$K^i = t\int T^{0i}d^3x - \int x^i T^{00}d^3x, \qquad (2.162)$$

where in the quantities $P^i = \int T^{0i}d^3x$ we recognize the conserved total momentum of the system, see (2.132). To provide an interpretation for the second term of K^i we define the position of the *center of mass of a relativistic system* as

$$\mathbf{x}_{\text{cm}}(t) = \frac{\int \mathbf{x}\,T^{00}d^3x}{\int T^{00}\,d^3x}, \qquad (2.163)$$

where $\varepsilon = \int T^{00}d^3x$ is the conserved total energy of the system. Notice that *formally* the definition (2.163) follows from the definition of the center of mass of a non-relativistic system, by replacing the mass density with the relativistic energy density T^{00}. The boost vector (2.162) can now be recast in the form

$$\mathbf{K} = t\,\mathbf{P} - \varepsilon\,\mathbf{x}_{\mathrm{cm}}(t),$$

and since it is time-independent we can evaluate it at $t = 0$ obtaining $\mathbf{K} = -\varepsilon\,\mathbf{x}_{\mathrm{cm}}(0)$. It follows that

$$\mathbf{x}_{\mathrm{cm}}(t) = \mathbf{x}_{\mathrm{cm}}(0) + \frac{\mathbf{P}}{\varepsilon}\,t.$$

We conclude, therefore, that the conservation of L^{0i} is equivalent to the assertion that the center of mass of an isolated system performs a *uniform linear motion*, with velocity

$$\mathbf{v}_{\mathrm{cm}} = \frac{\mathbf{P}}{\varepsilon}.$$

If, moreover, we assume that the constant vector P^{μ} is time-like, i.e. that it satisfies the inequality $P^{\mu}P_{\mu} = \varepsilon^2 - |\mathbf{P}|^2 > 0$, in a sense we can regard the whole system as a "particle" with mass $M = \sqrt{\varepsilon^2 - |\mathbf{P}|^2}$, four-momentum $P^{\mu} = (\varepsilon, \mathbf{P})$, and speed $v_{\mathrm{cm}} < 1$. Then there exists a privileged inertial frame , the center-of-mass reference frame, in which $P^{\mu} = (M, 0, 0, 0)$ and $v_{\mathrm{cm}} = 0$, i.e. a reference frame in which the center of mass is at rest. Notwithstanding the above interpretation, we must stress that the concept of *center of mass of a relativistic system*, as defined in (2.163), is not a relativistically invariant concept, since its coordinates $(t, \mathbf{x}_{\mathrm{cm}}(t))$ do not transform as a four-vector under Lorentz transformations. In other words, the center of mass of a relativistic system is represented by different points in different reference frames.

To summarize, from the fundamental equations of electrodynamics we were able to infer the existence of the *ten* conserved quantities P^{μ} and $L^{\alpha\beta}$ – as many as the parameters a^{μ} and $\omega^{\alpha\beta}$ describing the Poincaré group. In addition, we realized the presence of a further conservation law – that of the electric charge – in conjunction with a further one-parameter symmetry: gauge invariance. As mentioned earlier, these *coincidences* are due to the close relationship existing in nature between symmetry principles and conservation laws, a relationship that from a mathematical point of view is accomplished by *Noether's theorem*. In the following two chapters we will reanalyze the conservation laws derived in this chapter relying essentially on heuristic arguments, in the light of this powerful, even though somehow mysterious, theorem.

2.5 Problems

2.1 Using the identity $w^{\mu}u_{\mu} = 0$, show that the square of the four-acceleration $w^2 = w^{\mu}w_{\mu}$ satisfies the inequality

$$w^2 \leq 0.$$

Verify that in terms of the three-dimensional acceleration and velocity one has

$$w^2 = -\frac{a^2 - \dfrac{(\mathbf{a} \times \mathbf{v})^2}{c^2}}{c^4 \left(1 - \dfrac{v^2}{c^2}\right)^3}, \tag{2.164}$$

where we reintroduced the speed of light.
Hint: Consider the rest frame of the particle.

2.2 Show that the Bianchi identity can be written alternatively as in (2.39), (2.40) or (2.41).
Hint: The identities (1.37), (1.40) and (1.41) can prove useful.

2.3 Find all solutions of the equation

$$\left(x^2 - a^2\right) F(x) = 0, \quad a > 0, \tag{2.165}$$

for $F \in \mathcal{S}'(\mathbb{R})$, and show that every solution can be recast in the form

$$F(x) = f(x)\,\delta(x^2 - a^2), \tag{2.166}$$

for a convenient continuous function $f(x)$.

Solution: From Eq. (2.165) it follows that F can be "different from zero" at most for $x = \pm a$, i.e. its support is contained in the set $\{-a, a\}$, which is a set of *points*. From Theorem III of Sect. 2.3.1 on distributions with point-like support, it then follows that F is a *finite* linear combination of the distributions $\delta(x \pm a)$ and their derivatives. The linear combination

$$F_0(x) = c_1\,\delta(x - a) + c_2\,\delta(x + a),$$

with c_1 and c_2 arbitrary constants, is certainly a solution of (2.165), thanks to the identity (2.72). Conversely, regarding the first derivatives we notice that differentiating the identity $(x^2 - a^2)\,\delta(x \pm a) = 0$ we find

$$(x^2 - a^2)\,\delta'(x \pm a) = -2x\,\delta(x \pm a) = \pm 2a\,\delta(x \pm a) \neq 0.$$

Hence, the first derivatives of the δ-functions are not solutions of (2.165), and in the same way one proves that neither their higher derivatives are so. $F_0(x)$ is therefore the general solution of equation (2.165). To recast it in the form required by the problem, we resort to the identity (2.77) and multiply it by a generic continuous function $f(x)$

$$F(x) \equiv f(x)\,\delta(x^2 - a^2) = \frac{1}{2a}\left(f(a)\,\delta(x-a) + f(-a)\,\delta(x+a)\right).$$

Consequently, since the constants $f(a)$ and $f(-a)$ can take arbitrary values, $F_0(x)$ can always be recast in the form (2.166).

2.4 Show that a function $f : \mathbb{R}^D \to \mathbb{C}$ induces a regular distribution $F \in \mathcal{S}'(\mathbb{R}^D)$, defined by

$$F(\varphi) = \int f(x)\,\varphi(x)\,d^D x, \qquad (2.167)$$

if (a) its modulus $|f|$ is integrable in any ball $\mathcal{B} \subset \mathbb{R}^D$ – in particular if f has a finite number of integrable singularities – and if (b) it is asymptotically polynomially bounded.[11]

Hint: According to Theorem I of Sect. 2.3.1 one must show that $F(\varphi)$ satisfies the bound (2.61) for certain monomials $\mathcal{P}$ and $\mathcal{Q}$. For this purpose it is convenient to separate the integration domain $\mathbb{R}^D$ in (2.167) in a sufficiently large ball $\mathcal{B}$, and in its complement $\mathbb{R}^D \setminus \mathcal{B}$, and to use then the asymptotic properties (2.60) of φ (see also Problem 2.12).

2.5 *Birkhoff's theorem for electrodynamics*. Prove Birkhoff's theorem, enounced as follows. Consider a generic spherically symmetric, in general non-static, four-current

$$j^0(t, \mathbf{x}) = \rho(t, r), \qquad \mathbf{j}(t, \mathbf{x}) = \frac{\mathbf{x}}{r}\,j(t, r), \qquad r = |\mathbf{x}|, \qquad (2.168)$$

with compact spatial support, i.e.

$$j^\mu(t, \mathbf{x}) = 0, \quad \text{for } r > R, \quad \forall t.$$

Then the electromagnetic field in empty space, i.e. in the region $r > R$, is *static* and has the form

$$\mathbf{E} = \frac{Q\mathbf{x}}{4\pi r^3}, \qquad \mathbf{B} = 0,$$

where $Q = \int \rho(t, r)d^3x$ is the conserved total charge of the system. Conclude in particular that a spherically symmetric charge distribution – even though composed of *accelerated* charges – cannot emit electromagnetic waves, since the field in empty space is time-independent.

Hint: The spherical symmetry implies that the fields have the form $\mathbf{E} = \mathbf{x}f(t, r)$, $\mathbf{B} = \mathbf{x}g(t, r)$.

---

[11]A function f is said to be *asymptotically polynomially bounded* if there exist a number $L > 0$ and a positive polynomial $\mathcal{P}(x)$, such that for every x for which $\sqrt{(x^1)^2 + \cdots + (x^D)^2} > L$ the function f satisfies the bound $|f(x)| \leq \mathcal{P}(x)$.

2.6 *Invariant integrals*. Introduce the three-dimensional double tensor

$$H^{ij} = \int n^i n^j \, d\Omega, \tag{2.169}$$

where $n^i = x^i/r$, $r = |\mathbf{x}|$, and $d\Omega = \sin \vartheta \, d\vartheta \, d\varphi$ is the infinitesimal solid angle in three dimensions with $\int d\Omega = 4\pi$. The integrand in (2.169) depends hence only on the angles.

(a) Show that the integral (2.169) can be recast as

$$H^{ij} = \int \delta(r - 1) \, x^i x^j \, d^3 x. \tag{2.170}$$

(b) Prove that H^{ij} is an *invariant* tensor under the rotation group $SO(3)$, that is to say

$$R^i{}_m R^j{}_n H^{mn} = H^{ij}, \quad \forall \, R \in SO(3).$$

   Hint: Perform in (2.170) the change of variables $x^i = R^i{}_k y^k$.
(c) Knowing that the unique algebraically independent $SO(3)$-invariant tensors are δ^{ij} and ε^{ijk} – see Theorem II of Sect. 1.3.1 – conclude that $H^{ij} = k\delta^{ij}$ for some constant k. Determine k contracting both members of (2.169) with δ^{ij}.
(d) Along this line of reasoning establish the list of invariant integrals:

$$\int d\Omega = 4\pi,$$

$$\int n^i \, d\Omega = 0,$$

$$\int n^i n^j \, d\Omega = \frac{4\pi}{3} \, \delta^{ij},$$

$$\int n^i n^j n^k \, d\Omega = 0,$$

$$\int n^i n^j n^k n^l \, d\Omega = \frac{4\pi}{15} \left(\delta^{ij} \delta^{kl} + \delta^{ik} \delta^{jl} + \delta^{il} \delta^{jk} \right).$$

2.7 A particle of charge e and mass m is subject to a constant and uniform *external* electromagnetic field $F^{\mu\nu}$. Let $u^\mu(0)$ be the initial particle's four-velocity at proper time $s = 0$, subject to $u^2(0) = 1$.

(a) Show that in this case the Lorentz equation is equivalent to the first-order differential equation

$$\frac{dy^\mu}{ds} = u^\mu(s) = \left[e^{sA} \right]^\mu{}_\nu u^\nu(0),$$

   for a certain constant matrix $A^\mu{}_\nu$.
(b) Verify explicitly that $u^2(s) = 1$, $\forall \, s$. *Hint*: Observe that the exponential e^{sA} is an element of $SO(1,3)_c$, $\forall \, s$.

(c) Show that the scalar $w^2 = w^\mu(s)w_\mu(s)$ is independent of s and express it in terms of $F^{\mu\nu}$ and $u^\mu(0)$.

(d) It is known that, except when $E = B$ and simultaneously $\mathbf{E} \perp \mathbf{B}$, there exists always a reference frame in which the electric and magnetic fields are parallel and directed along the x axis: $\mathbf{E} = (E, 0, 0)$, $\mathbf{B} = (B, 0, 0)$. Show that in this reference frame the matrix A is block diagonal.

(e) Using the block-diagonal structure of A evaluate the exponential e^{sA} by expanding it in a Taylor series, and resumming it in terms of the functions sin, cos, sinh and cosh.

(f) Set $\mathbf{B} = 0$ and choose as initial velocity $\mathbf{v}_0 = (0, v_0, 0)$, that is

$$u^\mu(0) = \frac{1}{\sqrt{1 - v_0^2}}\,(1, 0, v_0, 0).$$

Determine $u^\mu(s)$ and $y^\mu(s)$ for every s, and finally the position vector $\mathbf{y}(t)$. Show in particular that for $v_0 = 0$ the particle performs the *hyperbolic motion*

$$\mathbf{y}(t) = \left(\sqrt{t^2 + \left(\frac{m}{eE}\right)^2}, 0, 0 \right).$$

2.8 *Ultraviolet divergences in non-relativistic electrodynamics*. Consider a system of N non-relativistic charges $\{e_r\}$, i.e. charges moving with speeds $v_r \ll c$, generating the electric and magnetic fields

$$\mathbf{E} = \sum_{r=1}^{N} \frac{e_r(\mathbf{x} - \mathbf{y}_r(t))}{4\pi|\mathbf{x} - \mathbf{y}_r(t)|^3}, \qquad \mathbf{B} = 0. \tag{2.171}$$

(a) Reintroducing the speed of light c in the Maxwell equations (2.24) and (2.25), verify that the fields (2.171) satisfy these equations, modulo terms of order $1/c$.

(b) Using the Maxwell equation $\nabla \cdot \mathbf{E} = \rho$, with ρ given in (2.109), and the relations

$$\mathbf{E} = -\nabla A^0, \qquad A^0 = \sum_{r=1}^{N} \frac{e_r}{4\pi|\mathbf{x} - \mathbf{y}_r(t)|},$$

show that the total energy of the electromagnetic field

$$\varepsilon_{em} = \frac{1}{2} \int \left(E^2 + B^2\right) d^3x \tag{2.172}$$

formally can be recast in the familiar form

$$\varepsilon_{em} = \frac{1}{2} \int A^0 \rho\, d^3x = \frac{1}{2} \sum_{r,s=1}^{N} \frac{e_r e_s}{4\pi|\mathbf{y}_r(t) - \mathbf{y}_s(t)|}. \tag{2.173}$$

Notice that both expressions (2.172) and (2.173) are *ultraviolet divergent*, because of the *self-interactions* of the charged particles. In Eq. (2.172) the divergences arise from the singular behavior of the electric field (2.171) near the position $\mathbf{y}_r(t)$ of each particle, i.e.

$$\mathbf{E} \sim \frac{e_r}{|\mathbf{x} - \mathbf{y}_r(t)|^2}.$$

This behavior produces in the integral (2.172) a non-integrable singularity proportional to e_r^2. Similarly, in Eq. (2.173) it is the term corresponding to $r = s$ to be infinite, which is again proportional to e_r^2. To retrieve for ε_{em} a finite result we may "renormalize away" this divergence, by restricting in (2.173) the sum to $r \neq s$.

(c) Introduce the non-relativistic kinetic energy of the particles and write the total *renormalized* energy as

$$\varepsilon = \sum_r \frac{1}{2} m_r v_r^2 + \sum_{r<s} \frac{e_r e_s}{4\pi |\mathbf{y}_r(t) - \mathbf{y}_s(t)|}. \tag{2.174}$$

(d) Show that in the presence of the fields (2.171) in the non-relativistic limit the Lorentz equations (2.23) become (see also (2.37))

$$m_r \mathbf{a}_r = \sum_{s=1}^{N} \frac{e_r^* e_s (\mathbf{y}_r(t) - \mathbf{y}_s(t))}{4\pi |\mathbf{y}_r(t) - \mathbf{y}_s(t)|^3}, \tag{2.175}$$

where in (2.23) momentarily we restored the *passive* charges $e_r^* = C e_r$, derived in (2.59) at the basis of *Newton's third law*. The term corresponding to $s = r$, proportional to e_r^2, represents the ultraviolet divergent self-force of the rth particle: As anticipated in Sect. 2.2.5, the Lorentz equation as it stands is ill-defined. To retrieve a finite result *renormalize* Eq. (2.175) by excluding from the sum the term corresponding to $s = r$.

(e) Verify explicitly that the total non-relativistic energy (2.174) is conserved if the particles satisfy the (renormalized version of the) Lorentz equations (2.175) with $C = 1$. Eventually, it is precisely the *energy-conservation paradigm* which justifies the, somehow pragmatic, renormalizations performed above. In Chap. 16 we will show how these rather improvised procedures can be generalized in a *systematic* way to the relativistic case, maintaining simultaneously Lorentz invariance as well as the conservation law $\partial_\mu T^{\mu\nu} = 0$.

2.9 Determine the general solution $y^\mu(\lambda)$ of the equation of motion of a free particle

$$\frac{d^2 y^\mu}{ds^2} = 0,$$

parameterizing the world line with a generic parameter λ, see (2.6). Verify that the general solution is determined only modulo a reparameterization.

2.10 Show that Eq. (2.36) can be recast in the form of Newton's second law

$$m\mathbf{a} = \mathbf{F}(\mathbf{y}, \mathbf{v}, t),$$

with $\mathbf{F}(\mathbf{y}, \mathbf{v}, t)$ given in (2.38).
Hint: Determine the quantity $\mathbf{v} \cdot \mathbf{a}$ by taking the scalar product of both sides of equation (2.36) with $\mathbf{v}$.

2.11 Prove that if an electromagnetic field $F^{\mu\nu}$ satisfies Eqs. (2.52) and (2.53) for every t and Eqs. (2.54) and (2.55) for $t = 0$, then it satisfies automatically Eqs. (2.54) and (2.55) for every t.
Hint: Evaluate the spatial divergence of equations (2.52) and (2.53).

2.12 Apply Theorem I of Sect. 2.3.1 to verify that the components of the three-dimensional tensor

$$G^{ij}(\mathbf{x}) = \frac{1}{r^3}\left(\delta^{ij} - 3\frac{x^i x^j}{r^2}\right),$$

appearing in (2.115), are distributions in $\mathcal{S}'(\mathbb{R}^3)$. Take into account that, by definition, in the resulting integral you must integrate first over the angles (*conditional convergence*).
Hint: Apply G^{ij} to a test function φ

$$G^{ij}(\varphi) = \int_{\mathbb{R}^3} G^{ij}(\mathbf{x})\,\varphi(\mathbf{x})\,d^3x, \qquad (2.176)$$

and separate the integration domain $\mathbb{R}^3$ in the two regions $r < 1$ and $r > 1$. In the region $r > 1$, where $G^{ij}(\mathbf{x})$ has no singularities, multiply and divide the integrand by r^{2n}, with n a sufficiently large positive integer. Using that the seminorm

$$\sup\nolimits_{\mathbf{x}\in\mathbb{R}^3} |\varphi(\mathbf{x})r^{2n}|$$

is finite, derive for the integral (2.176) in the region $r > 1$ a bound of the required type (2.61). In the region $r < 1$ write the test function in (2.176) as $(\varphi(\mathbf{x}) - \varphi(0)) + \varphi(0)$, and show – using the invariant integrals of Problem 2.6 – that the integral involving $\varphi(0)$ vanishes once you integrate over the angles. For the remaining term use the bound

$$|\varphi(\mathbf{x}) - \varphi(0)| = \left|\mathbf{x} \cdot \int_0^1 \boldsymbol{\nabla}\varphi(\lambda\mathbf{x})\,d\lambda\right| \le r \sum_{i=1}^3 \sup\nolimits_{\mathbf{x}\in\mathbb{R}^3} |\partial_i\varphi(\mathbf{x})|,$$

that cancels a power of r in the denominator of $G^{ij}(\mathbf{x})$. Using that in three dimensions $1/r^2$ is an integrable singularity, proceed as above to derive for (2.176) a bound of the type (2.61) also in the region $r < 1$.

2.13 Prove that the four linear functionals $j^\mu(\varphi)$ (2.103) define elements of $\mathcal{S}'(\mathbb{R}^4)$. *Hint*: Parameterize the world lines y_r^μ with the time coordinate y_r^0.

2.14 Prove the distributional limit in $\mathcal{S}'(\mathbb{R})$

$$\mathcal{S}' - \lim_{L \to \infty} \int_{-L}^{L} e^{-ikx} dk = 2\pi\delta(x),$$

showing that for all $\varphi \in \mathcal{S}(\mathbb{R})$ one has the ordinary limits

$$\lim_{L \to \infty} \int_{-\infty}^{\infty} dx\, \varphi(x) \int_{-L}^{L} e^{-ikx} dk = 2\pi\varphi(0).$$

Hint: Use the definite integral $\int_{-\infty}^{\infty} \frac{\sin y}{y}\, dy = \pi$.

2.15 Derive the transformation laws (2.30) of $\mathbf{E}$ and $\mathbf{B}$ under *parity* and *time reversal* from (2.29) (see the definitions (1.64) and (1.65)).

References

1. M. Reed, B. Simon, *Methods of Modern Mathematical Physics - I Functional Analysis* (Academic Press, New York, 1980)
2. W. Rudin, *Functional Analysis* (McGraw-Hill, New York, 1991)
3. E.G.P. Rowe, Structure of the energy tensor in the classical electrodynamics of point particles. Phys. Rev. D **18**, 3639 (1978)

Chapter 3
Variational Methods in Field Theory

As we saw, the fundamental equations of electrodynamics are invariant under Poincaré transformations and, in addition, they entail the conservation of the four-momentum and the four-dimensional angular momentum. At first sight these two aspects – relativistic invariance and the presence of conservation laws – do not seem to have anything to do with each other. *Noether's theorem* – which intertwines them in an inextricable way – relies, actually, heavily on a fundamental paradigm of theoretical physics, that we have not yet introduced: the *variational method*. In fact, the fundamental equations of electrodynamics own a property that at this stage of the presentation appears still hidden, namely, *they descend from a variational principle*: It is precisely this peculiar characteristic which ensures the above link between symmetries and conservation laws.

In general, the variational method allows to reformulate the dynamics of a theory in a compact and elegant way, providing a physically equivalent description of the theory. The crucial role this method plays in physics is underlined by the fact that all fundamental theories, from *Newtonian mechanics* to the *Standard Model of elementary particles*, *General Relativity*, and the more speculative *superstring theory*, can be deduced from a variational principle. In the absence of such a principle the – classical and quantum – internal consistency of these theories would be extremely difficult to control, and the validity of the main conservation laws would not be guaranteed. In this chapter we supply the basics of the variational method and establish its connection with Noether's theorem. Later, in Chap. 4, we will apply it to retrieve the fundamental equations of electrodynamics, together with its conservation laws.

The action. The key mathematical tool of the variational method is the *principle of least action*, also called *action principle*. This principle is based on a function of the dynamical variables of the system, the *Lagrangian L*, whose integral over time gives rise to the functional $I = \int L \, dt$, called *action*. The conceptual advantage of the method consists in the extreme "economy" employed in the construction of a physical theory: once the function L has been chosen, the principle of least

© Springer International Publishing AG, part of Springer Nature 2018
K. Lechner, *Classical Electrodynamics*, UNITEXT for Physics,
https://doi.org/10.1007/978-3-319-91809-9_3

action uniquely identifies the dynamics of the system. According to this principle the configurations that satisfy the equations of motion are, in fact, exactly those that make the action *stationary* – formally $\delta I = 0$ – under arbitrary variations of the dynamical variables. In the presence of symmetries Noether's theorem provides, in addition, the *explicit* form of the constants of motion in terms of L.

Relativistic invariance. In a relativistic theory the variational method is subject to an additional condition, namely the action must be *invariant under Poincaré transformations*, that is to say, a *four-scalar*. If I is the action in a reference frame K, and I' the corresponding action in K', then we must have

$$I' = I.$$

In this case the equations of motion associated with I satisfy automatically Einstein's *relativity principle*. In fact, schematically we have the double implications

eqs. of motion in K $\leftrightarrow$ $\delta I = 0$ $\leftrightarrow$ $\delta I' = 0$ $\leftrightarrow$ eqs. of motion in K'.

If the action is a scalar, the equations of motion have therefore automatically the same form in all reference frames.

Quantization. In theoretical physics the variational method plays a key role for a further important reason: it is the indispensable starting point for the *quantization* of whatever classical theory. In fact, the *canonical* quantization of a theory is based on a *Hamiltonian* which, in turn, descends from the Lagrangian by means of a Legendre transformation. In a relativistic theory, however, the canonical quantization does not represent a *manifestly covariant* process, simply because the Hamiltonian – being the fourth component of a four-vector – is not a scalar. On the other hand, there exists an alternative quantization method, Feynman's *functional integral* approach, which rests directly on the *action* and has the advantage of yielding a manifestly covariant quantum theory, provided the action is a relativistic invariant. In conclusion, if a theory is formulated in terms of a variational principle, the relativistic invariance of the classical action is automatically transferred to the corresponding quantum theory.

Fields and locality. We end these introductory remarks dwelling on a special characteristic of *relativistic* theories: the *locality* of the interactions. In non-relativistic physics particles interact through forces that exert an *action at a distance*. A particle of charge e_2, for example, exerts on a particle of charge e_1 the force

$$\mathbf{F} = \frac{e_1 e_2}{4\pi} \frac{\mathbf{y}_1 - \mathbf{y}_2}{|\mathbf{y}_1 - \mathbf{y}_2|^3},$$

which is transmitted *instantaneously*: if at a certain instant the charge e_2 starts moving, the charge e_1 perceives the effect of this displacement at the same instant. A *non-local* interaction of this type corresponds to a signal propagating at infinite speed and would, thus, be in conflict with the principles of special relativity. Conversely, in a relativistic theory the particles do not interact with each other *directly*, but rather by means of *fields*, and the interaction between fields and particles is a *contact* interaction, i.e. a *local* one. The Lorentz force

$$eF^{\mu\nu}(y)u_\nu$$

experienced by a relativistic charged particle depends, in fact, only on the value of the field at the particle's position y, and not on the values of the field at different points, or on the positions of the other particles. It follows, in particular, that the electromagnetic interaction between charged particles propagates at the speed of propagation of the electromagnetic field, i.e. at the speed of light. Hence, in relativistic theories the *locality* of the interactions is guaranteed by the *fields*, and the latter should be considered, in all respects, as independent dynamical *degrees of freedom*, at the same footing as the particles' coordinates: While in non-relativistic physics the concept of "field" is just *useful*, in a relativistic theory it is indispensable.

Finally, for comparison, we note that at in *quantum theory* the locality paradigm is realized by the virtual picture that the interaction between charged particles occurs via the emission and absorption of the *quanta* of the electromagnetic field, the photons, that, in turn, travel at the speed of light. Correspondingly, in the graphic representation of *quantum electrodynamics* via *Feynman diagrams*, the interaction between photons and charges occurs again *locally*, in so-called *vertices*, which symbolically represent the space-time points where the emission and absorption processes take place.

In summary, the formulation of a physical theory via the variational method follows the general scheme:

(1) identify the expression of the action;
(2) derive the equations of motion by means of the principle of least action;
(3) apply Noether's theorem to derive conservation laws.

As mentioned above, in a relativistic framework, and in particular in electrodynamics, generically we will have to deal with a system of point-like particles interacting with a system of fields. A physical system whose degrees of freedom consist of only fields is called a *field theory*. Hence, in general we will have to implement the variational method for a system of particles interacting with a field theory. In this chapter we present the variational method for a generic field theory which, in a sense, can be considered as a *Lagrangian* system with an *infinite* number of degrees of freedom. For this reason, in Sect. 3.1 we first recall how the method works in a Lagrangian system with a *finite* number of degrees of freedom, i.e. in *Newtonian mechanics*.

3.1 Principle of Least Action in Mechanics

We consider a mechanical – conservative and holonomic – system of N degrees of freedom, described by the Lagrangian coordinates $q_n(t)$, where $n = 1, \ldots, N$. We denote the coordinates collectively by $q = \{q_n(t)\}$, and their first derivatives by $\dot{q} = \{\dot{q}_n(t)\}$, where

$$\dot{q}_n(t) = \frac{dq_n(t)}{dt}.$$

We suppose furthermore that there exists a function of $2N$ variables – the Lagrangian $L(q, \dot{q})$ – such that the equations of motion of the underlying mechanical system are equivalent to the *Euler-Lagrange equations*

$$\frac{d}{dt} \frac{\partial L}{\partial \dot{q}_n} - \frac{\partial L}{\partial q_n} = 0, \qquad n = 1, \ldots, N. \tag{3.1}$$

Henceforth, we assume that the functions $q_n(t)$ and $L(q, \dot{q})$ are "sufficiently" regular, so that in particular equations (3.1) are well defined. A *prototypical* Lagrangian system is a system of M non-relativistic particles, with cartesian coordinates $\mathbf{y}_i(t)$, $i = 1, \ldots, M$, whose Lagrangian coordinates are thus given by the set $q = (\mathbf{y}_1, \ldots, \mathbf{y}_M) = \mathbf{y}$, and $N = 3M$. Denoting the interaction potential by $V(\mathbf{y})$ and the non-relativistic kinetic energy by $T(\dot{\mathbf{y}}) = \frac{1}{2} \sum_i m_i \dot{\mathbf{y}}_i \cdot \dot{\mathbf{y}}_i$, the Lagrangian of the system reads

$$L(\mathbf{y}, \dot{\mathbf{y}}) = T(\dot{\mathbf{y}}) - V(\mathbf{y}).$$

Equations (3.1) then take the familiar form

$$m_i \ddot{\mathbf{y}}_i = -\boldsymbol{\nabla}_i V(\mathbf{y}).$$

Turning to the general case, and fixing two boundary instants $t_1 < t_2$, we associate with the Lagrangian L the functional of the coordinates q, called *action*,

$$I[q] = \int_{t_1}^{t_2} L(q(t), \dot{q}(t)) \, dt. \tag{3.2}$$

The *principle of least action*, also known as *Hamilton's principle*, may then be enunciated as follows.

Theorem I (principle of least action in mechanics). The configuration q satisfies the Euler-Lagrange equations (3.1) in the interval (t_1, t_2), if and only if it is an *extremum* of the action $I[q]$ under arbitrary variations $\delta q = \{\delta q_n(t)\}$ vanishing at the boundaries:

$$\delta q_n(t_1) = 0 = \delta q_n(t_2), \quad \forall n.$$

Before proving the theorem we explain the terminology used in its statement. The *variations* $\delta q_n(t)$ denote N functions with the same regularity properties of the functions $q_n(t)$. We define the *variation* of a functional $I[q]$ around a configuration q under the variations δq as[1]

$$\delta I = \left. \frac{d}{d\alpha} I[q + \alpha \delta q] \right|_{\alpha=0}, \tag{3.3}$$

where α is a real parameter. Since the action (3.2) is given by the integral of a regular function L of q and $\dot{q}$, the definition (3.3) is equivalent to

$$\delta I = \lim_{\alpha \to 0} \frac{I[q + \alpha \delta q] - I[q]}{\alpha} = \left(I[q + \delta q] - I[q] \right)_{\text{lin}}. \tag{3.4}$$

With the last expression we understand the "Taylor expansion" of $I[q + \delta q] - I[q]$ in powers of $\delta q_n(t)$, truncated after the term linear in $\delta q_n(t)$. In practice to evaluate δI explicitly we will always proceed as specified in the last term of (3.4). Finally, we say that configuration q is an *extremum* of the functional $I[q]$ under certain variations δq, if $\delta I = 0$. Sometimes one also says that the action $I[q]$ is *stationary* at the configuration q, if $\delta I = 0$ for arbitrary variations δq vanishing at the boundaries.

Proof To prove the theorem we compute the variation δI for arbitrary variations δq, using (3.4),

$$\delta I = \left(I[q + \delta q] - I[q] \right)_{\text{lin}} = \int_{t_1}^{t_2} \left(L(q + \delta q, \dot{q} + \dot{\delta q}) - L(q, \dot{q}) \right)_{\text{lin}} dt$$

$$= \int_{t_1}^{t_2} \sum_n \left(\frac{\partial L}{\partial q_n} \delta q_n + \frac{\partial L}{\partial \dot{q}_n} \frac{d\delta q_n}{dt} \right) dt$$

$$= \int_{t_1}^{t_2} \sum_n \left(\frac{\partial L}{\partial q_n} \delta q_n + \frac{d}{dt} \left(\frac{\partial L}{\partial \dot{q}_n} \delta q_n \right) - \frac{d}{dt} \frac{\partial L}{\partial \dot{q}_n} \delta q_n \right) dt$$

$$= \int_{t_1}^{t_2} \sum_n \left(\frac{\partial L}{\partial q_n} - \frac{d}{dt} \frac{\partial L}{\partial \dot{q}_n} \right) \delta q_n \, dt + \sum_n \left. \frac{\partial L}{\partial \dot{q}_n} \delta q_n \right|_{t_1}^{t_2}.$$

Since $\delta q_n(t_1) = 0 = \delta q_n(t_2)$, the last sum vanishes. We conclude, therefore, that $\delta I = 0$ for arbitrary variations δq in the interval (t_1, t_2), if and only if in this interval the functions $q_n(t)$ satisfy the Euler-Lagrange equations (3.1). □

---

[1] δI is a functional of the $2N$ functions q and δq, and so, actually, it should be denoted by $\delta I[q, \delta q]$.

3.2   Principle of Least Action in Field Theory

A classical *field theory* is described by N functions of space and time $\varphi_r(t, \mathbf{x}) = \varphi_r(x)$, $r = 1, \ldots, N$, called *Lagrangian fields*, that we denote collectively by the symbol $\varphi = \{\varphi_r(x)\}$. These fields are supposed to thoroughly describe the system from a kinematical point of view, in the sense that every observable physical quantity can be expressed in terms of the fields φ_r, even though in general the fields themselves are not necessarily observables. In the case of electrodynamics, for example, the Lagrangian fields are not the electric and magnetic fields, but rather the four components of the vector potential A_0, A_1, A_2 and A_3, but, since these fields are not gauge invariant, they are, in fact, non directly observable. By means of the formal identification

$$\varphi_r(t, \mathbf{x}) \leftrightarrow q_{r,\mathbf{x}}(t),$$

and regarding the pair $(r, \mathbf{x})$ as a *discrete* index n, the set of fields $\varphi = \{\varphi_r(x)\}$ resembles, in a sense, a Lagrangian system with an *infinite* number of degrees of freedom. Therefore, also for a field theory our aim will be to derive the dynamics of the system from an action principle, starting from an action $I[\varphi]$ which is now a functional of the fields. In this case we will start from a Lagrangian *density* $\mathcal{L}$ – henceforth called simply *Lagrangian* – which in analogy to the finite-dimensional case should be a function of the fields φ and their time derivatives $\dot{\varphi} = \partial_0 \varphi$. However, if we want the action to be a relativistic invariant, then $\mathcal{L}$ must depend on all partial derivatives $\partial_\mu \varphi$, i.e.

$$\mathcal{L} = \mathcal{L}(\varphi(x), \partial\varphi(x)). \tag{3.5}$$

The *actual* Lagrangian $L(t)$ is then obtained by summing over all degrees of freedom, i.e. by integrating the Lagrangian density (3.5) over all spatial coordinates $\mathbf{x}$

$$L(t) = \int \mathcal{L}(\varphi(x), \partial\varphi(x))\, d^3x.$$

Finally, we define the action of the theory as

$$I[\varphi] = \int_{t_1}^{t_2} L(t)\, dt = \int_{t_1}^{t_2} \mathcal{L}(\varphi(x), \partial\varphi(x))\, d^4x. \tag{3.6}$$

Our aim is now to formulate a principle of least action relative to the action (3.6), analogous to the principle of least action for a system of finite number of degrees of freedom. As in that case we suppose that the functions $\varphi(x)$ and $\mathcal{L}(\varphi, \partial\varphi)$ are sufficiently regular, to make sure that our formal calculations retain their validity. In addition, we impose on these functions suitable asymptotic conditions. First of all the fields φ and their derivatives are required to vanish sufficiently fast at spatial infinity, so that, in particular, they fulfill the limit at fixed t

$$\lim_{|\mathbf{x}|\to\infty} \varphi_r(t, \mathbf{x}) = 0. \tag{3.7}$$

Moreover, we require that the function $\mathcal{L}(\varphi, \partial\varphi)$ in the limit for $\varphi \to 0$ decreases sufficiently fast, so that in the action (3.6) the integral in d^3x over the whole space $\mathbb{R}^3$ converges. The Euler-Lagrange equations relative to the Lagrangian $\mathcal{L}$ (3.5), analogous to (3.1), read

$$\partial_\mu \frac{\partial\mathcal{L}}{\partial(\partial_\mu \varphi_r)} - \frac{\partial\mathcal{L}}{\partial\varphi_r} = 0, \qquad r = 1, \ldots, N, \tag{3.8}$$

and they represent the *equations of motion* of the fields. We enunciate now the *principle of least action for the field theory* associated with the Lagrangian $\mathcal{L}$.

Theorem II (principle of least action in field theory). The field configuration φ satisfies the Euler-Lagrange equations (3.8) in the interval (t_1, t_2), if and only if it is an *extremum* of the action $I[\varphi]$ under arbitrary variations $\delta\varphi = \{\delta\varphi_r(x)\}$, vanishing at the boundaries:

$$\delta\varphi_r(t_1, \mathbf{x}) = 0 = \delta\varphi_r(t_2, \mathbf{x}), \ \forall r, \ \forall \mathbf{x}.$$

As in the case of a system of a finite number of degrees of freedom we say that a configuration φ is an extremum of the functional I under certain variations $\delta\varphi$, if the variation of the action

$$\delta I = \frac{d}{d\alpha} I[\varphi + \alpha\,\delta\varphi]\bigg|_{\alpha=0} = \lim_{\alpha\to 0} \frac{I[\varphi + \alpha\delta\varphi] - I[\varphi]}{\alpha} = \Big(I[\varphi + \delta\varphi] - I[\varphi]\Big)_{\text{lin}} \tag{3.9}$$

vanishes. It is understood that the allowed variations $\delta\varphi_r(x)$ entail the same regularity properties and the same asymptotic behaviors of the fields $\varphi_r(x)$.

Proof To prove the theorem we calculate the variation of the action (3.6) using the last expression of (3.9)

$$\delta I = \Big(I[\varphi + \delta\varphi] - I[\varphi]\Big)_{\text{lin}} = \int_{t_1}^{t_2} \Big(\mathcal{L}(\varphi + \delta\varphi, \partial\varphi + \partial\delta\varphi) - \mathcal{L}(\varphi, \partial\varphi)\Big)_{\text{lin}} d^4x$$

$$= \int_{t_1}^{t_2} \sum_r \left(\frac{\partial\mathcal{L}}{\partial\varphi_r} \delta\varphi_r + \frac{\partial\mathcal{L}}{\partial(\partial_\mu\varphi_r)} \partial_\mu\delta\varphi_r\right) d^4x.$$

Using Leibniz's rule we obtain

$$\delta I = \int_{t_1}^{t_2} \sum_r \left(\frac{\partial\mathcal{L}}{\partial\varphi_r} - \partial_\mu \frac{\partial\mathcal{L}}{\partial(\partial_\mu\varphi_r)}\right) \delta\varphi_r\, d^4x + \int_{t_1}^{t_2} \sum_r \partial_\mu \left(\frac{\partial\mathcal{L}}{\partial(\partial_\mu\varphi_r)} \delta\varphi_r\right) d^4x. \tag{3.10}$$

The second integral, whose integrand is a four-divergence, vanishes. In fact, applying the *fundamental theorem of calculus* to the time variable, and Gauss's theorem to the spatial divergence, with a spherical surface Γ_∞ placed at spatial infinity, we obtain:

$$\int_{t_1}^{t_2} \sum_r \partial_\mu \left(\frac{\partial \mathcal{L}}{\partial(\partial_\mu \varphi_r)} \, \delta\varphi_r \right) d^4x = \sum_r \int \frac{\partial \mathcal{L}}{\partial \dot{\varphi}_r} \, \delta\varphi_r \bigg|_{t_1}^{t_2} d^3x$$

$$+ \sum_r \int_{t_1}^{t_2} \left(\int_{\Gamma_\infty} \frac{\partial \mathcal{L}}{\partial(\partial_i \varphi_r)} \, \delta\varphi_r \, d\Sigma^i \right) dt.$$

The first term at the right-hand side vanishes because the variations $\delta\varphi_r(t, \mathbf{x})$ vanish at $t = t_1$ and $t = t_2$, whereas the second term vanishes thanks to the fact that at spatial infinity all fields decrease sufficiently fast. The variation (3.10) reduces thus to the first integral, and so δI vanishes for every choice of $\delta\varphi$, if and only if the fields satisfy the Euler-Lagrange equations in the interval (t_1, t_2). $\square$

3.2.1 Hypersurfaces in Minkowski Space-Time

In this section we introduce a few notions regarding hypersurfaces in four dimensions, which will prove useful later on.

Parameterizations of hypersurfaces. By definition a hypersurface Γ in a four-dimensional Minkowski space-time is a subset – more precisely a *submanifold* – of $\mathbb{R}^4$, of dimension three. In *parametric* form a hypersurface is described by four functions of three parameters

$$y^\mu(\boldsymbol{\lambda}), \tag{3.11}$$

where $\boldsymbol{\lambda}$ denotes the triple $\{\lambda^a\}$, with $a = 1, 2, 3$. Alternatively one may represent a hypersurface Γ in *implicit* form in terms of a single scalar function $f(x)$, through the relation

$$x^\mu \in \Gamma \qquad \Leftrightarrow \qquad f(x) = 0. \tag{3.12}$$

We can step from one representation to the other by inverting, for example, the spatial coordinates $\mathbf{y}(\boldsymbol{\lambda})$ of (3.11) to determine the parameters $\boldsymbol{\lambda}$ as functions of the spatial coordinates $\mathbf{x}$, i.e. by inverting the functions $\mathbf{x} = \mathbf{y}(\boldsymbol{\lambda}) \to \boldsymbol{\lambda}(\mathbf{x})$, and by setting then

$$f(x) = f(x^0, \mathbf{x}) = x^0 - y^0(\boldsymbol{\lambda}(\mathbf{x})).$$

This function satisfies, in fact, the identity

$$f(y(\boldsymbol{\lambda})) = 0. \tag{3.13}$$

We shall use one representation or the other, depending on the situation.

Hyperplanes. An important class of hypersurfaces are the *hyperplanes*, which in implicit form are described by a function of the kind

$$f(x) = M_\mu(x^\mu - x_*^\mu) = 0, \tag{3.14}$$

where M_μ and x_*^μ are constant vectors. The hyperplane corresponding to equation (3.14) passes through the point x_*^μ and is orthogonal to the vector M_μ.

Tangent and normal vectors. Given a generic point $P = y^\mu(\lambda) \in \Gamma$, the *tangent space* in P is defined as the three-dimensional vector space spanned by the three linearly independent four-vectors

$$U_a^\mu = \frac{\partial y^\mu(\lambda)}{\partial \lambda^a}, \qquad a = 1, 2, 3. \tag{3.15}$$

A generic vector U^μ tangent to Γ in P can thus be written as a linear combination of the basis vectors

$$U^\mu = \sum_{a=1}^{3} c^a U_a^\mu.$$

Since the space-time is four-dimensional, there exists a vector $N_\mu(\lambda)$ *normal* to Γ in P – unique modulo its normalization – characterized by the fact to be orthogonal to all tangent vectors

$$N_\mu U_a^\mu = 0, \quad \forall\, a. \tag{3.16}$$

If the hypersurface is given in the implicit form (3.12), by differentiating the identity (3.13) with respect to λ^a one obtains

$$0 = \frac{\partial f}{\partial x^\mu} \frac{\partial y^\mu}{\partial \lambda^a} = \frac{\partial f}{\partial y^\mu} U_a^\mu, \quad \forall\, a,$$

so that for N_μ one obtains the simple expression

$$N_\mu = \frac{\partial f}{\partial x^\mu}. \tag{3.17}$$

We distinguish three types of hypersurfaces.

Definition. A hypersurface Γ is called *space-like*, *time-like*, or *light-like*, if at every point $P \in \Gamma$ the normal vector N_μ is *time-like*, *space-like*, or *light-like*, respectively, that is to say

$$N^2 > 0, \qquad N^2 < 0, \qquad N^2 = 0.$$

Clearly, these characterizations are Lorentz invariant.

Space-like hypersurfaces. For a space-like hypersurface we have $N^2 > 0$, and correspondingly the tangent vectors are all space-like. To see this it is sufficient to exploit the property that, when $N^2 > 0$, for every fixed point $P \in \Gamma$ there exists a privileged inertial reference frame in which N_μ has the form $N_\mu = (N_0, 0, 0, 0)$. Since a generic tangent vector U^μ must obey the constraint $N_\mu U^\mu = 0$, it follows that $U^0 = 0$, and so in the privileged reference frame U^μ satisfies the Lorentz-invariant inequality

$$U^2 < 0, \tag{3.18}$$

which so holds in all reference frames. It is, moreover, possible to show that a hypersurface Γ is space-like, if and only if each pair of points x_1 and x_2 belonging to Γ obeys the inequality $(x_1 - x_2)^2 < 0$. This alternative characterization is easily verified in the case of the hyperplanes (3.14), for which (3.17) yields the constant normal vector

$$N_\mu = \frac{\partial f}{\partial x^\mu} = M_\mu.$$

In fact, if the points x_1 and x_2 belong to Γ, the conditions (3.14) imply that $M_\mu(x_1^\mu - x_*^\mu) = 0 = M_\mu(x_2^\mu - x_*^\mu)$, so that $(x_1^\mu - x_2^\mu)M_\mu = 0$. If $N^2 = M^2 > 0$, with the same reasoning that led to (3.18) one then concludes that the distance between the two points is space-like

$$(x_1 - x_2)^2 < 0.$$

It is easy to see that also the converse is true. Choosing for M_μ the time-like vector $M_\mu = (1, 0, 0, 0)$, one obtains the hyperplanes at constant time t_*

$$f(x) = M_\mu(x^\mu - x_*^\mu) = t - t_* = 0,$$

which correspond to the particular space-like hypersurfaces used in the action (3.6) to delimit the domain of the integral. The parametric form (3.11) of these hyperplanes is

$$y^0(\boldsymbol{\lambda}) = t_*, \qquad \mathbf{y}(\boldsymbol{\lambda}) = \boldsymbol{\lambda}. \tag{3.19}$$

Time-like hypersurfaces. For a time-like hypersurface the normal vector satisfies $N^2 < 0$, and in this case the tangent vectors can be space-like, time-like or light-like. If we consider, for example, the time-like hyperplane represented by the function $f(x) = z - z_* = 0$, corresponding to the vector $M_\mu = (0, 0, 0, 1) = N_\mu$, according to the conditions (3.16) the generic tangent vector has the form $U^\mu = (U^0, U^x, U^y, 0)$. The scalar product $U^\mu U_\mu = U^{02} - U^{x2} - U^{y2}$ can thus be positive, negative or zero. Another time-like hypersurface is represented by the function

$$f(x) = \frac{1}{2}\left(|\mathbf{x}|^2 - R^2\right) = 0, \quad N_\mu = \frac{\partial f}{\partial x^\mu} = (0, \mathbf{x}), \quad N^2 = -|\mathbf{x}|^2 = -R^2 < 0,$$

which describes a sphere of radius R, as time varies. In the limit for $R \to \infty$ this hypersurface tends to a *time-like hypersurface placed at spatial infinity*, a type of hypersurfaces that we will meet soon.

Gauss's theorem in four dimensions. Consider the integral of the four-divergence of a vector field $W^\mu(x)$ over a four-dimensional volume V whose boundary is the hypersurface $\Gamma = \partial V$. Parameterize Γ (locally) as in (3.11) and define the tangent vectors as in (3.15). Then we have the equality

$$\int_V \partial_\mu W^\mu \, d^4x = \int_\Gamma W^\mu d\Sigma_\mu, \tag{3.20}$$

where the *three-dimensional hypersurface element* is defined by

$$d\Sigma_\mu = \frac{1}{3!} \, \varepsilon_{\mu\alpha\beta\gamma} \, \varepsilon^{abc} U_a^\gamma U_b^\beta U_c^\alpha d^3\lambda = \varepsilon_{\mu\alpha\beta\gamma} \, U_1^\gamma U_2^\beta U_3^\alpha d^3\lambda, \tag{3.21}$$

ε^{abc} being the three-dimensional Levi-Civita tensor (1.21).

For the proof of the theorem we refer to a textbook of *Mathematical Analysis*. In what follows we want to give a geometrical interpretation of the second member of (3.20). The antisymmetry of the Levi-Civita tensor implies the identity

$$U_a^\mu \left(\varepsilon_{\mu\alpha\beta\gamma} \, U_1^\gamma U_2^\beta U_3^\alpha \right) = 0, \quad \forall \, a.$$

It follows that a normal vector is given by

$$N_\mu = \varepsilon_{\mu\alpha\beta\gamma} \, U_1^\gamma U_2^\beta U_3^\alpha = \frac{1}{3!} \, \varepsilon_{\mu\alpha\beta\gamma} \, \varepsilon^{abc} \, U_a^\gamma U_b^\beta U_c^\alpha, \tag{3.22}$$

so that we can rewrite the hypersurface element (3.21) as

$$d\Sigma_\mu = N_\mu d^3\lambda.$$

Gauss's theorem can therefore be rewritten as

$$\int_V \partial_\mu W^\mu d^4x = \int_\Gamma W^\mu N_\mu \, d^3\lambda.$$

Excluding the case of a light-like hypersurface Γ, $N^2 \neq 0$, we can rewrite $d\Sigma_\mu$ in a form that resembles the two-dimensional surface element $d\boldsymbol{\Sigma} = \mathbf{n} \, d\Sigma$ – where, we recall, $\mathbf{n}$ is the versor normal to the surface – since in this case the vector N_μ can be *normalized*. Taking the square of the vector (3.22), using the second relation in (1.40), and the determinant identity (1.69), we obtain in fact the identity

$$N^2 = N_\mu N^\mu = -\det g_{ab},$$

where $\det g_{ab}$ denotes the determinant of the *induced metric* on Γ

$$g_{ab} = U_a^\mu U_b^\nu \eta_{\mu\nu}. \tag{3.23}$$

We can now introduce a normal *versor*

$$n_\mu = \frac{N_\mu}{\sqrt{|N^2|}} = \frac{N_\mu}{\sqrt{g}}, \qquad g = |\det g_{ab}|,$$

obeying the relations

$$n^2 = 1, \text{ if } \Gamma \text{ is space-like;} \qquad n^2 = -1, \text{ if } \Gamma \text{ is time-like.}$$

In conclusion, the four-dimensional Gauss theorem can be recast in the *traditional* form

$$\int_V \partial_\mu W^\mu \, d^4x = \int_\Gamma W^\mu n_\mu \sqrt{g} \, d^3\lambda, \tag{3.24}$$

where $\sqrt{g} \, d^3\lambda$ represents the volume of the infinitesimal hypersurface element.[2] We conclude this section by illustrating the use of formula (3.24), for the case where a part of Γ is a space-like hyperplane Π with equation $t = t_*$, parameterized as in (3.19). For this hyperplane the definition (3.15) entails the tangent vectors

$$U_a^0 = 0, \qquad U_a^i = \delta_a^i,$$

so that formulas (3.22) and (3.23) yield the expressions

$$N_\mu = (1, 0, 0, 0), \qquad g_{ab} = -\delta_{ab}, \qquad n_\mu = N_\mu, \qquad N^2 = g = 1.$$

The contribution of the hyperplane Π to the integral at the second member of equation (3.24) then reduces to the expected result

$$\int_\Pi W^\mu n_\mu \sqrt{g} \, d^3\lambda = \int W^0(t_*, \boldsymbol{\lambda}) \, d^3\lambda.$$

3.2.2 Relativistic Invariance

So far we have made no assumptions about the invariance properties of the considered field theory. In this section we analyze some important aspects of the principle of least action, in the particular case of a *relativistic* field theory.

Principle of least action and manifest covariance. As explained in Chap. 1, in a relativistic field theory we expect the equations of motion to be *manifestly* covariant. In the variational framework manifest covariance is, actually, realized in a natural way, if the fields are organized into *tensor* multiplets, and the Lagrangian $\mathcal{L}$ is a *four-scalar*. In fact, in this case the Euler-Lagrange equations (3.8) are *automatically* manifestly covariant. In a relativistic theory we require, therefore, the Lagrangian to be invariant under the Poincaré transformations $x' = \Lambda x + a$, more precisely we demand the equality

---

[2]For further details on hypersurfaces, the properties of the induced metric (3.23), and the infinitesimal volume element $\sqrt{g} \, d^3\lambda$, see Sect. 19.3.2, where it is in particular shown that the latter is unaffected by arbitrary changes of the parameters $\boldsymbol{\lambda}$, as it must be.

$$\mathcal{L}(\varphi'(x'), \partial'\varphi'(x')) = \mathcal{L}(\varphi(x), \partial\varphi(x)) \tag{3.25}$$

to be fulfilled for all $\Lambda^\mu{}_\nu \in O(1,3)$, and for all a^μ. If this equality holds, we can ask whether the action (3.6) is a scalar, as postulated in the introduction of this chapter. Actually, from the expression (3.6) there emerges an obvious obstacle to the invariance of I: whereas the measure of the integral is invariant,

$$d^4x' = |\det \Lambda| \, d^4x = d^4x, \tag{3.26}$$

the integration domain is not, since the time variable is integrated over a finite interval. Nevertheless, it is not difficult to overcome this obstruction. In fact, it is sufficient to substitute in (3.6) the space-like hyperplanes $t = t_1$ and $t = t_2$, which bound the four-dimensional integration domain, with two generic infinitely extended and non-intersecting space-like *hypersurfaces* Γ_1 and Γ_2. In fact, a hyperplane at constant time is a particular space-like hypersurface, which after a Poincaré transformation is no longer a hyperplane at constant time, while remaining a space-like hypersurface. In order to overcome the above obstacle we, therefore, replace the expression (3.6) with the generalized action

$$I[\varphi] = \int_{\Gamma_1}^{\Gamma_2} \mathcal{L}(\varphi(x), \partial\varphi(x)) \, d^4x, \tag{3.27}$$

which, thanks to the relations (3.25) and (3.26), is indeed a relativistic invariant,

$$I' = \int_{\Gamma_1'}^{\Gamma_2'} \mathcal{L}(\varphi'(x'), \partial'\varphi'(x')) \, d^4x' = \int_{\Gamma_1}^{\Gamma_2} \mathcal{L}(\varphi(x), \partial\varphi(x)) \, d^4x = I.$$

We can now formulate a *manifestly covariant* principle of least action in field theory: the configuration $\{\varphi_r(x)\}$ satisfies the Euler-Lagrange equations (3.8) in the region between the hypersurfaces Γ_1 and Γ_2, if and only if it is an extremum of the action (3.27) under arbitrary variations $\{\delta\varphi_r(x)\}$, vanishing on the hypersurfaces Γ_1 and Γ_2

$$\delta\varphi_r(x)\big|_{\Gamma_1} = 0 = \delta\varphi_r(x)\big|_{\Gamma_2}. \tag{3.28}$$

The proof parallels the one of Theorem II above, apart from the four-divergences which must now be treated as in (3.30) and (3.31) below. Finally, the Lorentz-invariant version of the asymptotic condition (3.7) is

$$\lim_{x^2 \to -\infty} \varphi_r(x) = 0, \tag{3.29}$$

where $x^2 = x^\mu x_\mu = t^2 - |\mathbf{x}|^2$.

Equivalent Lagrangians and four-divergences. For a given Lagrangian $\mathcal{L}$ the N Euler-Lagrange equations of motion (3.8) are obviously uniquely determined, but frequently one has to face the inverse problem: given a set of N equations of motion for the fields $\{\varphi_r\}$, does there exist a Lagrangian $\mathcal{L}$ such that they can be recast in the form (3.8)? It is clear that for an arbitrary system of equations of motion – even though invariant under Poincaré transformations – in general there is no Lagrangian at all, such that they can be rewritten as in (3.8). On the other hand, if such a Lagrangian exists – as it happens for all *fundamental* physical theories – in general this Lagrangian is not uniquely determined. It is evident, for example, that the Lagrangians $\mathcal{L}$ and $\widehat{\mathcal{L}} = a\mathcal{L} + b$, with a and b real constants, give rise to the same Euler-Lagrange equations.

A less obvious indeterminacy is represented by the fact that the Lagrangians are determined up to *four-divergences*. The Lagrangians $\mathcal{L}$ and

$$\widehat{\mathcal{L}} = \mathcal{L} + \partial_\mu C^\mu(\varphi),$$

where $C^\mu(\varphi)$ are four arbitrary functions with the same regularity properties as $\mathcal{L}$, give, in fact, rise to identical Euler-Lagrange equations. To show this we compute the difference between the actions associated with the two Lagrangians, using Gauss's theorem (3.20),

$$\widehat{I} - I = \int_{\Gamma_1}^{\Gamma_2} \widehat{\mathcal{L}} \, d^4x - \int_{\Gamma_1}^{\Gamma_2} \mathcal{L} \, d^4x = \int_{\Gamma_1}^{\Gamma_2} \partial_\mu C^\mu d^4x = \int_{\partial V} C^\mu d\Sigma_\mu. \qquad (3.30)$$

In the last term V denotes the four-dimensional integration volume, whose boundary ∂V is composed of Γ_1, Γ_2, and of a time-like hypersurface Γ_∞ placed at spatial infinity. The difference between the new and old actions is then

$$\widehat{I} - I = \int_{\Gamma_2} C^\mu d\Sigma_\mu - \int_{\Gamma_1} C^\mu d\Sigma_\mu + \int_{\Gamma_\infty} C^\mu d\Sigma_\mu. \qquad (3.31)$$

Thanks to the asymptotic condition (3.29), the integral over Γ_∞ gives a vanishing contribution. The first two integrals are non-vanishing, while involving only the values of $\varphi_r(x)$ on Γ_1 and Γ_2. Thanks to the conditions (3.28) we obtain thus $\delta(\widehat{I} - I) = 0$, i.e.

$$\delta\widehat{I} = \delta I.$$

The actions $\widehat{I}$ and I give, hence, rise to the same Euler-Lagrange equations. To summarize, Lagrangians which differ by a four-divergence are *physically equivalent* and so, henceforth, we will identify them for all purposes.

Locality. We end this section by introducing a further restriction on the Lagrangians allowed for relativistic field theories. To the requirement of relativistic invariance we add, in fact, the one of *locality*, which parallels the requirement of a *contact interaction* between particles and fields discussed in the introduction to this chapter. In the case of a field theory, *locality* demands that the Lagrangian is formed by a finite sum of products of the fields and their derivatives, *evaluated at the same space-time point* x. We illustrate this requirement for a field theory described by two scalar Lagrangian fields $\varphi_1(x) = A(x)$ and $\varphi_2(x) = B(x)$. In this case we admit, for example, the Lagrangian

$$\mathcal{L}_1 = \frac{1}{2}\,\partial_\mu A(x)\partial^\mu A(x) + \frac{1}{2}\,\partial_\mu B(x)\partial^\mu B(x) - gA^2(x)B^2(x),$$

whereas we reject the Lagrangian

$$\mathcal{L}_2 = \frac{1}{2}\,\partial_\mu A(x)\partial^\mu A(x) + \frac{1}{2}\,\partial_\mu B(x)\partial^\mu B(x) - g_N \int A^2(x)\big((x-y)^2\big)^N B^2(y)d^4y, \tag{3.32}$$

although both are invariant under Poincaré transformations. In (3.32) g_N is a *coupling constant*, N is a positive integer, and $(x-y)^2 = (x^\mu - y^\mu)(x_\mu - y_\mu)$. In $\mathcal{L}_1$ the field $A(x)$ is in contact with the field $B(x)$, in the sense that it is evaluated at the same point x, while in $\mathcal{L}_2$ the field $A(x)$ is in contact with the field $B(y)$ for *every* value of y. In the Euler-Lagrange equations relative to $\mathcal{L}_2$ the "motion" of the field A at the point x is, therefore, influenced by the values of the field B in all points of space-time. Correspondingly, this Lagrangian generates a dynamics which is characterized by an *action at a distance*, and as such it is physically not acceptable. We stress, however, that $\mathcal{L}_2$ is a *scalar* under Poincaré transformations, and as such it gives rise to relativistically invariant equations of motion.

There is an additional reason that leads us to reject Lagrangians of the type (3.32). In fact, such Lagrangians cannot possess a *fundamental* character: The integer N appearing in the third term of (3.32) is arbitrary, and we could replace the term $g_N((x-y)^2)^N$ with an arbitrary function $f(x,y)$, as long as it is invariant under Poincaré transformations. It is not difficult to show that under certain regularity conditions such a function, expanded in a Taylor series, would have the general form

$$f(x,y) = \sum_{N=0}^{\infty} g_N\left((x-y)^2\right)^N, \tag{3.33}$$

depending thus on an *infinite* number of coupling constants g_N. Such a theory would have no predictive power at all, since an infinite number of experiments would be needed to determine the values of g_N. Nevertheless, Lagrangians of the type (3.32), with the replacement

$$g_N\left((x-y)^2\right)^N \to f(x,y),$$

are frequently employed to describe the dynamics of *effective theories*, that is to say, theories of limited validity which, nonetheless, in particular physical regimes, for example, at low or high energies, correctly reproduce the experimental results.

Quantum consistency. Not all Lagrangians possessing the properties imposed so far give rise to a classical dynamics which remains consistent if one takes quantum corrections into account. According to the paradigm of *relativistic quantum field theories*, classical Lagrangians $\mathcal{L}(\varphi, \partial\varphi)$ giving rise, for instance, via canonical quantization, to consistent quantum theories must be:

(1) invariant under Poincaré transformations;
(2) local expressions of the fields;
(3) polynomials of the fields and their derivatives of maximum order *four*.

These restrictions severely limit the form of the allowed Lagrangians and, along with other symmetry principles, often allow to determine them uniquely. Examples of such uniquely-fixed Lagrangians are the ones describing the electromagnetic, weak, and strong interactions. On the contrary, the Lagrangian describing the gravitational interaction – within the framework of General Relativity – meets requirements (1) and (2), but not requirement (3): due to the complicated self-interaction of the gravitational field, this Lagrangian is, in fact, *non-polynomial* in the field. This is the reason why – according to the current state of theoretical physics – *gravity* still appears in conflict with the laws of *quantum mechanics*, notwithstanding important recent advances accomplished in this research field, see e.g. [1–3].

3.2.3 Lagrangian for Maxwell's Equations

In this section we illustrate the variational method, retrieving the equations of the electromagnetic field through the principle of least action. In principle, our aim is thus to interpret equations (2.21) and (2.22) as the Euler-Lagrange equations relative to an appropriate Lagrangian. The first question to be addressed is the choice of the Lagrangian fields φ_r. Since equations (2.21) and (2.22) correspond altogether to *eight* equations, *a priori* we should hence introduce as many Lagrangian fields, namely eight. The natural choice $\varphi_r = F^{\mu\nu}$, which, by the way, would have the advantage of introducing only observable fields, is therefore precluded, for the Maxwell tensor corresponds not to eight, but only to six independent fields, i.e. $\mathbf{E}$ and $\mathbf{B}$, see in particular Problem 3.9. This strategy must hence be abandoned, and we must search for an alternative one. Such an alternative strategy consists in proceeding as anticipated in Sect. 2.2.4: we first solve the Bianchi identity in terms of a vector potential A_μ, setting

$$F_{\mu\nu} = \partial_\mu A_\nu - \partial_\nu A_\mu,$$

and then choose as Lagrangian fields the four components of this potential

$$\varphi_r = A_\mu, \tag{3.34}$$

with the identification $r = \mu = 0, 1, 2, 3$. According to this approach the principle of least action should give rise to the Maxwell equation

$$\partial_\mu F^{\mu\nu} - j^\nu = 0. \tag{3.35}$$

Henceforth, we assume that the current j^μ satisfies the continuity equation $\partial_\mu j^\mu = 0$, and that it does not depend on A^μ. Note that the choice of the four Lagrangian fields (3.34) is now consistent with the fact that the vector equation (3.35) corresponds to *four* equations. The problem is now reduced to find a Lagrangian $\mathcal{L} = \mathcal{L}(A, \partial A)$, so that its Euler-Lagrange equations

$$\partial_\mu \frac{\partial \mathcal{L}}{\partial(\partial_\mu A_\nu)} - \frac{\partial \mathcal{L}}{\partial A_\nu} = 0 \qquad (3.36)$$

are equivalent to (3.35). The Lagrangian we are searching for must be, first of all, a relativistic invariant. Since equations (3.35) involve only $F^{\mu\nu}$ and are hence gauge invariant, i.e. they are unaffected by the replacements

$$A'_\mu = A_\mu + \partial_\mu \Lambda, \qquad (3.37)$$

so must be $\mathcal{L}$, modulo four-divergences, see Sect. 3.2.2. To determine $\mathcal{L}$ we proceed heuristically relying on the structure of (3.35). The first term of this equation is linear in A_μ, while the second is independent of A_μ. Correspondingly, the Lagrangian must contain a term $\mathcal{L}_1$ quadratic in A_μ, and a term $\mathcal{L}_2$ linear in A_μ. Moreover, from the particular form of the two terms in (3.35) it follows that $\mathcal{L}_1$ must contain two derivatives, while $\mathcal{L}_2$ must contain none.

We determine first $\mathcal{L}_1$, which must be constructed with the derivatives of the vector potential. Gauge invariance forces $\mathcal{L}_1$ to depend on A_μ only through the gauge-invariant quantity $F^{\mu\nu}$, and so it must be quadratic in the latter. Actually, see (2.15), there exist only two independent relativistic invariants, i.e. Lorentz-invariant combinations, namely

$$F^{\mu\nu} F_{\mu\nu} \quad \text{and} \quad \varepsilon^{\mu\nu\rho\sigma} F_{\mu\nu} F_{\rho\sigma}.$$

Thanks to the Bianchi identity the second invariant amounts, however, to a four-divergence

$$\varepsilon^{\mu\nu\rho\sigma} F_{\mu\nu} F_{\rho\sigma} = 2\,\varepsilon^{\mu\nu\rho\sigma} \partial_\mu A_\nu F_{\rho\sigma} = 2\partial_\mu \left(\varepsilon^{\mu\nu\rho\sigma} A_\nu F_{\rho\sigma} \right) - 2A_\nu \left(\varepsilon^{\mu\nu\rho\sigma} \partial_\mu F_{\rho\sigma} \right)$$
$$= 2\partial_\mu \left(\varepsilon^{\mu\nu\rho\sigma} A_\nu F_{\rho\sigma} \right).$$

The invariant $\varepsilon^{\mu\nu\rho\sigma} F_{\mu\nu} F_{\rho\sigma}$ represents, therefore, an irrelevant contribution to the Lagrangian,[3] and consequently $\mathcal{L}_1$ must be proportional to $F^{\mu\nu} F_{\mu\nu}$.

Next we determine $\mathcal{L}_2$. This term must be linear in A_μ and involve the current j_μ. The only Lorentz-invariant combination linear in A_μ – not containing derivatives – that we can form with these two four-vectors is $\mathcal{L}_2 \propto A_\mu j^\mu$. This term is actually

---

[3]The invariant $\mathcal{L}_0 = \varepsilon^{\mu\nu\rho\sigma} F_{\mu\nu} F_{\rho\sigma} = -8\mathbf{E} \cdot \mathbf{B}$ is actually a *pseudoscalar* – in that under parity and time reversal it changes sign, see (1.68) – while the Lagrangian of electrodynamics is a *scalar*, see (4.17). The inclusion of $\mathcal{L}_0$ would therefore violate the invariance of electrodynamics under the complete Lorentz group $O(1, 3)$.

gauge invariant, modulo a four-divergence, once we take the continuity equation of the current into account. From (3.37) we derive in fact that the transformed term can be rewritten as

$$A'_\mu j^\mu = A_\mu j^\mu + \partial_\mu \Lambda j^\mu = A_\mu j^\mu + \partial_\mu(\Lambda j^\mu) - \Lambda \partial_\mu j^\mu = A_\mu j^\mu + \partial_\mu(\Lambda j^\mu).$$

To obtain the Maxwell equation with the correct relative normalizations we set

$$\mathcal{L} = \mathcal{L}_1 + \mathcal{L}_2, \qquad \mathcal{L}_1 = -\frac{1}{4} F^{\mu\nu} F_{\mu\nu}, \qquad \mathcal{L}_2 = -j^\nu A_\nu. \qquad (3.38)$$

With this choice of $\mathcal{L}$ it is easily verified that Eqs. (3.36) amount indeed to the Maxwell equation (3.35). Only $\mathcal{L}_2$ contributes to the derivative $\partial\mathcal{L}/\partial A_\nu$, which becomes

$$\frac{\partial\mathcal{L}}{\partial A_\nu} = -j^\nu.$$

On the other hand, only $\mathcal{L}_1$ contributes to the derivative $\partial\mathcal{L}/\partial(\partial_\mu A_\nu)$. To determine it explicitly it is convenient to evaluate the variation of $\mathcal{L}_1$ under an infinitesimal variation of ∂A,

$$\delta\mathcal{L}_1 = -\frac{1}{2} F^{\mu\nu}\delta F_{\mu\nu} = -\frac{1}{2} F^{\mu\nu}(\delta\partial_\mu A_\nu - \delta\partial_\nu A_\mu) = -F^{\mu\nu}\delta(\partial_\mu A_\nu),$$

giving

$$\frac{\partial\mathcal{L}}{\partial(\partial_\mu A_\nu)} = -F^{\mu\nu}. \qquad (3.39)$$

The Euler-Lagrange equations (3.36) reduce thus to

$$\partial_\mu \frac{\partial\mathcal{L}}{\partial(\partial_\mu A_\nu)} - \frac{\partial\mathcal{L}}{\partial A_\nu} = -\partial_\mu F^{\mu\nu} + j^\nu = 0, \qquad (3.40)$$

which is the Maxwell equation.

Vector potential and quantization. The Maxwell equation and the Bianchi identity have been formulated without any reference to the vector potential, or *gauge field*, A_μ and, as we shall see in Sect. 5.4.1, these equations can also be *solved* without introducing any potential. At the classical level the vector potential is, therefore, only a *useful* tool, while remaining, from a conceptual point of view, an *optional accessory*. Conversely, if we want to derive the electromagnetic field equations from a variational principle, as we have just seen, the introduction of a vector potential is *indispensable*. On the other hand, as we recalled at the beginning of the chapter, the variational principle – in turn – is the *imperative* starting point for the *quantization* of whatever classical theory. In summary, while at the classical level the use of the vector potential is optional, in quantum theory its presence as *fundamental* field is *unavoidable*. In *quantum electrodynamics*, this circumstance results in a series of problems, of non-trivial solution, related to the fact that the vector potential – although becoming a *self-adjoint operator* – not being gauge invariant cannot represent a physical observable.

A further distinctive feature arises in the description of the dynamics of the other three fundamental interactions, in that for them the vector potentials must be introduced already at the *classical* level. The reason is that the mediators of these interactions – the particles $W^{\pm}$, Z^0, the gluons, and the gravitons – unlike the photons, which mediate the electromagnetic interaction, are "charged", and hence subject to a *mutual* interaction. And it can be seen that the equations of motion of fields entailing mutual interactions involve already at the *classical* level the corresponding vector potentials, see also the discussion in the following section.

3.2.4   Gauge Bosons of Weak and Strong Interactions

In the particular case of a vanishing current, $j^{\mu} = 0$, the Lagrangian (3.38) gives rise to the equation of motion $\partial_{\mu} F^{\mu\nu} = 0$, describing the dynamics of the electromagnetic field in empty space, i.e. in the absence of charges. The Lagrangian

$$\mathcal{L}_1 = -\frac{1}{4} F^{\mu\nu} F_{\mu\nu} \tag{3.41}$$

describes thus a *free* gauge field A_{μ}. As seen in Sect. 3.2.3, the structure of this Lagrangian is determined essentially by symmetry principles, namely Lorentz invariance and gauge invariance. It is thus not surprising that also the *free* propagation of the mediators of the weak and strong interactions, subject to the same fundamental principles, is described by analogous Lagrangians. With the mediators of the weak interactions Z^0 and $W^{\pm}$ are associated the real gauge field Z^0_{μ} and the complex gauge field $W^{\pm}_{\mu} = W^{\mu}_1 \pm i\, W^{\mu}_2$, respectively, with the corresponding Maxwell tensors

$$F^0_{\mu\nu} = \partial_{\mu} Z^0_{\nu} - \partial_{\nu} Z^0_{\mu}, \qquad F^{\pm}_{\mu\nu} = \partial_{\mu} W^{\pm}_{\nu} - \partial_{\nu} W^{\pm}_{\mu}.$$

Likewise, to the eight mediators, gluons, of the strong interactions are associated the *gluon* vector potentials A^I_{μ} ($I = 1, \ldots, 8$), with the related Maxwell tensors

$$F^I_{\mu\nu} = \partial_{\mu} A^I_{\nu} - \partial_{\nu} A^I_{\mu}.$$

The Lagrangian describing the free propagation of all these fields is then given by

$$\mathcal{L}_0 = -\frac{1}{4}\left(F^{\mu\nu} F_{\mu\nu} + F^{0\mu\nu} F^0_{\ \mu\nu} + F^{+\mu\nu} F^-_{\ \mu\nu} + \sum_{I=1}^{8} F^{I\mu\nu} F^I_{\ \mu\nu} \right). \tag{3.42}$$

This Lagrangian is invariant under Lorentz transformations, under gauge transformations of A^μ, and under the gauge transformations of the mediators of the weak and strong interactions

$$W_\mu^\pm \to W_\mu^\pm + \partial_\mu \Lambda^\pm, \qquad Z_\mu^0 \to Z_\mu^0 + \partial_\mu \Lambda^0, \qquad A_\mu^I \to A_\mu^I + \partial_\mu \Lambda^I. \quad (3.43)$$

Massive vector bosons and gauge invariance. In Sect. 5.3 we will see that the general solution of Maxwell's equation in empty space $\partial_\mu F^{\mu\nu} = 0$ is a superposition of electromagnetic waves, propagating at the *speed of light*. Correspondingly, the associated mediators – the photons – are massless. According to the Lagrangian $\mathcal{L}_0$, assigning to all gauge fields the same dynamics, all mediators would thus have a vanishing mass. On the other hand, while photons and gluons are, in fact, massless particles, the vector bosons of the weak interactions are, actually, *massive.* Correspondingly, the Lagrangian $\mathcal{L}_0$ must be modified by the addition of a term $\mathcal{L}_m$, depending on $W_\mu^\pm$ and Z_μ^0, taking the non-vanishing masses m_W and m_Z of these particles into account. As we will see in Chap. 18, this term is given by the quadratic Lorentz-invariant polynomial

$$\mathcal{L}_m = \frac{1}{2\hbar^2} \left(m_W^2 W_\mu^+ W^{-\mu} + m_Z^2 Z_\mu^0 Z^{0\mu} \right), \quad (3.44)$$

where the presence of Planck's constant $\hbar$ is suggested by a dimensional analysis. The polynomial (3.44) entails, however, a fundamental drawback: it is not invariant under the gauge transformations (3.43) and, consequently, also the total Lagrangian $\mathcal{L}_0 + \mathcal{L}_m$ violates these fundamental symmetries. This conflict between gauge invariance and the fact that the mediators $W_\mu^\pm$ and Z^0 are massive has prevented for a long time the construction of a consistent *quantum field theory* of weak interactions. The problem has been overcome only when it was discovered that the violation of gauge invariance, introduced by the term (3.44), does not undermine the internal consistency of a quantum theory, provided that it occurs *spontaneously*, namely via the *condensation* of a *Higgs field*, see, for instance, reference [4]. In Chap. 18, devoted specifically to massive vector fields, we will nevertheless analyze in detail the effects of a term like $\mathcal{L}_m$ on the *classical* dynamics of these fields, some of them having, actually, a direct *quantum* mechanical counterpart.

Mutually interacting vector bosons. By construction the Lagrangian $\mathcal{L}_0 + \mathcal{L}_m$ describes the *free* propagation of the gauge fields involved. On the other hand, as we noted at the end of the preceding section, unlike the photons, the particles $W^\pm$ and Z^0, and the gluons are subject to *mutual* interactions. It can be seen that, to take these interactions into account, one must add to the Lagrangian $\mathcal{L}_0 + \mathcal{L}_m$ terms *cubic* and *quartic* in the potentials $W_\mu^\pm$, Z_μ^0 and A_μ^I, while no such terms appear for the electromagnetic potential A_μ: Since the photon is massless, as well as charge-less, in the absence of charged particles its dynamics is described by the simple quadratic Lagrangian (3.41). For further details on these issues we refer the reader to standard textbooks of the physics of elementary particles, as for instance [4, 5].

3.3   Noether's Theorem

In general, Noether's theorem states that to each one-parameter group of symmetries of a physical system corresponds a conservation law. The conservation of energy, for example, is associated with the invariance under time translations, and the conservation of the angular momentum is associated with the invariance under spatial rotations. It is worthwhile to clarify from the beginning that in the context of Noether's theorem the *invariance requirement* refers to a very specific circumstance. In the first place one could understand the invariance of the equations of motion. In general, however, as observed previously, this condition is too weak, as the invariance of these equations in no way guarantees the existence of constants of motion, an example being the *Lorentz-Dirac equation*, see Sect. 15.2.4. In fact, the validity of Noether's theorem is based on more restrictive assumptions:

- the equations of motion must descend from a variational principle;
- the corresponding action I must be invariant under the symmetry group.

As seen above, in field theory the action, in turn, is given in terms of the integral of a Lagrangian

$$I = \int \mathcal{L} \, d^4x.$$

For the theories considered in this book the invariance of I will always be a consequence of the separate invariance of $\mathcal{L}$ – modulo four-divergences, see Sect. 3.2.2 – and of the invariance of the measure d^4x. In the particular case of *internal symmetries*, for which, by definition, the space-time coordinates do not change, $x' = x$, we have trivially $d^4x' = d^4x$ (see also Problem 3.10). Similarly, for the Poincaré transformations $x' = \Lambda x + a$ we have $d^4x' = |\det \Lambda| \, d^4x = d^4x$.

Local conservation laws. A peculiar aspect of Noether's theorem in *field theory* is represented by the fact that it yields *local* conservation laws. We recall that a conservation law is called local if not only the total "charge" is conserved, but if, moreover, its conservation results from a *continuity equation*. In field theory Noether's theorem implies, in fact, for any one-parameter symmetry group, the existence of a four-current J^μ with vanishing four-divergence: $\partial_\mu J^\mu = 0$. Consequently, see Eqs. (2.120) and (2.121), the variation of the charge $Q_V = \int_V J^0 d^3x$ contained in a volume V is necessarily compensated by a *flux* $\mathbf{J}$ of charge across its boundary ∂V

$$\frac{dQ_V}{dt} = -\int_{\partial V} \mathbf{J} \cdot d\mathbf{\Sigma}. \tag{3.45}$$

In field theory it is therefore not possible that a charge, conserved *à la* Noether, disappears at a point and *instantaneously* appears at another point, without *flowing* from one point to the other. Finally, extending in (3.45) the volume V to the whole space, we retrieve that the *total* charge $Q = \int J^0 d^3x$ is time-independent, as would be required by a mere *global* conservation law.

For internal symmetries, as for instance gauge transformations, which do not involve transformations of space and time, the proof of Noether's theorem is quite simple, see e.g. Problem 3.10. Conversely, the Poincaré group originates precisely from transformations of the space-time coordinates, and so for this symmetry group the proof of the theorem is more involved. Nevertheless, given its conceptual importance and its phenomenological relevance in physics, below we prove this theorem for the Poincaré group in a *generic relativistic field theory*: We will see that the four-parameter subgroup of translations is associated with the conservation of the four-momentum P^μ, whereas the six-dimensional Lorentz subgroup is associated with the conservation of the six components of the angular momentum $L^{\alpha\beta}$.

3.3.1 Infinitesimal Poincaré Transformations

In the proof of Noether's theorem we will predominantly make us of the invariance of the Lagrangian under *infinitesimal* Poincaré transformations. In particular, we will need the explicit expressions of the infinitesimal variations of the fields, that is to say, of their variations at *first* order in the parameters $\omega^{\mu\nu}$ and a^μ (see (3.46) and (3.47) below). This preliminary section is dedicated to the determination of these variations. Up to now we denoted the N-tuple of Lagrangian fields generically by $\varphi = (\varphi_1, \ldots, \varphi_N)$. In a relativistic theory these fields must be organized into *multiplets* forming *tensors* under Poincaré transformations, i.e. scalar fields $\Phi(x)$, vector fields $A^\mu(x)$, rank-two tensor fields $B^{\mu\nu}(x)$ etc. Obviously, there can be several fields of the same rank in the theory. The index r of the set $\varphi = \{\varphi_r\}_{r=1}^N$ will denote all components of all multiplets. A generic Poincaré transformation has the form

$$x'^\mu = \Lambda^\mu{}_\nu x^\nu + a^\mu, \tag{3.46}$$

where, in this section, we shall assume that Λ belongs to the *proper* Lorentz group. We then have, see Sect. 1.4.2,

$$\Lambda^\mu{}_\nu = \left(e^\omega\right)^\mu{}_\nu, \qquad \omega^{\mu\nu} = -\omega^{\nu\mu}. \tag{3.47}$$

Under such a transformation the various fields transform according to their rank

$$\Phi'(x') = \Phi(x), \quad A'^\mu(x') = \Lambda^\mu{}_\nu A^\nu(x), \quad B'^{\mu\nu}(x') = \Lambda^\mu{}_\alpha \Lambda^\nu{}_\beta B^{\alpha\beta}(x) \text{ etc.} \tag{3.48}$$

Since these transformations are linear in the fields $\varphi_r(x)$, we can write them collectively as

$$\varphi'_r(x') = \mathcal{M}_r{}^s \varphi_s(x), \tag{3.49}$$

where $\mathcal{M}_r{}^s$ is an x-independent $N \times N$ matrix depending on the six parameters $\omega^{\mu\nu}$. Henceforth, the summation over repeated indices r and s is understood.

Total variations and form variations. For what follows it is convenient to distinguish two types of field variations: the *total variations* $\overline{\delta}\varphi_r$, and what we call the *form variations* $\delta\varphi_r$, defined by

$$\overline{\delta}\varphi_r = \varphi_r'(x') - \varphi_r(x), \tag{3.50}$$

$$\delta\varphi_r = \varphi_r'(x) - \varphi_r(x), \tag{3.51}$$

respectively. We now want to evaluate these variations under infinitesimal Poincaré transformations, that is to say, at first order in $\omega^{\mu\nu}$ and a^μ. By definition, the infinitesimal Poincaré transformations are composed of infinitesimal Lorentz transformations, see (3.47),

$$\Lambda^\mu{}_\nu = \delta^\mu{}_\nu + \omega^\mu{}_\nu, \tag{3.52}$$

and of infinitesimal translations. The defining relation (3.46) then yields for the infinitesimal Poincaré transformations of the coordinates[4]

$$\delta x^\mu = x'^\mu - x^\mu = (\delta^\mu{}_\nu + \omega^\mu{}_\nu)x^\nu + a^\mu - x^\mu = \omega^\mu{}_\nu x^\nu + a^\mu. \tag{3.53}$$

On the other hand, from the transformation laws (3.48) – using the relations (3.50) and (3.52), and retaining only terms of first order in $\omega^{\mu\nu}$ – we find for the infinitesimal total variations of the fields

$$\overline{\delta}\Phi = \Phi'(x') - \Phi(x) = 0, \tag{3.54}$$

$$\overline{\delta}A^\mu = A'^\mu(x') - A^\mu(x) = (\delta^\mu{}_\nu + \omega^\mu{}_\nu)A^\nu(x) - A^\mu(x) = \omega^\mu{}_\nu A^\nu(x), \tag{3.55}$$

$$\overline{\delta}B^{\mu\nu} = B'^{\mu\nu}(x') - B^{\mu\nu}(x) = (\delta^\nu{}_\alpha + \omega^\mu{}_\alpha)(\delta^\nu{}_\beta + \omega^\nu{}_\beta)B^{\alpha\beta}(x) - B^{\mu\nu}(x)$$
$$= \omega^\mu{}_\alpha B^{\alpha\nu}(x) + \omega^\nu{}_\beta B^{\mu\beta}(x). \tag{3.56}$$

As we see, by construction the infinitesimal variations $\overline{\delta}\varphi_r$ are linear in the parameters $\omega^{\mu\nu}$, as well as in the fields φ_r themselves, see also (3.49). These variations have, hence, the general structure

$$\overline{\delta}\varphi_r = \frac{1}{2}\omega_{\alpha\beta}\Sigma^{\alpha\beta}{}_r{}^s\varphi_s, \tag{3.57}$$

where the numerical quantities $\Sigma^{\alpha\beta}{}_r{}^s$ are antisymmetric in α and β, as is $\omega_{\alpha\beta}$,

$$\Sigma^{\alpha\beta}{}_r{}^s = -\Sigma^{\beta\alpha}{}_r{}^s. \tag{3.58}$$

The explicit expressions of these quantities can easily be read off from the field variations (3.54)–(3.56). For a scalar field Φ, for example, we have simply $\Sigma^{\alpha\beta}{}_1{}^1 = 0$, while for a vector field, i.e. $\varphi_r = A_\mu$, Eq. (3.55) gives

---

[4]To keep the notation simple we maintain for the infinitesimal variations of the fields and the coordinates the same symbol δ as for the finite ones.

$$\Sigma^{\alpha\beta}{}_r{}^s = \delta^\alpha_r \eta^{\beta s} - \delta^\beta_r \eta^{\alpha s}. \tag{3.59}$$

We now move on to the determination of the infinitesimal *form variations* of the fields. Adding and subtracting in the definition (3.51) the same term, and using the relations (3.53) and (3.57), we obtain

$$\delta\varphi_r = \varphi'_r(x) - \varphi'_r(x') + \varphi'_r(x') - \varphi_r(x) = \varphi'_r(x) - \varphi'_r(x + \delta x) + \overline{\delta}\varphi_r \tag{3.60}$$

$$= -\delta x^\nu \partial_\nu \varphi'_r + \overline{\delta}\varphi_r = -\delta x^\nu \partial_\nu \varphi_r + \overline{\delta}\varphi_r \tag{3.61}$$

$$= -\delta x^\nu \partial_\nu \varphi_r + \frac{1}{2}\,\omega_{\alpha\beta}\Sigma^{\alpha\beta}{}_r{}^s\,\varphi_s, \tag{3.62}$$

where we considered only terms of first order in $\omega^{\mu\nu}$ and a^μ. In particular, in line (3.61) we used the fact that the difference between $\varphi'_r(x)$ and $\varphi_r(x)$ is of first order in $\omega^{\mu\nu}$ and a^μ, so that the difference between $\delta x^\nu \partial_\nu \varphi'_r$ and $\delta x^\nu \partial_\nu \varphi_r$ is of second order in these parameters and drops out.

3.3.2 Noether's Theorem for the Poincaré Group

In field theory Noether's theorem, applied to the Poincaré group, is enounced as follows.

Theorem III (Noether's theorem in field theory). Consider a field theory whose dynamics follows from the action $I = \int \mathcal{L}\, d^4x$, relative to a certain Lagrangian $\mathcal{L}$. Then, if $\mathcal{L}$ is invariant under translations, the four-momentum is locally conserved, the *canonical* energy-momentum tensor being given by (3.71), while, if $\mathcal{L}$ is invariant under Lorentz transformations, the four-dimensional angular momentum is locally conserved, the *canonical* angular momentum density tensor being given by (3.72). These conservation laws hold, provided that the fields satisfy the Euler-Lagrange equations (3.8).

Translation invariance. Before proving the theorem, to gain a better understanding of the meaning of translation invariance of $\mathcal{L}$ we consider a class of Lagrangians which are slightly more general than the ones considered so far, i.e. Lagrangians of the form

$$\mathcal{L}(\varphi(x), \partial\varphi(x), x). \tag{3.63}$$

We thus allow $\mathcal{L}$ to exhibit also a generic *explicit* dependence on the space-time coordinate x. We recall that under a generic translation $x' = x + a$ the fields are invariant, namely $\varphi'_r(x') = \varphi_r(x)$. For the translated Lagrangian we obtain therefore

$$\mathcal{L}(\varphi'(x'), \partial'\varphi'(x'), x') = \mathcal{L}(\varphi(x), \partial\varphi(x), x + a).$$

Consequently, the translated Lagrangian equals the original one $\mathcal{L}(\varphi(x), \partial\varphi(x), x)$, only if $\mathcal{L}$ has no explicit dependence on x. We conclude, therefore, that a Lagrangian is translation invariant, if and only if it does not depend explicitly on x. Given the simplicity of this condition, this means that primarily we have to cope with the *Lorentz invariance* of $\mathcal{L}$. We proceed now to the proof of Theorem III.

Proof The first step in the proof of Noether's theorem consists in the evaluation of the variation of the Lagrangian under an arbitrary *finite* Poincaré transformation (see (3.46) and (3.48))

$$\Delta\mathcal{L} = \mathcal{L}(\varphi'(x'), \partial'\varphi'(x'), x') - \mathcal{L}(\varphi(x), \partial\varphi(x), x). \qquad (3.64)$$

For every fixed x this expression is a function of the parameters $\omega^{\mu\nu}$ and a^μ, and as such it can be expanded in a Taylor series around the values $\omega^{\mu\nu} = a^\mu = 0$. Since for vanishing values of these parameters $\Delta\mathcal{L}$ vanishes trivially, we obtain the expansion

$$\Delta\mathcal{L} = \delta\mathcal{L} + O(\omega^{\mu\nu}, a^\mu)^2, \qquad (3.65)$$

in which $\delta\mathcal{L}$ – the *infinitesimal* variation of the Lagrangian – denotes the terms of $\Delta\mathcal{L}$ which are linear in $\omega^{\mu\nu}$ and a^μ, while the last term denotes the rest of the Taylor series. If $\mathcal{L}$ is invariant under Poincaré transformations, the equation $\Delta\mathcal{L} = 0$ holds as an *identity*. In this case the expansion (3.65) – together with the *identity theorem of power series* – implies that

$$\delta\mathcal{L} = 0, \quad \forall\, \omega^{\mu\nu}, \ \forall\, a^\mu. \qquad (3.66)$$

Relying on (3.66), and assuming the validity of the Euler-Lagrange equations, below we will prove that certain tensors satisfy a continuity equation, meaning that the corresponding *charges* are locally conserved, as stated by Theorem III.

According to this strategy, we must first evaluate the infinitesimal variation $\delta\mathcal{L}$. For this purpose it is convenient to add and subtract from $\Delta\mathcal{L}$ in (3.64) the same term, and to evaluate then the expression so obtained retaining only the terms linear in the parameters:

$$\begin{aligned}
\delta\mathcal{L} &= [\mathcal{L}(\varphi'(x'), \partial'\varphi'(x'), x') - \mathcal{L}(\varphi'(x), \partial\varphi'(x), x)]_{\text{lin}} \\
&\quad + [\mathcal{L}(\varphi'(x), \partial\varphi'(x), x) - \mathcal{L}(\varphi(x), \partial\varphi(x), x)]_{\text{lin}}\,.
\end{aligned} \qquad (3.67)$$

The two term in the first line of (3.67) differ by the replacement $x \rightarrow x' = x + \delta x$, where δx is given in (3.53), while the two terms in the second line differ by the replacement $\varphi_r \rightarrow \varphi_r' = \varphi_r + \delta\varphi_r$, where $\delta\varphi_r$ is the *form variation* (3.51). Defining the *conjugate momenta*

$$\Pi^{\mu r}(x) = \frac{\partial\mathcal{L}}{\partial(\partial_\mu\varphi_r(x))} \qquad (3.68)$$

associated which each field φ_r, and noting that (3.53) implies

$$\partial_\mu \delta x^\mu = \eta_{\mu\nu}\omega^{\mu\nu} = 0,$$

we can rewrite the variation (3.67) as

$$
\begin{aligned}
\delta\mathcal{L} &= \delta x^\mu \partial_\mu \mathcal{L} + \delta\varphi_r \frac{\partial\mathcal{L}}{\partial\varphi_r} + \partial_\mu \delta\varphi_r \Pi^{\mu r} \\
&= \partial_\mu\left(\delta x^\mu \mathcal{L}\right) + \delta\varphi_r\left(\frac{\partial\mathcal{L}}{\partial\varphi_r} - \partial_\mu\Pi^{\mu r}\right) + \partial_\mu\left(\delta\varphi_r \Pi^{\mu r}\right) \qquad (3.69) \\
&= \partial_\mu\left(\delta x^\mu \mathcal{L} + \delta\varphi_r \Pi^{\mu r}\right) + \delta\varphi_r\left(\frac{\partial\mathcal{L}}{\partial\varphi_r} - \partial_\mu\Pi^{\mu r}\right).
\end{aligned}
$$

In this derivation we understand the identification $\mathcal{L} = \mathcal{L}(\varphi(x), \partial\varphi(x), x)$, and we consider all fields evaluated at x. We can rewrite the first two terms of the last line of (3.69) in a different form, using for the form variation of the fields (3.62)

$$\delta x^\mu \mathcal{L} + \delta\varphi_r \Pi^{\mu r} = \delta x_\nu\left(\eta^{\mu\nu}\mathcal{L} - \Pi^{\mu r}\partial^\nu\varphi_r\right) + \frac{1}{2}\,\Pi^{\mu r}\omega_{\alpha\beta}\Sigma^{\alpha\beta}{}_r{}^s\,\varphi_s. \qquad (3.70)$$

We define now the *canonical energy-momentum tensor* of the theory associated with the Lagrangian $\mathcal{L}$ as

$$\widetilde{T}^{\mu\nu} = \Pi^{\mu r}\partial^\nu\varphi_r - \eta^{\mu\nu}\mathcal{L}, \qquad (3.71)$$

and its *canonical angular momentum density tensor* as

$$\widetilde{M}^{\mu\alpha\beta} = x^\alpha \widetilde{T}^{\mu\beta} - x^\beta \widetilde{T}^{\mu\alpha} + \Pi^{\mu r}\Sigma^{\alpha\beta}{}_r{}^s\,\varphi_s, \qquad \widetilde{M}^{\mu\alpha\beta} = -\widetilde{M}^{\mu\beta\alpha}. \qquad (3.72)$$

Using these definitions, and (3.53), we can recast the expression (3.70) in the form

$$
\begin{aligned}
\delta x^\mu \mathcal{L} + \delta\varphi_r \Pi^{\mu r} &= -(a_\nu + \omega_{\nu\rho}\,x^\rho)\widetilde{T}^{\mu\nu} + \frac{1}{2}\,\Pi^{\mu r}\omega_{\alpha\beta}\Sigma^{\alpha\beta}{}_r{}^s\,\varphi_s \\
&= -a_\nu \widetilde{T}^{\mu\nu} + \frac{1}{2}\,\omega_{\alpha\beta}\widetilde{M}^{\mu\alpha\beta}.
\end{aligned}
$$

The infinitesimal variation (3.69) of the Lagrangian under a generic Poincaré transformation can finally be rewritten in terms of the above fundamental tensors as

$$\delta\mathcal{L} = -a_\nu \partial_\mu \widetilde{T}^{\mu\nu} + \frac{1}{2}\,\omega_{\alpha\beta}\partial_\mu\widetilde{M}^{\mu\alpha\beta} + \delta\varphi_r\left(\frac{\partial\mathcal{L}}{\partial\varphi_r} - \partial_\mu\Pi^{\mu r}\right). \qquad (3.73)$$

Before drawing the conclusions we are interested in, i.e. the proof of Noether's theorem, we present some significant applications of the general identity (3.73). Suppose, for example, that the Lagrangian is invariant under the one-parameter subgroup of time translations of the Poincaré group, namely $\Delta\mathcal{L} = 0$, see (3.64), for

$$t' = t + a^0, \qquad \mathbf{x}' = \mathbf{x}.$$

As seen above, this is equivalent to the assumption that $\mathcal{L}$ does not depend explicitly on t. In this case from the general expansion (3.65), in place of (3.66), we deduce that

$$\delta\mathcal{L} = 0, \quad \forall a^0 \in \mathbb{R}, \quad a^i = 0, \quad \omega^{\mu\nu} = 0.$$

If, moreover, we impose on the fields the Euler-Lagrange equations (3.8), which in the notation at hand read

$$\partial_\mu \Pi^{\mu r} - \frac{\partial\mathcal{L}}{\partial\varphi_r} = 0, \tag{3.74}$$

from (3.73) we deduce that $a_0 \partial_\mu \widetilde{T}^{\mu 0} = 0, \forall a_0 \in \mathbb{R}$, that is to say

$$\partial_\mu \widetilde{T}^{\mu 0} = 0.$$

We have, hence, retrieved the continuity equation for the *energy*. In particular, from (3.71) and (3.68) we obtain for the total energy the general expression

$$\varepsilon = \int \widetilde{T}^{00} d^3x = \int \left(\frac{\partial\mathcal{L}}{\partial\dot{\varphi}_r} \dot{\varphi}_r - \mathcal{L} \right) d^3x,$$

which parallels closely the analogous expression of the total energy in Newtonian mechanics

$$E = \sum_n \frac{\partial L}{\partial\dot{q}_n} \dot{q}_n - L.$$

In the same way from (3.73) we deduce that each of the ten parameters $\{a^\mu, \omega^{\alpha\beta}\}$ is associated with a current of vanishing four-divergence, and hence with a locally conserved quantity, if the Lagrangian is invariant under the corresponding one-parameter group: a^0 (time translations) corresponds to the conservation of energy, a^1 (translations along the x axis) corresponds to the conservation of the x component of momentum, ω^{12} (rotations about the z axis) corresponds to the conservation of the z component of the angular momentum, ω^{01} (special Lorentz transformations along the x axis) corresponds to the conservation of the x component of the boost generator, see Sect. 2.4.4, and so on. In particular, if the Lagrangian is invariant under the complete translation group, then the four-momentum is locally conserved, while if it is invariant under the complete Lorentz group, it is the four-dimensional angular momentum to be locally conserved.

Finally, if $\mathcal{L}$ is invariant under the complete Poincaré group, and the fields satisfy the Euler-Lagrange equations (3.74), the identity (3.73) reduces to

$$-a_\nu \partial_\mu \widetilde{T}^{\mu\nu} + \frac{1}{2} \omega_{\alpha\beta} \partial_\mu \widetilde{M}^{\mu\alpha\beta} = 0, \quad \forall a_\nu, \quad \forall \omega_{\alpha\beta}.$$

We thus obtain the ten continuity equations

$$\partial_\mu \widetilde{T}^{\mu\nu} = 0, \qquad \partial_\mu \widetilde{M}^{\mu\alpha\beta} = 0,$$

implying the ten constants of motion

$$\widetilde{P}^\mu = \int \widetilde{T}^{0\mu} d^3x, \qquad \widetilde{L}^{\alpha\beta} = \int \widetilde{M}^{0\alpha\beta} d^3x, \tag{3.75}$$

which represent the total four-momentum and the total four-dimensional angular momentum, respectively. This concludes the proof of Theorem III.   □

To stress, once more, the importance of this theorem, we recall that the field theories describing the four fundamental interactions satisfy Einstein's *relativity principle* and, moreover, descend from an *action principle*. Therefore, for these theories, Noether's theorem *automatically* ensures the conservation of the four-momentum and of the four-dimensional angular momentum.

On the canonical current densities. We end this section with some comments on the structure of the canonical currents (3.71) and (3.72). First of all. we notice that in general the canonical energy-momentum tensor (3.71) is not symmetric: $\widetilde{T}^{\mu\nu} \neq \widetilde{T}^{\nu\mu}$. In *special relativity* this circumstance, by itself, does not represent any problem. Vice versa, it can be seen that, according to the postulates of General Relativity, the existence of a *symmetric* energy-momentum tensor is an indispensable *prerequisite* for the coupling of whatever physical system to gravity.[5] Secondly, the expression (3.72) of the canonical angular momentum density is not *standard*, in the sense that it is not of the simple form

$$\mathcal{M}^{\mu\alpha\beta} = x^\alpha \widetilde{T}^{\mu\beta} - x^\beta \widetilde{T}^{\mu\alpha}, \tag{3.76}$$

as one might have expected at the basis of the general construction (2.156) of Chap. 2. On the other hand the tensor (3.76) could *anyhow* not be identified with the angular momentum density of the system, simply because in general it is not conserved. Actually, the two just mentioned *anomalies* affecting $\widetilde{T}^{\mu\nu}$ and $\widetilde{M}^{\mu\alpha\beta}$ are linked to each other, in that the four-divergence of $\mathcal{M}^{\mu\alpha\beta}$ equals precisely the antisymmetric part of the canonical energy-momentum tensor:

$$\partial_\mu \mathcal{M}^{\mu\alpha\beta} = \partial_\mu \big(x^\alpha \widetilde{T}^{\mu\beta} - x^\beta \widetilde{T}^{\mu\alpha}\big) = \partial_\mu x^\alpha \widetilde{T}^{\mu\beta} - \partial_\mu x^\beta \widetilde{T}^{\mu\alpha} = \widetilde{T}^{\alpha\beta} - \widetilde{T}^{\beta\alpha}.$$

Nonetheless $\widetilde{M}^{\mu\alpha\beta}$ reduces to $\mathcal{M}^{\mu\alpha\beta}$, if the quantities $\Sigma^{\alpha\beta}{}_r{}^s$ appearing in (3.72) are all vanishing. On the other hand, as we saw in Sect. 3.3.1, this happens only if all fields of the theory are *scalar* fields. In this last case, in turn, it is not difficult

---

[5]In General Relativity, Einstein's equations equal a certain two-index *symmetric* tensor, built with the *metric* $g_{\mu\nu}(x)$ and its derivatives, to the energy-momentum tensor. These equations would thus be inconsistent, would the latter not be a symmetric tensor, see Sect. 9.2.

to show that $\widetilde{T}^{\mu\nu}$ is always symmetric, see Problem 3.6, so that both anomalies disappear. In conclusion, the *anomaly* identified above concerning the canonical angular momentum density does not represent a conceptual *problem*, but rather a *naturalness* issue. Vice versa, would it not be possible to find a *symmetric* energy-momentum tensor, the field theory at hand would be incompatible with gravity, i.e. with the principles of General Relativity. The general solution to this basic problem will be provided in Sect. 3.4.

3.3.3 Canonical Energy-Momentum Tensor of the Electromagnetic Field

We exemplify the general expression of the canonical energy-momentum tensor (3.71) in the simple case of the *free* Maxwell field. The dynamics of this field is governed by the Lagrangian (3.38)

$$\mathcal{L}_1 = -\frac{1}{4}\, F^{\mu\nu} F_{\mu\nu}, \tag{3.77}$$

modulo the identification $\varphi_r = A_\alpha$. The conjugate momenta were already determined in (3.39)

$$\Pi^{\mu\alpha} = \frac{\partial \mathcal{L}_1}{\partial(\partial_\mu A_\alpha)} = -F^{\mu\alpha}. \tag{3.78}$$

Equation (3.71) then yields the expression

$$\widetilde{T}^{\mu\nu}_{\text{em}} = \Pi^{\mu\alpha}\partial^\nu A_\alpha - \eta^{\mu\nu}\mathcal{L}_1 = -F^{\mu\alpha}\partial^\nu A_\alpha + \frac{1}{4}\,\eta^{\mu\nu}F^{\alpha\beta}F_{\alpha\beta}. \tag{3.79}$$

The obtained tensor is affected by two *pathologies*: it is neither symmetric in μ and ν, nor is it gauge invariant. Even worse, the tensor (3.79) is in net disagreement with the tensor $T^{\mu\nu}_{\text{em}}$ in (2.137), recovered relying on heuristic arguments. We will address these problems in Sect. 3.4.1.

3.4 Symmetric Energy-Momentum Tensor

In this section we show that in a generic field theory, invariant under the *complete* Poincaré group and descending from an action principle, it is always possible to build a *symmetric* energy-momentum tensor, the construction being "canonical". This remarkable feature relies on the fact that, in reality, the energy-momentum tensor of a field theory is *not* uniquely determined. Suppose that the equations of motion of a field theory allow for the existence of a generic energy-momentum tensor $\widetilde{T}^{\mu\nu}$, satisfying the conservation equation

$$\partial_\mu \widetilde{T}^{\mu\nu} = 0. \tag{3.80}$$

Consider then a generic rank-three tensor $\phi^{\rho\mu\nu}$, which is antisymmetric in its first two indices

$$\phi^{\rho\mu\nu} = -\phi^{\mu\rho\nu}. \tag{3.81}$$

We can then introduce a new energy-momentum tensor $T^{\mu\nu}$ by setting

$$T^{\mu\nu} = \widetilde{T}^{\mu\nu} + \partial_\rho \phi^{\rho\mu\nu}. \tag{3.82}$$

Under mild fall-off assumptions at spatial infinity of the field $\phi^{\rho\mu\nu}$, the tensor $T^{\mu\nu}$ entails, in fact, the properties:

(1) $\partial_\mu T^{\mu\nu} = 0$;

(2) $P^\nu \equiv \int T^{0\nu} d^3x = \int \widetilde{T}^{0\nu} d^3x \equiv \widetilde{P}^\nu$.

Also the tensor $T^{\mu\nu}$ satisfies, hence, the continuity equation and, moreover, it gives rise to the same total four-momentum as $\widetilde{T}^{\mu\nu}$. To prove property (1) we compute the four-divergence of (3.82) as

$$\partial_\mu T^{\mu\nu} = \partial_\mu \widetilde{T}^{\mu\nu} + \partial_\mu \partial_\rho \phi^{\rho\mu\nu} = 0,$$

where the term $\partial_\mu \partial_\rho \phi^{\rho\mu\nu}$ vanishes, because a pair of antisymmetric indices is contracted with a pair of symmetric ones. To prove property (2) we use again (3.82) to compute the difference

$$P^\nu - \widetilde{P}^\nu = \int \left(T^{0\nu} - \widetilde{T}^{0\nu} \right) d^3x = \int \partial_\rho \phi^{\rho 0\nu} d^3x$$
$$= \int \partial_i \phi^{i0\nu} d^3x = \int_{\Gamma_\infty} \phi^{i0\nu} d\Sigma^i = 0.$$

To obtain the first term in the second line we used the identity $\phi^{00\nu} = 0$, implied by the antisymmetry of $\phi^{\rho\mu\nu}$ in its first two indices. In the last step we applied Gauss's theorem, with Γ_∞ a spherical surface at spatial infinity, assuming that as r tends to infinity $\phi^{\rho\mu\nu}$ decays more rapidly than $1/r^2$. Property (2) ensures, in particular, that the *Hamiltonian* of the system, represented by the component P^0, does not depend on the energy-momentum tensor one considers. $T^{\mu\nu}$ can, hence, be taken as energy-momentum of the system – on the same footing of $\widetilde{T}^{\mu\nu}$. Exploiting the freedom related to the choice of the tensor $\phi^{\rho\mu\nu}$, we prove now the following theorem.

Theorem IV. Consider a field theory whose dynamics follows from a Poincaré invariant Lagrangian $\mathcal{L}$. Then the energy-momentum tensor $T^{\mu\nu}$ given in (3.82) is *symmetric*, if $\widetilde{T}^{\mu\nu}$ is identified with the canonical energy-momentum tensor (3.71), and if the tensor $\phi^{\rho\mu\nu}$ is defined through Eqs. (3.87) and (3.83) below.

Proof Since by hypothesis the Lagrangian is invariant under the complete Poincaré group, we can apply Noether's theorem exploiting the existence of the *conserved* canonical tensors $\widetilde{T}^{\mu\nu}$ and $\widetilde{M}^{\mu\alpha\beta}$, given by (3.71) and (3.72), respectively. We recast the expression of the angular momentum tensor in the form

$$\widetilde{M}^{\mu\alpha\beta} = x^\alpha \widetilde{T}^{\mu\beta} - x^\beta \widetilde{T}^{\mu\alpha} + V^{\mu\alpha\beta}, \qquad V^{\mu\alpha\beta} = \Pi^{\mu r} \Sigma^{\alpha\beta}{}_r{}^s \varphi_s, \qquad (3.83)$$

where, thanks to (3.58), the tensor $V^{\mu\alpha\beta}$ is antisymmetric in its last two indices

$$V^{\mu\alpha\beta} = -V^{\mu\beta\alpha}. \qquad (3.84)$$

Using that both $\widetilde{M}^{\mu\alpha\beta}$ and $\widetilde{T}^{\mu\nu}$ satisfy the continuity equation we obtain

$$0 = \partial_\mu \widetilde{M}^{\mu\alpha\beta} = \partial_\mu x^\alpha \widetilde{T}^{\mu\beta} - \partial_\mu x^\beta \widetilde{T}^{\mu\alpha} + \partial_\mu V^{\mu\alpha\beta} = \widetilde{T}^{\alpha\beta} - \widetilde{T}^{\beta\alpha} + \partial_\mu V^{\mu\alpha\beta}, \qquad (3.85)$$

or, renaming the indices,

$$\partial_\rho V^{\rho\mu\nu} = \widetilde{T}^{\nu\mu} - \widetilde{T}^{\mu\nu}. \qquad (3.86)$$

The four-divergence $\partial_\rho V^{\rho\mu\nu}$ equals, hence, just the antisymmetric part of $\widetilde{T}^{\nu\mu}$. Nonetheless, we cannot simply identify $V^{\rho\mu\nu}$ with $\phi^{\rho\mu\nu}$, because the first is not antisymmetric in ρ and μ. Rather, the tensor $\phi^{\rho\mu\nu}$ defined by

$$\phi^{\rho\mu\nu} = \frac{1}{2} \left(V^{\rho\mu\nu} - V^{\mu\rho\nu} - V^{\nu\rho\mu} \right) \qquad (3.87)$$

satisfies the relations

$$\phi^{\rho\mu\nu} = -\phi^{\mu\rho\nu}, \qquad (3.88)$$

$$\partial_\rho \phi^{\rho\mu\nu} = \frac{1}{2} \left(\widetilde{T}^{\nu\mu} - \widetilde{T}^{\mu\nu} \right) - \frac{1}{2} \partial_\rho \left(V^{\mu\rho\nu} + V^{\nu\rho\mu} \right). \qquad (3.89)$$

Relation (3.88) follows from (3.87) and (3.84), and guarantees, we recall, that the new energy-momentum tensor $T^{\mu\nu}$ (3.82) is conserved. Equality (3.89) follows from Eqs. (3.86) and (3.87):

$$\partial_\rho \phi^{\rho\mu\nu} = \frac{1}{2} \left(\partial_\rho V^{\rho\mu\nu} - \partial_\rho V^{\mu\rho\nu} - \partial_\rho V^{\nu\rho\mu} \right)$$

$$= \frac{1}{2} \left(\widetilde{T}^{\nu\mu} - \widetilde{T}^{\mu\nu} \right) - \frac{1}{2} \partial_\rho \left(V^{\mu\rho\nu} + V^{\nu\rho\mu} \right).$$

Finally, with the aid of (3.89) we can recast the tensor (3.82) in the form

$$T^{\mu\nu} = \widetilde{T}^{\mu\nu} + \partial_\rho \phi^{\rho\mu\nu} = \frac{1}{2} \left(\widetilde{T}^{\nu\mu} + \widetilde{T}^{\mu\nu} \right) - \frac{1}{2} \partial_\rho \left(V^{\mu\rho\nu} + V^{\nu\rho\mu} \right), \qquad (3.90)$$

which is symmetric in μ and ν, so that the theorem is proven. $\qquad \square$

Taking advantage of the antisymmetry of $V^{\mu\nu\rho}$ in its last two indices, and resorting to the symmetrization convention (1.35), we can rewrite the new energy-momentum tensor (3.90) in a manifestly symmetric form, summarizing our results in the formulas

$$T^{\mu\nu} = \widetilde{T}^{(\mu\nu)} + \partial_\rho V^{(\mu\nu)\rho}, \qquad \partial_\mu T^{\mu\nu} = 0. \tag{3.91}$$

The tensor $T^{\mu\nu}$ is hence symmetric and conserved, and it gives rise to the same *total* four-momentum as the canonical energy-momentum tensor. Notice, however, that the four-momentum P_V^μ, contained in a finite volume V, *depends* on the chosen energy-momentum tensor. On the other hand, as observed previously, this four-momentum does not have tensorial character, i.e. P_V^μ is *not* a four-vector.

From the proof given above we draw, in particular, the following important conclusion: as seen in Sect. 3.3.2, the existence of a *conserved* energy-momentum tensor requires a field theory only to be translation invariant, while the existence of a *conserved and symmetric* energy-momentum tensor requires the theory, in addition, to be Lorentz invariant and, therefore, Poincaré invariant. In fact, the construction of the tensor $\phi^{\rho\mu\nu}$, with the required properties, relies in a crucial manner on the continuity equation

$$\partial_\mu \widetilde{M}^{\mu\alpha\beta} = 0,$$

see Eqs. (3.85)–(3.87), and the validity of this continuity equation is implied, in turn, by Lorentz invariance.

Poincaré group and General Relativity. We end this section with a comment about the *double* role played by Poincaré invariance, with regard to the gravitational interaction. In the first place, any theory that is invariant under the Poincaré group, within the framework of General Relativity, more specifically, according to the *equivalence principle*, admits a consistent so-called *minimal coupling* with the gravitational field. In the second place, we recall that the self-consistency of Einstein's equations, which govern the dynamics of the gravitational field, requires the existence of a *symmetric* energy-momentum tensor, whose existence is ensured, in turn, by Poincaré invariance. So we see that the internal coherence of the gravitational interaction of a physical system, eventually, is guaranteed by the *Poincaré invariance of the system in the absence of gravity*, in spite of the fact that gravity, actually, is based on a larger symmetry group than the Poincaré group, namely the group of *diffeomorphisms*. In four-dimensional space-time a diffeomorphism – sometimes also called *general coordinate transformation* – is a generic coordinate transformation from $\mathbb{R}^4$ to $\mathbb{R}^4$,

$$x^\mu \to x'^\mu(x), \tag{3.92}$$

invertible and of class C^∞ together with its inverse transformation. In a sense, diffeomorphisms constitute a generalization of the Poincaré transformations $x^\mu \to x'^\mu(x) = \Lambda^\mu{}_\nu x^\nu + a^\mu$. In conclusion, the importance of the Poincaré symmetry group stems not only from its role as *warrantor* of the covariance of the equations of motion, as well as of the main conservation laws of a theory, but also from its role as *guardian* of the internal consistency of General Relativity.

3.4.1 Symmetric Energy-Momentum Tensor of the Electromagnetic Field

We exemplify Theorem IV in the case of the free Maxwell field, whose Lagrangian is given in (3.77). The associated Euler-Lagrange equations are the Maxwell equations (3.40) in empty space

$$\partial_\mu F^{\mu\nu} = 0, \tag{3.93}$$

and the canonical energy-momentum tensor of this system has been derived in (3.79)

$$\widetilde{T}^{\mu\nu}_{\text{em}} = -F^{\mu\alpha} \partial^\nu A_\alpha + \frac{1}{4} \eta^{\mu\nu} F^{\alpha\beta} F_{\alpha\beta}.$$

Denoting the gauge fields indistinctly by A_r or A_α, we also recall the form of the conjugate momenta (3.78), and of the matrices $\Sigma^{\alpha\beta}{}_r{}^s$ (3.59) relative to a generic vector field:

$$\Pi^{\mu r} = -F^{\mu r}, \qquad \Sigma^{\alpha\beta}{}_r{}^s = \delta^\alpha_r \eta^{\beta s} - \delta^\beta_r \eta^{\alpha s}.$$

First we must determine the tensor $V^{\mu\alpha\beta}$ (3.83), antisymmetric in α and β,

$$V^{\mu\alpha\beta} = \Pi^{\mu r} \Sigma^{\alpha\beta}{}_r{}^s A_s = -F^{\mu\alpha} A^\beta + F^{\mu\beta} A^\alpha. \tag{3.94}$$

Next, the relation (3.87) yields for the tensor $\phi^{\rho\mu\nu}$, antisymmetric in ρ and μ, the expression (see Problem 3.7)

$$\phi^{\rho\mu\nu} = -F^{\rho\mu} A^\nu. \tag{3.95}$$

Computing its divergence we obtain

$$\partial_\rho \phi^{\rho\mu\nu} = -\partial_\rho F^{\rho\mu} A^\nu - F^{\rho\mu} \partial_\rho A^\nu = F^{\mu\alpha} \partial_\alpha A^\nu,$$

where we made use of the Euler-Lagrange equations (3.93). Finally, for the new energy-momentum tensor we retrieve the expression

$$\begin{aligned} T^{\mu\nu}_{\text{em}} &= \widetilde{T}^{\mu\nu}_{\text{em}} + \partial_\rho \phi^{\rho\mu\nu} = F^{\mu\alpha}(\partial_\alpha A^\nu - \partial^\nu A_\alpha) + \frac{1}{4} \eta^{\mu\nu} F^{\alpha\beta} F_{\alpha\beta} \\ &= F^{\mu\alpha} F_\alpha{}^\nu + \frac{1}{4} \eta^{\mu\nu} F^{\alpha\beta} F_{\alpha\beta}, \end{aligned} \tag{3.96}$$

which is *symmetric* in μ and ν as foreseen by Theorem IV, as well as gauge invariant, and moreover in perfect agreement with (2.137).

3.5 Standard Angular Momentum Density

We end this chapter proving the following theorem.

Theorem V. Consider a field theory whose dynamics follows from a Poincaré invariant Lagrangian $\mathcal{L}$. Then the *standard* angular momentum density

$$M^{\mu\alpha\beta} = x^\alpha T^{\mu\beta} - x^\beta T^{\mu\alpha},$$ (3.97)

in which $T^{\mu\nu}$ is the *symmetric* energy-momentum tensor (3.91), gives rise to the same *total* angular momentum as the *canonical* angular momentum density $\widetilde{M}^{\mu\alpha\beta}$ of equation (3.72).

Before proving the theorem we make some preliminary observations. First of all, thanks to the symmetry of $T^{\mu\nu}$, the tensor $M^{\mu\alpha\beta}$ (3.97) satisfies automatically the continuity equation $\partial_\mu M^{\mu\alpha\beta} = 0$, see the derivation (2.157). The proof of Theorem V closely parallels the one of Theorem IV of Sect. 3.4. In fact, also the angular momentum density is determined modulo the divergence of a four-tensor $\Lambda^{\mu\nu\alpha\beta}$, with certain symmetry properties. More precisely, this tensor must be antisymmetric in the first pair of indices, as well as in the second pair,

$$\Lambda^{\mu\nu\alpha\beta} = -\Lambda^{\nu\mu\alpha\beta} = -\Lambda^{\mu\nu\beta\alpha}.$$ (3.98)

Since by hypothesis $\mathcal{L}$ is Lorentz invariant, Noether's theorem, i.e. Theorem III of Sect. 3.3.2, ensures the existence of the conserved angular momentum density $\widetilde{M}^{\mu\alpha\beta}$ (3.72), and so we can define a new angular momentum density by setting

$$M^{\mu\alpha\beta} = \widetilde{M}^{\mu\alpha\beta} + \partial_\rho \Lambda^{\rho\mu\alpha\beta}.$$ (3.99)

This tensor entails, in fact, the properties:

(1) $M^{\mu\alpha\beta} = -M^{\mu\beta\alpha}$;

(2) $\partial_\mu M^{\mu\alpha\beta} = 0$;

(3) $L^{\alpha\beta} \equiv \int M^{0\alpha\beta} d^3x = \int \widetilde{M}^{0\alpha\beta} d^3x \equiv \widetilde{L}^{\alpha\beta}$.

Property (1) follows from the relations (3.99) and (3.98). Property (2) can be proven in the usual way, by evaluating the four-divergence

$$\partial_\mu M^{\mu\alpha\beta} = \partial_\mu \widetilde{M}^{\mu\alpha\beta} + \partial_\mu \partial_\rho \Lambda^{\rho\mu\alpha\beta} = 0,$$

where the first term vanishes thanks to Noether's theorem, and the term $\partial_\mu \partial_\rho \Lambda^{\rho\mu\alpha\beta}$ vanishes since a symmetric pair of indices contracts an antisymmetric one. To prove property (3) we compute the difference

$$L^{\alpha\beta} - \widetilde{L}^{\alpha\beta} = \int \left(M^{0\alpha\beta} - \widetilde{M}^{0\alpha\beta} \right) d^3x = \int \partial_\rho \Lambda^{\rho 0\alpha\beta} d^3x$$

$$= \int \partial_i \Lambda^{i0\alpha\beta} d^3x = \int_{\Gamma_\infty} \Lambda^{i0\alpha\beta} d\Sigma^i = 0.$$

We used that, thanks to the antisymmetry of $\Lambda^{\rho\mu\alpha\beta}$ in its first pair of indices, $\Lambda^{00\alpha\beta}$ is zero, and we applied Gauss's theorem assuming that $\Lambda^{\rho\mu\alpha\beta}$ decays sufficiently fast at spatial infinity. In conclusion, whenever $\Lambda^{\rho\mu\alpha\beta}$ satisfies the symmetry properties (3.98), the tensor (3.99) is conserved and gives rise to the same total angular momentum as the canonical tensor $\widetilde{M}^{\mu\alpha\beta}$.

Proof. To prove Theorem V it is now sufficient to find a tensor $\Lambda^{\rho\mu\alpha\beta}$, obeying the conditions (3.98), such that the second member of equation (3.99) equals the second member of equation (3.97). We begin by recalling the definition of the canonical angular momentum density (3.72)

$$\widetilde{M}^{\mu\alpha\beta} = x^\alpha \widetilde{T}^{\mu\beta} - x^\beta \widetilde{T}^{\mu\alpha} + V^{\mu\alpha\beta}, \qquad V^{\mu\alpha\beta} = \Pi^{\mu r} \Sigma^{\alpha\beta}{}_r{}^s \varphi_s, \qquad (3.100)$$

together with the relations (3.82) and (3.87) between the symmetric and canonical energy-momentum tensors

$$T^{\mu\nu} = \widetilde{T}^{\mu\nu} + \partial_\rho \phi^{\rho\mu\nu}, \qquad \phi^{\rho\mu\nu} = \frac{1}{2} \left(V^{\rho\mu\nu} - V^{\mu\rho\nu} - V^{\nu\rho\mu} \right). \qquad (3.101)$$

Replacing in the first equality of (3.100) $\widetilde{T}^{\mu\nu}$ with $T^{\mu\nu} - \partial_\rho \phi^{\rho\mu\nu}$, we obtain

$$\begin{aligned}
\widetilde{M}^{\mu\alpha\beta} &= x^\alpha T^{\mu\beta} - x^\beta T^{\mu\alpha} - x^\alpha \partial_\rho \phi^{\rho\mu\beta} + x^\beta \partial_\rho \phi^{\rho\mu\alpha} + V^{\mu\alpha\beta} \\
&= x^\alpha T^{\mu\beta} - x^\beta T^{\mu\alpha} - \partial_\rho \left(x^\alpha \phi^{\rho\mu\beta} - x^\beta \phi^{\rho\mu\alpha} \right) + \phi^{\alpha\mu\beta} - \phi^{\beta\mu\alpha} + V^{\mu\alpha\beta} \\
&= x^\alpha T^{\mu\beta} - x^\beta T^{\mu\alpha} - \partial_\rho \left(x^\alpha \phi^{\rho\mu\beta} - x^\beta \phi^{\rho\mu\alpha} \right).
\end{aligned}$$

$$(3.102)$$

To obtain the last line we used the definition of $\phi^{\rho\mu\nu}$ (3.101), which implies the identity

$$\phi^{\alpha\mu\beta} - \phi^{\beta\mu\alpha} = -V^{\mu\alpha\beta}.$$

Finally, choosing for the tensor $\Lambda^{\rho\mu\alpha\beta}$ the expression

$$\Lambda^{\rho\mu\alpha\beta} = x^\alpha \phi^{\rho\mu\beta} - x^\beta \phi^{\rho\mu\alpha}, \qquad (3.103)$$

satisfying the antisymmetry conditions (3.98), the equality (3.102) can be recast in the form

$$\widetilde{M}^{\mu\alpha\beta} = x^\alpha T^{\mu\beta} - x^\beta T^{\mu\alpha} - \partial_\rho \Lambda^{\rho\mu\alpha\beta}. \qquad (3.104)$$

Recalling the definition (3.97), we so retrieve equation (3.99). Property (3) above then implies the assertion of Theorem V.   □

3.6 Problems

3.1 Consider a real scalar field φ, describing a neutral spin-zero particle of mass m, with Lagrangian

$$\mathcal{L} = \frac{1}{2} \left(\partial_\mu \varphi \partial^\mu \varphi - m^2 \varphi^2 \right) - \frac{\lambda}{4!} \varphi^4,$$

where m and λ are real constants.

(a) Write the Euler-Lagrange equation relative to $\mathcal{L}$.
(b) Verify explicitly that the obtained equation is equivalent to the requirement that the action $I = \int_{t_1}^{t_2} \mathcal{L} \, d^4 x$ is stationary with respect to arbitrary variations of φ, vanishing at $t = t_1$ and $t = t_2$, see Sect. 3.1.

3.2 Consider a complex scalar field $\Phi = \varphi_1 + i\varphi_2$, describing a charged spin-zero particle of mass m, with Lagrangian

$$\mathcal{L} = \partial_\mu \Phi^* \partial^\mu \Phi - m^2 \Phi^* \Phi - \frac{\lambda}{4} \left(\Phi^* \Phi \right)^2,$$

where m and λ are real constants.

(a) Write the Euler-Lagrange equations relative to $\mathcal{L}$.
    Hint: Consider Φ and Φ^* as independent fields.
(b) For which values of λ and m the equations of motion of φ_1 and φ_2 are decoupled from each other?

3.3 Consider the Lagrangian $\mathcal{L}$ specified in (3.38), and the associated action

$$I = \int_{t_1}^{t_2} \mathcal{L} \, d^4 x.$$

(a) Determine the variation of I under arbitrary variations of the fields A_μ.
(b) Verify that the variation determined in (a) vanishes for arbitrary variations of A_μ, vanishing in $t = t_1$ and $t = t_2$, if and only if A_μ satisfies the Maxwell equation (3.35).

3.4 Consider the Lagrangian $\mathcal{L}$ of the real scalar field of Problem 3.1.

(a) Derive the explicit form of the associated canonical energy-momentum tensor (3.71), and analyze its symmetry properties.

(b) Write out the explicit expressions of the energy density and of the total energy. For which values of λ is the energy always positive definite?

3.5 Verify explicitly that the canonical energy-momentum tensor (3.79) of the free Maxwell field is conserved.

3.6 Show that in a field theory with only *scalar* fields the canonical energy-momentum tensor (3.71) is always symmetric in μ and ν.

Hint: Thanks to Lorentz invariance the Lagrangian can depend on $\partial_\mu \varphi_r$ only through the matrix $M_{rs} = \partial_\mu \varphi_r \partial^\mu \varphi_s$, which is symmetric in r and s.

3.7 Verify that for a free Maxwell field the tensor $\phi^{\rho\mu\nu}$ has the form (3.95).

Hint: Insert (3.94) in the general expression (3.87).

3.8 Determine the canonical angular momentum density $\widetilde{M}^{\mu\alpha\beta}$ for a free Maxwell field, using (3.83), (3.79) and (3.94). Verify that the tensor $\widetilde{M}^{\mu\alpha\beta} + \partial_\rho \Lambda^{\rho\mu\alpha\beta}$, with $\Lambda^{\rho\mu\alpha\beta}$ specified by (3.103) and (3.95), equals the *standard* angular momentum $M^{\mu\alpha\beta} = x^\alpha T_{\text{em}}^{\mu\beta} - x^\beta T_{\text{em}}^{\mu\alpha}$, see (3.96), as predicted by Theorem V and (3.104).

3.9 Consider a field theory described by the six Lagrangian fields $\varphi = \{\mathbf{E}, \mathbf{B}\}$, with Lagrangian

$$\mathcal{L} = \mathbf{E} \cdot \frac{\partial \mathbf{B}}{\partial t} + \frac{1}{2}\left(\mathbf{E} \cdot \nabla \times \mathbf{E} + \mathbf{B} \cdot \nabla \times \mathbf{B}\right) - \mathbf{j} \cdot \mathbf{B},$$

where $\mathbf{j}$ is treated as an external field, independent of $\mathbf{E}$ and $\mathbf{B}$. Compare the Euler-Lagrange equations following from $\mathcal{L}$ with Maxwell's equations (2.52)–(2.55).

3.10 *Theorem VI (Noether's theorem for internal symmetries)*. Consider the Lagrangian $\mathcal{L}$ for a complex scalar field of Problem 3.2.

(a) Verify that $\mathcal{L}$ is invariant under the continuous one-parameter group of *global* gauge transformations

$$\Phi'(x) = e^{i\Lambda}\Phi(x), \qquad \Phi^{*\prime}(x) = e^{-i\Lambda}\Phi^*(x),$$

where Λ is an x-independent real parameter. The set of phases $\{e^{i\Lambda}, \Lambda \in [0, 2\pi]\}$ forms a *unitary* group under multiplication, commonly denoted by $U(1)$.

(b) Verify that under a generic infinitesimal variation $\Phi \to \Phi + \delta\Phi$ one has

$$\delta\mathcal{L} = \left(\frac{\partial \mathcal{L}}{\partial \Phi} - \partial_\mu \frac{\partial \mathcal{L}}{\partial(\partial_\mu \Phi)}\right)\delta\Phi + \partial_\mu\left(\frac{\partial \mathcal{L}}{\partial(\partial_\mu \Phi)}\delta\Phi\right) + c.c.$$

(c) Prove Noether's theorem relative to the gauge group $U(1)$, showing in particular that the associated current is given by

$$J^{\mu} = i \left(\partial_{\mu} \Phi^* \Phi - \Phi^* \partial_{\mu} \Phi \right).$$

Hint: The infinitesimal gauge transformation of the field is $\delta \Phi = \Phi' - \Phi = i \Lambda \Phi$.

(d) Verify explicitly that J^{μ} is conserved, if Φ satisfies the Euler-Lagrange equation determined in Problem 3.2.

References

1. T. Ortín, *Gravity and Strings* (Cambridge University Press, Cambridge, 2004)
2. K. Becker, M. Becker, J.H. Schwarz, *String Theory and M-theory* (Cambridge University Press, Cambridge, 2007)
3. C. Rovelli, F. Vidotto, *Covariant Loop Quantum Gravity* (Cambridge University Press, Cambridge, 2015)
4. M.E. Peskin, D.V. Schroeder, *An Introduction to Quantum Field Theory* (Perseus Books Publishing, Reading, 1995)
5. T.-P. Cheng, L.-F. Li, *Gauge Theory of Elementary Particle Physics* (Clarendon Press, Oxford, 1984)

Chapter 4
Variational Methods in Electrodynamics

In this chapter we apply the variational principle to rederive the dynamics of *classical electrodynamics*, together with its conservation laws. This physical system represents a *field theory*, described by the fields $A_\mu(x)$, interacting with a system of charged particles, described by the world lines $y_r^\mu(\lambda_r)$. Before considering the coupled system, we establish the form of the action of a free relativistic particle.

4.1 Action for a Free Relativistic Particle

By definition, the action of a free relativistic particle must give rise to the equation of motion

$$\frac{dp^\mu}{ds} = 0, \tag{4.1}$$

where $p^\mu = mu^\mu$ is the four-momentum. Basically we are, hence, searching for the relativistic generalization of the Newtonian action for a free particle

$$I_0[\mathbf{y}] = \int_{t_a}^{t_b} \left(\frac{1}{2} m v^2 \right) dt, \qquad \mathbf{v} = \frac{d\mathbf{y}}{dt}, \tag{4.2}$$

where the Lagrangian coordinates are the three functions $\mathbf{y}(t)$. The first step in the formulation of a variational principle consists in the identification of the correct Lagrangian coordinates. Since we are searching for a relativistic action, the appropriate coordinates are not the $\mathbf{y}(t)$, but rather the four functions $y^\mu(\lambda)$ parameterizing the world line of the particle. Moreover, since the Eq. (4.1) that we want to derive is invariant under Poincaré transformations as well as under the world line reparameterizations $y^\mu(\lambda) \to y^\mu(\lambda(\lambda'))$, so must be the action $I[y]$ we are searching for. As a first step of the covariantization of the action I_0 (4.2) we replace the measure dt

© Springer International Publishing AG, part of Springer Nature 2018
K. Lechner, *Classical Electrodynamics*, UNITEXT for Physics,
https://doi.org/10.1007/978-3-319-91809-9_4

with the proper-time measure, which is invariant under Poincaré transformations as
well as under reparameterizations,

$$ds = \sqrt{\frac{dy^\mu}{d\lambda}\frac{dy_\mu}{d\lambda}}\, d\lambda. \tag{4.3}$$

In the non-relativistic limit the line element ds reduces in fact to cdt, see (2.7),
where we restored the speed of light. The action must hence have the form

$$I[y] = \int_a^b l(y,\dot{y})\, ds \tag{4.4}$$

for a certain Lagrangian l. In the integral (4.4) a and b are the delimiters of the
considered section of the world line, and we have introduced the shorthand notation

$$\dot{y}^\mu = \frac{dy^\mu}{d\lambda},$$

which will be used only in this chapter. Contrary to the "velocities" $\dot{y}^\mu$, the coordinates y^μ are *not* invariant under space-time translations – a subgroup of the Poincaré
group – and consequently the Lagrangian l can depend only on the formers. On the
other hand, the unique independent scalar that we can form with $\dot{y}^\mu$ is its square
$\dot{y}^\mu\dot{y}_\mu$, which, however, violates reparameterization invariance. Accordingly, l cannot
even depend on $\dot{y}^\mu$, and so it must be a *constant*. The action (4.4) reduces therefore
to the simple expression

$$I[y] = l\int_a^b ds, \tag{4.5}$$

which, from a geometric point of view, corresponds to the *length* of the arc of the
world line between a and b. Finally, to determine the constant l we impose that for
$v \ll c$, i.e. in the non-relativistic limit, the expression (4.5) reduces to the Newtonian
action (4.2). For this purpose we perform the non-relativistic expansion of the line
element (4.3), using (2.7),

$$ds = \sqrt{c^2 - v^2}\, dt = \left(1 - \frac{v^2}{2c^2} + O\left(\frac{v}{c}\right)^4\right) c\, dt,$$

and insert it in (4.5) keeping only the first two terms:

$$I[y] = lc(t_b - t_a) - \frac{l}{2c}\int_{t_a}^{t_b} v^2 dt.$$

The first term, being independent of the dynamical variables, is irrelevant, while the
second reduces, indeed, to (4.2), if we set $l = -mc$. In conclusion, the relativistic
action of a free particle is given by the expression

$$I[y] = -mc \int_a^b ds = -mc \int_a^b \sqrt{\frac{dy^\mu}{d\lambda} \frac{dy_\mu}{d\lambda}} \, d\lambda. \qquad (4.6)$$

Henceforth, we set the speed of light again equal to unity.

Derivation of the equations of motion. We determine now the world lines $y^\mu(\lambda)$ at which the action (4.6) is stationary for arbitrary variations of the coordinates[1]

$$\delta y^\mu(\lambda) = y'^\mu(\lambda) - y^\mu(\lambda),$$

provided they vanish at the boundaries:

$$\delta y^\mu(a) = 0 = \delta y^\mu(b). \qquad (4.8)$$

It is not difficult to show that these world lines are precisely those satisfying (4.1). For this purpose we compute first the variation of the action (4.6) under generic variations of the coordinates

$$\delta I = -m \int_a^b \frac{1}{2\sqrt{\frac{dy^\mu}{d\lambda} \frac{dy_\mu}{d\lambda}}} \delta \left(\frac{dy^\mu}{d\lambda} \frac{dy_\mu}{d\lambda} \right) d\lambda = -m \int_a^b \left(\frac{dy^\mu}{d\lambda} \frac{d\delta y_\mu}{d\lambda} \right) ds, \quad (4.9)$$

where we have used the identity

$$\sqrt{\frac{dy^\mu}{d\lambda} \frac{dy_\mu}{d\lambda}} = \frac{ds}{d\lambda}. \qquad (4.10)$$

Using (4.10) once more, and recalling the definition of the four-momentum $p^\mu = m \, dy^\mu/ds$, we can write (4.9) as

$$\delta I = -\int_a^b p^\mu \frac{d\delta y_\mu}{ds} \, ds.$$

Finally, integrating by parts we can recast this expression in the form

$$\delta I = -p^\mu \delta y_\mu \Big|_a^b + \int_a^b \frac{dp^\mu}{ds} \delta y_\mu \, ds.$$

---

[1]Alternatively, the action (4.6) could be considered as a functional of the Lagrangian coordinates $\mathbf{y}(t)$, in place of the $y^\mu(\lambda)$, in which case, choosing the parameter $\lambda = y^0/c = t$, it would become

$$I[\mathbf{y}] = -mc^2 \int \sqrt{1 - v^2/c^2} \, dt. \qquad (4.7)$$

The action (4.7) yields the three equations of motion $d\mathbf{p}/dt = 0$, where $\mathbf{p}$ is the relativistic momentum (2.11), which, although not *manifestly* covariant, are physically equivalent to the four equations (4.1).

Thanks to the boundary constraints (4.8) the first term is zero. Requiring that δI must vanish for otherwise arbitrary variations δy^μ, we retrieve the expected stationarity conditions

$$\frac{dp^\mu}{ds} = 0.$$

4.2 Action for the Electrodynamics of Point-Particles

We now consider a system of charged particles interacting with the electromagnetic field. Introducing, as usual, a vector potential A_μ through the field strength

$$F_{\mu\nu} = \partial_\mu A_\nu - \partial_\nu A_\mu,$$

the equations of motion of the system then are the Maxwell equation (2.22) for the vector potential, and the Lorentz equations (2.20) for the particles. In this section we want to rederive these equations from a variational principle. We need, thus, as starting point a Poincaré-invariant action $I[A, y]$, which must be a functional of the vector potential $A_\mu(x)$ and of the world lines of the particles $y \equiv \{y_r^\mu(\lambda_r)\}_{r=1}^N$. In Sect. 3.2.3 we derived the Maxwell equation from the Lagrangian (3.38), and, in addition, we already know the action (4.6) of a free particle. It is, therefore, natural to adopt as action for the coupled system the functional

$$I[A, y] = -\frac{1}{4} \int_{\Sigma_a}^{\Sigma_b} F^{\mu\nu} F_{\mu\nu} \, d^4x - \int_{\Sigma_a}^{\Sigma_b} A_\mu j^\mu \, d^4x - \sum_r m_r \int_{a_r}^{b_r} ds_r = I_1 + I_2 + I_3.$$

(4.11)

In this action the four-dimensional integrals are delimited by two non-intersecting space-like hypersurfaces Σ_a and Σ_b, and a_r and b_r are the intersection points of the rth world line γ_r with Σ_a and Σ_b, respectively.[2] We interpret I_1 as the term describing the free propagation of the electromagnetic field, I_3 as the one describing the free motion of the charges, and I_2 as the term describing the interaction between field and charges. The final justification of the action (4.11) derives, ultimately, from the fact that it yields the expected equations of motion, as shown below.

To set up the variational problem it is convenient to rewrite the action (4.11) in a different way. By inserting the explicit expression of the current (2.102), we can recast the interaction term in the form

$$I_2 = -\int_{\Sigma_a}^{\Sigma_b} A_\mu(x) \sum_r e_r \int \delta^4(x - y_r) \, dy_r^\mu \, d^4x = -\sum_r e_r \int_{a_r}^{b_r} A_\mu(y_r) \, dy_r^\mu.$$

(4.12)

---

[2]The world line γ_r intersects the hypersurfaces Σ_a and Σ_b at most once, because the first is time-like and the latter are space-like.

As in Sect. 4.1, we use for the derivatives of the coordinates the shorthand notation

$$\dot{y}_r^\mu = \frac{dy_r^\mu}{d\lambda_r}.$$

From Eqs. (4.6) and (4.12) we then obtain

$$I_2 + I_3 = -\sum_r \int_{a_r}^{b_r} \left(m_r\, ds_r + e_r A_\mu(y_r)\, dy_r^\mu \right) \tag{4.13}$$

$$= -\sum_r \int_{a_r}^{b_r} \left(m_r \sqrt{\dot{y}_r^\nu\, \dot{y}_{r\nu}} + e_r A_\nu(y_r)\, \dot{y}_r^\nu \right) d\lambda_r \tag{4.14}$$

$$= \sum_r \int_{a_r}^{b_r} L_r(y_r, \dot{y}_r)\, d\lambda_r, \tag{4.15}$$

where we have introduced the single-particle Lagrangians

$$L_r(y_r, \dot{y}_r) = -m_r \sqrt{\dot{y}_r^\nu\, \dot{y}_{r\nu}} - e_r A_\nu(y_r)\, \dot{y}_r^\nu. \tag{4.16}$$

Finally, from the formulas above we see that the action (4.11) can also be rewritten in the form

$$I[A, y] = \int_{\Sigma_a}^{\Sigma_b} \mathcal{L}\, d^4x,$$

if we introduce the total Lagrangian

$$\mathcal{L} = -\frac{1}{4} F^{\mu\nu} F_{\mu\nu} - A_\mu j^\mu - \sum_r m_r \int \delta^4(x - y_r)\, ds_r \tag{4.17}$$

$$= \mathcal{L}_1 + \mathcal{L}_2 + \mathcal{L}_3 = \mathcal{L}_1 + \sum_r \int L_r\, \delta^4(x - y_r)\, d\lambda_r. \tag{4.18}$$

Variational problem. According to the principle of least action we must search for the configurations of fields and world lines at which the action $I[A, y]$ is stationary under arbitrary variations δA^μ and δy_r^μ, subject to the boundary conditions

$$\delta A^\mu|_{\Sigma_a} = 0 = \delta A^\mu|_{\Sigma_b}, \qquad \delta y_r^\mu(a_r) = 0 = \delta y_r^\mu(b_r).$$

We consider the variations of the fields and of the world lines separately. Since I_3 is independent of A_μ, for what concerns the equations of motion of the vector potential the variational problem reduces to the action $I_1 + I_2 = \int (\mathcal{L}_1 + \mathcal{L}_2)\, d^4x$. From Theorem II of Sect. 3.2 we know, moreover, that the field configurations at which this action is stationary are those that satisfy the Euler-Lagrange equations relative to the Lagrangian $\mathcal{L}_1 + \mathcal{L}_2$. On the other hand, the latter have been derived in Sect. 3.2.3 and have been seen to coincide with the Maxwell equation, see (3.40).

Lorentz equations. It remains to impose on the action (4.11) the stationarity conditions relative to the world lines y_r^μ. Since I_1 is independent of the y_r^μ, in this instance the variation of (4.11) equals the one of the action $I_2 + I_3$. Computing the variation of the latter, using the techniques employed in Sect. 4.1, one finds indeed that the stationarity conditions amount to the Lorentz equations, see Problem 4.1. Here, we propose an alternative proof of this result, based on the Lagrangian method for a system with a finite number of degrees of freedom, described in Sect. 3.1. For this purpose we consider the action $I_2 + I_3$ written in the form (4.15), which, in turn, can be rewritten as a sum of N terms

$$I_2 + I_3 = \sum_r I[y_r], \qquad I[y_r] = \int_{a_r}^{b_r} L_r(y_r, \dot{y}_r) \, d\lambda_r,$$

where $I[y_r]$ is a functional only of the rth world line y_r^μ. Therefore, the action $I_2 + I_3$ will be stationary, if each action $I[y_r]$ is stationary under arbitrary variations of the y_r^μ, subject to the usual boundary conditions. On the other hand, $I[y_r]$ is the integral of the *mechanical* Lagrangian L_r (4.16). Theorem I of Sect. 3.1 then implies that the stationarity conditions for this action are equivalent to the Euler-Lagrange equations relative to L_r

$$\frac{d}{d\lambda_r} \frac{\partial L_r}{\partial \dot{y}_r^\mu} - \frac{\partial L_r}{\partial y_r^\mu} = 0. \tag{4.19}$$

It remains, thus, to evaluate the two terms of these equations. Omitting for simplicity the index r, from the Lagrangian (4.16) we obtain immediately

$$\frac{\partial L}{\partial y^\mu} = -e \, \partial_\mu A_\nu \dot{y}^\nu,$$

and

$$\frac{\partial L}{\partial \dot{y}^\mu} = -\frac{m \dot{y}_\mu}{\sqrt{\dot{y}^\nu \dot{y}_\nu}} - e A_\mu = -p_\mu - e A_\mu, \tag{4.20}$$

where $p_\mu = m dy^\mu/ds$ is the four-momentum, and we used the identity (4.10). Finally, the derivative of Eq. (4.20) is

$$\frac{d}{d\lambda} \frac{\partial L}{\partial \dot{y}^\mu} = -\frac{dp_\mu}{d\lambda} - e \, \dot{y}^\nu \partial_\nu A_\mu.$$

Eventually, the Euler-Lagrange equations (4.19) read

$$\frac{d}{d\lambda} \frac{\partial L}{\partial \dot{y}^\mu} - \frac{\partial L}{\partial y^\mu} = -\frac{dp_\mu}{d\lambda} + e\dot{y}^\nu (\partial_\mu A_\nu - \partial_\nu A_\mu)$$
$$= -\frac{ds}{d\lambda} \left(\frac{dp_\mu}{ds} - e F_{\mu\nu} u^\nu \right) = 0, \tag{4.21}$$

and they are, hence, equivalent to the Lorentz equation.

Parity and time reversal. In Sect. 2.2.2 we saw that the fundamental equations of electrodynamics are invariant under the *complete* Lorentz group $O(1,3)$ and, accordingly, so must be the related action. The action (4.11) is, indeed, *manifestly* invariant under the *proper* Lorentz group $SO(1,3)_c$. Hence, it remains to verify its invariance under the discrete symmetries $\mathcal{P}$ and $\mathcal{T}$, i.e. under parity and time reversal, see Sect. 1.4.3. We begin the analysis by observing that, while the proper time $ds = \sqrt{1-v^2}\, dt$ is a scalar under $\mathcal{P}$ and a *pseudoscalar* under $\mathcal{T}$, the integral $\int ds$, actually, is invariant under $\mathcal{P}$ as well as under $\mathcal{T}$. Furthermore, at the basis of the transformation laws (2.27), (2.29) and (2.48), under parity the fields j^μ, $F^{\mu\nu}$ and A^μ transform as tensors, and hence the action is manifestly invariant also under $\mathcal{P}$. Conversely, under time reversal the fields j^μ, $F^{\mu\nu}$ and A^μ transform as *pseudotensors*. Nevertheless, since in the action (4.11) these tensors appear multiplied by each other, and the product of two pseudotensors is a tensor, the action is invariant also under $\mathcal{T}$.

4.3   Noether's Theorem

In Sect. 4.2 we derived the fundamental equations of electrodynamics from the *Poincaré invariant* action (4.11). Accordingly, it is natural to expect that Noether's theorem, applied to translations and Lorentz transformations, respectively, yields the expressions of the conserved currents $T^{\mu\nu}$ (2.134) and $M^{\mu\alpha\beta}$ (2.156), previously derived by relying on heuristic arguments. The proof of the theorem follows essentially the scheme traced in Chap. 3 for a system of only fields. However, due to the presence of point-like defects, i.e. particles, from a technical point of view the proof will be slightly more involved. According to the strategy of Sect. 3.3, it is convenient to base the proof not directly on the action (4.11), but rather on the associated Lagrangian (4.18)

$$\mathcal{L} = \mathcal{L}_1 + \sum_r \int L_r \, \delta^4(x - y_r)\, d\lambda_r, \tag{4.22}$$

where

$$\mathcal{L}_1 = -\frac{1}{4} F^{\mu\nu} F_{\mu\nu}, \qquad L_r = -m_r \sqrt{\dot{y}_r^\nu \, \dot{y}_{r\nu}} - e_r A_\nu(y_r)\, \dot{y}_r^\nu.$$

For simplicity we will denote the functional dependencies of the Lagrangian (4.22) with $\mathcal{L}(A(x), y_r, x)$, omitting, hence, to indicate explicitly the dependencies on the derivatives $\partial_\nu A^\mu$ and $\dot{y}_r^\mu$. Formally, the Lagrangian $\mathcal{L}(A(x), y_r, x)$ carries thus also an *explicit* dependence on the coordinates x^μ – indicated by its third argument – due to the presence of the δ-functions $\delta^4(x - y_r)$ in (4.22). Nevertheless, as we will see in a moment, this Lagrangian preserves translation invariance. For the Poincaré transformations we adopt the same notations as in Sect. 3.3.1. The finite transformation laws are

$$x'^\mu = \Lambda^\mu{}_\nu x^\nu + a^\mu, \qquad y_r'^\mu = \Lambda^\mu{}_\nu y_r^\nu + a^\mu, \qquad A'^\mu(x') = \Lambda^\mu{}_\nu A^\nu(x),$$

and they reduce for infinitesimal transformations $\Lambda^\mu{}_\nu = \delta^\mu{}_\nu + \omega^\mu{}_\nu$ to

$$\delta x^\mu = x'^\mu - x^\mu = a^\mu + \omega^\mu{}_\nu x^\nu, \qquad \delta y_r^\mu = y_r'^\mu - y_r^\mu = a^\mu + \omega^\mu{}_\nu y_r^\nu. \quad (4.23)$$

The infinitesimal *form variation* of A^μ, see the definition (3.51), can be derived from the relations (3.55) and (3.61)

$$\delta A_\mu = A'_\mu(x) - A_\mu(x) = -\delta x^\nu \partial_\nu A_\mu + \overline{\delta} A_\mu = -\delta x^\nu \partial_\nu A_\mu + \omega_\mu{}^\nu A_\nu. \quad (4.24)$$

The invariance of the Lagrangian $\mathcal{L}$ (4.22) under finite Poincaré transformations is expressed by the identity

$$\Delta \mathcal{L} = \mathcal{L}(A'(x'), y_r', x') - \mathcal{L}(A(x), y_r, x) = 0. \tag{4.25}$$

The unique elements of (4.22) whose invariance must be checked explicitly are the δ-functions. Using the identity (2.82) we obtain

$$\delta^4(x' - y_r') = \delta^4(\Lambda x + a - (\Lambda y_r + a)) = \delta^4(\Lambda(x - y_r)) = \frac{\delta^4(x - y_r)}{|\det \Lambda|} = \delta^4(x - y_r).$$

This implies, in particular, that from now on the transformations of the δ-functions can be ignored. In analogy to the case of a theory with only fields, we rewrite the identity (4.25) in the form

$$\Delta \mathcal{L} = \big(\mathcal{L}(A'(x'), y_r', x') - \mathcal{L}(A'(x), y_r, x)\big) + \big(\mathcal{L}(A'(x), y_r, x) - \mathcal{L}(A(x), y_r, x)\big). \tag{4.26}$$

Considering – in the spirit of (3.65) and (3.66) – infinitesimal variations, the first two terms of (4.26) differ by the variations (4.23) of x and y_r, while the last two terms differ by the form variation of A^μ (4.24). In the first two terms of (4.26) it is convenient to use for $\mathcal{L}$ the expression (4.22), while in the last two terms it is more convenient to resort to the equivalent expression (4.17). In this way the *infinitesimal* version of (4.26) can be written as

$$\delta \mathcal{L} = \left(\delta x^\mu \partial_\mu \mathcal{L}_1 + \sum_r \int \delta L_r \, \delta^4(x - y_r) \, d\lambda_r\right) + \left(\frac{\partial \mathcal{L}}{\partial A_\nu} \delta A_\nu + \Pi^{\mu\nu} \partial_\mu \delta A_\nu\right), \tag{4.27}$$

where δL_r denotes the variation of L_r under the transformations δy_r (4.23), and we have introduced the usual conjugate momenta $\Pi^{\mu\nu} = \partial \mathcal{L}/\partial(\partial_\mu A_\nu) = -F^{\mu\nu}$. Using (4.17) we can recast the last two terms of (4.27) in the form

$$\frac{\partial \mathcal{L}}{\partial A_\nu} \delta A_\nu + \Pi^{\mu\nu} \partial_\mu \delta A_\nu = \partial_\mu \left(\Pi^{\mu\nu} \delta A_\nu \right) + \left(\frac{\partial \mathcal{L}}{\partial A_\nu} - \partial_\mu \Pi^{\mu\nu} \right) \delta A_\nu,$$

$$= \partial_\mu \left(\Pi^{\mu\nu} \delta A_\nu \right) + \left(\partial_\mu F^{\mu\nu} - j^\nu \right) \delta A_\nu,$$

where in the second line we recognize the Maxwell equation. In a similar way, when we evaluate δL_r, the Lorentz equations appear. In fact, we obtain

$$\delta L_r = \delta y_r^\nu \frac{\partial L_r}{\partial y_r^\nu} + \delta \dot{y}_r^\nu \frac{\partial L_r}{\partial \dot{y}_r^\nu} = \frac{d}{d\lambda_r} \left(\delta y_r^\nu \frac{\partial L_r}{\partial \dot{y}_r^\nu} \right) + \delta y_r^\nu \left(\frac{\partial L_r}{\partial y_r^\nu} - \frac{d}{d\lambda_r} \frac{\partial L_r}{\partial \dot{y}_r^\nu} \right)$$

$$= \frac{d}{d\lambda_r} \left(\delta y_r^\nu \frac{\partial L_r}{\partial \dot{y}_r^\nu} \right) + \frac{ds_r}{d\lambda_r} \left(\frac{dp_{r\nu}}{ds_r} - F_{\nu\mu} u_r^\mu \right) \delta y_r^\nu, \tag{4.28}$$

where, to obtain the last line, we have used the identity (4.21). The first term of (4.28) gives rise in (4.27) to the contribution

$$\sum_r \int \frac{d}{d\lambda_r} \left(\delta y_r^\nu \frac{\partial L_r}{\partial \dot{y}_r^\nu} \right) \delta^4(x - y_r)\, d\lambda_r = -\sum_r \int \delta y_r^\nu \frac{\partial L_r}{\partial \dot{y}_r^\nu} \frac{d}{d\lambda_r} \delta^4(x - y_r)\, d\lambda_r$$

$$+ \sum_r \left(\delta y_r^\nu \frac{\partial L_r}{\partial \dot{y}_r^\nu} \delta^4(x - y_r) \right) \Bigg|_{\lambda_r = -\infty}^{\lambda_r = +\infty}. \tag{4.29}$$

For every fixed x, in the limit for $\lambda_r \to \pm\infty$ the (symbolic) functions $\delta^4(x - y_r)$ vanish, so that the last term of (4.29) is zero. Regarding the first term we notice that

$$\frac{d}{d\lambda_r} \delta^4(x - y_r) = -\dot{y}_r^\mu \partial_\mu \delta^4(x - y_r) = -\partial_\mu \left(\dot{y}_r^\mu \delta^4(x - y_r) \right).$$

The expression (4.29) reduces therefore to the four-divergence

$$\sum_r \int \frac{d}{d\lambda_r} \left(\delta y_r^\nu \frac{\partial L_r}{\partial \dot{y}_r^\nu} \right) \delta^4(x - y_r)\, d\lambda_r = \partial_\mu \sum_r \int \dot{y}_r^\mu \delta y_r^\nu \frac{\partial L_r}{\partial \dot{y}_r^\nu} \delta^4(x - y_r)\, d\lambda_r.$$

Inserting these results in (4.27), and noticing that for the Poincaré transformations (4.23) we have $\partial_\mu \delta x^\mu = 0$, the infinitesimal variation (4.27) of the Lagrangian (4.17) of electrodynamics can eventually be recast in the form

$$\delta \mathcal{L} = \partial_\mu \left(\delta x^\mu \mathcal{L}_1 + \Pi^{\mu\nu} \delta A_\nu + \sum_r \int \dot{y}_r^\mu \delta y_r^\nu \frac{\partial L_r}{\partial \dot{y}_r^\nu} \delta^4(x - y_r)\, d\lambda_r \right)$$

$$+ (\partial_\mu F^{\mu\nu} - j^\nu) \delta A_\nu + \sum_r \int \left(\frac{dp_{r\nu}}{ds_r} - F_{\nu\mu} u_r^\mu \right) \delta y_r^\nu \delta^4(x - y_r)\, ds_r. \tag{4.30}$$

This formula has the structure expected by Noether's theorem: it equals the variation of the Lagrangian to the four-divergence of a certain four-vector, given in the first line of (4.30), modulo terms proportional to the equations of motion. It remains to write out this four-vector explicitly, inserting the Poincaré transformations (4.23) and (4.24). The first two terms can be recast in the form

$$
\begin{aligned}
\delta x^\mu \mathcal{L}_1 + \Pi^{\mu\nu} \delta A_\nu &= \delta x_\nu \left(\eta^{\mu\nu} \mathcal{L}_1 - \Pi^{\mu\alpha} \partial^\nu A_\alpha \right) + \Pi^{\mu\nu} \omega_{\nu\rho} A^\rho \\
&= -\delta x_\nu \widetilde{T}_{\mathrm{em}}^{\mu\nu} - F^{\mu\nu} \omega_{\nu\rho} A^\rho,
\end{aligned}
\tag{4.31}
$$

where we have retrieved the canonical energy-momentum tensor of the electromagnetic field $\widetilde{T}_{\mathrm{em}}^{\mu\nu}$ (3.79). In the third term of the first line of Eq. (4.30), thanks to the properties of the δ-function, we can replace δy_r^ν with δx^ν, see Eq. (4.23). Using the equality (4.20), and parameterizing the integral with the proper time ds_r, we can then rewrite this term in the form

$$
\begin{aligned}
&\sum_r \int \dot{y}_r^\mu \, \delta y_r^\nu \, \frac{\partial L_r}{\partial \dot{y}_r^\nu} \, \delta^4(x - y_r) \, d\lambda_r \\
&= -\delta x_\nu \sum_r \int u_r^\mu \left(p_r^\nu + e_r A^\nu(y_r) \right) \delta^4(x - y_r) \, ds_r = -\delta x_\nu \left(T_{\mathrm{p}}^{\mu\nu} + j^\mu A^\nu \right),
\end{aligned}
\tag{4.32}
$$

where we have retrieved the energy-momentum tensor of the particles $T_{\mathrm{p}}^{\mu\nu}$ (2.138). Summing up Eqs. (4.31) and (4.32), and inserting the variations $\delta x_\nu = a_\nu + \omega_{\nu\beta} x^\beta$, we can rewrite the four-vector appearing in the variation (4.30) in the form

$$
\begin{aligned}
&\delta x^\mu \mathcal{L}_1 + \Pi^{\mu\nu} \delta A_\nu + \sum_r \int \dot{y}_r^\mu \, \delta y_r^\nu \, \frac{\partial L_r}{\partial \dot{y}_r^\nu} \, \delta^4(x - y_r) \, d\lambda_r \\
&= -\delta x_\nu \left(T_{\mathrm{p}}^{\mu\nu} + \widetilde{T}_{\mathrm{em}}^{\mu\nu} + j^\mu A^\nu \right) - F^{\mu\nu} \omega_{\nu\rho} A^\rho = -a_\nu \widetilde{T}^{\mu\nu} + \frac{1}{2} \omega_{\alpha\beta} \widetilde{M}^{\mu\alpha\beta},
\end{aligned}
$$

where we have introduced the total *canonical* energy-momentum tensor and the total *canonical* angular momentum density of electrodynamics

$$
\widetilde{T}^{\mu\nu} = T_{\mathrm{p}}^{\mu\nu} + \widetilde{T}_{\mathrm{em}}^{\mu\nu} + j^\mu A^\nu,
\tag{4.33}
$$

$$
\widetilde{M}^{\mu\alpha\beta} = x^\alpha \widetilde{T}^{\mu\beta} - x^\beta \widetilde{T}^{\mu\alpha} - F^{\mu\alpha} A^\beta + F^{\mu\beta} A^\alpha.
\tag{4.34}
$$

In conclusion, the variation (4.27) of the total Lagrangian (4.17) under infinitesimal Poincaré transformations can be recast in the form

$$
\begin{aligned}
\delta \mathcal{L} = &- a_\nu \partial_\mu \widetilde{T}^{\mu\nu} + \frac{1}{2} \omega_{\alpha\beta} \partial_\mu \widetilde{M}^{\mu\alpha\beta} \\
&+ (\partial_\mu F^{\mu\nu} - j^\nu) \delta A_\nu + \sum_r \int \left(\frac{dp_{r\nu}}{ds_r} - F_{\nu\mu} u_r^\mu \right) \delta y_r^\nu \, \delta^4(x - y_r) \, ds_r,
\end{aligned}
\tag{4.35}
$$

to be compared with the analogous identity (3.73) for a pure field theory. On the other hand, since the Lagrangian (4.17) is invariant under Poincaré transformations, the variation $\delta \mathcal{L}$ (4.35) vanishes identically. We conclude, therefore, that if the fields and the particles satisfy their respective equations of motion, the currents $\widetilde{T}^{\mu\nu}$ and $\widetilde{M}^{\mu\alpha\beta}$ are conserved

$$\partial_\mu \widetilde{T}^{\mu\nu} = 0 = \partial_\mu \widetilde{M}^{\mu\alpha\beta}. \tag{4.36}$$

Symmetric energy-momentum tensor. We see, once more, that the obtained currents do not have the form anticipated in Sects. 2.4.3 and 2.4.4. In particular, $\widetilde{T}^{\mu\nu}$ is not symmetric, and $\widetilde{M}^{\mu\alpha\beta}$ is not of the standard form (2.156). Moreover, both currents – depending explicitly on the vector potential A^μ – violate *gauge invariance*. Apart from this, the *interference* term $j^\mu A^\nu$ in (4.33) does not admit a simple physical interpretation. Nevertheless, also in this instance we can resort to the general symmetrization procedure of the energy-momentum tensor developed in Sect. 3.4. We again set

$$T^{\mu\nu} = \widetilde{T}^{\mu\nu} + \partial_\rho \phi^{\rho\mu\nu}, \qquad \phi^{\rho\mu\nu} = -\phi^{\mu\rho\nu}, \tag{4.37}$$

where the tensor $\phi^{\rho\mu\nu}$ is tied to the tensor $V^{\mu\alpha\beta}$ by the relation (3.87). The latter can be determined by comparing Eq. (4.34) with the general expression (3.83). One finds

$$V^{\mu\alpha\beta} = -F^{\mu\alpha}A^\beta + F^{\mu\beta}A^\alpha,$$

as in the case of the *free* Maxwell field, see Eq. (3.94). Also the tensor $\phi^{\rho\mu\nu}$ coincides, thus, with the expression (3.95) relative to the free field

$$\phi^{\rho\mu\nu} = -F^{\rho\mu}A^\nu.$$

On the other hand, in the presence of particles the four-divergence of $\phi^{\rho\mu\nu}$ contains an additional term proportional to the four-current j^μ. Using the Maxwell equation we find indeed

$$\partial_\rho \phi^{\rho\mu\nu} = -\partial_\rho F^{\rho\mu}A^\nu - F^{\rho\mu}\partial_\rho A^\nu = -j^\mu A^\nu - F^{\alpha\mu}\partial_\alpha A^\nu.$$

Adding this expression to (4.33) one sees that the interference term $j^\mu A^\nu$ cancels out, and that the *symmetric* energy-momentum tensor $T_{em}^{\mu\nu}$ of the electromagnetic field shows up. Performing the computations, (4.37) yields in fact

$$T^{\mu\nu} = T_{em}^{\mu\nu} + T_p^{\mu\nu},$$

confirming hence the heuristic results (2.137) and (2.138) of Chap. 2.

Standard angular momentum. Similarly, according to the general prescription based on the relations (3.99) and (3.103), we can construct a standard angular momentum density by setting

$$M^{\mu\alpha\beta} = \widetilde{M}^{\mu\alpha\beta} + \partial_\rho \Lambda^{\rho\mu\alpha\beta}, \qquad \Lambda^{\rho\mu\alpha\beta} = x^\alpha \phi^{\rho\mu\beta} - x^\beta \phi^{\rho\mu\alpha}.$$

In the present case, using the first relation in (4.37), we obtain

$$\partial_\rho \Lambda^{\rho\mu\alpha\beta} = \phi^{\alpha\mu\beta} + x^\alpha \partial_\rho \phi^{\rho\mu\beta} - (\alpha \leftrightarrow \beta) = F^{\mu\alpha} A^\beta + x^\alpha (T^{\mu\beta} - \widetilde{T}^{\mu\beta}) - (\alpha \leftrightarrow \beta).$$

Adding this term to the expression (4.34) we find

$$M^{\mu\alpha\beta} = x^\alpha T^{\mu\beta} - x^\beta T^{\mu\alpha},$$

in agreement with the general definition (2.156).

In summary, we have retrieved the expected expressions for the energy-momentum tensor and for the angular momentum density of electrodynamics. What matters more than the – already known – results is the *systematic* method we employed to derive them, namely Noether's theorem. In Sect. 3.3 we gave a general proof of this theorem for a theory containing only *fields* as dynamical variables. In this section we proved it in a very different physical context, in which certain degrees of freedom are not distributed *continuously* in space, such as fields, but rather appear as point-like *defects*, i.e. as particles. In this chapter we have, indeed, exemplified a very fundamental aspect of Physics, namely that *Noether's theorem lasts at all levels*: it holds in classical physics as well as in quantum physics; it holds in theories whose degrees of freedom are described by fields, particles, strings and, more generally, by membranes of arbitrary spatial extension, so-called p-branes; it holds in *Newtonian mechanics* as well as in *special relativity*, as it holds true in *General Relativity* and in *superstring theory*.

4.4   Gauge Invariance and Charge Conservation

Until now we examined Noether's theorem relative to the Poincaré group, and to the corresponding conservation laws of the four-momentum and of the angular momentum (see, however, Problem 3.10). In electrodynamics there exists, however, a further locally conserved fundamental quantity – the *electric charge* – and if Noether's theorem has to be universally valid, also this conservation law should follow from a one-parameter group of symmetries. Actually, in electrodynamics there *is* a further fundamental symmetry, namely *gauge invariance*, which could be linked to the conservation of the charge. To analyze this hypothesis we notice, first of all, that the set of gauge transformations constitutes indeed a one-parameter group, with parameter Λ. If we set

$$A'_{1\mu} = A_\mu + \partial_\mu \Lambda_1,$$

we retrieve in fact the (abelian) composition law

$$A'_{2\mu} = A'_{1\mu} + \partial_\mu \Lambda_2 = A_\mu + \partial_\mu (\Lambda_1 + \Lambda_2).$$

In analyzing the change of the Lagrangian (4.17) under a generic gauge transformation $\delta A_\mu = A'_\mu - A_\mu = \partial_\mu \Lambda$ we obtain

$$\delta \mathcal{L} = -\partial_\mu \Lambda j^\mu = -\partial_\mu(\Lambda j^\mu) + \Lambda \partial_\mu j^\mu \to \Lambda \partial_\mu j^\mu,$$

where we took advantage of the fact that Lagrangians are defined modulo four-divergences, see Sect. 3.2.2. *Formally* we have thus established the link foreseen in general by Noether's theorem, namely that the invariance of the Lagrangian implies the local conservation of the charge

$$\delta \mathcal{L} = 0 \quad \Rightarrow \quad \partial_\mu j^\mu = 0. \tag{4.38}$$

Notwithstanding, the relation we have just established only partially overlaps with the main characteristics of Noether's theorem, as exemplified for the Poincaré group in Sects. 3.3.2 and 4.3. The first reason lies in the fact that the parameter $\Lambda(x)$ is not a *global* parameter, i.e. a coordinate-independent one, as required by Noether's theorem. Actually, if Λ is coordinate-independent, i.e. a constant, a gauge transformation reduces trivially to the identity map. In the second place, in the derivation (4.38) the equations of motion of electrodynamics played no role at all, while they were essential in the proof of the continuity equation (4.36). Accordingly, we know that the current (2.16) is *identically* conserved, whether the equations of motion are fulfilled or not. From the explicit check (2.105) one infers, in particular, that the identity $\partial_\mu j^\mu = 0$ eventually simply expresses the fact that the particles' world lines never begin and never end.

It should not be surprising that this *asymmetry* in the realization of the conservation laws associated with the Poincaré group and the group of gauge transformations – a peculiarity of *classical* electrodynamics – is due to the fact that in this theory the particles are assimilated to *point-like* defects. Actually, in this framework the charge conservation law itself becomes a *trivial* statement, in that it reduces simply to the counting of the particles contained in a given volume. Indeed, integrating the charge density (2.109) over a volume V, for the charge $Q_V(t)$ contained at the instant t in V we obtain

$$Q_V(t) = \int_V j^0(t, \mathbf{x}) \, d^3x = \sum_r e_r \int_V \delta^3(\mathbf{x} - \mathbf{y}_r(t)) \, d^3x = \sum_{r \in V} e_r,$$

where the last sum over r includes all particles contained at time t in the volume V. Conversely, it can be seen that once also the charged particles are represented by *fields* – on the same footing as the electromagnetic field – the conservation law of the electric charge follows precisely the Noether-scheme, that is to say: a) it descends from a continuous *global* one-parameter symmetry group, and b) it holds only if the equations of motion are fulfilled, as illustrated in Problem 3.10. In particular, in *quantum electrodynamics* the conservation of the electric charge occurs precisely according to this scheme.

4.5   Problems

4.1 Derive the Lorentz equations (2.20) imposing that the action (4.14) is stationary under arbitrary variations of the coordinates $\delta y_r^\mu(\lambda_r)$, vanishing at $\lambda_r = a_r$ and $\lambda_r = b_r$.

4.2 *Lagrangian of a system of non-relativistic charges.* Consider a system of N non-relativistic charges $\{e_r\}$, i.e. charges moving with speeds $v_r \ll c$, in the presence of an external electromagnetic field $F_{\mu\nu} = \partial_\mu A_\nu - \partial_\nu A_\mu$.

(a) Determine the non-relativistic limit of the action (4.13) and verify that the resulting Lagrangian of the system, neglecting the mutual interactions of the charges, admits the expansion

$$L(\mathbf{y}_r, \mathbf{v}_r, t) = \sum_{r=1}^{N} \left(\frac{1}{2} m_r v_r^2 - e_r \left(A^0(t, \mathbf{y}_r) - \frac{1}{c}\, \mathbf{v}_r \cdot \mathbf{A}(t, \mathbf{y}_r) \right) \right) + O\!\left(\frac{1}{c^2} \right).$$

(b) Determine the Lagrangian $L^*(\mathbf{y}_r, \mathbf{v}_r, t)$ of the system, which takes into account the mutual interactions of the charges.

   Hint: From Problem 2.8 one infers the answer

$$L^*(\mathbf{y}_r, \mathbf{v}_r, t) = L(\mathbf{y}_r, \mathbf{v}_r, t) - \frac{1}{4\pi} \sum_{r<s} \frac{e_r e_s}{|\mathbf{y}_r - \mathbf{y}_s|}.$$

Part II
Applications

Chapter 5
Electromagnetic Waves

In this chapter we start the search for exact solutions of Maxwell's equations and the analysis of their properties. The first class of solutions we will analyze are the *plane waves*, which form a particular complete set of solutions of Maxwell's equations in *empty space*, or in *vacuum*, that is to say, in the absence of charged sources throughout space

$$j^\mu(x) = 0$$

for all x. The phenomenological relevance of these solutions is self-evident: The major part of the energy provided by the Sun travels on top of electromagnetic waves, any signal propagating on the Earth surface via *ether* is composed of such waves, and almost all of the information that we gain about the Universe reaches the Earth via light signals emitted by cosmic objects, signals which consist of electromagnetic waves propagating throughout empty space and covering enormous distances. The Universe itself is pervaded by the so-called *cosmic microwave background* radiation – with excellent approximation isotropic and homogeneous – characterized by a *black–body* frequency spectrum with a temperature of about 2.73 °K. This radiation is a messenger of a primordial era in which the Universe was made up mainly of dissociated charged particles and radiation, in thermal equilibrium with each other. This balance was supported by collisions between the charged particles, mediated by the electromagnetic field. After the *recombination* of the charged particles, more precisely electrons and nuclei, into *neutral* atoms, approximately at the time of *last scattering*, the electromagnetic field decoupled from the charges and today manifests itself as an – *apparently* sourceless – background radiation.[1]

As mentioned above, in this chapter we analyze the properties of the electromagnetic waves as a *basis* of solutions of Maxwell's equations in vacuum, in the sense that every solution can be expressed as a linear superposition of these waves. In the

---

[1] According to Plank's law for the black-body spectrum a correct treatment of a low-intensity electromagnetic field, as the background radiation, of course requires a *quantum* analysis to take the particle-nature of the radiation into account: On average, a cm^3 of the Universe is, in fact, filled with about 400 photons.

© Springer International Publishing AG, part of Springer Nature 2018
K. Lechner, *Classical Electrodynamics*, UNITEXT for Physics,
https://doi.org/10.1007/978-3-319-91809-9_5

chapters to follow we shall instead deal with the solutions of Maxwell's equations in the presence of sources. In particular, we will determine the electromagnetic field generated by an arbitrary charge distribution confined to a finite region of space. Away from this region the field satisfies, in turn, Maxwell's equations in empty space, and correspondingly we will see that *far away* from this region, not surprisingly, it evolves once more into a superposition of plane waves. In order to properly address these issues, we must preliminarily determine the *kinematical* content of the electromagnetic field, i.e. we have to identify the independent variables which describe its *state* at every instant. In other words, we must identify the *physical degrees of freedom* involved in the time evolution of the electromagnetic field, since only then we shall be able to properly set the corresponding *Cauchy problem*. In the case at hand this problem includes the specification of a complete set of *independent* initial data, which through Maxwell's equations determine the electromagnetic field at any instant.

5.1 Degrees of Freedom

In the framework of field theory the concept of *degree of freedom* represents a generalization of the analogous concept in Newtonian mechanics, prototype of a Lagrangian system with a *finite* number of degrees of freedom. Before addressing this issue in a generic field theory, and then specifying the analysis to the electromagnetic field, we briefly recall the meaning of this basic concept in classical mechanics.

5.1.1 *Degrees of Freedom in Classical Mechanics*

In the context of Newtonian mechanics the concept of degree of freedom refers to the number of *Lagrangian* coordinates describing the system. A particle moving in a three-dimensional space, for example, carries three degrees of freedom in that, one would say, at any instant t its position is specified by the three coordinates $\mathbf{y}(t) = (x(t), y(t), z(t))$. We can actually analyze the system from a different point of view, asking the question: how many initial data, say at time $t = 0$, must we fix in order to predict the position of the particle at any time t? The answer – *six*, not *three* – is strictly tied to the *dynamics* of the system, specified by Newton's second law

$$m\ddot{\mathbf{y}} = \mathbf{F}(\mathbf{y}, \dot{\mathbf{y}}, t).$$

This equation corresponds indeed to a system of three *second-order* differential equations and admits, hence, a unique solution, once we impose the six initial data $\mathbf{y}(0)$ and $\dot{\mathbf{y}}(0)$. An equivalent manner to describe the dynamics of the system is offered by the *Hamiltonian* formalism, a framework complementary to the

Lagrangian formalism implicitly employed above, in which the position $\mathbf{y}(t)$ and the velocity $\mathbf{v}(t)$, spanning the *phase space*, are considered as independent variables. In this framework one imposes the six *first-order* differential equations

$$m\dot{\mathbf{v}} = \mathbf{F}(\mathbf{y}, \mathbf{v}, t), \qquad \dot{\mathbf{y}} = \mathbf{v}, \tag{5.1}$$

which admit a unique solution once the initial data $\mathbf{y}(0)$ and $\mathbf{v}(0)$ have been specified. In the Hamiltonian formalism the system appears, thus, as a system of *six* degrees of freedom. We realize so that the common statement "a particle carries three degrees of freedom" actually means "three degrees of freedom of the *second order*". Equivalently we could, in fact, say that the particle carries six degrees of freedom of the *first order*. The preference for the first convention – commonly adopted in physics – stems primarily from the peculiar relation existing between the degrees of freedom of a classical field and the particles associated to it at the *quantum* level, see the end of Sect. 5.1.2. Henceforth, with the term *degree of freedom* we will always understand *degree of freedom of the second order*, if not differently specified.

5.1.2   Degrees of Freedom in Field Theory

In field theory the fundamental variables are the fields, and from a mechanical point of view each field corresponds to a system of *infinite* degrees of freedom. Preserving the analogy with Newtonian mechanics based on Eq. (5.1) – while adapting the perspective – we give the following definition.

Definition. We say that a real field $\varphi(t, \mathbf{x})$ carries *one* degree of freedom if the equations of motion governing its dynamics are such that, once the data $\varphi(0, \mathbf{x})$ and $\partial_0\varphi(0, \mathbf{x})$ are known for every $\mathbf{x}$, they determine the field $\varphi(t, \mathbf{x})$ for any t and for any $\mathbf{x}$.

As prototype of such an equation we consider the partial differential equation for a scalar field

$$\Box\varphi = P(\varphi), \tag{5.2}$$

where

$$\Box = \partial_\mu\partial^\mu = \partial_0^2 - \nabla^2$$

is the *d'Alembertian* operator – a relativistic completion of the Laplacian – and $P(\varphi)$ is a polynomial in φ. Equation (5.2) is of second order in the time derivatives and we expect, hence, that it confers to φ *one* degree of freedom. To confirm this expectation explicitly we fix the initial data

$$\varphi(0, \mathbf{x}), \qquad \partial_0\varphi(0, \mathbf{x}), \tag{5.3}$$

and attempt to determine $\varphi(t, \mathbf{x})$ enforcing Eq. (5.2). Assuming that the solution is an analytic function of t we perform a Taylor expansion around $t = 0$

$$\varphi(t, \mathbf{x}) = \sum_{n=0}^{\infty} \frac{\partial_0^n \varphi(0, \mathbf{x})}{n!} t^n, \tag{5.4}$$

and try to determine its coefficients using (5.2). The coefficients relative to $n = 0$ and $n = 1$ correspond to the initial data (5.3). The coefficient relative to $n = 2$ can be determined evaluating (5.2) at $t = 0$

$$\partial_0^2 \varphi(0, \mathbf{x}) = \nabla^2 \varphi(0, \mathbf{x}) + P(\varphi(0, \mathbf{x})).$$

The second member of this equation is, in fact, known once $\varphi(0, \mathbf{x})$ is known for all $\mathbf{x}$. The coefficient relative to $n = 3$ can be obtained differentiating (5.2) with respect to time and evaluating the result at $t = 0$

$$\partial_0^3 \varphi(0, \mathbf{x}) = \nabla^2 \partial_0 \varphi(0, \mathbf{x}) + P'(\varphi(0, \mathbf{x})) \, \partial_0 \varphi(0, \mathbf{x}), \quad P'(\varphi) = \frac{dP(\varphi)}{d\varphi}.$$

Again, the second member of this equation is known once $\partial_0 \varphi(0, \mathbf{x})$ is known for all $\mathbf{x}$. Differentiating repeatedly (5.2) with respect to time one can determine all derivatives $\partial_0^n \varphi(0, \mathbf{x})$ in terms of the *spatial* partial derivatives of the initial data (5.3), and so the series (5.4) is uniquely determined. Vice versa, it is not difficult to show that the series obtained in this way satisfies Eq. (5.2). This "perturbative" proof can be easily generalized to cases where P is an arbitrary polynomial in φ and $\partial_\mu \varphi$, and where the second member of (5.2) contains an additional known term $j(x)$, independent of φ.

Degrees of freedom and particles. The number of degrees of freedom of a classical field theory has a direct counterpart in the corresponding *quantum* field theory. In general a quantum field theory describes a well-defined number of particles, which can, however, appear in different *species*. *Quantum electrodynamics*, for instance, describes the electron with spin $\pm\hbar/2$, the positron – its antiparticle – with spin $\pm\hbar/2$, and the photon with spin $\pm\hbar$. Hence, these particles appear altogether in six species. On the other hand, in semiclassical[2] electrodynamics a charged particle is described by the *Dirac equation* for a complex four-component wave function $\Psi(x)$, which corresponds to four complex first-order differential equations or, equivalently, to eight real first-order differential equations, and so it entails *four* degrees of freedom. In addition, as we will see in detail in Sect. 5.1.3, Maxwell's equations entail *two* degrees of freedom, so that the classical theory carries altogether six degrees of freedom. This example illustrates a fundamental general result regarding the number N of degrees of freedom of a classical field theory: the corresponding *quantum* field theory describes particles which appear altogether in precisely N *species*.

---

[2]For a massless electron Planck's constant $\hbar$ drops out from the Dirac equation and the theory can be considered as truly classical.

5.1.3   Cauchy Problem for Maxwell's Equations

In this section we formulate the Cauchy problem relative to the Maxwell equations (2.21) and (2.22). In doing so we will establish how many, and which ones, are the degrees of freedom associated with the electromagnetic field. For this purpose it is convenient to adopt the strategy outlined in Sect. 2.2.4, consisting in the solution of the Bianchi identity in terms of a vector potential A^μ, see (2.46). Accordingly, the system of equations we must solve can be written schematically as

$$\partial_\mu F^{\mu\nu} = j^\nu, \qquad F^{\mu\nu} = \partial^\mu A^\nu - \partial^\nu A^\mu, \qquad A^\mu \approx A^\mu + \partial^\mu \Lambda, \qquad (5.5)$$

where, we recall, the last relation signals that A^μ is defined modulo gauge transformations.

Asymptotic conditions. Before proceeding any further, we specify the class of vector potentials and currents that we consider *physically* acceptable. First of all we assume that the four-current is known and fulfils the conservation law $\partial_\mu j^\mu = 0$. Furthermore, we suppose that j^μ has *compact* spatial support, as it happens for all charge distributions realizable in nature. More precisely we require that

$$j^\mu(t, \mathbf{x}) = 0, \quad \text{for all } |\mathbf{x}| > l, \qquad (5.6)$$

where the radius l may depend on t, as it happens, for example, for a charged particle following an unbounded trajectory. Correspondingly, we will accept as *physical* solutions of Maxwell's equations only the vector potentials which for any fixed t vanish (sufficiently fast) at spatial infinity

$$\lim_{|\mathbf{x}| \to \infty} A^\mu(t, \mathbf{x}) = 0. \qquad (5.7)$$

Actually, in the particular case of an electromagnetic field in vacuum there seems to be no relation between condition (5.7) and the spatial distribution of the charges, simply because there are no charges at all. On the other hand, in general an electromagnetic field in vacuum represents a *mathematical* idealization of a real *physical* process. Such a field would, in fact, have been generated in the distant past by charges localized in a bounded region, far away from the considered "vacuum" zone, so that it is reasonable to assume that also in this case the vector potential vanishes at spatial infinity.[3] The condition (5.7) excludes, furthermore, a series of idealized solutions whose physical realization would require an *infinite energy*, and which, nonetheless, are frequently considered in theoretical physics since they can

---

[3]The cosmic microwave background radiation mentioned in the introduction to this chapter is supposed to fill the whole Universe and does, therefore, not satisfy (5.7). On the other hand, as we remarked previously, an appropriate description of this radiation requires a *quantum statistical* analysis of Maxwell's equations.

be treated analytically. Among these solutions we recall the constant and uniform electromagnetic fields $F^{\mu\nu}(x) = \mathcal{F}^{\mu\nu}$, with vector potential

$$A^{\mu}(x) = -\frac{1}{2}\,\mathcal{F}^{\mu\nu}x_{\nu},$$

the fields produced by infinitely extended charged wires and planes, or by currents flowing in infinite wires, and the plane waves themselves as they extend over the whole space.

We return now to the Maxwell equation (5.5), writing it in terms of the vector potential,

$$\partial_{\mu}\left(\partial^{\mu}A^{\nu} - \partial^{\nu}A^{\mu}\right) = \Box A^{\nu} - \partial^{\nu}(\partial_{\mu}A^{\mu}) = j^{\nu}. \tag{5.8}$$

Due to the presence of the second-order time derivatives, at first glance this system of equations seems to associate to A^{μ} *four* degrees of freedom. However, as anticipated in Sect. 2.2.5, this conclusion turns out to be wrong for several reasons.

A constraint. The first reason is that the four components of the vector equation (5.8) are not functionally independent. In fact, since the current is conserved, the four-vector defined by

$$G^{\nu} = \partial_{\mu}F^{\mu\nu} - j^{\nu} = \Box A^{\nu} - \partial^{\nu}(\partial_{\mu}A^{\mu}) - j^{\nu} \tag{5.9}$$

satisfies the *identity*

$$\partial_{\nu}G^{\nu} = 0 \quad \Leftrightarrow \quad \partial_{0}G^{0} = -\boldsymbol{\nabla}\cdot\mathbf{G}. \tag{5.10}$$

This means that the four components of the Maxwell equation (5.8), written as

$$G^{\nu} = 0,$$

are equivalent to the system

$$\mathbf{G}(t,\mathbf{x}) = 0, \quad \forall t,$$
$$G^{0}(0,\mathbf{x}) = 0.$$

In fact, once we have imposed the equation $\mathbf{G}(t,\mathbf{x}) = 0$ for all t, the identity (5.10) ensures that also the equation $\partial_{0}G^{0}(t,\mathbf{x}) = 0$ holds automatically for all t. The function $G^{0}(t,\mathbf{x})$ is thus time-independent and, consequently, it is sufficient to require its vanishing, say at $t = 0$. In conclusion, the $\nu = 0$ component of equation (5.8) represents a *constraint* on the initial data, rather than a *proper* equation of motion.

Gauge invariance and gauge-fixing. The second reason for which the above naive counting of the degrees of freedom is incorrect is associated with gauge invariance: since the vector potentials A^{μ} and $A^{\mu} + \partial^{\mu}\Lambda$ give rise to the same electromagnetic field $F^{\mu\nu}$, they are physically equivalent. Therefore, it is necessary to select among all vector potentials associated with a given $F^{\mu\nu}$ a unique *representative* or, as it is usually said, to enforce a *gauge-fixing* which determines A^{μ} uniquely. There exist obviously infinite ways of fixing the gauge – they are all physically equivalent – and the particular choice one makes is merely a matter of convenience, depending on the physical phenomenon one wants to analyze. Throughout this book we resort mainly to the *Lorenz gauge*[4]

$$\partial_{\mu}A^{\mu} = 0, \tag{5.11}$$

for its virtue of preserving Lorentz invariance.[5] To verify the consistency of this choice we must show that, starting from an arbitrary vector potential A^{μ}, it is always possible to perform a gauge transformation $A^{\mu} \rightarrow A^{\mu} + \partial^{\mu}\Lambda$, such that the new vector potential has vanishing four-divergence

$$\partial_{\mu}(A^{\mu} + \partial^{\mu}\Lambda) = 0.$$

As we see, it is sufficient to choose an arbitrary gauge function Λ such that

$$\Box\Lambda = -\partial_{\mu}A^{\mu},$$

an equation that we know to admit infinite solutions, see Sect. 5.1.2. In the Lorenz gauge (5.11) the Maxwell equation $G^{\mu} = 0$ simplifies to

$$G^{\mu} = \Box A^{\mu} - j^{\mu} = 0. \tag{5.12}$$

We shall use this form for the spatial components $\mathbf{G}$ of the equation while, for our purposes, for the component G^{0} it will be more convenient to resort to the original expression (5.9)

$$G^{0} = \Box A^{0} - \partial^{0}(\partial_{0}A^{0} + \partial_{i}A^{i}) - j^{0} = -\nabla^{2}A^{0} - \partial_{i}(\partial^{0}A^{i}) - j^{0} = 0. \tag{5.13}$$

Notice that – in line with the fact that this equation has to be interpreted as a constraint rather than a dynamical equation – G^{0} does not contain *second-order* time derivatives.

Residual gauge invariance. The Lorenz gauge does not determine the vector potential *uniquely*, but still allows for a subclass of gauge transformations called *residual gauge transformations*. In fact, assuming that A^{μ} satisfies the constraint $\partial_{\mu}A^{\mu} = 0$, a further gauge transformation $A^{\mu} \rightarrow A^{\mu} + \partial^{\mu}\Lambda$ gives rise to a vector potential satisfying the same constraint if

---

[4]The gauge-fixing condition (5.11) has been introduced by the Danish physicist Ludvig Valentin Lorenz (1829–1891), not to be confused with the Dutch physicist Hendrik Antoon Lorentz (1853–1928).

[5]Examples of non-covariant gauge-fixings used sometimes in the literature are the *Coulomb gauge* $\nabla \cdot \mathbf{A} = 0$, the *Weyl gauge* $A^{0} = 0$ and, more generally, the gauge $n_{\mu}A^{\mu} = 0$ where n_{μ} is a constant vector.

$$\partial_\mu (A^\mu + \partial^\mu \Lambda) = 0,$$

i.e. if the gauge function satisfies the wave equation $\Box \Lambda = 0$. The Lorenz gauge admits thus the residual gauge invariance

$$A'^\mu = A^\mu + \partial^\mu \Lambda, \qquad \Box \Lambda = 0. \tag{5.14}$$

Also the gauge-fixing of the residual gauge invariance can be accomplished in infinite different ways. We opt for the conditions

$$A^3(0, \mathbf{x}) = 0 = \partial_0 A^3(0, \mathbf{x}), \tag{5.15}$$

which can indeed be achieved through a convenient residual gauge transformation. To prove this we recall from Sect. 5.1.2 that the solution of the equation $\Box \Lambda = 0$ is completely determined by the initial conditions

$$\Lambda(0, \mathbf{x}) = \Phi_1(\mathbf{x}), \qquad \partial_0 \Lambda(0, \mathbf{x}) = \Phi_2(\mathbf{x}),$$

where $\Phi_1(\mathbf{x})$ and $\Phi_2(\mathbf{x})$ are arbitrary functions. Since under a generic gauge transformation A^3 transforms as

$$A'^3 = A^3 + \partial^3 \Lambda,$$

the transformed field satisfies the conditions (5.15) if we can choose $\Phi_1(\mathbf{x})$ and $\Phi_2(\mathbf{x})$ such that

$$A'^3(0, \mathbf{x}) = A^3(0, \mathbf{x}) + \partial^3 \Lambda(0, \mathbf{x}) = A^3(0, \mathbf{x}) - \partial_3 \Phi_1(\mathbf{x}) = 0,$$
$$\partial_0 A'^3(0, \mathbf{x}) = \partial_0 A^3(0, \mathbf{x}) + \partial^3 \partial_0 \Lambda(0, \mathbf{x}) = \partial_0 A^3(0, \mathbf{x}) - \partial_3 \Phi_2(\mathbf{x}) = 0.$$

It is thus sufficient to take for $\Phi_1(\mathbf{x})$ and $\Phi_2(\mathbf{x})$ primitive functions of $A^3(0, \mathbf{x})$ and $\partial_0 A^3(0, \mathbf{x})$, respectively, relative to the coordinate z. Further gauge transformations – preserving (5.11) and (5.15) – are forbidden, since they would require a z-independent gauge function $\Lambda(x)$, see Problem 5.10. In this case the transformed potentials $A^\mu + \partial^\mu \Lambda$ would, in fact, violate the asymptotic condition (5.7), since they would no longer vanish at infinity along the z-direction. The conditions (5.11) and (5.15) provide, therefore, a *complete* gauge-fixing.

Uniqueness of the solution. In virtue of the gauge-fixing conditions (5.11) and (5.15), Maxwell's equations finally reduce to the system of equations (see (5.12) and (5.13))

$$\Box \mathbf{A} = \mathbf{j}, \tag{5.16}$$
$$\nabla^2 A^0 = -\partial_i(\partial^0 A^i) - j^0, \quad \text{for } t = 0, \tag{5.17}$$
$$\partial_\mu A^\mu = 0, \tag{5.18}$$
$$A^3(0, \mathbf{x}) = 0 = \partial_0 A^3(0, \mathbf{x}). \tag{5.19}$$

This system admits a unique solution $A^\mu(x)$, once we fix the *physical* initial data

$$A^1(0, \mathbf{x}), \qquad \partial_0 A^1(0, \mathbf{x}), \qquad A^2(0, \mathbf{x}), \qquad \partial_0 A^2(0, \mathbf{x}). \qquad (5.20)$$

In fact, according to the initial conditions (5.19) and (5.20) the three Eqs. (5.16) determine $\mathbf{A}(t, \mathbf{x})$ for all t. Once $\mathbf{A}(t, \mathbf{x})$ is known, Eq. (5.17) uniquely determines $A^0(0, \mathbf{x})$ since, as we will show in Sect. 6.1, in the space of functions that vanish sufficiently fast for $|\mathbf{x}| \to \infty$, the Laplacian ∇^2 admits a unique inverse operator. Once $A^0(0, \mathbf{x})$ and $\mathbf{A}(t, \mathbf{x})$ are known, the condition (5.18) determines finally $A^0(t, \mathbf{x})$ for all t:

$$\partial_0 A^0 = -\boldsymbol{\nabla} \cdot \mathbf{A} \quad \Rightarrow \quad A^0(t, \mathbf{x}) = A^0(0, \mathbf{x}) - \int_0^t \boldsymbol{\nabla} \cdot \mathbf{A}(t', \mathbf{x}) \, dt'.$$

Our conclusion can also be stated differently: once we have set the physical initial conditions (5.20), Maxwell's equations determine the vector potential $A^\mu(x)$ *uniquely*, modulo gauge transformations.

According to our gauge-fixings, the physical fields are *represented* by the components A^1 and A^2. It is, however, obvious that different gauge-fixings lead to different representatives of the physical fields. What remains in any case invariant is the number of initial conditions – *four* – that we can choose freely, see Problem 5.3. It remains to derive the data (5.20) from the experimentally observable physical data, namely the electric and magnetic fields $\mathbf{E}(0, \mathbf{x})$ and $\mathbf{B}(0, \mathbf{x})$ at $t = 0$. This is, however, a purely *technical* problem in that, once the latter are known, the gauge-fixings (5.18) and (5.19) determine the initial data (5.20) uniquely.[6]

Degrees of freedom of the electromagnetic field. From the set of independent initial data (5.20) we infer that the electromagnetic field carries *two* degrees of freedom, and not four. The mechanism which removes from A^μ two degrees of freedom is essentially the following: one degree of freedom is eliminated by the Lorenz gauge, and the other by the residual gauge invariance in conjunction with the fact that the temporal component (5.13) of the Maxwell equation is a constraint. As mentioned above, which are the components of A^μ that appear as *physical* depends on the choice of gauge-fixing. Moreover, under a Lorentz transformation the components of A^μ transform into each other and, correspondingly, while the condition (5.18) is Lorentz invariant, the conditions (5.19) are not. However, this circumstance does not violate the relativistic invariance of the theory since, as we have seen, in any inertial reference frame the conditions (5.19) can be restored by performing a suitable gauge transformation. Finally, notice how the degrees-of-freedom analysis performed above matches with the results of the preliminary analysis of the kinematics of the electromagnetic field of Sect. 2.2.5, where we found four degrees of freedom of the *first order*. The fact that the electromagnetic field carries two degrees of free-

---

[6]The *six* initial data $\mathbf{E}(0, \mathbf{x})$ and $\mathbf{B}(0, \mathbf{x})$ are subject to the two constraints $\boldsymbol{\nabla} \cdot \mathbf{E}(0, \mathbf{x}) = j^0(0, \mathbf{x})$ and $\boldsymbol{\nabla} \cdot \mathbf{B}(0, \mathbf{x}) = 0$, so that they amount altogether only to *four* independent fields, as many as the fields in (5.20).

dom has important physical consequences: one is that at the classical level, as we shall see shortly, the electromagnetic waves are characterized by two independent polarization vectors, and another is that the photons, which make up these waves at the quantum level, appear in two different *variants*, or *species*, characterized by opposite *spin* and *helicity*.

5.2 The Wave Equation

Consider a real scalar field $\varphi(x)$ with Lagrangian

$$\mathcal{L} = \frac{1}{2}\,\partial_\mu\varphi\,\partial^\mu\varphi. \tag{5.21}$$

The Euler-Lagrange equation associated with $\mathcal{L}$

$$\partial_\mu\frac{\partial\mathcal{L}}{\partial(\partial_\mu\varphi)} - \frac{\partial\mathcal{L}}{\partial\varphi} = \partial_\mu\partial^\mu\varphi = \Box\,\varphi = 0 \tag{5.22}$$

is called *wave equation*. As this equation plays an important role in physics, especially in Electrodynamics, this section is devoted to a detailed analysis of its general solution. In particular we will find that the solution of Maxwell's equations in empty space is considerably facilitated by the knowledge of the general solution of the wave equation. In analogy to the asymptotic behavior of the electromagnetic vector potential (5.7), we will accept only solutions which satisfy the asymptotic condition

$$\lim_{|\mathbf{x}|\to\infty}\varphi(t,\mathbf{x}) = 0. \tag{5.23}$$

According to (3.68) and (3.71) the canonical energy-momentum tensor associated with the Lagrangian (5.21) is

$$T^{\mu\nu} = \frac{\partial\mathcal{L}}{\partial(\partial_\mu\varphi)}\,\partial^\nu\varphi - \eta^{\mu\nu}\mathcal{L} = \partial^\mu\varphi\partial^\nu\varphi - \frac{1}{2}\,\eta^{\mu\nu}\partial^\alpha\varphi\partial_\alpha\varphi. \tag{5.24}$$

Since φ is a scalar field, the canonical energy momentum tensor $\widetilde{T}^{\mu\nu}$ equals the symmetric one $T^{\mu\nu}$, see Sect. 3.3.2. The knowledge of (5.24) will be essential for the energy analysis of the solutions of the wave equation.

Wave equation and Fourier transform. Assuming that the space-time singularities of $\varphi(x)$ are at most of the distributional type, i.e. assuming that φ belongs to the space of distributions $\mathcal{S}' = \mathcal{S}'(\mathbb{R}^4)$, an efficient method to solve linear partial differential equations, like the wave equation, is offered by the *Fourier transform*: This map constitutes, in fact, a *bijection* from $\mathcal{S}'$ to itself, see Sect. 2.3.1, so that it does neither *loose* nor *add* any solutions. In symbolic notation the Fourier transform and its inverse are given by

$$\widehat{\varphi}(k) = \frac{1}{(2\pi)^2} \int e^{-ik\cdot x}\varphi(x)\,d^4x, \qquad \varphi(x) = \frac{1}{(2\pi)^2} \int e^{ik\cdot x}\widehat{\varphi}(k)\,d^4k, \quad (5.25)$$

respectively, where we have introduced the *dual* variable k^μ, also called *wave vector*, and the notation $k \cdot x = k^\mu x^\nu \eta_{\mu\nu}$. Among the properties of the Fourier transform we will frequently make use of the following ones.

(1) The Fourier transform of a *real* field $\varphi(x)$ satisfies the identity

$$\widehat{\varphi}^*(k) = \widehat{\varphi}(-k). \tag{5.26}$$

To prove it take the complex conjugate of the first relation in (5.25) and use that $\varphi^*(x) = \varphi(x)$.

(2) If we consider k^μ as a four-vector, that is to say, if we postulate that under a Lorentz transformation it transforms as

$$k'^\mu = \Lambda^\mu{}_\nu k^\nu,$$

then, since $\varphi(x)$ is a scalar field, i.e. $\varphi'(x') = \varphi(x)$, so is $\widehat{\varphi}(k)$:

$$\widehat{\varphi}'(k') = \widehat{\varphi}(k).$$

To prove it use the definition

$$\widehat{\varphi}'(k') = \frac{1}{(2\pi)^2} \int e^{-ik'\cdot x'}\varphi'(x')\,d^4x',$$

together with the relations $k' \cdot x' = k \cdot x$ and $d^4x' = d^4x$.

(3) The Fourier transform of a multiple derivative of φ amounts to

$$[\widehat{P(\partial_\mu)\varphi}](k) = P(ik_\mu)\,\widehat{\varphi}(k), \tag{5.27}$$

where $P(\partial_\mu)$ is an arbitrary polynomial of the partial derivatives. This relation is a particular case of the general identity (2.92).

With the aid of (5.27) the Fourier transform of $\Box\,\varphi$ becomes

$$\widehat{\Box\varphi}(k) = \widehat{\partial_\mu\partial^\mu\varphi}(k) = (ik_\mu)(ik^\mu)\,\widehat{\varphi}(k) = -k^2\widehat{\varphi}(k).$$

Since the Fourier transform represents a bijection from $\mathcal{S}'$ to itself, the wave equation goes over to the *equivalent* equation

$$k^2\widehat{\varphi}(k) = \left(\left(k^0\right)^2 - |\mathbf{k}|^2 \right)\widehat{\varphi}(k) = 0. \tag{5.28}$$

In the following we will denote the magnitude of the spatial wave vector $\mathbf{k}$ by the symbol

$$\omega = |\mathbf{k}|,$$

that is going to become the *frequency*. The Fourier transform has so converted the differential equation (5.22) in the *algebraic* equation (5.28) in the space of distributions, which is rather easy to solve. From (5.28) it follows in particular that the support of the distribution $\widehat{\varphi}(k)$ is (contained in) the *light cone* $k^2 = 0$, or

$$k^0 = \pm\omega,$$

so that it cannot be represented by a *function*. The solutions of (5.28) fall into two classes, which we analyze separately below.

Solutions of type I. We start the search for solutions of (5.28) looking, in the first place, at $\widehat{\varphi}(k)$ as a distribution in the variable k^0, while considering its dependence on $\mathbf{k}$ as a *parametric* dependence. Accordingly, the solutions – of type I – of Eq. (5.28) can be written in the form (see Problem 2.3)

$$\widehat{\varphi}_I(k) = \delta(k^2)f(k) = \delta\big((k^0)^2 - \omega^2\big)f(k), \tag{5.29}$$

where $f(k)$ is a complex *regular* function of k^μ, see below. Since $\widehat{\varphi}(k)$ is a scalar field, and $\delta(k^2)$ is Lorentz invariant, also $f(k)$ must be a scalar field. Moreover, the reality condition (5.26) imposes on this function the further constraint

$$f^*(k) = f(-k). \tag{5.30}$$

Using the distributional identity (2.77) the solution (5.29) can be rewritten as

$$\begin{aligned}
\widehat{\varphi}_I(k) &= \frac{1}{2\omega}\left(\delta(k^0 - \omega) + \delta(k^0 + \omega)\right)f(k^0, \mathbf{k}) \\
&= \frac{1}{2\omega}\left(\delta(k^0 - \omega)f(\omega, \mathbf{k}) + \delta(k^0 + \omega)f(-\omega, \mathbf{k})\right) \\
&= \frac{1}{2\omega}\left(\delta(k^0 - \omega)\,\varepsilon(\mathbf{k}) + \delta(k^0 + \omega)\,\varepsilon^*(-\mathbf{k})\right),
\end{aligned} \tag{5.31}$$

where we used (5.30) and defined the complex function of three variables

$$\varepsilon(\mathbf{k}) = f(\omega, \mathbf{k}).$$

Solutions of type II. The solutions (5.31) were derived under the hypothesis of a regular $f(k)$, leading to a regular function $\varepsilon(\mathbf{k})$. In addition, the δ-function in (5.29) – a "function" of k^0 – is ill-defined for $\mathbf{k} = 0$, which on the light cone corresponds to $k^\mu = 0$. Therefore, there could exist further solutions of (5.28), supported at the *point* $k^\mu = 0$, with weaker regularity properties than (5.31). In fact, from Theorem

III of Sect. 2.3.1 on distributions with point-like support, we know that such solutions are necessarily *finite* linear combinations of $\delta^4(k)$ and its derivatives

$$\widehat{\varphi}_{II}(k) = \sum_{n=1}^{N} C^{\mu_1 \cdots \mu_n} \, \partial_{\mu_1} \cdots \partial_{\mu_n} \delta^4(k), \tag{5.32}$$

where $C^{\mu_1 \cdots \mu_n}$ are arbitrary completely symmetric constant tensors. Performing the inverse Fourier transform of this expression – resorting again to (5.27) and using that the Fourier transform of $\delta^4(k)$ equals $1/(2\pi)^2$ – we obtain the polynomials[7]

$$\varphi_{II}(x) = \frac{1}{(2\pi)^2} \sum_{n=1}^{N} (-i)^n \, C^{\mu_1 \cdots \mu_n} x_{\mu_1} \cdots x_{\mu_n}. \tag{5.33}$$

This expression satisfies the wave equation $\Box \, \varphi_{II}(x) = 0$ if and only if the tensors $C^{\mu_1 \cdots \mu_n}$ are traceless

$$C_\nu{}^{\nu \mu_3 \cdots \mu_n} = 0. \tag{5.34}$$

There exist, indeed, infinitely many tensors satisfying these conditions. For $n = 2$, for example, in which case $\varphi_{II}(x)$ is a second-order polynomial, the general solution of (5.34) is of the form

$$C^{\mu\nu} = H^{\mu\nu} - \frac{1}{4} \eta^{\mu\nu} H^\rho{}_\rho,$$

where $H^{\mu\nu}$ is an arbitrary symmetric matrix. In conclusion, the functions $\varphi_{II}(x)$ specified by the relations (5.33) and (5.34) constitute a second class of solutions of the wave equation. However, being polynomials in x^μ, none of these functions vanish at spatial infinity, and so we do *not* admit them as *physical* solutions.

Returning to the solutions of type I (5.31) we evaluate their inverse Fourier transform according to (5.25). Performing in the integral involving $\varepsilon^*(-\mathbf{k})$ the change of variables $\mathbf{k} \to -\mathbf{k}$, we find that these solutions of the wave equation can be written as

$$\varphi(x) = \frac{1}{(2\pi)^2} \int \frac{d^3k}{2\omega} \int dk^0 \, e^{i(k^0 x^0 - \mathbf{k} \cdot \mathbf{x})} \left(\delta(k^0 - \omega) \, \varepsilon(\mathbf{k}) + \delta(k^0 + \omega) \, \varepsilon^*(-\mathbf{k}) \right)$$

$$= \frac{1}{(2\pi)^2} \int \frac{d^3k}{2\omega} \left(e^{ik \cdot x} \varepsilon(\mathbf{k}) + c.c. \right). \tag{5.35}$$

It is understood that the variable k^0 in the exponential $e^{ik \cdot x}$ of the last integral is defined by $k^0 = +\omega$. Finally, one may ask for which class of complex functions $\varepsilon(\mathbf{k})$ the expression (5.35) of $\varphi(x)$ satisfies the asymptotic condition (5.23). Stated

---

[7]As we have shown in Sect. 2.3.1, the inverse Fourier transform of a function, or distribution, can be computed performing its Fourier transform and changing eventually the sign of the argument.

differently, how *regular* must the function $\varepsilon(\mathbf{k})$ be in order that φ represents a *physically* acceptable solution? An answer to this question is provided by the following theorem.

Theorem I (Riemann-Lebesgue lemma). If $f : \mathbb{R}^D \to \mathbb{C}$ is a complex function belonging to the space $L^1(\mathbb{R}^D)$, see the definition (2.101), then its Fourier transform $\widehat{f}(k)$ is a *continuous* function such that

$$\lim_{|k| \to \infty} \widehat{f}(k) = 0.$$

For a fixed t, the integral in (5.35) amounts to the three-dimensional Fourier transform of the function

$$f(\mathbf{k}) = \frac{e^{i\omega t} \varepsilon(\mathbf{k})}{\omega}.$$

According to the Riemann-Lebesgue lemma, the function $\varphi(t, \mathbf{x})$ defined in Eq. (5.35) satisfies, thus, the asymptotic condition (5.23), if the function $|f(\mathbf{k})| = |\varepsilon(\mathbf{k})|/\omega$ is integrable on $\mathbb{R}^3$. In addition, in this case the solution $\varphi(t, \mathbf{x})$ is for all t a continuous function of $\mathbf{x}$. Obviously, the opposite is in general not true. Since a complex function can always be written as

$$\varepsilon(\mathbf{k}) = \varepsilon_1(\mathbf{k}) + i\,\varepsilon_2(\mathbf{k}),$$

the solution (5.35) of the wave equation is uniquely determined by the *two* real functions of three variables $\varepsilon_1(\mathbf{k})$ and $\varepsilon_2(\mathbf{k})$. This result matches with the fact that a real scalar field $\varphi(x)$ subject to the wave equation carries *one* degree of freedom, being likewise uniquely determined by the two real functions $\varphi(0, \mathbf{x})$ and $\partial_0 \varphi(0, \mathbf{x})$. In particular, as we will see in Sect. 5.2.2, the functions $\varepsilon_1(\mathbf{k})$ and $\varepsilon_2(\mathbf{k})$ can be written explicitly in terms of the initial data $\varphi(0, \mathbf{x})$ and $\partial_0 \varphi(0, \mathbf{x})$, and vice versa.

5.2.1 Elementary Scalar Waves

The general solution (5.35) can formally be considered as a *continuous* superposition of an infinite number of *elementary* waves with fixed wave vector k^μ and frequency ω, given by

$$\varphi_{\mathrm{el}}(x) = \varepsilon(\mathbf{k}) e^{ik \cdot x} + c.c., \qquad k^0 = \omega. \tag{5.36}$$

Below we report the main properties of these waves.

- The functions φ_{el} represent waves propagating at the *speed of light* along the direction of $\mathbf{k}$. Choosing $\mathbf{x} \parallel \mathbf{k}$ its *phase* can in fact be written as

$$k \cdot x = \omega t - \mathbf{k} \cdot \mathbf{x} = \omega(t - |\mathbf{x}|).$$

- The functions $\varphi_{\rm el}$ constitute *plane* waves, whose *phase planes* are orthogonal to **k**. For fixed t on such a plane the functions $\varphi_{\rm el}(t, \mathbf{x})$ take in fact the same value.
- The functions $\varphi_{\rm el}$ are *monochromatic* waves in that they possess a fixed frequency ω. Their *period* and *wavelength* are given by $T = 2\pi/\omega$ and $\lambda = 2\pi/\omega$, respectively.
- The functions $\varphi_{\rm el}$ represent *scalar* waves. With this we mean that the *polarization tensor* ε, that identifies its *amplitude*, is actually a Lorentz-scalar.
- The *energy content* of the elementary waves is codified by the energy-momentum tensor (5.24), whose evaluation requires the derivatives

$$\partial_\mu \varphi_{\rm el} = ik_\mu \varepsilon(\mathbf{k}) e^{ik\cdot x} + c.c. \tag{5.37}$$

We can recast them in a more compact form if we introduce the light-like vector

$$n^\mu = \frac{k^\mu}{\omega}, \quad n^0 = 1, \quad \mathbf{n} = \frac{\mathbf{k}}{\omega}, \quad n_\mu n^\mu = 0, \tag{5.38}$$

where the unit vector **n** identifies the propagation direction of the wave. Equation (5.37) then translate in the so-called *wave relations* (see Sect. 5.3.1)

$$\partial_\mu \varphi_{\rm el} = n_\mu \dot\varphi_{\rm el}. \tag{5.39}$$

It follows in particular that the quantity $\partial^\mu \varphi_{\rm el} \partial_\mu \varphi_{\rm el}$ vanishes. Substituting (5.39) in (5.24) we then find

$$T^{\mu\nu} = n^\mu n^\nu \dot\varphi_{\rm el}^2 = n^\mu n^\nu \omega^2 \left(2|\varepsilon|^2 - \varepsilon^2 e^{2ik\cdot x} - \varepsilon^{*2} e^{-2ik\cdot x}\right). \tag{5.40}$$

Taking the average of the energy-momentum tensor over time scales large with respect to the period T and over volume scales large with respect to the wavelength λ the (oscillating) exponentials drop out, and we obtain the simple expression

$$\langle T^{\mu\nu} \rangle = 2n^\mu n^\nu \omega^2 |\varepsilon|^2.$$

The mean energy density of the wave $\langle T^{00} \rangle = 2\omega^2 |\varepsilon|^2$ is hence proportional to the square of the amplitude, while its mean energy flux $\langle T^{0i} \rangle = 2\omega^2 |\varepsilon|^2 n^i$ is directed along the propagation direction. Finally, we evaluate the four-momentum P^μ contained in a small volume V, but large with respect to λ,

$$P^0 = \langle T^{00} \rangle V = 2\omega^2 |\varepsilon|^2 V, \qquad P^i = \langle T^{0i} \rangle V = 2\omega^2 |\varepsilon|^2 V n^i. \tag{5.41}$$

If we assume that the "radiation" in a volume V is associated with a *particle* with four-momentum P^μ, these expressions imply that the mass M of this particle vanishes:

$$M^2 = P^\mu P_\mu = \left(2\omega^2 |\varepsilon|^2 V\right)^2 \left(1 - |\mathbf{n}|^2\right) = 0.$$

This observation matches with the fact that in *quantum* field theory a real scalar field subject to the equation $\Box\, \varphi = 0$ is associated with a neutral and spinless *massless* particle.

5.2.2 Initial Value Problem

In general an *initial value problem*, or *Cauchy problem*, consists in the determination of the solution of a differential equation, or of a system of differential equations, subject to appropriate boundary conditions. In this section we want to determine explicitly the solution of the wave equation $\Box\, \varphi = 0$, subject to the initial data

$$\varphi(0, \mathbf{x}) = f(\mathbf{x}), \tag{5.42}$$

$$\partial_0 \varphi(0, \mathbf{x}) = h(\mathbf{x}). \tag{5.43}$$

In the light of the general solution (5.35) of the wave equation our task reduces to the determination of the complex function $\varepsilon(\mathbf{k})$ in terms of the real functions $f(\mathbf{x})$ and $h(\mathbf{x})$. For this purpose it is convenient to write the latter in terms of their Fourier transforms

$$f(\mathbf{x}) = \frac{1}{(2\pi)^{3/2}} \int d^3k \, e^{-i\mathbf{k}\cdot\mathbf{x}} \widehat{f}(\mathbf{k}), \quad h(\mathbf{x}) = \frac{1}{(2\pi)^{3/2}} \int d^3k \, e^{-i\mathbf{k}\cdot\mathbf{x}} \widehat{h}(\mathbf{k}), \tag{5.44}$$

and to compare them with the expression (5.35) and its time derivative, evaluated at $t = 0$,

$$f(\mathbf{x}) = \varphi(0, \mathbf{x}) = \frac{1}{(2\pi)^2} \int \frac{d^3k}{2\omega} \left(e^{-i\mathbf{k}\cdot\mathbf{x}} \, \varepsilon(\mathbf{k}) + c.c. \right), \tag{5.45}$$

$$h(\mathbf{x}) = \partial_0 \varphi(0, \mathbf{x}) = \frac{1}{(2\pi)^2} \int \frac{d^3k}{2\omega} \left(i\omega \, e^{-i\mathbf{k}\cdot\mathbf{x}} \, \varepsilon(\mathbf{k}) + c.c. \right). \tag{5.46}$$

From the comparison we recover the equations

$$\widehat{f}(\mathbf{k}) = \frac{1}{\sqrt{2\pi}} \frac{1}{2\omega} \left(\varepsilon(\mathbf{k}) + \varepsilon^*(-\mathbf{k}) \right),$$

$$\widehat{h}(\mathbf{k}) = \frac{1}{\sqrt{2\pi}} \frac{i}{2} \left(\varepsilon(\mathbf{k}) - \varepsilon^*(-\mathbf{k}) \right),$$

providing the relation we searched for

$$\varepsilon(\mathbf{k}) = \sqrt{(2\pi)} \left(\omega \widehat{f}(\mathbf{k}) - i \widehat{h}(\mathbf{k}) \right).$$

Inserting it in (5.35) we obtain

$$\varphi(x) = \frac{1}{(2\pi)^{3/2}} \int \frac{d^3k}{2\omega} \left(e^{ik\cdot x} \left(\omega \widehat{f}(\mathbf{k}) - i\,\widehat{h}(\mathbf{k}) \right) + c.c. \right). \tag{5.47}$$

Finally, we must invert the Fourier representations (5.44) to express $\widehat{f}(\mathbf{k})$ and $\widehat{h}(\mathbf{k})$ in terms of the known functions $f(\mathbf{x}) = \varphi(0, \mathbf{x})$ and $h(\mathbf{x}) = \partial_0 \varphi(0, \mathbf{x})$. Substituting the resulting expressions in (5.47) we find the formula solving our task, see Problem 5.1,

$$\varphi(t, \mathbf{x}) = \int \left(D(t, \mathbf{x} - \mathbf{y})\, \partial_0 \varphi(0, \mathbf{y}) + \partial_0 D(t, \mathbf{x} - \mathbf{y})\, \varphi(0, \mathbf{y}) \right) d^3y. \tag{5.48}$$

Antisymmetric kernel. In Eq. (5.48) we have introduced the *antisymmetric kernel* $D(x)$, a distribution in $\mathcal{S}'(\mathbb{R}^4)$, defined, in symbolic notation, by

$$D(t, \mathbf{x}) = \frac{1}{(2\pi)^3} \int \frac{d^3k}{2\omega i} \left(e^{ik\cdot x} - e^{-ik\cdot x} \right) = \frac{1}{(2\pi)^3} \int d^3k\, \frac{sen(\omega t)}{\omega}\, e^{i\mathbf{k}\cdot\mathbf{x}}, \tag{5.49}$$

where $k^0 = \omega$. The three-dimensional Fourier transforms appearing in these formulas have, in fact, to be understood in the distributional sense. Performing them explicitly one finds, see Problem 5.9,

$$D(t, \mathbf{x}) = \frac{1}{4\pi r} \left(\delta(t - r) - \delta(t + r) \right) = \frac{1}{2\pi} \varepsilon(t)\, \delta(x^2), \tag{5.50}$$

where $\varepsilon(\,\cdot\,)$ denotes the *sign* function, and $r = |\mathbf{x}|$. These equivalent representations yield the properties

$$\Box\, D(t, \mathbf{x}) = 0, \tag{5.51}$$

$$D(0, \mathbf{x}) = 0, \tag{5.52}$$

$$\partial_0 D(0, \mathbf{x}) = \delta^3(\mathbf{x}), \tag{5.53}$$

$$D(-t, \mathbf{x}) = -D(t, \mathbf{x}). \tag{5.54}$$

To show that $D(x)$ satisfies the wave equation (5.51) it is convenient to resort to the first formula in (5.49), swapping the derivatives with the integral sign and using that $k^2 = 0$. Property (5.52) follows from the second formula in (5.49). To prove (5.53) one can apply the time derivative to the second formula in (5.49)

$$\partial_0 D(t, \mathbf{x}) = \frac{1}{(2\pi)^3} \int \cos(\omega t)\, e^{i\mathbf{k}\cdot\mathbf{x}}\, d^3k,$$

evaluate it at $t = 0$ and use the formal identity (2.98). Property (5.54), following likewise from the second formula in (5.49), is the one that identifies $D(x)$ as *antisymmetric* kernel.

With the aid of (5.51)–(5.54) it can be explicitly verified that the expression (5.48) satisfies the wave equation and entails the correct initial data. First of all property (5.51) ensures that (5.48) satisfies the wave equation. Next, evaluating (5.48) at $t = 0$ and using (5.52) and (5.53) we retrieve $\varphi(0, \mathbf{x})$. Finally, by taking the time derivative of (5.48) and setting $t = 0$ we obtain

$$\partial_0 \varphi(0, \mathbf{x}) = \int \left(\partial_0 D(0, \mathbf{x} - \mathbf{y}) \, \partial_0 \varphi(0, \mathbf{y}) + \partial_0^2 D(0, \mathbf{x} - \mathbf{y}) \, \varphi(0, \mathbf{y}) \right) d^3 y.$$

Thanks to (5.53) the integral of the first term reduces to $\partial_0 \varphi(0, \mathbf{x})$. On the other hand, the integral of the second term vanishes, since (5.51) evaluated at $t = 0$ gives

$$\partial_0^2 D(0, \mathbf{x}) = \nabla^2 D(0, \mathbf{x}) = 0,$$

where in the last step we used (5.52).

5.2.3 Manifestly Invariant Cauchy Problem

In this section we present a manifestly Lorentz-invariant version of the solution (5.48), specifying the initial conditions for the field φ on an arbitrary space-like hypersurface Γ. On purpose, for simplicity, we disregard the regularity properties of the involved fields, so that the analysis will have, partially, formal character. We begin by showing that in addition to the properties (5.51)–(5.54) the antisymmetric kernel (5.50) is invariant under *proper* Lorentz transformations

$$D(\Lambda x) = D(x), \quad \forall \Lambda \in SO(1, 3)_c. \tag{5.55}$$

The factor $\delta(x^2)$ in (5.50) is manifestly invariant. It remains therefore to show that $\varepsilon(t)$ – the sign of t – restricted to the light cone $x^2 = 0$ is invariant under $SO(1, 3)_c$. This property follows from the following theorem.

Theorem II. Given a non space-like event $x^\mu = (t, \mathbf{x})$, that is to say, an event obeying the inequality

$$x^2 = t^2 - |\mathbf{x}|^2 \geq 0 \quad \Leftrightarrow \quad |t| \geq |\mathbf{x}|, \tag{5.56}$$

the sign of t is invariant under proper Lorentz transformations.

Proof We begin the proof recalling that if $\Lambda \in SO(1, 3)_c$, then $\Lambda^0{}_0 \geq 1$. In particular, the condition $\Lambda^\mu{}_\alpha \Lambda^\nu{}_\beta \eta^{\alpha\beta} = \eta^{\mu\nu}$ for $\mu = \nu = 0$ yields the relation

$$(\Lambda^0{}_0)^2 = 1 + |\mathbf{L}|^2, \quad L^i \equiv \Lambda^0{}_i. \tag{5.57}$$

On the other hand, for $t \neq 0$ the transformed time can be recast in the form

$$t' = \Lambda^0{}_0 t + \Lambda^0{}_i x^i = \Lambda^0{}_0 t + \mathbf{L} \cdot \mathbf{x} = \Lambda^0{}_0 t \left(1 + \frac{\mathbf{L} \cdot \mathbf{x}}{\Lambda^0{}_0 t} \right). \tag{5.58}$$

From (5.56) and (5.57) we derive the inequality

$$\left| \frac{\mathbf{L} \cdot \mathbf{x}}{\Lambda^0{}_0 t} \right| \le \frac{|\mathbf{L}| \cdot |\mathbf{x}|}{|t| \sqrt{1 + |\mathbf{L}|^2}} \le \frac{|\mathbf{L}|}{\sqrt{1 + |\mathbf{L}|^2}} < 1,$$

leading to

$$1 + \frac{\mathbf{L} \cdot \mathbf{x}}{\Lambda^0{}_0 t} > 0.$$

Since $\Lambda^0{}_0$ is positive, Eq. (5.58) then implies that t' and t have the same sign.   □

Relativistic invariance of the light cones. A corollary of Theorem II is that the *future light cone* L_+ and the *past light cone* L_-, that is to say, the two sets of four-vectors

$$L_+ = \{V^\mu \in \mathbb{R}^4 / V^2 \ge 0, \, V^0 > 0\}, \quad L_- = \{V^\mu \in \mathbb{R}^4 / V^2 \ge 0, \, V^0 < 0\},$$

are invariant under proper Lorentz transformations. We shall make extensive use of this important result in Chap. 6.

Property (5.55) allows to construct an invariant version of the Cauchy-solution (5.48), generalizing it to the case where the initial data of φ, or better its *boundary conditions*, are specified on a generic space-like hypersurface Γ, see Sect. 3.2.1 for the notation. In parametric form such a hypersurface is described by the four functions of three variables $y^\mu(\boldsymbol{\lambda})$, $\boldsymbol{\lambda} = (\lambda^1, \lambda^2, \lambda^3)$, and it is characterized by a time-like normal vector $n_\mu(\boldsymbol{\lambda})$ that can be normalized to unity, $n_\mu(\boldsymbol{\lambda}) n^\mu(\boldsymbol{\lambda}) = 1$. If we specify on Γ the values of φ and of its normal derivative setting

$$\varphi(y(\boldsymbol{\lambda})) = f(\boldsymbol{\lambda}), \quad n^\mu(\boldsymbol{\lambda}) \partial_\mu \varphi(y(\boldsymbol{\lambda})) = h(\boldsymbol{\lambda}), \tag{5.59}$$

the invariant version of (5.48) reads

$$\begin{aligned}
\varphi(x) &= \int_\Gamma \left(D(x-y) \, \partial^\mu \varphi(y) + \partial^\mu D(x-y) \, \varphi(y) \right) d\Sigma_\mu \\
&= \int_\Gamma \left(D(x-y) h(\boldsymbol{\lambda}) + n_\mu(\boldsymbol{\lambda}) \, \partial^\mu D(x-y) \, \varphi(y) \right) \sqrt{g(\boldsymbol{\lambda})} \, d^3\lambda,
\end{aligned} \tag{5.60}$$

where y^μ stands for $y^\mu(\boldsymbol{\lambda})$, and the surface element is given by $d\Sigma_\mu = n_\mu(\boldsymbol{\lambda}) \sqrt{g(\boldsymbol{\lambda})} \, d^3\lambda$, see Sect. 3.2.1. The integral (5.60) is certainly a solution of the wave equation, because the antisymmetric kernel is so. Furthermore, resorting to the covariant version of the identities (5.52) and (5.53)

$$D(x) = 0, \quad \text{for} \quad x^2 < 0,$$

$$\partial_\mu D(y(\boldsymbol{\lambda}') - y(\boldsymbol{\lambda})) = n_\mu(\boldsymbol{\lambda}) \frac{\delta^3(\boldsymbol{\lambda}' - \boldsymbol{\lambda})}{\sqrt{g(\boldsymbol{\lambda})}},$$

it can also be shown that (5.60) satisfies the boundary conditions (5.59). Instead of working out the details we provide an indirect prove of this property. Below we shall, in fact, show that the integral (5.60) is independent of the space-like hypersurface Γ. We can then choose for Γ the hyperplane $t = 0$, in which case

$$y^\mu(\boldsymbol{\lambda}) = (0, \lambda^1, \lambda^2, \lambda^3), \quad n^\mu(\boldsymbol{\lambda}) = (1, 0, 0, 0), \quad g(\boldsymbol{\lambda}) = 1,$$

and so (5.60) reduces to (5.48). Due to the uniqueness of the solution of the wave equation this amounts to an indirect proof of the validity of (5.59).

Independence of the invariant solution (5.60) ***from*** Γ. We introduce the vector field in the variables x and y

$$W^\mu(x, y) = D(x - y)\, \partial^\mu \varphi(y) + \partial^\mu D(x - y)\, \varphi(y), \tag{5.61}$$

and suppose that φ satisfies the wave equation. This vector field has then vanishing four-divergence with respect to y. In fact, leaving out the arguments we obtain

$$\frac{\partial}{\partial y^\mu} W^\mu(x, y) = -\partial_\mu D \partial^\mu \varphi + D\Box\varphi - \Box D\varphi + \partial_\mu D \partial^\mu \varphi = D\Box\varphi - \Box D\varphi = 0,$$

in that $\Box\varphi = 0 = \Box D$. Integrating this equation over a four-dimensional volume V whose boundary is composed of two non-intersecting space-like hypersurfaces Γ_1 and Γ_2, parameterized by $y_1^\mu(\boldsymbol{\lambda})$ and $y_2^\mu(\boldsymbol{\lambda})$, respectively, and of a time-like hypersurface Γ_∞ placed at spatial infinity, we obtain

$$\int_V \partial_\mu W^\mu(x, y)\, d^4 y = 0.$$

Gauss's theorem in four dimensions then yields

$$\int_{\Gamma_2} W^\mu(x, y_2)\, d\Sigma_\mu - \int_{\Gamma_1} W^\mu(x, y_1)\, d\Sigma_\mu + \int_{\Gamma_\infty} W^\mu(x, y_\infty)\, d\Sigma_\mu = 0.$$

Since at spatial infinity φ vanishes sufficiently fast the third integral is zero, and we conclude that

$$\int_{\Gamma_2} W^\mu(x, y_2)\, d\Sigma_\mu = \int_{\Gamma_1} W^\mu(x, y_1)\, d\Sigma_\mu.$$

Given the definition (5.61) of W^μ this proves that the integral (5.60) is independent of Γ.

5.3 Maxwell's Equations in Empty Space

The main goal of this section is to find the general physical solution of Maxwell's equations in the absence of sources, which according to (5.5) are equivalent to

$$\partial_\mu F^{\mu\nu} = 0, \qquad F^{\mu\nu} = \partial^\mu A^\nu - \partial^\nu A^\mu, \qquad A^\mu \approx A^\mu + \partial^\mu \Lambda. \tag{5.62}$$

An electromagnetic field satisfying these equations is called a *free field*, or also a *radiation field*. Our goal is, therefore, to find the explicit expression of a generic radiation field. Since the system (5.62) is linear in the fields, and we consider the field strength $F^{\mu\nu}$ as well as the vector potential A^μ as distributions, the most powerful technique for its solution is, once more, the *Fourier transform*.

The solution of the system (5.62) requires first of all an appropriate gauge-fixing condition. As exemplified in Sect. 5.1.3, a convenient choice is provided by the covariant *Lorenz gauge* $\partial_\mu A^\mu = 0$, giving rise to the residual gauge invariance (5.14). Accordingly, the system (5.62) reduces to the system of differential equations

$$\Box A^\mu = 0, \tag{5.63}$$
$$\partial_\mu A^\mu = 0, \tag{5.64}$$
$$A^\mu \approx A^\mu + \partial^\mu \Lambda, \qquad \Box \Lambda = 0. \tag{5.65}$$

The Fourier transforms of the vector potential and of the gauge parameter are

$$\widehat{A}^\mu(k) = \frac{1}{(2\pi)^2} \int e^{-ik\cdot x} A^\mu(x)\, d^4 x, \qquad \widehat{\Lambda}(k) = \frac{1}{(2\pi)^2} \int e^{-ik\cdot x} \Lambda(x)\, d^4 x.$$

The main difference between $\widehat{A}^\mu(k)$ and the Fourier transform $\widehat{\varphi}(k)$ of a scalar field is represented by the fact that under a Lorentz transformation $\widehat{A}^\mu(k)$ transforms as a *vector-field* (see Problem 5.2)

$$\widehat{A}'^\mu(k') = \Lambda^\mu{}_\nu \widehat{A}^\nu(k).$$

The Fourier transform maps the system (5.63)–(5.65) into the system of algebraic equations

$$k^2 \widehat{A}^\mu(k) = 0, \tag{5.66}$$
$$k_\mu \widehat{A}^\mu(k) = 0, \tag{5.67}$$
$$\widehat{A}^\mu(k) \approx \widehat{A}^\mu(k) + ik^\mu \widehat{\Lambda}(k), \qquad k^2 \widehat{\Lambda}(k) = 0. \tag{5.68}$$

The general solution of equation (5.66) can be determined as in the case of the scalar field, see (5.29) and (5.31), the only difference being that the *weight* function is now a complex four-vector $f^\mu(k)$:

$$\widehat{A}^\mu(k) = \delta(k^2)f^\mu(k) = \frac{1}{2\omega}\left(\delta(k^0 - \omega)\,\varepsilon^\mu(\mathbf{k}) + \delta(k^0 + \omega)\,\varepsilon^{*\mu}(-\mathbf{k})\right). \quad (5.69)$$

We have set

$$\varepsilon^\mu(\mathbf{k}) = f^\mu(\omega, \mathbf{k})$$

and we have introduced again the frequency $\omega = |\mathbf{k}|$. As opposed to the case of the scalar field, where $\varepsilon(\mathbf{k})$ was a four-scalar, $\varepsilon^\mu(\mathbf{k})$ is a four-vector called *polarization vector*. Equation (5.67) imposes on this vector the *transversality condition*

$$k_\mu \varepsilon^\mu = 0, \qquad\qquad (5.70)$$

with $k^0 = \omega$. Similarly, the general solution of the equation $k^2 \widehat{\Lambda}(k) = 0$ in (5.68) is

$$\widehat{\Lambda}(k) = \frac{1}{2\omega i}\left(\delta(k^0 - \omega)\lambda(\mathbf{k}) - \delta(k^0 + \omega)\lambda^*(-\mathbf{k})\right),$$

in which for notational reasons we introduced a factor $1/i$. The equivalence relation in (5.68) then implies that the polarization vectors, apart from being subject to the constraint (5.70), are defined modulo the residual gauge transformations

$$\varepsilon^\mu \to \varepsilon^\mu + k^\mu \lambda. \qquad\qquad (5.71)$$

Clearly, thanks to the relations $k^2 = 0 = k_\mu \varepsilon^\mu$, these transformations preserve the Lorenz gauge (5.70)

$$k_\mu(\varepsilon^\mu + k^\mu \lambda) = 0. \qquad\qquad (5.72)$$

We realize, therefore, that out of the four complex components of the polarization vector, only *two* are physically significant: one component is eliminated by the Lorenz gauge (5.70), and the other by means of the residual gauge transformation (5.71). Clearly, this counting reflects the fact that the electromagnetic field entails two physical degrees of freedom.

Performing the inverse Fourier transform of (5.69) we find the vector potential, representing the general physical solution of the system (5.63)–(5.65),

$$A^\mu(x) = \frac{1}{(2\pi)^2} \int \frac{d^3k}{2\omega}\left(e^{ik\cdot x}\varepsilon^\mu(\mathbf{k}) + c.c.\right). \qquad\qquad (5.73)$$

Consequently, for the electromagnetic field – *general solution of Maxwell's equations in empty space* – we find

$$F^{\mu\nu} = \partial^\mu A^\nu - \partial^\nu A^\mu = \frac{1}{(2\pi)^2} \int \frac{d^3k}{2\omega}\left(ie^{ik\cdot x}\left(k^\mu\varepsilon^\nu - k^\nu\varepsilon^\mu\right) + c.c.\right), \quad (5.74)$$

which is hence the sought expression of a generic radiation field. Notice that, while the potential (5.73) is still sensitive to the residual gauge transformation (5.71), the field (5.74) is instead *invariant*. Introducing for the integration variable k polar coordinates $(\omega, \varphi, \vartheta)$, with $d^3k = \omega^2 d\omega\, d\Omega$, we can rewrite (5.74) as

$$F^{\mu\nu}(t, \mathbf{x}) = \frac{i}{2(2\pi)^2} \int_0^\infty d\omega\, \omega\, e^{i\omega t} \int d\Omega \left(e^{-i\mathbf{k}\cdot\mathbf{x}} \left(k^\mu \varepsilon^\nu - k^\nu \varepsilon^\mu \right) \right) + c.c., \quad (5.75)$$

which can be recast in the form

$$F^{\mu\nu}(t, \mathbf{x}) = \frac{1}{\sqrt{2\pi}} \int_{-\infty}^\infty e^{i\omega t} F^{\mu\nu}(\omega, \mathbf{x})\, d\omega. \quad (5.76)$$

So we see that the Fourier transform of $F^{\mu\nu}(x)$ with respect to the time variable – i.e. the quantity $F^{\mu\nu}(\omega, \mathbf{x})$ – represents the *weight* with which a frequency ω appears in the superposition of elementary waves composing the generic *radiation field* (5.74). We will resort to this *duality property* when we will analyze the energy content of electromagnetic radiation "frequency by frequency", i.e. when performing its *spectral analysis*, see Chap. 11.

5.3.1 Elementary Electromagnetic Waves

From (5.73) we infer that the vector potential associated with the general solution of Maxwell's equation in empty space amounts to a superposition of *elementary* electromagnetic waves, with a fixed wave vector k^μ subject to $k^2 = 0$, given by

$$A_{\text{el}}^\mu(x) = \varepsilon^\mu e^{ik\cdot x} + c.c., \quad k^0 = \omega, \quad k_\mu \varepsilon^\mu = 0, \quad \varepsilon^\mu \approx \varepsilon^\mu + k^\mu \lambda. \quad (5.77)$$

From Sect. 5.2 we know that these waves are *plane* and *monochromatic*, and that they propagate at the *speed of light*. However, these waves are no longer *scalar* waves in that the polarization tensor ε^μ is now a vector.

Wave relations. To derive the additional characteristics implied by the vectorial nature of these waves, i.e. properties (1)–(5) below, it is convenient to derive first a compact expression for the derivatives of A_{el}^μ. To keep the notation simple, henceforth we write A^μ instead of A_{el}^μ. Differentiating (5.77) with respect to x we find

$$\partial_\mu A^\nu = ik_\mu \varepsilon^\nu e^{ik\cdot x} + c.c. \quad (5.78)$$

As in the case of the scalar waves we introduce the light-like vector

$$n^\mu = \frac{k^\mu}{\omega}, \quad n^0 = 1, \quad \mathbf{n} = \frac{\mathbf{k}}{\omega}, \quad (5.79)$$

where the unit vector **n** denotes the propagation direction of the wave. Equations (5.77)–(5.79) then imply the *wave relations*

$$\partial_\mu A^\nu = n_\mu \dot{A}^\nu, \qquad n_\mu \dot{A}^\mu = 0, \qquad n^\mu n_\mu = 0. \tag{5.80}$$

These relations are preserved by the residual gauge transformations (5.71), which for an elementary wave (5.77) amount to the replacements

$$A^\mu \rightarrow A^\mu + n^\mu \varphi, \tag{5.81}$$

where φ is an arbitrary elementary scalar wave (5.36). We shall base the proof of properties (1), (2) and (5) below on the wave relations (5.80), rather than on the explicit expressions (5.77), for a reason that will be clarified at the end of the section.

(1) Transversality. The elementary electromagnetic waves (5.77) are *transverse*, that is to say, the related electric and magnetic fields are orthogonal to their propagation direction

$$\mathbf{n} \cdot \mathbf{E} = 0 = \mathbf{n} \cdot \mathbf{B}. \tag{5.82}$$

To prove (5.82) we determine the electromagnetic field of the wave using (5.80),

$$F^{\mu\nu} = \partial^\mu A^\nu - \partial^\nu A^\mu = n^\mu \dot{A}^\nu - n^\nu \dot{A}^\mu, \tag{5.83}$$

and rewrite the relation $n_\mu \dot{A}^\mu = 0$ as

$$\dot{A}^0 = \mathbf{n} \cdot \dot{\mathbf{A}}. \tag{5.84}$$

Equation (5.83) then gives

$$E^i = F^{i0} = n^i \dot{A}^0 - \dot{A}^i = (\mathbf{n} \cdot \dot{\mathbf{A}})\, n^i - \dot{A}^i, \tag{5.85}$$

$$B^i = -\frac{1}{2}\, \varepsilon^{ijk} F^{jk} = -\varepsilon^{ijk} n^j \dot{A}^k, \tag{5.86}$$

which by inspection satisfy Eqs. (5.82).

(2) Relations between **E** *and* **B**. The electric and magnetic fields have the same magnitude and are orthogonal to each other

$$E = B, \qquad \mathbf{E} \cdot \mathbf{B} = 0. \tag{5.87}$$

To prove these relations it is convenient to use the relativistic invariants (2.15)

$$\varepsilon^{\alpha\beta\gamma\delta} F_{\alpha\beta} F_{\gamma\delta} = -8\, \mathbf{E} \cdot \mathbf{B}, \qquad F^{\alpha\beta} F_{\alpha\beta} = 2\left(B^2 - E^2\right).$$

In fact, inserting the expressions (5.83) one finds that both invariants vanish: the first for the antisymmetry of the Levi-Civita tensor, and the second thanks to the wave relations. Properties (1) and (2) can be summarized in the formulas

$$\mathbf{B} = \mathbf{n} \times \mathbf{E}, \qquad \mathbf{n} \cdot \mathbf{E} = 0. \tag{5.88}$$

The magnetic field is, hence, uniquely determined by the electric field, and vice versa.

(3) Two physical polarization states. As anticipated after Eq. (5.72), for every fixed $\mathbf{k}$ there exist two linearly independent physical polarization states. To analyze them more in detail we choose as z axis the propagation direction of the wave, so that the wave vector becomes

$$k^\mu = (\omega, 0, 0, \omega).$$

The condition $k_\mu \varepsilon^\mu = \omega(\varepsilon^0 - \varepsilon^3) = 0$ then implies that $\varepsilon^\mu = (\varepsilon^0, \varepsilon^1, \varepsilon^2, \varepsilon^0)$. The polarization vector can therefore be considered as a superposition of the non-physical *longitudinal* state

$$\varepsilon_L^\mu = (\varepsilon^0, 0, 0, \varepsilon^0)$$

and of the two physical *transverse* states

$$\varepsilon_T^\mu = (0, \varepsilon^1, \varepsilon^2, 0). \tag{5.89}$$

Under a residual gauge transformation (5.71) the transverse states are invariant, while the longitudinal one changes according to

$$\varepsilon'^\mu = \varepsilon^\mu + \lambda k^\mu = (\varepsilon^0 + \lambda\omega, \varepsilon^1, \varepsilon^2, \varepsilon^0 + \lambda\omega) = \varepsilon_T^\mu + \left(1 + \frac{\lambda\omega}{\varepsilon^0}\right)\varepsilon_L^\mu. \tag{5.90}$$

Hence, we can always eliminate ε_L^μ choosing $\lambda = -\varepsilon^0/\omega$, a choice that amounts to the residual gauge-fixing $\varepsilon'^0 = 0$. However, sometimes it is convenient to keep the longitudinal state alive, in that its *virtual* presence can be exploited to control the correctness of certain computations: The *observable* quantities must, in fact, be insensitive to the presence of the longitudinal state, and so they must be invariant under residual gauge transformations. As an example we verify the invariance of the electric and magnetic fields (5.85) and (5.86), certainly observable quantities. Performing the transformation (5.81), i.e. sending $\mathbf{A} \to \mathbf{A} + \varphi\mathbf{n}$, we obtain in fact

$$E^i \to E^i + (\mathbf{n} \cdot \dot\varphi\,\mathbf{n})\,n^i - \dot\varphi\,n^i = E^i,$$
$$B^i \to B^i - \varepsilon^{ijk}n^j n^k \dot\varphi = B^i.$$

(4) Linear, circular and elliptical polarization. Inserting (5.77) in (5.85) we find that the electric field of an elementary wave has the general form

$$\mathbf{E}(t, \mathbf{x}) = \mathcal{E}\, e^{ik \cdot x} + \mathcal{E}^* e^{-ik \cdot x} = \cos(k \cdot x)\, \mathbf{V}_1 + \sin(k \cdot x)\, \mathbf{V}_2, \qquad (5.91)$$

where

$$\mathcal{E} = \frac{1}{2}\, (\mathbf{V}_1 - i\mathbf{V}_2)$$

is an arbitrary complex vector orthogonal to $\mathbf{n}$, and $\mathbf{V}_1$ and $\mathbf{V}_2$ arbitrary real vectors, likewise orthogonal to $\mathbf{n}$. The expression (5.91) of $\mathbf{E}$ depends, thus, on four arbitrary real parameters, which are in one-to-one correspondence with the two complex physical polarizations (5.89) of the vector potential (5.77). The *polarization properties* of an elementary wave are related to the constraints satisfied by the complex vector $\mathcal{E}$ or, equivalently, to the links existing between the vectors $\mathbf{V}_1$ and $\mathbf{V}_2$. A wave is said to be *linearly polarized*, if the direction of $\mathbf{E}$ does not change with time, i.e. if

$$\mathbf{V}_1 \parallel \mathbf{V}_2 \qquad \Leftrightarrow \qquad \mathcal{E} = e^{i\gamma}\mathbf{U}, \ \text{with } \gamma \text{ and } \mathbf{U} \text{ real.} \qquad (5.92)$$

A wave is said to be *circularly polarized*, if at fixed $\mathbf{x}$ as t varies the tip of $\mathbf{E}$ describes a circle, i.e. if

$$\mathbf{V}_1 \perp \mathbf{V}_2 \ \ \text{and} \ \ |\mathbf{V}_1| = |\mathbf{V}_2| \qquad \Leftrightarrow \qquad \mathbf{n} \times \mathcal{E} = \pm i\,\mathcal{E}. \qquad (5.93)$$

If we choose the z axis along $\mathbf{n}$, this condition amounts to $\mathcal{E}^x = \pm i\,\mathcal{E}^y$, see Problem 5.5. A circular polarization is said to be *clockwise* (*counter-clockwise*), if the tip of $\mathbf{E}$ runs along the circle clockwise (counter-clockwise) or, equivalently, if the vector $\mathbf{V}_2 \times \mathbf{V}_1$ is parallel (antiparallel) to $\mathbf{n}$. Finally, if $\mathcal{E}$ is a generic complex vector orthogonal to $\mathbf{n}$, the wave is said to be *elliptically polarized*. Actually, in this case the wave does not possess any particular polarization properties. Finally, if we introduce the quantities

$$\mathbf{W}_1 = \cos\alpha \mathbf{V}_1 - \sin\alpha \mathbf{V}_2, \ \ \mathbf{W}_2 = \sin\alpha \mathbf{V}_1 + \cos\alpha \mathbf{V}_2, \ \ \tan 2\alpha = \frac{2\mathbf{V}_1 \cdot \mathbf{V}_2}{V_2^2 - V_1^2},$$

the electric field (5.91) can be recast in the form

$$\mathbf{E}(t, \mathbf{x}) = \cos(k \cdot x + \alpha)\mathbf{W}_1 + \sin(k \cdot x + \alpha)\mathbf{W}_2, \quad \text{where } \mathbf{W}_1 \perp \mathbf{W}_2. \ (5.94)$$

In this representation the wave is linearly polarized if either $\mathbf{W}_1$ or $\mathbf{W}_2$ vanish, while it is circularly polarized if these vectors have the same magnitude, $W_1 = W_2$. From (5.94) we infer, furthermore, that in general the tip of the electric field describes an *ellipse*, whence the terminology *elliptical polarization*. If we direct the x axis along $\mathbf{W}_1$ and the y axis along $\mathbf{W}_2$, the relations (5.94) imply, in fact, that the components of $\mathbf{E}$ obey the equation of an ellipse

$$\frac{(E^x)^2}{W_1^2} + \frac{(E^y)^2}{W_2^2} = 1.$$

(5) Energy and momentum. The energy and momentum content of a generic electromagnetic field is codified by the energy-momentum tensor (2.137)

$$T_{\text{em}}^{\mu\nu} = F^\mu{}_\alpha F^{\alpha\nu} + \frac{1}{4}\eta^{\mu\nu}F^{\alpha\beta}F_{\alpha\beta}.$$

For the elementary waves (5.77), using (5.83) and the wave relations, and taking into account that the invariant $F^{\alpha\beta}F_{\alpha\beta}$ vanishes, we obtain the simple expression

$$T_{\text{em}}^{\mu\nu} = (n^\mu \dot{A}_\alpha - n_\alpha \dot{A}^\mu)(n^\alpha \dot{A}^\nu - n^\nu \dot{A}^\alpha) = -n^\mu n^\nu (\dot{A}^\alpha \dot{A}_\alpha). \tag{5.95}$$

Since the energy-momentum tensor is an observable quantity, this expression must be invariant under the residual gauge transformations (5.81), a property which is easily verified thanks to the wave relations. Eliminating from (5.95) A^0 via (5.84), we obtain an expression involving only the spatial components of the vector potential

$$T_{\text{em}}^{\mu\nu} = n^\mu n^\nu \left(|\dot{\mathbf{A}}|^2 - (\mathbf{n}\cdot\dot{\mathbf{A}})^2 \right). \tag{5.96}$$

Choosing as z axis the propagation direction we have $\mathbf{n} = (0,0,1)$, and (5.96) simplifies further to

$$T_{\text{em}}^{\mu\nu} = n^\mu n^\nu \left((\dot{A}_1)^2 + (\dot{A}_2)^2 \right), \tag{5.97}$$

an expression which involves only the two transverse (physical) components A^1 and A^2, while the longitudinal (non-physical) component A^3 dropped out, as expected. In fact, as the residual gauge invariance (5.81) guarantees that the energy-momentum tensor is invariant under the transformations

$$A^1 \to A^1, \qquad A^2 \to A^2, \qquad A^3 \to A^3 + \varphi,$$

its independence of A^3 is automatic. Finally, comparing (5.97) with the right hand side of (5.40), and redoing the analysis performed after that formula, we conclude that in quantum field theory each of the two physical polarization states of the elementary wave is associated with a massless particle, i.e. with a *transverse* photon. We will return to the physical meaning of these two components in Sect. 5.3.3, in connection with the concept of *helicity*. Recalling Eqs. (5.85) and (5.87), the tensor (5.96) can also be recast in the form

$$T_{\text{em}}^{\mu\nu} = n^\mu n^\nu E^2 = \frac{1}{2}n^\mu n^\nu \left(E^2 + B^2 \right). \tag{5.98}$$

Similarly, using Eqs. (5.88) the Poynting vector of an elementary wave becomes

$$\mathbf{S} = \mathbf{E}\times\mathbf{B} = \mathbf{E}\times(\mathbf{n}\times\mathbf{E}) = E^2\mathbf{n}. \tag{5.99}$$

From (5.98) we retrieve thus the general expressions $T_{em}^{00} = \frac{1}{2}(E^2 + B^2)$ and $T_{em}^{0i} = S^i$. In particular, the energy flux and the momentum density of an elementary wave – both represented by the vector $\mathbf{S}$ – are directed along the propagation direction of the wave, as expected. Another important physical quantity of a wave phenomenon is the *intensity* $\mathcal{I}$, which is defined as the mean energy flowing per unit time across a unit surface orthogonal to the propagation direction. The intensity of an elementary wave is thus given by

$$\mathcal{I} = \overline{\mathbf{n} \cdot \mathbf{S}} = \overline{E^2}. \tag{5.100}$$

Electromagnetic field in the wave zone. We conclude this section with the caveat that properties (1)–(5) hold for the *elementary* waves (5.77), but not for a generic radiation field (5.74) which is a *superposition* of such waves. Nonetheless, as we will see in Sect. 8.1, the wave relations (5.80) hold true for a generic electromagnetic field in the so-called *wave zone*, that is to say, at large distances from the charges which generated it. Since to derive properties (1), (2) and (5) we used only these relations, these specific properties will hold also for a generic electromagnetic field in the wave zone. We refer here in particular to formulas (5.85), (5.88) and (5.98), giving the electric field, the magnetic field and the energy-momentum tensor in terms of only the spatial components of the vector potential:

$$\mathbf{E} = \mathbf{n} \times (\mathbf{n} \times \dot{\mathbf{A}}) = -\dot{\mathbf{A}} + (\mathbf{n} \cdot \dot{\mathbf{A}})\mathbf{n}, \tag{5.101}$$

$$\mathbf{B} = \mathbf{n} \times \mathbf{E}, \quad \mathbf{n} \cdot \mathbf{E} = 0, \tag{5.102}$$

$$T_{em}^{\mu\nu} = n^\mu n^\nu E^2. \tag{5.103}$$

We emphasize this point since we shall see that the energy analysis of a generic *radiation process* does not require the knowledge of the exact electromagnetic field, but only its (asymptotic) behavior in the wave zone. For this purpose we will therefore be allowed to use the simple formulas (5.101)–(5.103), and consequently the energy analysis will be significantly facilitated.

5.3.2 Elementary Gravitational Waves

In Sect. 5.3.3 we will analyze a characteristic property of elementary waves, called *helicity*. To provide a better understanding of this concept we will compare electromagnetic waves with scalar and gravitational ones. For this reason, in the present section we anticipate from Chap. 9, in particular from Sect. 9.3, a few results regarding the latter. According to the theory of General Relativity, on one hand the gravitational potential created by a generic physical system has the form of a symmetric tensor $H_{\mu\nu}(x)$, and on the other hand the space-time *curvature* is described by a symmetric tensor $g_{\mu\nu}(x)$ – the *Riemannian metric* – replacing the Minkowski metric $\eta_{\mu\nu}$. In particular, in a curved space-time the interval between two events whose coordinates differ by dx^μ becomes

$$ds^2 = dx^\mu dx^\nu g_{\mu\nu}(x).$$

The theory establishes, furthermore, that the two fields are tied together by the relation

$$g_{\mu\nu}(x) = \eta_{\mu\nu} + H_{\mu\nu}(x) - \frac{1}{2}\eta_{\mu\nu}H_\rho{}^\rho(x), \tag{5.104}$$

linking the gravitational action to the deformation of space-time. The field $H_{\mu\nu}(x)$ quantifies, hence, the deviation of the metric $g_{\mu\nu}(x)$ of a curved space-time from the metric $\eta_{\mu\nu}$ of a flat one. A key ingredient of General Relativity is its invariance under the *diffeomorphisms* (3.92), a group of local transformations which, in a sense, represent the "gauge invariance of the gravitational interaction". In Chap. 9 we shall see that in the *weak-field* approximation, in which by definition

$$|H_{\mu\nu}(x)| \ll 1, \quad \forall\,\mu,\nu,$$

the general solution of *Einstein's equations* in empty space is a superposition of plane and monochromatic elementary *gravitational waves*, propagating at the speed of light, of the form

$$H_{el}^{\mu\nu}(x) = \varepsilon^{\mu\nu}e^{ik\cdot x} + c.c., \quad k^2 = 0, \quad k^0 = \omega. \tag{5.105}$$

As usual we have introduced the frequency $\omega = |\mathbf{k}|$. The waves (5.105) are characterized by a symmetric complex polarization *tensor* $\varepsilon^{\mu\nu}$, subject to the *harmonic* gauge-fixing of diffeomorphisms and to the residual gauge invariance

$$k_\mu\varepsilon^{\mu\nu} = 0, \quad \varepsilon^{\mu\nu} \approx \varepsilon^{\mu\nu} + \lambda^\mu k^\nu + \lambda^\nu k^\mu - \eta^{\mu\nu}\lambda_\rho k^\rho, \tag{5.106}$$

respectively, see Sect. 9.3. As for the electromagnetic waves (5.77), the residual gauge transformations in (5.106) – involving now *four* complex gauge parameters λ^μ, instead of one – preserve the harmonic gauge. In fact, thanks to $k^2 = 0$ we have

$$k_\mu\varepsilon'^{\mu\nu} = k_\mu(\varepsilon^{\mu\nu} + \lambda^\mu k^\nu + \lambda^\nu k^\mu - \eta^{\mu\nu}\lambda_\rho k^\rho) = k_\mu\varepsilon^{\mu\nu} + k^2\lambda^\nu = 0.$$

Two physical polarization states. The symmetric tensor $\varepsilon^{\mu\nu}$ has ten independent components, which are, however, subject to four gauge-fixing conditions as well as to four residual gauge transformations. The gravitational waves (5.105) are thus characterized by $10 - 4 - 4 = 2$ physical polarization states. To determine them explicitly we must fix the residual gauge invariance, as in the case of the electromagnetic waves, see (5.90). In the case at hand it is convenient to impose the four conditions

$$\varepsilon^{0i} = 0, \quad \varepsilon^{jj} = 0. \tag{5.107}$$

To prove their consistency we perform a residual gauge transformation (5.106) and impose the system of four equations in the four unknowns λ^μ

$$\varepsilon'^{0i} = \varepsilon^{0i} + \lambda^i \omega + \lambda^0 k^i = 0,$$
$$\varepsilon'^{jj} = \varepsilon^{jj} + 2\lambda^j k^j + 3(\lambda^0 \omega - \lambda^j k^j) = \varepsilon^{jj} + 3\lambda^0 \omega - \lambda^j k^j = 0,$$

admitting, in fact, the unique solution

$$\lambda^0 = -\frac{1}{4\omega}\left(\varepsilon^{jj} + \frac{k^i}{\omega}\varepsilon^{i0}\right), \qquad \lambda^i = -\frac{1}{4\omega^2}\left(4\omega\varepsilon^{i0} - \left(\varepsilon^{jj} + \frac{k^j}{\omega}\varepsilon^{j0}\right)k^i\right).$$

Thanks to the constraints (5.107), the gauge conditions $k_\mu \varepsilon^{\mu\nu} = 0$ yield for $\nu = 0$ and $\nu = i$ the further relations

$$k_\mu \varepsilon^{\mu 0} = \omega\varepsilon^{00} + k_j \varepsilon^{j0} = \omega\varepsilon^{00} = 0 \qquad \Rightarrow \qquad \varepsilon^{00} = 0,$$
$$k_\mu \varepsilon^{\mu i} = \omega\varepsilon^{0i} + k_j \varepsilon^{ji} = k_j \varepsilon^{ji} = 0 \qquad \Rightarrow \qquad k^i \varepsilon^{ij} = 0.$$

The completely gauge-fixed polarization tensor is hence characterized by the relations

$$\varepsilon^{00} = 0, \qquad \varepsilon^{0i} = 0, \qquad \varepsilon^{jj} = 0, \qquad k^i \varepsilon^{ij} = 0, \qquad (5.108)$$

which identify the so-called *transverse-traceless* gauge. If we choose as z axis the propagation direction of the wave, where $\mathbf{k} = (0, 0, \omega)$, the conditions (5.108) imply that the only non-vanishing components of $\varepsilon^{\mu\nu}$ are

$$\varepsilon^{12} = \varepsilon^{21}, \qquad \varepsilon^{11} = -\varepsilon^{22}.$$

In this gauge the polarization tensor has the simple form

$$\varepsilon^{\mu\nu} = \begin{pmatrix} 0 & 0 & 0 & 0 \\ 0 & \varepsilon^{11} & \varepsilon^{12} & 0 \\ 0 & \varepsilon^{12} & -\varepsilon^{11} & 0 \\ 0 & 0 & 0 & 0 \end{pmatrix}, \qquad (5.109)$$

and the two physical polarization states of the gravitational wave (5.105) are represented by the components ε^{12} and ε^{11}. Notice that the components ε^{12} and $\frac{1}{2}(\varepsilon^{11} - \varepsilon^{22})$ are invariant under the residual gauge transformations (5.106). The presence of two physical polarization states in the gravitational waves reflects the fact that Einstein's equations confer on the gravitational field *two* degrees of freedom, see Chap. 9.

5.3.3 Helicity

The *helicity* of a classical wave is intimately related to a physical quantity which plays a fundamental role in quantum physics: the *spin*.[8] More precisely, it can be seen that the spin of the particles describing a given elementary wave at the quantum level is proportional to the helicity of the wave, the proportionality constant being Planck's constant $\hbar$. Below we discuss the helicity of the scalar, electromagnetic, and gravitational waves

$$\varphi = \varepsilon\, e^{ik\cdot x} + c.c., \tag{5.110}$$

$$A^\mu = \varepsilon^\mu e^{ik\cdot x} + c.c., \quad k_\mu \varepsilon^\mu = 0, \quad \varepsilon^\mu \approx \varepsilon^\mu + \lambda k^\mu, \tag{5.111}$$

$$H^{\mu\nu} = \varepsilon^{\mu\nu} e^{ik\cdot x} + c.c., \quad k_\mu \varepsilon^{\mu\nu} = 0, \quad \varepsilon^{\mu\nu} \approx \varepsilon^{\mu\nu} + \lambda^\mu k^\nu + \lambda^\nu k^\mu - \eta^{\mu\nu}\lambda_\rho k^\rho. \tag{5.112}$$

Helicity and rotations. The notion of helicity is related to the properties of the polarizations tensors $\varepsilon(\mathbf{k})$, $\varepsilon^\mu(\mathbf{k})$ and $\varepsilon^{\mu\nu}(\mathbf{k})$ under spatial rotations. Under a generic Lorentz transformation $\Lambda^\mu{}_\nu$ these tensors transform according to the rules

$$\varepsilon'(\mathbf{k}') = \varepsilon(\mathbf{k}), \quad \varepsilon'^\mu(\mathbf{k}') = \Lambda^\mu{}_\nu\, \varepsilon^\nu(\mathbf{k}), \quad \varepsilon'^{\mu\nu}(\mathbf{k}') = \Lambda^\mu{}_\alpha \Lambda^\nu{}_\beta\, \varepsilon^{\alpha\beta}(\mathbf{k}), \tag{5.113}$$

where

$$k'^\mu = \Lambda^\mu{}_\nu k^\nu.$$

From now on we consider a fixed wave vector $\mathbf{k}$. An elementary wave is then completely characterized by its *complex* polarization tensor, subject to the gauge-fixing conditions indicated in (5.110)–(5.112), respectively. Let V_i, with $i = 1, 2, 3$, denote the complex linear spaces of the three polarization tensors, constrained by the corresponding gauge-fixing conditions. The dimensions d_i of these spaces are

$$d_1 = 1, \qquad d_2 = 4 - 1 = 3, \qquad d_3 = 10 - 4 = 6.$$

Consider now the $U(1)$ subgroup of the Lorentz group constituted by the spatial rotations of a generic angle φ around the direction of $\mathbf{k}$. Denoting a generic element of this $U(1)$ by $\Lambda^\mu{}_\nu(\varphi)$ we have in particular

$$\Lambda^\mu{}_\nu(\varphi_1)\Lambda^\nu{}_\rho(\varphi_2) = \Lambda^\mu{}_\rho(\varphi_1 + \varphi_2).$$

Under such a transformation the vector $\mathbf{k}$ and its magnitude $k^0 = |\mathbf{k}|$ remain clearly invariant

$$k'^\mu = \Lambda^\mu{}_\nu(\varphi)\, k^\nu = k^\mu.$$

---

[8]The term *helicity* is also in use in quantum physics, where it denotes the projection of the spin of a particle on the direction of its momentum.

This means that in (5.113) only the components of the polarization tensors transform, while their argument $\mathbf{k}' = \mathbf{k}$ remains invariant. Consequently, the transformed
polarization tensors still satisfy the gauge-fixing conditions of (5.110)–(5.112) –
with the same k^μ – and so they still belong to V_i. This means that each space V_i
hosts a *representation* of $U(1)$, which in general is, however, *reducible*. According
to a known theorem of group theory, all complex *irreducible* representations of an
abelian Lie group G are *one-dimensional*, with representation space $\mathbb{C}$. Moreover, in
the case of $G = U(1)$ in each irreducible representation an element $\Lambda^\mu{}_\nu(\varphi) \in U(1)$
acts on an element $\mathcal{E} \in \mathbb{C}$ according to the rule

$$\mathcal{E} \to \mathcal{E}' = e^{in\varphi}\mathcal{E}, \tag{5.114}$$

where n is a fixed real number. It must therefore be possible to decompose the
spaces V_i of the polarization tensors into d_i one-dimensional subspaces, hosting an
irreducible representation of $U(1)$ of the type (5.114). Thus, each of these subspaces
identifies a – physical or non-physical – polarization state of the wave, which is
uniquely associated with a real number n, called *helicity*. The important result mentioned at the beginning of the section is that each state of helicity n at the quantum
level corresponds to a particle of *spin* $n\hbar$.

To facilitate the explicit decomposition in irreducible representations it is convenient to choose as z axis the propagation direction of the wave, so that $k^\mu =
(\omega, 0, 0, \omega)$. The matrix $\Lambda^\mu{}_\nu(\varphi)$ then corresponds to a rotation of an angle φ around
the z axis

$$\Lambda^\mu{}_\nu(\varphi) = \begin{pmatrix} 1 & 0 & 0 & 0 \\ 0 & \cos\varphi & \sin\varphi & 0 \\ 0 & -\sin\varphi & \cos\varphi & 0 \\ 0 & 0 & 0 & 1 \end{pmatrix}. \tag{5.115}$$

To reduce the representations (5.113) of $U(1)$ in one-dimensional representations
we must find linear combinations $\mathcal{E}$ of the components of the polarization tensors, so
that the transformation rules (5.113) assume the *diagonal* form (5.114). We perform
now this reduction explicitly for the three types of waves.

Scalar waves. For the scalar waves (5.110) we have $d_1 = 1$. Under an arbitrary
Lorentz transformation, and hence also under $\Lambda^\mu{}_\nu(\varphi)$, we have $\varepsilon' = \varepsilon$. This representation is already one-dimensional, and (5.114) holds with $\mathcal{E} = \varepsilon$ and $n = 0$. The
scalar waves entail thus a unique (physical) polarization state of helicity *zero*.

Electromagnetic waves. In this case we have $d_2 = 3$, since the condition $k_\mu \varepsilon^\mu =
0$ implies that ε^μ has three independent components: the two transverse physical
polarizations ε^1 and ε^2, and the longitudinal non-physical component $\varepsilon^0 = \varepsilon^3$. From
(5.113) and (5.115) we derive the transformation laws

$$\varepsilon'^0 = \varepsilon^0,$$
$$\varepsilon'^1 = \cos\varphi\,\varepsilon^1 + \sin\varphi\,\varepsilon^2,$$
$$\varepsilon'^2 = -\sin\varphi\,\varepsilon^1 + \cos\varphi\,\varepsilon^2,$$
$$\varepsilon'^3 = \varepsilon^3.$$

The longitudinal component possesses, thus, zero helicity. The two remaining transformation laws are diagonalized by the linear combinations

$$\mathcal{E}_\pm = \varepsilon^1 \mp i\varepsilon^2,$$

in that

$$\mathcal{E}'_\pm = \varepsilon'^1 \mp i\varepsilon'^2 = \cos\varphi\,\varepsilon^1 + \sin\varphi\,\varepsilon^2 \mp i\left(-\sin\varphi\,\varepsilon^1 + \cos\varphi\,\varepsilon^2\right) = e^{\pm i\varphi}\mathcal{E}_\pm.$$

Comparing with (5.114) we conclude that an electromagnetic wave is associated with a non-physical polarization state of helicity $n = 0$, and with two physical polarization states of helicity $n = \pm 1$. The latter correspond to electromagnetic waves of circular clockwise and counter-clockwise polarizations, see Problem 5.5.

Gravitational waves. For the gravitational waves (5.112) the gauge-fixing $k_\mu \varepsilon^{\mu\nu} = 0$ implies that the tensor $\varepsilon^{\mu\nu}$ has $d_3 = 6$ independent components, two physical ones and four non-physical ones. For the sake of brevity we analyze only the two physical ones ε^{12} and ε^{11}, see (5.109). From the third relation of (5.113) we find that under a rotation around the z axis the component ε^{11} transforms as

$$\varepsilon'^{11} = \Lambda^1{}_1(\varphi)\Lambda^1{}_1(\varphi)\,\varepsilon^{11} + 2\Lambda^1{}_2(\varphi)\Lambda^1{}_1(\varphi)\,\varepsilon^{12} + \Lambda^1{}_2(\varphi)\Lambda^1{}_2(\varphi)\,\varepsilon^{22}$$
$$= \cos^2\varphi\,\varepsilon^{11} + 2\sin\varphi\cos\varphi\,\varepsilon^{12} - \sin^2\varphi\,\varepsilon^{11}$$
$$= \cos 2\varphi\,\varepsilon^{11} + \sin 2\varphi\,\varepsilon^{12}.$$

Similarly we find

$$\varepsilon'^{12} = -\sin 2\varphi\,\varepsilon^{11} + \cos 2\varphi\,\varepsilon^{12}.$$

As in the case of the electromagnetic waves these transformation laws are diagonalized by the combinations

$$\mathcal{E}_\pm = \varepsilon^{11} \mp i\varepsilon^{12},$$

leading to

$$\mathcal{E}'_\pm = e^{\pm 2i\varphi}\mathcal{E}_\pm.$$

The two physical polarization states of a gravitational wave entail, thus, helicity $n = \pm 2$. In conclusion, gravitational and electromagnetic waves share the propagation velocity as well as the number of physical states, but differ by their helicity.

Finally, at the basis of the general connection between spin and helicity recalled above, we infer that at the quantum level a real scalar field, whose dynamics is governed by the Lagrangian (5.21), is associated with a neutral massless particle of spin zero, that the electromagnetic field consists of massless particles of spin $\pm\hbar$, the photons, and that the gravitational field – *if* there exists a consistent quantum theory of gravity – will consist of massless particles of spin $\pm 2\hbar$, the gravitons.

Different bases of solutions. In this section we have analyzed a particular complete basis of Maxwell's equations in empty space – the plane waves – and we have investigated their most salient properties. We add here an additional, last, property, not less significant than the other ones and, probably, the most characteristic one: under a Lorentz transformation each element of the basis goes over to another element of the basis, that is to say, under a Lorentz transformation the plane wave (5.77) remains a plane wave. It is, however, clear that the plane-wave basis – even though particularly relevant in a relativistic theory – is not the unique basis of physical interest. Another important complete system of solutions of Maxwell's equations in empty space is represented by the so-called *spherical waves*, a system that appears useful for a systematic *multipole expansion* of the electromagnetic radiation. We will not dwell on the details of this system as we shall resort to the multipole expansion mainly in the non-relativistic limit, see Sect. 8.3, where only the dipole and quadrupole terms are of physical relevance. For more details on spherical waves see, for example, Sects. 9.6 and 9.7 of the textbook [1].

5.4 Cauchy Problem for the Radiation Field

In this section we address the Cauchy problem for the free electromagnetic field in the light of formula (5.74), i.e. the general solution of Maxwell's equation in vacuum. On this occasion, in Sect. 5.4.1 we rederive it resorting to an alternative method, having the advantage of being *manifestly* gauge invariant.

5.4.1 Radiation Field and Manifest Gauge Invariance

As pointed out in Sect. 3.2.3, the introduction of the vector potential is unavoidable if one wants to derive the laws of electrodynamics from a variational principle, the latter being, in turn, the indispensable starting point for the quantization of the theory. On the other hand, the approaches that, in addition to the electromagnetic field, involve explicitly the vector potential, entail the disadvantage of violating *manifest* gauge invariance. In the context of *classical* electrodynamics the introduction of the vector potential constitutes, actually, only a matter of *convenience*, as it can make the analysis of certain phenomena easier. We saw, for example, that the introduction of the vector potential, together with the Fourier transform, allowed to solve Maxwell's equations in vacuum in a simple way. However, it is important to realize that it is possible – not only in principle, but also in practice – to solve the fundamental equations of classical electrodynamics, in full generality, in terms of only the electromagnetic field $F^{\mu\nu}$. The obvious advantages of such an approach are that one never introduces *non-physical* elements, and that gauge invariance is manifest at all stages. To illustrate this alternative framework, below we again solve Maxwell's equations in vacuum by relying only on the electromagnetic field. In this perspective we must solve the system of equations

$$\partial_\mu F^{\mu\nu} = 0, \tag{5.116}$$

$$\partial_{[\mu} F_{\nu\rho]} = \frac{1}{3}\left(\partial_\mu F_{\nu\rho} + \partial_\nu F_{\rho\mu} + \partial_\rho F_{\mu\nu}\right) = 0, \tag{5.117}$$

where we included again the Bianchi identity. We prove first of all that all components of the Maxwell tensor must satisfy the wave equation. Applying to the Bianchi identity the derivative operator ∂^μ we obtain the equation

$$\frac{1}{3}\left(\Box F_{\nu\rho} + \partial_\nu \partial^\mu F_{\rho\mu} + \partial_\rho \partial^\mu F_{\mu\nu}\right) = 0.$$

Thanks to (5.116) the second and third terms vanish, so that we remain indeed with the wave equations

$$\Box F^{\mu\nu} = 0. \tag{5.118}$$

Notice that these equations *follow* from Eqs. (5.116) and (5.117), but they do *not* imply them. Notwithstanding this, from Sect. 5.2 we know the general solution of (5.118), see (5.35),

$$F^{\mu\nu} = \frac{1}{(2\pi)^2} \int \frac{d^3k}{2\omega} \left(e^{ik\cdot x} f^{\mu\nu}(\mathbf{k}) + c.c.\right), \tag{5.119}$$

where $f^{\mu\nu}(\mathbf{k})$ is an arbitrary antisymmetric complex tensor. To impose on (5.119) Eqs. (5.116) and (5.117) we must evaluate the partial derivatives

$$\partial_\rho F^{\mu\nu} = \frac{i}{(2\pi)^2} \int \frac{d^3k}{2\omega} \left(e^{ik\cdot x} k_\rho f^{\mu\nu}(\mathbf{k}) + c.c.\right).$$

Enforcing (5.116) and (5.117), and applying the inverse Fourier transform with respect to the variable $\mathbf{x}$, these differential equations turn into the algebraic ones ($k^2 = 0$, $k^0 = \omega$)

$$k_\mu f^{\mu\nu} = 0, \tag{5.120}$$

$$k_{[\mu} f_{\nu\rho]} = 0. \tag{5.121}$$

The general solution of (5.121) is

$$f_{\mu\nu} = k_\mu \beta_\nu - k_\nu \beta_\mu, \tag{5.122}$$

where $\beta^\mu = \beta^\mu(\mathbf{k})$ is an arbitrary complex four-vector. Then Eq. (5.120) imposes on β^μ the constraint

$$k_\mu f^{\mu\nu} = k_\mu(k^\mu \beta^\nu - k^\nu \beta^\mu) = k^2 \beta^\nu - k^\nu(k_\mu \beta^\mu) = -k^\nu(k_\mu \beta^\mu) = 0,$$

that is to say

$$k_\mu \beta^\mu = 0. \tag{5.123}$$

However, there are different vectors β^μ leading to the same solution (5.119). In fact, the vectors β^μ and $\beta^\mu + \lambda k^\mu$ give rise to the same tensor $f^{\mu\nu}$ and, moreover, both satisfy (5.123). Finally, the field (5.119), with $f^{\mu\nu}$ specified by the relations (5.122) and (5.123), agrees with the solution (5.74) derived previously, via the identification

$$\beta^\mu = i\varepsilon^\mu.$$

Correspondingly, the replacement of β^μ with $\beta^\mu + \lambda k^\mu$ reflects the residual gauge invariance (5.71).

5.4.2 Initial Value Problem

We address now the initial value problem for the free electromagnetic field. In the light of Eqs. (5.118) and (5.119), from the analysis of Sect. 5.2.2 we know already how to write $F^{\mu\nu}(x)$ in terms of the initial data $F^{\mu\nu}(0, \mathbf{x})$ and $\partial_0 F^{\mu\nu}(0, \mathbf{x})$, and of the antisymmetric kernel $D(x)$, see (5.48),

$$F^{\mu\nu}(x) = \int \big(D(t, \mathbf{x} - \mathbf{y})\, \partial_0 F^{\mu\nu}(0, \mathbf{y}) + \partial_0 D(t, \mathbf{x} - \mathbf{y})\, F^{\mu\nu}(0; \mathbf{y}) \big)\, d^3 y. \tag{5.124}$$

On the other hand, the time derivatives $\partial_0 F^{\mu\nu}(0, \mathbf{x})$ are tied to the initial values of the fields $F^{\mu\nu}(0, \mathbf{x})$ through Eqs. (5.116) and (5.117), that is to say

$$\frac{\partial \mathbf{E}}{\partial t} = \boldsymbol{\nabla} \times \mathbf{B}, \qquad \frac{\partial \mathbf{B}}{\partial t} = -\boldsymbol{\nabla} \times \mathbf{E}.$$

Evaluating these equations at $t = 0$, and inserting them into (5.124), we obtain the formulas we sought for

$$\mathbf{E}(t, \mathbf{x}) = \int \big(D(t, \mathbf{x} - \mathbf{y})\boldsymbol{\nabla} \times \mathbf{B}(0, \mathbf{y}) + \partial_0 D(t, \mathbf{x} - \mathbf{y})\, \mathbf{E}(0, \mathbf{y}) \big)\, d^3 y,$$

$$\mathbf{B}(t, \mathbf{x}) = \int \big(-D(t, \mathbf{x} - \mathbf{y})\boldsymbol{\nabla} \times \mathbf{E}(0, \mathbf{y}) + \partial_0 D(t, \mathbf{x} - \mathbf{y})\, \mathbf{B}(0, \mathbf{y}) \big)\, d^3 y, \tag{5.125}$$

expressing the fields at a generic time t in terms of the initial data $\mathbf{E}(0, \mathbf{x})$ and $\mathbf{B}(0, \mathbf{x})$. Moreover, since $D(x)$ obeys the wave equation, it is easy to verify that the expressions (5.125) satisfy Maxwell's equations with vanishing sources (2.52)-(2.55), if the initial data satisfy the physical constraints

$$\nabla \cdot \mathbf{E}(0, \mathbf{x}) = 0 = \nabla \cdot \mathbf{B}(0, \mathbf{x}).$$

From Sect. 2.2.5 we already know, in fact, that the knowledge of only four components of the electromagnetic field at $t = 0$, for instance, $E^1(0, \mathbf{x})$, $E^2(0, \mathbf{x})$, $B^1(0, \mathbf{x})$ and $B^2(0, \mathbf{x})$, is sufficient to determine the fields for all t. We end this section underlining the most salient properties of the solutions (5.124) and (5.125).

Manifest covariance. From Sect. 5.2.3 we know how to set (5.124) in a manifestly covariant form, see (5.60). For a given space-like hypersurface Γ, parameterized by $y^\mu(\boldsymbol{\lambda})$ and with normal unit vector $n^\rho(\boldsymbol{\lambda})$, on which we specify the values of $F^{\mu\nu}$ and of $n_\rho \partial^\rho F^{\mu\nu}$, the covariant version of (5.124) is

$$F^{\mu\nu}(x) = \int_\Gamma \left(D(x - y)\, \partial^\rho F^{\mu\nu}(y) + \partial^\rho D(x - y)\, F^{\mu\nu}(y) \right) d\Sigma_\rho, \qquad (5.126)$$

where $d\Sigma_\rho = n_\rho \sqrt{g}\, d^3\lambda$, see Sect. 3.2.1. Also in this case it can be shown that the values of the derivatives $n_\rho \partial^\rho F^{\mu\nu}(y)$ on Γ are determined by the values of the six fields $F^{\mu\nu}(y)$ on Γ, and that, in addition, only four out of the latter are independent.

Causality. A fundamental characteristic of the antisymmetric kernel (5.50) is that its support is the light cone, i.e. $D(t, \mathbf{x})$ is different from zero only for $t = \pm|\mathbf{x}|$. This property ensures, in fact, that a generic radiation field – and not only the elementary waves – propagates at the speed of light. We illustrate this fundamental feature with an example. Suppose that the initial fields $\mathbf{E}(0, \mathbf{x})$ and $\mathbf{B}(0, \mathbf{x})$ are different from zero only inside a sphere S_L of radius L centered at the origin, so that in formulas (5.125) the integrals over $\mathbf{y}$ are restricted to the region $|\mathbf{y}| < L$. Consider now a point P lying outside S_L with coordinate $\mathbf{x}_0$. Since $D(t, \mathbf{x} - \mathbf{y}) = 0$ for $|\mathbf{x} - \mathbf{y}| \neq \pm t$, Eqs. (5.125) imply that at a time $t > 0$ the field at P is different from zero only if there exist some $\mathbf{y}$ such that

$$t = |\mathbf{x}_0 - \mathbf{y}|, \quad |\mathbf{y}| < L.$$

This implies that the *first* signal arrives in P at the time $t_0 = |\mathbf{x}_0| - L$, while at all instants preceding t_0 the field at P is zero. Since the distance of P from the sphere S_L is $|\mathbf{x}_0| - L$, we conclude that the radiation fields propagates at the speed of light.

Huygens's principle. The kernel $D(t, \mathbf{x})$ (5.50) is invariant under spatial rotations:

$$D(t, R\mathbf{x}) = D(t, \mathbf{x}), \quad \forall R \in SO(3).$$

Consequently, in the light of formulas (5.125), the electromagnetic field propagates in all directions in an *isotropic* way. In addition, $D(t, \mathbf{x} - \mathbf{y})$ is proportional to the δ-function $\delta(t^2 - |\mathbf{x} - \mathbf{y}|^2)$. If we perform in the three-dimensional integrals (5.125) the shift $\mathbf{y} \rightarrow \mathbf{y} + \mathbf{x}$, and use this δ-function to integrate over the radial variable $|\mathbf{y}|$, the fields $\mathbf{E}(t, \mathbf{x})$ and $\mathbf{B}(t, \mathbf{x})$ are expressed as *surface* integrals over the sphere of radius t centered in $\mathbf{x}$. These fields result thus from the superposition of all *signals* emitted from this sphere and joining $\mathbf{x}$. Both the above characteristics – *isotropy*, and *integrals over surfaces* as secondary sources – form the basis of *Huygens's principle*. If one applies the same reasoning to the covariant formula (5.126), involving an integral over a generic three-dimensional space-like hypersurface Γ, after the integration over the δ-function one remains with a residual surface integral over a generic closed surface – no longer a spherical one – corresponding to the general form of Huygens's principle.

5.4.3   Time-Reversal Invariance

Property (5.54) asserts that $D(x)$ is antisymmetric in the time variable

$$D(-t, \mathbf{x}) = -D(t, \mathbf{x}), \qquad \frac{\partial}{\partial t^*} D(t^*, \mathbf{x})\bigg|_{t^*=-t} = \frac{\partial}{\partial t} D(t, \mathbf{x}). \qquad (5.127)$$

This feature is by no means a coincidence being, rather, strictly related to the *time-reversal invariance* of electrodynamics. Returning for a moment to the simpler case of the wave equation for a scalar field $\Box\, \varphi = 0$, and to the corresponding Lagrangian $\mathcal{L} = \frac{1}{2}\, \partial_\mu \varphi \partial^\mu \varphi$, we see that both are invariant under time reversal. As t goes into $t^* = -t$, the scalar field transforms in fact according to the rule $\varphi^*(t^*, \mathbf{x}) = \varphi(t, \mathbf{x})$, or $\varphi^*(t, \mathbf{x}) = \varphi(-t, \mathbf{x})$. Consequently, if $\varphi(x)$ is a solution of the wave equation, so is $\varphi^*(x)$. More precisely, replacing in the solution (5.48) t with $-t$, and using (5.127), we see that $\varphi^*(x)$ is a solution of the wave equation with the correct initial data $\varphi^*(0, \mathbf{x}) = \varphi(0, \mathbf{x})$ and $\partial_0 \varphi^*(0, \mathbf{x}) = -\partial_0 \varphi(0, \mathbf{x})$.

Similarly, the equations of electrodynamics are invariant under time reversal, and in Sect. 2.2.2 we have already determined the transformation laws (2.30) of the fields. From (2.31) and (2.32) we know in particular that if the fields $\mathbf{E}(t, \mathbf{x})$ and $\mathbf{B}(t, \mathbf{x})$ are solutions of Maxwell's equations, so are the fields $\mathbf{E}^*(t, \mathbf{x}) = \mathbf{E}(-t, \mathbf{x})$ and $\mathbf{B}^*(t, \mathbf{x}) = -\mathbf{B}(-t, \mathbf{x})$. Replacing in the solutions (5.125) t with $-t$, and using again (5.127), we see that actually the fields $\mathbf{E}^*(x)$ and $\mathbf{B}^*(x)$ satisfy Maxwell's equations with the correct initial data $\mathbf{E}^*(0, \mathbf{x}) = \mathbf{E}(0, \mathbf{x})$ and $\mathbf{B}^*(0, \mathbf{x}) = -\mathbf{B}(0, \mathbf{x})$.

In conclusion, if the fields $\mathbf{E}$ and $\mathbf{B}$ represent a radiation field, a *free* electromagnetic field, observed in nature, then also the fields $\mathbf{E}^*$ and $\mathbf{B}^*$ represent a radiation field which can exist in nature, and vice versa. In contrast to this, later on we will find that in the presence of sources, where the fields are no longer *free*, this correspondence will no longer hold true: If the couple $(\mathbf{E}, \mathbf{B})$ is an electromagnetic field observable in nature, the couple $(\mathbf{E}^*, \mathbf{B}^*)$ obtained from the former by time reversal, in general will no longer be so, although it satisfies Maxwell's equations. In other words, in electrodynamics time-reversal invariance is, eventually, violated *spontaneously*, see Sect. 6.2.3.

5.5   Relativistic Doppler Effect

In Sect. 5.3.3 we saw that in the transition from one inertial reference frame to another an elementary wave remains an elementary wave. However, its polarization, propagation direction, and frequency change. In this section we investigate its change in frequency. For this purpose we consider a source which in its rest frame emits monochromatic light signals of *proper* wavelength λ_0 and of frequency $\omega_0 = 2\pi/\lambda_0$. We want to determine the frequency of the signal in a reference frame K where the source moves with constant velocity $\mathbf{v}$. Denote by K^* the inertial reference frame in which the source is at rest. In K^* its four-velocity and the wave vector are then given by

$$u^{*\mu} = (1, 0, 0, 0), \qquad k^{*\mu} = (\omega_0, \mathbf{k}_0),$$

where $\omega_0 = |\mathbf{k}_0|$. In the reference frame K the analogous quantities are

$$u^{\mu} = \left(\frac{1}{\sqrt{1 - v^2}}, \frac{\mathbf{v}}{\sqrt{1 - v^2}} \right), \qquad k^{\mu} = (\omega, \mathbf{k}).$$

Denoting by α the angle between the propagation direction of the wave and the velocity of the source, both measured in K, we may use the relativistic invariance of the scalar product $u_{\mu} k^{\mu}$ to write

$$\omega_0 = u^*_\mu k^{*\mu} = u_\mu k^\mu = \frac{\omega - \mathbf{v} \cdot \mathbf{k}}{\sqrt{1 - v^2}} = \frac{\omega - \omega v \cos \alpha}{\sqrt{1 - v^2}}.$$

The frequency and wavelength in K are then given by the relations

$$\omega = \frac{\sqrt{1 - v^2}}{1 - v \cos \alpha} \omega_0, \qquad \lambda = \frac{1 - v \cos \alpha}{\sqrt{1 - v^2}} \lambda_0, \qquad (5.128)$$

which represent the *relativistic Doppler effect*. In the particular case of a source that approaches the observer head-on (recedes from the observer) we have $\alpha = 0$ ($\alpha = \pi$), and so, restoring the velocity of light, we obtain

$$\lambda = \frac{1 \mp v/c}{\sqrt{1 - v^2/c^2}} \lambda_0. \qquad (5.129)$$

These expressions can be compared with the non-relativistic Doppler effect formulas

$$\lambda_{nr} = (1 \mp v/v_p) \lambda_0,$$

where v_p represents the propagation velocity of the signal. As the source moves with velocities v small with respect to the speed of light, formally the relativistic formulas (5.129) reduce to the non-relativistic ones, if we set $v_p = c$.

Cosmological redshift and expansion of the Universe. We end the section with an important application of the relativistic Doppler effect, the *cosmological redshift*. For sources receding from the observer, (5.129) gives for the relative variation of the wavelength

$$z = \frac{\lambda - \lambda_0}{\lambda_0} = \sqrt{\frac{1 + v/c}{1 - v/c}} - 1 > 0. \qquad (5.130)$$

Correspondingly, as the recession velocity v increases, the wavelengths increase and the frequencies decrease, a phenomenon that is known as *redshift*, since the spectral lines of the visible spectrum move towards its red end. This effect plays an important role in several branches of physics, in particular in cosmology. Through a systematic analysis of the redshift of the radiation emitted from a group of galaxies E. Hubble in 1929 discovered the expansion of the Universe. The galaxies he observed had small velocities compared to the velocity of light, of the order of $v \sim 3000$ km/s, and hence the relative increase of the wavelengths was rather small. For $v/c \ll 1$ the redshift formula (5.130) gives in fact

$$z \approx \frac{v}{c} \sim 10^{-2}.$$

On the other hand, today we know galaxies with large values of z, of order unity or larger. For the galaxy GN-z11, for example, the data collected by the *Hubble Space Telescope* permitted in 2015 [2] to infer a redshift of $z = 11.1$, for which (5.130) gives a *nominal*[9] recession velocity of $v = 0.986\,c$. Very precise measurements of the cosmological redshift in type Ia *supernovae* permitted to draw new and revolutionary conclusions about the current status of the Universe. They revealed, in fact, that not only the Universe is expanding, but that the expansion velocity is increasing, i.e. the Universe is *accelerating*. According to General Relativity an accelerating Universe requires, in turn, a non-vanishing and positive *cosmological constant*, a circumstance that has enriched modern cosmology by a series of new theoretical problems, which are still waiting for convincing solutions.

---

[9]Due to the non-stationary, and curved, metric of the expanding Universe the physical interpretation of this velocity requires, however, some care, see e.g. [3, 4].

5.6 Problems

5.1 Carrying out the inverse Fourier transform of the relations (5.44), and using the definitions (5.42) and (5.43), show that (5.47) can be recast in the form (5.48).

5.2 Assuming that $A^\mu(x)$ is a vector field and that under Lorentz transformations the wave vector transforms as $k'^\mu = \Lambda^\mu{}_\nu k^\nu$, prove that also its Fourier transform

$$\widehat{A}^\mu(k) = \frac{1}{(2\pi)^2} \int e^{-ik\cdot x} A^\mu(x)\, d^4 x$$

transforms as a vector field.

5.3 Relying on the *Weyl* gauge-fixing $A^0 = 0$, verify that the electromagnetic field entails two physical degrees of freedom. Proceed as follows:

(a) impose initial conditions on A^1 and A^2 and on their time derivatives at $t = 0$;
(b) determine the form of the residual gauge transformations;
(c) enforcing the equation $G^0 = \partial_\mu F^{\mu 0} - j^0 = 0$ at $t = 0$, and using the residual gauge transformations, determine the initial conditions for A^3 and $\partial_0 A^3$ at $t = 0$;
(d) address the same problem for the *axial* gauge-fixing $A^3 = 0$.

5.4 Consider the general solution of Maxwell's equations in vacuum (5.74).

(a) Derive the general expressions for the radiation fields

$$\mathbf{E}(t, \mathbf{x}) = \frac{1}{2(2\pi)^2} \int \left(i e^{ik\cdot x} \left((\mathbf{n} \cdot \boldsymbol{\varepsilon})\, \mathbf{n} - \boldsymbol{\varepsilon} \right) + c.c. \right) d^3 k,$$

$$\mathbf{B}(t, \mathbf{x}) = \frac{1}{2(2\pi)^2} \int \left(i e^{ik\cdot x} \left(\boldsymbol{\varepsilon} \times \mathbf{n} \right) + c.c. \right) d^3 k,$$

where $\boldsymbol{\varepsilon} = \boldsymbol{\varepsilon}(\mathbf{k})$ is a complex three-dimensional vector field.
(b) Check explicitly that these fields satisfy Maxwell's equations (2.52)–(2.55) in empty space, in addition to the wave equations

$$\Box\, \mathbf{E} = 0 = \Box\, \mathbf{B}.$$

(c) Assuming that the initial fields $\mathbf{E}(0, \mathbf{x})$ and $\mathbf{B}(0, \mathbf{x})$ are known, determine the vector field $\mathbf{V}(\mathbf{k}) = \boldsymbol{\varepsilon} - (\mathbf{n} \cdot \boldsymbol{\varepsilon})\, \mathbf{n}$, and hence $\mathbf{E}(t, \mathbf{x})$ and $\mathbf{B}(t, \mathbf{x})$ for all t.
Hint: See Sect. 5.2.2.
(d) Is the vector field $\boldsymbol{\varepsilon}(\mathbf{k})$ determined uniquely?

5.5 Consider the vector potential of an elementary wave

$$A^\mu = \varepsilon^\mu e^{ik\cdot x} + c.c.$$

with wave vector $k^\mu = (\omega, 0, 0, \omega)$ and with a generic polarization vector $\varepsilon^\mu = (\varepsilon^0, \varepsilon^1, \varepsilon^2, \varepsilon^0)$.

(a) Determine the fields $\mathbf{E}$ and $\mathbf{B}$ and verify that they are gauge-invariant, i.e. independent of ε^0, and that they satisfy the transversality conditions $E^z = 0 = B^z$.

(b) Introduce the *complex* electric field $E = E^x + iE^y$. Show that

$$E = -i\omega \left(\mathcal{E}_- \, e^{ik \cdot x} - \mathcal{E}_+^* \, e^{-ik \cdot x} \right), \tag{5.131}$$

where the coefficients $\mathcal{E}_\pm = \varepsilon^1 \mp i\varepsilon^2$ represent the helicity eigenstates.

(c) Show that for $\mathcal{E}_+ = 0$ ($\mathcal{E}_- = 0$) the wave is clockwise (counter-clockwise) circularly polarized. Compare the corresponding expressions of $\mathbf{E}$ with formulas (5.91) and (5.93).

(d) Show that the wave is linearly polarized if the relation $\mathcal{E}_-^* = e^{i\gamma} \mathcal{E}_+$ holds for a real γ.

5.6 Show that the energy-momentum tensor of the elementary wave (5.77), averaged over time scales large with respect to the period T, amounts to

$$\langle T_{\mathrm{em}}^{\mu\nu} \rangle = -2k^\mu k^\nu \varepsilon^{*\alpha} \varepsilon_\alpha.$$

Verify the inequality $\langle T_{\mathrm{em}}^{00} \rangle \geq 0$.

5.7 Consider the *spherical* scalar wave

$$\Phi(t, \mathbf{x}) = \frac{1}{r} f(t - r), \quad r = |\mathbf{x}|,$$

where f is an arbitrary function.

(a) Show that Φ satisfies the wave equation $\Box \Phi = 0$ for all $r \neq 0$.
    Hint: Write the Laplacian in polar coordinates

$$\nabla^2 = \frac{1}{r} \frac{\partial^2}{\partial r^2} r + \frac{1}{r^2} L^2, \tag{5.132}$$

where L^2 is a differential operator – the square of the quantum mechanical angular momentum – involving only the polar angles φ and ϑ.

(b) Explain the mathematical reason for why Φ is not a solution of the wave equation in whole space, and give a physical interpretation of the failure.

5.8 Consider the one-dimensional wave equation

$$\left(\partial_t^2 - \partial_x^2 \right) \Phi(t, x) = 0.$$

(a) Resorting to the Fourier-transform method prove that its general solution has the form

$$\Phi(t,x) = f(t-x) + g(t+x),$$

where f and g are arbitrary functions.

(b) Express $\Phi(t,x)$ in terms of the initial data $F(x) = \Phi(0,x)$ and $G(x) = \dot{\Phi}(0,x)$.

5.9 Show that the kernel $D(t,\mathbf{x})$ (5.49) can be recast in the form (5.50).
Hint: Perform the integral over d^3k in (5.49) in polar coordinates and use the rotation invariance of $D(t,\mathbf{x})$ to set $\mathbf{x} = (0,0,r)$. Use the representation (2.98) of the δ-function.

5.10 Consider the four-dimensional scalar wave equation $\Box \Lambda = 0$. Show that if one imposes the z-independent initial condition $\Lambda(0,x,y,z) = f(x,y)$ and $\partial_0\Lambda(0,x,y,z) = g(x,y)$, its solution $\Lambda(t,\mathbf{x})$ is z-independent for all t.
Hint: Solve the wave equation using the perturbative method of Sect. 5.1.2.

References

1. J.D. Jackson, *Classical Electrodynamics* (Wiley, New York, 1998)
2. P.A. Oesch et al., A remarkably luminous galaxy at $z = 11.1$ measured with Hubble Space Telescope grism spectroscopy. Astr. Phys. J. **819**, 129 (2016)
3. E.F. Bunn, D.W. Hogg, The kinematic origin of the cosmological redshift. Am. J. Phys. **77**, 688 (2009)
4. S. Braeck, O. Elgaroy, A physical interpretation of Hubble's law and the cosmological redshift from the perspective of a static observer. Gen. Rel. Grav. **44**, 2603 (2012)

Chapter 6
Generation of Electromagnetic Fields

In Chap. 5 we saw that an electromagnetic field satisfying Maxwell's equations in vacuum, a radiation field, is a linear superposition of elementary electromagnetic waves. In this chapter we address another fundamental problem of classical electrodynamics: the determination of the electromagnetic field generated by an arbitrary distribution of moving charges, which amounts to solve Maxwell's equations in the presence of a generic four-current j^μ. As first application of the general solution we will derive the electromagnetic field generated by a particle in uniform linear motion, distinguishing the cases of massive and massless particles which create, actually, fields with radically different characteristics. Next, in Chap. 7, we shall apply the solution to determine the electromagnetic field created by a particle in arbitrary motion. To solve Maxwell's equations in the presence of charges we will resort to a standard approach: the *Green function method*. Already in the simple case of a particle in uniform linear motion this method reveals its efficiency, deriving from its high degree of flexibility, as well as its limits, deriving from its failure if applied to the *light-like* trajectories of massless particles. We will solve this problem, in full generality, in Chap. 17, resorting to an alternative method.

In the presence of charges the electromagnetic field satisfies the equations

$$\partial_\mu F^{\mu\nu} = j^\nu, \qquad F^{\mu\nu} = \partial^\mu A^\nu - \partial^\nu A^\mu,$$

which in Lorenz gauge become

$$\Box A^\mu = j^\mu, \tag{6.1}$$

$$\partial_\mu A^\mu = 0. \tag{6.2}$$

As will become clear later on, in this case the residual gauge transformations 5.14 play no role. Equations (6.1) and (6.2) form a system of linear nonhomogeneous differential equations. Correspondingly, its general solution is obtained by adding to

© Springer International Publishing AG, part of Springer Nature 2018
K. Lechner, *Classical Electrodynamics*, UNITEXT for Physics,
https://doi.org/10.1007/978-3-319-91809-9_6

a particular solution, A_{ret}^{μ}, the general solution, A_{in}^{μ}, of the associated homogeneous system:

$$A^{\mu} = A_{ret}^{\mu} + A_{in}^{\mu}. \tag{6.3}$$

The potential A_{in}^{μ} is hence the general solution of the system

$$\Box A_{in}^{\mu} = 0, \qquad \partial_{\mu} A_{in}^{\mu} = 0,$$

and it corresponds thus to a radiation field, generic superposition of electromagnetic waves. Being unrelated to the source j^{μ}, this vector potential represents the *external* field. Conversely, the vector potential A_{ret}^{μ} represents the field *causally* generated by the four-current j^{μ} through Eqs. (6.1) and (6.2), whose solution we will address in the following sections. The subscripts *in* and *ret* stand for *incoming* and *retarded*, respectively. This terminology is related to the convention that the radiation associated with A_{in}^{μ}, which is superposed on the *retarded* electromagnetic field associated with A_{ret}^{μ}, without disturbing it, *enters* from infinity. Ultimately, this interpretation originates from the spontaneous violation of time-reversal invariance occurring in classical electrodynamics, see Sect. 6.2.3. In the remainder of this chapter we will ignore the radiation field A_{in}^{μ} and, correspondingly, denote A_{ret}^{μ} simply by A^{μ}. As mentioned above, an efficient technique for solving partial differential equations like (6.1) is represented by the Green function method. Before applying this method to (6.1), in the next section we illustrate it in the case of a simpler, though physically relevant, equation.

6.1   Green Function Method: Poisson Equation

Consider the three-dimensional *Poisson equation* in the unknown F

$$-\nabla^2 F(\mathbf{x}) = \varphi(\mathbf{x}), \tag{6.4}$$

in which φ is a known term. For concreteness we shall assume that

$$F \in \mathcal{S}'(\mathbb{R}^3) = \mathcal{S}', \qquad \varphi \in \mathcal{S}(\mathbb{R}^3) = \mathcal{S},$$

although the solutions we will find maintain their validity also if φ belongs to a particular subset of $\mathcal{S}'$, see below. If we interpret F as the electrostatic potential A^0, and φ as the charge density j^0, the *source* of A_0, Eq. (6.4) becomes the fundamental equation of *electrostatics*. Inspired by this interpretation we add the physical condition that F vanishes, sufficiently fast, at infinity

$$\lim_{|\mathbf{x}| \to \infty} F(\mathbf{x}) = 0. \tag{6.5}$$

Of course, in general it does not make sense to impose on a *distribution* an asymptotic condition like (6.5). Notwithstanding, we will see that in the case at hand, for $\varphi \in \mathcal{S}$, all solutions F of (6.4) are *regular* distributions, i.e. *functions*, and consequently the condition (6.5) is well posed. In particular, we will show that under this asymptotic condition the Poisson equation admits a *unique* solution. Nonetheless, in Sect. 6.1.3 we shall also analyze the *general* solution of the Poisson equation, irrespective of the validity of (6.5).

6.1.1   Particular Solution

Since the Poisson equation is a linear nonhomogeneous differential equation, its general solution is obtained by adding to a *particular* solution the general solution of the associated homogenous equation, namely the *Laplace equation* $\nabla^2 F = 0$. Obviously, the particular solution is not unique, but we can circumscribe it by adding further requirements. We begin with the observation that Eq. (6.4) is *jointly* linear in F and φ, in the sense that a solution relative to the source $\varphi = \varphi_1 + \varphi_2$ can be obtained by adding the solutions F_1 and F_2, relative to φ_1 and φ_2. Ignoring for a moment the regularity properties of the quantities involved, it is then natural to assume that the value of F at a point $\mathbf{x}$ depends *linearly* on the values $\varphi(\mathbf{y})$ of the source at *all* points $\mathbf{y}$. In other words, we assume that for every fixed $\mathbf{x}$ the number $F(\mathbf{x})$ defines a "linear functional" $f_\mathbf{x}$ on the space of functions φ, such that

$$F(\mathbf{x}) = f_\mathbf{x}(\varphi).   \tag{6.6}$$

In symbolic notation the functional $f_\mathbf{x}$ is associated with a function of two variables $f_\mathbf{x}(\mathbf{y}) = g(\mathbf{x}, \mathbf{y})$, so that the relation (6.6) can be written in the form of the integral

$$F(\mathbf{x}) = \int g(\mathbf{x}, \mathbf{y})\, \varphi(\mathbf{y})\, d^3 y.   \tag{6.7}$$

Euclidean invariance. To constrain the form of the function $g(\mathbf{x}, \mathbf{y})$ we return to the electrostatic interpretation of the Poisson equation, requiring that its solutions are invariant under the *Euclidean group* of transformations

$$\mathbf{x} \rightarrow \mathbf{x}' = R\mathbf{x} + \mathbf{a}, \quad R \in O(3), \quad \mathbf{a} \in \mathbb{R}^3.$$

Under such a transformation the electrostatic potential and the charge density remain, indeed, invariant

$$F'(\mathbf{x}') = F(\mathbf{x}), \qquad \varphi'(\mathbf{x}') = \varphi(\mathbf{x}).$$

On the other hand, in the transformed coordinate system Eq. (6.7) reads[1]

$$F'(\mathbf{x}') = \int g(\mathbf{x}', \mathbf{y}')\, \varphi'(\mathbf{y}')\, d^3y'. \tag{6.8}$$

Equating this expressions to (6.7), since $d^3y' = d^3y$ we see that the function $g(x, y)$ must obey the constraint

$$g(\mathbf{x}', \mathbf{y}') = g(\mathbf{x}, \mathbf{y}) \tag{6.9}$$

for all elements of the Euclidean group, and for all $\mathbf{x}$ and $\mathbf{y}$. Choosing in (6.9) $R = 1$ and $\mathbf{a} = -\mathbf{y}$, we obtain

$$g(\mathbf{x} - \mathbf{y}, 0) = g(\mathbf{x}, \mathbf{y}) \equiv g(\mathbf{x} - \mathbf{y}).$$

Setting in (6.9) $\mathbf{a} = 0$, we find that the function $g(\,\cdot\,)$ must be rotation invariant: $g(R\mathbf{x}) = g(\mathbf{x})$, for all $R \in O(3)$. Thus, $g(\mathbf{x})$ can depend on the vector $\mathbf{x}$ only through its magnitude $|\mathbf{x}|$. Eventually, the *ansatz* (6.7) assumes the form

$$F(\mathbf{x}) = \int g(\mathbf{x} - \mathbf{y})\, \varphi(\mathbf{y})\, d^3y. \tag{6.10}$$

Recalling the definition of the convolution (2.69), this integral can be recast also in the form

$$F = g * \varphi. \tag{6.11}$$

Once F is written as in (6.11), since φ belongs to $\mathcal{S}$, F belongs to $\mathcal{S}'$ whenever g belongs to $\mathcal{S}'$. More precisely, as recalled in Sect. 2.3.1, the convolution between a generic distribution and a test function is a *regular* distribution, i.e. a function, belonging to O_M.

Green function. Given the representation (6.11), the Poisson equation translates now into an equation for g. Inserting (6.11) in (6.4), and using the properties of the convolution (2.70) and (2.78), we find

$$-\nabla^2 F = -\nabla^2 (g * \varphi) = -\nabla^2 g * \varphi = \varphi.$$

Since φ is arbitrary, g must thus satisfy the *kernel equation*

$$-\nabla^2 g(\mathbf{x}) = \delta^3(\mathbf{x}). \tag{6.12}$$

Clearly one arrives to the same conclusion if, proceeding formally, one swaps in (6.10) the derivatives with the integral sign:

---

[1]The function $g(\cdot, \cdot)$ must be the same in all Cartesian coordinate systems, because otherwise an observer would be able to distinguish one coordinate system from another, in contrast to Euclidean invariance. In other words, we must have $g(\mathbf{x}', \mathbf{y}') = g(\mathbf{x}, \mathbf{y})$, rather than $g'(\mathbf{x}', \mathbf{y}') = g(\mathbf{x}, \mathbf{y})$.

$$-\nabla^2 F(\mathbf{x}) = -\int \nabla^2 g(\mathbf{x} - \mathbf{y})\, \varphi(\mathbf{y})\, d^3y = \varphi(\mathbf{x}) \;\Rightarrow\; -\nabla^2 g(\mathbf{x} - \mathbf{y}) = \delta^3(\mathbf{x} - \mathbf{y}).$$

Equation (6.12) identifies g as the *Green function* of the Laplacian, sometimes also called *propagator*, or *integral kernel*, of the partial differential equation. Notice that, would we not adopt the distributional framework, the basic equation (6.12) would be meaningless. The *Green function method* requires first to solve the kernel equation (6.12) for g, and then to write the solution of the original equation in the well-defined form (6.11). The latter, for a sufficiently regular kernel g, can be rewritten in the integral form (6.10). The efficiency of the method relies on the fact that the solution of Eq. (6.4) – which a priori should be solved separately for each source φ – is reduced to the solution of a single equation, the kernel equation (6.12).

Inverse of the Laplacian. The relations (6.11) and (6.12) allow for an operatorial interpretation of the Green function. As any other integral kernel, g induces in fact a linear operator $\mathcal{O}_g$ in the space of functions, defined by

$$\mathcal{O}_g : \varphi \to \mathcal{O}_g \varphi = g * \varphi.$$

In light of the identity

$$\left(-\nabla^2 \mathcal{O}_g\right) \varphi = -\nabla^2 (g * \varphi) = -(\nabla^2 g) * \varphi = \delta^3 * \varphi = \varphi \quad \leftrightarrow \quad -\nabla^2 \mathcal{O}_g = 1,$$

the integral operator $\mathcal{O}_g$ represents, hence, an inverse of the differential operator $-\nabla^2$. For this reason it is sometimes said that the kernel g constitutes an *inverse of the Laplacian*, and accordingly one introduces the formal notation

$$g = \frac{1}{-\nabla^2}.$$

Green function and particular solution. Now that the search for a particular solution of the Poisson equation is reduced to the kernel equation (6.12), according to the additional requirements above we must first solve the system

$$-\nabla^2 g(\mathbf{x}) = \delta^3(\mathbf{x}), \qquad g(\mathbf{x}) = g(|\mathbf{x}|), \qquad g \in \mathcal{S}'. \tag{6.13}$$

Thanks to the identity (2.119), a solution of this system is the Green function[2]

$$g(\mathbf{x}) = \frac{1}{4\pi |\mathbf{x}|}. \tag{6.14}$$

Inserting this expression in (6.10), we obtain as particular solution of the Poisson equation

---

[2]From the form of the general solution (6.27) of the Laplace equation it is not difficult to infer that the *general* solution of the system (6.13) is $g(\mathbf{x}) = 1/4\pi|\mathbf{x}| + C$, where C is a constant.

$$F(\mathbf{x}) = \frac{1}{4\pi} \int \frac{\varphi(\mathbf{y})}{|\mathbf{x} - \mathbf{y}|} \, d^3y, \tag{6.15}$$

which, in particular, reproduces the correct expression for the electrostatic potential created by the charge density $\varphi(\mathbf{x})$. Moreover, the solution (6.15) satisfies also the asymptotic condition (6.5). More precisely, we have the limit

$$\lim_{|\mathbf{x}| \to \infty} |\mathbf{x}| F(\mathbf{x}) = \frac{1}{4\pi} \lim_{|\mathbf{x}| \to \infty} \int \frac{|\mathbf{x}| \, \varphi(\mathbf{y})}{|\mathbf{x} - \mathbf{y}|} \, d^3y = \frac{1}{4\pi} \int \varphi(\mathbf{y}) \, d^3y = \frac{Q}{4\pi},$$

where the total *charge* Q is finite, because φ belongs to $\mathcal{S}$. It follows that $F(\mathbf{x})$ entails the asymptotic behavior

$$F(\mathbf{x}) \to \frac{Q}{4\pi |\mathbf{x}|}, \quad \text{as } |\mathbf{x}| \to \infty. \tag{6.16}$$

We end the section observing that the solution (6.15) may hold true even for some particular φ not belonging to $\mathcal{S}$, but rather to $\mathcal{S}'$. Consider, for example, the source corresponding to the charge density of a system of point-like particles

$$\varphi(\mathbf{x}) = \sum_{r=1}^{N} e_r \, \delta^3(\mathbf{x} - \mathbf{y}_r). \tag{6.17}$$

In this case the integral (6.15) remains, in fact, meaningful if we extend formally the symbolic rule (2.81) from a test function to less regular functions, obtaining thereby the known Coulomb potential

$$F(\mathbf{x}) = \frac{1}{4\pi} \sum_{r=1}^{N} e_r \int \frac{\delta^3(\mathbf{y} - \mathbf{y}_r)}{|\mathbf{x} - \mathbf{y}|} \, d^3y = \sum_{r=1}^{N} \frac{e_r}{4\pi |\mathbf{x} - \mathbf{y}_r|}. \tag{6.18}$$

This potential still entails the correct asymptotic behavior (6.16)

$$F(\mathbf{x}) \to \frac{\sum_{r=1}^{N} e_r}{4\pi |\mathbf{x}|}, \quad \text{as } |\mathbf{x}| \to \infty.$$

6.1.2 General Validity

As long as φ belongs to $\mathcal{S}$, the particular solution of the Poisson equation represented by the convolution (6.11) is a well-defined distribution F. Moreover, in this case it can be written in the integral form (6.15). On the other hand, in most physical applications φ does not belong to $\mathcal{S}$, being actually much more singular than a test function. In electrostatics, e.g., φ can be a singular macroscopic charge density, as the ones corresponding to charge distributions on surfaces or wires, or the even more singular charge density (6.17) of point-particles. In these cases φ does not belong to $\mathcal{S}$, but rather to $\mathcal{S}'$, and therefore, since in general the convolution of two distributions is ill-defined, a priori Eq. (6.11) is meaningless. However, if one of the two distributions is of *compact* support, their convolution can be given a precise meaning as a distribution, as we will now show.

Generalized convolution of distributions. Consider in a generic space $\mathbb{R}^D$ a distribution $g \in \mathcal{S}'$ and a test function $\varphi \in \mathcal{S}$, together with their convolution, in turn belonging to $\mathcal{S}'$,

$$A = g * \varphi. \tag{6.19}$$

Performing the Fourier transform of this relation, from the general rule (2.100) we find

$$\widehat{A}(k) = (2\pi)^{D/2}\, \widehat{g}(k)\, \widehat{\varphi}(k). \tag{6.20}$$

Since the Fourier transform maps $\mathcal{S}'$ into $\mathcal{S}'$, $\widehat{A}(k)$ is of course a distribution. In the case at hand it is easy to check it explicitly, in that the second member of (6.20) is the product of an element $\widehat{g} \in \mathcal{S}'$ and of an element $\widehat{\varphi} \in \mathcal{S}$. However, if φ is not a test function, but a *distribution of compact support*, Theorem IV of Sect. 2.3.1 ensures that $\widehat{\varphi}$ belongs to O_M, and correspondingly the right hand side of (6.20) still defines a distribution, see Sect. 2.3.1. In this case we can, thus, define the *generalized convolution* $g * \varphi$, as the distribution corresponding to the inverse Fourier transform of the right hand side of (6.20). It is easy to see that this definition preserves the basic rules (2.70) and (2.100).

The introduction of the generalized convolution allows, in turn, to generalize the Green function method to sources φ, which are distributions of compact support. In the particular case of the Poisson equation (6.4), as first step we must then evaluate analytically the Fourier transform $\widehat{g}(\mathbf{k})$ of the Green function (6.14). To do it quickly we proceed *formally*, i.e. we evaluate the Fourier transform by means of an integral, which is, actually, diverging. Resorting to rotation invariance to set $\mathbf{k} = (0, 0, k)$, and going over to polar coordinates, we obtain

$$
\begin{aligned}
\widehat{g}(\mathbf{k}) &= \frac{1}{(2\pi)^{3/2}} \int d^3x\, e^{-i\mathbf{k}\cdot\mathbf{x}}\, \frac{1}{4\pi|\mathbf{x}|} \\
&= \frac{1}{4\pi(2\pi)^{3/2}} \int_0^\infty r^2 dr \int_0^{2\pi} d\varphi \int_{-1}^1 d\cos\vartheta\, e^{-irk\cos\vartheta}\, \frac{1}{r} \\
&= \frac{i}{2(2\pi)^{3/2}k} \int_0^\infty dr\, \left(e^{-ikr} - e^{ikr}\right) = \frac{i}{2(2\pi)^{3/2}k} \int_{-\infty}^\infty dx\, e^{-ikx}\, \varepsilon(x) \\
&= \frac{i\,\widehat{\varepsilon}(k)}{2(2\pi)k},
\end{aligned}
$$

where in the last step we introduced the Fourier transform $\widehat{\varepsilon}(k)$ of the sign function $\varepsilon(x)$. The former can be expressed in terms of the principal value, see Eqs. (2.85) and (2.95),

$$\widehat{\varepsilon}(k) = -i\sqrt{\frac{2}{\pi}}\,\mathcal{P}\frac{1}{k}.$$

Since k is positive we obtain thus

$$\widehat{g}(\mathbf{k}) = \frac{1}{(2\pi)^{3/2}|\mathbf{k}|^2}. \tag{6.21}$$

Equation (6.20) then yields for the Fourier transform $\widehat{F}(\mathbf{k})$ of a particular solution $F(\mathbf{x})$ of the Poisson equation

$$\widehat{F}(\mathbf{k}) = \frac{\widehat{\varphi}(\mathbf{k})}{|\mathbf{k}|^2} \in \mathcal{S}'. \tag{6.22}$$

The expressions (6.21) and (6.22) satisfy, indeed, the algebraic equations obtained from the Fourier transform of Eqs. (6.12) and (6.4). If we choose, e.g., as source $\varphi(\mathbf{x})$ the distribution of compact support (6.17) we obtain

$$\widehat{\varphi}(\mathbf{k}) = \frac{1}{(2\pi)^{3/2}} \sum_{r=1}^{N} e_r\, e^{-i\mathbf{k}\cdot\mathbf{y}_r},$$

which clearly belongs to O_M, since it is of class C^{∞} and all its derivatives are bounded by a constant. In this case, the distributional inverse Fourier transform of the function (6.22) amounts, indeed, to the Coulomb potential (6.18).

In conclusion, whenever φ is a distribution of compact support, the formal expression of F (6.11), *defined* as the inverse Fourier transform of the right hand side of Eq. (6.22), belongs to $\mathcal{S}'$ and solves Eq. (6.4). If, moreover, F is not too singular, so that it can be expressed in the integral form (6.15), its asymptotic behavior is (6.16) with Q given by

$$Q = \int \varphi(\mathbf{y})\, d^3y = (2\pi)^{3/2}\,\widehat{\varphi}(0),$$

a finite number because $\widehat{\varphi}(\mathbf{k})$ belongs to O_M. Therefore, since all charge densities φ realized in nature are of compact support, and their most singular behavior is (6.17), the Green function method provides a *physically* as well as *mathematically* well-defined solution of the Poisson equation.

6.1.3 Uniqueness of the Solution and Laplace Equation

To check whether (6.11) is the unique *physical* solution of the Poisson equation, i.e. a solution satisfying (6.5), we must first of all analyze the general solution $F_0 \in \mathcal{S}'$ of the Laplace equation

$$\nabla^2 F_0(\mathbf{x}) = 0. \tag{6.23}$$

This equation admits, indeed, infinite linearly independent solutions, but none of them vanishes at infinity, in contrast with (6.5). To prove it we apply the Fourier transform to Eq. (6.23), mapping it into the algebraic equation

$$|\mathbf{k}|^2 \widehat{F}_0(\mathbf{k}) = 0. \tag{6.24}$$

The support of $\widehat{F}_0(\mathbf{k})$ is, hence, the origin $\mathbf{k} = 0$. Theorem III of Sect. 2.3.1 on distributions with point-like support then implies that $\widehat{F}_0(\mathbf{k})$ is a *finite* linear combination of $\delta^3(\mathbf{k})$ and its derivatives,

$$\widehat{F}_0(\mathbf{k}) = \sum_{n=1}^{N} C^{i_1 \cdots i_n} \partial_{i_1} \cdots \partial_{i_n} \delta^3(\mathbf{k}), \tag{6.25}$$

where $C^{i_1 \cdots i_n}$ are completely symmetric constant tensors. Inserting (6.25) in (6.24) we find that these tensors must be traceless (see the analogous problem for the wave equation in Sect. 5.2)

$$\delta_{i_1 i_2} C^{i_1 \cdots i_n} = 0. \tag{6.26}$$

Performing the inverse Fourier transform of (6.25) we obtain eventually the, infinitely many, solutions of the Laplace equation

$$F_0(\mathbf{x}) = \frac{1}{(2\pi)^{3/2}} \sum_{n=1}^{N} (-i)^n C^{i_1 \cdots i_n} x^{i_1} \cdots x^{i_n}. \tag{6.27}$$

However, since the functions (6.27) are polynomials, none of them vanishes at infinity, unless $F_0 = 0$. It follows that the convolution (6.11), with g given in (6.14), is the unique solution of the Poisson equation vanishing at infinity.

Green function method: general case. The Green function method generalizes naturally to a linear partial differential equation, in a D-dimensional space, of the form

$$P(\partial)F = \varphi, \tag{6.28}$$

where $P(\partial)$ is an arbitrary polynomial of the derivatives. By definition, the Green function g associated with this operator must satisfy the distributional kernel equation

$$P(\partial)\, g(x) = \delta^D(x),$$

and correspondingly a particular solution of Eq. (6.28) is given by

$$F = g * \varphi. \tag{6.29}$$

In fact, using the properties of the convolution (2.70) and (2.78), we find

$$P(\partial)\, F = P(\partial)\, (g * \varphi) = P(\partial)\, g * \varphi = \delta^D * \varphi = \varphi.$$

A *sufficient* condition for F in (6.29) to be a well-defined distributional solution of (6.28) is that φ is a distribution of compact support. If it is not so, in general the Green function method may fail to provide a solution to Eq. (6.28), in that the convolution in (6.29), even if considered as a *generalized* one, could be ill-defined. For explicit examples see Sects. 6.2.4 and 7.2.2.

6.2 The Field Generated by a Generic Current

In the presence of a generic current Maxwell's equations amount to

$$\Box A^\mu = j^\mu, \tag{6.30}$$

$$\partial_\mu A^\mu = 0. \tag{6.31}$$

As explained at the beginning of the chapter, we are not searching for the general solution of this system, but rather for the field generated *causally* by the current j^μ. This means that we are searching for a *specific* particular solution. Accordingly, since (6.30) is a linear nonhomogeneous partial differential equation, it is natural to resort again to the Green function method to solve the system. There is, however, a crucial difference between (6.30) and the Poisson equation (6.4): the source $j^\mu(x)$, being a function of space as well as time, is *not* a distribution of compact support, in contrast to the three-dimensional electrostatic charge density $\varphi(\mathbf{x})$. In this case, as observed at the end of the preceding section, the Green function method is not guaranteed to provide a solution of Eq. (6.30). Nevertheless, we can pursue the method *formally*, resorting to the symbolic notation, disregarding, momentarily, the question whether our calculations are mathematically meaningful. We postpone the answer to this fundamental question to Sect. 6.2.4, where we will see that it depends crucially on the regularity properties of j^μ. Below we first address the solution of Eq. (6.30), and subsequently we deal with the Lorenz gauge (6.31).

Poincaré invariance. The fact that Maxwell's equations (6.30) and (6.31) live in four-dimensional Minkowski space-time, whereas the Poisson equation (6.4) lives in three-dimensional space, leads to a further important distinctive feature between the two: while the symmetry group of the latter is the *Euclidean group*, the symmetry group of the former is the *Poincaré group*. As in the case of the Poisson equation, due to the joined linearity of Eq. (6.30) in A^μ and j^μ, we search for a solution of the form

$$A^\mu(x) = \int G(x, y) j^\mu(y) \, d^4y, \tag{6.32}$$

where the *Green function* $G(x, y)$ is a function of the four-dimensional coordinates x^μ and y^μ. The form of this function is, however, severely constrained by Poincaré invariance. In fact, if we perform a Poincaré transformation

$$x' = \Lambda x + a,$$

in the new inertial frame the solution (6.32) reads[3]

$$A'^\mu(x') = \int G(x', y') j'^\mu(y') \, d^4y'. \tag{6.33}$$

According to the transformation laws

$$A'^\mu(x') = \Lambda^\mu{}_\nu A^\nu(x), \qquad j'^\mu(y') = \Lambda^\mu{}_\nu j^\nu(y), \qquad d^4y' = d^4y,$$

from (6.33) we find

$$A^\mu(x) = \int G(x', y') j^\mu(y) \, d^4y.$$

Comparing this equation with (6.32) we deduce that G is subject to the invariance constraint[4]

$$G(\Lambda x + a, \Lambda y + a) = G(x, y), \quad \forall \Lambda \in SO(1,3)_c, \quad \forall a \in \mathbb{R}^4. \tag{6.34}$$

Choosing $\Lambda = 1$ and $a = -y$, we obtain

$$G(x - y, 0) = G(x, y) \equiv G(x - y). \tag{6.35}$$

Setting in (6.34) $a = 0$, and keeping Λ generic, we then conclude that the function $G(x)$ must be invariant under proper Lorentz transformations:

---

[3] As in the case of the Poisson equation, $G(x, y)$ must not be considered as a scalar *field* in x and y, but rather as an *invariant function* of x and y, with a fixed functional dependence. This function must be the same in all inertial frames, because otherwise two currents with the same functional dependence on the coordinates in two different inertial frames would give rise to vector potentials with different functional dependencies, in contrast to the *relativity principle*. In other words, we must have $G(x', y') = G(x, y)$, rather than $G'(x', y') = G(x, y)$.

[4] The reason why we consider the *proper* Lorentz group $SO(1,3)_c$, rather than $O(1,3)$, will become clear soon.

$$G(\Lambda x) = G(x), \quad \forall \Lambda \in SO(1,3)_c. \tag{6.36}$$

Thanks to (6.35), the *ansatz* (6.32) now reads

$$A^\mu(x) = \int G(x-y)j^\mu(y)\,d^4y, \tag{6.37}$$

which corresponds, once more, to the convolution between the Green function and the source

$$A^\mu = G * j^\mu. \tag{6.38}$$

Finally, inserting (6.37) in the Maxwell equation (6.30), we find

$$\Box A^\mu(x) = \int \Box G(x-y)j^\mu(y)\,d^4y = j^\mu(x).$$

Imposing that this condition holds for all currents, we derive the kernel equation

$$\Box G(x) = \delta^4(x), \tag{6.39}$$

which identifies $G(x)$ as a Green function of the *d'Alembertian*. The search for a particular solution of (6.30) has, hence, been reduced to the solution of the kernel equation (6.39), subject to the constraint (6.36).

Time ordering. As we will see in the next section, the conditions (6.36) and (6.39) do not determine the Green function uniquely. Hence, we add a further physical requirement, concerning the *causal* propagation of the electromagnetic field: the potential $A^\mu(x)$ at a point x cannot depend on the values $j^\mu(y)$ of the current at points y occurring chronologically *after* x, that is to say, at points y for which $y^0 > x^0$. In light of the relation (6.37) this implies that $G(x-y)$ must vanish whenever $y^0 > x^0$, i.e.

$$G(x) = 0, \quad \forall x^0 < 0.$$

In the following we will see that with this additional requirement the conditions (6.36) and (6.39) admit, indeed, a unique solution. The resulting Green function is called *retarded* Green function and is frequently denoted by G_{ret}, while the corresponding vector potential is called *retarded* potential and denoted by

$$A_{\text{ret}}^\mu = G_{\text{ret}} * j^\mu. \tag{6.40}$$

This terminology stems, in reality, from quantum field theory, where, for technical reasons, one introduces in addition to G_{ret} the *advanced* Green function G_{adv}, satisfying in addition to (6.36) and (6.39) the specular condition

$$G(x) = 0, \quad \forall x^0 > 0. \tag{6.41}$$

With this Green function one associates the *advanced* potential

$$A^\mu_{\mathrm{adv}} = G_{\mathrm{adv}} * j^\mu, \tag{6.42}$$

still satisfying Maxwell's equations (6.30) and (6.31), so that the general solution (6.3) could be recast also in the form $A^\mu = A^\mu_{\mathrm{adv}} + A^\mu_{\mathrm{out}}$, where A^μ_{out} is an *outgoing* radiation potential different from A^μ_{in}. However, since A_{adv} does not respect causality, this solution plays no role in our classical analysis.

6.2.1   Retarded Green Function

The retarded Green function is defined by the conditions

$$\Box\, G(x) = \delta^4(x), \tag{6.43}$$

$$G(\Lambda x) = G(x), \quad \forall\, \Lambda \in SO(1,3)_c, \tag{6.44}$$

$$G(x) = 0, \quad \forall\, x^0 < 0. \tag{6.45}$$

Before facing the solution of this system we remark that a function $G(x)$ satisfying conditions (6.44) and (6.45) vanishes automatically for all x^μ, such that $x^2 < 0$, i.e. for all space-like x^μ. To prove this property we choose the z axis along the direction of $\mathbf{x}$, so that $x^\mu = (x^0, 0, 0, x^3)$. The condition $x^2 < 0$ is then equivalent to the inequality $|x^3| > |x^0|$. It is then easy to see that there exists a proper Lorentz transformation, more precisely, a special Lorentz transformation along the z axis, so that for the transformed vector $x' = \Lambda x$ one has $x'^0 < 0$. The conditions (6.44) and (6.45) then imply that $G(x) = G(x') = 0$. We postpone the interpretation of this result to Sect. 6.2.2, where we analyze the causality properties of a generic Green function.

Uniqueness. Before addressing the explicit solution of the system (6.43)–(6.45) we prove that, if it admits solutions, the solution is unique. This is equivalent to prove that the homogeneous equation associated with (6.43), i.e. the wave equation

$$\Box\, F = 0, \tag{6.46}$$

has no solution satisfying (6.44) and (6.45). We shall first determine all solutions of Eq. (6.46) satisfying the Lorentz-invariance condition (6.44), showing then that none of them satisfies the causality constraint (6.45). To solve Eq. (6.46) it is convenient to consider its Fourier transform. The condition (6.44), i.e. $F(\Lambda x) = F(x)$, then implies that also the Fourier transform $\widehat{F}(k)$ is Lorentz invariant. In fact, performing in the integral defining $\widehat{F}(k)$ the change of variables $x = \Lambda y$, we obtain

$$\widehat{F}(\Lambda k) = \frac{1}{(2\pi)^2} \int e^{-i\Lambda k \cdot x} F(x) \, d^4x = \frac{1}{(2\pi)^2} \int e^{-i\Lambda k \cdot \Lambda y} F(\Lambda y) \, d^4y$$

$$= \frac{1}{(2\pi)^2} \int e^{-i\, k \cdot y} F(y) \, d^4y = \widehat{F}(k).$$

We must, thus, solve the system

$$k^2 \widehat{F}(k) = 0, \qquad \widehat{F}(\Lambda k) = \widehat{F}(k), \ \ \forall \Lambda \in SO(1,3)_c.$$

The wave equation $k^2 \widehat{F}(k) = 0$ implies that the support of $\widehat{F}(k)$ is a subset of the light cone $k^2 = 0$, and the invariance under $SO(1,3)_c$ requires its support to be Lorentz invariant. Therefore, the support of $\widehat{F}(k)$ must be $I)$ either the whole light cone $k^0 = \pm|\mathbf{k}|$, or $II)$ the point $k^\mu = 0$. The corresponding two types of solutions of the wave equation have been analyzed in Sect. 5.2, where we have found (see (5.29) and (5.32))

$$\widehat{F}_I(k) = \delta(k^2) \, f(k), \tag{6.47}$$

$$\widehat{F}_{II}(k) = \sum_{n=1}^{N} C^{\mu_1 \cdots \mu_n} \, \partial_{\mu_1} \cdots \partial_{\mu_n} \delta^4(k), \quad C_\nu{}^{\nu \mu_3 \cdots \mu_n} = 0. \tag{6.48}$$

We must now select those solutions that are invariant under proper Lorentz transformations. Concerning the solutions of type I, due to rotation invariance – implied by Lorentz invariance – on the light cone the function $f(k)$ can depend on k^μ only through the component $k^0 = \pm|\mathbf{k}|$. However, the only two functions of k^0 which are Lorentz invariant on the light cone are the constant and the sign function $\varepsilon(k^0)$. Taking the reality condition $\widehat{F}^*(k) = \widehat{F}(-k)$ into account, we obtain thus the two linearly independent solutions

$$\widehat{F}_1(k) = \delta(k^2), \qquad \widehat{F}_2(k) = i\varepsilon(k^0)\, \delta(k^2). \tag{6.49}$$

Concerning the solutions of type II, Lorentz invariance requires the coefficients $C^{\mu_1 \cdots \mu_n}$ to be *invariant tensors*, see Sect. 1.3.1. The odd-rank tensors must then vanish, while the even-rank ones must be proportional to symmetrized products of Minkowski metrics:

$$C^{\mu_1 \cdots \mu_n} = A_n \, \eta^{(\mu_1 \mu_2} \cdots \eta^{\mu_{n-1}\mu_n)},$$

where the A_n are constants. But, according to the constraint in (6.48), these tensors must also be traceless

$$C_\nu{}^{\nu \mu_3 \cdots \mu_n} = A_n \, \frac{n+2}{n-1} \, \eta^{(\mu_3 \mu_4} \cdots \eta^{\mu_{n-1}\mu_n)} = 0,$$

and so all A_n with $n \neq 0$ must vanish. For $n = 0$ we obtain a third independent solution

$$\widehat{F}_3(k) = \delta^4(k). \tag{6.50}$$

Notice that, formally, all three solutions can be retrieved from the type I solutions (5.31) of the wave equation, by setting $\varepsilon(\mathbf{k}) = 1$, i, $\omega\delta^3(\mathbf{k})$, respectively. Finally, to impose the condition (6.45), we must know the inverse Fourier transforms[5] of the distributions (6.49) and (6.50)

$$F_1(x) = -\frac{1}{\pi}\mathcal{P}\frac{1}{x^2}, \qquad F_2(x) = -\varepsilon(x^0)\,\delta(x^2), \qquad F_3(x) = \frac{1}{(2\pi)^2}. \tag{6.51}$$

The composite principal value appearing in F_1 refers to the variable x^0 of $x^2 = (x^0)^2 - |\mathbf{x}|^2$, see (2.86). As we see, all three solutions are invariant under $SO(1,3)_c$ – as must be by construction – but none of them satisfies (6.45). The retarded Green function, if it exists, is therefore unique.

Determination of the retarded Green function. We begin the solution of the system (6.43)–(6.45) observing that (6.44) requires the Green function to be invariant under spatial rotations:

$$G(t, R\mathbf{x}) = G(t, \mathbf{x}), \quad \forall R \in SO(3).$$

This means that G can depend on $\mathbf{x}$ only through its length $r = |\mathbf{x}|$, so that we may set

$$G(t, \mathbf{x}) = G(t, r).$$

Consider now the three-dimensional space without the origin, $\mathbf{x} \neq 0$, where (6.43) reduces to the wave equation $\Box\, G = 0$. In this region we are allowed to use polar coordinates and we can, thus, write the Laplacian as in Eq. (5.132) of Problem 5.7. Since G does not depend on the angles, we obtain therefore

$$\Box G = \left(\frac{\partial^2}{\partial t^2} - \frac{1}{r}\frac{\partial^2}{\partial r^2}r\right)G = \frac{1}{r}\left(\frac{\partial^2}{\partial t^2} - \frac{\partial^2}{\partial r^2}\right)(rG) = 0. \tag{6.52}$$

This means that the product rG must satisfy the one-dimensional wave equation, whose general solution is known, see Problem 5.8. Consequently, G must be of the form

$$G(t, r) = \frac{1}{r}\left(f(t-r) + g(t+r)\right),$$

where f and g are arbitrary functions. However, since according to the causality condition (6.45) G must vanish for all $t < 0$, we conclude that $g = 0$. In fact, since r is positive, for $t < 0$ the sum $t + r$ can become any real number, while the differ-

---

[5]The inverse Fourier transforms (6.51) can be retrieved from Eqs. (18.59)–(18.61) of Sect. 18.4, taking the limit for $M \to 0$.

ence $t - r$ takes only negative values. Therefore, the function g must vanish for all values of its argument, while the function f can be different from zero for positive arguments. We have thus

$$G = \frac{1}{r} f(t - r). \tag{6.53}$$

Finally, to determine f we impose that the expression (6.53) satisfies the kernel equation (6.43) in the sense of distributions

$$\frac{\partial^2 G}{\partial t^2} - \nabla^2 G = \delta^3(\mathbf{x})\,\delta(t). \tag{6.54}$$

Denoting the derivative of f with respect to its argument with a *prime*, we obtain

$$\frac{\partial^2 G}{\partial t^2} = \frac{1}{r} f''(t - r). \tag{6.55}$$

To evaluate $\nabla^2 G$ we must proceed with more caution, since the factor $1/r$ is singular at $r = 0$. Nevertheless, we can apply Leibniz's rule, if we suppose that $f(t - r)$ is regular at $r = 0$, a property that we will verify *a posteriori*,

$$\nabla^2 G = \left(\nabla^2 \frac{1}{r}\right) f(t - r) + \frac{1}{r} \nabla^2 f(t - r) + 2\left(\nabla \frac{1}{r}\right) \cdot \nabla f(t - r). \tag{6.56}$$

When acting on functions which are regular at $r = 0$, the Laplacian can again be written as in Eq. (5.132), and so we obtain

$$\nabla^2 f(t - r) = \frac{1}{r}\frac{\partial^2}{\partial r^2}(r f(t - r)) = f''(t - r) - \frac{2}{r} f'(t - r).$$

Moreover, we have

$$\nabla \frac{1}{r} = -\frac{\mathbf{x}}{r^3}, \quad \nabla f(t - r) = -\frac{\mathbf{x}}{r} f'(t - r),$$

together with the distributional identity, see (2.119),

$$\nabla^2 \frac{1}{r} = -4\pi\delta^3(\mathbf{x}).$$

Inserting these relations in (6.56) we see that the first derivatives of f cancel out, and we remain with

$$\nabla^2 G = -4\pi\delta^3(\mathbf{x}) f(t) + \frac{1}{r} f''(t - r). \tag{6.57}$$

With (6.55) and (6.57) the kernel equation (6.54) reduces to

$$\frac{\partial^2 G}{\partial t^2} - \nabla^2 G = 4\pi \delta^3(\mathbf{x}) f(t) = \delta^3(\mathbf{x}) \delta(t),$$

so that we must have

$$f(t) = \frac{\delta(t)}{4\pi}.$$

Correspondingly, Eq. (6.53) yields the retarded Green function

$$G_{\text{ret}}(x) = \frac{1}{4\pi r} \delta(t - r). \tag{6.58}$$

This expression satisfies the causality condition (6.45), but, at first glance, it does not seem to be Lorentz invariant. Nonetheless, the identity (2.77) implies the decomposition

$$\delta(x^2) = \delta(t^2 - r^2) = \frac{1}{2r} \big(\delta(t - r) + \delta(t + r) \big),$$

so that, thanks to the equality $H(t)\,\delta(t + r) = H(-r)\,\delta(t + r) = 0$, we can recast the expression (6.58) in the manifestly Lorentz-invariant form

$$G_{\text{ret}}(x) = \frac{1}{2\pi} H(x^0)\,\delta(x^2). \tag{6.59}$$

In fact, on the light cone $x^2 = 0$ the sign of x^0 is invariant under $SO(1,3)_c$, see Theorem II of Sect. 5.2.3. Notice also that $G_{\text{ret}}(x)$ vanishes for $x^2 < 0$, as anticipated at the beginning of this section. Similarly, for the advanced Green function, satisfying (6.43), (6.44) and (6.41), one obtains

$$G_{\text{adv}}(x) = \frac{1}{4\pi r} \delta(t + r) = \frac{1}{2\pi} H(-x^0)\,\delta(x^2). \tag{6.60}$$

We have, hence, found two Green functions satisfying (6.43) and (6.44), both belonging to $\mathcal{S}'$, see Problem 6.1 In particular, they satisfy the equations $\Box\, G_{\text{ret}} = \delta^4(x) = \Box\, G_{\text{adv}}$. A priori we could, thus, have chosen as Green function any combination of the form

$$G_a = a\, G_{\text{ret}} + (1 - a)\, G_{\text{adv}}, \qquad \Box\, G_a = \delta^4(x), \tag{6.61}$$

with a an arbitrary real number, but the causality condition (6.45) selects, eventually, the value $a = 1$. We end the section noticing a simple relation existing between the advanced and retarded Green functions, and the antisymmetric kernel D (5.50):

$$D = G_{\text{ret}} - G_{\text{adv}}. \tag{6.62}$$

This relation implies immediately that D satisfies the characteristic equation

$$\Box D = 0, \tag{6.63}$$

which identifies it as the *propagator* of a free field, see the solutions (5.48) and (5.125).

6.2.2 Retarded Vector Potential

Henceforth, we denote the retarded Green function G_{ret} simply with the symbol G. Inserting its expression (6.59) in (6.37) we obtain the retarded vector potential, in a manifestly covariant form,

$$A^\mu(x) = \frac{1}{2\pi} \int H(x^0 - y^0)\, \delta\left((x-y)^2\right) j^\mu(y)\, d^4y. \tag{6.64}$$

Using the equivalent expression (6.58) we can integrate over y^0, obtaining the alternative representation

$$A^\mu(t, \mathbf{x}) = \frac{1}{4\pi} \int \left(\int \frac{1}{|\mathbf{x}-\mathbf{y}|}\, \delta(t - y^0 - |\mathbf{x}-\mathbf{y}|)\, j^\mu(y^0, \mathbf{y})\, dy^0 \right) d^3y$$

$$= \frac{1}{4\pi} \int \frac{1}{|\mathbf{x}-\mathbf{y}|}\, j^\mu(t - |\mathbf{x}-\mathbf{y}|, \mathbf{y})\, d^3y. \tag{6.65}$$

In the following we will make use of both representations (6.64) and (6.65): The former has the advantage of being manifestly Lorentz invariant, whereas the latter involves one integration less. It remains to verify that the vector potential (6.64) satisfies the Lorenz gauge (6.31). For this purpose it is convenient to return to the abstract notation (6.38) and to use the property (2.70) of the convolution. In this way we obtain

$$\partial_\mu A^\mu = \partial_\mu(G * j^\mu) = G * \partial_\mu j^\mu = 0,$$

where in the conclusion we used current conservation. Notice that the potential $A^\mu = G * j^\mu$ fulfils the Lorenz gauge $\partial_\mu A^\mu = 0$, independently of the form of the Green function $G(x)$.

Green functions and relativistic causality. We analyze now the causal structure of the retarded potential (6.64). We derived the expression (6.64) requiring that the Green function vanishes at negative times, ensuring so that future events cannot influence past events. A priori, such a requirement is in apparent conflict with relativity, since in general the time ordering between two events is not preserved by a Lorentz transformation, not even by a *proper* one. To preserve the time ordering we must impose the further constraint that the two events can influence each other only if they are *time-like* or *light-like* separated. In fact, according to *relativistic causality* an event y can influence an event x only if both inequalities

$$(x - y)^2 \geq 0, \qquad x^0 \geq y^0, \tag{6.66}$$

are satisfied. The events x define the *future light cone* L_+ of y, a set that we know indeed to be invariant under proper Lorentz transformations, see Sect. 5.2.3. If x and y satisfy the conditions (6.66), their time ordering is, hence, the same in all reference frames. Correspondingly, a generic *relativistic causal* Green function must satisfy the conditions

$$G(x) = 0, \quad \forall\, x^0 < 0, \tag{6.67}$$
$$G(x) = 0, \quad \forall\, x^2 < 0. \tag{6.68}$$

In other words, *the support of the Green function must be contained in the future light cone* L_+. The retarded Green function (6.59) satisfies the conditions (6.67) and (6.68), and is thus a causal relativistic Green function, and moreover its support belongs to the *boundary* of the light cone. Correspondingly, the potential $A^\mu(x)$ (6.64) at a point x is causally connected only to points y of the current which are light-like separated from x and belong to its past. From a physical point of view this means that in the electromagnetic field the information propagates at the speed of light *from the charged particles to the observation point*, and not vice versa. We will come back to this issue in Sect. 6.2.3.

Retarded time and delay. It is instructive to compare the expression (6.65) with the solution (6.15) of the Poisson equation. We report the latter in its electrostatic version, turning on the time,

$$A^0(t, \mathbf{x}) = \frac{1}{4\pi c} \int \frac{1}{|\mathbf{x} - \mathbf{y}|}\, j^0(t, \mathbf{y})\, d^3 y. \tag{6.69}$$

To compare (6.65) properly with (6.69), we rewrite the former restoring the speed of light:

$$A^\mu(t, \mathbf{x}) = \frac{1}{4\pi c} \int \frac{1}{|\mathbf{x} - \mathbf{y}|}\, j^\mu \left(t - \frac{|\mathbf{x} - \mathbf{y}|}{c}, \mathbf{y} \right) d^3 y. \tag{6.70}$$

As we see, the unique difference between the two solutions is the appearance in (6.70) of the *delay* $|\mathbf{x} - \mathbf{y}|/c$ in the time argument of the current, a delay that equals the time spent by a light signal to travel from the point $\mathbf{y}$, where the charge is located, to the observation point $\mathbf{x}$, where the electromagnetic field is evaluated. Correspondingly, at a time t the field at a point $\mathbf{x}$ depends on the value of the current at a point $\mathbf{y}$ not at the same time t, but rather at the *retarded time* $t' = t - |\mathbf{x} - \mathbf{y}|/c$. Conversely, in the non-relativistic potential (6.69) one assumes implicitly an *instantaneous* action at a distance, propagating at infinite speed, so that no delay occurs.

6.2.3 Spontaneous Violation of Time-Reversal Invariance

The time reversal transformation is a discrete element of the Lorentz group $O(1,3)$, not belonging to the proper group $SO(1,3)_c$, sending t in $-t$, while leaving the spatial coordinates invariant, see Sect. 1.4.3. It is therefore represented by a matrix $\mathcal{T}^\mu{}_\nu$ with elements

$$\mathcal{T}^0{}_0 = -1, \qquad \mathcal{T}^i{}_j = \delta^i{}_j, \qquad \mathcal{T}^\mu{}_\nu = 0, \ \text{if } \mu \neq \nu. \qquad (6.71)$$

If a physical system is invariant under time reversal, any process recorded by video camera and then screened in a reversed chronological order appears as *realistic* as the original one, meaning that it appears as a process that could, actually, occur in nature. By recording, e.g., an elastic collision between two particles, if their interaction respects time-reversal invariance, its projection on a screen in a reversed order shows a – generally different – collision, which may nevertheless occur in nature. An unbiased spectator would not be able to recognize, therefore, whether the projection is in the initial chronological order, or in the reversed one. Before discussing the spontaneous violation of time-reversal invariance in classical electrodynamics, we briefly recall how this symmetry is realized in classical mechanics.

Time reversal in classical mechanics. Consider Newton's second law for a particle subject to a generic position- and velocity-dependent force

$$m\mathbf{a} = \mathbf{F}(\mathbf{y}, \mathbf{v}). \qquad (6.72)$$

We analyze now the conditions under which this equation maintains its form under the inversion of the time axis $t \to t^* = -t$, inducing the transformations

$$\mathbf{y}^*(t^*) = \mathbf{y}(t), \quad \mathbf{v}^*(t^*) = \frac{d\mathbf{y}^*}{dt^*} = -\mathbf{v}(t), \quad \mathbf{a}^*(t^*) = \frac{d^2\mathbf{y}^*}{dt^{*2}} = \mathbf{a}(t). \qquad (6.73)$$

Since under these replacements the first member of (6.72) remains the same, if we want this equation to maintain its form, the force must satisfy the condition

$$\mathbf{F}(\mathbf{y}, -\mathbf{v}) = \mathbf{F}(\mathbf{y}, \mathbf{v}). \qquad (6.74)$$

This condition is certainly satisfied by a velocity-independent force. However, Eq. (6.74) holds true also if the force depends on the velocity vector $\mathbf{v}$ only through its magnitude and its direction, but not through its *orientation*.[6] An important consequence of time-reversal invariance is that, if the trajectory $\mathbf{y}(t)$ satisfies Newton's second law, this law is satisfied also by the trajectory $\mathbf{y}^*(t) = \mathbf{y}(-t)$. The trajectories $\mathbf{y}(t)$ and $\mathbf{y}^*(t)$ correspond, obviously, to the same *orbits*, but in the case of $\mathbf{y}^*(t)$ the orbit is traced *backwards* in time, with all velocities reversed. If the

---

[6]Examples of forces of this kind are $\mathbf{F}_1 = v^2\mathbf{b}$, and $\mathbf{F}_2 = (\mathbf{b} \cdot \mathbf{v})\,\mathbf{v}$, where $\mathbf{b}$ is a constant vector.

force depends, conversely, also on the orientation of $\mathbf{v}$, then Newton's law violates *explicitly* time-reversal invariance. In this case it happens that, if the trajectory $\mathbf{y}(t)$ obeys Eq. (6.72), the trajectory $\mathbf{y}^*(t)$ in general does not. A force that violates time-reversal invariance is, for instance, the viscous force $\mathbf{F} = -k\mathbf{v}$. In this case the solution $\mathbf{y}(t) = e^{-kt/m}\mathbf{y}(0)$ describes a motion with an exponentially damped velocity, while the trajectory $\mathbf{y}^*(t)$ describes a motion with an exponentially increasing velocity, which does *not* satisfy Eq. (6.72). Finally, if the force is velocity-independent and *conservative*, $\mathbf{F} = -\nabla V(\mathbf{y})$, Newton's second law can be deduced from the Lagrangian $L = \frac{1}{2}mv^2 - V(\mathbf{y})$, which, in turn, is time-reversal invariant.

From these considerations we also infer that, if $\mathbf{F}$ is a time-reversal invariant force, and $\mathbf{y}(t)$ is the solution of Eq. (6.72) relative to the initial data $\mathbf{y}(0) = \mathbf{y}_0$ and $\mathbf{v}(0) = \mathbf{v}_0$, then $\mathbf{y}^*(t) = \mathbf{y}(-t)$ is the solution of the same equation relative to the initial data $\mathbf{y}^*(0) = \mathbf{y}_0$ and $\mathbf{v}^*(0) = -\mathbf{v}_0$. The crucial point is, however, the following: since the initial data $\{\mathbf{y}_0, -\mathbf{v}_0\}$ are arbitrary, $\mathbf{y}^*(t)$ describes a *physical* motion, i.e. a motion which can occur in nature. In other words, in Newtonian mechanics no *spontaneous violation* of time-reversal invariance occurs.

Spontaneous violation of time-reversal invariance in electrodynamics. As shown in Sect. 2.2.2, the fundamental equations of electrodynamics (2.20)–(2.22) are invariant under time reversal. It follows that, if the configuration Σ specified by

$$\{\mathbf{y}_r(t), j^0(t, \mathbf{x}), \mathbf{j}(t, \mathbf{x}), \mathbf{E}(t, \mathbf{x}), \mathbf{B}(t, \mathbf{x}), A^0(t, \mathbf{x}), \mathbf{A}(t, \mathbf{x})\}$$

solves these equations, so does the configuration Σ^* specified by (see the transformation laws (2.28), (2.30) and (2.49))

$$\{\mathbf{y}_r(-t), j^0(-t, \mathbf{x}), -\mathbf{j}(-t, \mathbf{x}), \mathbf{E}(-t, \mathbf{x}), -\mathbf{B}(-t, \mathbf{x}), A^0(-t, \mathbf{x}), -\mathbf{A}(-t, \mathbf{x})\}.$$

In particular, the Poynting vectors $\mathbf{S} = \mathbf{E} \times \mathbf{B}$ of the two configurations are linked by the relation

$$\mathbf{S}^*(t, \mathbf{x}) = \mathbf{E}^*(t, \mathbf{x}) \times \mathbf{B}^*(t, \mathbf{x}) = \mathbf{E}(-t, \mathbf{x}) \times (-\mathbf{B}(-t, \mathbf{x})) = -\mathbf{S}(-t, \mathbf{x}). \tag{6.75}$$

Consider now a pair $\{j^\mu, A^\mu\}$ satisfying the fundamental equations of electrodynamics and inducing, hence, a solution Σ. The vector potential A^μ is then given by (6.64), to which we must add, in case, the vector potential A_{in}^μ of the external field. We may think, for example, that j^μ is the current associated with an electron in a *clockwise* circular motion. Considering the corresponding solution Σ^*, represented by the pair $\{A^{*\mu}, j^{*\mu}\}$, we may then ask for the relationship existing between $A^{*\mu}$ and $j^{*\mu}$. The question is of concrete physical relevance in that, if j^μ is a current realizable in nature, so is the current $j^{*\mu}$. In the example above $j^{*\mu}$ corresponds, in fact, to an electron in a *counter-clockwise* circular motion. One could therefore conclude that the electron in counter-clockwise motion generates the vector potential $A^{*\mu}$, in that the pair $\{A^{*\mu}, j^{*\mu}\}$ – we repeat – satisfies indeed Maxwell's equations. However, this conclusion is wrong: The *physical* vector potential generated by $j^{*\mu}$ is *not* $A^{*\mu}$. In fact, it is not difficult to establish the relationship between $A^{*\mu}$ and $j^{*\mu}$ explicitly. From the way the potentials and the sources appear in Σ and Σ^* we see that it suffices to analyze the relationship between the time components $j^{*0}(t, \mathbf{x}) = j^0(-t, \mathbf{x})$ and $A^{*0}(t, \mathbf{x}) = A^0(-t, \mathbf{x})$. The potential $A^{*0}(t, \mathbf{x})$ is obtained simply replacing in (6.64) the variable $x^0 = t$ with $-t$

$$
A^{*0}(t, \mathbf{x}) = \frac{1}{2\pi} \int H(-t - y^0)\, \delta\left((-t - y^0)^2 - |\mathbf{x} - \mathbf{y}|^2\right) j^0(y^0, \mathbf{y})\, d^4 y
$$

$$
= \frac{1}{2\pi} \int H(-t + y^0)\, \delta\left((x - y)^2\right) j^{*0}(y^0, \mathbf{y})\, d^4 y,
$$

where in the first integral we performed the change of variable $y^0 \to -y^0$. Recalling the definition of the advanced Green function (6.60), we recognize that this relation can be recast in the form $A^{*0} = G_{\mathrm{adv}} * j^{*0}$. Proceeding in the same way for the components A^{*i}, eventually we find the relation

$$
A^{*\mu} = G_{\mathrm{adv}} * j^{*\mu}, \tag{6.76}
$$

meaning that $A^{*\mu}$ is the *advanced* potential (6.42) generated by $j^{*\mu}$. Therefore, the solution $\{A^{*\mu}, j^{*\mu}\}$ conflicts manifestly with relativistic causality. In fact, according to formula (6.40), the *physical* vector potential created by the current $j^{*\mu}$ is the *retarded* potential

$$
A^{*\mu}_{\mathrm{ret}} = G_{\mathrm{ret}} * j^{*\mu},
$$

which is different from $A^{*\mu}$. Hence, we recognize that in classical electrodynamics time-reversal invariance is violated *spontaneously*, in the sense that it is violated by the physical *solutions*.

From a physical point of view, the choice between the vector potentials $A^{*\mu}_{\mathrm{ret}}$ and $A^{*\mu}$ is intimately related to an electromagnetic phenomenon of fundamental importance: an accelerated charge in nature always *emits* radiation, rather than *absorbing* it. To be concrete, we return to our example of the electron moving on a circle, anticipating a general result of Chap. 7: $A^{*\mu}_{\mathrm{ret}}$ represents the electromagnetic field created by the electron in counter-clockwise motion *emitting* radiation, as observed in nature, while $A^{*\mu}$ represents the electromagnetic field generated by the electron still in counter-clockwise motion, but *absorbing* radiation, a phenomenon that *never* occurs in nature. In particular, from the relation (6.75) we see that the Poynting vectors associated with the configurations Σ and Σ^* are one the opposite of the other. This means that if in the first case the radiation is outgoing, in the second case it is incoming, in contrast to what occurs in nature.[7] It is thus as if the radiation would choose for its propagation a preferred *arrow of time*, that, mysteriously, coincides with the direction of time in which our Universe is expanding. gg

---

[7]In Sect. 6.3 we will see that, if the particle is in *uniform* linear motion, actually the potential $A^{*\mu}_{\mathrm{ret}}$ equals $A^{*\mu}$. In this case the particle, not being accelerated, does neither emit nor absorb radiation, and the retarded and advanced solutions coincide.

Ultimately, it is precisely the spontaneous violation of time-reversal invariance to be responsible for all radiation phenomena occurring in nature: If we would require the solutions of Maxwell's equations to respect time-reversal symmetry – choosing in Eq. (6.61) the value $a = 1/2$ instead of $a = 1$, and obtaining thus the time-reversal-preserving Green function $G = \delta(x^2)/4\pi$ – the same amount of radiation would be emitted and absorbed in all physical processes. We will analyze the, in a sense dramatic, effect of this symmetry violation on the *internal* consistency of classical electrodynamics in Chap. 15.

6.2.4 General Validity

The retarded solution A^μ (6.40) of Maxwell's equations, in integral form (6.65), has been derived disregarding the regularity properties of the quantities involved. Actually, since the retarded Green function (6.58) is a distribution, the vector potential (6.40) would be well defined if j^μ is a test function belonging to S or, according to the definition of the generalized convolution based on (6.20), if j^μ is a distribution of compact support; unfortunately, both alternatives amount to highly non-physical four-currents. On the other hand, the most singular currents appearing in nature are the *microscopic* point-particle currents of the form (2.16). Since any *macroscopic* current appearing in classical electrodynamics, being an average of the former, is less singular than (2.16), it is sufficient to restrict our attention to the δ-like currents (2.16).

Fourier transform approach. As in the case of the Poisson equation, we try to make sense of the solution (6.40) considering its Fourier transform

$$\widehat{A}^\mu(k) = (2\pi)^2 \widehat{G}(k) \, \widehat{j}^\mu(k). \tag{6.77}$$

We need thus the Fourier transform $\widehat{G}(k)$ of the retarded Green function (6.59), which for this purpose we rewrite as

$$G(x) = \frac{1}{4\pi} \left(\delta(x^2) + \varepsilon(x^0) \, \delta(x^2) \right).$$

Using the Fourier transforms (6.49) of the distributions F_1 and F_2 in (6.51), we then obtain

$$\widehat{G}(k) = -\frac{1}{(2\pi)^2} \left(P\frac{1}{k^2} + i\pi\varepsilon(k^0) \, \delta(k^2) \right), \tag{6.78}$$

an expression that, by construction, satisfies the Fourier transform of the kernel equation (6.43)

$$-k^2 \widehat{G}(k) = \frac{1}{(2\pi)^2}.$$

So the relation (6.77) becomes

$$\widehat{A}^\mu(k) = -\left(\mathcal{P}\frac{1}{k^2} + i\pi\varepsilon(k^0)\,\delta(k^2)\right)\widehat{j}^\mu(k). \tag{6.79}$$

Whether the expression at second member defines a distribution depends, clearly, on the specific form of the Fourier transform $\widehat{j}^\mu(k)$ of the four-current, which in general belongs, however, not to O_M. Nevertheless, if the second member of (6.79) *is* a distribution, we can *define* the potential $A^\mu(x)$ as its inverse Fourier transform. As we will show in Sect. 7.2.2, for massive particles on strictly time-like trajectories (see (7.3)), via (6.64) the current (2.16) gives rise to the celebrated *distribution-valued* Liénard-Wiechert potentials, obeying Maxwell's equations. This provides an indirect proof that for those trajectories the right hand side of (6.79) is a distribution. On the other hand, it is less known that for massless particles on light-like trajectories the second member of (6.79) in general does *not* define a distribution, and that a fortiori it does not provide a solution to Maxwell's equations. In other words, for massless charges in general the Green function method fails. Below we illustrate the failure of the Green function method – when applied to massless charges – for the prototypical case of a particle in *uniform linear* motion. For this simple world line we provide a solution to the problem in Sect. 6.3.2, while we postpone the more involved solution of Maxwell's equations for *accelerated* massless charges to Chap. 17.

Uniform linear motion. The world line of a particle in uniform linear motion can be written as $y^\mu(\lambda) = \lambda u^\mu$, where λ is an arbitrary parameter, so that its current assumes the simple form

$$j^\mu(x) = e\,u^\mu \int \delta^4(x - \lambda u)\,d\lambda. \tag{6.80}$$

Its Fourier transform can be computed to be

$$\begin{aligned}
\widehat{j}^\mu(k) &= \frac{eu^\mu}{(2\pi)^2}\int d^4x\,e^{-ik\cdot x}\int \delta^4(x - \lambda u)\,d\lambda \\
&= \frac{eu^\mu}{(2\pi)^2}\int e^{-i(k\cdot u)\lambda}\,d\lambda = \frac{eu^\mu}{2\pi}\,\delta(u\cdot k),
\end{aligned}$$

an expression that, obviously, is a distribution, although it does not belong to O_M. Nevertheless, we can write the product (6.79)

$$\widehat{A}^\mu(k) = -\frac{eu^\mu}{2\pi}\left(\mathcal{P}\frac{1}{k^2} + i\pi\varepsilon(k^0)\,\delta(k^2)\right)\delta(u\cdot k). \tag{6.81}$$

We analyze now separately the cases of massless and massive particles.

Time-like trajectories. A massive particle follows a time-like trajectory, i.e. a trajectory with $v < 1$, so that we can normalize the four-velocity to $u^2 = 1$. We can then choose its rest frame, where $u^\mu = (1, 0, 0, 0)$ and $\delta(u \cdot k) = \delta(k^0)$. The second term at the right hand side of (6.81) is then zero, since (see (2.72))

$$\varepsilon(k^0)\, \delta(k^2)\, \delta(u \cdot k) = \frac{1}{2|\mathbf{k}|}\left(\delta(k^0 - |\mathbf{k}|) - \delta(k^0 + |\mathbf{k}|)\right)\delta(k^0) = 0. \qquad (6.82)$$

Equation (6.81) then yields the vector potential

$$\widehat{A}^\mu(k) = -\frac{eu^\mu}{2\pi}\left(\mathcal{P}\frac{1}{k^2}\right)\delta(k^0) = \frac{eu^\mu}{2\pi|\mathbf{k}|^2}\,\delta(k^0),$$

which defines indeed a distribution. Its inverse Fourier transform is then well defined and can be easily evaluated explicitly. Equations (6.14) and (6.21) yield, in fact, the known Coulomb potential

$$A^\mu(x) = \frac{eu^\mu}{4\pi|\mathbf{x}|}. \qquad (6.83)$$

Light-like trajectories. A massless particle follows a light-like trajectory, i.e. a trajectory with $v = 1$, so that $u^2 = 0$. We can then choose a reference frame where $u^\mu = (1, 0, 0, 1)$, so that $\delta(u \cdot k) = \delta(k^0 - k^3)$. In this case in the place of (6.82) we obtain

$$\varepsilon(k^0)\, \delta(k^2)\, \delta(u \cdot k) \overset{!}{=} \varepsilon(k^0)\, \delta\left((k^1)^2 + (k^2)^2\right)\delta(k^0 - k^3). \qquad (6.84)$$

Inserting this expression in (6.81) we would thus end up with the vector potential

$$\widehat{A}^\mu(k) = \frac{eu^\mu}{2\pi}\left(\frac{1}{(k^1)^2 + (k^2)^2} - i\pi\varepsilon(k^0)\,\delta\left((k^1)^2 + (k^2)^2\right)\right)\delta(k^0 - k^3),$$
$$(6.85)$$

which is definitely not a distribution. In fact, none of the two terms between parentheses is a distribution: the first, because it is non-integrable in the plane $k^1 = k^2 = 0$, and the second, because the argument of the δ-function does not have simple zeros, see (2.76). The reason for which the expression (6.85) is not a distribution is essentially that, in this case, the two distributions involved in the product (6.77) are *too* singular. Similarly, if – without resorting to the Fourier transform – one substitutes the current (6.80) in the formal solution (6.64), one obtains the vector potential, see Problem 6.4,

$$A^\mu(x) = \frac{eu^\mu}{4\pi}\frac{H(u \cdot x)}{u \cdot x}, \qquad (6.86)$$

which is likewise *not* a distribution, since it is non-integrable in the region $u \cdot x = t - z = 0$. So we see that in the case of a light-like trajectory in general the Green function method fails. There remains the question of *which specific feature* of such a trajectory causes the problem: As we will show in Sect. 7.2.2, the failure of the Green function method is due to the fact that the particle's speed approaches the speed of light on a *straight line* in the *infinite past*. Consequently, the method fails also for so-called *hyperbolic motions*, see Sect. 7.1, which are actually *time-like* for all finite t.

Conversely, as observed above, for massive charges on strictly time-like trajectories the product (6.79) defines always a distribution and, correspondingly, also the integral representations (6.64) and (6.65) are well defined. Henceforth, for such trajectories we will always resort to these representations.

6.3 Field of a Particle in Uniform Linear Motion

In this section we determine the electromagnetic field generated by a generic charged particle in uniform linear motion. While for a massive particle we can resort to the general solution (6.64), for a massless one we must follow a different path. Actually, in the first case we could derive the field also by means of a Lorentz transformation from the rest frame of the particle, where the fields are given by

$$\mathbf{E} = \frac{e\mathbf{x}}{4\pi r^3}, \qquad \mathbf{B} = 0,$$

to the laboratory frame, see Problem 6.2. This approach entails, however, the drawback of violating *manifest* Lorentz invariance. In the second case this approach could, anyhow, not be applied, because a massless particle does not admit a rest frame.

6.3.1 The Field of a Massive Particle

Choosing appropriately the origin of the coordinate system, the world line of a massive particle in uniform linear motion, with speed $v < 1$, can be written as $y^\mu(s) = su^\mu$, where the constant four-velocity u^μ is subject to the constraint $u^2 = 1$, and s is the proper time. Correspondingly, the four-current (2.16) simplifies to

$$j^\mu(y) = e\,u^\mu \int \delta^4(y - su)\,ds. \tag{6.87}$$

Retarded potential. To determine the vector potential generated by the particle we insert the current (6.87) in (6.64), obtaining

$$A^\mu(x) = \frac{eu^\mu}{2\pi} \int d^4y \int H(x^0 - y^0)\, \delta\left((x-y)^2\right) \delta^4(y - su)\, ds$$

$$= \frac{eu^\mu}{2\pi} \int H(x^0 - su^0)\, \delta(f(s))\, ds. \tag{6.88}$$

We have introduced the function of s

$$f(s) = (x - su)^2 = x^2 - 2s(ux) + s^2, \qquad (ux) = u^\mu x_\mu,$$

where the dependence on x is understood. To evaluate the integral (6.88) we write out the distribution $\delta(f(s))$ applying the rule (2.76), which requires to determine the zeros of $f(s)$. Being a second order polynomial, $f(s)$ has the two zeros

$$s_\pm = (ux) \mp \sqrt{(ux)^2 - x^2}, \qquad f(s_\pm) = 0, \tag{6.89}$$

both real, since the discriminant $(ux)^2 - x^2$ is always greater than or equal to zero. To see it rapidly we use the Lorentz invariance of the discriminant, to evaluate it in the rest frame of the particle, where $u^\mu = (1, 0, 0, 0)$,

$$(ux)^2 - x^2 = (x^0)^2 - \left((x^0)^2 - |\mathbf{x}|^2\right) = |\mathbf{x}|^2 \geq 0. \tag{6.90}$$

The rule (2.76) then yields

$$\delta(f(s)) = \frac{\delta(s - s_+)}{|f'(s_+)|} + \frac{\delta(s - s_-)}{|f'(s_-)|}. \tag{6.91}$$

Since $f'(s) = 2(s - ux)$, we have furthermore

$$|f'(s_\pm)| = 2\sqrt{(ux)^2 - x^2}.$$

Inserting these results in (6.88) we obtain

$$A^\mu(x) = \frac{eu^\mu}{4\pi\sqrt{(ux)^2 - x^2}} \int \left[H\left(x^0 - s_+u^0\right) \delta(s - s_+) \right. \tag{6.92}$$
$$\left. + H\left(x^0 - s_-u^0\right) \delta(s - s_-) \right] ds.$$

The evaluation of the remaining integral requires the knowledge of the signs of the quantities $x^0 - s_\pm u^0$. To determine these signs we use again a covariance argument. Define the four-vectors

$$V_\pm^\mu = x^\mu - s_\pm u^\mu,$$

which by construction belong to the light cone, $V_\pm^2 = 0$. Consequently, the signs of their time components $V_\pm^0 = x^0 - s_\pm u^0$ are Lorentz invariant, so that we can determine them, once more, in the rest frame of the particle. In this frame we have, see Eqs. (6.89) and (6.90),

$$s_\pm = x^0 \mp |\mathbf{x}| \quad \Rightarrow \quad V_\pm^0 = x^0 - s_\pm u^0 = \pm |\mathbf{x}|.$$

We conclude, therefore, that in any reference frame we have the inequalities

$$x^0 - s_+ u^0 > 0, \qquad x^0 - s_- u^0 < 0.$$

This means that $H(x^0 - s_+ u^0) = 1$, and $H(x^0 - s_- u^0) = 0$, so that (6.92) yields the manifestly covariant potential

$$A^\mu(x) = \frac{e u^\mu}{4\pi \sqrt{(ux)^2 - x^2}}. \tag{6.93}$$

From the above calculation it is obvious that, had we used the advanced Green function (6.60) – by replacing in Eq. (6.88) $H(x^0)$ with $H(-x^0)$ – we still would have obtained the same expression (6.93): As anticipated in Sect. 6.2.3, in the case of a particle in uniform linear motion, no spontaneous violation of time-reversal invariance occurs. In particular, for a static particle, for which $u^\mu = (1, 0, 0, 0)$, (6.93) returns the usual Coulomb potential

$$A^0 = \frac{e}{4\pi |\mathbf{x}|}, \qquad A^i = 0. \tag{6.94}$$

Differentiating equation (6.93) with respect to x^ν we obtain

$$\partial^\nu A^\mu = \frac{e}{4\pi} \frac{x^\nu - u^\nu(ux)}{((ux)^2 - x^2)^{3/2}} u^\mu, \tag{6.95}$$

so that for the electromagnetic field we derive the covariant expression

$$F^{\mu\nu} = \partial^\mu A^\nu - \partial^\nu A^\mu = \frac{e}{4\pi} \frac{x^\mu u^\nu - x^\nu u^\mu}{((ux)^2 - x^2)^{3/2}}. \tag{6.96}$$

If, instead, in Eq. (6.95) we contract the indices, we confirm that the vector potential obeys the Lorenz gauge $\partial_\mu A^\mu = 0$, as it must by construction.

Electric and magnetic fields. From (6.96) we read off the electric and magnetic fields

$$E^i = F^{i0} = \frac{e}{4\pi} \frac{x^i u^0 - x^0 u^i}{((ux)^2 - x^2)^{3/2}} = \frac{eu^0}{4\pi} \frac{x^i - v^i t}{((ux)^2 - x^2)^{3/2}}, \tag{6.97}$$

$$B^k = -\frac{1}{2} \varepsilon^{kij} F^{ij} = -\frac{e}{8\pi} \frac{\varepsilon^{kij}(x^i u^j - x^j u^i)}{((ux)^2 - x^2)^{3/2}} = \frac{eu^0}{4\pi} \frac{\varepsilon^{kij} v^i x^j}{((ux)^2 - x^2)^{3/2}}$$

$$= \frac{eu^0}{4\pi} \frac{\varepsilon^{kij} v^i (x^j - v^j t)}{((ux)^2 - x^2)^{3/2}} = \varepsilon^{kij} v^i E^j. \tag{6.98}$$

By inspection we see that these fields are tied together by the relation

$$\mathbf{B} = \frac{\mathbf{v}}{c} \times \mathbf{E}, \tag{6.99}$$

where we restored the velocity of light. In each point the magnetic field is, hence, a simple function of the electric field, so that it is sufficient to analyze the properties of the latter. In particular, the magnetic field is suppressed by a factor v/c with respect to the electric field, in agreement with the fact that the former represents a relativistic effect. To analyze the electric field in more detail it is convenient to introduce the vector

$$\mathbf{R} = \mathbf{x} - \mathbf{v}t,$$

connecting the observation point $\mathbf{x}$ at time t with the position $\mathbf{y}(t) = \mathbf{v}t$ occupied by the particle at the same time. Carrying out the algebra we then find

$$(ux)^2 - x^2 = \frac{R^2 + (\mathbf{v} \cdot \mathbf{R})^2 - v^2 R^2}{1 - v^2}, \tag{6.100}$$

so that the field (6.97) can be rewritten as

$$\mathbf{E} = \frac{e(1 - v^2)\mathbf{R}}{4\pi(R^2 + (\mathbf{v} \cdot \mathbf{R})^2 - v^2 R^2)^{3/2}}. \tag{6.101}$$

Introducing the angle ϑ between $\mathbf{v}$ and $\mathbf{R}$, we can recast this expression in the form

$$\mathbf{E} = \frac{1 - v^2}{(1 - v^2 \sin^2 \vartheta)^{3/2}} \mathbf{E}_{nr}, \tag{6.102}$$

where $\mathbf{E}_{nr}$ is the non-relativistic Coulomb field

$$\mathbf{E}_{nr} = \frac{e\mathbf{R}}{4\pi R^3}. \tag{6.103}$$

We see thus that at fixed time, $\mathbf{E}$ decays at large distances as $1/r^2$, where $r = |\mathbf{x}|$. The electromagnetic field (6.96) maintains, hence, asymptotically the inverse-square-law behavior of the non-relativistic Coulomb field:

$$F^{\mu\nu} \sim \frac{1}{r^2}, \quad \text{for} \quad r \to \infty. \tag{6.104}$$

Furthermore, $\mathbf{E}$ is a *radial* field, as is $\mathbf{E}_{nr}$, but it is no longer spherically symmetric in that its magnitude E depends on the direction. In fact, for $\mathbf{R}$ orthogonal ($\vartheta = \pi/2$) and parallel ($\vartheta = 0, \pi$) to $\mathbf{v}$, Eq. (6.102) yields for the corresponding magnitudes

$$E_\perp = \frac{1}{\sqrt{1 - v^2}} E_{nr} > E_{nr}, \tag{6.105}$$

$$E_\| = (1 - v^2) E_{nr} < E_{nr}, \tag{6.106}$$

respectively. Hence, along the trajectory of the particle the intensity of the field is reduced with respect to E_{nr}, in both directions, while in the plane passing through the particle's position and orthogonal to the trajectory, it is enhanced with respect to E_{nr}. In particular, for velocities approaching the speed of light, i.e. in the *ultrarelativistic* limit, $E_\|$ vanishes, while $E_\perp$ diverges. In practice, for very high velocities the electromagnetic field is essentially zero in all directions, except for values of ϑ close to $\pi/2$, where it becomes very intense. In light of the relation (6.99) the magnitude B of the magnetic field has characteristics similar to those of E. However, whereas the field lines of $\mathbf{E}$ are radial straight lines, according to (6.99) the field lines of $\mathbf{B}$ are concentric circles surrounding the trajectory.

6.3.2 The Field of a Massless Particle

In light of the ultrarelativistic behavior of the field (6.102) we expect that the electromagnetic field produced by a massless particle in uniform linear motion entails singularities so strong, that they can make sense only in the framework of distributions. It appears, therefore, natural to address the solution of Maxwell's equations by means of an appropriate distributional *limiting procedure*. The trajectory of a massless particle in uniform linear motion can be written as $\mathbf{y}(t) = t\mathbf{n}$, where the unit vector $\mathbf{n}$ identifies the propagation direction. Since in this case the proper time s of the particle is not defined, we parameterize its world line $y^\mu(\lambda)$ with a generic parameter λ. Introducing a light-like vector n^μ we then have

$$y^\mu(\lambda) = \lambda n^\mu, \qquad n^\mu = (1, \mathbf{n}), \qquad n^2 = 0.$$

In this case the current (2.16) reduces to

$$\mathcal{J}^\mu(x) = e n^\mu \int \delta^4(x - \lambda n) \, d\lambda = e \, n^\mu \delta^3(\mathbf{x} - t\mathbf{n}), \tag{6.107}$$

and the field $\mathcal{F}^{\mu\nu}$ generated by the particle must obey the Maxwell equations

$$\partial_\mu \mathcal{F}^{\mu\nu} = \mathcal{J}^\nu, \qquad \partial_{[\mu}\mathcal{F}_{\nu\rho]} = 0. \tag{6.108}$$

Limiting procedure. To derive the solution $\mathcal{F}^{\mu\nu}$ of the system (6.108) we start from the field (6.96) created by a massive particle, moving with a constant speed $v < 1$ in the same direction $\mathbf{n}$ as the massless one. Setting in (6.87) $\mathbf{v} = v\mathbf{n}$, we then obtain for its current

$$j^\mu(x) = eu^\mu \int \delta^4(x - su)\, ds = e\,(1, v\mathbf{n})\, \delta^3(\mathbf{x} - v t\mathbf{n}). \tag{6.109}$$

By construction, this current and the field $F^{\mu\nu}$ (6.96) satisfy the Maxwell equations

$$\partial_\mu F^{\mu\nu} = j^\nu, \qquad \partial_{[\mu}F_{\nu\rho]} = 0. \tag{6.110}$$

The basis of our limiting procedure is the existence of the obvious distributional limit

$$\mathcal{S}' - \lim_{v\to 1} j^\mu = \mathcal{J}^\mu, \tag{6.111}$$

meaning, in fact, that the ordinary limits

$$\lim_{v\to 1} j^\mu(\varphi) = \lim_{v\to 1} e\,(1, v\mathbf{n}) \int \varphi(t, v t\mathbf{n})\, dt = e\,(1, \mathbf{n}) \int \varphi(t, t\mathbf{n})\, dt = \mathcal{J}^\mu(\varphi)$$

hold for all test functions $\varphi \in \mathcal{S} = \mathcal{S}(\mathbb{R}^4)$. *If* also the distributional limit for $v \to 1$ of $F^{\mu\nu}$ exists, then we can take the distributional limits of both equations in (6.110) and – resorting to Theorem II of Sect. 2.3.1, stating that distributional limits commute with distributional derivatives – swap at their first members the limits with the derivatives. The resulting system of equations is then precisely (6.108). In conclusion, if the limit

$$\mathcal{F}^{\mu\nu} = \mathcal{S}' - \lim_{v\to 1} F^{\mu\nu} \tag{6.112}$$

exists, then the so-defined field $\mathcal{F}^{\mu\nu}$ satisfies *automatically* the Maxwell equations (6.108). We emphasize that it is crucial to perform the limit (6.112) in the sense of distributions. In fact, from Eqs. (6.99) and (6.102) we see that the *pointwise* limit for $v \to 1$ of the tensor (6.96) *vanishes* almost everywhere.

Limit of the vector potential. Although it is not too difficult to evaluate the limit (6.112) directly, see Problem 6.5, it is more instructive to address its determination in an indirect way, starting not from the field (6.96), but rather from the associated potential (6.93)

$$A^\mu(x) = \frac{eu^\mu}{4\pi\sqrt{(ux)^2 - x^2}}, \qquad u^\mu = \frac{(1, v\mathbf{n})}{\sqrt{1 - v^2}}. \tag{6.113}$$

If this potential admits a distributional limit, we could, in fact, write

$$\mathcal{S}' - \lim_{v \to 1} F^{\mu\nu} = \partial^\mu \left(\mathcal{S}' - \lim_{v \to 1} A^\nu \right) - \partial^\nu \left(\mathcal{S}' - \lim_{v \to 1} A^\mu \right),$$

again because limits commute with derivatives. Performing – preliminarily – the *pointwise* limit of the potential (6.113), we obtain, actually, the finite expression

$$\lim_{v \to 1} A^\mu(x) = \frac{e}{4\pi} \lim_{v \to 1} \frac{(1, v\mathbf{n})}{\sqrt{(t - v\mathbf{n} \cdot \mathbf{x})^2 - (1 - v^2)x^2}} = \frac{e(1, \mathbf{n})}{4\pi |t - \mathbf{n} \cdot \mathbf{x}|} . \quad (6.114)$$

However, being non-integrable at $t = \mathbf{n} \cdot \mathbf{x}$, the limiting function (6.114) is *not* a distribution. It is, actually, not difficult to see that also the distributional limit for $v \to 1$ of the potential (6.113) does not exist. There arises, hence, naturally the question whether or not $F^{\mu\nu}$ admits a limit in $\mathcal{S}'$. The answer could still be affirmative if, so to say, the "part of A^μ" which diverges for $v \to 1$ in the distributional sense, does not contribute to $F^{\mu\nu}$: In other words, this part should be a *pure gauge*, i.e. of the form $\partial_\mu \Lambda$.

A gauge transformation. Since the field strength $F^{\mu\nu}$ (6.96) is supposed to admit a distributional limit, whereas its potential A^μ (6.113) does not, we try to cure the latter by means of an appropriate gauge transformation. Consider the gauge function

$$\Lambda(x) = \frac{e}{4\pi} \ln \left| (ux) - \sqrt{(ux)^2 - x^2} \right| \in \mathcal{S}'. \quad (6.115)$$

A simple calculation then shows that the transformed potential – completely equivalent to (6.113), although no longer obeying the Lorenz gauge – is given by

$$\widetilde{A}^\mu = A^\mu + \partial^\mu \Lambda = \frac{e}{4\pi} \left(1 + \frac{(ux)}{\sqrt{(ux)^2 - x^2}} \right) \frac{x^\mu}{x^2}. \quad (6.116)$$

Of course, the associated field strength $\partial^\mu \widetilde{A}^\nu - \partial^\nu \widetilde{A}^\mu$ still equals the field strength $F^{\mu\nu}$ (6.96). In (6.116) with $1/x^2$ we understand the composite principal value $\mathcal{P}(1/x^2)$ with respect to x^0, see (2.86). It is not difficult to show that the potential $\widetilde{A}^\mu$ admits a limit for $v \to 1$ in the distributional sense, and that, moreover, this limit coincides with its pointwise limit. Using the expression of the four-velocity (6.113) we obtain the pointwise limit

$$\lim_{v \to 1} \frac{(ux)}{\sqrt{(ux)^2 - x^2}} = \lim_{v \to 1} \frac{t - v\mathbf{n} \cdot \mathbf{x}}{\sqrt{(t - v\mathbf{n} \cdot \mathbf{x})^2 - (1 - v^2)x^2}} = \frac{(nx)}{|(nx)|} = \varepsilon(nx),$$

where $\varepsilon(\cdot)$ is the sign function, and $(nx) = t - \mathbf{n} \cdot \mathbf{x}$. The potential (6.116) admits, hence, the distributional limit

$$\mathcal{A}^\mu = \mathcal{S}' - \lim_{v \to 1} \widetilde{A}^\mu = \frac{ex^\mu}{2\pi x^2} H(nx), \tag{6.117}$$

where $1/x^2$ stands again for $\mathcal{P}(1/x^2)$.

Electromagnetic field. The existence of the limit (6.117) implies the existence of the sought limit (6.112). More precisely, the rule (2.73) for the derivative of the Heaviside function yields

$$\mathcal{F}^{\mu\nu} = \mathcal{S}' - \lim_{v \to 1} F^{\mu\nu} = \mathcal{S}' - \lim_{v \to 1} \left(\partial^\mu \widetilde{A}^\nu - \partial^\nu \widetilde{A}^\mu \right) = \partial^\mu \mathcal{A}^\nu - \partial^\nu \mathcal{A}^\mu \tag{6.118}$$

$$= \frac{e(n^\mu x^\nu - n^\nu x^\mu)}{2\pi x^2} \delta(nx). \tag{6.119}$$

For the electric and magnetic fields we then gain the expressions

$$\mathcal{E} = -\frac{e(\mathbf{x} - \mathbf{n}t)}{2\pi x^2} \delta(nx), \qquad \mathcal{B} = \mathbf{n} \times \mathcal{E}, \qquad \mathbf{n} \cdot \mathcal{E} = 0. \tag{6.120}$$

In particular, their "magnitudes" $\mathcal{E}$ and $\mathcal{B}$ are equal. As we see, at any instant the fields are different from zero only on the plane orthogonal to the trajectory and passing through the particle's position at that instant. If the particle moves, for instance, along the z axis, formulas (6.120) reduce to

$$\mathcal{E} = \frac{e(x, y, 0)}{2\pi(x^2 + y^2)} \delta(z - t), \tag{6.121}$$

$$\mathcal{B} = \frac{e(-y, x, 0)}{2\pi(x^2 + y^2)} \delta(z - t). \tag{6.122}$$

In this case, at time t the fields are non-vanishing on the xy plane located at $z = t$, where they are *very intense*, i.e. proportional to a δ-function. We recall that, by construction, the above fields satisfy Maxwell's equations (6.108). It is, for example, easy to verify that the fields (6.121) and (6.122) obey the equations, see Problem 6.3,

$$\nabla \cdot \mathcal{E} = \mathcal{J}^0(x) = e\,\delta(x)\,\delta(y)\,\delta(z - t), \qquad \nabla \cdot \mathcal{B} = 0. \tag{6.123}$$

Finally, as anticipated above, the field (6.119) can also be derived directly from the solution (6.96) of Maxwell's equations for $v < 1$, by taking its distributional limit for $v \to 1$, see Problem 6.5.

Vector potential in the Lorenz gauge. There remains the question whether or not a massless particle admits a potential $\mathcal{A}'^\mu$ subject to the Lorenz gauge $\partial_\mu \mathcal{A}'^\mu = 0$. Stated differently: is it possible to find a gauge function Λ such that $\partial_\mu \mathcal{A}'^\mu = \partial_\mu(\mathcal{A}^\mu + \partial^\mu \Lambda) = 0$, where $\mathcal{A}^\mu$ is the potential (6.117)? The answer is affirmative in that the gauge-transformed potential

$$\mathcal{A}'^\mu = \mathcal{A}^\mu - \partial^\mu \left(\frac{e}{4\pi} H(nx) \ln |x^2| \right) = -\frac{en^\mu}{4\pi} \ln |x^2| \, \delta(nx) \qquad (6.124)$$

satisfies, indeed, the condition $\partial_\mu \mathcal{A}'^\mu = 0$. Notice that $\mathcal{A}'^\mu$, a distribution, has nothing to do with the *formal* Green-function potential (6.86), which is not a distribution. In conclusion, also for massless particles in uniform linear motion Maxwell's equations (6.108) can be solved in Lorenz gauge, in which case they assume the familiar form $\Box \mathcal{A}'^\mu = \mathcal{J}^\mu$, $\partial_\mu \mathcal{A}'^\mu = 0$. However, although the latter are of the standard form (6.30) and (6.31), they cannot be solved – we repeat – by means of the Green function method: The resulting expressions of the potential – (6.86) in coordinate space and (6.85) in momentum space – are in fact meaningless, and, a fortiori, they are not solutions.

Shock waves. Fields of the form (6.120) are called *shock waves* in that the field is different from zero only on a plane which, in the case at hand, moves at the speed of light. Their main characteristic is that a test charge experiences an effect only if this plane hits it, changing instantaneously its momentum by a finite amount. Suppose, e.g., that the shock-wave plane created by a massless particle of charge e moving along the z axis hits at time $t = 0$ a massive particle of charge e^*, situated at the position $\mathbf{b} = (x, y, 0)$ with non-relativistic velocity $\mathbf{v} = (v_x, v_y, v_z)$. In this case, in the Lorentz equation

$$\frac{d\mathbf{p}}{dt} = e^* (\boldsymbol{\mathcal{E}} + \mathbf{v} \times \boldsymbol{\mathcal{B}}) \qquad (6.125)$$

the magnetic field can be neglected. Inserting the electric field (6.121) in (6.125), and integrating the resulting equation between two instants preceding and ensuing the collision, we find that the momentum of the particle changes by

$$\Delta \mathbf{p} = \int_{-t}^{t} \frac{d\mathbf{p}}{dt'} \, dt' \simeq e^* \int_{-t}^{t} \boldsymbol{\mathcal{E}} \, dt' = e^* \int_{-t}^{t} \frac{e\mathbf{b}}{2\pi b^2} \, \delta(z(t') - t') \, dt'$$

$$= \frac{e^* e \mathbf{b}}{2\pi b^2 (1 - v_z)} \simeq \frac{e^* e \mathbf{b}}{2\pi b^2 c},$$

where, to underline the relativistic origin of the effect, in the final result we restored the velocity of light. Above we have used also the identity

$$\delta(z(t) - t) = \frac{\delta(t)}{1 - \dot{z}(0)} = \frac{\delta(t)}{1 - v_z},$$

following from the general rule (2.76), in that by hypothesis $z(0) = 0$. The collision causes, hence, a kick on e^* along the plane of the shock wave towards e, if the signs of the charges are opposite, and a kick in the opposite direction, if their signs are equal. However, according to our current knowledge, in electrodynamics shock waves represent a mathematical idealization, and not a phenomenon realized in nature, as no massless charged particle has ever been observed. On the other hand,

Einstein's equations of General Relativity imply that the *gravitational* field generated by a massless particle moving at the speed of light is, again, of the shock-wave
type [1, 2]. However, in this case these solutions are *physical*, because a massless
particle, like the photon, possesses a non-vanishing *energy*, which in General Relativity plays actually the role of the *gravitational* charge. Therefore, in a classical
world such a particle would indeed create a gravitational field of the shock-wave
type, and so the mathematical idealization would correspond to a real physical phenomenon.

6.4 Problems

6.1 Prove that the retarded Green function (6.59) defines a distribution in $\mathcal{S}'(\mathbb{R}^4)$.

6.2 Consider a particle of charge e moving in the laboratory frame K with constant
velocity v along the z axis. Recall that in the inertial frame K', where the particle is
at rest at the position $\mathbf{x}' = 0$, the vector potential has the form

$$A'^\mu(x') = \frac{e}{4\pi|\mathbf{x}'|}\,(1,0,0,0).$$

(a) Determine the Lorentz transformation $\Lambda^\mu{}_\nu$ connecting the coordinates of a
    generic event in K with the corresponding coordinates in K'.
(b) Determine the form of $A^\mu(x)$ in K using the fact that the vector potential is a
    four-vector, and compare the result with (6.93).

6.3 Verify that the fields (6.121) and (6.122) satisfy the Maxwell equations (6.123),
proving in particular the distributional identity in two dimensions

$$\boldsymbol{\nabla}\cdot\frac{\mathbf{x}}{r^2} = 2\pi\,\delta^2(\mathbf{x}),$$

where $\mathbf{x} = (x,y)$ and $r = \sqrt{x^2+y^2}$. Conclude that the Green function associated
with the two-dimensional Laplacian is given by the *logarithm*:

$$\nabla^2\left(\frac{1}{2\pi}\ln r\right) = (\partial_x^2+\partial_y^2)\left(\frac{1}{2\pi}\ln\sqrt{x^2+y^2}\right) = \delta^2(\mathbf{x}).$$

6.4 Consider the current (6.107) of a massless charged particle in uniform motion
along the z axis

$$\mathcal{J}^\mu(t,\mathbf{x}) = e\,(1,0,0,1)\,\delta(x)\,\delta(y)\,\delta(z-t).$$

Show that the vector potential A^μ obtained by applying *formally* the integral representation (6.64) is given by (6.86), where $u^\mu = (1,0,0,1)$.

6.5 Derive the shock-wave field (6.119) as the distributional limit for $v \to 1$ of the field (6.96), evaluating for a generic test function $\varphi \in \mathcal{S}$ the limit

$$\mathcal{F}^{\mu\nu}(\varphi) = \lim_{v \to 1} F^{\mu\nu}(\varphi). \tag{6.126}$$

Proceed as indicated below.

(a) Consider the velocity $\mathbf{v} = (0, 0, v)$ and, using (6.100), write the field (6.96) in the form

$$F^{\mu\nu}(t, \mathbf{x}) = \frac{e}{4\pi} \frac{(1 - v^2)(x^\mu v^\nu - x^\nu v^\mu)}{[(z - vt)^2 + (1 - v^2)(x^2 + y^2)]^{3/2}}, \quad v^\mu = (1, 0, 0, v). \tag{6.127}$$

(b) In the integral

$$F^{\mu\nu}(\varphi) = \int F^{\mu\nu}(t, \mathbf{x}) \, \varphi(t, \mathbf{x}) \, d^4x$$

perform the change of variables $z \to \sqrt{1 - v^2}\, z + vt$.

(c) To evaluate the limit (6.126) swap the limit for $v \to 1$ with the integral sign, and then perform the integral over z.

References

1. P.C. Aichelburg, R.U. Sexl, On the gravitational field of a massless particle. Gen. Rel. Grav. **2**, 303 (1971)
2. J. Groah, B. Temple, Shock-wave solutions of the Einstein equations with perfect fluid sources: existence and consistency by a locally inertial glimm scheme. Memoirs AMS **172**, No. 813 (2004)

Chapter 7
Liénard-Wiechert Fields

As second basic application of the general solution (6.64) of Maxwell's equations we determine the electromagnetic field generated by a particle following a generic *strictly* time-like trajectory, see condition (7.3). This field plays a fundamental role in classical electrodynamics and is named after its discoverers, A.-M. Liénard (1898) [1] and E.J. Wiechert (1900) [2]. The solution of Maxwell's equations for light-like trajectories, not achievable with the Green function method, will be addressed in Chap. 17.

In general, a particle moving on a generic trajectory has a non-vanishing acceleration, and generates an electromagnetic field whose characteristics differ substantially from those of the *Coulomb field* (6.96) generated by a particle in uniform linear motion. The most significant differences between the two fields can be summarized as follows. For an accelerated particle the field (6.96) undergoes a deformation, while preserving its asymptotic $1/r^2$-behavior at large distances. In addition to this field there appears a new field – generated by the acceleration of the particle – which at large distances decreases more weakly, namely as $1/r$, and dominates hence over the Coulomb field. This particular and, in a sense, more intense asymptotic $1/r$-behavior is at the origin of all electromagnetic radiation processes: radiation – a core phenomenon of classical electrodynamics – is thus produced by *accelerated charges*.

7.1 World Lines and Asymptotic Conditions

We begin with a few general considerations concerning the particles' trajectories we will admit. Of course, due to causality, the speeds of all particles are bounded from above by the speed of light, $v(t) \leq 1$. In this chapter we deal with massive particles, and so at any *finite* time t their speeds must satisfy the strict inequality $v(t) < 1$. Nevertheless, for $t \to \pm\infty$ in principle the speed of a massive particle

© Springer International Publishing AG, part of Springer Nature 2018
K. Lechner, *Classical Electrodynamics*, UNITEXT for Physics,
https://doi.org/10.1007/978-3-319-91809-9_7

could approach the speed of light. This happens, for example, to a particle subject to a uniform constant infinitely extended electric field, in which case the particle undergoes a uniformly accelerated motion, also called *hyperbolic motion*, see Problem 2.7. If the electric field E is parallel to the z axis, the world line of the associated hyperbolic motion, parameterized by the proper time s, is given by

$$y^\mu(s) = (k\sinh(s/k),\, 0,\, 0,\, k\cosh(s/k)), \tag{7.1}$$

where $k = m/eE$. This world line corresponds to the trajectory

$$\mathbf{y}(t) = \left(0,\, 0,\, \sqrt{t^2 + k^2}\right), \quad \mathbf{v}(t) = \left(0,\, 0,\, \frac{t}{\sqrt{t^2 + k^2}}\right), \quad \lim_{t\to\pm\infty} v(t) = 1. \tag{7.2}$$

As we will see in Sect. 7.2.2, for trajectories like (7.2) which become *asymptotically* light-like in the *infinite past* the Green function method fails, as it fails for generic light-like trajectories approaching a straight line in the infinite past. For motions of this kind we will present the solution of Maxwell's equations in Sect. 17.5, in the context of massless particles. In the present chapter, to exclude world lines like (7.1), we impose on the trajectories a slightly stricter constraint than $v(t) < 1$, namely we require that there exists a maximum speed v_M such that

$$v(t) < v_M < 1, \quad \forall t. \tag{7.3}$$

Trajectories of this kind are called *strictly* time-like, in that they are time-like for *all* $-\infty \leq t \leq \infty$. The condition (7.3) ensures, in particular, the inequality

$$\sqrt{1 - v^2(t)} \geq \sqrt{1 - v_M^2}, \tag{7.4}$$

so that the proper time

$$s(t) = \int_0^t \sqrt{1 - v^2(t')}\, dt' \tag{7.5}$$

satisfies the asymptotic conditions

$$\lim_{t\to\pm\infty} s(t) = \pm\infty. \tag{7.6}$$

According to (7.4) and (7.5) $s(t)$ is a strictly increasing function of time, and consequently the conditions (7.6) guarantee that we can use the parameters s and t interchangeably during the whole time evolution. The trajectories realized in nature are basically of two types, *bounded* motions and *unbounded* ones, and both satisfy the inequality (7.3), see below.

Unbounded motions. By definition the four-velocity of an unbounded motion admits the finite limits

$$\lim_{t\to\pm\infty} u^\mu = u_\pm^\mu. \tag{7.7}$$

The existence of these limits is equivalent to the assumption that the velocity $\mathbf{v}(t)$ for $t \to \pm\infty$ tends to limit vectors $\mathbf{v}_\pm$, whose magnitudes satisfy the strict inequalities $v_\pm < 1$. Consequently, there exists a speed $v_M < 1$ for which $v(t)$ satisfies the bound (7.3). From a physical point of view the conditions (7.7) are motivated by the fact that the force fields realized in nature fundamentally extend only over finite regions of space, tending rapidly to zero at large distances. In this case the Lorentz equations (2.35) and (2.36) then imply that for $t \to \pm\infty$ the four-momentum p^μ admits *finite* limits. Since $u^\mu = p^\mu/m$, this means that the four-velocity admits the limits (7.7). Common examples of unbounded motions are the open trajectories of scattering experiments, and the trajectories of particles incoming from infinity and outgoing to infinity, crossing a finite region of space filled with an electromagnetic field. Conversely, the hyperbolic motion (7.2) violates the conditions (7.3) as well as (7.7), although it entails the limits (7.6). In fact, the time component of the world line (7.1) is equivalent to the relation $s(t) = k \operatorname{arcsinh}(t/k)$, and so the proper time has for $t \to \pm\infty$ the asymptotic behaviors $s(t) \sim \pm k \ln |t|$, implying (7.6).

Bounded motions. By definition a bounded motion satisfies the constraints

$$v(t) < v_M < 1, \quad |\mathbf{y}(t)| < l, \quad \forall t.$$

This class of motions concerns particles confined to a finite region of space, as e.g. the electrons in an antenna, or the charged particles circling in a *synchrotron*, see Chap. 12. In the first case the particles are subject to an oscillating electric force, but at the same time they dissipate energy via Joule heating and through radiation. As a result, their energy remains finite and correspondingly their speed remains strictly less than the speed of light. Similarly, in the synchrotron, along some stretches of the cycle, in addition to the magnetic field there are electric fields, so-called *resonant cavities*, which accelerate the particles, thereby increasing their energy. However, asymptotically this increase is compensated by the energy loss due to radiation and other dissipative effects, so that the maximum speeds are again strictly less than the speed of light, although often very close to it. Unless otherwise stated, henceforth, all trajectories considered are supposed to correspond to one of the above two types of motions.

7.2 Liénard-Wiechert Vector Potential

In this section we determine the vector potential generated by a particle in arbitrary motion, proceeding in the same way as in Sect. 6.3.1. The four-current of a particle with charge e and world line $y^\mu(s)$ is given by

$$j^\mu(y) = e \int u^\mu(s)\, \delta^4(y - y(s))\, ds. \tag{7.8}$$

Inserting this expression in the general solution (6.64) we find the vector potential

$$A^\mu(x) = \frac{e}{2\pi} \int d^4 y \int u^\mu(s)\, H(x^0 - y^0)\, \delta((x - y)^2)\, \delta^4(y - y(s))\, ds \quad (7.9)$$

$$= \frac{e}{2\pi} \int u^\mu(s)\, H(x^0 - y^0(s))\, \delta\left((x - y(s))^2\right) ds \qquad\qquad (7.10)$$

$$= \frac{e}{2\pi} \int u^\mu(s)\, H(x^0 - y^0(s))\, \delta(f(s))\, ds. \qquad\qquad\qquad (7.11)$$

We introduced the function of s

$$f(s) = (x - y(s))^2 = (x^0 - y^0(s))^2 - |\mathbf{x} - \mathbf{y}(s)|^2, \qquad (7.12)$$

where the dependence on the observation point $x = (x^0, \mathbf{x})$ is understood. As in the case of a uniform linear motion, to evaluate $\delta(f(s))$ we must determine the zeros of the function f. Below, in Sect. 7.2.1, we show that – if the world lines correspond to bounded or unbounded motions according to the definitions of Sect. 7.1 – also in the case at hand $f(s)$ possess two zeros $s_\pm(x) \equiv s_\pm$, satisfying the inequalities

$$x^0 - y^0(s_+) > 0, \qquad x^0 - y^0(s_-) < 0. \qquad (7.13)$$

Applying (2.76), and noticing that

$$f'(s) = -2(x^\mu - y^\mu(s)) u_\mu(s), \qquad\qquad (7.14)$$

we can rewrite the integrand of (7.11) in the form

$$H(x^0 - y^0(s))\, \delta(f(s)) = H(x^0 - y^0(s)) \left(\frac{\delta(s - s_+)}{|f'(s_+)|} + \frac{\delta(s - s_-)}{|f'(s_-)|} \right)$$

$$= H(x^0 - y^0(s_+)) \frac{\delta(s - s_+)}{|f'(s_+)|} + H(x^0 - y^0(s_-)) \frac{\delta(s - s_-)}{|f'(s_-)|}$$

$$= \frac{\delta(s - s_+)}{|f'(s_+)|} = \frac{\delta(s - s_+)}{2(x - y(s_+)) u(s_+)}, \qquad (7.15)$$

where we introduced the shorthand notation $(x - y(s)) u(s) = (x^\mu - y^\mu(s)) u_\mu(s)$. In the last line above we used the fact that the scalar product $(x - y(s_+)) u(s_+)$ is positive. To show this we evaluate the scalar product in the inertial frame where the particle at the proper time s_+ is at rest, so that $u^\mu(s_+) = (1, 0, 0, 0)$. From the first inequality in (7.13) we then obtain

$$(x - y(s_+)) u(s_+) = x^0 - y^0(s_+) > 0.$$

Inserting the expression (7.15) in the integral (7.11) we obtain the famous *Liénard-Wiechert potential*

$$A^\mu(x) = \frac{e}{4\pi} \frac{u^\mu(s)}{(x - y(s))u(s)} \bigg|_{s=s_+(x)}, \tag{7.16}$$

where the function $s_+(x)$ is uniquely determined by the implicit relations

$$(x - y(s))^2 = 0, \qquad x^0 - y^0(s) > 0, \tag{7.17}$$

which are equivalent to the single equation

$$x^0 - y^0(s) = |\mathbf{x} - \mathbf{y}(s)|. \tag{7.18}$$

Retarded time. To clarify the physical meaning of the proper "retarded" time $s_+(x)$ we parameterize the world line with the time coordinate $t' = y^0(s)$, so that the world line now reads

$$y^\mu(t') = (t', \mathbf{y}(t')).$$

Then Eq. (7.18) translates in the *delay equation* for t'

$$t - t' = \frac{1}{c} |\mathbf{x} - \mathbf{y}(t')|, \tag{7.19}$$

where we reintroduced the speed of light. The (unique) solution of this equation defines the *retarded time* $t'(t, \mathbf{x})$. As we see, this time is determined in such a way that the position $\mathbf{y}(t')$ of the particle at the time t' is connected by means of a *future light-like* signal to the event $(t, \mathbf{x})$, where the potential is evaluated. In this – in a sense non-covariant – perspective the Liénard-Wiechert potential (7.16) assumes the form (see (2.10))

$$A^\mu(x) = \frac{e}{4\pi} \frac{\left(1, \frac{\mathbf{v}(t')}{c}\right)}{|\mathbf{x} - \mathbf{y}(t')| - (\mathbf{x} - \mathbf{y}(t')) \cdot \frac{\mathbf{v}(t')}{c}}, \tag{7.20}$$

where in the denominator we used the delay equation (7.19). As one sees, the potential at the point $x = (t, \mathbf{x})$ depends not on the kinematical variables $\mathbf{y}$ and $\mathbf{v}$ at time t, but on the values of these variables at the retarded time t'. Therefore, with respect to the non-relativistic potential

$$A^\mu_{\mathrm{nr}}(x) = \frac{e}{4\pi} \frac{(1, 0, 0, 0)}{|\mathbf{x} - \mathbf{y}(t)|}$$

the potential (7.20) entails *explicit* relativistic corrections, represented by the factors $\mathbf{v}(t')/c$, in addition to *implicit* relativistic corrections, due to the appearance of the retarded time $t'(t, \mathbf{x}) = t + O(1/c)$. More precisely, solving the implicit equation (7.19) perturbatively one finds that the retarded time entails the expansion in powers of $1/c$

$$t'(t, \mathbf{x}) = t - \frac{|\mathbf{x} - \mathbf{y}(t)|}{c} - \frac{(\mathbf{x} - \mathbf{y}(t)) \cdot \mathbf{v}(t)}{c^2} + O\left(\frac{1}{c^3}\right). \qquad (7.21)$$

7.2.1 The Zeros of the Function $f(s)$

Theorem I. If the motions associated with the world line $y^\mu(s)$ are either *bounded* or *unbounded*, as specified in Sect. 7.1, for any x^μ not belonging to the world line the function

$$f(s) = (x - y(s))^2 = (x^0 - y^0(s))^2 - |\mathbf{x} - \mathbf{y}(s)|^2 \qquad (7.22)$$

has two real zeros $s_\pm$, satisfying the inequalities

$$x^0 - y^0(s_+) > 0, \qquad x^0 - y^0(s_-) < 0. \qquad (7.23)$$

Proof The function f entails the limits

$$\lim_{s \to \pm\infty} f(s) = +\infty. \qquad (7.24)$$

For bounded motions these limits are obvious in that for $s \to \pm\infty$ we have (see (7.6) and the comments thereafter)

$$y^0(s) = t(s) \to \pm\infty,$$

whereas in the same limit the spatial coordinate $\mathbf{y}(s)$ remains bounded. In the case of unbounded motions, for $s \to \pm\infty$ the four-velocities converge to the limits $u_\pm^\mu$, see (7.7), so that the world line has the asymptotic form $y^\mu(s) \to su_\pm^\mu$. Correspondingly, the function (7.22) entails the asymptotic behavior

$$f(s) \to x^2 - 2\left(x_\mu u_\pm^\mu\right) s + s^2 \to +\infty, \qquad \text{for } s \to \pm\infty.$$

The limits (7.24) imply that the function $f(s)$ has at least an extremum – in particular at least a minimum – and so its derivative has at least one zero. Choose an arbitrary extremum $s = a$. From the derivative (7.14) we obtain

$$f'(a) = -2(x^\mu - y^\mu(a))u_\mu(a) = 0. \qquad (7.25)$$

This relation implies that

$$f(a) < 0.$$

In fact, since the quantities $f(s)$ and $f'(s)$ are Lorentz scalars, we can evaluate them in an arbitrary inertial frame. We choose the rest frame of the particle at proper time

$s = a$, so that $u^\mu(a) = (1, 0, 0, 0)$. Equations (7.22) and (7.25) then imply

$$0 = f'(a) = -2(x^0 - y^0(a)) \quad \Rightarrow \quad f(a) = -|\mathbf{x} - \mathbf{y}(a)|^2 < 0.$$

All maxima and minima of $f(s)$ lie, therefore, in the lower half-plane. This information, together with (7.24), allows us to conclude that f possesses precisely two zeros $s_\pm$, and we choose $s_+ < s_-$. In fact, if f had more than two zeros, it would have at least an extremum in the upper half-plane. At s_+ the function $f(s)$ switches from positive to negative values, and at s_- it switches from negative to positive ones. Correspondingly, we have

$$f'(s_+) < 0, \qquad f'(s_-) > 0.$$

If, using (7.14), we evaluate these inequalities in the rest frame of the particle at the proper times s_+ and s_-, respectively, we obtain the inequalities (7.23). However, since $f(s_\pm) = 0$, the vectors $x^\mu - y^\mu(s_\pm)$ belong to the light cone and, consequently, the sign of $x^0 - y^0(s_\pm)$ is a relativistic invariant. It follows that the inequalities (7.23) hold in all reference frames. $\qquad\square$

Hyperbolic motion. If the conditions (7.3) and, for unbounded motions, (7.7) are not satisfied, in general Theorem I does not apply. In this case it can happen that for some x the delay equation (7.19), equivalent to the conditions (7.17), admits no solution. An important example is the hyperbolic motion (7.1), which violates (7.3) and (7.7), and also (7.24). In fact, for the world line (7.1) the conditions (7.17) admit no solution for s, if x^μ belongs to the set (see Problem 7.2)

$$\Sigma = \{x^\mu \in \mathbb{R}^4 / t + z < 0\}. \tag{7.26}$$

Although for hyperbolic motions the Green function method fails, this feature – arising from a causality argument – suggests that for every fixed t in the region $z < -t$ the electromagnetic field produced by a particle in hyperbolic motion along the z axis is zero. This expectation will be confirmed in Sect. 17.5.

7.2.2 General Validity

Since for a microscopic current like (7.8) it is not a priori guaranteed that the Green function method provides a solution to Maxwell's equations – not even for strictly time-like trajectories – the validity of the method must be checked a posteriori. A necessary condition for the *formal* steps we carried out to be meaningful, is that the so-derived Liénard-Wiechert potential (7.16) defines a distribution. The analysis of this section aims to determine the conditions under which (7.16) is, actually, a distribution and to identify, in turn, the trajectories for which the Green function method necessarily fails.

Since the functional dependence on x of the potential A^μ (7.16) is implicit – in general it is not possible to solve the delay equation (7.19) analytically – to detect its potential singularities we need to apply it to a test function $\varphi \in \mathcal{S}(\mathbb{R}^4)$. To evaluate the corresponding quantities

$$A^\mu(\varphi) = \int A^\mu(x)\,\varphi(x)\,d^4x$$

it is preferable to return to the intermediate expression (7.10)

$$
\begin{aligned}
A^\mu(\varphi) &= \frac{e}{2\pi} \int d^4x \int u^\mu(s)\,H(x^0 - y^0(s))\,\delta\left((x - y(s))^2\right) \varphi(x)\,ds \\
&= \frac{e}{2\pi} \int d^4x \int u^\mu(s)\,H(x^0)\,\delta\left(x^2\right) \varphi(x + y(s))\,ds \\
&= \frac{e}{2\pi} \int dx^0\,d^3x \int u^\mu(s)\frac{1}{2r}\,\delta(x^0 - r)\,\varphi\left(x^0 + y^0(s), \mathbf{x} + \mathbf{y}(s)\right) ds \\
&= \frac{e}{4\pi} \int d^3x \int u^\mu(s)\,\frac{1}{r}\,\varphi\left(y^0(s) + r, \mathbf{x} + \mathbf{y}(s)\right) ds \\
&= \frac{e}{4\pi} \int \frac{v^\mu(t)}{r}\,\varphi(t + r, \mathbf{x} + \mathbf{y}(t))\,d^3x\,dt, \\
&= \frac{e}{4\pi} \int \frac{v^\mu(t - r)}{r}\,\varphi(t, \mathbf{x} + \mathbf{y}(t - r))\,d^3x\,dt, \qquad (7.27)
\end{aligned}
$$

where in the last two lines we have parameterized the trajectory with time, in place of proper time. We have set $r = |\mathbf{x}|$ and we have introduced the vector $v^\mu(t) = (1, \mathbf{v}(t))$. For A^μ to be a distribution it is, first of all, necessary that the four-dimensional integral appearing in (7.27) converges for all test functions φ. Since $1/r$ is integrable in d^3x in the neighborhood of $\mathbf{x} = 0$, the convergence is guaranteed in all *finite* regions of the four-dimensional space-time. This means that we have only to worry about the behavior of the integrand at infinity, i.e. for $x^\mu = (t, \mathbf{x}) \to \infty$, where the convergence of the integral (7.27) *should* be guaranteed by the fast decay of $\varphi(x)$ in all directions, see (2.60). Moreover, since the absolute value of all components of $v^\mu(t - r)$ is less than 1, this vector plays no role in this respect and can be disregarded.

Suppose now first that the, time-like or light-like, trajectory $\mathbf{y}(t)$ is bounded: $|\mathbf{y}(t)| < l$ for all t. In this case, for $x^\mu \to \infty$ all four arguments of the test function in (7.27) go to infinity, since in this limit $(t, \mathbf{x} + \mathbf{y}(t - r)) \approx (t, \mathbf{x})$. This means that, thanks to the fast decay of φ in all directions, for an *arbitrary bounded motion* the integral (7.27) converges. Next consider an unbounded, time-like or light-like, trajectory which in the infinite past, i.e. for $t \to -\infty$, tends to a straight line L with constant speed v_∞. Taking as z axis the line L, then for $t \to -\infty$ the trajectory entails the asymptotic behavior

$$\mathbf{y}(t) \approx (0, 0, v_\infty t), \quad v_\infty \leq 1. \qquad (7.28)$$

In this case, as $t \to \pm\infty$ the test function in (7.27) still goes rapidly to zero, in that the absolute value of its time argument goes to infinity. On the other hand, considering the limit for $\mathbf{x} \to \infty$, from (7.28) we derive for the spatial argument of the test function in (7.27) the behavior

$$\mathbf{x} + \mathbf{y}(t - r) \approx (x, y, tv_\infty + z - v_\infty r).$$

In particular, if $\mathbf{x}$ goes to infinity along the *positive* z axis this behavior becomes

$$\mathbf{x} + \mathbf{y}(t - r) \approx (x, y, tv_\infty + (1 - v_\infty)z).$$

Therefore, if $v_\infty < 1$ the spatial argument of the test function goes to infinity in all directions and so the integral (7.27) converges. Conversely, if $v_\infty = 1$, the spatial argument of the test function does not vanish as $\mathbf{x}$ goes to infinity along the positive z axis, becoming rather independent of z, and, consequently, in general the integral (7.27) *diverges*.

It is a simple exercise to show that in the cases considered above, once the integral (7.27) expressing $A^\mu(\varphi)$ converges, it can also be bounded by a finite sum of semi-norms, as required by (2.61). This means that in these cases the Liénard-Wiechert potential (7.16) *is* a distribution. We conclude, therefore, that (a) for bounded time-like and light-like trajectories and (b) for unbounded time-like trajectories which in the infinite past tend to a straight line with asymptotic speed less than the speed of light, the Green function method gives rise to a distribution-valued vector potential. Vice versa, for unbounded trajectories which in the infinite past tend to a straight line with asymptotic speed equal to the speed of light, the Green function method does not yield a distribution-valued vector potential. This method works thus, in particular, for all strictly time-like trajectories, whereas it fails for all light-like trajectories as well as for all time-like asymptotically light-like trajectories – like the hyperbolic motion (7.2) – which in the infinite past tend to a straight line.

7.3 Liénard-Wiechert Fields

Henceforth, we consider again strictly time-like, bounded or unbounded, trajectories, for which the potential (7.16) is a distribution and solves Maxwell's equations.[1] We determine now the associated electromagnetic field $F^{\mu\nu}$. In the following for simplicity we denote the retarded proper time $s_+(x)$ – a function of x – with the symbol s. Apart from the four-velocity $u^\mu(s) = dy^\mu(s)/ds$, we introduce also the four-acceleration $w^\mu(s) = du^\mu(s)/ds$, and the vector field

$$L^\mu(x) = x^\mu - y^\mu(s), \tag{7.29}$$

---

[1] Of course, the analysis of Sect. 7.2.2 only proves that for strictly time-like trajectories the Liénard-Wiechert potential (7.16) is a distribution, but not that it satisfies Maxwell's equations: Being a *historical* result, we take it for granted.

which depends on x also through s. The conditions (7.17) can then be recast in the equivalent form

$$L_\alpha L^\alpha = 0, \qquad L^0 > 0. \tag{7.30}$$

In this way the potential (7.16) and the Maxwell tensor $F^{\mu\nu} = \partial^\mu A^\nu - \partial^\nu A^\mu$ can be written as

$$A^\mu = \frac{eu^\mu}{4\pi(uL)}, \qquad F^{\mu\nu} = \frac{e}{4\pi(uL)}\left(\partial^\mu u^\nu - \frac{\partial^\mu(uL)u^\nu}{(uL)} - (\mu \leftrightarrow \nu)\right). \tag{7.31}$$

We have introduced for the scalar product of two four-vectors the notation

$$a^\mu b_\mu = (ab). \tag{7.32}$$

To evaluate the derivatives appearing in (7.31) we must determine the partial derivatives of s with respect to x^μ. To this end we differentiate the constraint (7.30) with respect to x^μ

$$0 = L^\alpha \partial_\mu L_\alpha = L^\alpha \partial_\mu(x_\alpha - y_\alpha(s)) = L^\alpha\left(\eta_{\alpha\mu} - \frac{\partial s}{\partial x^\mu}\frac{dy_\alpha}{ds}\right) = L_\mu - (uL)\frac{\partial s}{\partial x^\mu},$$

leading to

$$\frac{\partial s}{\partial x^\mu} = \frac{L_\mu}{(uL)}. \tag{7.33}$$

For the derivatives appearing in (7.31) we then obtain

$$\partial^\mu u^\nu = \frac{\partial s}{\partial x_\mu}\frac{du^\nu}{ds} = \frac{L^\mu w^\nu}{(uL)}, \tag{7.34}$$

$$\partial_\mu L_\nu = \eta_{\mu\nu} - \frac{\partial s}{\partial x^\mu}\frac{dy_\nu}{ds} = \eta_{\mu\nu} - \frac{L_\mu u_\nu}{(uL)}, \tag{7.35}$$

$$\partial_\mu(uL) = (\partial_\mu u^\nu)L_\nu + u^\nu\partial_\mu L_\nu = \frac{(wL)}{(uL)}L_\mu + u^\nu\left(\eta_{\mu\nu} - \frac{L_\mu u_\nu}{(uL)}\right) \tag{7.36}$$

$$= \frac{(wL) - 1}{(uL)}L_\mu + u_\mu. \tag{7.37}$$

Substituting these expressions in the second formula in (7.31) the Maxwell tensor becomes

$$F^{\mu\nu} = \frac{e}{4\pi(uL)^3}\left(L^\mu u^\nu + L^\mu((uL)w^\nu - (wL)u^\nu) - (\mu \leftrightarrow \nu)\right), \tag{7.38}$$

which is the *Liénard-Wiechert field* in manifestly covariant form.

7.3.1   Velocity and Acceleration Fields

We begin the analysis of the electromagnetic field (7.38) by studying its behavior at large distances from the particle. To this end it is convenient to separate the terms appearing in $F^{\mu\nu}$ in two classes, according to their dependence on the variable, see (7.29) and (7.30),

$$R \equiv L^0 \doteq |\mathbf{x} - \mathbf{y}(s)|. \tag{7.39}$$

We introduce also the light-like vector

$$m^\mu = \frac{L^\mu}{R}, \qquad m^\mu m_\mu = 0,$$

whose components are

$$m^0 = 1, \qquad \mathbf{m} = \frac{\mathbf{x} - \mathbf{y}(s)}{|\mathbf{x} - \mathbf{y}(s)|}, \qquad |\mathbf{m}| = 1.$$

Substituting in (7.38) the vector L^μ with Rm^μ, we can recast the Liénard-Wiechert field as sum of two terms, *the velocity field* $F_v^{\mu\nu}$ and the *acceleration field* $F_a^{\mu\nu}$:

$$F^{\mu\nu} = F_v^{\mu\nu} + F_a^{\mu\nu}, \tag{7.40}$$

$$F_v^{\mu\nu} = \frac{e}{4\pi(um)^3 R^2}\,(m^\mu u^\nu - m^\nu u^\mu), \tag{7.41}$$

$$F_a^{\mu\nu} = \frac{e}{4\pi(um)^3 R}\Big(m^\mu\big((um)w^\nu - (wm)u^\nu\big)\big) - (\mu \leftrightarrow \nu)\Big). \tag{7.42}$$

In $F_a^{\mu\nu}$ we have included the terms proportional to $1/R$, and in $F_v^{\mu\nu}$ those proportional to $1/R^2$. The former is linear in the four-acceleration w^μ, while the latter is independent of w^μ. Let us now analyze the behavior of these fields at large distances from the particle. To this end we suppose that the particle is confined to a finite region of space, $|\mathbf{y}(s)| < l$ for all s, and we consider the field at a point $\mathbf{x}$ far away from this region, $|\mathbf{x}| \gg l$. Setting $|\mathbf{x}| = r$, we then have the asymptotic identification

$$\frac{1}{R} = \frac{1}{|\mathbf{x} - \mathbf{y}|} \to \frac{1}{r}, \qquad \text{for} \quad r \gg l.$$

Supposing that the four-vectors u^μ and w^μ are bounded, we see that at large distances from the particle the acceleration field decreases as

$$F_a^{\mu\nu} \sim \frac{1}{r}, \tag{7.43}$$

while the velocity field decreases as

$$F_v^{\mu\nu} \sim \frac{1}{r^2}. \tag{7.44}$$

In particular, at large distances from the particle the acceleration field dominates over the velocity field, so that the total field decreases as $F^{\mu\nu} \sim 1/r$. Notice that this behavior is in sharp contrast with the asymptotic behavior (6.104) of the field of a particle in uniform linear motion. In order to analyze more closely the velocity field, we rewrite it in the form

$$F_v^{\mu\nu} = \frac{e}{4\pi(uL)^3} \left(L^\mu u^\nu - L^\nu u^\mu \right). \tag{7.45}$$

It is easy to see that for a uniform linear motion this field reduces to (6.96). In fact, for $y^\mu(s) = su^\mu$ we have $L^\mu = x^\mu - su^\mu$, so that by inserting the retarded time (6.89) we obtain the relations

$$L^\mu u^\nu - L^\nu u^\mu = x^\mu u^\nu - x^\nu u^\mu,$$
$$(uL) = u^\mu(x_\mu - su_\mu) = (ux) - s_+(x) = \sqrt{(ux)^2 - x^2}.$$

The field $F_v^{\mu\nu}$ represents hence a *deformation* of the field (6.96) of a particle in uniform linear motion, inheriting in particular its asymptotic behavior $1/r^2$. For this reason frequently $F_v^{\mu\nu}$ is also called *Coulomb field*. Conversely, the acceleration field $F_a^{\mu\nu}$ – generated, in fact, by the acceleration of the particle – constitutes a new dynamical effect which is responsible for the *radiation* phenomenon, see Sect. 7.4.

7.3.2 Electric and Magnetic Fields

We now write out the electric and magnetic fields associated with the Maxwell tensor (7.38). According to Eqs. (7.40)–(7.42) these fields split, in turn, in velocity fields – independent of the acceleration and proportional to $1/R^2$ – and in acceleration fields – linear in the acceleration and proportional to $1/R$ –

$$\mathbf{E} = \mathbf{E}_v + \mathbf{E}_a, \tag{7.46}$$
$$\mathbf{B} = \mathbf{B}_v + \mathbf{B}_a. \tag{7.47}$$

As first step we write out the four-acceleration in terms of the spatial acceleration $\mathbf{a}$

$$w^\mu = \frac{du^\mu}{ds} = \frac{(\mathbf{a} \cdot \mathbf{v})\,u^\mu}{(1-v^2)^{3/2}} + \frac{(0,\mathbf{a})}{1-v^2}.$$

When inserting this expression in the term $(um)w^\nu - (wm)u^\nu$, the contribution proportional to u^μ cancels out. Using also the relation

$$(um) = \frac{1 - \mathbf{v} \cdot \mathbf{m}}{\sqrt{1 - v^2}},$$

from (7.41) and (7.42) we derive the Liénard-Wiechert electric and magnetic fields

$$\mathbf{E}_v = \frac{e}{4\pi R^2} \frac{\left(1 - \frac{v^2}{c^2}\right)\left(\mathbf{m} - \frac{\mathbf{v}}{c}\right)}{\left(1 - \frac{\mathbf{v} \cdot \mathbf{m}}{c}\right)^3}, \qquad \mathbf{B}_v = \mathbf{m} \times \mathbf{E}_v,$$

$$ \tag{7.48}$$

$$\mathbf{E}_a = \frac{e}{4\pi R c^2} \frac{\mathbf{m} \times \left(\left(\mathbf{m} - \frac{\mathbf{v}}{c}\right) \times \mathbf{a}\right)}{\left(1 - \frac{\mathbf{v} \cdot \mathbf{m}}{c}\right)^3}, \qquad \mathbf{B}_a = \mathbf{m} \times \mathbf{E}_a,$$

where we reintroduced the velocity of light. It is important to keep in mind that the kinematical variables $\mathbf{y}$, $\mathbf{v}$ and $\mathbf{a}$ appearing in the above formulas are evaluated not at the time t, but at the retarded time $t'(t, \mathbf{x})$ defined by (7.19). Equation (7.48) imply first of all the relation between $\mathbf{E}$ and $\mathbf{B}$

$$\mathbf{B} = \mathbf{m} \times \mathbf{E}.$$

The total electric and magnetic fields are, therefore, at each point orthogonal to each other. From the expressions of $\mathbf{E}_v$ and $\mathbf{B}_v$ we read off the further equality

$$\mathbf{B}_v = \frac{\mathbf{v}}{c} \times \mathbf{E}_v,$$

implying that the magnetic velocity field is suppressed by a factor of v/c with respect to the electric velocity field, exactly as in the case of the uniform linear motion, see Eq. (6.99). Conversely, the magnetic and electric acceleration fields have *equal* magnitude, in that they obey the relations

$$\mathbf{m} \cdot \mathbf{E}_a = 0, \qquad \mathbf{B}_a = \mathbf{m} \times \mathbf{E}_a \qquad \Rightarrow \qquad B_a = E_a. \tag{7.49}$$

Finally, notice that with respect to the velocity field $\mathbf{E}_v$, the fields $\mathbf{E}_a$ and $\mathbf{B}_a$ carry a prefactor $1/c^2$: The acceleration fields represent, therefore, a genuinely *relativistic* effect.

Large distance behavior of the field of a system of particles. We conclude this section with an important generalization. Since the general solution (6.64) of Maxwell's equations is *linear* in the current, and the current (2.16) of a system of point-particles corresponds to the sum of currents

$$j^\mu = \sum_{r=1}^{N} j_r^\mu,$$

the asymptotic behaviors (7.43) and (7.44) of the Liénard-Wiechert field extend automatically to the field generated by an arbitrary bounded system of charged particles, i.e. a system of particles for which $|\mathbf{y}_r(s)| < l$ for all s and for all $r = 1, \ldots, N$. More precisely, the total electromagnetic field can still be written as a sum of two contributions, $F^{\mu\nu} = F_v^{\mu\nu} + F_a^{\mu\nu}$, which at large distances, i.e. for $r \gg l$, decrease again as

$$F_v^{\mu\nu} \sim \frac{1}{r^2}, \qquad F_a^{\mu\nu} \sim \frac{1}{r}. \tag{7.50}$$

Furthermore, at the *asymptotic* level also the relations (7.49) can be generalized. In fact, at large distances the unit vectors $\mathbf{m}_r$ loose their dependence on the position $\mathbf{y}_r(s)$ of the particle,

$$\mathbf{m}_r = \frac{\mathbf{x} - \mathbf{y}_r(s)}{|\mathbf{x} - \mathbf{y}_r(s)|} \rightarrow \frac{\mathbf{x}}{r} = \mathbf{n}, \quad \text{for} \quad r \rightarrow \infty, \tag{7.51}$$

coinciding thus with the unit vector $\mathbf{n}$ identifying the asymptotic direction in which the field is evaluated. For a system of particles from the relations (7.49) – via linearity – we deduce hence the *asymptotic* relations

$$\mathbf{n} \cdot \mathbf{E}_a = 0, \qquad \mathbf{B}_a = \mathbf{n} \times \mathbf{E}_a, \qquad B_a = E_a, \qquad \text{for} \quad r \rightarrow \infty. \tag{7.52}$$

Finally, since also the *macroscopic* currents – as the ones associated with the electrons in an antenna or in an electric circuit – are linear superpositions, or averages, of the currents (2.16) of point-like charges, the asymptotic relations (7.50) and (7.52) hold also for the electromagnetic fields generated by those currents.

7.4 Radiation from Accelerated Charged Particles

Now that we have an analytic expression of the electromagnetic field generated by a charged particle in arbitrary motion, we are in a position to understand the mechanism by which accelerated charges emit or absorb energy, and more generally four-momentum, *through* their field. Notice that the quantity we are interested in is not the four-momentum the particles exchange with the field, but rather the four-momentum the system *field + particles* exchanges with the *environment*, since this is the physical quantity detected experimentally. By an abuse of language, that we shall adopt too, normally one refers to this four-momentum as the *four-momentum emitted by the particles*.

Four-momentum emission. Consider a system of charged particles generating an electromagnetic field according to Maxwell's equations. As seen in Sect. 2.4.3, the four-momentum transport of such a system is quantified by the energy-momentum tensor of *only* the electromagnetic field

$$T_{\text{em}}^{\mu\nu} = F^\mu{}_\alpha F^{\alpha\nu} + \frac{1}{4} \eta^{\mu\nu} F^{\alpha\beta} F_{\alpha\beta}.$$

Considering the exchanged four-momentum as "positive" when it is *ceded* – as, henceforth, we will always do – at the basis of Eq. (2.151) the four-momentum ceded by the system across a closed surface Γ per unit time is given by the surface integral

$$\frac{dP^\mu}{dt} = \int_\Gamma T_{\text{em}}^{i\mu} \, d\Sigma^i. \tag{7.53}$$

However, the four-momentum can be considered as truly *emitted*, i.e. *definitively* ceded by the system to the environment, only if subsequently it is not reabsorbed. In other words, the emitted four-momentum is the one that manages to escape to infinity.[2] This means that in Eq. (7.53) we must choose for Γ a sphere of radius r, and then let r tend to infinity. Writing the surface element of the sphere as $d\Sigma = \mathbf{n}\,r^2 d\Omega$, where $d\Omega$ is the infinitesimal solid angle and $\mathbf{n}$ the outgoing unit vector normal to the sphere, for the four-momentum emitted per unit time we obtain thus

$$\frac{dP^\mu}{dt} = r^2 \int T_{\text{em}}^{i\mu} \, n^i d\Omega \Big|_{r\to\infty}. \tag{7.54}$$

From this expression we can eventually read off the four-momentum emitted in the direction $\mathbf{n}$ per unit solid angle per unit time

$$\frac{d^2 P^\mu}{dt d\Omega} = r^2 \left(T_{\text{em}}^{i\mu} \, n^i \right)\big|_{r\to\infty}. \tag{7.55}$$

Equation (7.55) constitutes the basis for the analysis of the energy- and momentum-emission of a generic system of charged particles. As one sees, to evaluate its right hand side it is sufficient to select from $T_{\text{em}}^{i\mu}$ the terms which for $r \to \infty$ decrease as $1/r^2$, i.e. – since $T_{\text{em}}^{\mu\nu}$ is quadratic in $F^{\mu\nu}$ – to select from $F^{\mu\nu}$ the terms which decrease as $1/r$. In light of the asymptotic behaviors (7.50), this implies that only the acceleration field contributes to the second member of Eq. (7.55). From this analysis we draw, thus, a two-fold general conclusion: *in a system of charged particles it is only the acceleration field to cause emission of four-momentum, and to quantify the latter it is sufficient to evaluate the former at large distances from the charged particles.* Henceforth, in Eq. (7.55) the limit for $r \to \infty$ will always be understood.

Radiation formula. We now investigate more closely the emission of energy via radiation. The $\mu = 0$ component of Eq. (7.55) yields for the energy $\varepsilon = P^0$ emitted per unit time per unit solid angle, i.e. for the *power* $\mathcal{W} = d\varepsilon/dt$ emitted per unit solid angle, the expression (remember that T_{em}^{i0} equals the Poynting vector S^i)

---

[2]At the quantum level this means that we consider as *emitted* only those photons which are able to reach spatial infinity, without being reabsorbed by the charged particles.

$$\frac{d\mathcal{W}}{d\Omega} = \frac{d^2\varepsilon}{dt d\Omega} = r^2 (\mathbf{n} \cdot \mathbf{S}).$$
(7.56)

According to the above analysis, it is sufficient to keep in the Poynting vector the radiation fields,

$$\mathbf{S} = \mathbf{E} \times \mathbf{B} \quad \rightarrow \quad \mathbf{E}_a \times \mathbf{B}_a,$$

and, in addition, the latter must be evaluated at large distances from the particles. We can, therefore, resort to the general asymptotic relations (7.52), obtaining for the Poynting vector the simple asymptotic formula

$$\mathbf{S} = \mathbf{E}_a \times \mathbf{B}_a = \mathbf{E}_a \times (\mathbf{n} \times \mathbf{E}_a) = E_a^2 \, \mathbf{n}.$$
(7.57)

The vector $\mathbf{S}$ has, hence, the same direction and the same orientation as $\mathbf{n}$, meaning that the energy flux is always radially *outgoing* to infinity: The energy is, therefore, always *emitted* by the charged particles, and never *absorbed*. Notice that, if in Eq. (6.64) in place of the retarded Green function G_{ret} we had used the advanced one G_{adv}, the energy flux would have been always *incoming* from infinity. Clearly, this *asymmetry* is a manifestation of the spontaneous violation of time-reversal invariance discussed in Sect. 6.2.3. Inserting the Poynting vector (7.57) in Eq. (7.56), and reintroducing the speed of light, eventually for the angular distribution of the emitted power we obtain the simple expression

$$\frac{d\mathcal{W}}{d\Omega} = c \, r^2 E_a^2.$$
(7.58)

This equation represents the fundamental *radiation formula*: it ties the radiated energy to the magnitude of the electric acceleration field, evaluated at large distances from the charges. Notice, in particular, that since $\mathbf{E}_a$ decreases as $1/r$, in the limit for $r \rightarrow \infty$, which is always understood, Eq. (7.58) yields always a *finite* result.

Large distance relations. In the case of a single particle the asymptotic electric radiation field $\mathbf{E}_a$ assumes a relatively simple form. In fact, performing in the fields (7.48) the asymptotic identifications $\mathbf{m} \rightarrow \mathbf{n}$ and $R \rightarrow r$, we obtain

$$\mathbf{E}_a = \frac{e}{4\pi r c^2} \frac{\mathbf{n} \times \left(\left(\mathbf{n} - \frac{\mathbf{v}}{c} \right) \times \mathbf{a} \right)}{\left(1 - \frac{\mathbf{v} \cdot \mathbf{n}}{c} \right)^3}.$$
(7.59)

In particular we have, see Problem 7.1,

$$\mathbf{E}_a = 0, \ \forall \mathbf{n} \quad \Leftrightarrow \quad \mathbf{a} = 0,$$

meaning that the presence of a non-vanishing emitted energy is inseparably tied to the occurrence of a non-vanishing acceleration of the charged particle. The expression (7.59) looks simpler than it is in reality, in that the kinematic variables appearing therein are evaluated at the retarded time $t'(t, \mathbf{x})$, determined by Eq. (7.19)

$$t - t' = \frac{1}{c}|\mathbf{x} - \mathbf{y}(t')|. \tag{7.60}$$

In the asymptotic field (7.59), also this equation must be considered at large distances from the particle. If the particle is confined to a sphere S_l of radius l centered at the origin, the second member of (7.60) must be evaluated for $r = |\mathbf{x}| \gg l > |\mathbf{y}(t')|$. Setting $\mathbf{y}(t') = \mathbf{y}$, and using the expansion

$$|\mathbf{x} - \mathbf{y}| = r\left|\mathbf{n} - \frac{\mathbf{y}}{r}\right| = r\sqrt{1 - 2\frac{\mathbf{n}\cdot\mathbf{y}}{r} + \frac{y^2}{r^2}}$$

$$= r\left(1 - \frac{\mathbf{n}\cdot\mathbf{y}}{r} + O\left(\frac{y^2}{r^2}\right)\right) = r - \mathbf{n}\cdot\mathbf{y} + O\left(\frac{y^2}{r}\right), \tag{7.61}$$

at the asymptotic level Eq. (7.60) becomes

$$t' = t - \frac{r}{c} + \frac{\mathbf{n}\cdot\mathbf{y}(t')}{c}. \tag{7.62}$$

At large distances the retarded time t' can, thus, be interpreted as sum of the *macroscopic* retarded time $t - r/c$, and of the *microscopic* delay $\mathbf{n}\cdot\mathbf{y}(t')/c$. The former represents the retarded instant at which the electromagnetic signal must leave the center of the sphere S_l, in order to arrive at the instant t at the observation point $\mathbf{x}$. This instant is independent of the particle's motion as well as of the propagation direction $\mathbf{n}$ of the signal. Conversely, the microscopic term represents an additional delay dependent on $\mathbf{n}$, caused by the motion $\mathbf{y}(t')$ of the particle inside the sphere S_l. In Sect. 8.3.1 we will see that in the non-relativistic limit this term is negligible.

Acceleration field as radiation field. The acceleration field $F_a^{\mu\nu}$ in a sense resembles a radiation field, i.e. a solution of Maxwell's equations in vacuum, see Chap. 5. A first indication pointing in this direction comes from the observation that in the complement the world line of a particle the total field (7.40) satisfies, actually, the equations of a radiation field

$$\partial_\mu F^{\mu\nu} = 0 = \partial_{[\mu}F_{\nu\rho]}. \tag{7.63}$$

Since $F^{\mu\nu} = F_v^{\mu\nu} + F_a^{\mu\nu}$, and $F_v^{\mu\nu}$ decreases as $1/r^2$, Eqs. (7.63) imply that the field $F_a^{\mu\nu}$ – decreasing as $1/r$ – satisfies these equations asymptotically,[3] i.e. modulo terms of order $1/r^2$

$$\partial_\mu F_a^{\mu\nu} = O\left(\frac{1}{r^2}\right), \qquad \partial_{[\mu}F_{a\,\nu\rho]} = O\left(\frac{1}{r^2}\right).$$

Accordingly, we expect that at large distances from the particle the acceleration field behaves as a radiation field, amounting in particular to a *superposition of ele-*

---

[3] Actually, it can be shown that the acceleration and velocity fields both satisfy the Bianchi identity exactly: $\partial_{[\mu}F_{a\,\nu\rho]} = 0 = \partial_{[\mu}F_{v\,\nu\rho]}$.

mentary waves. If this is true, from the expression of the Poynting vector (7.57) – formally identical to the expression (5.99) of the Poynting vector of an elementary wave – we conclude that the waves composing $F_a^{\mu\nu}$ propagate along the radial outgoing direction. In Chap. 8, where we analyze in detail the asymptotic properties of a generic acceleration field, we will actually confirm these general predictions. For the characteristics just described the field $F_a^{\mu\nu}$ is often also called *radiation field*.

7.4.1 Non-relativistic Limit and Larmor's Formula

As a basic application of the formulas of the preceding section we determine the total power

$$W = \int \frac{dW}{d\Omega} \, d\Omega \tag{7.64}$$

emitted by a non-relativistic particle, $v \ll c$, in all directions. To be precise, we want to evaluate the power (7.64) at lowest order in $1/c$, which, as we will see in a moment, corresponds to the order $W \sim 1/c^3$. In general, to determine $dW/d\Omega$ we must insert the asymptotic Liénard-Wiechert field (7.59) in the radiation formula (7.58). Since in the present case we are interested in the lowest order in $1/c$, due to the presence of the prefactor $1/c^2$ the field (7.59) simplifies to

$$\mathbf{E}_a = \frac{e}{4\pi r c^2} \, \mathbf{n} \times (\mathbf{n} \times \mathbf{a}), \qquad E_a^2 = \frac{e^2 \, |\mathbf{n} \times \mathbf{a}|^2}{16\pi^2 r^2 c^4}. \tag{7.65}$$

In these expressions the acceleration is still evaluated at the exact asymptotic retarded time $t'(t, \mathbf{x})$, determined by Eq. (7.62). However, as anticipated above, in the non-relativistic limit the microscopic delay $\mathbf{n} \cdot \mathbf{y}(t')/c$ is negligible, see Sect. 8.3.1, so that t' reduces simply to

$$t' = t - \frac{r}{c}.$$

Equations (7.58) and (7.65) yield thus the formula

$$\frac{dW}{d\Omega}(t, r, \mathbf{n}) = \frac{e^2}{16\pi^2 c^3} \left| \mathbf{n} \times \mathbf{a}\left(t - \frac{r}{c}\right) \right|^2, \tag{7.66}$$

whose correct interpretation is the following: the expression at second member represents the energy emitted by a non-relativistic particle per unit time per unit solid angle, *detected* at time t at a large distance r from the particle in the direction $\mathbf{n}$. Accordingly, the acceleration appearing there is evaluated at the retarded time $t - r/c$.

Larmor's formula. In Eq. (7.66) the acceleration does no longer depend on the angles, and so the total power (7.64) can be calculated explicitly. Choosing, for fixed t and r, as z axis the direction of $\mathbf{a}$, and using the relations

$$|\mathbf{n} \times \mathbf{a}|^2 = a^2 \sin^2 \vartheta, \qquad d\Omega = \sin \vartheta \, d\vartheta d\varphi,$$

Equations (7.64) and (7.66) give for the total power the expression

$$\mathcal{W} = \frac{e^2 a^2}{16\pi^2 c^3} \int_0^{2\pi} d\varphi \int_0^{\pi} \sin^3 \vartheta \, d\vartheta.$$

Performing the integrals we obtain the celebrated *Larmor formula* (1897)

$$\mathcal{W} = \frac{e^2 a^2}{6\pi c^3}, \tag{7.67}$$

which gives the energy emitted per unit time by a non-relativistic particle of charge e, having acceleration $\mathbf{a}$. We stress that in this formula the power $\mathcal{W}$ – detected at time t at a distance r from the particle – involves at its second member the acceleration at the time $t - r/c$. Precisely because radiation propagates at the speed of light, Larmor's formula can also be *interpreted* saying that if at an instant t the particle has acceleration $\mathbf{a}$ at that instant it emits radiation with total power $e^2 a^2/6\pi c^3$. We will come back to this interpretation in Sect. 10.1, in the context of the relativistic generalization of (7.67). We postpone the investigation of the physical implications of Larmor's formula to Chap. 8, where we retrieve it by means of a more systematic method.

7.5 Non-relativistic Expansion of Potentials and Fields

In this section, for future reference, we perform the non-relativistic expansion of the Liénard-Wiechert fields, formally corresponding to a Taylor series expansion in powers of $1/c$. This expansion is, hence, physically reliable, whenever the speed of the charged particle is much less than the speed of light, $v \ll c$. From the form of the Lorentz force $e(\mathbf{E} + \mathbf{v} \times \mathbf{B}/c)$ we see that if we expand $\mathbf{E}$ up to the powers $1/c^n$, it is sufficient to expand $\mathbf{B}$ up to the powers $1/c^{n-1}$. In this section we perform the expansion of the Liénard-Wiechert fields (7.46)–(7.48) up to the powers corresponding to $n = 3$.

Expansion of the Liénard-Wiechert potential. From a technical point of view the expansion of the fields (7.48) is complicated by the presence of the retarded time $t'(t, \mathbf{x})$, which, in turn, must be expanded in powers of $1/c$, see (7.21). For practical purposes it is more convenient first to expand the Liénard-Wiechert potential (7.20), and then to use the relations

$$\mathbf{E} = -\nabla A^0 - \frac{1}{c}\frac{\partial \mathbf{A}}{\partial t}, \qquad \mathbf{B} = \nabla \times \mathbf{A}, \tag{7.68}$$

to derive the expansions of the fields. Correspondingly, we must expand the scalar potential A^0 up to terms of order $1/c^3$, and the vector potential $\mathbf{A}$ up to terms of order $1/c^2$. Moreover, instead of expanding the implicit representation (7.20) it is preferable to expand the equivalent integral representation (6.70)

$$A^\mu(t, \mathbf{x}) = \frac{1}{4\pi c} \int \frac{1}{|\mathbf{x} - \mathbf{z}|} j^\mu\left(t - \frac{|\mathbf{x} - \mathbf{z}|}{c}, \mathbf{z}\right) d^3 z, \tag{7.69}$$

in which the current is given by

$$j^\mu(t, \mathbf{x}) = e V^\mu(t)\, \delta^3(\mathbf{x} - \mathbf{y}(t)), \qquad V^\mu(t) = (c, \mathbf{v}(t)). \tag{7.70}$$

Expanding the integral (7.69) in powers of $1/c$ up to terms of order $1/c^3$, taking into account that V^μ is of order c, we obtain

$$\begin{aligned}
A^\mu(t, \mathbf{x}) &= \frac{1}{4\pi c} \int \left(\frac{j^\mu(t, \mathbf{z})}{|\mathbf{x} - \mathbf{z}|} - \frac{1}{c} \frac{\partial j^\mu(t, \mathbf{z})}{\partial t} \right. \\
&\qquad \left. + \frac{1}{2c^2} |\mathbf{x} - \mathbf{z}| \frac{\partial^2 j^\mu(t, \mathbf{z})}{\partial t^2} - \frac{1}{6c^3} |\mathbf{x} - \mathbf{z}|^2 \frac{\partial^3 j^\mu(t, \mathbf{z})}{\partial t^3} \right) d^3 z \\
&= \frac{1}{4\pi c} \left(\int \frac{j^\mu(t, \mathbf{z})}{|\mathbf{x} - \mathbf{z}|} d^3 z - \frac{1}{c} \frac{\partial}{\partial t} \int j^\mu(t, \mathbf{z})\, d^3 z \right. \\
&\qquad \left. + \frac{1}{2c^2} \frac{\partial^2}{\partial t^2} \int |\mathbf{x} - \mathbf{z}|\, j^\mu(t, \mathbf{z})\, d^3 z - \frac{1}{6c^3} \frac{\partial^3}{\partial t^3} \int |\mathbf{x} - \mathbf{z}|^2 j^\mu(t, \mathbf{z})\, d^3 z \right) \\
&= \frac{e}{4\pi c} \left(\frac{V^\mu}{R} - \frac{1}{c} \frac{\partial V^\mu}{\partial t} + \frac{1}{2c^2} \frac{\partial^2}{\partial t^2} (R V^\mu) - \frac{1}{6c^3} \frac{\partial^3}{\partial t^3} (R^2 V^\mu) \right). \tag{7.71}
\end{aligned}$$

We use the notation[4]

$$\mathbf{R} = \mathbf{x} - \mathbf{y}(t), \qquad R = |\mathbf{x} - \mathbf{y}(t)|, \qquad \widehat{\mathbf{R}} = \frac{\mathbf{R}}{R}.$$

Inserting in (7.71) for V^μ the expression (7.70), we obtain the expansion of the Liénard-Wiechert potentials at the required orders

$$A^0 = \frac{e}{4\pi} \left(\frac{1}{R} + \frac{1}{2c^2} \frac{\partial^2 R}{\partial t^2} - \frac{1}{6c^3} \frac{\partial^3 R^2}{\partial t^3} \right), \tag{7.72}$$

$$\mathbf{A} = \frac{e}{4\pi} \left(\frac{\mathbf{v}}{cR} - \frac{\mathbf{a}}{c^2} \right). \tag{7.73}$$

Notice that in A^0 the term of order $1/c$ dropped out.

---

[4] In this section R has a different meaning from the same symbol of Sect. 7.3, where $R = |\mathbf{x} - \mathbf{y}(t')|$, see (7.39).

Expansion of the Liénard-Wiechert fields. To compute the electric field we need the partial derivatives

$$-\boldsymbol{\nabla} A^0 = \frac{e}{4\pi}\left(\frac{\mathbf{R}}{R^3} - \frac{1}{2c^2}\frac{\partial^2 \widehat{\mathbf{R}}}{\partial t^2} - \frac{1}{3c^3}\frac{d\mathbf{a}}{dt}\right), \tag{7.74}$$

$$-\frac{1}{c}\frac{\partial \mathbf{A}}{\partial t} = -\frac{e}{4\pi}\left(\frac{1}{c^2}\frac{\partial}{\partial t}\left(\frac{\mathbf{v}}{R}\right) - \frac{1}{c^3}\frac{d\mathbf{a}}{dt}\right). \tag{7.75}$$

By adding these expressions, with the aid of the derivatives

$$\frac{\partial R}{\partial t} = -\widehat{\mathbf{R}}\cdot\mathbf{v}, \qquad \frac{\partial \widehat{\mathbf{R}}}{\partial t} = \frac{(\widehat{\mathbf{R}}\cdot\mathbf{v})\widehat{\mathbf{R}} - \mathbf{v}}{R}, \tag{7.76}$$

we obtain the intermediate result

$$\mathbf{E} = \frac{e}{4\pi}\left(\frac{\mathbf{R}}{R^3} - \frac{1}{2c^2}\frac{\partial}{\partial t}\left(\frac{\mathbf{v} + (\widehat{\mathbf{R}}\cdot\mathbf{v})\widehat{\mathbf{R}}}{R}\right) + \frac{2}{3c^3}\frac{d\mathbf{a}}{dt}\right). \tag{7.77}$$

The remaining derivative can be evaluated using again the formulas in (7.76). Finally, to compute the magnetic field it is sufficient to evaluate the curl of (7.73). In this way we obtain the expansions

$$\mathbf{E} = \frac{e}{4\pi}\left(\frac{\mathbf{R}}{R^3} - \frac{1}{2c^2R}\left(\mathbf{a} + (\widehat{\mathbf{R}}\cdot\mathbf{a})\widehat{\mathbf{R}} + \frac{(3(\widehat{\mathbf{R}}\cdot\mathbf{v})^2 - v^2)\widehat{\mathbf{R}}}{R}\right)\right.$$
$$\left. + \frac{2}{3c^3}\frac{d\mathbf{a}}{dt}\right) + O\left(\frac{1}{c^4}\right), \tag{7.78}$$
$$\mathbf{B} = \frac{e}{4\pi c}\,\mathbf{v}\times\frac{\mathbf{R}}{R^3} + O\left(\frac{1}{c^3}\right).$$

In the expression of $\mathbf{E}$, at lowest order we recognize the Coulomb field. The terms of order $1/c^2$ represent relativistic corrections of the *kinetic* type to this field, see Sect. 15.4.5. Conversely, the term of order $1/c^3$ traces back to *radiation*, as we will see in detail in Sect. 15.4. In the expression of $\mathbf{B}$ the term of order $1/c^2$ is absent, because in (7.73) the term of order $1/c^2$ is proportional to the acceleration $\mathbf{a}$, which is $\mathbf{x}$-independent. From the formulas above we also read off the general relation

$$\mathbf{B} = \frac{\mathbf{v}}{c}\times\mathbf{E} + O\left(\frac{1}{c^3}\right). \tag{7.79}$$

We stress that the expressions (7.78) represent the non-relativistic expansions of the Liénard-Wiechert fields (7.46)–(7.48). We end the chapter with the caveat that the expansion in powers of $1/c$ and the expansion for large $r = |\mathbf{x}|$ are operations which do not commute with each other. As an example compare the limit for large r of the $1/c^2$-terms of the electric field in (7.78) – in which case one has the identifications

$R \approx r$ and $\widehat{\mathbf{R}} \approx \mathbf{n}$ – with the electric field (7.65) obtained by performing first the expansion of the Liénard-Wiechert field for large r, and successively its expansion in powers of $1/c$.

7.6   Problems

7.1 Show that the asymptotic acceleration field (7.59) vanishes in all directions $\mathbf{n}$, if and only if $\mathbf{a} = 0$.

7.2 Consider the world line (7.1) of a hyperbolic motion

$$y^\mu(s) = (k\sinh(s/k),\, 0,\, 0,\, k\cosh(s/k)), \quad k > 0.$$

Show that the retarded-time conditions (7.17) do not admit solutions for s, if x^μ belongs to the set $\Sigma = \{x^\mu \in \mathbb{R}^4 / x^0 + x^3 < 0\}$.
Hint: The equation $(x - y(s))^2 = 0$ can be recast in the equivalent form

$$\left(x^0 + x^3 - ke^{s/k}\right)\left(x^0 + x^3 - ke^{s/k} - 2(x^0 - k\sinh(s/k))\right) + \left(x^1\right)^2 + \left(x^2\right)^2 = 0,$$

and the condition $x^0 > y^0(s)$ amounts to $x^0 > k\sinh(s/k)$.

References

1. A.-M. Liénard, Champ électrique et magnétique produit par une charge électrique concentrée en un point et animée d'un mouvement quelconque. L'Éclair. Électr. **16**, 5, 53, 106 (1898)
2. E.J. Wiechert, Elektrodynamische Elementargesetze. Ann. der Phys. **309**, 667 (1901)

Chapter 8
Electromagnetic Radiation

One of the main conclusions of Chap. 7 was that a system of charged particles in general emits electromagnetic radiation. In the first place, we determined the electromagnetic field generated by a particle in arbitrary motion – the Liénard-Wiechert field – realizing that, if the particle is accelerated, this field includes a *radiation field*, which at large distances decreases as $1/r$ and transports energy and momentum. In the second place, we found that the determination of the radiated four-momentum does not require the knowledge of the *exact* Liénard-Wiechert field, but only its shape at large distances from the particle. Via the superposition principle, we then extended these *qualitative* properties to generic charge distributions. In this chapter we perform a systematic *quantitative* analysis of the radiation emitted by a generic system of charges in arbitrary motion, represented by a generic four-current j^μ. One of our main purposes is the determination of the four-momentum radiated by the system per unit time per unit solid angle, see (7.55),

$$\frac{d^2 P^\mu}{dt d\Omega} = r^2 \left(T_{\text{em}}^{i\mu} n^i \right).$$ (8.1)

In the (understood) limit for $r \to \infty$, in the second member of this equation contribute only the fields – and hence the potentials – which at large distances decrease as $1/r$. Accordingly, in Sect. 8.1 first we perform a detailed analysis of the potential (6.65) at large distances from the charges – in the so-called *wave zone* – then we derive the associated electromagnetic field and, finally, we proceed to the evaluation of the emitted four-momentum (8.1).

8.1 The Electromagnetic Field in the Wave Zone

We begin with a few specifications about the nature of the four-currents we will consider. Of course, in the first place, the currents must be conserved, $\partial_\mu j^\mu = 0$. In the second place, we assume that our currents are of compact spatial support

© Springer International Publishing AG, part of Springer Nature 2018
K. Lechner, *Classical Electrodynamics*, UNITEXT for Physics,
https://doi.org/10.1007/978-3-319-91809-9_8

$$j^\mu(t, \mathbf{x}) = 0, \quad \forall r > l, \quad \forall t, \tag{8.2}$$

where $r = |\mathbf{x}|$, meaning that the charges forming j^μ always move inside a sphere S_l of radius l. The limitation to currents of this kind has its physical motivation in the fact that the charge distributions appearing in nature are commonly confined to finite regions of space. Nevertheless, most of our results extend also to processes involving *unconfined* charges, as e.g. the elastic scattering of particles incoming from and outgoing to infinity, see Sect. 8.4.3, or the *beta decay*, see Sect. 11.4.3.

Spectral decomposition of currents. The currents appearing in nature are essentially of two types, according to their dependence on time: *aperiodic* currents and *periodic* ones. An aperiodic current admits a Fourier transform in the time variable, corresponding to the *spectral decomposition*

$$j^\mu(t, \mathbf{x}) = \frac{1}{\sqrt{2\pi}} \int_{-\infty}^{\infty} e^{i\omega t} j^\mu(\omega, \mathbf{x}) \, d\omega, \tag{8.3}$$

in which the Fourier transform $j^\mu(\omega, \mathbf{x})$ represents the *continuous* weight with which the *frequency* ω contributes to the current. Since $j^\mu(x)$ is real, the weights with frequency ω and $-\omega$ are linked by the relation

$$j^\mu(-\omega, \mathbf{x}) = j^{\mu*}(\omega, \mathbf{x}),$$

so that in the following we shall consider only *positive* frequencies. Examples of processes corresponding to aperiodic currents are the elastic collisions between two charged particles, and the deflection of charged particles crossing a finite region of space where an electromagnetic field is present. For a periodic current of period T, obeying the relation $j^\mu(t + T, \mathbf{x}) = j^\mu(t, \mathbf{x})$ for all t and for all $\mathbf{x}$, the spectral decomposition (8.3) is replaced by the Fourier series[1]

$$j^\mu(t, \mathbf{x}) = \sum_{N=-\infty}^{\infty} e^{iN\omega_0 t} j_N^\mu(\mathbf{x}), \qquad j_N^{\mu*}(\mathbf{x}) = j_{-N}^\mu(\mathbf{x}), \tag{8.4}$$

where $\omega_0 = 2\pi/T$ is called the *fundamental frequency*. In this case the Fourier coefficient $j_N^\mu(\mathbf{x})$ represents the *discrete* weight with which the frequency

$$\omega_N = N\omega_0$$

---

[1] In the space of distributions the decomposition (8.4) can be regarded as a particular case of the representation (8.3), corresponding to the choice

$$j^\mu(\omega, \mathbf{x}) = \sqrt{2\pi} \sum_{N=-\infty}^{\infty} \delta(\omega - \omega_N) \, j_N^\mu(\mathbf{x}).$$

contributes to the current. Examples of periodic currents are the macroscopic current corresponding to the oscillating electron flow in an antenna, and the current associated with a particle circling in a synchrotron. Occasionally, we will also make reference to *monochromatic* currents – i.e. currents with a fixed frequency ω – of the form

$$j^\mu(t, \mathbf{x}) = e^{i\omega t} j^\mu(\omega, \mathbf{x}) + c.c. \tag{8.5}$$

Any current can, indeed, be thought of as a – discrete or continuous – superposition of monochromatic ones. The terminology *frequency* for the variable ω derives from the fact that a monochromatic current generates an electromagnetic field which in the wave zone assumes the form of a monochromatic wave with the *same* frequency as the current, see Sect. 8.1.2.

Vector potential in the wave zone. From Sect. 6.2.2 we know that a current $j^\mu(x)$ generates the vector potential (6.65)

$$A^\mu(x) = \frac{1}{4\pi} \int \frac{1}{|\mathbf{x} - \mathbf{y}|} j^\mu(t - |\mathbf{x} - \mathbf{y}|, \mathbf{y}) \, d^3y. \tag{8.6}$$

Being interested in this potential at large distances from the charges, we expand it in powers of $1/r$. Since the current is zero outside S_l, see (8.2), the integral in (8.6) can be restricted to the region $y < l$. Accordingly, in the integrand we can resort to the expansions, see (7.61),

$$|\mathbf{x} - \mathbf{y}| = r - \mathbf{n} \cdot \mathbf{y} + O\left(\frac{y^2}{r}\right), \qquad \mathbf{n} = \frac{\mathbf{x}}{r}, \tag{8.7}$$

$$\frac{1}{|\mathbf{x} - \mathbf{y}|} = \frac{1}{r} + O\left(\frac{y}{r^2}\right). \tag{8.8}$$

Inserting them in (8.6) we obtain the *potential in the wave zone*[2]

---

[2]In the literature one sometimes defines the *wave zone* as the region

$$r \gg \lambda, \qquad r \gg l, \qquad r \gg \omega l^2, \tag{8.9}$$

where $\lambda = 2\pi/\omega$ is the wavelength, and ω is a generic frequency present in the currents (8.3) or (8.4). The first condition ensures that the concept of wavelength is meaningful. The second and third one make sure that the expression (8.10) maintains its validity also for *finite* values of r. In fact, since $y < l$, the second condition ensures the validity of the expansions (8.7) and (8.8). The third condition ensures, instead, that the large-distance expansion of the current in (8.6) can be truncated at the lowest order, leading to (8.10). In fact, to derive (8.10), in the time argument of the current in (8.6) we have neglected a term of the order y^2/r, which in the expansion of the current would give rise to a contribution of the type $(y^2/r) \, \partial_0 j^\mu$. Considering the monochromatic current (8.5), schematically we have $\partial_0 j^\mu \sim \omega j^\mu$, so that the contribution $(y^2/r) \, \partial_0 j^\mu \sim (\omega y^2/r) j^\mu$ becomes negligible with respect to j^μ, if $\omega y^2/r < \omega l^2/r \ll 1$, i.e. if the third condition in (8.9) holds.

$$A^\mu(x) = \frac{1}{4\pi r} \int j^\mu(t - r + \mathbf{n} \cdot \mathbf{y}, \mathbf{y}) \, d^3y, \tag{8.10}$$

which – by definition – is the potential truncated at the first power in $1/r$. In the time argument of the current we recognize the macroscopic retarded time $t - r$, together with the microscopic delay $\mathbf{n} \cdot \mathbf{y}$.

Wave relations. To derive the main properties of the electromagnetic field in the wave zone we resort to the *wave relations* (5.80)

$$\partial_\mu A^\nu = n_\mu \dot{A}^\nu, \quad n_\mu \dot{A}^\mu = 0, \quad n^\mu n_\mu = 0, \tag{8.11}$$

which hold also for the potential in the wave zone (8.10) – modulo terms of order $1/r^2$ – as we now show. For this purpose we must, in the first place, specify the light-like four-vector n^μ

$$n^0 = 1, \qquad \mathbf{n} = \frac{\mathbf{x}}{r}, \qquad n^\mu n_\mu = 0.$$

The unit vector $\mathbf{n}$ identifies the propagation direction of the "wave", which equals in each point the radial outgoing direction. To prove the first relation in (8.11) we need to evaluate the derivative with respect to x^i of the integrand in (8.10), i.e. of

$$j^\mu(t - r + \mathbf{n} \cdot \mathbf{y}, \mathbf{y}). \tag{8.12}$$

Omitting the arguments of this current, we obtain

$$\partial_i j^\mu = \partial_i (t - r + \mathbf{n} \cdot \mathbf{y}) \, \partial_0 j^\mu = -\frac{x^i}{r} \, \partial_0 j^\mu + O\left(\frac{1}{r}\right) = n_i \partial_0 j^\mu + O\left(\frac{1}{r}\right).$$

Modulo terms of order $1/r$, the derivatives of the current (8.12) can therefore be recast in the form

$$\partial_\nu j^\mu = n_\nu \partial_0 j^\mu.$$

For the derivatives of (8.10) we obtain thus, modulo terms of order $1/r^2$,

$$\partial_\mu A^\nu = \frac{1}{4\pi r} \int \partial_\mu j^\nu d^3y \ = \frac{n_\mu}{4\pi r} \int \partial_0 j^\nu d^3y \ = n_\mu \partial_0 \frac{1}{4\pi r} \int j^\nu d^3y = n_\mu \partial_0 A^\nu,$$

which is the first relation in (8.11). The second relation follows from the first in that A^μ, by construction, satisfies the Lorenz gauge $\partial_\mu A^\mu = 0$. This relation allows, in turn, to determine $\dot{A}^0$ in terms of $\dot{\mathbf{A}}$

$$\dot{A}^0 = \mathbf{n} \cdot \dot{\mathbf{A}}. \tag{8.13}$$

Once we have established the validity of the wave relations (8.11), we conclude that the *electromagnetic field in the wave zone*, i.e. the field truncated at terms of order $1/r$, shares with the elementary waves (5.77) the basic properties (5.101)–(5.103):

$$\mathbf{E} = \mathbf{n} \times (\mathbf{n} \times \dot{\mathbf{A}}) = -\dot{\mathbf{A}} + (\mathbf{n} \cdot \dot{\mathbf{A}})\mathbf{n}, \tag{8.14}$$

$$\mathbf{B} = \mathbf{n} \times \mathbf{E}, \quad \mathbf{n} \cdot \mathbf{E} = 0, \quad E = B, \tag{8.15}$$

$$T_{\text{em}}^{\mu\nu} = n^\mu n^\nu E^2. \tag{8.16}$$

In this way we confirm, in particular, that the asymptotic properties (7.52) of the field of a system of charged particles hold true for the field generated by a generic conserved four-current j^μ. Moreover, since the potential (8.10) entails the asymptotic behavior $\dot{\mathbf{A}} \sim 1/r$, the relations (8.14) and (8.15) imply for the electromagnetic field the expected asymptotic behavior $F^{\mu\nu} \sim 1/r$.

8.1.1   Four-Momentum Emission

Since in the four-momentum emission formula (8.1) the right hand side must be evaluated in the limit for $r \to \infty$, we can use the wave-zone expressions (8.10), (8.14) and (8.15). Inserting the energy-momentum tensor (8.16) in formula (8.1), we then obtain for the emitted four-momentum the general formula

$$\frac{d^2 P^\mu}{dt d\Omega} = r^2 n^\mu E^2. \tag{8.17}$$

Denoting the radiated power by $\mathcal{W} = d\varepsilon/dt$, from this formula we derive for the energy and momentum radiated per unit time per unit solid angle in the direction $\mathbf{n}$ the expressions

$$\frac{d^2\varepsilon}{dt d\Omega} = \frac{d\mathcal{W}}{d\Omega} = r^2 E^2 = r^2 |\mathbf{n} \times \dot{\mathbf{A}}|^2 = r^2 \dot{A}^i \dot{A}^j (\delta^{ij} - n^i n^j), \tag{8.18}$$

$$\frac{d^2\mathbf{P}}{dt d\Omega} = \frac{d\mathcal{W}}{d\Omega} \mathbf{n}. \tag{8.19}$$

According to Eq. (8.19), giving the momentum flux of the electromagnetic field in terms of its energy flux, the momentum $\Delta\mathbf{P}$ and the energy $\Delta\varepsilon$ of a small amount of radiation are linked by the relations

$$\Delta\mathbf{P} = \mathbf{n}\Delta\varepsilon, \quad (\Delta\varepsilon)^2 - |\Delta\mathbf{P}|^2 = 0, \tag{8.20}$$

holding also for the elementary waves (5.77), see Eqs. (5.41). Considering that at the quantum level radiation corresponds to a flux of *photons*, the relations (8.20) state that these particles are *massless*, and that far away from the source they propagate in the *radial* direction. The most significant result of this section is Eq. (8.18), a refined version of the basic radiation formula (7.58). In fact, in fact, the power radiated by a generic charge distribution in terms of only the *spatial* components of the potential in the wave zone (8.10).

8.1.2 Monochromatic Currents and Elementary Waves

As seen above, the electromagnetic field $F^{\mu\nu}$ in the wave zone shares several properties with the elementary electromagnetic waves (5.77). This circumstance is not accidental, since outside the sphere S_l, i.e. for $|\mathbf{x}| > l$, this field actually satisfies Maxwell's equations for a free field. Therefore, although $F^{\mu\nu}$ is not a free field in the *whole* space, the farer we go from the charged sources, the closer its structure should be to that of a *true* free field. Correspondingly, we expect that in the wave zone the field becomes essentially a superposition of elementary waves, and we will actually see that it shares several properties with these waves. However, it is also clear that in general the asymptotic field is not formed by a *single* elementary wave.

To decompose the electromagnetic field in the wave zone in elementary waves, we resort to the spectral decompositions of the current (8.3) and (8.4), and observe that the potential in the wave zone (8.10) is linear in the current. Accordingly, it is sufficient to analyze the potential generated by a *monochromatic* current of frequency ω. Inserting the corresponding expression (8.5) in (8.10) we obtain the potential

$$
\begin{aligned}
A^\mu(x) &= \frac{1}{4\pi r} \int e^{i\omega(t-r+\mathbf{n}\cdot\mathbf{y})} j^\mu(\omega, \mathbf{y})\, d^3y + c.c. \\
&= e^{i\omega(t-r)} \frac{1}{4\pi r} \int e^{i\omega\mathbf{n}\cdot\mathbf{y}} j^\mu(\omega, \mathbf{y})\, d^3y + c.c. \\
&= e^{ik\cdot x} \varepsilon^\mu + c.c.
\end{aligned}
\tag{8.21}
$$

We have introduced the wave vector k^μ with components

$$
k^0 = \omega, \qquad \mathbf{k} = \omega\mathbf{n},
$$

obeying the relations

$$
k^2 = 0, \qquad k \cdot x = k^\mu x_\mu = \omega(t - r),
$$

and the polarization vector

$$
\varepsilon^\mu = \frac{1}{4\pi r} \int e^{i\omega\mathbf{n}\cdot\mathbf{y}} j^\mu(\omega, \mathbf{y})\, d^3y.
\tag{8.22}
$$

We see, therefore, that a monochromatic current generates a field which in the wave zone reduces *formally* to an elementary wave propagating in the radial direction $\mathbf{n}$, having the wave and polarization vectors specified above. In particular, a current of frequency ω generates a wave with the *same* frequency ω.

Plane waves and spherical waves. The expression (8.21) does, however, not represent a true *plane* wave. In fact, both the wave vector and the polarization vector entail a residual dependence on the position $\mathbf{x} = r\mathbf{n}$, i.e. on the propagation direction $\mathbf{n}$ as well as on the distance r from the charged sources. In particular, the polarization vector (8.22) decreases as $1/r$, an asymptotic behavior which is dictated by energy conservation. To see this we recall that the Poynting vector associated with the wave (8.21), averaged over time, is given by (see Problem 5.6)

$$\langle \mathbf{S} \rangle = -2\omega^2 \varepsilon^{*\mu} \varepsilon_\mu \mathbf{n} = 2\omega^2 |\mathbf{n} \times \boldsymbol{\varepsilon}|^2 \mathbf{n}, \tag{8.23}$$

where, to obtain the second expression, we have used the constraint $k_\mu \varepsilon^\mu = 0$, yielding $\varepsilon^0 = \mathbf{n} \cdot \boldsymbol{\varepsilon}$. Since the quantity $\langle \mathbf{S} \rangle$ is proportional to $1/r^2$, the energy crossing per unit time the spherical cap of an infinitesimal cone subtending a solid angle $d\Omega$, namely the quantity

$$\langle \mathbf{S} \rangle \cdot \mathbf{n}\, r^2 d\Omega = 2\omega^2 r^2 |\mathbf{n} \times \boldsymbol{\varepsilon}|^2 d\Omega,$$

is independent of r. This means that the energy flows to infinity remaining *conserved*.

Besides this dependence on r, the polarization vector (8.22) entails also a dependence on the emission direction $\mathbf{n}$, through the exponential $e^{i\omega \mathbf{n} \cdot \mathbf{y}}$. According to these residual functional dependencies on r and $\mathbf{n}$, the potential (8.21) corresponds, properly speaking, to a superposition of *spherical* waves,[3] rather than to a plane wave. Nevertheless, in a region of spatial extension L small compared to r,

$$L \ll r,$$

the vectors $\mathbf{k}$ and ε^μ are practically constant, and the potential (8.21) behaves with high accuracy as a plane wave. To see this more in detail we observe that inside such a region the relative variations of r and $\mathbf{n}$ are bounded by

$$\frac{\Delta r}{r} < \frac{L}{r}, \qquad |\Delta \mathbf{n}| < \frac{L}{r}. \tag{8.24}$$

It follows that the relative variation of the wave vector $\mathbf{k} = \omega \mathbf{n}$ is bounded by

$$\frac{|\Delta \mathbf{k}|}{\omega} = |\Delta \mathbf{n}| < \frac{L}{r}. \tag{8.25}$$

---

[3]For more details on spherical waves see e.g. Ref. [1].

Similarly, for the variation of the polarization vector (8.22) we obtain

$$\Delta\varepsilon^\mu = \frac{1}{4\pi r}\int\left(-\frac{\Delta r}{r}+i\omega\Delta\mathbf{n}\cdot\mathbf{y}\right)e^{i\omega\mathbf{n}\cdot\mathbf{y}}j^\mu(\omega,\mathbf{y})\,d^3y. \qquad (8.26)$$

Comparing this expression, in turn, with (8.22), and taking into account that in the integral (8.26) the variable y is bounded from above by l, for the relative variation of ε^μ we obtain the bound

$$\left|-\frac{\Delta r}{r}+i\omega\Delta\mathbf{n}\cdot\mathbf{y}\right| < \frac{L}{r}+\omega l\,|\Delta\mathbf{n}| < (1+\omega l)\frac{L}{r}, \qquad (8.27)$$

where we used again the bounds (8.24). Since ωl is of the order of magnitude of the typical speeds $v < 1$ of the charged particles composing the current (8.5), see Sect. 8.3.1, in a region of spatial extension L the relative variation of the polarization vector is hence at most

$$\left|-\frac{\Delta r}{r}+i\omega\Delta\mathbf{n}\cdot\mathbf{y}\right| \sim \frac{L}{r}. \qquad (8.28)$$

As an example we consider the radiation emitted by the Sun and observed on the surface of the Earth. In this case r equals the distance between the Earth and the Sun, $r \approx 1.5\cdot 10^8$ km, and L equals at most the diameter of the Earth, $L \approx 1.2\cdot 10^4$ km. According to (8.25) and (8.28), on the surface of the Earth the wave vector and the polarization vector are, hence, subject to very small relative variations, at most of the order $L/r \sim 10^{-4}$. When the radiation emitted by the Sun reaches the Earth, in practice it appears thus as a superposition of plane waves.

Generic currents. Since the wave-zone potential (8.10) is *linear* in j^μ, the above conclusions – derived for monochromatic currents – admit straightforward generalizations to the generic currents (8.3) and (8.4), superpositions of the former. In the general case the field in the wave zone is, thus, a superposition of monochromatic (locally) plane waves, and the frequencies present in the radiation are a *subset of the frequencies present in the current*. In fact, it may happen that for certain ω's the integral (8.22) for some directions $\mathbf{n}$ vanishes. In particular, a system of charged particles in periodic motion with the same period $T = 2\pi/\omega_0$ is associated with a periodic current of the type (8.4). Such a system emits, therefore, radiation with frequencies belonging to the *discrete* set

$$\omega_N = N\omega_0, \quad N = 1,2,3,\ldots.$$

Vice versa, a system of charged particles in arbitrary motion is associated with an aperiodic current of the type (8.3), and such a system emits, hence, radiation with a *continuous* spectrum of frequencies.

8.2 Radiation of a Linear Antenna

According to the radiation formulas (8.18) and (8.19), the analysis of the four-momentum emitted by a generic charge distribution via radiation has been reduced to the determination of the spatial components $\mathbf{A}$ of the wave-zone potential (8.10). Unfortunately, the integral appearing in (8.10) can be rarely evaluated analytically, and in general one must resort to a perturbative method, as e.g. the *multipole expansion*, see Sect. 8.3. One of the rare physical systems for which the integral (8.10) can be calculated exactly is represented by the *linear antenna*. For concreteness, let us consider an antenna of length L disposed along the z axis, fed at its center by an alternating-current generator of frequency ω. Without entering into details, we give the (idealized) form of the associated spatial current

$$\mathbf{j}(t, \mathbf{y}) = I\, \delta(y^1)\, \delta(y^2) \sin\left(\omega\left(\frac{L}{2} - |y^3|\right)\right) \cos(\omega t)\, \mathbf{u}, \qquad (8.29)$$

where for future reference we set

$$I = \frac{I_0}{\sin\left(\frac{\omega L}{2}\right)}. \qquad (8.30)$$

It is understood that $\mathbf{j}(t, \mathbf{y})$ vanishes for $|y^3| > L/2$. The current (8.29) vanishes at its boundary, at $y^3 = \pm L/2$, and for any time t it has a maximum at the *gap*, i.e. at $y^3 = 0$, where the antenna is fed. I_0 has the dimension of a *current*, in the sense of charge divided by time, and it equals the peak amplitude of the current at the gap. The vector $\mathbf{u} = (0, 0, 1)$ is the versor of the z axis. Comparing the currents (8.5) and (8.29) we see that the latter is a *monochromatic* current of *fundamental* frequency $\omega_0 = \omega$, and consequently the antenna emits only radiation with frequency ω and wavelength $\lambda = 2\pi/\omega$.

To compute the spatial potential in the wave zone, we must insert the current (8.29) in (8.10)

$$\mathbf{A} = \frac{I\mathbf{u}}{4\pi r} \int_{-L/2}^{L/2} dy^3 \int dy^1 \int dy^2\, \delta(y^1)\, \delta(y^2)$$
$$\sin\left(\omega\left(\frac{L}{2} - |y^3|\right)\right) \cos\left(\omega(t - r + \mathbf{n} \cdot \mathbf{y})\right). \qquad (8.31)$$

Once we have integrated out the δ-functions in the variables y^1 and y^2, amounting to set $y^1 = y^2 = 0$, we can replace $\mathbf{n} \cdot \mathbf{y}$ with $n^1 y^1 + n^2 y^2 + n^3 y^3 = n^3 y^3 = y^3 \cos\vartheta$, where ϑ is the angle between $\mathbf{n}$ and the z axis. In this way (8.31) becomes

$$\mathbf{A} = \frac{I\mathbf{u}}{4\pi r} \int_{-L/2}^{L/2} dy^3 \sin\left(\omega\left(\frac{L}{2} - |y^3|\right)\right) \cos\left(\omega\left(t - r + y^3 \cos\vartheta\right)\right).$$

The remaining integral over y^3 is elementary and leads to

$$\mathbf{A} = \frac{I \cos\left(\omega(t - r)\right)}{2\pi r \omega \sin^2 \vartheta} \left(\cos\left(\frac{\omega L}{2} \cos\vartheta \right) - \cos\frac{\omega L}{2} \right) \mathbf{u}. \qquad (8.32)$$

The spatial potential is hence always parallel to the z axis, as is its time derivative $\dot{\mathbf{A}}$. From Eq. (8.14) it follows then that the electric field $\mathbf{E}$ in the wave zone lies in the plane spanned by the z axis and by the propagation direction $\mathbf{n}$, being, of course, orthogonal to $\mathbf{n}$. Since this direction is time-independent, the radiation emitted by the antenna is *linearly* polarized. To determine the *angular distribution* of the radiated power we must calculate the time derivative of the potential (8.32) and insert it in the radiation formula (8.18). Since $\dot{\mathbf{A}}$ is parallel to the z axis, we obtain

$$\frac{d\mathcal{W}}{d\Omega} = r^2 \left| \mathbf{n} \times \dot{\mathbf{A}} \right|^2 = r^2 \left| \dot{\mathbf{A}} \right|^2 \sin^2 \vartheta. \qquad (8.33)$$

The time derivative of (8.32) amounts to the replacement

$$\cos(\omega(t - r)) \;\rightarrow\; -\omega \sin(\omega(t - r)).$$

Since the emitted power is usually observed over time scales large with respect to the period $T = 2\pi/\omega$, the physically meaningful quantity is the average over a period of the emitted power (8.33), which amounts to the further replacement

$$\sin^2(\omega(t - r)) \;\rightarrow\; \overline{\sin^2(\omega(t - r))} = \frac{1}{2}.$$

In this way, Eqs. (8.32) and (8.33) yield for the angular distribution of the mean power emitted by the antenna[4]

$$\frac{d\overline{\mathcal{W}}}{d\Omega} = \frac{I_0^2}{8\pi^2} \left(\frac{\cos\left(\frac{\omega L}{2} \cos\vartheta \right) - \cos\frac{\omega L}{2}}{\sin\left(\frac{\omega L}{2} \right) \sin\vartheta} \right)^2. \qquad (8.34)$$

Since this angular distribution is invariant under the replacement $\vartheta \to \pi - \vartheta$, reflecting the fact that the antenna has no preferred orientation, it is sufficient to analyze its shape in the interval $\vartheta \in [0, \pi/2]$. The existence of directions ϑ of maximum or minimum energy emission depends heavily on the value of the ratio

$$\frac{\omega L}{2} = \frac{\pi L}{\lambda}.$$

In place of a systematic analysis of the angular distribution (8.34), in the following we confine our attention to some physically interesting cases. In general we see, however, that $d\overline{\mathcal{W}}/d\Omega$ vanishes for $\vartheta = 0$, i.e. in the direction parallel to the

---

[4]For $L = n\lambda = 2\pi n/\omega$, with n integer, the normalization of I in (8.30) must be abandoned.

antenna, while it has a maximum for $\vartheta = \pi/2$, i.e. in the plane orthogonal to the antenna, if only $L/\lambda \neq 2n$, with n integer. Moreover, it can be seen that if $L \leq \lambda$ the angular distribution (8.34) has no other extrema, while if $L > \lambda$ there exist further directions in which the emitted power has a maximum, or vanishes.

From formulas (8.34) and (8.19) we can compute the total mean momentum emitted by the antenna per unit time in all directions. Since the angular distribution (8.34) is invariant under $\vartheta \to \pi - \vartheta$, and independent of the azimuthal angle, the mean power emitted in the directions n and $-$n is the same. We obtain, therefore,

$$\frac{d\mathbf{P}}{dt} = \int \frac{d\overline{W}}{d\Omega}\, \mathbf{n}\, d\Omega = 0. \tag{8.35}$$

The antenna does, hence, emit no overall momentum. Qualitatively, linear antennas fall into two classes: *long* antennas, with lengths of the order $L \sim \lambda$, and *short* antennas, with length $L \ll \lambda$. We shall analyze short antennas in detail in Sect. 8.4.1, in the context of the *dipole approximation*, while below we consider a typical example of a long antenna.

Half-wave antenna, radiation resistance, and radiation efficiency. Particularly interesting cases of long antennas are *half-wave* antennas, of length $L = \lambda/2$, and *full-wave* antennas, of length $L = \lambda$. For a half-wave antenna we have $\omega L/2 = \pi L/\lambda = \pi/2$. In this case the angular distribution (8.34) reduces to

$$\frac{d\overline{W}}{d\Omega} = \frac{I_0^2}{8\pi^2} \frac{\cos^2\left(\frac{\pi}{2}\cos\vartheta\right)}{\sin^2\vartheta}, \tag{8.36}$$

which has a unique maximum at $\vartheta = \pi/2$ and a unique minimum at $\vartheta = 0$, where it vanishes. To analyze the *radiation efficiency* of the antenna we compute the total mean power $\overline{W}$ emitted in all directions, integrating (8.36) over the solid angle $d\Omega = \sin\vartheta\, d\vartheta\, d\varphi$

$$\overline{W} = \int \frac{d\overline{W}}{d\Omega}\, d\Omega = \frac{I_0^2}{4\pi} \int_0^\pi \frac{\cos^2\left(\frac{\pi}{2}\cos\vartheta\right)}{\sin\vartheta}\, d\vartheta.$$

The remaining integral can be evaluated only numerically and is equal to 1.22. We obtain thus

$$\overline{W} = 0.097\, I_0^2 = \frac{1}{2}\, I_0^2 R_{\text{rad}}^{(1/2)}. \tag{8.37}$$

We have introduced the *radiation resistance* of the half-wave antenna

$$R_{\text{rad}}^{(1/2)} = 0.194, \tag{8.38}$$

not to be confused with its *ohmic resistance* R_{ohm}. To switch to MKS units we must multiply (8.38) with the *wave resistance of free space*

$$R_0 = \sqrt{\frac{\mu_0}{\varepsilon_0}} = \frac{1}{c\varepsilon_0} \approx 377\,\text{ohm}, \qquad (8.39)$$

leading to the radiation resistance

$$R_{\text{rad}}^{(1/2)} = 0.194\,R_0 \approx 73\,\text{ohm}. \qquad (8.40)$$

For a *full-wave* antenna, proceeding similarly, one finds

$$R_{\text{rad}}^{(1)} \approx 201\,\text{ohm}.$$

It can be seen that these values are typically much larger than the ohmic resistance of the antenna,

$$R_{\text{rad}} \gg R_{\text{ohm}},$$

meaning that most part of the energy supplied by the generator is radiated in the form of electromagnetic waves, while only a small fraction is dissipated via ohmic heating. Therefore, a long antenna has in general a high radiation efficiency. Conversely, as we will see in Sect. 8.4.1, a short antenna entails a low radiation efficiency, i.e. in that case $R_{\text{rad}} \lesssim R_{\text{ohm}}$.

8.3   Radiation in the Non-relativistic Limit

The analysis of the four-momentum of the radiation produced by a generic system of charges has been reduced to the calculation of the potential in the wave zone (8.10), which we rewrite reintroducing the velocity of light

$$A^\mu(x) = \frac{1}{4\pi rc} \int j^\mu\left(t - \frac{r}{c} + \frac{\mathbf{n}\cdot\mathbf{y}}{c}, \mathbf{y}\right) d^3y. \qquad (8.41)$$

As observed above, in general this three-dimensional integral cannot be evaluated analytically, and so it is necessary to resort to an approximation method. If the charged particles present in the current j^μ have velocities small with respect to the velocity of light, there is a reliable systematic perturbative method known as *multipole expansion*.

8.3.1   Multipole Expansion

By definition, the multipole expansion of the integral (8.41) amounts to a Taylor series expansion of the integrand $j^\mu\left(t - r/c + \mathbf{n}\cdot\mathbf{y}/c, \mathbf{y}\right)$ around the macroscopic retarded time $T = t - r/c$, in powers of the microscopic delay $\mathbf{n}\cdot\mathbf{y}/c$

$$A^\mu(x) = \frac{1}{4\pi rc} \int \left(j^\mu(T,\mathbf{y}) + \frac{\mathbf{n}\cdot\mathbf{y}}{c}\,\partial_t j^\mu(T,\mathbf{y}) + \frac{(\mathbf{n}\cdot\mathbf{y})^2}{2c^2}\,\partial_t^2 j^\mu(T,\mathbf{y}) + \cdots \right) d^3y.$$
(8.42)

The first term is called *dipole term*, the second *quadrupole term*, the third *sestupole term*, and so on. Being a power series in $1/c$, the expansion (8.42) constitutes a *non-relativistic* approximation scheme. To show, more in detail, that it yields sensible results if the speeds of the charged particles involved in j^μ are small with respect to the speed of light, let us suppose that the particles move with a characteristic velocity v. They then spend a characteristic time l/v to cross the sphere S_l of radius l, inside which they are confined. This means that j^μ varies appreciably over time scales of the order $\tau = l/v$. Therefore, the Taylor expansion of the current $j^\mu(T + \mathbf{n}\cdot\mathbf{y}/c, \mathbf{y})$ in powers of $\mathbf{n}\cdot\mathbf{y}/c$ is sensible, if

$$\left|\frac{\mathbf{n}\cdot\mathbf{y}}{c}\right| \ll \tau.$$
(8.43)

On the other hand, since $y < l$, the microscopic delay is at most

$$\left|\frac{\mathbf{n}\cdot\mathbf{y}}{c}\right| \sim \frac{l}{c} = \frac{v}{c}\tau,$$
(8.44)

and so the condition (8.43) translates into

$$\frac{v}{c}\tau \ll \tau \quad \Leftrightarrow \quad v \ll c.$$

The expansion (8.42) gives, thus, physically meaningful results, if during the radiation process the speeds of the charges remain much smaller than the speed of light. An alternative way to understand the meaning of the multipole expansion (8.42) consists in analyzing the potential (8.41) *frequency by frequency*, i.e. by considering a monochromatic current of fixed frequency ω (see Eq. (8.5))

$$j^\mu(t,\mathbf{x}) = e^{i\omega t} j^\mu(\omega,\mathbf{x}) + c.c.$$
(8.45)

In this case the characteristic time of the current is $\tau = 1/\omega$, and correspondingly the charges have a characteristic speed $v = l/\tau = \omega l$. On the other hand, the multiple time derivatives of the current (8.45) schematically amount to

$$\partial_t^N j^\mu \sim \omega^N j^\mu.$$

According to (8.44), the Nth term of the expansion (8.42) can then be written, again schematically, as

$$\frac{1}{N!}\frac{(\mathbf{n}\cdot\mathbf{y})^N}{c^N}\,\partial_t^N j^\mu(T,\mathbf{y}) \simeq \frac{1}{N!}\frac{(\omega l)^N}{c^N}\,j^\mu(T,\mathbf{y}) = \frac{1}{N!}\left(\frac{v}{c}\right)^N j^\mu(T,\mathbf{y}).$$

The series (8.42) resembles, hence, an expansion in powers of v/c, suitable if $v \ll c$.

8.4   Electric Dipole Radiation

In this section we perform a systematic analysis of the radiation in the *dipole approximation*, which amounts to keep in the expansion of the potential (8.42) only the first term

$$\mathbf{A}(t, \mathbf{x}) = \frac{1}{4\pi rc} \int \mathbf{j}\left(t - \frac{r}{c}, \mathbf{y}\right) d^3y. \tag{8.46}$$

The resulting electromagnetic field is called *electric dipole field*, or simply *dipole field*, and the associated radiation is called *dipole radiation*. When the speeds of the involved charges are non-relativistic, this approximation in general yields accurate values for the radiated four-momentum. If, instead, one requires a higher degree of precision – or if the dipole field is zero – then it is necessary to keep in (8.42) also the next term, corresponding to the *quadrupole field*. In fact, as we will see in Sect. 8.5, the energy radiated by the quadrupole field is suppressed with respect to the energy radiated by the dipole field by a factor $(v/c)^2$.

Electric dipole moment. The integral (8.46) can be recast in a simpler form, if we introduce for a generic four-current $j^\mu = (c\rho, \mathbf{j})$ the *electric dipole moment*

$$\mathbf{D}(t) = \int \mathbf{y}\rho(t, \mathbf{y}) \, d^3y. \tag{8.47}$$

The reason is that its time derivative equals precisely the integral appearing in (8.46)

$$\dot{\mathbf{D}}(t) = \int \mathbf{j}\,(t, \mathbf{y}) \, d^3y. \tag{8.48}$$

In fact, thanks to the conservation equation $\dot{\rho} = -\partial_k j^k$, we obtain

$$\dot{D}^i(t) = \int y^i \dot{\rho} \, d^3y = -\int y^i \partial_k j^k \, d^3y = -\int \left(\partial_k(y^i j^k) - j^i\right) d^3y = \int j^i \, d^3y.$$

In the last step we have applied Gauss's theorem to a sphere placed at spatial infinity, using that for $y > l$ the current $\mathbf{j}\,(t, \mathbf{y})$ vanishes. The *dipole potential* (8.46) can then be rewritten as

$$\mathbf{A}(t, \mathbf{x}) = \frac{1}{4\pi rc} \, \dot{\mathbf{D}}\left(t - \frac{r}{c}\right). \tag{8.49}$$

A peculiar aspect of this formula – characteristic for the dipole approximation – is that it expresses the spatial potential in terms of only the charge density ρ, without explicitly involving the spatial current $\mathbf{j}$.

The scalar potential A^0. Although the analysis of the radiation emitted by a system does not require the scalar potential A^0, it is worthwhile to point out that in the non-relativistic expansion (8.42) the components $\mathbf{A}$ and A^0 must be treated on a different footing. In fact, the time component j^0 of the current is related to the charge density by $j^0 = c\rho$, whereas its spatial components are independent of c. Recall, in this regard, that for a single charged particle we have $\mathbf{j} = \rho\mathbf{v}$. It follows that, if we truncate the expansion (8.42) of $\mathbf{A}$ at the order $1/c^N$, for consistency in the expansion of A^0 we must keep also the order $1/c^{N+1}$. In particular, if we want to determine A^0 in dipole approximation, it is necessary to keep in the expansion (8.42) also the term linear in $\mathbf{n} \cdot \mathbf{y}/c$. Setting $\rho(t - r/c, \mathbf{y}) = \rho$ we obtain, thus,

$$A^0(t, \mathbf{x}) = \frac{1}{4\pi rc} \left(\int c\rho \, d^3y + \frac{1}{c}\mathbf{n} \cdot \partial_t \int \mathbf{y} \, c\rho \, d^3y \right) = \frac{1}{4\pi r} \left(Q + \frac{1}{c}\mathbf{n} \cdot \dot{\mathbf{D}} \right), \tag{8.50}$$

where $Q = \int \rho(t, \mathbf{y}) \, d^3y$ is the total conserved charge of the system, and $\mathbf{D}$ is evaluated at the time $t - r/c$. In the first, time-independent, term we recognize the classical Coulomb potential, while the second, time-dependent, term represents a relativistic correction. Notice that only with this correction the four-potential specified by formulas (8.49) and (8.50) satisfies the wave relation (8.13), as well as the Lorenz-gauge $\partial_\mu A^\mu = \frac{1}{c}\dot{A}^0 + \nabla \cdot \mathbf{A} = 0$, modulo terms of order $1/r^2$.

Four-momentum emission. Equations (8.14), (8.15) and (8.49) yield for the electric and magnetic fields in the wave zone the simple expressions

$$\mathbf{E} = -\frac{1}{4\pi rc^2} \left(\ddot{\mathbf{D}} - (\mathbf{n} \cdot \ddot{\mathbf{D}})\mathbf{n} \right), \qquad \mathbf{B} = -\frac{\mathbf{n} \times \ddot{\mathbf{D}}}{4\pi rc^2}, \tag{8.51}$$

where, henceforth, we understand the dependence of the dipole moment $\mathbf{D}$ on the retarded time $t - r/c$. From the first equation in (8.51) we see that, at time t, at a distance r from the source along a direction $\mathbf{n}$, the electric field lies in the plane spanned by the vectors $\ddot{\mathbf{D}}(t - r/c)$ and $\mathbf{n}$. Inserting the dipole potential (8.49) in the radiation formula (8.18), we obtain for the angular distribution of the power radiated by a generic *non-relativistic* charged source the equivalent expressions

$$\frac{dW}{d\Omega} = \frac{1}{16\pi^2 c^3} \ddot{D}^i \ddot{D}^j (\delta^{ij} - n^i n^j) = \frac{\sin^2\vartheta \, |\ddot{\mathbf{D}}|^2}{16\pi^2 c^3} = \frac{|\mathbf{n} \times \ddot{\mathbf{D}}|^2}{16\pi^2 c^3}, \tag{8.52}$$

where ϑ is the angle between the vectors $\ddot{\mathbf{D}}$ and $\mathbf{n}$. So we realize that the dipole radiation for fixed t and r has a very simple *universal* angular distribution: it vanishes in the direction of the vector $\ddot{\mathbf{D}}(t - r/c)$, and it has a maximum in the plane perpendicular to this vector, with no other maxima or minima in between. Finally, to determine the total power emitted by the system in all directions, we integrate (8.52) over the solid angle. With the aid of the invariant integrals of Problem 2.6 we obtain the basic formula for the total power of the dipole radiation

$$W = \int \frac{dW}{d\Omega}\, d\Omega = \frac{1}{16\pi^2 c^3}\, \ddot{D}^i \ddot{D}^j \int \left(\delta^{ij} - n^i n^j\right) d\Omega$$

$$= \frac{1}{16\pi^2 c^3}\, \ddot{D}^i \ddot{D}^j \left(4\pi\delta^{ij} - \frac{4\pi}{3}\, \delta^{ij}\right) = \frac{|\ddot{\mathbf{D}}|^2}{6\pi c^3}. \tag{8.53}$$

Conversely, the *total* momentum emitted by the system in all directions is *zero*. In fact, by integrating the radiation formula (8.19) over the angles we obtain

$$\frac{d\mathbf{P}}{dt} = \frac{1}{c}\int \mathbf{n}\, \frac{dW}{d\Omega}\, d\Omega = \frac{1}{16\pi^2 c^4}\, \ddot{D}^i \ddot{D}^j \int \mathbf{n}\left(\delta^{ij} - n^i n^j\right) d\Omega = 0, \tag{8.54}$$

where the conclusion stems from the vanishing of the invariant integrals of an odd number of $\mathbf{n}$'s. This result is actually a consequence of the invariance of the angular distribution of the radiated energy (8.52) under the replacement $\mathbf{n} \to -\mathbf{n}$. In fact, since the energies emitted in the directions $\mathbf{n}$ and $-\mathbf{n}$ are equal, according to (8.19) the momenta emitted in the two directions are *opposite* and, hence, they cancel each other.

In conclusion, in the non-relativistic *dipole approximation*, a generic charged source radiates energy with instantaneous power given by (8.53), whereas the *total* radiated momentum is zero. In this approximation the radiated energy is proportional to $1/c^3$, while the momentum radiated in a fixed direction is proportional to $1/c^4$, as can be seen from Eq. (8.54). It follows that the *exact* radiated total momentum starts with terms of order $1/c^5$.

Systems of charged particles and bremsstrahlung. As particular case we consider a system of non-relativistic charged particles. The charge density of this system is given by

$$\rho(t, \mathbf{y}) = \sum_r e_r\, \delta^3(\mathbf{y} - \mathbf{y}_r(t)), \tag{8.55}$$

so that Eq. (8.47) yields the dipole moment

$$\mathbf{D}(t) = \int \mathbf{y} \sum_r e_r\, \delta^3(\mathbf{y} - \mathbf{y}_r(t))\, d^3 y = \sum_r e_r\, \mathbf{y}_r(t), \quad \ddot{\mathbf{D}} = \sum_r e_r\, \mathbf{a}_r, \tag{8.56}$$

where $\mathbf{a}_r$ is the acceleration of the rth particle. According to (8.53) this system emits therefore dipole radiation with instantaneous total power

$$W = \frac{1}{6\pi c^3} \left|\sum_r e_r\, \mathbf{a}_r\right|^2, \tag{8.57}$$

an equation that generalizes Larmor's formula (7.67) to a system of particles. It is important to realize that what appears in (8.57) is *not* the sum of the individually emitted powers $\sum_r e_r^2 |\mathbf{a}_r|^2 /6\pi c^3$. In fact, the electromagnetic field satisfies the superposition principle and so it obeys the laws of interference: If $\mathbf{E}_r$ denotes the asymptotic field of the rth particle, the emitted power (8.18) reads indeed

$$\frac{dW}{d\Omega} = cr^2 \left| \sum_r \mathbf{E}_r \right|^2, \tag{8.58}$$

rather than $dW/d\Omega = cr^2 \sum_r |\mathbf{E}_r|^2$. In particular, in the non-relativistic limit the field $\mathbf{E}_r$ of each particle has the simple form (8.59), and in this case it is immediately seen that by integrating equation (8.58) over the angles one retrieves the result (8.57). The radiation emitted by a charged particle subject to a temporary, or long-term, acceleration is commonly called *bremsstrahlung*, i.e. *braking radiation*. Equation (8.57) quantifies hence the instantaneous amount of bremsstrahlung, summed over angles, emitted by an arbitrary system of non-relativistic charged particles. In the next sections we will analyze several interesting examples of non-relativistic bremsstrahlung, by resorting repeatedly to this equation.

Bremsstrahlung from a single particle. Let us analyze more closely the radiation emitted by a single charged particle, in which case $\ddot{\mathbf{D}} = e\mathbf{a}$. The fields in the wave zone (8.51) then take the simple form

$$\mathbf{E} = -\frac{e}{4\pi r c^2} (\mathbf{a} - (\mathbf{n} \cdot \mathbf{a})\,\mathbf{n}), \qquad \mathbf{B} = \frac{e\,(\mathbf{n} \times \mathbf{a})}{4\pi r c^2}, \tag{8.59}$$

where we must keep in mind that the acceleration $\mathbf{a}$ is evaluated at the time $t - r/c$. These fields differ radically from the fields generated at large distances by a particle in uniform linear motion, see Eqs. (6.102) and (6.99) in the limit $v \ll c$. In particular, the field $\mathbf{E}$ is no longer radial, being rather orthogonal to the radial direction $\mathbf{n}$, and it belongs to the plane spanned by the vectors $\mathbf{n}$ and $\mathbf{a}$. For the angular distribution of the radiated power Eq. (8.52) yields

$$\frac{dW}{d\Omega} = \frac{e^2 |\mathbf{n} \times \mathbf{a}|^2}{16\pi^2 c^3}. \tag{8.60}$$

Therefore, the particle emits no radiation in the direction of the acceleration, whereas the intensity of the radiation has a maximum in the plane *orthogonal* to the acceleration. We anticipate that this angular distribution is peculiar for the radiation emitted by a *non-relativistic* particle. In fact, in Sect. 10.3 we will see that the shape of the angular distribution of the radiation emitted by an ultrarelativistic particle is radically different from (8.60). Finally, notice that Eqs. (8.59) and (8.60) perfectly match with Eqs. (7.65) and (7.66) of Sect. 7.4.1, derived starting from the exact Liénard-Wiechert field.

Vanishing of the electric dipole radiation. We mention some important cases in which a system of charged particles emits no dipole radiation. Apart from the trivial case of a system of particles in uniform linear motion, moving hence far away from each other, the dipole radiation is absent for an *isolated* system of particles for which the ratio $e_r/m_r = \gamma$ is independent of r. In this case the dipole moment can, in fact, be written as

$$\mathbf{D} = \sum_r e_r \, \mathbf{y}_r = \gamma \sum_r m_r \, \mathbf{y}_r,$$

and, since the total momentum $\sum_r m_r \, \mathbf{v}_r$ of an isolated non-relativistic system is conserved, it follows that

$$\ddot{\mathbf{D}} = \gamma \frac{d}{dt} \left(\sum_r m_r \, \mathbf{v}_r \right) = 0.$$

In particular, in any physical process involving only a single species of charged particles, as e.g. during the collision of two identical particles, no dipole radiation is emitted. Other important accelerated systems of charges which do not emit dipole radiation are the *spherically symmetric* charge distributions. This is a direct consequence of *Birkhoff's theorem* for electrodynamics, see Problem 2.5, ensuring that an isotropic charge distribution generates an electromagnetic field which in empty space is necessarily static, and a static field cannot support radiation. Equation (8.52) of our non-relativistic analysis matches, indeed, with this exact statement, since in this case $\mathbf{D}$ vanishes identically. To see it we observe that for an isotropic charge distribution the charge density depends only on t and on the radius $y = |\mathbf{y}|$, $\rho(t, \mathbf{y}) = \rho(t, y)$. Switching to polar coordinates, setting $\mathbf{y} = y\mathbf{n}$ and $d^3y = y^2 dy d\Omega$, for the dipole moment of such a system we the obtain

$$\mathbf{D}(t) = \int \mathbf{y}\rho(t, y) \, d^3y = \left(\int_0^\infty y^3 \rho(t, y) \, dy \right) \left(\int \mathbf{n} \, d\Omega \right) = 0, \qquad (8.61)$$

where the conclusion follows from the vanishing of the integral of $\mathbf{n}$ over the angles. As observed previously, when the dipole radiation is absent, it is the quadrupole term of the multipole expansion (8.42) to dominate the series. Of course, for a spherically symmetric system the radiation vanishes at all orders of the expansion.

Summary. We end the section collecting the expressions for the vector potential in different regimes, and the associated angular distributions of the radiated power.

• Exact potential:

$$A^\mu(x) = \frac{1}{4\pi c} \int \frac{1}{|\mathbf{x} - \mathbf{y}|} \, j^\mu \left(t - \frac{|\mathbf{x} - \mathbf{y}|}{c}, \mathbf{y} \right) d^3y.$$

Locally radiated power:

$$\frac{d\mathcal{W}}{d\Omega} = cr^2 \mathbf{n} \cdot (\mathbf{E} \times \mathbf{B}).$$

- Potential in the wave zone:

$$A^\mu(x) = \frac{1}{4\pi rc} \int j^\mu \left(t - \frac{r}{c} + \frac{\mathbf{n} \cdot \mathbf{y}}{c}, \mathbf{y}\right) d^3y.$$

Exact radiated power:

$$\frac{d\mathcal{W}}{d\Omega} = \frac{r^2}{c} \left|\mathbf{n} \times \dot{\mathbf{A}}\right|^2.$$

- Potential in the wave zone in dipole approximation:

$$\mathbf{A}(x) = \frac{1}{4\pi rc} \int \mathbf{j} \left(t - \frac{r}{c}, \mathbf{y}\right) d^3y = \frac{\dot{\mathbf{D}}}{4\pi rc}.$$

Radiated power in dipole approximation:

$$\frac{d\mathcal{W}}{d\Omega} = \frac{\left|\mathbf{n} \times \ddot{\mathbf{D}}\right|^2}{16\pi^2 c^3}.$$

Total radiated power in dipole approximation:

$$\mathcal{W} = \frac{\left|\ddot{\mathbf{D}}\right|^2}{6\pi c^3}.$$

8.4.1 Radiation of a Short Linear Antenna

As first application of the dipole approximation we investigate the radiation emitted by a short linear antenna, that is to say, a linear antenna which is much shorter than the wavelength of the emitted radiation,

$$L \ll \lambda \quad \leftrightarrow \quad \frac{\omega L}{c} \ll 1. \tag{8.62}$$

Actually, in Sect. 8.2 we derived an exact formula for the angular distribution of the energy radiated by an antenna of *arbitrary* length, see Eq. (8.34). Consequently, the analysis of this section will also allow us to discuss the validity of the dipole approximation in a concrete example. Our starting point is again the spatial current density (8.29)

$$\mathbf{j}(t, \mathbf{y}) = I_0 \, \delta(y^1) \, \delta(y^2) \frac{\sin\left(\frac{\omega}{c}\left(\frac{L}{2} - |y^3|\right)\right)}{\sin\left(\frac{\omega L}{2c}\right)} \cos(\omega t) \, \mathbf{u}, \tag{8.63}$$

where we have reintroduced the speed of light. To find out whether under the hypothesis (8.62) the dipole approximation is, indeed, reliable we must check the validity of (8.43). The characteristic time scale of the current (8.63) is its period $\tau = 2\pi/\omega$ and, since $|\mathbf{y}|$ equals at most $L/2$, the condition (8.43) translates into

$$\left|\frac{\mathbf{n} \cdot \mathbf{y}}{c}\right| \sim \frac{L}{c} \ll \tau = \frac{2\pi}{\omega},$$

which is precisely (8.62). This means that the radiation of the antenna can be analyzed accurately in terms of the dipole formulas of Sect. 8.4, involving only the dipole moment (8.47). To evaluate the latter we need the charge density $\rho(t, \mathbf{y})$ of the antenna, which can be derived from the spatial current (8.63) via four-current conservation,

$$\dot{\rho} = -\partial_i j^i = -\frac{\partial j^3}{\partial y^3} = \omega I_0\, \delta(y^1)\, \delta(y^2)\, \frac{\cos\left(\frac{\omega}{c}\left(\frac{L}{2} - |y^3|\right)\right)}{c \sin\left(\frac{\omega L}{2c}\right)}\, \cos(\omega t)\, \varepsilon(y^3),$$

where $\varepsilon(\cdot)$ denotes the sign function. The sought charge density is thus given by

$$\rho(t, \mathbf{y}) = I_0\, \delta(y^1)\, \delta(y^2)\, \frac{\cos\left(\frac{\omega}{c}\left(\frac{L}{2} - |y^3|\right)\right)}{c \sin\left(\frac{\omega L}{2c}\right)}\, \sin(\omega t)\, \varepsilon(y^3).$$

For the dipole moment we the obtain

$$\mathbf{D} = \int \mathbf{y}\rho(t, \mathbf{y})\, d^3y = I_0 \sin(\omega t) \int \mathbf{y}\, \delta(y^1)\, \delta(y^2)\, \frac{\cos\left(\frac{\omega}{c}\left(\frac{L}{2} - |y^3|\right)\right)}{c \sin\left(\frac{\omega L}{2c}\right)}\, \varepsilon(y^3)d^3y$$

$$= 2I_0 \sin(\omega t)\mathbf{u} \int_0^{\frac{L}{2}} y^3 \frac{\cos\left(\frac{\omega}{c}\left(\frac{L}{2} - y^3\right)\right)}{c \sin\left(\frac{\omega L}{2c}\right)}\, dy^3 = \frac{2I_0 c\, \sin(\omega t)\left(1 - \cos\left(\frac{\omega L}{2c}\right)\right)}{\omega^2 \sin\left(\frac{\omega L}{2c}\right)}\, \mathbf{u}.$$

Since by hypothesis $\omega L/c \ll 1$, this expression simplifies to

$$\mathbf{D} = \frac{I_0 L}{2\omega}\, \sin(\omega t)\, \mathbf{u}, \qquad \ddot{\mathbf{D}} = -\frac{\omega I_0 L}{2}\, \sin(\omega t)\, \mathbf{u}.$$

Equation (8.52) then yields the instantaneous radiated power

$$\frac{d\mathcal{W}}{d\Omega} = \frac{(\omega I_0 L)^2 \sin^2(\omega(t - r/c))}{64\pi^2 c^3}\, \sin^2 \vartheta, \qquad (8.64)$$

where ϑ is the angle between $\mathbf{n}$ and the z axis. Averaging this expression over a period we obtain the mean power

$$\frac{d\overline{\mathcal{W}}}{d\Omega} = \frac{(I_0 \omega L)^2}{128\pi^2 c^3}\, \sin^2 \vartheta, \qquad (8.65)$$

to be compared with the exact mean power (8.34). Indeed, it is easy to verify that for $\omega L/c \ll 1$ the expression (8.34) actually reduces to (8.65). Furthermore, the angular distribution of the radiated energy (8.65) entails the typical dipole behavior: it has a maximum in the plane orthogonal to the antenna, at $\vartheta = \pi/2$, and it vanishes in the direction of the antenna, at $\vartheta = 0$. From a more close comparison between the shapes of (8.65) and (8.34) it emerges that, as long as $L \leq \lambda$, the latter has actually a *form* very similar to the former: a unique maximum at $\vartheta = \pi/2$, and a unique zero at $\vartheta = 0$.

To analyze the differences between the emitted powers of short and long antennas more quantitatively, we compare the corresponding total powers. Integrating equation (8.65) over the angles we find

$$\overline{W} = \int \frac{d\overline{W}}{d\Omega}\, d\Omega = \frac{(\omega L)^2}{48\pi c^3}\, I_0^2 = \frac{1}{2c}\, I_0^2\, R_{\text{rad}}^s, \qquad (8.66)$$

having introduced the radiation resistance of the short antenna

$$R_{\text{rad}}^s = \frac{(\omega L)^2}{24\pi c^2} = \frac{\pi}{6}\left(\frac{L}{\lambda}\right)^2. \qquad (8.67)$$

Switching to MKS units this formula reads

$$R_{\text{rad}}^s = \frac{\pi}{6}\left(\frac{L}{\lambda}\right)^2 R_0 = 197\left(\frac{L}{\lambda}\right)^2 \text{ohm},$$

where $R_0 = 377$ ohm is the wave resistance of free space (8.39). Choosing, for example, in line with the working hypothesis (8.62), $L = \lambda/25$, we obtain the radiation resistance

$$R_{\text{rad}}^s = 0.32\,\text{ohm},$$

which is much smaller than the radiation resistance of the half-wave antenna $R_{\text{rad}}^{(1/2)} = 73$ ohm, see (8.40). The relevant point is, however, that the *ohmic* resistance R_{ohm}^s of a short antenna is in general of the same order, or also sensibly larger, than its radiation resistance

$$R_{\text{ohm}}^s \gtrsim R_{\text{rad}}^s.$$

This means that a short antenna has in general a *low* radiation efficiency. Finally, one may ask which value we would have obtained for the radiated power of the half-wave antenna if – erroneously – we would have applied the dipole approximation. The result would have been (8.66) with $L = \lambda/2 = \pi c/\omega$, i.e.

$$\overline{W} = \frac{\pi I_0^2}{48c} = 0.065\,\frac{I_0^2}{c},$$

instead of the exact answer (8.37), i.e. $\overline{\mathcal{W}} = 0.097 \, I_0^2/c$. Thus, despite the wrong number, we would have obtained the correct order of magnitude of the radiated power.

8.4.2 Thomson Scattering

The scattering of electromagnetic radiation by charged particles is a process playing a central role in the physics of fundamental interactions. In classical physics, in the simplest case this process is described by an electromagnetic wave hitting a free charged particle at rest, and is called *Thomson scattering*. In quantum physics, the same process is described in terms of collisions between the photons of the incident wave and the charged particle, and is called *Compton scattering*. The basic characteristics of Thomson scattering can be summarized as follows. A charged particle hit by an *incident* wave of frequency ω is subject to the Lorentz force and starts to oscillate, mainly in the direction of the electric field of the incident wave and with the same frequency ω. Being accelerated, the particle, in turn, emits electromagnetic radiation in all directions, but with an *anisotropic* angular distribution. If the speed of the particle is much smaller than the speed of light, this *scattered* radiation has the same frequency ω as the incident wave, and is linearly polarized.

Incident wave. We take as incident wave a linearly polarized plane wave of frequency ω, propagating in the direction $\mathbf{u}$. The electric and magnetic fields of the wave are then of the form

$$\boldsymbol{\mathcal{E}} = \boldsymbol{\mathcal{E}}_0 \cos(\omega t - \mathbf{k} \cdot \mathbf{x}), \qquad \boldsymbol{\mathcal{B}} = \mathbf{u} \times \boldsymbol{\mathcal{E}}, \qquad \mathbf{u} \cdot \boldsymbol{\mathcal{E}}_0 = 0, \qquad (8.68)$$

where the "amplitude" $\boldsymbol{\mathcal{E}}_0$ of the electric field is a real vector, and the spatial wave vector is given by $\mathbf{k} = \omega \mathbf{u}/c$. The intensity $\mathcal{I}$ of the incident wave, i.e. the mean energy crossing a unit surface orthogonal to $\mathbf{u}$ per unit time, is then given by (see (5.100))

$$\mathcal{I} = c\,\overline{\mathcal{E}^2} = c\,\mathcal{E}_0^2\,\overline{\cos^2(\omega t - \mathbf{k} \cdot \mathbf{x})} = \frac{c\,\mathcal{E}_0^2}{2}. \qquad (8.69)$$

Henceforth, we assume that $\mathcal{I}$ is sufficiently small, in order that the induced speed of the particle is much smaller than c, see (8.73). In this case it is, hence, appropriate to analyze the radiation scattered by the particle by means of the dipole approximation. Let us now investigate the effect of the wave (8.68) when it hits a non-relativistic particle of mass m and charge e, which is thus subject to the Lorentz equation

$$m\mathbf{a} = e\left(\boldsymbol{\mathcal{E}} + \frac{\mathbf{v}}{c} \times \boldsymbol{\mathcal{B}}\right). \qquad (8.70)$$

Since $v \ll c$ and $\mathcal{B} = \mathcal{E}$, the magnetic field can be neglected and (8.70) reduces to

$$m\ddot{\mathbf{y}}(t) = e\,\boldsymbol{\mathcal{E}}_0 \cos\left(\omega t - \mathbf{k} \cdot \mathbf{y}(t)\right). \tag{8.71}$$

Taking as z axis the direction of $\mathbf{u}$, and as x axis the direction of $\boldsymbol{\mathcal{E}}_0$, in which case we have $\mathbf{u} = (0,0,1)$, $c\mathbf{k} = (0,0,\omega)$ and $\boldsymbol{\mathcal{E}}_0 = (\mathcal{E}_0,0,0)$, Eq. (8.71) admits the *stationary* solution

$$x(t) = -\frac{e\mathcal{E}_0}{m\omega^2}\cos(\omega t), \quad y(t) = 0, \quad z(t) = 0. \tag{8.72}$$

The particle oscillates, hence, in the direction of the electric field, with the same frequency ω as the incident wave. In particular, its maximum speed is $v_M = e\mathcal{E}_0/m\omega$. The reliability of the dipole approximation then requires that

$$\frac{e\mathcal{E}_0}{m\omega} \ll c \quad \Leftrightarrow \quad \mathcal{E}_0 \ll \frac{m\omega c}{e}, \tag{8.73}$$

so that the intensity (8.69) of the incident wave is subject to the limitation $\mathcal{I} \ll m^2\omega^2 c^3/e^2$.

Scattered radiation. The acceleration of the trajectory (8.72) can be recast in the simple form

$$\mathbf{a}(t) = \frac{e\boldsymbol{\mathcal{E}}_0}{m}\cos(\omega t). \tag{8.74}$$

Formulas (8.59) and (8.60) then yield for the radiation field and for the radiated power the expressions

$$\mathbf{E} = -\frac{e^2}{4\pi m r c^2}\left(\boldsymbol{\mathcal{E}}_0 - (\mathbf{n}\cdot\boldsymbol{\mathcal{E}}_0)\,\mathbf{n}\right)\cos\left(\omega\left(t - \frac{r}{c}\right)\right), \quad \mathbf{B} = \mathbf{n}\times\mathbf{E}, \tag{8.75}$$

$$\frac{dW}{d\Omega} = \frac{e^4}{16\pi^2 m^2 c^3}\left(\mathcal{E}_0^2 - (\mathbf{n}\cdot\boldsymbol{\mathcal{E}}_0)^2\right)\cos^2\left(\omega\left(t - \frac{r}{c}\right)\right). \tag{8.76}$$

The *scattered* radiation field (8.75) has thus the same frequency as the incident wave, but it propagates radially in all directions. In particular, from its angular distribution (8.76) we see that its intensity has a maximum in the plane (containing the charge) orthogonal to $\boldsymbol{\mathcal{E}}_0$, i.e. for $\mathbf{n} \perp \boldsymbol{\mathcal{E}}_0$, while it vanishes in the direction of $\boldsymbol{\mathcal{E}}_0$, i.e. for $\mathbf{n} \parallel \boldsymbol{\mathcal{E}}_0$. Furthermore, from (8.75) we see that the electric field lies in the plane spanned by the vectors $\boldsymbol{\mathcal{E}}_0$ and $\mathbf{n}$, and so its direction is *constant* in time. The scattered radiation is, thus, *linearly* polarized.

Unpolarized incident radiation. In many physically interesting situations the incident radiation is *unpolarized*, an important example being *natural* light. In these cases the incident radiation is an equiprobable superposition – a statistical mixture – of all polarizations $\boldsymbol{\mathcal{E}}_0$ orthogonal to $\mathbf{k}$. For an incident radiation of this kind we must average the radiated power (8.76) over all vectors $\boldsymbol{\mathcal{E}}_0$ orthogonal to $\mathbf{k}$, subject to the constraint $\mathcal{E}_0^2 = 2\mathcal{I}/c$. To perform this average we write out the term in (8.76) which depends on the direction of $\boldsymbol{\mathcal{E}}_0$. Choosing again $\mathbf{u}=(0,0,1)$, we have $\mathcal{E}_{0z}=0$, and so

$$(\mathbf{n} \cdot \boldsymbol{\mathcal{E}}_0)^2 = (n_x \mathcal{E}_{0x} + n_y \mathcal{E}_{0y})^2 = n_x^2 \mathcal{E}_{0x}^2 + n_y^2 \mathcal{E}_{0y}^2 + 2 n_x n_y \mathcal{E}_{0x} \mathcal{E}_{0y}.$$

To perform the average of this expression we use the relations

$$\overline{\mathcal{E}_{0x}^2} = \overline{\mathcal{E}_{0y}^2} = \frac{1}{2}\mathcal{E}_0^2, \qquad \overline{\mathcal{E}_{0x}\mathcal{E}_{0y}} = 0.$$

Denoting the angle between $\mathbf{n}$ and the z axis by ϑ, we have also $n_z = \cos\vartheta$ and $n_x^2 + n_y^2 = \sin^2\vartheta$. In this way we then obtain

$$\overline{\mathcal{E}_0^2 - (\mathbf{n} \cdot \boldsymbol{\mathcal{E}}_0)^2} = \mathcal{E}_0^2 - \frac{1}{2}\sin^2\vartheta\,\mathcal{E}_0^2 = \frac{1}{2}\left(1 + \cos^2\vartheta\right)\mathcal{E}_0^2.$$

Considering, in addition, the time average of the radiated power (8.76), we must also perform the replacement $\cos^2(\omega(t - r/c)) \to 1/2$. For unpolarized incident radiation, (8.76) eventually yields for the average angular distribution of the scattered radiation

$$\frac{d\overline{\mathcal{W}}}{d\Omega} = \frac{e^4 \mathcal{E}_0^2}{64\pi^2 m^2 c^3}\left(1 + \cos^2\vartheta\right). \tag{8.77}$$

The scattered radiation has, hence, a maximum in the propagation direction $\mathbf{u}$ of the incident wave – for both orientations $\vartheta = 0$ and $\vartheta = \pi$ – in agreement with the fact that for an arbitrarily polarized incident wave the particle oscillates always in the plane orthogonal to $\mathbf{u}$. Recall, in this regard, that a non-relativistic particle admits predominantly in the directions orthogonal to its acceleration. Finally, to compute the total radiated power we integrate (8.77) over the angles. With the aid of the integral

$$\int (1 + \cos^2\vartheta)\,d\Omega = \int_0^{2\pi} d\varphi \int_0^\pi \sin\vartheta\,d\vartheta\,(1 + \cos^2\vartheta) = \frac{16\pi}{3} \tag{8.78}$$

we find

$$\overline{\mathcal{W}} = \int \frac{d\overline{\mathcal{W}}}{d\Omega}\,d\Omega = \frac{e^4 \mathcal{E}_0^2}{12\pi m^2 c^3}. \tag{8.79}$$

Of course, the same expression can be obtained by inserting the acceleration (8.74) in Larmor's formula (7.67), and by performing its average over time. The resulting radiated power turns out to be independent of the polarization of the incident wave, so that the final average over polarizations is trivial.

Thomson cross section. From an experimental point of view in general the physically most relevant quantities regarding a scattering process are the *differential cross* section $d\sigma/d\Omega$, and the total *cross section* σ. In the case at hand $d\sigma/d\Omega$ is defined as the ratio between the energy emitted per unit time per unit solid angle in a direction **n**, and the incoming energy crossing a unit surface per unit time, i.e. the incident intensity $\mathcal{I}$. Similarly, σ is defined as the ratio between the total energy emitted in all directions per unit time, and the incident intensity $\mathcal{I}$. For Thomson scattering, from Eqs. (8.69) and (8.77) we find

$$\frac{d\sigma}{d\Omega} = \frac{1}{\mathcal{I}}\frac{d\overline{W}}{d\Omega} = \frac{1+\cos^2\vartheta}{2}r_0^2, \tag{8.80}$$

where we have introduced the *classical radius* r_0 of a charged particle, which for an electron is equal to

$$r_0 = \frac{e^2}{4\pi mc^2} = 2.8\cdot 10^{-13}\,\text{cm}. \tag{8.81}$$

Notice that this radius is much smaller than both the *Bohr radius* r_B and the *Compton wavelength* λ_C of the electron:

$$r_B = \frac{4\pi\hbar^2}{me^2} = 5.3\cdot 10^{-9}\,\text{cm}, \qquad \lambda_C = \frac{\hbar}{mc} = 3.8\cdot 10^{-11}\,\text{cm}. \tag{8.82}$$

To compute the total cross section we integrate equation (8.80) over the angles, using again (8.78),

$$\sigma = \int \frac{d\sigma}{d\Omega}\,d\Omega = \frac{\overline{W}}{\mathcal{I}} = \frac{8\pi}{3}r_0^2. \tag{8.83}$$

This cross section is called *Thomson cross section* and, evidently, it has the dimensions of an area. By its very definition, σ can be interpreted as the surface *offered* by the particle as target to the incident wave, or also as the *image* of the particle captured by the incident wave: It is precisely the fact that the Thomson cross section is proportional to r_0^2 to confer to r_0 the interpretation of *classical electron radius*.

Four-momentum balance and self-force. We end this section with an analysis of the four-momentum conservation mechanism in a Thomson-scattering experiment. According to Eq. (8.83), the scattering process can be interpreted as follows: only the fraction of the – conceptually infinitely extended – incident radiation that hits the surface σ gets *scattered*, whereas the rest propagates undisturbed by the presence of the particle, and forms the *transmitted* radiation. We consider the four-momentum balance separately for the transmitted radiation, the scattered radiation, and the particle. Obviously, the initial and final four-momenta of the transmitted radiation are equal. Also the particle conserves, on average, its four-momentum in that it performs a stationary motion. Concerning the scattered radiation, by definition its energy flux before the collision with the particle is $\mathcal{I}\sigma$, while after the collision it is $\overline{W}$: Thanks to the equality $\overline{W} = \mathcal{I}\sigma$, its energy is thus conserved. Conversely, according to Eqs. (5.103), (8.69) and (8.83), the momentum flux of the scattered radiation in the z direction before the collision is

$$\frac{dP^z}{dt} = \overline{T_{\text{em}}^{zz}}\,\sigma = \overline{\mathcal{E}^2}\,\sigma = \frac{4\pi}{3}\,r_0^2\,\mathcal{E}_0^2,$$

while after the collision it is represented by the fields (8.75), entailing a *vanishing* total momentum flux. In fact, in dipole approximation the radiation emitted by a charged particle carries zero total momentum, see Eq. (8.54). If the total momentum of the system is to be conserved, we must conclude that the missing momentum flux has been transferred to the particle. Consequently, the latter must be acted on by a mean forward-force equal to

$$\mathbf{f} = \frac{d\mathbf{P}}{dt} = \frac{4\pi}{3}\,r_0^2\,\mathcal{E}_0^2\,\mathbf{u} = \frac{e^4\mathcal{E}_0^2}{12\pi m^2 c^4}\,\mathbf{u}, \tag{8.84}$$

which should be added to the right hand side of the equation of motion (8.70). Notice that $\mathbf{f}$ is of order $1/c^4$, i.e. of the same order as the momentum flux *locally* associated with dipole radiation, see Eq. (8.54).

The picture which emerges is thus the following. The Lorentz force $e(\mathcal{E} + \mathbf{v} \times \mathcal{B}/c)$, originating the whole process, causes the particle to perform an oscillatory motion. Being accelerated, the particle emits electromagnetic radiation which, in turn, causes the particle a boost in the forward direction, represented by the force $\mathbf{f}$. This force originates thus from the "interaction" between the particle and the field created by the particle itself, and correspondingly it is called *self-force* or *radiation reaction force*. Notice that $\mathbf{f}$ – a relativistic effect – does not arise from the Lorentz force $e\mathbf{v} \times \mathcal{B}/c$, which in (8.70) has indeed been neglected. More precisely, from the trajectory (8.72) and the fields (8.68) we find

$$e\,\frac{\mathbf{v}}{c} \times \mathcal{B} = \frac{e^2\mathcal{E}_0^2}{m\omega c}\,\sin(\omega t)\cos(\omega t)\,\mathbf{u}, \tag{8.85}$$

a correction which is quadratic in $\mathcal{E}_0$ and of order $1/c$, while $\mathbf{f}$ is quadratic in $\mathcal{E}_0$ but of order $1/c^4$. Actually, the time average of the force (8.85) vanishes, so that it produces no *net* effect at all.

From the point of view of first principles, represented by the fundamental equations (2.20)–(2.22), the force $\mathbf{f}$ must appear automatically once, in place of the approximate equation of motion (8.70), we consider the *complete* Lorentz equation (2.37)

$$\frac{d\mathbf{p}}{dt} = e\left(\mathcal{E} + \frac{\mathbf{v}}{c} \times \mathcal{B}\right) + \mathbf{f}_{\text{self}}, \qquad \mathbf{f}_{\text{self}} = e\left(\mathbf{E} + \frac{\mathbf{v}}{c} \times \mathbf{B}\right), \tag{8.86}$$

where $\mathbf{E}$ and $\mathbf{B}$ are the exact Liénard-Wiechert fields generated by the particle. In Eq. (8.86) these fields are evaluated at the position of the particle, so that the *microscopic* self-force $\mathbf{f}_{\text{self}}$ introduced above, actually, represents a force imputable to self-interaction. A detailed analysis shows that $\mathbf{f}_{\text{self}}$, indeed, *induces* the force $\mathbf{f}$ (8.84), see Problem 15.2. The problematic aspect, however, is that $\mathbf{f}_{\text{self}}$ as it stands is *infinite*: As mentioned several times, the electromagnetic field of a particle evaluated at the particle's position is always divergent. At this point, in principle we have two alternatives. If we keep in Eq. (8.86) the term $\mathbf{f}_{self}$, the equation predicts the correct self-force $\mathbf{f}$, but is by itself meaningless in that its second member diverges. Conversely, if we disregard in (8.86) the term $\mathbf{f}_{self}$, the equation is finite, but violates four-momentum conservation in that it does not predict the presence of $\mathbf{f}$. Obviously, both choices are unsatisfactory. We postpone a systematic treatment of the problematic self-interaction phenomenon – which includes in particular the replacement of the Lorentz equation with the *Lorentz-Dirac* equation – to Chap. 15.

Quantum mechanical aspects. *Thomson scattering* as presented above disregards the quantum nature of the interaction between radiation and charges. At the quantum level, the scattering process of radiation of frequency ω by free charged particles is, in fact, realized through collisions between incident photons of energy $\hbar\omega$ and particles considered at rest, a process known as *Compton scattering*. The scattered radiation, in turn, is then composed of photons propagating in all directions. As long as the wavelength of the incident radiation is much larger than the Compton wavelength of the charged particle, $\lambda \gg \lambda_C = \hbar/mc$, i.e. $\hbar\omega \ll mc^2$, quantum effects are negligible, and so the description of the process in terms of the classical Thomson scattering is appropriate. Vice versa, if λ is of the order of λ_C, the incident photons cede part of their energy to the charged particle and emerge, hence, from the collision with a smaller frequency, i.e. with a wavelength λ' larger than λ. In fact, imposing four-momentum conservation one derives the known Compton effect formula

$$\lambda' = \lambda + 2\pi(1 - \cos\vartheta)\,\lambda_C,$$

where the *scattering angle* ϑ is the angle between the incident and outgoing photons, having thus the same meaning as in the classical equation (8.80). For $\lambda \gg \lambda_C$ the incoming and outgoing photons have, thus, almost the same energy $\hbar\omega$ – independently from the scattering angle – and in this limit the number of photons in the ingoing and outgoing radiations is given simply by the energy divided by $\hbar\omega$. In this case the cross sections (8.80) and (8.83) equal the number of outgoing photons per unit time, divided the number of incident photons per unit time per unit area. The main differences between the classical and quantum pictures of the scattering process can be summarized as follows.

- At the quantum level the energy of the incident radiation is not absorbed and emitted *continuously* by means of electromagnetic waves, as assumed in the Thomson scattering, but rather under the form of discrete light *quanta*, i.e. photons.
- The energy of the outgoing photons is smaller than the energy of the incoming ones, while in the classical treatment the outgoing radiation has the same frequency as the incoming one.
- It can be seen that the Thomson cross section (8.83) is subject to quantum corrections, which at first order in $\hbar$ result in the modified formula

$$\sigma_q = \frac{8\pi}{3}\, r_0^2 \left(1 - 4\pi \frac{\lambda_C}{\lambda}\right).$$

8.4.3   Bremsstrahlung from Coulomb Interaction

We investigate now the radiation generated in the interaction between two non-relativistic charged particles, a prototype of non-relativistic bremsstrahlung. We will mainly be interested in the determination of the radiated energy. Since in the non-relativistic limit the interaction between two charged particles is governed by the Coulomb force $\mathbf{F} = e_1 e_2 \mathbf{r}/4\pi r^3$, the corresponding orbits are conics: ellipses, hyperbolas, parabolas. Of course, the knowledge of the explicit form of the orbit facilitates the determination of the radiated energy. We consider an isolated system of two particles with masses m_1 and m_2 and charges e_1 and e_2. The position vectors $\mathbf{r}_1$ and $\mathbf{r}_2$, the relative position $\mathbf{r} = \mathbf{r}_1 - \mathbf{r}_2$, and the position of the center of mass $\mathbf{r}_{cm}$ are then connected by the known relations

$$\mathbf{r}_1 = \mathbf{r}_{cm} + \frac{m_2}{m_1 + m_2}\,\mathbf{r}, \qquad \mathbf{r}_2 = \mathbf{r}_{cm} - \frac{m_1}{m_1 + m_2}\,\mathbf{r}. \tag{8.87}$$

After the separation of the center-of-mass motion, the dynamics of the system is then governed by the equations of motion

$$\mu\,\ddot{\mathbf{r}} = \mathbf{F} = \frac{\alpha \mathbf{r}}{r^3}, \qquad \ddot{\mathbf{r}}_{cm} = 0, \tag{8.88}$$

where we have set
$$\alpha = \frac{e_1 e_2}{4\pi}, \qquad \mu = \frac{m_1 m_2}{m_1 + m_2},$$

μ being the reduced mass of the system.

Cinematics of conics. As the Coulomb force is central and spherically symmetric, the relative motion is a planar motion, and the total energy ε and angular momentum $\mathbf{L}$ are conserved quantities. Introducing in the plane of the orbit the polar coordinates r and φ, the constants of motion assume the form

$$\varepsilon = \frac{1}{2}\mu v^2 + \frac{\alpha}{r}, \qquad L = \mu r^2 \dot{\varphi}, \tag{8.89}$$

where $\mathbf{v} = \dot{\mathbf{r}}$ is the relative velocity. If the energy is negative, $\varepsilon < 0$, and so necessarily $\alpha < 0$, the orbit is an ellipse with equation

$$r(\varphi) = \frac{(1 - e^2)a}{1 + e\cos\varphi}, \tag{8.90}$$

where the major semi-axis a and the eccentricity $e < 1$ are given by

$$a = \left|\frac{\alpha}{2\varepsilon}\right|, \qquad e = \sqrt{1 + \frac{2\varepsilon L^2}{\mu\alpha^2}}. \tag{8.91}$$

We also recall that the period of the orbit can be expressed as

$$T = 2\pi \sqrt{\frac{\mu a^3}{|\alpha|}}, \tag{8.92}$$

and that the angular momentum can be recast in the form

$$L = \sqrt{\mu a |\alpha|} \sqrt{1 - e^2}.$$

If the energy is positive, $\varepsilon > 0$, the orbits are hyperbolas with equation

$$r(\varphi) = \frac{(e^2 - 1)\, a}{\pm 1 + e \cos \varphi}, \tag{8.93}$$

where the plus sign corresponds to an *attractive* Coulomb potential, $\alpha < 0$, and the minus sign to a *repulsive* one, $\alpha > 0$. The parameters a and e are still given by Eq. (8.91), but now $e > 1$. Since hyperbolas are open orbits, the constants of motion can also be expressed in terms of the *impact parameter* b and of the *asymptotic speed* v_∞

$$\varepsilon = \frac{1}{2} \mu v_\infty^2, \qquad L = \mu b v_\infty. \tag{8.94}$$

Furthermore, from Eq. (8.93) it follows that for hyperbolic orbits the angular variable φ is subject to the limitations

$$-\varphi_\infty < \varphi < \varphi_\infty, \qquad \cos \varphi_\infty = \mp \frac{1}{e}. \tag{8.95}$$

Radiated energy. We compute now the energy radiated via bremsstrahlung. The total radiated power is given by Eq. (8.53) in terms of the dipole moment $\mathbf{D}$ of the system

$$W = \frac{|\ddot{\mathbf{D}}|^2}{6\pi c^3}. \tag{8.96}$$

To evaluate the latter we insert the relations (8.87) in (8.56)

$$\mathbf{D} = e_1 \mathbf{r}_1 + e_2 \mathbf{r}_2 = (e_1 + e_2)\, \mathbf{r}_{\mathrm{cm}} + \mu \left(\frac{e_1}{m_1} - \frac{e_2}{m_2} \right) \mathbf{r}.$$

Taking the second derivative of this expression, and using the equations of motion (8.88), we find

$$\ddot{\mathbf{D}} = \mu \left(\frac{e_1}{m_1} - \frac{e_2}{m_2} \right) \ddot{\mathbf{r}} = \left(\frac{e_1}{m_1} - \frac{e_2}{m_2} \right) \frac{\alpha \mathbf{r}}{r^3}.$$

For the instantaneous power (8.96) we then obtain[5]

$$W = \frac{\alpha^2}{6\pi c^3} \left(\frac{e_1}{m_1} - \frac{e_2}{m_2} \right)^2 \frac{1}{r^4}. \tag{8.97}$$

We see that the dipole radiation is absent if the particles have the same ratio e/m, in particular if they are identical particles, as shown in general in Sect. 8.4. To determine the energy radiated along a finite arc of the orbit we must integrate Eq. (8.97) between the corresponding instants t_1 and t_2. To evaluate the resulting integral it is convenient to switch from the integration variable t to the angular variable φ, resorting to the conservation of the angular momentum (8.89). Writing

$$dt = \frac{\mu r^2}{L} \, d\varphi,$$

and denoting the angles corresponding to t_1 and t_2 by φ_1 and φ_2, for the energy radiated during this time interval we then obtain

$$\Delta \varepsilon = \int_{t_1}^{t_2} W \, dt = \frac{\mu \alpha^2}{6\pi L c^3} \left(\frac{e_1}{m_1} - \frac{e_2}{m_2} \right)^2 \int_{\varphi_1}^{\varphi_2} \frac{1}{r^2} \, d\varphi. \tag{8.98}$$

Inserting in the integrand the polar equations (8.90) and (8.93) we obtain integrals which can be computed analytically. Below we investigate separately the energy radiated from elliptical and hyperbolic orbits.

Elliptical orbits. If the relative orbit is elliptical, both particles perform *periodic* motions with the same period T (8.92). From Sect. 8.1.2 we then know that the system emits radiation with the discrete frequencies $\omega_N = 2\pi N/T$, with N integer. In this case, as the total radiated energy is infinite, the physically meaningful quantity is the mean power $\overline{W}$. Averaging Eq. (8.98) over a period, and inserting the polar equation (8.90), we find

$$\overline{W} = \frac{1}{T} \int_0^T W \, dt = \frac{\mu \alpha^2}{6\pi L T c^3} \left(\frac{e_1}{m_1} - \frac{e_2}{m_2} \right)^2 \int_0^{2\pi} \frac{1}{r^2} \, d\varphi$$

$$= \frac{\mu \alpha^2}{6\pi L T c^3} \left(\frac{e_1}{m_1} - \frac{e_2}{m_2} \right)^2 \frac{1}{a^2(1-e^2)^2} \int_0^{2\pi} (1 + e \cos\varphi)^2 d\varphi.$$

Substituting the kinematical relations above and using the integral

$$\int_0^{2\pi} (1 + e \cos\varphi)^2 d\varphi = \pi \left(2 + e^2 \right),$$

---

[5]Equation (8.97) represents the radiated power observed at a time t at a large distance r^* from the particles, if the radius r appearing there is evaluated at the macroscopic retarded time $t - r^*/c$. Vice versa, if r is evaluated at the time t, the equation yields the fraction of energy emitted at time t *which reaches infinity*. We will come back to this point in Chap. 10.

we eventually obtain the mean radiated power

$$\overline{W} = \frac{\alpha^2}{12\pi a^4 c^3} \left(\frac{e_1}{m_1} - \frac{e_2}{m_2} \right)^2 \frac{2 + e^2}{(1 - e^2)^{5/2}}. \tag{8.99}$$

Correspondingly, during a cycle the bremsstrahlung carries away the energy

$$\Delta \varepsilon_c = T \overline{W}. \tag{8.100}$$

Therefore, if the total energy is to be conserved, the mechanical energy ε in (8.89) of the two-particle system during any cycle must decrease by the amount $T \overline{W}$. We come so to the conclusion that bremsstrahlung necessarily breaks up elliptical orbits, deforming them into instable spiraling trajectories. The force *directly* responsible for this phenomenon should be, once more, the radiation reaction force f_{self} introduced in (8.86). In Sect. 8.4.4 we will quantify the energy loss (8.100) for a system of fundamental importance in the history of quantum mechanics, namely the hydrogen atom, showing that according to the laws of classical electrodynamics this force would cause the electron to crash into the nucleus in a fraction of a second.

Hyperbolic orbits. If the relative orbit is hyperbolic, both particles perform *aperiodic* motions and the system emits radiation with a continuous spectrum of frequencies. This process corresponds to the collision of two particles incoming from infinity, deflecting each other, and then outgoing again to infinity. In the initial and final asymptotic states the particles have vanishing acceleration and, correspondingly, we expect the total energy $\Delta \varepsilon$ radiated during the whole process to be finite. To calculate it we must set in Eq. (8.98) $t_1 = -\infty$ and $t_2 = \infty$, corresponding to $\varphi_1 = -\varphi_\infty$ and $\varphi_2 = \varphi_\infty$, see (8.95). Inserting the polar equation (8.93) in (8.98) we find

$$\Delta \varepsilon = \int_{-\infty}^{\infty} W \, dt = \frac{\mu \alpha^2}{6\pi L c^3} \left(\frac{e_1}{m_1} - \frac{e_2}{m_2} \right)^2 \frac{1}{a^2 (e^2 - 1)^2} \int_{-\varphi_\infty}^{\varphi_\infty} (\pm 1 + e \cos \varphi)^2 d\varphi$$

$$= \frac{\mu \alpha^2}{6\pi L c^3} \left(\frac{e_1}{m_1} - \frac{e_2}{m_2} \right)^2 \frac{1}{a^2 (e^2 - 1)^2} \left((2 + e^2) \varphi_\infty \pm 3 \sqrt{e^2 - 1} \right). \tag{8.101}$$

To express $\Delta \varepsilon$ in terms of the asymptotic speed v_∞ and of the impact parameter b, it is convenient to introduce the dimensionless parameter (see (8.91) and (8.94))

$$\gamma = \frac{\mp 1}{\sqrt{e^2 - 1}} = \frac{\alpha}{\mu v_\infty^2 b}, \tag{8.102}$$

and to rewrite the relation between φ_∞ and e in (8.95) in the form

$$\varphi_\infty = \frac{\pi}{2} - \arctan \gamma.$$

For an attractive (repulsive) Coulomb potential we have $\alpha < 0$ ($\alpha > 0$) and $\gamma < 0$ ($\gamma > 0$). Working out the algebra the radiated energy (8.101) then takes the form

$$\Delta\varepsilon = \frac{\mu^3 v_\infty^5}{6\pi\alpha c^3} \left(\frac{e_1}{m_1} - \frac{e_2}{m_2}\right)^2 \left[(3\gamma^2 + 1)\left(\frac{\pi}{2} - \arctan\gamma\right) - 3\gamma\right]\gamma^3. \quad (8.103)$$

For large impact parameters, corresponding to small values of γ, $\Delta\varepsilon$ goes rapidly to zero. In fact, in the limit for $\gamma \to 0$ the expression (8.103) reduces to

$$\Delta\varepsilon \approx \frac{\mu^3 v_\infty^5}{12\alpha c^3} \left(\frac{e_1}{m_1} - \frac{e_2}{m_2}\right)^2 \gamma^3 = \frac{\alpha^2}{12 v_\infty c^3} \left(\frac{e_1}{m_1} - \frac{e_2}{m_2}\right)^2 \frac{1}{b^3}, \quad (8.104)$$

i.e. $\Delta\varepsilon \sim 1/b^3$. Obviously, this behavior follows from the fact that for large impact parameters the particles perform nearly uniform linear motions. Consequently, their acceleration is small during the whole process, and so they emit only a small amount of bremsstrahlung.

Small impact parameters. As opposed to the previous case, the intensity of the emitted bremsstrahlung is expected to reach a maximum in head-on collisions, corresponding to the limits

$$b \to 0, \qquad \gamma \to \pm\infty.$$

In this case the emitted energy (8.103) entails two different behaviors, depending on whether the potential is attractive or repulsive. In the attractive case, $\alpha < 0$, the parameter γ tends to $-\infty$, and so both terms between square brackets in Eq. (8.103) tend to plus infinity. The radiated energy is thus infinite, in agreement with the fact that the acceleration diverges when the particles collide. However, in this case also the speeds of the particles tend to infinity, and so the (non-relativistic) dipole approximation breaks down. Vice versa, in the repulsive case we have $\alpha > 0$, and so the particles reach the finite minimum distance (Eq. (8.89) with $\varepsilon = \mu v_\infty^2/2$ and $v = 0$)

$$r_{\min} = \frac{2\alpha}{\mu v_\infty^2}.$$

Correspondingly, the total radiated energy $\Delta\varepsilon^*$ is finite. In fact, since $\alpha > 0$, as $b \to 0$ the parameter γ in (8.102) tends to $+\infty$. Performing carefully the limit of the expression (8.103) we find, indeed, the finite value

$$\Delta\varepsilon^* = \lim_{\gamma \to +\infty} \Delta\varepsilon = \frac{2\mu^3 v_\infty^5}{45\pi\alpha c^3} \left(\frac{e_1}{m_1} - \frac{e_2}{m_2}\right)^2. \quad (8.105)$$

To analyze the back-reaction of this energy loss on the dynamics of the system, we assume for simplicity that one of the two particles is much heavier than the other, $m_2 \gg m_1 = m$, and that their charges are equal, as happens, for instance, in the collision of a proton and a positron. In this case the process amounts to a light particle colliding with a heavy one, practically at rest, and we have $\mu \approx m$ and $\alpha = e^2/4\pi$. The radiated energy (8.105) then reduces to

$$\Delta\varepsilon^* \approx \frac{8mv_\infty^5}{45c^3}.$$

In the same limit the mechanical energy in (8.94) becomes $\varepsilon \approx mv_\infty^2/2$, so that the fractional energy decrease of the system – due to bremsstrahlung – amounts to

$$\frac{\Delta\varepsilon^*}{\varepsilon} \approx \frac{16}{45}\left(\frac{v_\infty}{c}\right)^3. \tag{8.106}$$

In the case at hand we are considering non-relativistic particles, $v_\infty \ll c$, and so $\Delta\varepsilon^*/\varepsilon \ll 1$. This means that in the non-relativistic limit the energy loss due to bremsstrahlung is completely negligible, even under the most *favorable* condition of a head-on collision. In Chap. 10, where we investigate the characteristics of bremsstrahlung emitted by ultrarelativistic particles, we will come to drastically different conclusions: when accelerated particles have speeds close to the speed of light, radiation effects can cause considerable energy losses, even for interactions of the Coulomb type.

8.4.4 Radiation of the Classical Hydrogen Atom

In this section we briefly illustrate the phenomenological picture that would arise for a hydrogen atom, if its dynamics would be governed by the laws of classical physics. We concentrate our analysis on the ground state, which classically is described by an electron circling at a radius r around the proton, with a constant velocity $v \ll c$. In this case we can thus apply our general results of Sect. 8.4.3, relative to elliptic orbits with vanishing eccentricity. Since the proton is much heavier than the electron we have $m_2 \gg m_1 = m$, and furthermore in this case $\alpha = -e^2/4\pi$ and $a = r$. For a circular motion we have in addition $mv^2/r = e^2/4\pi r^2$, and so the first relation in (8.89) allows to write the energy of the electron and its angular velocity in the form

$$\varepsilon = -\frac{e^2}{8\pi r} = -\frac{1}{2}mv^2, \qquad \omega = \frac{v}{r} = \sqrt{\frac{e^2}{4\pi mr^3}} = c\sqrt{\frac{r_0}{r^3}} = \frac{me^4}{(4\pi)^2\hbar^3}. \tag{8.107}$$

We have introduced the classical electron radius r_0 (8.81) and, to get the last expression, we have identified r with the Bohr radius r_B (8.82).

Emitted frequencies. The electron performs a periodic motion with period $T = 2\pi/\omega$, and its acceleration $\mathbf{a}(t)$ is a *trigonometric* periodic function of the form $\mathbf{a}(t) = a_0(\cos(\omega t), \sin(\omega t), 0)$. According to (8.59), its non-relativistic radiation fields in the wave zone $\mathbf{E}$ and $\mathbf{B}$ are then linear combinations of the functions $\cos(\omega t - kR)/R$ and $\sin(\omega t - kR)/R$, where R is the distance from the hydrogen atom, and $k = \omega/c$. This means that these fields are composed of a *single* monochromatic wave of frequency ω, and so the classical hydrogen atom would emit solely radiation of the fundamental frequency ω. As we will see in Chap. 12, a *relativistic* particle in uniform circular motion would, instead, emit radiation with frequencies $\omega_N = N\omega$, for all positive integers N. These predictions are, in any case, in sharp contrast with the quantum mechanical Rydberg formula, predicting the emission spectrum

$$\omega_{MN} = \left(\frac{1}{N^2} - \frac{1}{M^2}\right) \frac{me^4}{2(4\pi)^2\hbar^3} = \left(\frac{1}{N^2} - \frac{1}{M^2}\right) \frac{\omega}{2},$$

where N and M are positive integers.

Energy loss and collapse. We proceed now to the investigation of the physical implications of the emitted radiation, starting from the evaluation of the radiated energy. Setting in (8.99) the eccentricity to zero, and identifying a with r, we obtain for the emitted power the equivalent expressions

$$W = \left(\frac{e^2}{4\pi}\right)^2 \frac{e^2}{6\pi m^2 r^4 c^3} = \frac{e^2 c}{6\pi} \frac{r_0^2}{r^4} = \frac{e^2(\omega^2 r)^2}{6\pi c^3}, \tag{8.108}$$

in agreement with Larmor's formula (7.67). Since the total energy must be conserved, the mechanical energy of the atom (8.107) must decrease continuously according to the equation

$$\frac{d\varepsilon}{dt} = -W.$$

Furthermore, since ε is proportional to $-1/r$, the energy decrease implies also a decrease of the radius of the orbit. In fact, from Eqs. (8.107) and (8.108) we derive the fractional decrease

$$\frac{1}{r}\frac{dr}{dt} = -\frac{1}{\varepsilon}\frac{d\varepsilon}{dt} = \frac{W}{\varepsilon} = -\frac{4c}{3}\frac{r_0^2}{r^3} \approx -2 \cdot 10^{10}\, s^{-1}, \tag{8.109}$$

where we have inserted the numerical values (8.81) and (8.82). This means that in the very small time span of 10^{-10} s the radius of the orbit would reduce to about half of its value, and consequently the hydrogen atom would collapse in a fraction of a second. Notice, however, that in light of the relation $v^2 = e^2/4\pi mr$, as $r \to 0$ the speed of the electron would tend to infinity, a pathology which obviously is due to the non-relativistic nature of our analysis, which breaks down when the radius becomes to small. It is, nevertheless, interesting to compute the fractional decrease of the atom's energy during a cycle

$$\frac{\Delta\varepsilon}{\varepsilon} = \frac{TW}{\varepsilon} = \frac{2\pi W}{\omega\varepsilon} = \frac{8\pi}{3}\left(\frac{r_0}{r}\right)^{3/2} \approx 3 \cdot 10^{-6}, \tag{8.110}$$

which, actually, is a very small fraction. What, eventually, makes the classical hydrogen atom collapse in a very short time is the brevity of a cycle

$$T = \frac{2\pi r}{c} \sqrt{\frac{r}{r_0}} \approx 1.5 \cdot 10^{-16} \text{s}.$$

Finally, we observe that the classical velocity of the electron has the value

$$\frac{v}{c} = \omega r = \sqrt{\frac{r_0}{r}} \approx 0.7 \cdot 10^{-2},$$

so that a posteriori it was correct to address the problem relying on the non-relativistic dipole approximation.

We end this section with a caveat about the general validity of our analysis. From a quantitative point of view, the analysis performed above is in fact only valid until the radius of the orbit does not vary appreciably. Since the radius, actually, is not constant, the acceleration to be inserted in Larmor's formula varies with time in an unknown way, and so does the emitted power. This means that the equation of motion of the electron should be adapted to take into account the energy loss by means of Larmor's formula, which, in turn, involves the acceleration of the electron. So we realize that to address the problem of the classical hydrogen atom in a rigorous way, in principle we must regard the Maxwell and Lorentz equations as a system of *coupled* differential equations, an extremely difficult mathematical problem which, ultimately, can be dealt with only by numerical methods. Furthermore, as mentioned above, from a certain point onwards the dynamics can no longer be addressed in terms of a non-relativistic approximation. Despite these technical – and to a certain extent conceptual – difficulties it is, nevertheless, clear that the main qualitative conclusions of our analysis hold true.

8.5 Electric Quadrupole and Magnetic Dipole Radiations

When the electric dipole radiation is absent, meaning that the second derivative of the dipole moment vanishes, $\ddot{\mathbf{D}} = 0$, the dominant term of the expansion (8.42) of the potential in the wave zone is the one linear in $\mathbf{n} \cdot \mathbf{y}/c$. This term gives rise to the *electric quadrupole* and *magnetic dipole* radiations, and below we want to determine the energy content of these radiations. Thanks to the general formula (8.18), for this purpose it is again sufficient to determine the spatial components of the four-potential.

8.5.1   Vector Potential at Order $1/c^2$

Considering in the multiple expansion (8.42) also the term linear in $\mathbf{n} \cdot \mathbf{y}/c$, and understanding the functional dependence of the current $\mathbf{j}$ on the argument $(t - r/c, \mathbf{y})$, up to terms of order $1/c^2$ the spatial potential reads

$$
\begin{aligned}
A^i &= \frac{1}{4\pi rc} \int \left(j^i + \frac{1}{c} \left(n^k y^k \right) \partial_t j^i \right) d^3 y \\
&= \frac{1}{4\pi rc} \left(\dot{D}^i + \frac{1}{c} n^k \partial_t \int \left(\frac{1}{2} \left(y^k j^i - y^i j^k \right) + \frac{1}{2} \left(y^k j^i + y^i j^k \right) \right) d^3 y \right) \\
&= \frac{1}{4\pi rc} \left(\dot{D}^i - \frac{1}{c} \dot{M}^{ik} n^k + \frac{1}{2c} n^k \partial_t \int \left(y^k j^i + y^i j^k \right) d^3 y \right).
\end{aligned}
\tag{8.111}
$$

We have introduced the three-dimensional antisymmetric time-dependent tensor

$$
M^{ik} = \frac{1}{2} \int \left(y^i j^k - y^k j^i \right) d^3 y,
\tag{8.112}
$$

linked to the *magnetic dipole moment*

$$
\mathbf{M} = \frac{1}{2} \int \mathbf{y} \times \mathbf{j} \, d^3 y
\tag{8.113}
$$

by the relations

$$
M^i = \frac{1}{2} \varepsilon^{ijk} M^{jk}, \qquad M^{ij} = \varepsilon^{ijk} M^k.
\tag{8.114}
$$

For a system of charged point-particles the current density is given by

$$
\mathbf{j}(t, \mathbf{y}) = \sum_r e_r \, \mathbf{v}_r(t) \, \delta^3(\mathbf{y} - \mathbf{y}_r(t)),
$$

and in this case we find the simple expression

$$
\mathbf{M} = \frac{1}{2} \sum_r e_r \, \mathbf{y}_r \times \mathbf{v}_r.
\tag{8.115}
$$

To evaluate the last integral in (8.111) it is convenient to introduce the *electric quadrupole moment* D^{ij}, and its *reduced* trace-less version $\mathcal{D}^{ij}$, defined by

$$
D^{ij} = \int y^i y^j \rho \, d^3 y, \qquad \mathcal{D}^{ij} = D^{ij} - \frac{1}{3} \delta^{ij} D^{kk}, \qquad \mathcal{D}^{ii} = 0.
\tag{8.116}
$$

In fact, the integral in (8.111) can be expressed in terms of D^{ij} by means of the identity

$$\dot{D}^{ij} = \int \left(y^i j^j + y^j j^i \right) d^3 y, \tag{8.117}$$

which resembles the analogous identity (8.48) satisfied by the electric dipole moment. As in the case of the latter, the relation (8.117) can be proved through an integration by parts, using the continuity equation $\dot{\rho} = -\partial_k j^k$, together with the fact that $\mathbf{j}$ has compact support:

$$\dot{D}^{ij} = \int y^i y^j \dot{\rho} \, d^3 y = -\int y^i y^j \left(\partial_k j^k \right) d^3 y = \int \partial_k \left(y^i y^j \right) j^k d^3 y$$

$$= \int \left(\delta_k^i y^j + y^i \delta_k^j \right) j^k d^3 y = \int \left(y^i j^j + y^j j^i \right) d^3 y.$$

In this way we can recast the expansion (8.111) in the form

$$A^i = \frac{1}{4\pi r} \left(\frac{1}{c} \dot{D}^i + \frac{1}{2c^2} \left(\ddot{D}^{ij} - 2 \dot{M}^{ij} \right) n^j \right), \tag{8.118}$$

where it is understood that all multipole moments are evaluated at the retarded time $t - r/c$. Up to terms of order $1/c^2$, the potential in the wave zone $\mathbf{A}$ results, hence, in a superposition of an *electric dipole* term, an *electric quadrupole* term, and a *magnetic dipole* term, the last two being suppressed by a factor $1/c$ with respect to the first. Although not required for the determination of the emitted power, we also report the expansion up to terms of order $1/c^2$ of the time component of the potential (8.42). Recalling that $j^0 = c\rho$, we obtain the expansion (compare with formula (8.50))

$$A^0 = \frac{1}{4\pi r} \int \left(\rho + \frac{1}{c} \left(n^i y^i \right) \dot{\rho} + \frac{1}{2c^2} \left(n^i y^i \right) \left(n^j y^j \right) \ddot{\rho} \right) d^3 y$$

$$= \frac{1}{4\pi r} \left(Q + \frac{1}{c} \mathbf{n} \cdot \dot{\mathbf{D}} + \frac{1}{2c^2} n^i n^j \ddot{D}^{ij} \right), \tag{8.119}$$

where all multipole moments are evaluated at the time $t - r/c$. Notice that formulas (8.118) and (8.119) satisfy the wave relations (8.13).

8.5.2 Total Radiated Power

In the presence of the magnetic dipole and electric quadrupole radiations the angular distribution (8.18) of the emitted energy entails a rather complicated dependence on the direction $\mathbf{n}$ of the radiation, the reason being that $\mathbf{n}$ appears now not only in the projector $\Lambda^{ij} = \delta^{ij} - n^i n^j$, but also in the vector potential $\mathbf{A}$. Nonetheless, it is still possible to derive a rather simple expression for the total radiated power

$$\mathcal{W} = \frac{r^2}{c} \int \dot{A}^i \dot{A}^j \left(\delta^{ij} - n^i n^j \right) d\Omega. \tag{8.120}$$

Inserting the potential (8.118) we find

$$\mathcal{W} = \frac{1}{16\pi^2 c^3} \int \left(\ddot{D}^i + \frac{1}{2c} \left(\dddot{D}^{ik} - 2\ddot{M}^{ik} \right) n^k \right) \cdot$$
$$\left(\ddot{D}^j + \frac{1}{2c} \left(\dddot{D}^{jl} - 2\ddot{M}^{jl} \right) n^l \right) \left(\delta^{ij} - n^i n^j \right) d\Omega. \tag{8.121}$$

The integrals over the angles can be evaluated, as usual, resorting to the invariant integrals of Problem 2.6. These integrals produce products of the Kronecker symbols δ^{ij}, contracting the multipole moments. The terms of order $1/c^3$ give rise to the electric dipole power (8.53). The terms of order $1/c^4$ drop out, since they involve integrals of an *odd* number of $\mathbf{n}$'s. Concerning the terms of order $1/c^5$, the (mixed) contractions between the tensors $\dddot{D}^{ij}$ and $\ddot{M}^{kl}$ drop out, because the former is symmetric, whereas the latter is antisymmetric. Therefore, eventually in the integral (8.121) only the diagonal terms give non-vanishing contributions:

$$\mathcal{W} = \frac{\left| \ddot{\mathbf{D}} \right|^2}{6\pi c^3} + \frac{1}{64\pi^2 c^5} \left(\dddot{D}^{ik} \dddot{D}^{jl} + 4\ddot{M}^{ik} \ddot{M}^{jl} \right) \int \left(n^k n^l \delta^{ij} - n^k n^l n^i n^j \right) d\Omega$$
$$= \frac{\left| \ddot{\mathbf{D}} \right|^2}{6\pi c^3} + \frac{1}{64\pi^2 c^5} \left(\dddot{D}^{ik} \dddot{D}^{jl} + 4\ddot{M}^{ik} \ddot{M}^{jl} \right) \cdot$$
$$\left(\frac{4\pi}{3} \delta^{kl} \delta^{ij} - \frac{4\pi}{15} \left(\delta^{kl} \delta^{ij} + \delta^{ki} \delta^{lj} + \delta^{kj} \delta^{il} \right) \right).$$

The computation of the remaining contractions is facilitated by the fact that M^{ij} is antisymmetric, while D^{ij} is symmetric. Moreover, the relations (8.114) and (8.116) imply the identities

$$\ddot{M}^{ij} \ddot{M}^{ij} = 2 \left| \ddot{\mathbf{M}} \right|^2, \qquad \dddot{D}^{ij} \dddot{D}^{ij} - \frac{1}{3} \dddot{D}^{ii} \dddot{D}^{jj} = \dddot{\mathcal{D}}^{ij} \dddot{\mathcal{D}}^{ij}. \tag{8.122}$$

In this way, we obtain for the total radiated power the expression

$$\mathcal{W} = \frac{\left| \ddot{\mathbf{D}} \right|^2}{6\pi c^3} + \frac{\left| \ddot{\mathbf{M}} \right|^2}{6\pi c^5} + \frac{\dddot{\mathcal{D}}^{ij} \dddot{\mathcal{D}}^{ij}}{80\pi c^5}. \tag{8.123}$$

The appearance of the *reduced* quadrupole moment $\mathcal{D}^{ij}$ is a consequence of Birkhoff's theorem, as we will explain in a moment. As anticipated, the contributions to the emitted energy of the electric quadrupole and the magnetic dipole radiations are suppressed by a factor $1/c^2$ with respect to the electric dipole radiation. Furthermore, there are no corrections of order $1/c^4$, although contributions of this order are present in the angular distribution $d\mathcal{W}/d\Omega$ (8.18) of the radiated energy.

Sestupole radiation. We add an important comment concerning the correct use of formula (8.123). Returning to the non-relativistic expansion (8.42) of the potential in the wave zone, we write it schematically as

$$\mathbf{A} = \frac{1}{c}\,\mathbf{A}_1 + \frac{1}{c^2}\,\mathbf{A}_2 + \frac{1}{c^3}\,\mathbf{A}_3 + \cdots, \tag{8.124}$$

where with $\mathbf{A}_N$ we understand the $2N$-pole term, including also the correspondent magnetic contributions. Equation (8.118) represents the first two terms of this expansion. In particular, $\mathbf{A}_N$ contains $N - 1$ factors of $\mathbf{n}$. Inserting the expansion (8.124) in the total radiated power (8.120), we obtain for $\mathcal{W}$ a power series in $1/c$. However, since the integral over the angles of an odd number of $\mathbf{n}$'s is zero, only products of the type $\dot{A}_N^i \dot{A}_M^j$ with $M + N$ even survive the integration. Therefore, the total radiated power (8.120) is given by

$$\mathcal{W} = r^2 \int \left(\frac{1}{c^3}\,\dot{A}_1^i \dot{A}_1^j + \frac{1}{c^5}\left(\dot{A}_2^i \dot{A}_2^j + 2\dot{A}_1^i \dot{A}_3^j\right) + O\!\left(\frac{1}{c^7}\right) \right)\left(\delta^{ij} - n^i n^j\right) d\Omega.$$

Thus, we see that to correctly determine the radiated power up to terms of order $1/c^5$, to (8.123) we must add the terms coming from the product $\dot{A}_1^i \dot{A}_3^j$, involving the *sestupole* radiation $\mathbf{A}_3$. However, in the particular case in which the electric dipole radiation is absent,

$$\dot{\mathbf{A}}_1 = 0 \quad \Leftrightarrow \quad \ddot{\mathbf{D}} = 0,$$

the product $\dot{A}_1^i \dot{A}_3^j$ vanishes, and so Eq. (8.123) gives, indeed, the correct power up to terms of order $1/c^5$. Only in this case the formula (8.123) is, hence, of practical utility; otherwise, in general one must take into account also the sestupole radiation.

Vanishing of the magnetic dipole radiation. Let us now investigate some important cases in which the contributions of order $1/c^5$ in (8.123) vanish. Thanks to angular momentum conservation in an isolated system, the magnetic dipole radiation vanishes for a system of isolated charges, for which the ratio $e_r/m_r = \gamma$ is the same for all particles. In fact, inserting the relation $e_r = \gamma m_r$ in the definition (8.115) of the magnetic dipole moment, we obtain

$$\mathbf{M} = \frac{\gamma}{2} \sum_r \mathbf{y}_r \times m_r \mathbf{v}_r = \frac{\gamma}{2}\,\mathbf{L},$$

where $\mathbf{L}$ is the conserved total angular momentum of the system. It follows that $\dot{\mathbf{M}} = 0$. Similarly, the magnetic dipole radiation is absent in a generic isolated two-particle system, if one chooses as origin its center of mass. In fact, in this case we have $m_1 \mathbf{r}_1 + m_2 \mathbf{r}_2 = 0$, and $\mathbf{p}_1 = -\mathbf{p}_2$, and so the magnetic dipole moment (8.115) can be recast in the form

$$\mathbf{M} = \frac{1}{2}\left(e_1\,\mathbf{r}_1 \times \mathbf{v}_1 + e_2\,\mathbf{r}_2 \times \mathbf{y}_2\right) = \frac{1}{2}\left(\frac{e_1}{m_1^2} + \frac{e_2}{m_2^2}\right)\frac{m_1 m_2}{m_1 + m_2}\,\mathbf{L},$$

where $\mathbf{L} = \mathbf{r}_1 \times \mathbf{p}_1 + \mathbf{r}_2 \times \mathbf{p}_2$ is the total angular momentum. It follows again that $\dot{\mathbf{M}}$ is zero.

Birkhoff's theorem and spherically symmetric charge distributions. According to Birkhoff's theorem, a spherically symmetric charge distribution in empty space generates a *static* electromagnetic field, see Problem 2.5. Therefore, a system of this kind cannot emit radiation and, consequently, the radiated power $\mathcal{W}$ and all multipole moments appearing in (8.123) must vanish. To verify these predictions of the theorem we recall that for an isotropic system the four-current $j^\mu(t, \mathbf{y})$ assumes the general form (2.168), i.e. $\rho = \rho(t, y), \mathbf{j} = \mathbf{y}j(t, y)/y$. In Eq. (8.61) we have already shown that in this case the electric dipole moment $\mathbf{D}$ is zero. Similarly, the magnetic dipole moment $\mathbf{M}$ (8.113) vanishes, because $\mathbf{j}$ is parallel to $\mathbf{y}$. To verify that also the reduced electric quadrupole moment (8.116) vanishes, we introduce polar coordinates, finding

$$\mathcal{D}^{ij} = \int\left(y^i y^j - \frac{\delta^{ij}}{3}\,y^2\right)\rho\,d^3 y = \left(\int_0^\infty y^4 \rho\,dy\right)\int\left(n^i n^j - \frac{\delta^{ij}}{3}\right)d\Omega = 0,$$
(8.125)

where the conclusion follows from the vanishing of the last integral. In this regard we emphasize that, although an isotropic system does not emit radiation, and its electromagnetic field in empty space is a *static* Coulomb field, the four-potential in the wave zone of Eqs. (8.118) and (8.119) is by no means static. In fact, since M^{ij} and D^i vanish, and ρ depends only on t and y, it is immediately seem that in this case the expansions (8.118) and (8.119) reduce to the non-static expressions

$$A^0 = \frac{1}{4\pi r}\left(Q + \ddot{f}\left(t - \frac{r}{c}\right)\right), \qquad \mathbf{A} = \frac{\mathbf{n}}{4\pi r}\,\ddot{f}\left(t - \frac{r}{c}\right),$$
(8.126)

where the function f is defined by

$$f(t) = \frac{2\pi}{3c^2}\int_0^\infty y^4 \rho(t, y)\,dy.$$

Equations (8.126) are still in agreement with the vanishing of the radiated power (8.18) in all directions, since $\mathbf{A}$ is parallel to $\mathbf{n}$. On the other hand, since $F^{\mu\nu}$ is a static Coulomb field, exploiting gauge invariance it should be possible to find also a static four-potential. It is, indeed, not difficult to recognize that the gauge transformation that transform (8.126) in a static four-potential, is the one corresponding to the gauge function

$$\Lambda(x) = -\frac{c}{4\pi r}\,\dot{f}\left(t - \frac{r}{c}\right).$$
(8.127)

In fact, remembering that $\partial^0 = \frac{1}{c} \partial/\partial t$ and that $\partial^i = -\partial/\partial x^i$, from (8.126), modulo terms of order $1/r^2$ we find (recall that A^μ is a potential in the wave-zone)

$$A'^\mu = A^\mu + \partial^\mu \Lambda = \left(\frac{Q}{4\pi r}, 0, 0, 0 \right). \tag{8.128}$$

Furthermore, from Problem 5.7 we know that a function of the type $g(t - r/c)/r$ in the region $r \neq 0$ obeys the wave equation. Therefore, the gauge function (8.127) satisfies the equation $\Box \Lambda = 0$, and so (8.128) represents a *residual* gauge transformation, preserving thus the Lorenz gauge. In fact, the transformed potential (8.128) still obeys $\partial_\mu A'^\mu = 0$.

8.6 Problems

8.1 Synchrotron radiation in the non-relativistic limit. The radiation created by a charged particle in a uniform circular motion is generically called *synchrotron radiation*. Consider a particle of charge e and mass m, which in the presence of a constant uniform magnetic field B parallel to the z axis performs a circular motion of radius R and of angular velocity $\omega = eB/mc$, the non-relativistic *cyclotron frequency*. Suppose that the motion is non-relativistic, i.e. $v = \omega R \ll c$.

(a) Determine the electric field $\mathbf{E}(t, \mathbf{x})$ generated by the particle in the wave zone.
   Hint: See Eqs. (8.59).
(b) For every fixed time t, determine the directions $\mathbf{n}$ in which the intensity of the emitted radiation has maxima and minima.
   Hint: See Eq. (8.60).
(c) Show that the time average of the angular distribution of the radiated power is given by

$$\frac{d\overline{W}}{d\Omega} = \frac{e^2 \omega^4 R^2}{32\pi^2 c^3} (1 + \cos^2 \vartheta), \tag{8.129}$$

   where ϑ is the angle between the z axis and the emission direction $\mathbf{n}$.
   Hint: Apply Eq. (8.60) writing $|\mathbf{n} \times \mathbf{a}|^2 = a^2 - (\mathbf{n} \cdot \mathbf{a})^2$, and use the time averages

$$\overline{a_x^2} = \overline{a_y^2} = \frac{a^2}{2}, \qquad \overline{a_x a_y} = 0.$$

(d) Suppose that the particle is constrained to move without friction on a ring of radius R. Taking into account the energy loss via radiation, determine the decreasing particle's speed $v(t)$ as a function of time. Assume that

$$|\dot{v}| \ll \frac{v^2}{R},$$

so that in Larmor's formula the tangential acceleration can be neglected. Verify the validity of this hypothesis a posteriori.

Hint: Equate the time derivative of the kinetic energy $mv^2/2$ to minus the emitted power, as given by Larmor's formula.

8.2 Consider a light charged particle performing a uniform circular motion around a heavy one, under the same hypotheses as in Sect. 8.4.4.

(a) Determine the speed $v(t)$ and the period $T(t)$ of the light particle as functions of time.
   Hint: See Eq. (8.109).
(b) Discuss the validity of the result.

8.3 *Spherically symmetric charge distribution*. Consider a spherically symmetric charge distribution, associated with a four-current j^μ of the form (2.168)

$$j^0(t, \mathbf{y}) = \rho(t, y), \qquad j^i(t, \mathbf{y}) = \frac{y^i}{y}\, j(t, y), \qquad y = |\mathbf{y}|. \tag{8.130}$$

Verify explicitly that such a charge distribution does not emit radiation, as predicted by Birkhoff's theorem, by showing that the radiated power (8.18) vanishes for all $\mathbf{n}$.

Solution: It is sufficient to show that the spatial components of the potential in the wave zone (8.10) are of the form

$$\mathbf{A}(t, \mathbf{x}) = g(t, r)\, \mathbf{n} \tag{8.131}$$

for some function g, where $r = |\mathbf{x}|$ and $\mathbf{n} = \mathbf{x}/r$. For this purpose we recall a theorem on three-dimensional vector functions.

Theorem. Let $f(\mathbf{x}, \mathbf{y})$ be a rotation invariant function of two three-dimensional variables, that is to say, a function $f : \mathbb{R}^3 \times \mathbb{R}^3 \to \mathbb{R}$ satisfying the equality

$$f(R\mathbf{x}, R\mathbf{y}) = f(\mathbf{x}, \mathbf{y}), \tag{8.132}$$

for all matrices $R \in SO(3)$. Then the vector function $\mathbf{F} : \mathbb{R}^3 \to \mathbb{R}^3$ defined by

$$\mathbf{F}(\mathbf{x}) = \int \mathbf{y} f(\mathbf{x}, \mathbf{y})\, d^3y \tag{8.133}$$

is of the form

$$\mathbf{F}(\mathbf{x}) = g(r)\, \mathbf{n}, \tag{8.134}$$

for some scalar function g. Notice that the condition (8.132) is equivalent to the assumption that f depends on $\mathbf{x}$ and $\mathbf{y}$ only through the invariants $|\mathbf{x}|$, $|\mathbf{y}|$ and $\mathbf{x} \cdot \mathbf{y}$.

Proof Performing in the integral (8.133) the change of variables $\mathbf{y} \to R\mathbf{y}$, and replacing $\mathbf{x}$ with $R\mathbf{x}$, thanks to (8.132) we deduce that $\mathbf{F}$ is a *rotation covariant* function, that is to say,

$$\mathbf{F}(R\mathbf{x}) = R\mathbf{F}(\mathbf{x}), \quad \forall R \in SO(3).$$

$\mathbf{F}(\mathbf{x})$ is then necessarily of the form (8.134). □

If the current is of the form (8.130), the integral for $\mathbf{A}(t, \mathbf{x})$ in (8.10) is for every fixed t of the form (8.133), with $f(\mathbf{x}, \mathbf{y})$ a function satisfying (8.132). The above theorem then ensures that the spatial potential has the functional dependence (8.131). Notice that in general the function $g(t, r)$ is non-vanishing, as well as time-dependent, see, for instance, Eqs. (8.126). However, if the four-current is of the form (8.130), the structure of the integral (8.10) implies that g is of the particular form $g(t, r) = f(t - r)/r$, and in this case there exists a residual gauge transformation which eliminates $\mathbf{A}(t, \mathbf{x})$, see Eqs. (8.126)–(8.128).

8.4 Using the covariant expression (7.38) of the Liénard-Wiechert field, verify that the energy-momentum tensor of the electromagnetic field in the wave zone, generated by a particle, reduces to the form

$$T_{\text{em}}^{\mu\nu} = n^\mu n^\nu E^2,$$

in agreement with the general formula (8.16).

Solution: In the wave zone the field (7.38) reduces to the acceleration field (7.42), that we rewrite in the form (remember the notation (7.32))

$$F^{\mu\nu} \to F_a^{\mu\nu} = \frac{e}{4\pi(um)^3 R}(m^\mu \Delta^\nu - m^\nu \Delta^\mu), \quad \Delta^\mu = (um)w^\mu - (wm)u^\mu.$$

With the aid of the relations

$$m_\mu m^\mu = 0 = m_\mu \Delta^\mu, \quad \Delta^2 = (um)^2 w^2 + (wm)^2,$$

we then find

$$T_{\text{em}}^{\mu\nu} = F^{\mu\rho}F_\rho{}^\nu + \frac{1}{4}\eta^{\mu\nu}F^{\rho\sigma}F_{\rho\sigma} = -\frac{e^2\Delta^2}{16\pi^2(um)^6 R^2}m^\mu m^\nu. \tag{8.135}$$

Moreover, in the wave zone one has the identifications $R \to r$ and $m^\mu \to n^\mu$. Finally, a simple algebraic calculation shows that the coefficient of $m^\mu m^\nu$ in (8.135) equals indeed E_a^2, see (7.59).

8.5 *Non-relativistic bremsstrahlung in a Coulomb field*. A non-relativistic electron incoming from infinity grazes a nucleus of charge Ze remaining always at a large distance, so that its trajectory $\mathbf{y}(t)$ deviates only little from a uniform linear motion. Consider the nucleus at rest during the whole process. Denoting the asymptotic velocity of the electron by $\mathbf{v}$, where $v \ll c$, and its impact parameter by b, its distance from the nucleus as a function of time has then the approximate form

$$r(t) = |\mathbf{y}(t)| \simeq \sqrt{b^2 + v^2 t^2}. \tag{8.136}$$

(a) Considering that the acceleration of the electron is given by

$$\mathbf{a} = -\frac{Ze^2}{4\pi m} \frac{\mathbf{y}}{r^3},$$

show that during the scattering process it radiates the energy

$$\Delta\varepsilon(v, b) = \frac{e^6 Z^2}{192\pi^2 m^2 v c^3 b^3}. \tag{8.137}$$

Compare this result with the exact expression (8.103).

(b) Suppose now that a beam of electrons with speed v hits the nucleus of charge Ze. Show that the *radiation cross section* $\chi(v)$, defined as the ratio between the radiated power $\mathcal{W}_{\mathrm{rad}}$ and the incident electron flux j, is related to $\Delta\varepsilon(v, b)$ by the general formula

$$\chi(v) = \int_0^\infty \Delta\varepsilon(v, b)\, 2\pi b\, db, \qquad \mathcal{W}_{\mathrm{rad}} = \chi(v)\, j. \tag{8.138}$$

Hint: The *incident flux* denotes the number of incoming particles crossing a unit area per unit time. Notice that $\chi(v)$ has the dimensions of energy times area.

(c) In the case at hand the integral (8.138) diverges for $b \to 0$. On the other hand, one must keep in mind that the above computation of $\Delta\varepsilon(v, b)$ is valid only for large b, and that, moreover, at small distances quantum effects can no longer be ignored. Actually, quantum mechanics provides a natural *cut-off* for b in terms of the uncertainty principle, suggesting to estimate the minimum distance D between the electrons and the nucleus via $D \cdot mv \sim \hbar$, i.e. $D \sim \hbar/mv$. We can thus provide an estimate for the radiation cross section, by replacing the lower limit of integration in (8.138) with $b \sim D$. Inserting (8.137), in this way we obtain

$$\chi(v) \sim \frac{e^6 Z^2}{96\pi m^2 v c^3} \int_{\hbar/mv}^\infty \frac{db}{b^2} = \frac{e^6 Z^2}{96\pi m \hbar c^3}. \tag{8.139}$$

This estimate reproduces, indeed, the correct order of magnitude of the radiation cross section as computed in quantum mechanics. However, it can be seen that for a non-relativistic electron beam, as the one at hand, colliding, for instance, with a solid, the radiative energy loss (8.139) is suppressed by a factor $(v/c)^2$ with respect to the *collisional* energy loss, see [1] for details. This means that in this case, like in the repulsive Coulomb interaction, see Eq. (8.106), the radiation phenomenon entails appreciable physical implications only in the ultrarel-ativistic limit, i.e. if the incident electrons have speeds of the order of c.

8.6 Consider a non-relativistic charged particle in uniform circular motion with trajectory $\mathbf{y}(t) = (R\cos(\omega t), R\sin(\omega t), 0)$, as in Problem 8.1.

(a) Show that the magnetic dipole radiation is absent.
   Hint: According to Eq. (8.115), the derivative of the magnetic dipole moment of a system of particles is equal to

$$\dot{\mathbf{M}} = \frac{1}{2} \sum_r e_r \, \mathbf{y}_r \times \mathbf{a}_r.$$

(b) Determine the emitted power of the electric quadrupole radiation and compare it with the corresponding power of the electric dipole radiation.
   Hint: Use Eq. (8.123), and notice that Eqs. (8.55) and (8.116) yield for the electric quadrupole moment of a system of particles the expression

$$D^{ij}(t) = \sum_r e_r \, y_r^i(t) \, y_r^j(t). \tag{8.140}$$

(c) Show that the electric quadrupole radiation has frequency 2ω.
   Hint: Analyze the dependence on the variable $\omega(t - r/c) = \omega t - kr$ of the second term in (8.118).

8.7 Investigate the collision between two *identical* non-relativistic charged particles, subject to the mutual Coulomb interaction, in their center of mass frame. Provide an estimate for the instantaneous radiated power $\mathcal{W}_{\mathrm{id}}$, and compare it with the instantaneous power $\mathcal{W}_{\mathrm{opp}}$ radiated in the collision between two non-relativistic particles of the same mass, but with opposite charges. Suppose that the asymptotic speeds and the impact parameters in the two collisions are the same.

Hint: Consider Eq. (8.123) and remember that a system of identical particles does not emit electric dipole radiation. In the center of mass frame one has $\mathbf{y}_1 = -\mathbf{y}_2 = \mathbf{y}$, and so Eqs. (8.56) and (8.140) imply the qualitative behaviors

$$\dddot{D}^i \sim ea, \qquad \dddot{D}^{ij} \sim e\,(va + y\dot{a}).$$

Moreover, for a Coulomb force the acceleration behaves as $a \sim 1/y^2$, and consequently $\dot{a} \sim v/y^3 \sim av/y$. Equation (8.123) then yields the relation, of general validity, $\mathcal{W}_{\mathrm{id}} \sim (v/c)^2\, \mathcal{W}_{\mathrm{opp}}$.

8.8 *Large distance expansion of the Liénard-Wiechert potential*. Consider a charged particle following a generic trajectory $\mathbf{y}(t)$, with four-current

$$j^0(t, \mathbf{x}) = e\,\delta^3(\mathbf{x} - \mathbf{y}(t)), \qquad \mathbf{j}(t, \mathbf{x}) = e\,\mathbf{v}(t)\,\delta^3(\mathbf{x} - \mathbf{y}(t)). \tag{8.141}$$

(a) Show that in this case the potential in the wave zone (8.10) reduces to

$$A^\mu = \frac{e}{4\pi r} \cdot \frac{(1, \mathbf{v}(t')/c)}{1 - \mathbf{n} \cdot \mathbf{v}(t')/c}, \tag{8.142}$$

where t' is determined by the implicit equation $t' = t - (r - \mathbf{n} \cdot \mathbf{y}(t'))/c$. Compare the expression (8.142) with the exact Liénard-Wiechert potential (7.20).
Hint: Insert the sources (8.141) in (8.10), and integrate over the δ-function resorting to the identity (2.83). If $\mathbf{a}$ and $\mathbf{b}$ are two arbitrary vectors, one has the determinant identity $\det(\delta^{ij} + a^i b^j) = 1 + \mathbf{a} \cdot \mathbf{b}$.

(b) Determine the electric field associated with the potential (8.142), and compare the result with the field (7.59).
Hint: Since the potential (8.142) is valid only at first order in in $1/r$, for consistency in the field the terms of order $1/r^2$ must be neglected.

(c) Derive the non-relativistic expansions of the four-potential (8.142)

$$A^0 = \frac{e}{4\pi r}\left(1 + \frac{1}{c}\mathbf{n} \cdot \mathbf{v} + \frac{1}{c^2}\left((\mathbf{n} \cdot \mathbf{v})^2 + (\mathbf{n} \cdot \mathbf{y})(\mathbf{n} \cdot \mathbf{a})\right)\right) + O\left(\frac{1}{c^3}\right), \tag{8.143}$$

$$\mathbf{A} = \frac{e\mathbf{v}}{4\pi rc} + O\left(\frac{1}{c^2}\right), \tag{8.144}$$

where the kinematical quantities $\mathbf{y}$, $\mathbf{v}$ and $\mathbf{a}$ are evaluated at the time $t - r/c$.

(d) Compute the electric field associated with the potentials (8.143), (8.144), and compare the result with the non-relativistic asymptotic electric field (7.65).

8.9 Show that for the aperiodic current (8.3) the potential in the wave zone (8.10) can be recast in the form

$$A^\mu(t, \mathbf{x}) = \frac{1}{2(2\pi)^{3/2}r} \int e^{i\omega(t-r)} \left(\int e^{i\omega \mathbf{n} \cdot \mathbf{y}} j^\mu(\omega, \mathbf{y})\, d^3y\right) d\omega, \tag{8.145}$$

and interpret the result in terms of elementary waves.

Reference

1. J.D. Jackson, *Classical Electrodynamics* (Wiley, New York, 1998)

Chapter 9
Gravitational Radiation

One of the purposes of this chapter is a close comparison of the basic properties of the electromagnetic and gravitational radiations. For concreteness, we will consider both radiations mainly in the non-relativistic limit, i.e. we shall assume that they are generated by objects moving with speeds much smaller than the speed of light. In this case the multipole approximation yields sensible, analytic, results, and so we will have at our disposal sufficiently explicit formulas, allowing for a detailed qualitative as well as quantitative comparison. For obvious reasons we report a few predictions of General Relativity without derivations, providing, however, where possible, heuristic arguments.

Although the existence of gravitational waves has been confirmed only recently via a *direct* detection by the LIGO observatories [1], in the past there was little doubt that any accelerated body emits such waves, not least because Einstein's equations predict not only their existence, but also their precise generation mechanism. Furthermore, already before this fundamental experimental breakthrough, there were incontestable *indirect* signals of the existence of these waves, the first one coming from the binary pulsar PSR B1913+16, discovered in 1974 by R.A. Hulse and J.H. Taylor [2], which earned its discoverers the 1993 Nobel Prize for the first indirect detection of gravitational waves, in perfect quantitative agreement with the theoretical predictions of General Relativity. Rather recently, the observations made on the double pulsar PSR J0737-3039A/B, discovered in 2003 [3], provided a second compelling indirect proof of the reality of gravitational radiation, so confirming General Relativity as a fundamental paradigm of nature.

9.1 Gravitational Versus Electromagnetic Waves

The electromagnetic and gravitational radiations share several fundamental properties, but they entail also basic distinctions. Below we collect the principal analogies and differences between the two, thereby anticipating some of the main conclusions of this chapter.

© Springer International Publishing AG, part of Springer Nature 2018
K. Lechner, *Classical Electrodynamics*, UNITEXT for Physics,
https://doi.org/10.1007/978-3-319-91809-9_9

- Whereas electromagnetic waves are *exact* solutions of Maxwell's equations, gravitational waves satisfy Einstein's equations only in the *weak-field* limit. This approximation is more than sensible for gravitational waves traveling across the Universe, since their amplitude decreases as the inverse of the distance.
- Both types of waves are transverse and possess two physical polarization states, and they propagate at the speed of light carrying energy and momentum.
- Whereas the electromagnetic waves have helicity ± 1, the gravitational ones have helicity ± 2. Accordingly, contrary to photons which have spin $\pm \hbar$, gravitons – if they exist – have spin $\pm 2\hbar$.
- In the same way as the source of the electromagnetic field is the four-current j^μ of the charge distribution, the source of the gravitational field is the energy-momentum tensor $T^{\mu\nu}$ of the matter system. Correspondingly, just like a generic accelerated charge emits electromagnetic radiation, a generic accelerated body emits gravitational radiation.
- According to Birkhoff's theorem, which holds in General Relativity as well as in electrodynamics, a spherically symmetric system does emit neither electromagnetic nor gravitational waves.
- There exists no gravitational dipole radiation, neither of the electric nor of the magnetic type, and consequently in the non-relativistic limit this radiation is dominated by the *quadrupole term*. Correspondingly, the intensity of the gravitational radiation is suppressed with respect to the electromagnetic one by a relativistic factor $(v/c)^2$, so that the former is in general substantially weaker than the latter. This strong suppression is at the origin of the extreme difficulties involved in the direct detection of gravitational waves.
- The fundamental local symmetry of General Relativity, with quit the same role as *gauge invariance* in electrodynamics, is represented by the *diffeomorphisms* (3.92), which we can write equivalently in the form

$$x^\mu \to x'^\mu(x) = x^\mu - \Lambda^\mu(x). \tag{9.1}$$

This symmetry group depends on the four local arbitrary *gauge parameters* $\Lambda^\mu(x)$, and it can be seen that in General Relativity it implies the local conservation of *four-momentum*, in the same way as in electrodynamics gauge invariance implies the local conservation of the *charge*, see Sect. 4.4. However, from an algebraic formal point of view there is a basic difference between gauge transformations and diffeomorphisms: the former form an abelian symmetry group, while the latter form a non-abelian one. In this sense General Relativity is closer to the *non-abelian* gauge theories describing the weak and strong interactions. Finally, on one hand diffeomorphisms are linked in a geometric way to special relativity in that, in a sense, they generalize the Poincaré transformations $x'^\mu = \Lambda^\mu{}_\nu x^\nu + a^\mu$, and on the other hand they constitute the mathematical device realizing the phenomenological cornerstone of General Relativity, i.e. the *equivalence principle*. Both these properties are peculiar to the symmetry group of the gravitational interaction, having no counterpart in the symmetries of the other fundamental interactions.

9.2 The Equations of the Weak Gravitational Field

In this section we present a heuristic derivation of the equations of motion governing the dynamics of a low-intensity gravitational field. On one hand we will rely on covariance arguments, and on the other hand we will take advantage from the close analogy existing between electromagnetism and gravity in the non-relativistic regime. The field equations (9.14) we shall derive in this way coincide with Einstein's equations in the *weak-field* limit, subject to the *harmonic* gauge-fixing of diffeomorphisms, see below. The structure of these equations parallels closely the structure of Maxwell's equations in Lorenz gauge, so that, relying on our experience with the latter, we will have no difficulty in solving the former.

We start from the simple observation that at the non-relativistic level the gravitational and electromagnetic interactions have, indeed, an identical structure. In fact, the corresponding quasi-static forces between two bodies of charges e_1 and e_2 and masses m_1 and m_2 are given by

$$\mathbf{F}_{\text{em}} = \frac{e_1 e_2}{4\pi r^3}\,\mathbf{r}, \tag{9.2}$$

$$\mathbf{F}_{\text{gr}} = -G\frac{m_1 m_2}{r^3}\,\mathbf{r}, \tag{9.3}$$

where G is Newton's constant. We recall that the minus sign in Eq. (9.3) signals that the gravitational force between masses is attractive, whereas the electrostatic force between charges of the same sign is repulsive. Correspondingly, the electric and gravitational scalar potentials, which in the one-body case are given by

$$\varphi_{\text{em}} = \frac{e}{4\pi r}, \qquad \varphi_{\text{gr}} = -\frac{Gm}{r}, \tag{9.4}$$

in general satisfy the Poisson equations

$$-\nabla^2 \varphi_{\text{em}} = \rho_{\text{e}}, \tag{9.5}$$

$$\nabla^2 \varphi_{\text{gr}} = 4\pi G \rho_{\text{m}}, \tag{9.6}$$

where ρ_{e} is the charge density and ρ_{m} the mass density. Equations (9.5) and (9.6) violate Lorentz invariance, but at least in the case of electromagnetism we know how to modify equation (9.5) to restore it. In the first place, we must covariantize the Laplacian operator $-\nabla^2$ replacing it with the d'Alembertian operator $\Box = -\nabla^2 + \partial_0^2$, and so we obtain

$$\Box \varphi_{\text{em}} = \rho_{\text{e}}. \tag{9.7}$$

In the second place, we must find out the four-dimensional tensors which combine the non-relativistic physical quantities involved in Eq. (9.7). In this regard we recall that the charge density is the time component of the four-current (2.102), i.e. $\rho_{\text{e}} = j^0/c$. Correspondingly, the electrostatic potential must be identified with the time component of a certain four-vector A^μ, i.e. $\varphi_{em} = A^0$. Enforcing Lorentz invariance, (9.7) leads us then to postulate the equations

$$\Box A^\mu = \frac{1}{c} j^\mu. \tag{9.8}$$

Finally, local charge conservation, encoded by the continuity equation $\partial_\mu j^\mu = 0$, forces the vector potential to satisfy the constraint

$$\partial_\mu A^\mu = 0. \tag{9.9}$$

In this way, we have actually retrieved Maxwell's equations in Lorenz gauge (6.1) and (6.2).

Let us now follow a similar procedure to derive from the Poisson equation (9.6) a relativistic equation for the gravitational field. Again we start by replacing Eq. (9.6) with

$$\Box \varphi_{gr} = -4\pi G \rho_m. \tag{9.10}$$

To find out the tensor associated with the potential φ_{gr} we must, in turn, find a tensor which has as one of its components the mass density. In special relativity the mass is a form of energy and we must hence expect that in a relativistic theory of gravitation the gravitational field is generated not by the mass, but by the *energy* of the system. This hypothesis is reinforced by the observation that photons are deflected when passing trough a gravitational field. In fact, according to Newton's principle of *action and reaction*, also photons must then create a gravitational field. But since photons, although possessing energy, have no mass, we conclude that ultimately it is the energy that creates the gravitational field. This leads us to replace in Eq. (9.10) the mass density with the energy density, which is nothing else than the 00 component of the energy-momentum tensor

$$\rho_m \rightarrow \frac{1}{c^2} T^{00}. \tag{9.11}$$

Notice that for a system of non-relativistic particles T^{00} reduces indeed to the mass density multiplied by c^2, see Eq. (2.139). With the replacement (9.11), if Eq. (9.10) has to become Lorentz invariant, we are forced to consider φ_{gr} as the 00 component of a two-index tensor $H^{\mu\nu}$, the "gravitational potential". Conventionally we set

$$\varphi_{gr} = \frac{1}{4} H^{00}, \tag{9.12}$$

so that $H^{\mu\nu}$ has the same dimensions of φ_{gr}, namely those of a velocity squared. Equation (9.10) translates now into

$$\Box H^{00} = -\frac{16\pi G}{c^2} T^{00}, \tag{9.13}$$

which admits a natural covariantization in terms of the ten independent equations

$$\Box H^{\mu\nu} = -\frac{16\pi G}{c^2} T^{\mu\nu}. \tag{9.14}$$

As shown in Sect. 3.4, a generic relativistic system whose dynamics follows from an action principle allows for a *symmetric* energy-momentum tensor. If we choose for $T^{\mu\nu}$ this "standard" tensor, then Eq. (9.14) implies that also the gravitational potential $H^{\mu\nu}$ must be *symmetric*. Finally, the energy-momentum conservation law $\partial_\mu T^{\mu\nu} = 0$, analogous to the continuity equation for the electric charge $\partial_\mu j^\mu = 0$, imposes on $H^{\mu\nu}$ the further constraint

$$\partial_\mu H^{\mu\nu} = 0. \tag{9.15}$$

Condition (9.15) can be seen to correspond to the *harmonic* gauge-fixing of the diffeomorphism invariance (9.1), see Sect. 9.3, in the same way as the Lorenz-condition (9.9) fixes the gauge invariance in electrodynamics. Comparing Eqs. (9.8) and (9.14) we conclude, hence, that the source of the gravitational field is the energy-momentum tensor, in the same way as the four-current is the source of the electro-magnetic field. Apart from this basic *physical* difference, the *formal* structure of the system of differential equations (9.14), (9.15) is identical to that of Maxwell's equations (9.8), (9.9).

9.2.1   Weak-Field Limit and Einstein's Equations

The gravitational field equations (9.14), (9.15) represent a *minimal* covariantization of the non-relativistic equation (9.6), in the sense that they realize Lorentz invariance in the simplest way. It can, actually, be seen that, once one imposes on Einstein's equations the harmonic gauge (9.15), they reduce to (9.14) only in the weak-field limit, that is to say, if the gravitational potential satisfies the *weak-field conditions*

$$|H^{\mu\nu}| \ll c^2, \quad \text{for all } \mu, \nu. \tag{9.16}$$

Notice that these conditions are dimensionally consistent in that $H^{\mu\nu}$ has the dimensions of a velocity squared. The dimensionless field $H^{\mu\nu}/c^2$ is sometimes called *gravitational strain*, because it provides a measure for the relative gravitational stretches $\Delta l/l$ of the length and time scales, see Eqs. (9.20) and (9.21) below.

Non-linearity of Einstein's equations. A simple physical argument, based again on the non-relativistic analogy between gravity and electromagnetism, allows us to infer that Eq. (9.14) can describe the dynamics of the gravitational field only in an approximate manner. It starts from the observation that, according to (9.14), all "matter" would indeed create a gravitational field, by means of the associated energy-momentum tensor $T^{\mu\nu}$. For instance, if the matter system is a system of charged particles, in which case $T^{\mu\nu} = T^{\mu\nu}_{\text{em}} + T^{\mu\nu}_{\text{p}}$, the particles as well as the electromagnetic field would create a gravitational field. Nonetheless, Eq. (9.14) misses a fundamental physical effect. In fact, in the same way as the electric field $\mathbf{E} = -\boldsymbol{\nabla}\varphi_{\text{em}}$ entails an energy density proportional to $E^2 = (\partial\varphi_{\text{em}})^2$, the gravitational field $\boldsymbol{\mathcal{G}} = -\boldsymbol{\nabla}\varphi_{\text{gr}}$ entails an energy density proportional to $\mathcal{G}^2 = (\partial\varphi_{\text{gr}})^2$. However, while the former is indeed present in the second member of equation (9.14) – in the term $T^{\mu\nu}_{\text{em}}$ – there is no contribution proportional to $(\partial\varphi_{\text{gr}})^2$. Therefore, to describe the dynamics of the gravitational field correctly, we must complete Eq. (9.14) by adding to its right hand side the *energy-momentum tensor of the gravitation field* itself, a tensor that we call $T^{\mu\nu}_{\text{gr}}$. We replace hence (9.14) with the modified field equations

$$\Box H^{\mu\nu} = -\frac{16\pi G}{c^2}\,\mathbb{T}^{\mu\nu}, \qquad \mathbb{T}^{\mu\nu} = T^{\mu\nu} + T^{\mu\nu}_{\text{gr}}, \qquad \partial_\mu \mathbb{T}^{\mu\nu} = 0, \qquad (9.17)$$

while we maintain the harmonic gauge (9.15). Notice that, since in general the matter system exchanges energy and momentum with the gravitational field, only the *total* energy-momentum tensor $\mathbb{T}^{\mu\nu}$ is guaranteed to satisfy the continuity equation. To find out, at least, the qualitative expression of $T^{\mu\nu}_{\text{gr}}$, we again resort to the analogy between gravity and electromagnetism. From the Newtonian forces (9.2) and (9.3) we deduce that masses and charges are related to each other via the identification $e \leftrightarrow \sqrt{4\pi G}\, m$, so that, according to the Poisson equations (9.5) and (9.6), the electric and gravitational potentials are related to each other via $\varphi_{\text{em}} \leftrightarrow \varphi_{\text{gr}}/\sqrt{4\pi G}$. It follows that in General Relativity the role of A_μ is played by the field $H_{\mu\nu}/\sqrt{4\pi G}$. Correspondingly, since $T^{\mu\nu}_{\text{em}}$ is quadratic in ∂A, the tensor $T^{\mu\nu}_{\text{gr}}$ should be of the form

$$T^{\mu\nu}_{\text{gr}} \sim \frac{1}{4\pi G}\,\partial H \partial H. \qquad (9.18)$$

However, since H/c^2 is a dimensionless field, the tensor $T^{\mu\nu}_{\text{gr}}$ could contain relativistic corrections in which the term (9.18) is multiplied by an arbitrary power of H/c^2. In fact, according to General Relativity, the complete tensor $T^{\mu\nu}_{\text{gr}}$ contains an *infinite* number of such terms, being schematically of the form

$$T^{\mu\nu}_{\text{gr}} \sim \frac{1}{4\pi G} \sum_{N=0}^{\infty} \frac{1}{c^{2N}}\,\partial\partial H^{N+2}. \qquad (9.19)$$

In this way, Eqs. (9.17) indeed represent the exact Einstein equations in the harmonic gauge (9.15). The precise form of the terms of the series (9.19) is fixed by (i) the continuity equation $\partial_\mu \mathbb{T}^{\mu\nu} = 0$, and by (ii) the invariance of Einstein's equations under the diffeomorphisms (9.1).

We discover so a fundamental difference between Maxwell's equations in electrodynamics, and Einstein's equations in General Relativity: while the former are *linear* in A_μ, just because the electromagnetic field carries no *electric charge*, the latter are highly *non-linear* in $H^{\mu\nu}$, because the gravitational field carries *four-*

momentum. However, if the gravitational field is of so low intensity that it does not affect its own propagation pattern, i.e. if it satisfies the bound (9.16), then in Eqs. (9.17) the gravitational energy-momentum tensor $T_{\text{gr}}^{\mu\nu}$, codifying the non-linearities, can be neglected, and in this case the dynamics of the field is described with high accuracy by the linear equations (9.14). In all experiments conducted so far to reveal the presence of gravitational waves via *indirect* methods, see Sect. 9.5.1, the weak-field conditions (9.16) were satisfied during the whole radiation process, and so the use of Eqs. (9.14) was fully justified. Similarly, the interferometric ground-based experiments [1, 4] which recently led to a *direct* detection of gravitational waves revealed extremely small peak amplitudes, of the order $H^{\mu\nu}/c^2 \sim 10^{-21}$. Conversely, the gravitational field involved in the *emission* process of these waves, namely the merging of two massive relativistic black holes, was extremely intense, of the order $H^{\mu\nu}/c^2 \sim 1$, see Sect. 9.5. This means that, whereas the weak-field limit is in any case perfectly appropriate to analyze gravitational wave signals detected on Earth, it is not always suitable to analyze their production processes, where gravity can be in a *strong-field* regime.

The way matter curves space-time. Let us briefly explain in which sense, in the presence of matter, space-time becomes *curved*. As anticipated in Sect. 5.3.2, in the presence of a gravitational field the interval between two events assumes the form

$$ds^2 = dx^\mu dx^\nu g_{\mu\nu}(x), \tag{9.20}$$

where $g_{\mu\nu}(x)$ represents the *metric* of a curved space-time. The metric is related to the gravitational potential $H_{\mu\nu}(x)$ – generated by a matter system according to Einstein's equations (9.17) – by the relations (5.104), which we rewrite reintroducing the velocity of light

$$g_{\mu\nu}(x) = \eta_{\mu\nu} + \frac{1}{c^2}\left(H_{\mu\nu}(x) - \frac{1}{2}\eta_{\mu\nu}H^\rho{}_\rho(x)\right). \tag{9.21}$$

In their original form, Einstein's equations (9.17) are second-order partial differential equations for the matrix $g_{\mu\nu}(x)$ and its inverse $g^{\mu\nu}(x)$, and if written in terms of these matrices, rather than in terms of $H^{\mu\nu}(x)$, they become, actually, *polynomial* equations. As observed above, in the weak-field limit Einstein's equations reduce to the linear equations (9.14). The latter can be solved explicitly for $H^{\mu\nu}(x)$, see Eqs. (9.30) below, once the dynamics of the matter system is known, i.e. once the components of $T^{\mu\nu}(x)$ are known functions of space and time, and so the relations (9.21) determine the metric at any space-time point in terms of the matter distribution. As we see from these relations, the potential $H^{\mu\nu}(x)$ quantifies the displacement of the metric $g_{\mu\nu}(x)$ of the curved space-time from the Minkowski metric $\eta_{\mu\nu}$ of the *flat* space-time. If the conditions (9.16) hold, $g_{\mu\nu}(x)$ deviates only little from the flat metric. In particular, in the absence of matter, $T^{\mu\nu}(x) = 0$, and of gravitational radiation, we have $H^{\mu\nu}(x) = 0$ and $g_{\mu\nu}(x) = \eta_{\mu\nu}$.

Alternative gravitational potentials. As seen above, in the context of General Relativity the two-index tensor $H^{\mu\nu}$ plays a twofold role: on one hand it codifies the dynamics of the gravitational field by means of equations (9.17), and, on the other hand, it curves the space-time by means of the metric (9.21). If we ignore this second role – of more geometric nature – then *a priori* the gravitational interaction could be mediated also by tensors of different rank. A first alternative tensor is related to the possibility of considering ρ_{m} as the time component of the conserved *mass* four-current, which for a system of particles of masses m_r reads

$$J_{\mathrm{m}}^{\mu} = \sum_r m_r \int u_r^{\mu}\, \delta^4(x - y_r)\, ds_r, \qquad \partial_{\mu} J_{\mathrm{m}}^{\mu} = 0. \tag{9.22}$$

We have indeed

$$J_{\mathrm{m}}^0 = c \sum_r m_r\, \delta^3(\mathbf{x} - \mathbf{y}_r) = c\rho_{\mathrm{m}}.$$

According to Eq. (9.10) we should then consider the field φ_{gr} as the time component of a four-vector field $\mathcal{A}^{\mu}$, i.e. $\varphi_{\mathrm{gr}} = \mathcal{A}^0$. In this way we would obtain the system of equations

$$\Box \mathcal{A}^{\mu} = -\frac{4\pi G}{c}\, J_{\mathrm{m}}^{\mu}, \qquad \partial_{\mu} \mathcal{A}^{\mu} = 0.$$

However, this system would conflict with two fundamental experimental facts. In the first place, we would have constructed a relativistic theory of gravity *completely* analogous to electrodynamics, in sharp contrast with the fact that the former conceives only positive "charges", the masses, while the latter predicts charges of both signs. In the second place, in light of the current (9.22), what would be conserved is the total *mass* of the system, rather than its energy, in contrast with special relativity.

A second alternative would be to consider φ_{gr} as a relativistic *scalar* field. In this case we could resort to the fact that in the non-relativistic limit the components T^{ij} of the energy-momentum tensor of a system of particles are negligible with respect to the energy density T^{00}, see Eq. (2.139), so that $T^{\mu}{}_{\mu} = T^{00} - T^{ii} \approx T^{00} \approx c^2 \rho_{\mathrm{m}}$. In the non-relativistic limit the mass density ρ_{m} coincides, thus, with the trace of $T^{\mu\nu}/c^2$, and so in place of (9.14) we could conceive the Lorentz-invariant equation

$$\Box \varphi_{\mathrm{gr}} = -\frac{4\pi G}{c^2}\, T^{\mu}{}_{\mu}. \tag{9.23}$$

However, since the electromagnetic energy-momentum tensor $T_{\mathrm{em}}^{\mu\nu}$ (2.137) has a vanishing trace, see Eq. (2.148), according to the field equation (9.23) an electromagnetic field would never generate a gravitational field, in contrast with the observed bending of light rays passing close to a massive body.

9.3 Generation of Gravitational Radiation

As noticed above, in the weak-field limit, Einstein's equations in harmonic gauge

$$\Box H^{\mu\nu} = -\frac{16\pi G}{c^2}\,T^{\mu\nu}, \qquad \partial_\mu H^{\mu\nu} = 0, \qquad \partial_\mu T^{\mu\nu} = 0, \qquad (9.24)$$

have the same structure as Maxwell's equations in Lorenz gauge. We can thus solve the former with the same techniques used to solve the latter. Below we the analyze the basic properties of the solutions of the system (9.24), focusing our attention on the radiation phenomena.

Gravitational waves. In empty space, where $T^{\mu\nu} = 0$, Eqs. (9.24) reduce to

$$\Box H^{\mu\nu} = 0, \qquad \partial_\mu H^{\mu\nu} = 0. \qquad (9.25)$$

The most general solution of this system is a superposition of the *elementary gravitational waves*

$$H^{\mu\nu} = \varepsilon^{\mu\nu} e^{ik\cdot x} + c.c., \qquad k_\mu \varepsilon^{\mu\nu} = 0, \qquad k^2 = 0, \qquad (9.26)$$

where $\varepsilon^{\mu\nu}$ is a symmetric polarization tensor. The gravitational potentials (9.26) represent monochromatic plane waves propagating at the speed of light. To determine the *physical* polarization states associated with these waves we need to know how the diffeomorphisms (9.1) act on $H^{\mu\nu}$. This information is provided by General Relativity which predicts that under a generic coordinate transformation of the type (9.1) – involving the four "gauge" parameters $\Lambda^\mu(x)$ – in the weak-field limit the gravitational potential changes according to

$$H'^{\mu\nu} = H^{\mu\nu} + \partial^\mu \Lambda^\nu + \partial^\nu \Lambda^\mu - \eta^{\mu\nu}\partial_\rho \Lambda^\rho, \qquad (9.27)$$

a transformation that resembles the gauge transformation (2.44) of the electromagnetic vector potential. We now address the question which *residual* coordinate transformations preserve the harmonic gauge $\partial_\mu H^{\mu\nu} = 0$. From (9.27) we obtain

$$\partial_\mu H'^{\mu\nu} = \partial_\mu H^{\mu\nu} + \Box \Lambda^\nu = 0,$$

so that the parameters must satisfy the wave equations

$$\Box \Lambda^\nu = 0. \qquad (9.28)$$

Notice that the transformations (9.27), subject to the constraints (9.28), resemble the residual gauge transformations (5.14) of electrodynamics. Thanks to (9.28), the transformed potential $H'^{\mu\nu}$ (9.27) still satisfies Einstein's equations (9.24). As a consequence of (9.27) and (9.28), the polarization tensor of the wave (9.26) is determined modulo the transformations

$$\varepsilon'^{\mu\nu} = \varepsilon^{\mu\nu} + k^{\mu}\lambda^{\nu} + k^{\nu}\lambda^{\mu} - \eta^{\mu\nu}k_{\rho}\lambda^{\rho}, \tag{9.29}$$

where λ^{μ} is an arbitrary complex vector. The solutions of equations (9.25) represented by the relations (9.26) and (9.29) – derived for the first time by A. Einstein in 1916 – correspond exactly to the elementary waves (5.112) analyzed in Sects. 5.3.2 and 5.3.3, where we found, in particular, that these waves possess *two* physical polarization states of helicity ± 2.

Asymptotic potential and generation of gravitational radiation. In the presence of a matter source of compact support, represented by a conserved energy-momentum tensor $T^{\mu\nu}(t, \mathbf{x})$, the system (9.24) admits the exact causal solution, see Eqs. (6.30), (6.38), (6.58) and (6.65),

$$H^{\mu\nu} = G_{ret} * \left(-\frac{16\pi G}{c^2} T^{\mu\nu} \right) = -\frac{4G}{c^2} \int \frac{1}{|\mathbf{x} - \mathbf{y}|} T^{\mu\nu} \left(t - \frac{|\mathbf{x} - \mathbf{y}|}{c}, \mathbf{y} \right) d^3 y. \tag{9.30}$$

Thanks to the property of the convolution (2.70), and to the continuity equation $\partial_{\mu}T^{\mu\nu} = 0$, the potential (9.30) automatically satisfies the harmonic-gauge condition (9.15). As in the case of electrodynamics, to explore the presence of radiation we must analyze the behavior of $H^{\mu\nu}$ in the wave zone, i.e. at large distances from the source. Repeating the asymptotic analysis for large $r = |\mathbf{x}|$ carried out in Sect. 8.1, we find that modulo terms of order $1/r^2$ the potential (9.30) reduces to

$$H^{\mu\nu}(t, \mathbf{x}) = -\frac{4G}{rc^2} \int T^{\mu\nu} \left(t - \frac{r}{c} + \frac{\mathbf{n} \cdot \mathbf{y}}{c}, \mathbf{y} \right) d^3 y, \tag{9.31}$$

where $\mathbf{n} = \mathbf{x}/r$. At large distance from the source, $H^{\mu\nu}(t, \mathbf{x})$ decreases thus as $1/r$, as it must happen for a potential describing a *radiation field*. Setting $n^{\mu} = (1, \mathbf{n})$, as in Sect. 8.1 it can be shown that modulo terms of order $1/r^2$ the potential (9.31) satisfies the *wave relations*

$$\partial_{\rho}H^{\mu\nu} = \frac{1}{c} n_{\rho}\dot{H}^{\mu\nu}, \qquad n_{\mu}\dot{H}^{\mu\nu} = 0, \qquad n^2 = 0. \tag{9.32}$$

Finally, repeating the analysis of Sect. 8.1.2 we find that the potential in the wave zone (9.31) is a superposition of elementary gravitational waves of the form (9.26).

Energy emission. A quantitative analysis of the radiated energy associated with the wave-zone potential (9.31) requires the precise expression of the gravitational energy-momentum tensor $T_{\mathrm{gr}}^{\mu\nu}$ (9.19), provided by General Relativity. However, as in electrodynamics, for the evaluation of the emitted energy it is sufficient to retain in $T_{\mathrm{gr}}^{\mu\nu}$ the terms of order $1/r^2$, see (9.34) below. Consequently, in the series (9.19) only the terms relative to $N = 0$ give a non-vanishing contribution to the emitted energy, meaning that the quadratic approximation (9.18) of $T_{\mathrm{gr}}^{\mu\nu}$ is exactly what is needed for this purpose. The precise form of the terms quadratic in $H^{\mu\nu}$, provided by General Relativity, greatly simplifies if one takes into account the wave relations (9.32). The resulting expression of the gravitational energy-momentum tensor is

$$T_{\text{gr}}^{\mu\nu} = \frac{1}{32\pi G} \left(\partial_\mu H^{\alpha\beta} \partial_\nu H_{\alpha\beta} - \frac{1}{2} \partial_\mu H^\alpha{}_\alpha \partial_\nu H^\beta{}_\beta \right). \tag{9.33}$$

Once $T_{\text{gr}}^{\mu\nu}$ is known, the angular distribution of the radiated power $dW_{\text{gr}}/d\Omega$, and the total radiated power W_{gr}, can now be determined in complete analogy with the electromagnetic case (see the $\mu = 0$ component of equation (7.55)

$$\frac{dW_{\text{gr}}}{d\Omega} = cr^2 \left(T_{\text{gr}}^{0i} \, n^i \right), \qquad W_{\text{gr}} = \int \frac{dW_{\text{gr}}}{d\Omega} \, d\Omega, \qquad r \to \infty. \tag{9.34}$$

In Sect. 9.4 we will explicitly evaluate formulas (9.31),(9.33) and (9.34) for an arbitrary *non-relativistic* matter system, deriving, in particular, for the total power W_{gr} the compact, universal, expression (9.58), known as *quadrupole formula*.

9.3.1 Heuristic Derivation of the Quadrupole Formula

Before moving on to the explicit evaluation of the formulas above, in this section we determine W_{gr} for a non-relativistic matter system using heuristic arguments, again based on the analogy with electrodynamics. These arguments allow to derive the quadrupole formula modulo an overall coefficient, but, at the same time, they lead to a deeper insight into its physical meaning. We start by emphasizing an important aspect of the "fundamental" equations (9.24) regarding the continuity equation $\partial_\mu T^{\mu\nu} = 0$, which we supposed to be satisfied by the energy-momentum tensor of an *isolated* matter system. In reality, in light of the exact equation (9.17), this continuity equation is violated. In fact, from (9.33) we find qualitatively

$$\partial_\mu T^{\mu\nu} = -\partial_\mu T_{\text{gr}}^{\mu\nu} \sim \frac{1}{G} \partial H \partial^2 H \sim H^2, \tag{9.35}$$

precisely because the system emits four-momentum via radiation. However, since our aim is the computation of the emitted power W_{gr} (9.34), which is quadratic in H, in the calculation of $H^{\mu\nu}$ via (9.30) we can neglect terms of order H^2, and assume, hence, that $\partial_\mu T^{\mu\nu}$ vanishes. In other words, for the calculation of the radiated energy it is sufficient to consider the dynamics of the matter system at the unperturbed level, i.e. neglecting the *gravitational radiation reaction force*, and so the four-momentum of the matter system can be considered as conserved.

Let us now return to the expression (8.123) of the total electromagnetic power radiated by a generic non-relativistic charge distribution

$$\mathcal{W}_{\text{em}} = \frac{\left|\ddot{\mathbf{D}}\right|^2}{6\pi c^3} + \frac{\left|\dddot{\mathbf{M}}\right|^2}{6\pi c^5} + \frac{\dddot{\mathcal{D}}^{ij}\dddot{\mathcal{D}}^{ij}}{80\pi c^5}. \tag{9.36}$$

Let us suppose that the system is formed by a certain number of particles with charges e_r and masses m_r. The formal analogy between the forces (9.2) and (9.3) suggests then to estimate the gravitational power $\mathcal{W}_{\text{gr}}$ radiated by the same system, by making in Eq. (9.36) the replacements

$$e_r \rightarrow \sqrt{4\pi G}\, m_r, \tag{9.37}$$

i.e. by replacing the charge density ρ_{e} with the mass density ρ_{m} according to

$$\rho_{\text{e}} \rightarrow \sqrt{4\pi G}\, \rho_{\text{m}}.$$

Recalling the definition of the multipole moments (8.58), (8.115) and (8.116), these replacements lead to the estimate

$$\mathcal{W}_{\text{gr}} \approx G\left(\frac{2\left|\dot{\mathbf{P}}\right|^2}{3c^3} + \frac{\left|\ddot{\mathbf{L}}\right|^2}{6c^5} + \frac{\dddot{\mathcal{P}}^{ij}\dddot{\mathcal{P}}^{ij}}{20c^5}\right), \tag{9.38}$$

where $\mathbf{P} = \sum_r m_r \mathbf{v}_r$ is the total momentum of the system, $\mathbf{L} = \sum_r \mathbf{y}_r \times m_r \mathbf{v}_r$ is its total angular momentum, and $\mathcal{P}^{ij}$ is its *reduced gravitational quadrupole moment*

$$\mathcal{P}^{ij} = P^{ij} - \frac{1}{3}\delta^{ij}P^{kk}, \qquad P^{ij} = \frac{1}{c^2}\int y^i y^j\, T^{00}\, d^3 y. \tag{9.39}$$

Recall that in the non-relativistic limit the energy density T^{00} is identical with $c^2\rho_{\text{m}}$. However, since the system is isolated, its momentum as well as its angular momentum do not vary with time,

$$\dot{\mathbf{P}} = 0, \qquad \dot{\mathbf{L}} = 0,$$

and so in (9.38) both dipole-moment contributions vanish! Although the above argument made no explicit reference to it, the vanishing of the gravitational dipole radiations is intimately related with the *equivalence principle*, stating that the *gravitational charge* of a body, namely its gravitational mass M_r, equals its inertial mass m_r. To clarify this relationship we recall from Sects. 8.4 and 8.5.2 a result that we proved in electrodynamics: an isolated system of particles with charges e_r and inertial masses m_r does emit neither electric nor magnetic dipole radiation, if the ratio e_r/m_r does not depend on r. In the gravitational case, once according to (9.37) we have replaced the electric charges e_r with the gravitational charges $\sqrt{4\pi G}\, M_r$, thanks to the equivalence principle the ratio e_r/m_r takes the form

$$\frac{e_r}{m_r} \rightarrow \frac{\sqrt{4\pi G}\, M_r}{m_r} = \sqrt{4\pi G},$$

and is thus independent of r for *all* particles. Therefore, no matter system of massive particles of whatever kind generates gravitational dipole radiations. According to the above argument, the gravitational power radiated by a non-relativistic system reduces to the last term in (9.38). As we will see in Sect. 9.4, General Relativity indeed confirms our heuristic result – apart from a factor 4. In fact, we will derive the fundamental *quadrupole formula* for gravitational radiation

$$\mathcal{W}_{\text{gr}} = \frac{G}{5c^5} \, \ddot{\mathcal{P}}^{ij} \ddot{\mathcal{P}}^{ij}, \tag{9.40}$$

first obtained by A. Einstein in 1918. This formula represents in every respect the gravitational counterpart of the analogous result (8.53) of electrodynamics

$$\mathcal{W}_{\text{em}} = \frac{|\ddot{\mathbf{D}}|^2}{6\pi c^3}.$$

In fact, both formulas give the dominant term of the total energy emitted by a non-relativistic system per unit time, via radiation. If one operates the identification $e \leftrightarrow \sqrt{4\pi G}\, m$, one realizes that, for dimensional reasons, the intensity of the gravitational radiation is suppressed by a factor $(v/c)^2$ with respect to the electromagnetic one.

Birkhoff's theorem. The appearance of the *reduced* quadrupole moment (9.39) in Eq. (9.40) is a consequence of Birkhoff's theorem, see Problem 2.5. This theorem holds also in General Relativity, for which, originally, it has been proven. It states that the gravitational field produced, via Einstein's equations, by a spherically symmetric system in empty space is *static*. Therefore, such a system cannot emit gravitational waves and, consequently, the exact emitted power $\mathcal{W}_{\text{gr}}$ must vanish identically. It is not difficult to realize that the quadrupole formula (9.40) is in agreement with this theorem. In fact, for a spherically symmetric system we have $T^{00}(t, \mathbf{y}) = T^{00}(t, y)$, and so the argument used in (8.125) to prove the vanishing of $\mathcal{D}^{ij}$ applies equally well to $\mathcal{P}^{ij}$. It follows that for a spherically symmetric system $\mathcal{P}^{ij}$ vanishes identically, and so $\mathcal{W}_{\text{gr}} = 0$.

9.4 Energy of the Quadrupole Radiation

We now derive the quadrupole formula (9.40) from Eqs. (9.31), (9.33) and (9.34). We consider first a matter system formed by bodies of arbitrary speeds, and only subsequently we perform the non-relativistic expansion. We begin by rewriting Eq. (9.33) using the wave relations (9.32)

$$T_{\text{gr}}^{\mu\nu} = \frac{n^\mu n^\nu}{32\pi G c^2} \left(\dot{H}^{\alpha\beta} \dot{H}_{\alpha\beta} - \frac{1}{2} \left(\dot{H}^\alpha{}_\alpha \right)^2 \right). \tag{9.41}$$

Notice the formal analogy between this formula and the corresponding expression of the electromagnetic energy-momentum tensor (5.95)

$$T_{\text{em}}^{\mu\nu} = -\frac{n^\mu n^\nu}{c^2} \left(\dot{A}^\alpha \dot{A}_\alpha \right).$$

The angular distribution of the radiated power (9.34) then becomes

$$\frac{dW_{\text{gr}}}{d\Omega} = \frac{r^2}{32\pi Gc} \left(\dot{H}^{\alpha\beta} \dot{H}_{\alpha\beta} - \frac{1}{2} (\dot{H}^\alpha{}_\alpha)^2 \right). \tag{9.42}$$

As in the electromagnetic case, we now express the quantity between parentheses in terms of only the spatial components H^{ij} of the potential. For this purpose we recall from (9.32) the wave relations $n_\mu \dot{H}^{\mu\nu} = 0$, that in component form read

$$\dot{H}^{00} - n^i \dot{H}^{i0} = 0,$$
$$\dot{H}^{i0} - n^k \dot{H}^{ki} = 0.$$

Inserting the second relation in the first, we can express all components of $\dot{H}^{\mu\nu}$ in terms of only $\dot{H}^{ij}$:

$$\dot{H}^{00} = n^i n^j \dot{H}^{ij}, \tag{9.43}$$
$$\dot{H}^{0i} = n^k \dot{H}^{ki}. \tag{9.44}$$

For future reference we observe that these relations, being equivalent to $n_\mu \dot{H}^{\mu\nu} = 0$, are equivalent to the identity $\partial_\mu H^{\mu\nu} = 0$, which, in turn, is satisfied by virtue of the continuity equation $\partial_\mu T^{\mu\nu} = 0$, see the relation (9.30). Ultimately, the relations (9.43), (9.44) are thus consequences of four-momentum conservation, in the same vein as in electrodynamics the relation $\dot{A}^0 = n^i \dot{A}^i$ in (8.13) is a consequence of charge conservation. Inserting Eqs. (9.43) and (9.44) in (9.42), we obtain for the angular distribution of the radiated power (see Problem 9.1)

$$\frac{dW_{\text{gr}}}{d\Omega} = \frac{r^2}{32\pi Gc} \dot{H}^{ij} \dot{H}^{lm} \Lambda^{ijlm}, \tag{9.45}$$

$$\Lambda^{ijlm} = \delta^{il}\delta^{jm} - \frac{1}{2}\delta^{ij}\delta^{lm} - 2\delta^{il}n^j n^m + \delta^{ij}n^l n^m + \frac{1}{2}n^i n^j n^l n^m, \tag{9.46}$$

to be compared with the analogous expression (8.18) of the electromagnetic power

$$\frac{dW_{\text{em}}}{d\Omega} = \frac{r^2}{c} \dot{A}^i \dot{A}^j \Lambda^{ij}, \qquad \Lambda^{ij} = \delta^{ij} - n^i n^j. \tag{9.47}$$

Non-relativistic approximation. If the bodies of the matter system are non-relativistic, with speeds $v \ll c$, in the potential (9.31) we can neglect the microscopic delay

$\mathbf{n} \cdot \mathbf{y}/c$, like we did in electrodynamics. The spatial components of the gravitational potential then take the simple form

$$H^{ij} = -\frac{4G}{rc^2} \int T^{ij}\left(t - \frac{r}{c}, \mathbf{y}\right) d^3y, \qquad (9.48)$$

which is independent of the emission direction $\mathbf{n}$. In this limit the components H^{ij} are related in a simple way to the quadrupole moment P^{ij} (9.39), in virtue of the identity

$$\int T^{ij}\, d^3y = \frac{1}{2}\, \ddot{P}^{ij}. \qquad (9.49)$$

To prove (9.49) we resort, once more, to the continuity equation $\partial_\mu T^{\mu\nu} = 0$, which in component form reads

$$\frac{1}{c}\, \dot{T}^{00} = -\partial_k T^{k0}, \qquad (9.50)$$

$$\frac{1}{c}\, \dot{T}^{0k} = -\partial_m T^{mk}. \qquad (9.51)$$

Substituting Eq. (9.51) in the time derivative of Eq. (9.50), we derive the further relation

$$\frac{1}{c^2}\, \ddot{T}^{00} = -\frac{1}{c}\, \partial_k \dot{T}^{k0} = \partial_k \partial_m T^{km}. \qquad (9.52)$$

Inserting this expression for $\ddot{T}^{00}$ in the second time derivative of the quadrupole moment (9.39), and integrating twice by parts, we finally obtain

$$\ddot{P}^{ij} = \frac{1}{c^2} \int y^i y^j\, \ddot{T}^{00}\, d^3y = \int y^i y^j\, \partial_k \partial_m T^{km}\, d^3y = \int \partial_m \partial_k \left(y^i y^j\right) T^{km}\, d^3y$$

$$= \int \left(\delta_k^i \delta_m^j + \delta_k^j \delta_m^i\right) T^{km}\, d^3y = 2 \int T^{ij}\, d^3y,$$

which is Eq. (9.49). In the non-relativistic limit the gravitational potential in the wave zone is, thus, related to the second derivative of the quadrupole moment by the formula

$$H^{ij}(t, \mathbf{x}) = -\frac{2G}{rc^2}\, \ddot{P}^{ij}\left(t - \frac{r}{c}\right), \qquad (9.53)$$

to be compared with the electromagnetic counterpart (8.49). There are two basic, related, differences: the dominant term of the gravitational radiation is a *quadrupole* term, as predicted by our heuristic argument of Sect. 9.3.1, while the dominant term of the electromagnetic radiation is a *dipole* term. Correspondingly, the potential (9.53) is suppressed with respect to the potential (8.49) by a relativistic factor $1/c$. Substituting the potential (8.49) in the radiation formula (9.45) we obtain

$$\frac{d\mathcal{W}_{\mathrm{gr}}}{d\Omega} = \frac{G}{8\pi c^5}\, \dddot{P}^{ij}\, \dddot{P}^{lm}\, \Lambda^{ijlm}. \tag{9.54}$$

In light of the rather involved form of the tensor (9.46), in general the intensity of the gravitational radiation depends in a rather complicated way on the emission direction **n**. However, it is still possible to derive a compact expression for the total power, see Problem 9.2,

$$\mathcal{W}_{\mathrm{gr}} = \frac{G}{8\pi c^5}\, \dddot{P}^{ij}\, \dddot{P}^{lm} \int \Lambda^{ijlm}\, d\Omega \tag{9.55}$$

$$= \frac{G}{8\pi c^5}\, \dddot{P}^{ij}\, \dddot{P}^{lm}\, \frac{2\pi}{15} \left(11\delta^{il}\delta^{jm} + \delta^{im}\delta^{jl} - 4\,\delta^{ij}\delta^{lm}\right) \tag{9.56}$$

$$= \frac{G}{5c^5} \left(\dddot{P}^{ij}\,\dddot{P}^{ij} - \frac{1}{3}\,\dddot{P}^{ii}\,\dddot{P}^{jj}\right) \tag{9.57}$$

$$= \frac{G}{5c^5}\, \dddot{\mathcal{P}}^{ij}\, \dddot{\mathcal{P}}^{ij}, \tag{9.58}$$

which is the quadrupole formula (9.40). In the last line we used the identity (8.122) applied to the reduced quadrupole moment (9.39).

9.4.1 Vanishing of the Gravitational Dipole Radiation

In Sect. 9.3.1 we gave a heuristic argument for the disappearance of the gravitational dipole radiation. The analysis carried out indicated that the reason for this vanishing is to be sought in the equivalence principle, as well as in the conservation of the total four-momentum of an isolated system. We now want to test this hypothesis by performing the non-relativistic expansion of the potential (9.31). As usual, the non-relativistic expansion of the integral (9.31) amounts to a Taylor series expansion in powers of the microscopic delay $\mathbf{n} \cdot \mathbf{y}/c$. To take into account the dipole and quadrupole radiations we must expand $H^{\mu\nu}$ up to the powers $(\mathbf{n} \cdot \mathbf{y}/c)^2$. However, since different components of the matter energy-momentum tensor $T^{\mu\nu}$ have different relativistic behaviors, schematically

$$T^{00} \sim Mc^2, \qquad T^{0i} \sim Mcv, \qquad T^{ij} \sim Mvv,$$

the expansions of the components H^{00}, H^{0i} and H^{ij} must be truncated at different powers of $\mathbf{n} \cdot \mathbf{y}/c$. Modulo terms of order $1/c^3$, from the potential (9.31) so we obtain the expansions

$$H^{00} = -\frac{4G}{rc^2} \int \left(T^{00} + \frac{1}{c}\, n^k y^k\, \dot{T}^{00} + \frac{1}{2c^2}\, n^k n^l y^k y^l\, \ddot{T}^{00}\right) d^3y, \tag{9.59}$$

$$H^{0i} = -\frac{4G}{rc^2} \int \left(T^{0i} + \frac{1}{c}\, n^k y^k\, \dot{T}^{0i}\right) d^3y, \tag{9.60}$$

$$H^{ij} = -\frac{4G}{rc^2} \int T^{ij} d^3y, \tag{9.61}$$

where it is understood that $T^{\mu\nu}$ is evaluated at the point $(t - r/c, \mathbf{y})$. In order to rewrite these expressions in a more compact form, first of all we introduce the *conserved* total four-momentum of the matter system

$$P^\mu = \int T^{0\mu} d^3y = (Mc^2, c\mathbf{P}),$$

whose components appear in the first terms of the potentials (9.59) and (9.60). Here M denotes the total mass of the system, defined as its energy divided by c^2, and $\mathbf{P}$ is its conserved momentum. In the third term of (9.59) we recognize the second time derivative of the quadrupole moment P^{ij} (9.39). The second term of (9.59) can be expressed in terms of $\mathbf{P}$, using (9.50),

$$\frac{1}{c} \int y^k \dot{T}^{00} d^3y = -\int y^k \partial_i T^{i0} d^3y = \int (\partial_i y^k) T^{i0} d^3y = \int T^{k0} d^3y = cP^k.$$

Similarly, the second term of (9.60) can be expressed in terms of P^{ij}, using (9.51),

$$\frac{1}{c} \int y^k \dot{T}^{0i} d^3y = -\int y^k \partial_j T^{ji} d^3y = \int (\partial_j y^k) T^{ji} d^3y = \int T^{ki} d^3y = \frac{1}{2} \ddot{P}^{ki}.$$

In this way the potentials (9.59)–(9.61) can be recast in the form

$$H^{00} = -\frac{4G}{r} \left(M + \frac{1}{c} n^k P^k + \frac{1}{2c^2} n^k n^l \ddot{P}^{kl} \right), \tag{9.62}$$

$$H^{0i} = -\frac{4G}{r} \left(\frac{1}{c} P^i + \frac{1}{2c^2} n^k \ddot{P}^{ki} \right), \tag{9.63}$$

$$H^{ij} = -\frac{2G}{rc^2} \ddot{P}^{ij}, \tag{9.64}$$

where the quantities M, $\mathbf{P}$ and P^{ij} are of *zero* order in $1/c$.

Equations (9.62)–(9.64) are the non-relativistic expansions of the gravitational potential in the wave zone, up to order $1/c^2$, which parallel thus the expansions (8.118) and (8.119) of the electromagnetic potential A^μ. As in that case, the dipole radiations correspond to the terms of order $1/c$, which in the potentials (9.62)–(9.64) are proportional to $\mathbf{P}$, and non-vanishing. However – and this is the key observation – what appears in the gravitational energy-momentum tensor (9.33), as well as in the radiated power (9.42), is not the *potential* $H^{\mu\nu}$, but rather the *field* $\partial_\rho H^{\mu\nu} = n_\rho \dot{H}^{\mu\nu}/c$, see the wave relations (9.32). But, since $\mathbf{P}$ is a constant of motion, the dipole radiation drops out from the fields (9.62)–(9.64), once one considers their derivatives $\dot{H}^{\mu\nu}$. Similarly, the Newtonian potential $-4MG/r$, a term of order zero in $1/c$, drops out from H^{00} once one considers the field $\dot{H}^{00}$.

In conclusion, thanks to *four-momentum conservation* the field $\dot{H}^{\mu\nu}$ contains only terms of order $1/c^2$, proportional to the third derivative of P^{ij}, and it is thus a quadrupole radiation. Although it was not explicitly involved, in this analysis the equivalence principle acted implicitly under the surface. In fact, due to Lorentz invariance, the time component M of the four-momentum P^{μ} is, by definition, the *inertial* mass, the same that appears in the momentum $\mathbf{P} = M\mathbf{V}/\sqrt{1-V^2}$. On the other hand, the same M appears in the scalar potential (9.62), where thoroughly it plays the role of the *gravitational* mass. Would these two quantities not be the same, the whole above construction would be self-contradictory. Finally, using the expressions (9.62)–(9.64) it is easy to verify that the identities (9.43) and (9.44), needed to express the radiated power (9.42) in terms of only the quadrupole moment, hold precisely in virtue of the constancy of P^{μ}.

9.5 Predictions of General Relativity and Experimental Tests

The quadrupole formula (9.58) quantifies the energy carried away by the gravitational radiation emitted by a non-relativistic system – in the weak-field approximation – in terms of its quadrupole moment. The latter involves, in turn, the energy density of the system T^{00}, which in the non-relativistic limit is dominated by the mass density multiplied by c^2. If the system is formed by a certain number of particles of Masses M_r with trajectories $\mathbf{y}_r(t)$, or, more generally, by a certain number of rigid bodies with negligible rotational motions, the energy density has thus the simple form

$$T^{00} = \sum_r M_r\, c^2\, \delta^3(\mathbf{y} - \mathbf{y}_r).$$

The quadrupole moment (9.39) then takes the simple form

$$P^{ij} = \frac{1}{c^2} \int y^i y^j\, T^{00}\, d^3y = \sum_r M_r \int y^i y^j\, \delta^3(\mathbf{y} - \mathbf{y}_r)\, d^3y = \sum_r M_r\, y_r^i y_r^j.$$

$$\text{(9.65)}$$

Differentiating this expression three times with respect to time, subtracting its trace, and inserting the resulting expression in (9.58), it is straightforward to determine the energy radiated by the system per unit time. For the reasons explained above, under *ordinary* physical circumstances the emitted power is very small, proportional to $1/c^5$, and so its effects are difficult to observe. In fact, an experimental test of formula (9.58) requires a system of bodies undergoing *large accelerations* and/or having *large masses*. In particular, a system of bodies in uniform linear motion with constant velocities $\mathbf{v}_r$ emits no gravitational radiation at all. In fact, in this case the second derivative of the quadrupole moment (9.65) is given by

$$\ddot{P}^{ij} = 2\sum_r M_r v_r^i v_r^j,$$

so that its third derivative $\dddot{P}^{ij}$ vanishes. Today, there exist two viable experimental methods to probe the presence of gravitational radiation, furnishing a direct and indirect evidence, respectively.

Strong gravitational fields and direct detection. A direct detection of gravitational waves aims to measure the deformation of space and time caused on Earth by the gravitational potential $H^{\mu\nu}(x)$, via the metric (9.21). The experimental techniques for measurements of this kind employ *gravitational antennas* or *laser interferometers*. The former probe the presence of gravitational waves through small changes of the shape of heavy cylindrical resonant bars, while the latter exploit the phase shifts induced in the interference pattern of light rays by the oscillating space-time geometry generated by the wave.

The most promising sources of gravitational radiation are of astrophysical origin. A potential type of *continuous* sources of gravitational waves are isolated pulsars – spinning neutron stars endowed with strong magnetic fields – which emit gravitational radiation if they are in non-axisymmetric rotation.[1] Continuous gravitational radiation is also produced by two stars orbiting each other on quasi-stationary trajectories, in which case the acceleration is mainly of the centripetal type. *Transient* gravitational-wave signals are instead produced by supernovae, caused by the collapse and subsequent explosion of a star, if the collapse – resulting in a neutron star or a black hole – takes place in an anisotropic way. Similarly, intense gravitational radiation can be produced by the merger of massive compact *binary* systems, formed by white dwarfs, neutron stars, or black holes. The gravitational-wave signal GW150914 observed by the laser interferometer LIGO [1] has, in fact, been produced during the final stage of the merger of two spinning black holes, of masses $M_1 = 36 M_\odot$ and $M_2 = 29 M_\odot$, where $M_\odot$ is the solar mass, which occurred at a distance of about $d_{\text{obs}} = 1.2 \cdot 10^{22}$ km from our galaxy. Since the mass of the final black hole was determined to be $62 M_\odot$, the energy of about three solar masses was carried away by gravitational radiation. The measured peak amplitude of the signal was $H_{\text{obs}}^{\mu\nu}/c^2 \sim 10^{-21}$, an extremely *weak* field, and so the *observed* gravitational potential was with high precision a superposition of elementary waves of the form (9.26). Vice versa, at emission the gravitational field was very strong. To obtain an estimate of its magnitude we assume that the peak amplitude originated from an emission just outside the *horizons* of the black holes, with *Schwarzschild radii* $d_{\text{em}} \approx 2GM_1/c^2 \approx 2GM_2/c^2 \approx 100$ km. Since the gravitational potential decreases as $1/r$, see Eq. (9.31), we derive for the peak amplitude *at emission* the crude estimate

$$\frac{H_{\text{em}}^{\mu\nu}}{c^2} \sim \frac{d_{\text{obs}}}{d_{\text{em}}} \frac{H_{\text{obs}}^{\mu\nu}}{c^2} \sim 0.1,$$

---

[1] The energy-momentum tensor of a rigid body in axisymmetric rotation is constant in time, and generates, therefore, a time-independent gravitational field.

which corresponds to an extremely intense gravitational field. However, the crucial point is that in the final phase of the merger the gravitational potential $H_{\text{fin}}^{\mu\nu}/c^2$ created by, say, black hole 1 and experienced by black hole 2 was in a strong-field regime as well. In fact, if we estimate the closest distance of the black holes before the merger by the sum of their Schwarzschild radii $R = 2G(M_1 + M_2)/c^2 \approx 200\,\text{km}$, we obtain (according to (9.12) and (9.4) we take $H^{\mu\nu}(r) \sim 4GM/r$)

$$\frac{H_{\text{fin}}^{\mu\nu}}{c^2} \sim \frac{4GM_1}{R} \sim 1.$$

Correspondingly, just before their fusion the speeds of the black holes became very large, of the order $v/c \sim 0.5$. This implies that during the emission process the non-relativistic weak-field approximation breaks down, and thus the theoretical predictions had to be based on the exact Einstein equations. However, due to their highly non-linear structure, in the relativistic strong-field regime it is difficult to derive from Einstein's equations the form of the gravitational signal entering the detector, with sufficient accuracy. Moreover, the analysis of the LIGO data involved an average of a large number of such *template waveforms*, and so, taking the statistical errors into account, the detected signal allowed only to derive 90% credible intervals for the relevant physical parameters of the source, including the component masses, the final black hole spin, and the distance and sky position of the source. Eventually, these data allowed to firmly establish the absence of deviations between the observations and General Relativity, whereas a more quantitative comparison remained problematic.

Weak gravitational fields and indirect detection. Contrary to direct, earth-based, detection experiments, indirect detection methods of gravitational radiation allow for much more stringent *quantitative* tests of the predictions of General Relativity. If a quasi-periodic binary system has small masses, or is subject to small accelerations, then it emits a continuous low-intensity gravitational radiation of fixed frequency, see Problem 9.3, that can be too weak to be detected on Earth. On the other hand, if the total energy is to be conserved, gravitational radiation causes necessarily an energy loss of the binary system. Although the *instantaneous* radiated power can be very small, if the dynamics of the system lasts for a sufficiently long time it can happen that the persistent energy loss causes in the system *cumulative* astrophysical effects, so large that they become observable from Earth. These effects include very small variations of the velocities and of the form of the orbits, which otherwise would be periodic, and small changes in the period. The velocities of the components of such a binary systems are, typically, non-relativistic, and the gravitational fields they produce are weak. Therefore, in these cases the expressions of the gravitational potential (9.62)–(9.64) and of the emitted energy (9.58) are sensible, and one can use them to make precise quantitative predictions about the changes in the dynamics of the system induced by the emission of gravitational radiation. Consequently, a comparison between these predictions of General Relativity and astrophysical Earth-based observations can furnish an *indirect* evidence of gravitational waves, and allow, in particular, to perform high-precision tests of the theory. In the next section we describe the first indirect detection of gravitational waves, based on a binary system discovered in 1974 [2].

9.5.1 The Binary Pulsar PSR B1913+16

The pulsar PSR B1913+16 and its unseen companion star – probably another neutron star – rotate around each other on quasi-elliptic orbits with periastron and apastron separations of $d_{per} = 746600$ km and $d_{ap} = 3153600$ km, respectively, with an orbital period of $T = 7.75$ h. For comparison, we recall that the solar radius is $R_\odot = 695700$ km. The radii of both stars are estimated to be of the order of 10 km. The pulsar is spinning rapidly around one of its axes with *spin period* $\tau \approx 59$ ms, and correspondingly it emits electromagnetic pulses at a frequency of $1/\tau \approx 17$ Hz. The observation of these pulses, in particular of the oscillations of τ due to the Doppler effect caused by the orbital motion, allowed to perform a series of very precise measurements of the kinematics of the system. In fact, a characteristic feature of *isolated* pulsars is that the time τ between two consecutive pulses is constant in time, with a precision that sometimes rivals the accuracy of atomic clocks. In this way it was possible to determine several physical parameters of the binary system with very high accuracy [5, 6]. For instance, the masses of the pulsar and its companion are

$$M_1 = 1.4414(2)M_\odot, \qquad M_2 = 1.3867(2)M_\odot, \tag{9.66}$$

respectively, the eccentricity of the pulsar orbit is

$$e = 0.6171338(4), \tag{9.67}$$

and the period is

$$T = 0.322997448930(4) \, \text{days.} \tag{9.68}$$

Among the observations made of this system, which allowed in particular to verify several predictions of General Relativity with high precision, the most relevant experimental fact is that the orbital period decays in time, although at a very slow rate. Observations carried out over three decades have indeed shown that the period decreases with a constant systematic decay rate given by [6]

$$\left(\frac{dT}{dt}\right)_{\text{obs}} = -(2.4056 \pm 0.0051) \cdot 10^{-12}. \tag{9.69}$$

Notice that in a solar year the period (9.68) of 7.75 hours decreases by only $7.6 \cdot 10^{-5}$ s.

Orbital period decay. Let us now analyze the effects of the emitted gravitational radiation on the dynamics of a binary system which, in a sense, parallel the effects of the electromagnetic radiation causing the instability of the classical hydrogen atom, see Sect. 8.4.4. From the data above we deduce that the gravitational potential generated by, say, the companion and felt by the pulsar is at most of the order $H^{\mu\nu}/c^2 \sim 4GM_2/d_{per}c^2 \sim 10^{-5}$, so that we can safely work in the weak-field approximation.[2] Furthermore, for the speeds of the stars we obtain $v/c \sim 2\pi r_{ave}/cT \sim 0.7 \cdot 10^{-3}$, where for the average radius we took $r_{ave} = (d_{per} + d_{ap})/4 \approx 10^6$ km. This implies that also the non-relativistic framework is perfectly suitable.

We shall analyze the period decay in the simplified case of circular orbits with constant radius r, considering two stars of the same mass, $M_1 = M_2 = M$. We then have $\mathbf{y}_1 = -\mathbf{y}_2 = \mathbf{y}$, and $r = |\mathbf{y}|$. To evaluate the radiated power (9.58) we first compute the quadrupole moment (9.65)

$$P^{ij} = M\left(y_1^i y_1^j + y_2^i y_2^j\right) = 2M y^i y^j. \tag{9.70}$$

Introducing the velocity $\mathbf{v} = d\mathbf{y}/dt$, and the centripetal acceleration $\mathbf{a} = -v^2\mathbf{y}/r^2$, for its third derivative we obtain

$$\dddot{P}^{ij} = -\frac{8Mv^2}{r^2}\left(y^i v^j + y^j v^i\right). \tag{9.71}$$

The identity $\mathbf{y} \cdot \mathbf{v} = 0$ then implies that the trace $\dddot{P}^{ii}$ vanishes, and therefore the third derivative of the reduced quadrupole moment (9.39) just equals (9.71)

$$\dddot{\mathcal{P}}^{ij} = \dddot{P}^{ij}.$$

The quadrupole formula (9.58) then yields for the energy emitted by the system per unit time

$$\mathcal{W}_{gr} = \frac{128 G M^2 v^6}{5 r^2 c^5}. \tag{9.72}$$

To quantify the effects of this energy loss on the system we set

$$\mathcal{W}_{gr} = -\frac{d\varepsilon}{dt}, \tag{9.73}$$

---

[2]Pulsars are very compact stars with radii of a few kilometers and masses of around one solar mass. This means that at their surface the gravitational field is, actually, rather intense, of the order $H^{\mu\nu}/c^2 \sim 4MG/Rc^2 \sim 1$, and so the *internal* gravitational field is in a *strong-field* regime. In this case an *exact*, although implicit, solution of Einstein's equations (9.17) can be obtained by replacing in (9.30) $T^{\mu\nu}$ with the exact energy-momentum tensor $\mathbb{T}^{\mu\nu}$, see Eqs. (9.17) and (9.19). The non-relativistic wave-zone potential is then again of the form (9.53), where the quadrupole moment is now given by (9.39), with the total energy density $\mathbb{T}^{00}$ in place of T^{00}. Eventually, this amounts thus merely to the identification of Mc^2 with the total *energy* of the pulsar. See also Ref. [7], written in resolution of the *quadrupole-formula controversy*.

where ε is the non-relativistic total energy of the unperturbed binary system. From the Newtonian equation

$$\frac{Mv^2}{r} = \frac{GM^2}{(2r)^2} \tag{9.74}$$

we obtain

$$\varepsilon = 2 \cdot \frac{1}{2} M v^2 - \frac{GM^2}{2r} = -\frac{GM^2}{4r}. \tag{9.75}$$

On the other hand, since the period can be written as

$$T = \frac{2\pi r}{v} = \frac{4\pi r^{3/2}}{\sqrt{MG}}, \tag{9.76}$$

Equation (9.75) can be recast in the equivalent form

$$\varepsilon = -\left(\frac{\pi^2 M^5 G^2}{4}\right)^{1/3} T^{-2/3} = KT^{-2/3},$$

where K is a constant. A decrease in energy $d\varepsilon$ implies hence the period decrease

$$dT = -\frac{3T}{2\varepsilon} d\varepsilon.$$

Consequently, from Eqs. (9.72) and (9.73) it follows that the period decays in time according to the rule

$$\frac{dT}{dt} = -\frac{3T}{2\varepsilon}\frac{d\varepsilon}{dt} = -\frac{48\pi}{5c^5}\left(\frac{4\pi MG}{T}\right)^{5/3}, \tag{9.77}$$

where we have eliminated r and v in favor of the period T, using Eqs. (9.74) and (9.76). Finally, as shown in [8], the elliptic structure of the orbits modifies the second member of equation (9.77) by the eccentricity-dependent factor

$$f(e) = \frac{1 + 73e^2/24 + 37e^4/96}{(1 - e^2)^{7/2}},$$

and the fact that the stars have different masses results in the replacement of M with the *chirp mass* $(M_1 M_2)^{3/5}/\left(\frac{1}{2}(M_1 + M_2)\right)^{1/5}$. For two stars of arbitrary masses on elliptic orbits the period decay (9.77) becomes thus

$$\frac{dT}{dt} = -\frac{192\pi f(e)}{5c^5}\left(\frac{2\pi G}{T}\right)^{5/3}\frac{M_1 M_2}{(M_1 + M_2)^{1/3}}. \tag{9.78}$$

Substituting at second member the experimental values (9.66)–(9.68), the period decay predicted by General Relativity turns out to be

$$\left(\frac{dT}{dt}\right)_{\text{GR}} = -(2.40242 \pm 0.00002) \cdot 10^{-12}.$$

The observed decay (9.69) is thus perfectly in agreement with the hypothesis of emission of gravitational radiation, as predicted by General Relativity. One obtains, in fact,

$$\frac{\left(\frac{dT}{dt}\right)_{\text{obs}}}{\left(\frac{dT}{dt}\right)_{\text{GR}}} = 1.0013 \pm 0.0021.$$

Theory and experiment agree, hence, to within about 0.2%.

9.6 Problems

9.1 Show that the tensor Λ^{ijlm} has the form (9.46), by inserting Eqs. (9.43) and (9.44) in Eq. (9.42).

9.2 Verify that the integral over the angles in Eq. (9.55) gives (9.56).
Hint: Use the invariant integrals of Problem 2.6.

9.3 Consider a binary system formed by two identical non-relativistic stars of mass M, rotating around each other with speed v on circular orbits of radius r, as in Sect. 9.5.1.

(a) Determine the energy fraction $\Delta\varepsilon/\varepsilon$ dissipated by the system via gravitational radiation during a period, and compare it with the corresponding fraction dissipated by the classical hydrogen atom, see Eq. (8.110).
(b) Determine the frequencies present in the emitted radiation.
Hint: Investigate the time dependence of the gravitational potential (9.53), using the explicit expression of the quadrupole moment (9.70). The answer is that the system emits only radiation of frequency $2v/r$, i.e. *twice* the frequency of the motion.

9.4 Introduce for a generic matter system the traceless *reduced* gravitational potential

$$\mathcal{H}^{ij} = H^{ij} - \frac{1}{3}\delta^{ij}H^{kk}, \qquad \mathcal{H}^{kk} = 0, \tag{9.79}$$

where the wave-zone potential H^{ij} is given in (9.31).

(a) Show that in terms of this potential the angular distribution of the radiated power (9.45) can be recast in the form

$$\frac{d\mathcal{W}_{\mathrm{gr}}}{d\Omega} = \frac{r^2}{32\pi Gc}\,\dot{\mathcal{H}}^{ij}\,\dot{\mathcal{H}}^{lm}\,\Sigma^{ijlm}, \tag{9.80}$$

$$\Sigma^{ijlm} = \delta^{il}\delta^{jm} - 2\,\delta^{il}n^j n^m + \frac{1}{2}\,n^i n^j n^l n^m, \tag{9.81}$$

and provide an interpretation of this result.

Hint: Consider the non-relativistic limit and recall Birkhoff's theorem, see Sect. 9.3.1.

(b) Verify that the radiated power (9.80) is positive for all directions **n**.

Hint: Use rotation invariance to choose the emission direction $\mathbf{n} = (0,0,1)$.

(c) Show that for a *spherically symmetric* system the radiated power (9.80) vanishes for all directions **n**. Consider that in general the components of the matter system have arbitrary speeds, so that you cannot resort to formulas (9.48) and (9.53).

Hint: The spatial components of the energy-momentum tensor of a spherically symmetric system are of the form $T^{ij}(t,\mathbf{y}) = \delta^{ij}\,a(t,y) + y^i y^j\,b(t,y)$. Equation (9.31) then implies that the spatial components of the potential in the wave zone are of the form $H^{ij}(t,\mathbf{x}) = \delta^{ij}f(t,r) + n^i n^j g(t,r)$, see Problem 8.3.

References

1. B.P. Abbott et al., Observation of gravitational waves from a binary black hole merger. Phys. Rev. Lett. **116**, 061102 (2016)
2. R.A. Hulse, J.H. Taylor, Discovery of a pulsar in a binary system. Astrophys. J. Lett. **195**, L51 (1975)
3. M. Kramer et al., Tests of general relativity from timing the double pulsar. Science **314**, 97 (2006)
4. B.P. Abbott et al., GW151226: observation of gravitational waves from a 22-solar-mass binary black hole coalescene. Phys. Rev. Lett. **116**, 241103 (2016)
5. J.M. Weisberg, J.H. Taylor, Observations of post-newtonian timing effects in the binary pulsar PSR 1913+16. Phys. Rev. Lett. **52**, 1348 (1984)
6. J.M. Weisberg, J.H. Taylor, Relativistic binary pulsar B1913+16: thirty years of observations and analysis. ASP Conf. Ser. **328**, 25 (2005)
7. T. Damour, Gravitational radiation reaction in the binary pulsar and the quadrupole formula controversy. Phys. Rev. Lett. **51**, 1019 (1983)
8. P.C. Peters, J. Mathews, Gravitational radiation from point masses in a Keplerian orbit. Phys. Rev. **131**, 435 (1963)

Chapter 10
Radiation in the Ultrarelativistic Limit

Modern high-energy physics experiments commonly employ charged particles of very high velocities, frequently extremely close to the speed of light. To gain such ultrarelativistic velocities, the energies of the particles must be increased far beyond their rest masses. In addition, if one wants to confine the particles in a bounded region, for instance, in an *accumulation ring*, their trajectories must be bent. During both these processes the particles are subject to acceleration and so they emit electromagnetic radiation, thus dissipating part of the accumulated energy. The evaluation of the energy radiated by ultrarelativistic charged particles can no longer be based on the multipole expansion, legitimate only in the non-relativistic limit, and requires appropriate computational tools yielding exact results. One of the basic tools of this kind is the *relativistic Larmor formula*, which allows, in particular, to compute the energy dissipated via radiation in high-energy particle accelerators.

In Chaps. 7 and 8 we have developed the basic instruments for the analysis of the radiation emitted by a generic charge distribution. In particular, we have seen that the evaluation of the radiated four-momentum, see Eqs. (7.55) and (8.17),

$$\frac{d^2 P^\mu}{dt d\Omega} = r^2 \left(T_{\text{em}}^{\mu i} n^i \right) = r^2 n^\mu E^2, \qquad n^\mu = (1, \mathbf{n}), \qquad (10.1)$$

requires only the knowledge of the electric field $\mathbf{E}$ in the wave zone. In the following we shall mainly deal with the radiation emitted by a single charged particle, and so for $\mathbf{E}$ we can use the asymptotic Liénard-Wiechert field (7.59)

$$\mathbf{E} = \frac{e}{4\pi r} \frac{\mathbf{n} \times ((\mathbf{n} - \mathbf{v}) \times \mathbf{a})}{(1 - \mathbf{v} \cdot \mathbf{n})^3}, \qquad (10.2)$$

where the kinematical variables are evaluated at the retarded time t', determined by the implicit equation (7.62)

© Springer International Publishing AG, part of Springer Nature 2018
K. Lechner, *Classical Electrodynamics*, UNITEXT for Physics,
https://doi.org/10.1007/978-3-319-91809-9_10

$$t = t' + r - \mathbf{n} \cdot \mathbf{y}(t').$$ (10.3)

Differentiation this equation with respect to t', keeping r and $\mathbf{n}$ fixed, we obtain the important relation

$$\frac{dt}{dt'} = 1 - \mathbf{n} \cdot \mathbf{v}(t'),$$ (10.4)

which will be frequently used in the following. Inserting the field (10.2) in the radiation formula (10.1), one obtains a rather complicated expression for the angular distribution of the radiated four-momentum. Nevertheless, in Sect. 10.1 we will be able to derive a simple formula for the *total* four-momentum dP_{rad}^{μ}/ds radiated by the particle per unit proper time in all directions, see Eq. (10.7), which constitutes the relativistic generalization of Larmor's formula (7.67). In Sect. 10.3 we will then perform a qualitative analysis of the angular distribution of the radiation emitted by a generic ultrarelativistic charged particle.

10.1   Relativistic Generalization of Larmor's Formula

We consider a charged particle in arbitrary motion with world line $y^{\mu}(s)$. To calculate the four-momentum dP_{rad}^{μ}/ds it radiates in all directions we should insert the electric field (10.2) in the radiation formula (10.1), then integrate the resulting expression over the angles, and finally multiply the result by $u^0 = dt/ds$. This is an instructive, although slightly involved, calculation, that we will carry out explicitly in Sect. 10.1.2, leading to the relativistic Larmor formula (10.7). In Sect. 10.1.1 below we present an alternative, quicker, derivation of this formula, based on a covariance argument.

10.1.1   A Covariance Argument

Let us return momentarily to the energy and momentum radiated per unit time by a non-relativistic charged particle, Eqs. (8.54) and (8.57),

$$\frac{d\varepsilon}{dt} = \frac{e^2 a^2 (t - r)}{6\pi}, \qquad \frac{d\mathbf{P}}{dt} = 0.$$

We recall that this four-momentum is detected at a time t at a distance r from the particle, and correspondingly the acceleration is evaluated at the retarded time $t - r$. It is precisely this circumstance that allows us to interpret the expression

$$\frac{dP_{\text{rad}}^{\mu}}{dt} = \frac{e^2 a^2 (t)}{6\pi} (1, 0, 0, 0)$$ (10.5)

as the fraction of four-momentum emitted by the particle at time t, which *escapes to infinity*. Let us now consider a particle with arbitrary velocity $\mathbf{v}(t)$. Since we are in the presence of a single particle, we can replace the coordinate time t with the proper time s of the particle, and ask what is the four-momentum dP^{μ}_{rad}/ds it radiates per unit proper time. In the following we will assume that this quantity is a *four-vector*,[1] as suggested by the notation. To establish a connection with the formula (10.5) we consider for a fixed proper time s the inertial frame K^* in which at this time the particle is at rest, being thus strictly non-relativistic. According to (10.5), in K^* the radiated four-momentum is then given by

$$\frac{dP^{*\mu}_{\text{rad}}}{ds} = \frac{e^2 a^{*2}}{6\pi} u^{*\mu}, \qquad u^{*\mu} = (1,0,0,0), \tag{10.6}$$

where $u^{*\mu}$ is the four-velocity of the particle in K^*. We have set $dt^* = ds$, because $\mathbf{v}^* = 0$. Moreover, in K^* the four-acceleration at the instant s takes the form $w^{*\mu} = (0, \mathbf{a}^*)$, so that

$$w^{*2} = w^{*\mu} w^*_{\mu} = -a^{*2}.$$

Equation (10.6) can then be recast in the equivalent form

$$\frac{dP^{*\mu}_{\text{rad}}}{ds} = -\frac{e^2 w^{*2}}{6\pi} u^{*\mu}.$$

Since this equation equals two four-vectors it holds in *all* inertial frames. We have thus proven the *relativistic Larmor formula*

$$\frac{dP^{\mu}_{\text{rad}}}{ds} = -\frac{e^2 w^2}{6\pi} u^{\mu}. \tag{10.7}$$

We emphasize that this formula does not yield the *total* four-momentum emitted by the particle at the proper time s, but only that fraction that escapes to infinity. Multiplying Eq. (10.7) with u^0, and using that $d/ds = u^0 d/dt$, we can rewrite it in the equivalent form

$$\frac{dP^{\mu}_{\text{rad}}}{dt} = -\frac{e^2 w^2}{6\pi u^0} u^{\mu} = -\frac{e^2 w^2}{6\pi} \left(1, \frac{\mathbf{v}}{c}\right), \tag{10.8}$$

where we have restored the velocity of light. The time component of this equation yields for the energy $\varepsilon_{\text{rad}} = c P^0_{\text{rad}}$ radiated by a relativistic particle per unit time – the radiated power – the formula

---

[1] If dP^{μ}_{rad}/ds would equal the *total* four-momentum loss of the particle at the proper time s, then surely these four quantities would form a four-vector. Actually, in Sect. 15.2.4 we will see that at the same time s the particle exchanges with the electromagnetic field a further fraction of four-momentum – the *Schott term*, see Eq. (15.44) – which is, however, separately Lorentz invariant. Therefore, the assumption made in the text is justified a posteriori.

$$W = \frac{d\varepsilon_{\text{rad}}}{dt} = -\frac{ce^2 w^2}{6\pi}, \tag{10.9}$$

where the explicit expression of w^2 in terms of the spatial acceleration $\mathbf{a}$ is given in (2.164). Equation (10.9) represents the relativistic generalization of Larmor's original formula (7.67)

$$W_{\text{nr}} = \frac{e^2 a^2}{6\pi c^3}.$$

Notice that the second member of equation (10.9) is a Lorentz scalar, although in general the *power* is not a relativistic invariant. In the case at hand the Lorentz invariance of W is a consequence of the fact that dP_{rad}^μ / ds is proportional to u^μ. The spatial components of equation (10.8) tell us, instead, that the radiation emitted by a relativistic charged particle entails the non-vanishing momentum flow

$$\frac{d\mathbf{P}_{\text{rad}}}{dt} = -\frac{e^2 w^2 \mathbf{v}}{6\pi c} = \frac{d\varepsilon_{\text{rad}}}{dt} \frac{\mathbf{v}}{c^2}. \tag{10.10}$$

As the radiated energy (10.9) is a quantity of order $1/c^3$, this relation tells us in particular that the momentum radiated by the particle in all directions is a quantity of order $1/c^5$. This is in agreement with our general non-relativistic analysis of the emitted radiation of Sect. 8.4, where we found that the total momentum carried by the dipole radiation – a priori a quantity of order $1/c^4$ – vanishes identically, see Eq. (8.54). Henceforth, we set again $c = 1$.

Radiated power for $\mathbf{a} \parallel \mathbf{v}$ *and for* $\mathbf{a} \perp \mathbf{v}$. To properly compare the radiated power (10.9) with the non-relativistic one W_{nr} we express the former in terms of the spatial acceleration $\mathbf{a}$ using Eq. (2.164)

$$W = \frac{e^2}{6\pi} \frac{a^2 - (\mathbf{a} \times \mathbf{v})^2}{(1 - v^2)^3}. \tag{10.11}$$

For small speeds, $v \ll 1$, W obviously reduces to W_{nr}. Conversely, for ultrarelativistic speeds, $v \approx 1$, assuming equal accelerations the factor $1/(1 - v^2)^3$ implies that $W \gg W_{\text{nr}}$. Let us consider in more detail linear motions, for which $\mathbf{a} \parallel \mathbf{v}$, and motions with only centripetal acceleration, where $\mathbf{a} \perp \mathbf{v}$. In these cases equation (10.11) yields the emitted powers

$$W_\parallel = \frac{e^2 a^2}{6\pi} \frac{1}{(1 - v^2)^3}, \qquad W_\perp = \frac{e^2 a^2}{6\pi} \frac{1}{(1 - v^2)^2}. \tag{10.12}$$

Assuming equal accelerations, for ultrarelativistic particles we would thus have $W_\parallel \gg W_\perp$, so that during a linear motion a particle would radiate much more energy than during a motion with purely centripetal acceleration. However, this analysis does not take into account the accelerations that can be reached in practice for the two types of motion, and, moreover, it does not relate the amount of emitted energy to the energy possessed by the particle. In high-energy accelerators, for instance, the situation is actually *reversed*. In fact, in Sect. 10.2 we will see that the effects of radiative energy loss are much more pronounced in circular accelerators, than in linear ones.

10.1.2 Derivation from First Principles

We now derive the relativistic Larmor formula (10.7) from the fundamental radiation formula (10.1). We shall base the derivation of dP^μ_{rad}/ds on the evaluation of the total four-momentum ΔP^μ emitted by the particle along the whole trajectory. To make sure that this quantity is finite, we will suppose that the particle is accelerated only during a finite time interval (or that its acceleration vanishes sufficiently fast for $t \to \pm\infty$). In this way, the particle emits radiation only during a finite time interval, and so the emitted four-momentum ΔP^μ is certainly finite. To evaluate ΔP^μ in a manifestly covariant way, it is more convenient to start from the first expression in (10.1), using for $T^{\mu\nu}_{\text{em}}$ the asymptotic form (8.135) (remember the notation (7.32))

$$T^{\mu\nu}_{\text{em}} = -\frac{e^2 \left((un)^2 w^2 + (wn)^2 \right)}{16\pi^2 (un)^6 r^2} \, n^\mu n^\nu.$$

In this way we obtain

$$\frac{d^2 P^\mu}{dt d\Omega} = -\frac{e^2 \left((un)^2 w^2 + (wn)^2 \right)}{16\pi^2 (un)^6} \, n^\mu. \tag{10.13}$$

Obviously, one obtains the same result if one inserts the electric field (10.2) in the second expression in (10.1). To determine ΔP^μ we must integrate equation (10.13) over the angles and over all times

$$\Delta P^\mu = -\frac{e^2}{16\pi^2} \int d\Omega \int_{-\infty}^{\infty} dt \, n^\mu \left(\frac{w^2}{(un)^4} + \frac{(wn)^2}{(un)^6} \right). \tag{10.14}$$

The integrand of equation (10.14) depends in a complicated way on t and on the angles $\mathbf{n}$, because the kinematical variables u^μ and w^μ are evaluated at the retarded time $t'(t, \mathbf{x})$, specified by Eq. (10.3). Therefore, to simplify the integral it is convenient to change integration variable from t to the proper time s. In fact, for every fixed $\mathbf{x}$ there exists a one-to-one correspondence between t and t', Eq. (10.3), and also a one-to-one correspondence between t' and s, Eq. (7.5). Correspondingly, using the relation (10.4), we can express dt in terms of ds

$$dt = \frac{dt'}{ds} \frac{dt}{dt'} \, ds = u^0 (1 - \mathbf{n} \cdot \mathbf{v}) \, ds = (un) \, ds,$$

so that the emitted four-momentum (10.14) becomes

$$\Delta P^\mu = -\frac{e^2}{16\pi^2} \int_{-\infty}^\infty ds \int d\Omega\, n^\mu \left(\frac{w^2}{(un)^3} + \frac{(wn)^2}{(un)^5} \right). \tag{10.15}$$

Now that u^μ and w^μ are evaluated at s, which is an independent variable, the integral over the angles can be performed analytically. Below we compute this integral explicitly, resorting to a few mathematical techniques which are frequently used in theoretical physics.

Integration over the angles. The integrand in equation (10.15) depends on the angle-independent *parameters* u^μ and w^μ, which are subject to the kinematical constraints $u^2 = 1$ and $(uw) = 0$. To perform the angular integral it is convenient to consider u^μ and w^μ as *arbitrary* vectors, subject to no constraint at all. Eventually, we will recover the integral we are interested in enforcing in the final result again these constraints. Considering, hence, u_μ as a free variable, we can rewrite the integrand of equation (10.15) as a gradient with respect to u_μ

$$n^\mu \left(\frac{w^2}{(un)^3} + \frac{(wn)^2}{(un)^5} \right) = -\frac{1}{2} \frac{\partial}{\partial u_\mu} \left(\frac{w^2}{(un)^2} + \frac{1}{2} \frac{(wn)^2}{(un)^4} \right).$$

Taking the derivative with respect to u_μ out of the angular integral we then obtain

$$\Delta P^\mu = \frac{e^2}{32\pi^2} \int_{-\infty}^\infty ds\, \frac{\partial}{\partial u_\mu} \int d\Omega \left(\frac{w^2}{(un)^2} + \frac{1}{2} \frac{(wn)^2}{(un)^4} \right).$$

In this way we remain with a *single* integral over the angles. We can further simply the latter using the identity

$$\frac{(wn)^2}{(un)^4} = w_\alpha w_\beta \frac{n^\alpha n^\beta}{(un)^4} = \frac{1}{6} w_\alpha w_\beta \frac{\partial^2}{\partial u_\alpha \partial u_\beta} \frac{1}{(un)^2},$$

and taking the derivatives with respect to u^α and u^β out of the integral

$$\Delta P^\mu = \frac{e^2}{32\pi^2} \int_{-\infty}^\infty ds\, \frac{\partial}{\partial u_\mu} \left(\left(w^2 + \frac{1}{12} w_\alpha w_\beta \frac{\partial^2}{\partial u_\alpha \partial u_\beta} \right) \int \frac{d\Omega}{(un)^2} \right). \tag{10.16}$$

We have, thus, reduced the integral over the angles to the evaluation of a single elementary integral, and to the calculation of a few derivatives. Resorting to spatial rotation invariance we can take as four-velocity $u^\mu = (u^0, 0, 0, u^3)$, thereby obtaining

$$\int \frac{d\Omega}{(un)^2} = 2\pi \int_0^\pi \frac{\sin\vartheta\, d\vartheta}{(u^0 - u^3 \cos\vartheta)^2} = \frac{4\pi}{(u^0)^2 - (u^3)^2} = \frac{4\pi}{u^2}.$$

The integral (10.16) then becomes

$$\Delta P^\mu = \frac{e^2}{8\pi} \int_{-\infty}^{\infty} ds \, \frac{\partial}{\partial u_\mu} \left(\left(w^2 + \frac{1}{12} w_\alpha w_\beta \frac{\partial^2}{\partial u_\alpha \partial u_\beta} \right) \frac{1}{u^2} \right). \tag{10.17}$$

The evaluation of the remaining derivatives leads to

$$\frac{\partial}{\partial u_\mu} \left(\left(w^2 + \frac{1}{12} w_\alpha w_\beta \frac{\partial^2}{\partial u_\alpha \partial u_\beta} \right) \frac{1}{u^2} \right) =$$
$$\left(\frac{2}{3} - 2u^2 \right) \frac{w^2 u^\mu}{(u^2)^3} + \left(\frac{4}{3} u^2 w^\mu - 4(uw)u^\mu \right) \frac{(uw)}{(u^2)^4} = -\frac{4}{3} w^2 u^\mu, \tag{10.18}$$

where in the second expression, valid for arbitrary u^μ and w^μ, we have again enforced the constraints $u^2 = 1$ and $(uw) = 0$. Using (10.18) in (10.16) we conclude that the total four-momentum radiated by the particle during its entire history is given by the integral along its world line

$$\Delta P^\mu = -\frac{e^2}{6\pi} \int_{-\infty}^{\infty} w^2 u^\mu \, ds. \tag{10.19}$$

Formula (10.19) entails the following physical interpretation: the four-momentum emitted by the particle along the whole trajectory is composed of a "sum" of infinite individual contributions, each one referred to an emission instant s and to a corresponding proper-time interval Δs, given by

$$\Delta P^\mu_{\text{rad}}(s) = -\frac{e^2 w^2(s)}{6\pi} u^\mu(s) \Delta s. \tag{10.20}$$

This conclusion is equivalent to the relativistic Larmor formula (10.7).

Instantaneous emission of four-momentum. Now let us compare the formula for the total radiated four-momentum (10.19) with the four-momentum dP^μ/ds *instantaneously* radiated by the particle at the time s. Equation (10.19) certainly implies the identification

$$\Delta P^\mu = \int_{-\infty}^{\infty} \frac{dP^\mu}{ds} \, ds = -\frac{e^2}{6\pi} \int_{-\infty}^{\infty} w^2 u^\mu \, ds. \tag{10.21}$$

However, Eq. (10.21) does not imply the equality

$$\frac{dP^\mu}{ds} = -\frac{e^2 w^2}{6\pi} u^\mu,$$

but only the existence of a four-vector $G^\mu(s)$ subject to the constraint

$$\int_{-\infty}^{\infty} G^\mu(s) \, ds = 0, \tag{10.22}$$

so that

$$\frac{dP^\mu}{ds} = -\frac{e^2 w^2(s)}{6\pi}\, u^\mu(s) + G^\mu(s). \tag{10.23}$$

This equality tells us that the four-momentum instantaneously radiated by the particle at the time s during an interval Δs is composed of two contributions: the first is given by the term (10.20) *flowing to infinity*, and the second is given by the four-momentum $G^\mu(s)\Delta s$. According to the sum rule (10.22), as s varies along the world line the latter four-momenta, actually, are *emitted* and *absorbed* along the trajectory, eventually summing up to zero. In this way, via Eqs. (10.22) and (10.23) we retrieve the interpretation of the relativistic Larmor formula given after Eq. (10.7), see also footnote 1 in Sect. 10.1.1. In Chap. 15 we will find that the four-vector $G^\mu(s)$ is indeed non-vanishing, being actually identical with the *Schott term* in Eq. (15.44).

10.2   Energy Loss in High-Energy Particle Accelerators

We now apply the relativistic Larmor formula to the evaluation of the radiative energy loss of the particle beams of high-energy accelerators, where the trajectories of the particles are determined by external electric and magnetic fields. First we derive a general formula for the power emitted via bremsstrahlung by a charged particle moving in a generic external electromagnetic field $F^{\mu\nu}$. Then we apply this formula to quantify the effects of the radiative energy loss in high-energy particle accelerators. We will find that, whereas in linear accelerators these effects are completely negligible, in circular accelerators the radiative energy loss can become the *dominant* dynamical factor, which significantly limits the maximum energies attainable. We start from the kinematical relation

$$w^\mu = \frac{1}{m}\frac{dp^\mu}{ds},$$

which permits to express the emitted power (10.9) in terms of dp^μ/dt as

$$\mathcal{W} = -\frac{e^2}{6\pi m^2}\frac{dp^\mu}{ds}\frac{dp_\mu}{ds} = \frac{e^2}{6\pi m^2(1-v^2)}\left(\left|\frac{d\mathbf{p}}{dt}\right|^2 - \left(\frac{d\varepsilon}{dt}\right)^2\right). \tag{10.24}$$

If the particle moves under the influence of an electromagnetic field $F^{\mu\nu}$ it is subject to the Lorentz equation

$$\frac{dp^\mu}{ds} = eF^{\mu\nu}u_\nu. \tag{10.25}$$

In this case the emitted power (10.24) can be expressed in terms of the fields and of the velocity $\mathbf{v}$ of the particle. In fact, thanks to the spatial Lorentz equation (2.36) and to the work-energy theorem (2.35), which are equivalent to (10.25), Eq. (10.24) takes the form

$$W = \frac{e^4}{6\pi m^2} \frac{\left|\mathbf{E} + \mathbf{v} \times \mathbf{B}\right|^2 - (\mathbf{v} \cdot \mathbf{E})^2}{1 - v^2}. \tag{10.26}$$

This formula gives the emitted power in terms of the external fields evaluated on the trajectory $\mathbf{y}(t)$ of the particle, which, in turn, should be determined by solving the Lorentz equation (10.25). Therefore, the expression (10.26) is particularly useful when the former can be solved exactly, as happens, for instance, for a uniform constant field $F^{\mu\nu}$. However, we must keep in mind that, proceeding in this way, we neglect the effects of the radiation on the form of the trajectory, i.e. the effects of the *radiation reaction force*. Correspondingly, the expression (10.26) of the emitted power yields significant predictions only if the energy loss, derived in this way, is small with respect to the energy possessed by the particle, because then it induces only minor deformations of the trajectory.

Light versus heavy charged particles. We end these introductory considerations with a general observation concerning the masses of the charged particles employed by particle accelerators. For this purpose we rewrite the emitted power (10.26) in terms of the particle's energy $\varepsilon = m/\sqrt{1 - v^2}$

$$W = \frac{e^4 \varepsilon^2}{6\pi m^4} \left(\left|\mathbf{E} + \mathbf{v} \times \mathbf{B}\right|^2 - (\mathbf{v} \cdot \mathbf{E})^2\right).$$

From the high powers of m in the denominator we see that for fixed external fields, and for the same attained energy ε, in the ultrarelativistic limit a *light* particle radiates much more energy than a *heavy* one. The physical origin of this fact is essentially that, according to Newton's second law, for a fixed applied force a light particle has a larger acceleration than a heavy one. An important physical consequence is that, since the mass of the proton is about two thousand times that of the electron, from the point of view of the energy dissipated via radiation the accelerators of protons and antiprotons, as for instance the LHC (Large Hadron Collider) and the Tevatron, are far more convenient than the accelerators of electrons and positrons, as for instance LEP (Large Electron-Positron Collider), or the possible future accelerators ILC (International Linear Collider) and CEPC (Circular Electron Positron Collider).

10.2.1 Linear Accelerators

We now analyze the effects of the radiative energy loss in linear accelerators. In these accelerators the particles move under the influence of a constant electric field $\mathbf{E}$ parallel to their direction of motion, which supplies them with the *external* power, see Eq. (2.35),

$$W_{\text{ex}} = \frac{d\varepsilon}{dt} = evE. \tag{10.27}$$

Setting in Eq. (10.26) $\mathbf{B} = 0$, for the dissipated power we then obtain

$$\mathcal{W} = \frac{e^4 E^2}{6\pi m^2}.$$ (10.28)

At first sight this formula seems to conflict with the expression of $\mathcal{W}_\parallel$ in (10.12), in that the relativistic factors $1/\sqrt{1 - v^2}$ seem to have disappeared. However, the contradiction is only apparent because for a one-dimensional motion the Lorentz equation

$$m \frac{d}{dt} \left(\frac{v}{\sqrt{1 - v^2}} \right) = eE$$

can be recast in the equivalent form (see Eq. (2.38))

$$a = \frac{dv}{dt} = \left(\sqrt{1 - v^2} \right)^3 \frac{eE}{m}.$$

Equations (10.12) and (10.28) give thus the same result. To analyze the significance of the dissipated power (10.28) we compare it with the power (10.27) furnished by the external force:

$$\frac{\mathcal{W}}{\mathcal{W}_{\text{ex}}} = \frac{e^3 E}{6\pi m^2 v} = \frac{2r_0}{3mv} \frac{d\varepsilon}{dx},$$ (10.29)

where

$$\frac{d\varepsilon}{dx} = \frac{1}{v} \frac{d\varepsilon}{dt} = eE$$ (10.30)

is the energy furnished per unit space interval by the external field, and $r_0 = e^2/4\pi m$ is the classical radius of the particle. For an ultrarelativistic particle, $v \approx 1$, the ratio (10.29) reduces to

$$\frac{\mathcal{W}}{\mathcal{W}_{\text{ex}}} = \frac{2r_0}{3m} \frac{d\varepsilon}{dx} = \frac{e^3 E}{6\pi m^2}.$$ (10.31)

Therefore, the radiative energy loss is significant only in the presence of external fields which are so strong, that the particle gains an energy of the order of magnitude of its mass m, during a small time interval in which it travels a distance of the order of magnitude of its classical radius. However, the electrostatic fields available experimentally are much smaller and hardly exceed the value $E \sim 100\,\text{MV/m}$, for which Eq. (10.30) gives

$$\frac{d\varepsilon}{dx} \approx 100 \, \frac{\text{MeV}}{\text{m}}.$$ (10.32)

On the other hand, for a given electric field E the ratio (10.31) is largest for the lightest known charged particle – the electron – for which $m \approx 0.5\,\text{MeV}$ and $r_0 \approx 3 \cdot 10^{-15}\,\text{m}$, see Eq. (8.81). In this case, even for the maximum field (10.32) the ratio (10.31) reduces to the extremely small value

$$\frac{\mathcal{W}}{\mathcal{W}_{ex}} \approx 4 \cdot 10^{-13}.$$

For a proton one would obtain the even smaller ratio $\mathcal{W}/\mathcal{W}_{ex} \sim 10^{-19}$. In linear high-energy accelerators the radiative energy loss is, hence, completely negligible.

10.2.2 Circular Accelerators

In a circular accelerator a charged particle performs a uniform circular motion under the influence of a constant magnetic field $\mathbf{B}$. Since $\mathbf{E}$ is zero, in this case the Lorentz equation (2.36) takes the form

$$\frac{d\mathbf{u}}{dt} = \mathbf{u} \times \left(\frac{e}{m} \sqrt{1 - v^2} \, \mathbf{B} \right),$$

from which one reads the relativistic *cyclotron frequency*

$$\omega_0 = \frac{eB}{m} \sqrt{1 - v^2} = \frac{eB}{\varepsilon}. \tag{10.33}$$

Notice that ω_0 is obtained from the non-relativistic cyclotron frequency $\omega_{nr} = eB/m$, by replacing the mass of the particle with its energy $\varepsilon = m/\sqrt{1 - v^2}$. For $\mathbf{E} = 0$ Eq. (10.26) yields for the emitted power the expression

$$\mathcal{W} = \frac{e^4}{6\pi m^2} \frac{v^2 B^2}{1 - v^2} = \frac{e^2}{6\pi} \frac{v^2 \omega_0^2}{(1 - v^2)^2}, \tag{10.34}$$

to be compared with the non-relativistic Larmor formula

$$\mathcal{W}_{nr} = \frac{e^2 a^2}{6\pi}, \qquad a = \frac{veB}{m}.$$

To investigate the effects of the emitted radiation we compute the energy $\Delta\varepsilon$ dissipated during a cycle, i.e. during a period $T = 2\pi/\omega_0$. If the accumulation ring has a radius R we have $\omega_0 = v/R$ and $T = 2\pi R/v$, so that Eq. (10.34) yields the energy loss per cycle

$$\Delta\varepsilon = T\mathcal{W} = \frac{e^2}{3R} \frac{v^3}{(1 - v^2)^2} = \frac{e^2 v^3 \varepsilon^4}{3Rm^4}.$$

If the speed of the particle is close to the speed of light, in the numerator one can set $v \to 1$. In this way we obtain for the energy loss per cycle the celebrated *accelerator radiation formula*

$$\Delta\varepsilon = \frac{e^2}{3R}\left(\frac{\varepsilon}{m}\right)^4. \tag{10.35}$$

This formula imposes, indeed, severe restrictions on the technical characteristics of circular accelerators aiming to accumulate particles of a given energy ε. In particular, for a fixed particle's energy the effect of the emitted radiation is small for *large* accumulation rings, and for *heavy* particles.

High-energy synchrotrons. Since during any cycle a charged particle in a circular accelerator looses the energy (10.35), to maintain the particles in stable orbits at constant energy, along the accumulation ring one must dispose electric fields, so-called radio-frequency *resonant cavities*, which periodically compensate the energy loss. Circular accelerators of this kind, in which electric and magnetic fields act in a synchronized way, are called *synchrotrons*, see also Chap. 12. As an example we evaluate the energy loss per cycle of the Cornell Synchrotron, which was active between the years 1968 and 1979. This accelerator had an instantaneous magnetic bending radius of about $R = 100\,\mathrm{m}$, and accelerated electrons up to an energy of $\varepsilon = 10\,\mathrm{GeV}$, the highest energy attained at that time in any electron synchrotron. With these values the accelerator radiation formula (10.35) gives the energy dissipated per cycle as

$$\Delta\varepsilon \approx 8.9\,\mathrm{MeV}, \qquad \frac{\Delta\varepsilon}{\varepsilon} \sim 10^{-3},$$

whereas the resonant cavities were able to supply a maximum energy of 10.5 MeV per cycle. At an energy of $10\,\mathrm{GeV}$ the Cornell Synchrotron operated, hence, close to its technological limits. As second example we consider the accelerator LEP, which was active at the CERN laboratory near Geneva from 1989 to 2000, which accumulated electrons and positrons, circulating in opposite directions. The radius of the storage ring was $R = 4.3\,\mathrm{km}$, and the maximum energy per particle was about $\varepsilon = 100\,\mathrm{GeV}$. In this case equation (10.35) gives

$$\Delta\varepsilon \approx 2\,\mathrm{GeV}, \qquad \frac{\Delta\varepsilon}{\varepsilon} \sim 2\cdot 10^{-2},$$

corresponding to an energy loss of about 2% during each cycle. Since the particles traveled practically at the speed of light, in one second they performed about 11.000 cycles, and so, in the absence of resonance cavities, all their energy would have been radiated away in the fraction of a second. In fact, at LEP the number of resonant cavities installed along the storage ring to compensate this energy loss grew in time, and in its finale stage there were 344 such cavities. The primary technological limit of LEP was, indeed, its *radiation damping*, eventually limiting the maximum energy of the particles to 104.5 GeV. As last example we analyze the effect of the radiation damping in the LHC at CERN, which realizes collisions between two proton beams with a design energy of $\varepsilon = 7\,\mathrm{TeV}$ per particle. The beams circulate in the same storage ring of LEP, with a radius $R = 4.3\,\mathrm{km}$. Since the mass of a proton is about two thousand times that of an electron, in this case the energy (10.35) radiated per cycle is very small:

$$\Delta\varepsilon \approx 3\,\text{keV}, \qquad \frac{\Delta\varepsilon}{\varepsilon} \sim 5 \cdot 10^{-10}.$$

Within one hour, for instance, during which the protons circle about $4 \cdot 10^7$ times around the ring, their energy would just diminish by about 2%. Correspondingly, the LHC needs only a small number of resonant cavities, just 8 per beam.

Apart from the technological problems caused by bremsstrahlung, the potentialities of circular accelerators are severely constrained also by the strong magnetic fields needed to bend the particles' trajectories. More precisely, since the cyclotron frequency ω_0 in (10.33) is equal to v/R, in the limit for $v \to 1$ the required magnetic field becomes proportional to the energy

$$B = \frac{\varepsilon}{eR}.$$

In particular, as the LHC reaches energies about 70 times larger than those of LEP, and both accelerators operate in the same storage ring, the former needs magnetic fields which are about 70 times stronger than those employed by LEP. Since the magnetic fields produced by the even most powerful superconducting electromagnets are subject to technological limitations, to further increase the energy of the particles it only remains to resort to storage rings of increasing radii.

10.3   Angular Distribution in the Ultrarelativistic Limit

We now perform a *qualitative* analysis of the angular distribution of the radiation emitted by an ultrarelativistic charged particle, $v \approx 1$. For comparison we recall the angular distribution (8.60) of the radiation generated by a generic non-relativistic particle, $v \ll 1$,

$$\frac{d\mathcal{W}}{d\Omega} = \frac{e^2 |\mathbf{n} \times \mathbf{a}|^2}{16\pi^2} = \frac{e^2}{16\pi^2}\,|\mathbf{a}|^2 \sin^2\vartheta, \tag{10.36}$$

where ϑ is the angle between the particle's acceleration $\mathbf{a}$ and the direction $\mathbf{n}$ of the radiation. In the non-relativistic limit the radiated power has thus a "continuous" angular distribution, with a zero in the direction of the acceleration, and a maximum in the plane orthogonal it. In particular, the radiated power (10.36) is independent of the direction of the particle's velocity. As we will show in this section, the angular distribution of the radiation emitted by an ultrarelativistic particle has a quite different shape, being sharply peaked along a preferred direction. We begin the analysis by inserting the asymptotic Liénard-Wiechert field (10.2) in the radiation formula (8.18), obtaining thus for the power radiated by a single particle the general expression, valid for arbitrary velocities,

$$\frac{d\mathcal{W}}{d\Omega} = \frac{e^2}{16\pi^2}\,\frac{\left|\mathbf{n} \times ((\mathbf{n} - \mathbf{v}) \times \mathbf{a})\right|^2}{(1 - \mathbf{v} \cdot \mathbf{n})^6}. \tag{10.37}$$

In the non-relativistic limit, i.e. for $v \approx 0$, Eq. (10.37) obviously reduces to Eq. (10.36). Vice versa, in the limit for $v \to 1$ the dependence of the second member of equation (10.37) on the emission direction $\mathbf{n}$ is dominated by the factor $1/(1 - \mathbf{v} \cdot \mathbf{n})^6$. For non-relativistic speeds this factor is close to unity for all $\mathbf{n}$, whereas for ultrarelativistic ones, $v \approx 1$, it becomes very large in the flight direction of the particle, i.e. for $\mathbf{n} \approx \mathbf{v}/v$. In fact, for $\mathbf{n} = \mathbf{v}/v$ we have $1 - \mathbf{v} \cdot \mathbf{n} = 1 - v$. To analyze the effect of this factor more closely we recast Eq. (10.37) as the product of two factors

$$\frac{dW}{d\Omega} = \frac{e^2}{16\pi^2} \left| \frac{\mathbf{n} \times ((\mathbf{n} - \mathbf{v}) \times \mathbf{a})}{1 - \mathbf{v} \cdot \mathbf{n}} \right|^2 \cdot \frac{1}{(1 - \mathbf{v} \cdot \mathbf{n})^4}, \qquad (10.38)$$

and distinguish two kinematical situations.

Generic acceleration. Let us consider an instant at which the particle's velocity and acceleration form a generic *non-vanishing* angle. Choosing the direction $\mathbf{n} = \mathbf{v}/v$ we then have[2]

$$\frac{\mathbf{n} - \mathbf{v}}{1 - \mathbf{v} \cdot \mathbf{n}} = \mathbf{n}, \qquad (10.39)$$

and, consequently, in the flight direction the first factor of Eq. (10.38) becomes velocity-independent

$$\left| \frac{\mathbf{n} \times ((\mathbf{n} - \mathbf{v}) \times \mathbf{a})}{1 - \mathbf{v} \cdot \mathbf{n}} \right|^2 = |\mathbf{n} \times \mathbf{a}|^2. \qquad (10.40)$$

Vice versa, the second factor of (10.38) in the direction $\mathbf{n} = \mathbf{v}/v$ assumes the value $1/(1 - v)^4$, which for $v \approx 1$ is very large. We see, thus, that the radiation emitted by an ultrarelativistic particle in generic motion has a pronounced peak at its instantaneous flight direction. Let us now estimate the angular aperture of the cone, coaxial with the flight direction, in which the predominant part of the radiation is emitted. The corresponding directions $\mathbf{n}$ must be such that the term $1 - \mathbf{v} \cdot \mathbf{n}$ remains close to its minimum, i.e.

$$1 - \mathbf{v} \cdot \mathbf{n} \sim 1 - v, \qquad (10.41)$$

so that the factor $1/(1 - \mathbf{v} \cdot \mathbf{n})^4$ in Eq. (10.38) remains close to its maximum $1/(1 - v)^4$. Denoting the small angle between the vectors $\mathbf{n}$ and $\mathbf{v}$ by α, we obtain the expansion

---

[2]With a more accurate analysis, left as exercise, one can prove the general inequalities

$$1 \le \left| \frac{\mathbf{n} - \mathbf{v}}{1 - \mathbf{v} \cdot \mathbf{n}} \right| \le \frac{1}{\sqrt{1 - v^2}}.$$

If α denotes the angle between $\mathbf{v}$ and $\mathbf{n}$, the lower bound is attained for $\alpha = 0$ and $\alpha = \pi$, while the upper bound is attained for $\sin \alpha = \sqrt{1 - v^2}$. Therefore, if $v \approx 1$, the magnitude of the vector $(\mathbf{n} - \mathbf{v})/(1 - \mathbf{v} \cdot \mathbf{n})$ becomes, actually, very large for $\alpha \approx \sqrt{1 - v^2}$. This means that the estimates (10.39) and (10.40) in reality are *lower bounds*.

$$1 - \mathbf{v} \cdot \mathbf{n} = 1 - v \cos \alpha \approx 1 - v \left(1 - \frac{\alpha^2}{2} \right) \approx 1 - v + \frac{\alpha^2}{2},$$

which implies that (10.41) remains valid as long as α does not exceed a value of the order $\sqrt{1 - v}$. Equivalently, since $1 - v = (1 - v^2)/(1 + v) \approx (1 - v^2)/2$, the relation (10.41) holds true as long as α remains below the value

$$\alpha \sim \sqrt{1 - v^2}. \tag{10.42}$$

In conclusion, an ultrarelativistic charged particle in generic motion emits radiation mainly along its flight direction, and the predominant part of the radiation is emitted inside a narrow cone centered around this direction, having an angular aperture of the order $\alpha \sim \sqrt{1 - v^2}$. An ultrarelativistic electron circling in a synchrotron, for example, emits radiation via a highly collimated beam spiraling in the plane of the orbit, resembling a pulsating flash. Notice that this emission pattern has little to do with the angular shape of the radiation emitted by a *non-relativistic* synchrotron, see Eq. (8.129) of Problem 8.1.

Linear motion. If a particle performs a linear motion, as happens, for instance, in a linear accelerator, $\mathbf{a}$ is always parallel to $\mathbf{v}$ and so Eq. (10.38) reduces to

$$\frac{d\mathcal{W}}{d\Omega} = \frac{e^2}{16\pi^2} \frac{|\mathbf{n} \times \mathbf{a}|^2}{(1 - \mathbf{v} \cdot \mathbf{n})^6} = \frac{e^2}{16\pi^2} \frac{a^2 \sin^2 \alpha}{(1 - v \cos \alpha)^6}, \tag{10.43}$$

where α is again the angle between $\mathbf{n}$ and $\mathbf{v}$. In this case, as $d\mathcal{W}/d\Omega$ vanishes at $\alpha = 0$, the particle emits no radiation in the flight direction. However, analyzing the function of α appearing in (10.43), it is not difficult to see that in the ultrarelativistic limit $d\mathcal{W}/d\Omega$ has a pronounced peak at $\alpha \sim \sqrt{1 - v^2}$, see Problem 10.3. Therefore, also in this case the predominant part of the radiation is emitted inside a cone coaxial with the linear trajectory, with an angular aperture of the order $\alpha \sim \sqrt{1 - v^2}$.

Emitted versus observed energy. We end this section with a comment concerning the physical interpretation of the single-particle radiation formula (10.37). By its very definition, this formula yields the energy of the radiation which at a fixed time t crosses the sphere of radius r per unit time dt in the direction $\mathbf{n}$. This radiation originates from the position of the particle at the retarded time t', satisfying the equation $t = t' + r - \mathbf{n} \cdot \mathbf{y}(t')$. The energy emitted by the particle between the instants $t' = \tau_1$ and $t' = \tau_2$ is thus given by the integral

$$\frac{d\varepsilon}{d\Omega} = \int_{\tau_1 + r - \mathbf{n} \cdot \mathbf{y}(\tau_1)}^{\tau_2 + r - \mathbf{n} \cdot \mathbf{y}(\tau_2)} \frac{d\mathcal{W}}{d\Omega} \, dt = \int_{\tau_1}^{\tau_2} \frac{d\mathcal{W}}{d\Omega} (1 - \mathbf{n} \cdot \mathbf{v}) \, dt',$$

where we have used (10.4). Consequently, the energy $\mathcal{W}'$ emitted by the particle per unit acceleration time dt' is given by

$$\frac{dW'}{d\Omega} = \frac{d^2\varepsilon}{dt'd\Omega} = (1 - \mathbf{n} \cdot \mathbf{v})\frac{dW}{d\Omega}. \tag{10.44}$$

In other words, $dW/d\Omega$ represents the energy detected by a distant observer, whereas $dW'/d\Omega$ represents the energy emitted by the particle. Finally, we can introduce a third quantity, namely the energy W^0 emitted by the particle per unit proper time ds

$$\frac{dW^0}{d\Omega} = \frac{dt'}{ds}\frac{dW'}{d\Omega} = \frac{1}{\sqrt{1 - v^2}}\frac{dW'}{d\Omega} = \frac{1 - \mathbf{n} \cdot \mathbf{v}}{\sqrt{1 - v^2}}\frac{dW}{d\Omega}.$$

It follows that

$$\frac{dW}{d\Omega} = \frac{\sqrt{1 - v^2}}{1 - \mathbf{n} \cdot \mathbf{v}}\frac{dW^0}{d\Omega}.$$

In this formula one recognizes the proportionality factor of the relativistic Doppler effect, see Eq. (5.128), which relates the inverse of the emission-time interval, i.e. the proper frequency $\omega_0 = 1/\Delta s$ of a light source moving with constant velocity $\mathbf{v}$, to the inverse of the reception-time interval, i.e. the frequency $\omega = 1/\Delta t$ measured by a static observer.

10.4   Problems

10.1 Show that the total energy radiated by a particle of charge e, mass m and arbitrarily large speed v, passing at a large impact parameter b in the vicinity of a static nucleus of charge Ze, is given by

$$\Delta\varepsilon(v, b) = \frac{e^6 Z^2 \left(1 - \frac{v^2}{4c^2}\right)}{192\pi^2 m^2 v c^3 b^3 \left(1 - \frac{v^2}{c^2}\right)}.$$

Compare this result with the expression (8.137) of Problem 8.5.
Hint: Apply the general formula (10.26), using that for a large impact parameter the particle's trajectory deviates only slightly from a uniform linear motion.

10.2 An electromagnetic wave with electric field

$$\mathbf{E}(t, \mathbf{x}) = (E_0 \cos(\omega(t - z)), E_0 \sin(\omega(t - z)), 0)$$

hits a charged particle which as a consequence performs a *relativistic* motion.

(a) Verify that the incident wave is circularly polarized.
     Hint: See Eqs. (5.91) and (5.93).
(b) Determine the magnetic field $\mathbf{B}(t, \mathbf{x})$ of the wave.

(c) Verify that in the presence of a circularly polarized wave, the stationary motions of the particle are uniform circular motions, and determine their speed and radius.

   Hint: By definition, in a stationary motion the average velocity of the particle is zero, $\overline{\mathbf{v}} = 0$. These motions depend, thus, on three independent parameters.

(d) Assuming that the particle performs a stationary motion calculate the total radiated power.

   Hint: Use Eq. (10.26).

10.3 Analyze the angular distribution of the radiation emitted by an ultrarelativistic particle in linear motion, see Eq. (10.43). Determine in particular the directions of maximum and minimum intensity. Compare the answers with the non-relativistic angular distribution (10.36).

Chapter 11
Spectral Analysis

In the previous chapters we have developed a series of tools for the determination of the four-momentum carried by the radiation generated by a generic charge distribution. In particular, we have derived explicit formulas for the total energy radiated per unit time, as well as for the angular distribution of the emitted radiation. For some systems we have also been able to determine the *frequencies* entailed by the radiation. So we have seen that the linear antenna emits radiation of a fixed frequency, that in the Thomson scattering the radiation emitted by a non-relativistic electron has the same frequency as the incident wave, and that a charge in uniform circular motion emits dipole radiation with the same frequency of its orbital motion, as well as quadrupole radiation with twice this frequency, see Sect. 8.4.4 and Problem 8.6. Similarly, the frequency of the *gravitational* radiation emitted by a non-relativistic periodic matter system is twice its orbital frequency, see Problem 9.3.

In general, the electromagnetic radiation emitted by a relativistic system is spread over a wide range of frequencies, called its *spectrum*. For many radiating physical systems – from molecules to pulsars – the spectrum constitutes a kind of *genetic code*, which makes the system uniquely recognizable. The relevant physical quantity is the energy $\Delta\varepsilon$ of the radiation with frequencies contained in the small interval $[\omega, \omega + \Delta\omega]$, or, equivalently, the *spectral weight* $\Delta\varepsilon/\Delta\omega$ of the frequency ω. The investigation of the properties of this quantity is called *spectral analysis*, which is the main topic of this chapter. A further, no less important, physical quantity characterizing the radiation of a system is the direction of the associated electric and magnetic fields, that is to say, its *polarization*. The general framework for a systematic analysis of this quantity will be presented in Sect. 11.2. In general, the combined experimental observation of these two quantities – the spectral weights and the polarization – furnishes important information on the structure of the physical system emitting the radiation, allowing, at times, to identify it unambiguously.

© Springer International Publishing AG, part of Springer Nature 2018
K. Lechner, *Classical Electrodynamics*, UNITEXT for Physics,
https://doi.org/10.1007/978-3-319-91809-9_11

11.1   Fourier Analysis and General Results

The general solution of Maxwell's equations in empty space is a radiation field – superposition of monochromatic elementary waves – and accordingly the temporal Fourier transform of the electromagnetic field corresponds to a *frequency* decomposition, see Eq. (5.76). As shown in Sect. 8.1.2, also the field generated by a generic charge distribution in the *wave zone* is a radiation field, and so its temporal Fourier transform amounts, once more, to a frequency decomposition. One of the basic results of Sect. 8.1.2 was that a monochromatic current of frequency ω generates a monochromatic wave of the same frequency. More specifically, the temporal Fourier transform $\mathbf{E}(\omega, \mathbf{x})$ of the electric field generated in the wave zone by a generic, say, aperiodic system can be directly linked to the Fourier coefficients $\mathbf{j}(\omega, \mathbf{x})$ of the corresponding current $\mathbf{j}(t, \mathbf{x})$, see Eq. (8.3). In fact, inserting the current (8.3) in the potential in the wave zone (8.10), and using the wave relation (8.14), we find the continuous Fourier decomposition of the electric field

$$\mathbf{E}(t, \mathbf{x}) = \frac{1}{\sqrt{2\pi}} \int_{-\infty}^{\infty} e^{i\omega t}\, \mathbf{E}(\omega, \mathbf{x})\, d\omega, \tag{11.1}$$

where the Fourier coefficients, representing the amplitudes of the waves of frequency ω, are given by

$$\mathbf{E}(\omega, \mathbf{x}) = \frac{i\omega e^{-i\omega r}}{4\pi r}\, \mathbf{n} \times \left(\mathbf{n} \times \int e^{i\omega \mathbf{n} \cdot \mathbf{y}}\, \mathbf{j}(\omega, \mathbf{y})\, d^3 y \right). \tag{11.2}$$

Vice versa, for a periodic charge distribution the current admits the discrete decomposition (8.4), and the electric field is periodic. In this case the continuous Fourier representation (11.1) is thus replaced by a *Fourier series*, see below. As we will see shortly, the Fourier coefficients $\mathbf{E}(\omega, \mathbf{x})$ of the electric field in the wave zone are the main ingredients of the spectral analysis of an arbitrary radiation field. For a generic charge distribution with current $\mathbf{j}(t, \mathbf{x})$ this analysis could be based on the general formula (11.2) for $\mathbf{E}(\omega, \mathbf{x})$, or on its discrete counterpart, which link these coefficients directly to the current. However, in the particular case of the radiation of a single particle, on which we mainly focus in this chapter, it is more convenient to compute these coefficients by starting directly from the Liénard-Wiechert fields. Nevertheless, we shall retrieve the general representation (11.2) in Sect. 11.5 – see Eq. (11.139) – which is devoted specifically to the spectral analysis of the radiation emitted by *generic* charge distribution.

Fourier analysis. Let $\mathbf{E}(t, \mathbf{x})$ be the electric field generated by a generic system in the wave zone. To simplify the notation, in the following we will omit to indicate explicitly the dependence on the spatial coordinate, writing hence $\mathbf{E}(t)$ in place of $\mathbf{E}(t, \mathbf{x})$. We shall adopt analogous conventions for the temporal Fourier transforms, and Fourier series, of the electric field. As recalled above, an *aperiodic* charge distribution creates an electric field $\mathbf{E}(t)$, which admits a Fourier transform $\mathbf{E}(\omega)$. These functions obey the standard relations

$$\mathbf{E}(t) = \frac{1}{\sqrt{2\pi}} \int_{-\infty}^{\infty} e^{i\omega t}\, \mathbf{E}(\omega)\, d\omega, \tag{11.3}$$

$$\mathbf{E}(\omega) = \frac{1}{\sqrt{2\pi}} \int_{-\infty}^{\infty} e^{-i\omega t}\, \mathbf{E}(t)\, dt, \tag{11.4}$$

$$\int_{-\infty}^{\infty} \left|\mathbf{E}(t)\right|^2 dt = \int_{-\infty}^{\infty} \left|\mathbf{E}(\omega)\right|^2 d\omega = 2 \int_{0}^{\infty} \left|\mathbf{E}(\omega)\right|^2 d\omega. \tag{11.5}$$

We have introduced the shorthand notation for the scalar product between a complex vector and its complex conjugate

$$\left|\mathbf{E}(\omega)\right|^2 = \mathbf{E}^*(\omega) \cdot \mathbf{E}(\omega).$$

The spectrum of the emitted frequencies, that is to say, the set of ω for which $\mathbf{E}(\omega)$ is different from zero, is hence a *continuous* subset of $\mathbb{R}$.

A *periodic* charge distribution, with period T and *fundamental frequency* $\omega_0 = 2\pi/T$, generates instead a periodic field $\mathbf{E}(t)$, admitting thus a Fourier-series representation with Fourier coefficients $\mathbf{E}_N$. The relations analogous to (11.3)–(11.5) are

$$\mathbf{E}(t) = \sum_{N=-\infty}^{\infty} e^{iN\omega_0 t}\, \mathbf{E}_N, \tag{11.6}$$

$$\mathbf{E}_N = \frac{1}{T} \int_{0}^{T} e^{-iN\omega_0 t}\, \mathbf{E}(t)\, dt, \tag{11.7}$$

$$\frac{1}{T} \int_{0}^{T} \left|\mathbf{E}(t)\right|^2 dt = \sum_{N=-\infty}^{\infty} \left|\mathbf{E}_N\right|^2 = 2 \sum_{N=1}^{\infty} \left|\mathbf{E}_N\right|^2, \tag{11.8}$$

where Eq. (11.8) is known as *Parseval's identity*. The Fourier series (11.6) gives the electric field as a linear superposition of infinite trigonometric functions, called *harmonics*, each one corresponding to an elementary wave of amplitude $\mathbf{E}_N$, and of frequency

$$\omega_N = N\omega_0.$$

For a periodic charge distribution the spectrum of the emitted frequencies is hence a *discrete* subset of $\mathbb{R}$. The term relative to $N = 1$ in Eq. (11.6) is called the *fundamental* harmonic, and the ones relative to $|N| > 1$ are called *higher* harmonics. In Eq. (11.8) we have intentionally omitted the Fourier coefficient relative to $N = 0$, because $\mathbf{E}_0$ vanishes identically. In fact, Eq. (8.14) implies that the electric field in the wave zone can be written as the total time derivative

$$\mathbf{E}(t) = \frac{\partial}{\partial t}\left(\mathbf{n} \times (\mathbf{n} \times \mathbf{A}(t))\right),$$

where also the vector potential in the wave zone $\mathbf{A}(t) = \mathbf{A}(t, \mathbf{x})$ is a periodic function of time, see Eq. (8.10). The formula for the Fourier coefficients (11.7) for $N = 0$ then yields

$$\mathbf{E}_0 = \frac{1}{T} \int_0^T \mathbf{E}(t) \, dt = \frac{1}{T} \mathbf{n} \times (\mathbf{n} \times (\mathbf{A}(T) - A(0))) = 0. \qquad (11.9)$$

To write the third expression in Eqs. (11.5) and (11.8) we took advantage from the fact that the electric field is real, so that its Fourier coefficients obey the relations, see property (5.26),

$$\mathbf{E}^*(\omega) = \mathbf{E}(-\omega), \qquad \mathbf{E}_N^* = \mathbf{E}_{-N}. \qquad (11.10)$$

These relations allow us to introduce only *positive* frequencies. In particular, henceforth, we shall assume $N > 0$. From the general relation (11.2) we see that for an aperiodic charge distribution the Fourier coefficients $\mathbf{E}(\omega) = \mathbf{E}(\omega, \mathbf{x}) = \mathbf{E}(\omega, r\mathbf{n})$ depend on r only through the factor $e^{-i\omega r}/r$, and hence they are essentially functions of ω and of the emission direction $\mathbf{n}$. Similarly, for a periodic charge distribution the Fourier coefficient $\mathbf{E}_N$ for a fixed N is essentially a function of $\mathbf{n}$. Returning to the basic radiation formula (8.18)

$$\frac{d\mathcal{W}}{d\Omega} = \frac{d^2\varepsilon}{dt d\Omega} = r^2 |\mathbf{E}(t)|^2, \qquad (11.11)$$

we now want to express the *spectral weights* of the radiation in terms of the Fourier coefficients $\mathbf{E}(\omega)$ and $\mathbf{E}_N$, for aperiodic and periodic charge distributions, respectively, see Eqs. (11.12) and (11.14) below.

Aperiodic charge distributions. In general an aperiodic system is composed of a set of charges performing *unbounded* motions, see Sect. 7.1, so that their accelerations vanish rapidly for $t \to \pm\infty$. This implies that the total energy $d\varepsilon/d\Omega$ radiated per unit solid angle between $t = -\infty$ and $t = \infty$ is finite. From Eqs. (11.11) and (11.5) we then derive the sum rule

$$\frac{d\varepsilon}{d\Omega} = \int_{-\infty}^{\infty} \frac{d\mathcal{W}}{d\Omega} \, dt = r^2 \int_{-\infty}^{\infty} |\mathbf{E}(t)|^2 dt = 2r^2 \int_0^{\infty} |\mathbf{E}(\omega)|^2 d\omega.$$

From this relation we conclude that the energy radiated in the direction $\mathbf{n}$ per unit solid angle and per unit frequency interval – the *continuous spectral weight* of the radiation – is given by

$$\frac{d^2\varepsilon}{d\omega d\Omega} = 2r^2 |\mathbf{E}(\omega)|^2. \qquad (11.12)$$

Periodic charge distributions. For a periodic system of charges the energy emitted between $t = -\infty$ and $t = \infty$ is infinite. In this case the physically meaningful quantity is the *mean power*, more precisely, the energy $d\overline{\mathcal{W}}/d\Omega$ radiated per unit solid angle during a period, divided by the period T. From Eqs. (11.11) and (11.8) we now obtain the sum rule

$$\frac{d\overline{W}}{d\Omega} = \frac{1}{T}\int_0^T \frac{dW}{d\Omega}\,dt = \frac{r^2}{T}\int_0^T \left|\mathbf{E}(t)\right|^2 dt = 2r^2 \sum_{N=1}^\infty \left|\mathbf{E}_N\right|^2. \qquad (11.13)$$

From this rule we conclude that the mean power of the radiation with frequency $\omega_N = N\omega_0$ emitted in the direction n per unit solid angle – the *discrete spectral weight* of the radiation – is given by

$$\frac{dW_N}{d\Omega} = 2r^2\left|\mathbf{E}_N\right|^2, \quad N = 1, 2, 3, \ldots. \qquad (11.14)$$

Integrating Eqs. (11.13) and (11.14) over the total solid angle, we find for the total mean power $\overline{W}$ the sum rule

$$\overline{W} = \sum_{N=1}^\infty W_N, \qquad W_N = 2r^2\int \left|\mathbf{E}_N\right|^2 d\Omega, \qquad (11.15)$$

where W_N is the total mean power of the radiation with frequency $N\omega_0$. Formulas (11.12) and (11.14) yield simple expressions for the spectral weights, or the *spectral distributions*, of a generic radiation field, and they represent hence the basis for the spectral analysis of a generic radiation process. It is important to keep in mind that in all above equations the symbol $\mathbf{E}(\cdot) = \mathbf{E}(\cdot, \mathbf{x})$ does not stand for the *exact* electric field, but for the electric field in the *wave zone*.

11.2 Polarization

The physical observable called *polarization* generically refers to the direction of the electric field, which in the wave zone lies in the plane orthogonal to the propagation direction n. Therefore, the vector $\mathbf{E}(t)$, henceforth denoted simply by $\mathbf{E}$, can be decomposed along two directions orthogonal to n, which are identified by two unit vectors $\mathbf{e}_p$ subject to the constraints $(p, q = 1, 2)$

$$\mathbf{n}\cdot\mathbf{e}_p = 0, \qquad \mathbf{e}_p\cdot\mathbf{e}_q = \delta_{pq}. \qquad (11.16)$$

The choice of $\mathbf{e}_1$ and $\mathbf{e}_2$, in general, depends on the underlying geometry of the system that emits the radiation. If in an experiment, in addition to the frequency, also the polarization states of the radiation are observed, to compare the observations with the theoretical predictions we need a formula for the spectral weights of the radiation with polarization parallel to a given direction. To derive such a formula we start again from the radiation formula (11.11), and insert a completeness relation in the modulus squared of the electric field

$$|\mathbf{E}|^2 = (\mathbf{e}_1 \cdot \mathbf{E})^2 + (\mathbf{e}_2 \cdot \mathbf{E})^2 + (\mathbf{n} \cdot \mathbf{E})^2 = (\mathbf{e}_1 \cdot \mathbf{E})^2 + (\mathbf{e}_2 \cdot \mathbf{E})^2. \qquad (11.17)$$

The angular distribution of the radiation (11.11) can then be written as the sum of two terms, representing the radiation with polarization parallel to $\mathbf{e}_1$ and $\mathbf{e}_2$, respectively,

$$\frac{d\mathcal{W}}{d\Omega} = r^2 \left((\mathbf{e}_1 \cdot \mathbf{E})^2 + (\mathbf{e}_2 \cdot \mathbf{E})^2\right) = \frac{d\mathcal{W}^1}{d\Omega} + \frac{d\mathcal{W}^2}{d\Omega}. \qquad (11.18)$$

In general, the angular distribution of the power of the radiation with polarization parallel to an arbitrary direction $\mathbf{e}_p$ is hence given by

$$\frac{d\mathcal{W}^p}{d\Omega} = r^2 (\mathbf{e}_p \cdot \mathbf{E})^2. \qquad (11.19)$$

In the same way, from Eqs. (11.12) and (11.14) we find that the spectral weights of the radiation polarized parallel to $\mathbf{e}_p$ are given by

$$\frac{d^2\varepsilon^p}{d\omega d\Omega} = 2r^2 \left|\mathbf{e}_p \cdot \mathbf{E}(\omega)\right|^2, \qquad (11.20)$$

$$\frac{d\mathcal{W}_N^p}{d\Omega} = 2r^2 \left|\mathbf{e}_p \cdot \mathbf{E}_N\right|^2, \qquad (11.21)$$

for a continuous and discrete spectrum of frequencies, respectively.

Linear polarization. We recall that radiation is said to be linearly polarized if the direction of $\mathbf{E}$ is constant in time, see the condition (5.92). Radiation propagating in a certain direction $\mathbf{n}$ is then linearly polarized, say along $\mathbf{e}_1$, if $\mathbf{E} \parallel \mathbf{e}_1$ for all t and for all $\mathbf{x}$ or, equivalently, if

$$|\mathbf{e}_1 \cdot \mathbf{E}| = |\mathbf{E}|, \qquad \mathbf{e}_2 \cdot \mathbf{E} = 0. \qquad (11.22)$$

For a generic polarization direction $\mathbf{e}_p$ forming with $\mathbf{e}_1$ the angle φ_p, Eqs. (11.18) and (11.19) then imply the relation

$$\frac{d\mathcal{W}^p}{d\Omega} = \cos^2 \varphi_p \frac{d\mathcal{W}}{d\Omega}. \qquad (11.23)$$

The spectral weight $d\mathcal{W}^p/d\Omega$ has thus a maximum for $\mathbf{e}_p \parallel \mathbf{e}_1$, where it equals the total spectral weight, and a minimum for $\mathbf{e}_p \perp \mathbf{e}_1$, where it is zero. Notice, however, that if the radiation is not linearly polarized, there exists no choice of the directions $\mathbf{e}_1$ and $\mathbf{e}_2$ for which Eqs. (11.22) and (11.23) hold. Furthermore, it can happen that the radiation is linearly polarized only for certain frequencies, say, in the case of a periodic system, for the frequency $N\omega_0$. In this case the conditions (11.22) are replaced by $|e_1 \cdot \mathbf{E}_N| = |\mathbf{E}_N|$ and $e_2 \cdot \mathbf{E}_N = 0$, and the spectral weights relative to e_p are given by

$$\frac{dW_N^p}{d\Omega} = \cos^2 \varphi_p \frac{dW_N}{d\Omega}. \tag{11.24}$$

Completely analogous results hold for aperiodic systems.

Circular polarization. Let us now analyze the case of circular polarization. For concreteness we consider the periodic field (11.6) and select the harmonic with frequency $N\omega_0$

$$\mathbf{E} = e^{iN\omega_0 t} \mathbf{E}_N + c.c. \tag{11.25}$$

We recall from Sect. 5.3.1 that a generic monochromatic wave (5.91) is circularly polarized, if and only if its polarization vector $\mathcal{E}$ satisfies either one of the conditions (5.93). For the wave (11.25) this implies that the Fourier coefficient $\mathbf{E}_N$ must obey either one of the constraints

$$\mathbf{n} \times \mathbf{E}_N = \pm i\mathbf{E}_N. \tag{11.26}$$

To find a simple criterion allowing to check whether radiation is circularly polarized, we consider an arbitrary pair of orthogonal unit vectors e_1 and e_2, and recall the relation $e_2 = \pm \mathbf{n} \times e_1$. The conditions (11.26) for circular polarization then imply the equality

$$\left| e_1 \cdot \mathbf{E}_N \right| = \left| e_1 \cdot (\mathbf{n} \times \mathbf{E}_N) \right| = \left| (\mathbf{n} \times e_1) \cdot \mathbf{E}_N \right| = \left| e_2 \cdot \mathbf{E}_N \right|. \tag{11.27}$$

Since e_1 is arbitrary, from the sum rule $|e_1 \cdot \mathbf{E}_N|^2 + |e_2 \cdot \mathbf{E}_N|^2 = |\mathbf{E}_N|^2$ it then follows that for a circularly polarized wave the quantity $|e_p \cdot \mathbf{E}_N|$ is independent of e_p, and so we have

$$\left| e_p \cdot \mathbf{E}_N \right| = \frac{1}{\sqrt{2}} \left| \mathbf{E}_N \right|, \quad \text{for all } e_p. \tag{11.28}$$

It is easy to show that also the opposite is true, namely that the condition (11.28) implies that the complex vector $\mathbf{E}_N$ satisfies either one of the constraints (11.26). In conclusion, a wave of the form (11.25) is circularly polarized, if and only if the polarization vector $\mathbf{E}_N$ satisfies the conditions (11.28). Finally, from the form of the polarized spectral weights (11.21) and of the total spectral weights (11.14), we see that the conditions (11.28) are equivalent to the equalities

$$\frac{dW_N^p}{d\Omega} = \frac{1}{2} \frac{dW_N}{d\Omega}, \quad \text{for all } e_p. \tag{11.29}$$

This means that an arbitrarily oriented linear polarizer filters of a circularly polarized wave *half* of the total radiation.

Elliptical polarization. We consider again the monochromatic wave (11.25), recalling that for a generic complex polarization vector $\mathbf{E}_N$ the wave is said to be elliptically polarized. For this case in Sect. 5.3.1 we have seen that there exist two principal axes $\mathbf{e}_1$ and $\mathbf{e}_2$ – the axes of the ellipse described by the electric field – such that the polarization vector in (11.25) can be recast in the form, see Eq. (5.94),

$$\mathbf{E}_N = e^{i\alpha_N} (E_{N1}\,\mathbf{e}_1 + iE_{N2}\,\mathbf{e}_2), \tag{11.30}$$

where α_N, E_{N1} and E_{N2} are real numbers. For a generic polarization direction $\mathbf{e}_p$ forming with, say $\mathbf{e}_1$, the angle φ_p, the spectral weights (11.21) then take the form

$$\frac{dW_N^p}{d\Omega} = 2r^2 \left(\cos^2\varphi_p\, E_{N1}^2 + \sin^2\varphi_p\, E_{N2}^2\right) = \cos^2\varphi_p \frac{dW_N^1}{d\Omega} + \sin^2\varphi_p \frac{dW_N^2}{d\Omega}. \tag{11.31}$$

It is easy to recognize Eqs. (11.24) and (11.29) as particular cases of this formula. Completely analogous results hold for aperiodic systems.

11.3   Spectral Analysis in the Non-relativistic Limit

In the non-relativistic limit the continuous and discrete spectral weights (11.12) and (11.14) take particularly simple forms. In fact, in this limit the electric field in the wave zone can be expressed in terms of the dipole moment $\mathbf{D}(t)$ (8.47) as, see Eq. (8.51),

$$\mathbf{E}(t) = \frac{1}{4\pi r}\,\mathbf{n} \times \left(\mathbf{n} \times \ddot{\mathbf{D}}(t - r)\right). \tag{11.32}$$

We perform the spectral analysis separately for aperiodic and periodic systems.

Aperiodic systems. The dipole moment (8.47) of an aperiodic system is a generic function of time, so that we can introduce its Fourier transform

$$\mathbf{D}(\omega) = \frac{1}{\sqrt{2\pi}} \int_{-\infty}^{\infty} e^{-i\omega t}\,\mathbf{D}(t)\, dt. \tag{11.33}$$

Equation (11.32) then allows to express the Fourier coefficients of the electric field in terms of those of the dipole moment as

$$\mathbf{E}(\omega) = -\frac{\omega^2 e^{-i\omega r}}{4\pi r}\,\mathbf{n} \times (\mathbf{n} \times \mathbf{D}(\omega)), \tag{11.34}$$

where we used the property (2.92) for the Fourier transform of the derivatives of a function. Equation (11.12) then yields the angular spectral weights

$$\frac{d^2\varepsilon}{d\omega d\Omega} = \frac{\omega^4}{8\pi^2} \left|\mathbf{n} \times \mathbf{D}(\omega)\right|^2. \tag{11.35}$$

Integrating both sides of this equation over the solid angle – proceeding in the same way as in Eqs. (8.52) and 8.53 – we find the total spectral weights

$$\frac{d\varepsilon}{d\omega} = \frac{\omega^4}{8\pi^2} \int |\mathbf{n} \times \mathbf{D}(\omega)|^2 d\Omega = \frac{\omega^4}{3\pi} |\mathbf{D}(\omega)|^2. \tag{11.36}$$

Finally, integrating this equation over all frequencies we obtain for the total radiated energy the expression

$$\Delta\varepsilon = \frac{1}{3\pi} \int_0^\infty \omega^4 |\mathbf{D}(\omega)|^2 d\omega.$$

Characteristic frequencies of non-relativistic particles. We consider a single non-relativistic charged particle performing a generic *aperiodic* motion $\mathbf{y}(t)$, under the influence of a force $\mathbf{F}(t)$. In this case the dipole moment is given by $\mathbf{D}(t) = e\mathbf{y}(t)$, and so $\ddot{\mathbf{D}}(t) = e\mathbf{a}(t)$, where $\mathbf{a}(t)$ is the acceleration of the particle. The Fourier transform of this relation reads $-\omega^2 \mathbf{D}(\omega) = e\mathbf{a}(\omega)$, where $\mathbf{a}(\omega)$ denotes the Fourier transform of $\mathbf{a}(t)$. The total spectral weights (11.36) then take the simple form

$$\frac{d\varepsilon}{d\omega} = \frac{e^2}{3\pi} |\mathbf{a}(\omega)|^2. \tag{11.37}$$

Let us denote by $\mathcal{T}$ the characteristic time scale of the force $\mathbf{F}(t)$ acting on the particle, meaning that the force varies significantly during a time interval of the order of $\mathcal{T}$. In the simplest case $\mathbf{F}(t)$ is markedly different from zero only during a time $\mathcal{T}$, where it is practically constant. Since the acceleration is linked to the acceleration by Newton's second law $\mathbf{a}(t) = \mathbf{F}(t)/m$, according to the Fourier-transform uncertainty inequality[1] the function $|\mathbf{a}(\omega)|$ is then significantly different from zero only for values of ω not exceeding $1/\mathcal{T}$. We have, therefore, proved the following general result: if a non-relativistic particle in aperiodic motion is acted on by a force varying with a characteristic time scale $\mathcal{T}$, then most of the emitted radiation has the characteristic frequencies

$$\omega \lesssim \frac{1}{\mathcal{T}}. \tag{11.38}$$

Periodic systems. The dipole moment (8.47) of a charge distribution performing a periodic motion is likewise periodic, with the same period T, and it can thus be expanded in a Fourier series whose Fourier coefficients are

$$\mathbf{D}_N = \frac{1}{T} \int_0^T e^{-iN\omega_0 t} \mathbf{D}(t)\, dt. \tag{11.39}$$

---

[1]We refer here to the known fact that the *widths* Δx and Δk of a function $f(x)$ and of its Fourier transform $\widehat{f}(k)$ are related by the uncertainty inequality $\Delta x \Delta k \geq 1/2$.

Equations (11.7) and (11.32) then allow to express the Fourier coefficients of the electric field in terms of those of the dipole moment,

$$\mathbf{E}_N = -\frac{(N\omega_0)^2 e^{-iN\omega_0 r}}{4\pi r} \, \mathbf{n} \times (\mathbf{n} \times \mathbf{D}_N), \tag{11.40}$$

so that Eq. (11.14) gives the spectral weights

$$\frac{d\mathcal{W}_N}{d\Omega} = \frac{(N\omega_0)^4}{8\pi^2} \left| \mathbf{n} \times \mathbf{D}_N \right|^2. \tag{11.41}$$

Integrating equation (11.41) over the solid angle we obtain the total power radiated with frequency $N\omega_0$

$$\mathcal{W}_N = \frac{(N\omega_0)^4}{3\pi} \left| \mathbf{D}_N \right|^2. \tag{11.42}$$

Characteristic frequencies of non-relativistic particles. Again we consider a non-relativistic charged particle, this time with a periodic trajectory $\mathbf{y}(t)$. Since we have still $\mathbf{D}(t) = e\mathbf{y}(t)$ and $\ddot{\mathbf{D}}(t) = e\mathbf{a}(t)$, expanding both members of the latter equation in a Fourier series we obtain the relations

$$- (N\omega_0)^2 \mathbf{D}_N = e\mathbf{a}_N, \tag{11.43}$$

and, from (11.40),

$$\mathbf{E}_N = \frac{e\, e^{-iN\omega_0 r}}{4\pi r} \, \mathbf{n} \times (\mathbf{n} \times \mathbf{a}_N), \tag{11.44}$$

where $\mathbf{a}_N$ is the Nth Fourier coefficient of the acceleration. It is convenient to recast this coefficient in the form

$$\mathbf{a}_N = \frac{1}{T} \int_0^T e^{-iN\omega_0 t} \, \mathbf{a}(t) \, dt = \frac{\sqrt{2\pi}}{T} \, \mathbf{C}(N\omega_0), \tag{11.45}$$

where $\mathbf{C}(\omega)$ is the Fourier transform of the vector function

$$\mathbf{C}(t) = \mathbf{a}(t) \, \chi_{[0,T]}(t), \tag{11.46}$$

and $\chi_{[0,T]}$ is the characteristic function of the interval $[0,T]$, see (2.75). For $t \in [0,T]$ the function $\mathbf{C}(t)$ coincides, hence, with the acceleration $\mathbf{a}(t)$, while in the complement of this interval it is zero. Using Eqs. (11.43) and (11.45) the spectral weights (11.42) can be written as

$$\mathcal{W}_N = \frac{e^2 \left| \mathbf{a}_N \right|^2}{3\pi} = \frac{2e^2 \left| \mathbf{C}(N\omega_0) \right|^2}{3T^2}. \tag{11.47}$$

In light of the relation (11.46), and considering that $\mathbf{a}(t) = \mathbf{F}(t)/m$, if the characteristic time of the force $\mathbf{F}(t)$ coincides with the period T of the motion, according to the Fourier-transform uncertainty inequality the function $\mathbf{C}(\omega)$ is significantly different from zero only for $\omega \lesssim 1/T = \omega_0/2\pi$. Consequently, the quantity $\mathbf{C}(N\omega_0)$ governing the spectral weight (11.47) is significantly different from zero only if $N\omega_0 \lesssim \omega_0/2\pi$, that is to say, if N is of order *unity*. This means that in this case the particle emits radiation predominantly via the fundamental harmonic, and via the first lowest ones. More generally, if the force varies more rapidly, say with a characteristic time scale of the order T/K with $K > 1$, then $\mathbf{C}(N\omega_0)$ is significantly different from zero for $N\omega_0 \lesssim K/T = K\omega_0/2\pi$. In this case the particle emits thus radiation mainly via the harmonics with frequencies

$$\omega_N = N\omega_0, \qquad \text{with } N \lesssim K. \tag{11.48}$$

Simple harmonic motion. As particular case we consider a system of non-relativistic charged particles performing the *simple harmonic motions*

$$\mathbf{y}_r(t) = \sin(\omega_0 t)\mathbf{b}_r + \cos(\omega_0 t)\mathbf{c}_r, \tag{11.49}$$

with the same period $T = 2\pi/\omega_0$ for all r. Examples are uniform circular motions, and oscillatory motions on straight lines. Such a system emits only radiation via the fundamental harmonic, with frequency ω_0. In fact, in this case the Fourier coefficients of the dipole moment $\mathbf{D}(t) = \sum_r e_r \mathbf{y}_r(t)$ are given by

$$\mathbf{D}_N = \sum_r e_r \mathbf{y}_{rN}, \tag{11.50}$$

where the $\mathbf{y}_{rN}$ are the Fourier coefficients of the trajectories $\mathbf{y}_r(t)$. Since the latter are of the form (11.49), all coefficients $\mathbf{y}_{rN}$ vanish, apart from $\mathbf{y}_{r\pm1}$. Consequently, the only non-vanishing Fourier coefficients (11.50) of the dipole moment are $\mathbf{D}_1$ and $\mathbf{D}_{-1} = \mathbf{D}_1^*$, and so the unique non-vanishing spectral weight in (11.42) is $\mathcal{W}_1$. This means that the system emits only radiation with the fundamental frequency ω_0. Notice that, although the radiation is monochromatic in all directions, in general it is *elliptically* polarized, see Problem 11.5. We emphasize that the results of this section hold only in the non-relativistic limit. In particular, if the particles performing the simple harmonic motions (11.49) have arbitrarily high speeds, they emit higher multipole radiations of all frequencies $N\omega_0$, see Chap. 12.

11.3.1 Continuous Spectrum of Frequencies and Infrared Catastrophe

We begin to illustrate the general results derived above with the analysis of the radiation emitted by a non-relativistic charged particle, crossing a finite region where a constant and uniform electric field $\mathbf{E}$ is present. In this case the acceleration of the particle is different from zero only for a finite time, and so it performs an aperiodic motion. Therefore, it emits radiation with a *continuous* spectrum of frequencies. In the following we want to analyze the shape of the spectral distributions (11.20) and (11.37) and, in particular, to compare the latter with the general prediction of the characteristic frequencies (11.38). Without loss of generality, we can assume that the particle enters the zone of the electric field at the time $t = -\mathcal{T}$ and that it leaves it at the time $t = \mathcal{T}$. During this time interval its acceleration is

$$\mathbf{a} = \frac{e\mathbf{E}}{m},$$

whereas for $|t| > \mathcal{T}$ it is zero. To compute the spectral weights (11.37) we must evaluate the Fourier transform

$$\mathbf{a}(\omega) = \frac{1}{\sqrt{2\pi}} \int_{-\infty}^{\infty} e^{-i\omega t}\, \mathbf{a}(t)\, dt = \frac{\mathbf{a}}{\sqrt{2\pi}} \int_{-\mathcal{T}}^{\mathcal{T}} e^{-i\omega t}\, dt = \sqrt{\frac{2}{\pi}}\, \frac{\sin(\omega \mathcal{T})\, \mathbf{a}}{\omega}, \tag{11.51}$$

giving

$$\frac{d\varepsilon}{d\omega} = \frac{2e^2 a^2}{3\pi^2}\, \frac{\sin^2(\omega \mathcal{T})}{\omega^2}. \tag{11.52}$$

As a function of the frequency $d\varepsilon/d\omega$ has a maximum for $\omega = 0$, and it vanishes for the first time for $\omega = \pi/\mathcal{T}$. Then, for $\omega \gtrsim 1/\mathcal{T}$ it decreases rapidly, in agreement with the general estimate (11.38). In this case it is also easy to evaluate the polarized spectral angular distribution (11.20). To calculate the Fourier transform $\mathbf{E}(\omega)$ of the field in the wave zone we resort to formula (11.34), use the relation between Fourier transforms $\mathbf{D}(\omega) = -e\mathbf{a}(\omega)/\omega^2$, and substitute the result (11.51)

$$\mathbf{E}(\omega) = \frac{\sqrt{2}\, e \sin(\omega \mathcal{T})\, e^{-i\omega r}}{4\pi^{3/2} r \omega} \big((\mathbf{n} \cdot \mathbf{a})\, \mathbf{n} - \mathbf{a}\big). \tag{11.53}$$

Inserting this expression in the general formula (11.20), taking into account that $\mathbf{e}_p$ is orthogonal to $\mathbf{n}$, we obtain the polarized spectral weights

$$\frac{d^2 \varepsilon^p}{d\omega d\Omega} = \frac{e^2 \sin^2(\omega \mathcal{T})(\mathbf{e}_p \cdot \mathbf{a})^2}{4\pi^3 \omega^2}. \tag{11.54}$$

Although the dependence on the emission direction $\mathbf{n}$ formally dropped out, the expression (11.54) still depends on $\mathbf{n}$ through the unit vector $\mathbf{e}_p$ which, in fact, must be orthogonal to the former. To study the behavior of these spectral weights, for a fixed n we chose as basis for the polarization directions the vector e_1 contained in the plane spanned by a and n and orthogonal to n, and the vector e_2 orthogonal to e_1, a and n. We can then decompose e_p according to $e_p = \cos \varphi_p \, e_1 + \sin \varphi_p \, e_2$. Denoting the angle formed by the vectors a and n by β, we then obtain

$$e_p \cdot a = (e_1 \cdot a) \cos \varphi_p = a \sin \beta \cos \varphi_p.$$

For a given n the spectral weights (11.54) hence take the form

$$\frac{d^2 \varepsilon^p}{d\omega d\Omega} = \left(\frac{e^2 a^2 \sin^2 \beta \sin^2 (\omega T)}{4\pi^3 \omega^2} \right) \cos^2 \varphi_p. \tag{11.55}$$

Comparing this expression with Eq. (11.23), or rather with its continuous version, we conclude that for any frequency ω and for any emission direction n the radiation is *linearly* polarized, with polarization parallel to the vector e_1 specified above. Of course, this conclusion matches with the explicit form of the polarization vector of the electric field (11.53), see also the condition (5.92).

Total radiated energy and ultraviolet divergences. To compute the total energy $\Delta \varepsilon$ radiated during the deflection process we can integrate equation (11.52) over all frequencies, using the integral

$$\int_0^\infty \left(\frac{\sin x}{x} \right)^2 dx = \frac{\pi}{2}, \tag{11.56}$$

or, alternatively, apply Larmor's formula $\mathcal{W} = e^2 a^2 / 6\pi$. The result is

$$\Delta \varepsilon = \int_0^\infty \frac{d\varepsilon}{d\omega} \, d\omega = \int_{\mathcal{T}}^{\mathcal{T}} \mathcal{W} \, dt = \frac{e^2 a^2 \mathcal{T}}{3\pi} = \frac{e^2 |\Delta v|^2}{12\pi \mathcal{T}}, \tag{11.57}$$

where Δv denotes the difference between the final and initial velocities

$$\Delta v = v_f - v_i = 2\mathcal{T} a. \tag{11.58}$$

The energy $\Delta \varepsilon$ can be calculated also by integrating the spectral weights (11.55), for $\varphi_p = 0$, over all angles and over all frequencies, see Problem 11.3. Equation (11.57) establishes a direct link between the radiated energy and its cause, i.e. the change of the velocity. Let us now analyze the behavior of the spectral distribution (11.52) in the limit in which the duration $2\mathcal{T}$ of the acceleration tends to zero, while keeping Δv fixed. In this limit the process degenerates in an *instantaneous* deflection, entailing an infinite acceleration. Expressing the spectral distribution (11.52) in terms of the variation of the velocity (11.58) we obtain the finite limiting distribution

$$\lim_{T \to 0} \frac{d\varepsilon}{d\omega} = \lim_{T \to 0} \left(\frac{e^2 |\Delta \mathbf{v}|^2}{6\pi^2} \frac{\sin^2(\omega T)}{\omega^2 T^2} \right) = \frac{e^2 |\Delta \mathbf{v}|^2}{6\pi^2}. \tag{11.59}$$

The resulting spectrum is thus *flat*, in the sense that all frequencies appear in the radiation with the same weight, a result that formally is still in agreement with the general estimate (11.38) if we send $T \to 0$. However, as in this limit the instantaneous acceleration becomes infinite, the total radiated energy (11.57) for $T \to 0$ *diverges*. Clearly, the same result is obtained if one integrates the flat spectral distribution (11.59) over all frequencies. We see, therefore, that the schematization of collision- and deflection-processes of charged particles as *instantaneous* processes, frequently used in theoretical physics for the conceptual as well as technical simplifications it entails, actually, is physically inconsistent, because the radiated energy of these idealized processes is infinite.

In elementary particle physics, an infinity of the type encountered above is called an *ultraviolet divergence*. By definition, a divergence of this kind is due to the singular behavior of a physical quantity for small distances, or for small times. In the case above this physical quantity is the total emitted energy, which diverges as the time $2T$ of the deflection goes to zero or, equivalently, as the spatial region where the deflection occurs shrinks to zero. It is a general rule that divergences of this kind – via *Fourier duality* – entail also a singular behavior of the physical quantity for large energies, or for large momenta. In fact, in the frequency domain, in Eq. (11.59) above we saw that the divergence of the total emitted energy is due to the large "ultraviolet" frequencies, which in a quantum mechanical picture correspond indeed to photons with high energies. Of course, the ultraviolet divergence of the total energy (11.57) for $T \to 0$ is non-physical, as in nature there exist no instantaneous processes. A specular variant of these divergences, namely the infrared ones, is presented below.

Infrared catastrophe. We conclude the analysis of this prototypical radiation process – entailing a continuous frequency spectrum – illustrating a fundamental *quantum mechanical* phenomenon known as *infrared catastrophe*. We recall that in quantum theory electromagnetic radiation of frequency ω is composed of photons of energy $\hbar\omega$. We can then ask, in basis of the spectral distribution (11.52), how many photons are emitted during the whole deflection process with frequencies contained in the interval ω and $\omega + d\omega$. The answer is[2]

$$\frac{dN}{d\omega} = \frac{1}{\hbar\omega} \frac{d\varepsilon}{d\omega} = \frac{2e^2 a^2}{3\pi^2 \hbar} \frac{\sin^2(\omega T)}{\omega^3}.$$

Therefore, the number of emitted photons with frequencies contained in the interval $[\omega_1, \omega_2]$ is given by the integral

---

[2]On general grounds, we are allowed to analyze the electromagnetic radiation by means of classical instruments, that is to say, ignoring quantum effects, as long as the involved wavelengths are much longer than the Compton wavelength, $\lambda \gg \lambda_C = \hbar/mc$, i.e. $\omega \ll mc^2/\hbar$. Since the infrared catastrophe concerns frequencies tending to zero, our classical analysis is, therefore, anyhow valid.

$$N(\omega_1, \omega_2) = \int_{\omega_1}^{\omega_2} \frac{dN}{d\omega}\, d\omega = \frac{2e^2 a^2}{3\pi^2 \hbar} \int_{\omega_1}^{\omega_2} \frac{\sin^2(\omega \mathcal{T})}{\omega^3}\, d\omega. \qquad (11.60)$$

In particular, the number of *hard* photons, that is to say, photons of high frequencies, is finite, since the integral $N(\omega_1, \infty)$ is finite for every $\omega_1 > 0$. Conversely, in the limit for $\omega \to 0$ the integrand of Eq. (11.60) behaves as

$$\frac{\sin^2(\omega \mathcal{T})}{\omega^3} \quad \to \quad \frac{\mathcal{T}^2}{\omega},$$

so that the number $N(0, \omega_2)$ of photons with frequencies below ω_2 *diverges* logarithmically for any $\omega_2 > 0$. We see therefore that, although the total radiated energy (11.57) is finite, the particle radiates an *infinite* number of *soft* photons, i.e. photons whose frequencies tend to zero. This phenomenon is called *infrared catastrophe*, in that it entails the presence of an infinite number of photons with wavelength tending to infinity. However, only a finite number of these photons can be observed experimentally, because any experimental apparatus – of whatever kind – has a finite sensitivity, meaning that it can detect only photons with energies above a certain threshold.

General analysis. The infrared catastrophe is a *universal* phenomenon – caused just by the non-vanishing (aperiodic) acceleration of the particle – which is in particular independent of the form of the force acting on the particle. This phenomenon accompanies, therefore, *whatsoever* collision process involving charged particles. To show it we consider a generic deflection process during which the velocity of a charged particle experiences a non-vanishing variation $\Delta \mathbf{v} = \mathbf{v}_f - \mathbf{v}_i$. In the limit for $\omega \to 0$ the Fourier transform $\mathbf{a}(\omega)$ of the particle's acceleration then tends to the non-vanishing finite value

$$\lim_{\omega \to 0} \mathbf{a}(\omega) = \lim_{\omega \to 0} \frac{1}{\sqrt{2\pi}} \int_{-\infty}^{\infty} e^{-i\omega t} \mathbf{a}(t)\, dt = \frac{1}{\sqrt{2\pi}} \int_{-\infty}^{\infty} \mathbf{a}(t)\, dt = \frac{\Delta \mathbf{v}}{\sqrt{2\pi}}. \qquad (11.61)$$

From the general expression of the spectral distribution (11.37) we then derive that in the limit for $\omega \to 0$ the number of emitted photons per unit frequency interval behaves as

$$\frac{dN}{d\omega} = \frac{1}{\hbar\omega} \frac{d\varepsilon}{d\omega} = \frac{e^2 |\mathbf{a}(\omega)|^2}{3\pi\hbar\omega} \quad \to \quad \frac{e^2 |\Delta \mathbf{v}|^2}{6\pi^2\hbar} \frac{1}{\omega}. \qquad (11.62)$$

Consequently, the number of soft photons $N(0, \omega_2) = \int_0^{\omega_2} (dN/d\omega)\, d\omega$ again diverges logarithmically for any $\omega_2 > 0$, unless the velocity variation $\Delta \mathbf{v}$ vanishes.

Infrared catastrophe in the fundamental interactions. The infrared catastrophe is intimately tied to the fact that the mediator of the electromagnetic interaction, the photon, has *zero mass*, so that its energies $\hbar\omega$ can reach arbitrarily low values. Since the total energy emitted in *whatsoever* physical process is finite, this phenomenon can, therefore, not take place in the weak interactions, because their mediators – the particles $W^{\pm}$ and Z^0 – have a non-vanishing mass. Conversely, at first glance the infrared catastrophe occurs both in the gravitational as well as in the strong interactions, since their mediators, the gravitons and gluons, respectively, are massless particles. However, in the strong interactions, due to the *color confinement* paradigm, the soft gluons radiated by the quarks during a scattering process rapidly form bound states, leading to hadronic, massive, particles, and so they do not manifest themselves asymptotically as *free* particles.

Being a low energy phenomenon, the infrared catastrophe, analyzed by us at the semiclassical level, recurs also in *quantum field theory*, where it causes a series of problems of both technical and conceptual character: in particular, it causes *divergent* transition amplitudes and cross sections. In *quantum electrodynamics*, as well as in *quantum chromodynamics*, the field theory that describes the strong interactions, the problem of these *infrared* divergences has found an, at least, *pragmatic* solution, allowing to compare the predictions of these theories with the experimental data [1, 2]. On the other hand, in quantum chromodynamics this problem is still waiting for a *conceptual* solution, related with the finiteness of a *complete* set of physical observables, see, for instance, Refs. [3, 4]. The difficulty of this theory lies in the fact that each of the infinite soft gluons emitted by a quark has a non-vanishing color *charge*, and it emits therefore, in turn, an infinite number of soft gluons, and so on. Conversely, a soft photon has no electric charge and so it cannot emit further photons. This is the reason for why, eventually, infrared divergences are easier to control in quantum electrodynamics, than in quantum chromodynamics.

Finally, contrary to what one might think, the infrared divergences entailed by the gravitational and electromagnetic interactions have a quite similar structure. In fact, in the first place, as we saw in Chap. 9, the gravitational analogue of the electric charge – so to say, the *gravitational* charge – is the mass, which in a relativistic theory, in turn, must be replaced with the energy. In the second place, in the same way as any accelerated charged particle emits electromagnetic radiation, any accelerated body emits gravitational radiation. In a quantum theory of gravity this radiation is composed of gravitons, in particular of an infinite number of *soft* gravitons, which, being gravitationally charged, i.e. having a non-vanishing energy, emit, in turn, an infinite number of soft gravitons, and so on. However, as we recalled above, in a relativistic theory of gravity the gravitational coupling constant is not the mass, but rather the energy, and since a soft graviton has an energy which tends to zero, the probability of emission of additional soft gravitons is, therefore, strongly suppressed. For this reason, the soft gravitons that cause infrared problems are only the *primary* ones, namely those emitted directly by the accelerated body. Eventually, these gravitons are completely analogous to the soft photons emitted by a charged particle in quantum electrodynamics, and therefore they can be controlled in quite the same way, see Ref. [5].

11.3.2   Bessel and Neumann Functions

Bessel functions of integer order. In the following we will frequently encounter the Bessel functions of integer order N, defined by

$$J_N(x) = \frac{1}{2\pi} \int_0^{2\pi} e^{i(Ny - x\sin y)}\, dy = \frac{1}{\pi} \int_0^\pi \cos(Ny - x\sin y)\, dy. \quad (11.63)$$

In this section we list some of their basic properties, referring for a more exhaustive list to textbooks of special functions, see, for instance, Refs. [6, 7]. The Bessel functions satisfy the parity relations

$$J_N(-x) = J_{-N}(x) = (-)^N J_N(x), \quad (11.64)$$

and the integral formulas

$$\frac{1}{2\pi} \int_0^{2\pi} e^{i(Ny - x\sin y)} \cos y\, dy = \frac{N}{x} J_N(x), \quad (11.65)$$

$$\frac{1}{2\pi} \int_0^{2\pi} e^{i(Ny - x\sin y)} \sin y\, dy = i J_N'(x), \quad (11.66)$$

where the *prime* denotes the derivative with respect to x. Property (11.65) follows from the identity

$$\int_0^{2\pi} \frac{d}{dy}\, e^{i(Ny - x\sin y)}\, dy = 0,$$

whereas property (11.66) is implied by the definition (11.63). Similarly, one can prove the recursion relations for the derivatives of the Bessel functions

$$J_N'(x) = \frac{1}{2}\left(J_{N-1}(x) - J_{N+1}(x)\right).$$

We have the asymptotic behaviors

$$J_N(x) \approx \sqrt{\frac{2}{\pi x}} \cos\left(x - \frac{\pi}{4}(2N+1)\right), \quad \text{for } x \to +\infty, \quad N \text{ fixed}, \quad (11.67)$$

$$J_N(x) \approx \frac{1}{N!}\left(\frac{x}{2}\right)^N, \quad \text{for } x \to 0, \quad N \geq 0 \text{ fixed}, \quad (11.68)$$

$$J_N(x) \approx \frac{1}{N!}\left(\frac{x}{2}\right)^N, \quad \text{for } N \to \infty, \quad x \text{ fixed}, \quad (11.69)$$

where the formal identity of the behaviors (11.68) and (11.69) has to be considered as a coincidence. The Bessel function J_N satisfies the differential equation

$$x^2 J_N'' + x J_N' + \left(x^2 - N^2\right) J_N = 0. \tag{11.70}$$

Neumann functions of integer order. Equation (11.70) is a second-order differential equation, and as such for any fixed N it admits two linearly independent solutions. One solution is J_N, and another independent solution is the *Neumann function* of order N, also called Bessel function of the *second type*,

$$Y_N(x) = \frac{1}{\pi} \int_0^\pi \sin\left(x \sin y - N y\right) dy - \frac{1}{\pi} \int_0^\infty \left(e^{Ny} + (-)^N e^{-Ny}\right) e^{-x \sinh y} dy,$$

which is defined for $x > 0$. These functions share some properties with the Bessel functions, as for instance the identities

$$Y_{-N}(x) = (-)^N Y_N(x), \qquad Y_N'(x) = \frac{1}{2}\left(Y_{N-1}(x) - Y_{N+1}(x)\right), \tag{11.71}$$

while they differ by others. In particular, they entail the distinctive asymptotic behaviors

$$Y_N(x) \approx \sqrt{\frac{2}{\pi x}} \sin\left(x - \frac{\pi}{4}\left(2N + 1\right)\right), \quad \text{for } x \to +\infty, \quad N \text{ fixed,} \tag{11.72}$$

and the behaviors for $x \to 0$

$$Y_N(x) \approx \begin{cases} -\dfrac{(N-1)!}{\pi}\left(\dfrac{2}{x}\right)^N, & \text{for } N \geq 1, \\ \dfrac{2}{\pi} \ln x, & \text{for } N = 0. \end{cases} \tag{11.73}$$

Notice that, unlike the Bessel functions, the Neumann functions are singular at the point $x = 0$. Moreover, the functions J_N admit an analytic continuation in the whole complex plane, while the analytically-continued functions Y_N have a *branch cut* along the real negative semi-axis. For further properties of the Neumann functions we refer the reader to [6, 7].

11.3.3 Discrete Spectrum of Frequencies: Anharmonic Oscillator

We illustrate the qualitative features of the radiation generated by a non-relativistic periodic system discussed above, in the case of a charged particle oscillating with period T along a finite arc of a circle of radius R, representing thus an anharmonic oscillator. For concreteness we choose the trajectory, lying in the xy plane,

$$\mathbf{y}(t) = (R \cos \varphi(t), R \sin \varphi(t), 0), \qquad \varphi(t) = \Phi \sin(\omega_0 t), \tag{11.74}$$

where $\omega_0 = 2\pi/T$ is the fundamental frequency, and $\Phi < \pi/2$ is the angular amplitude of the motion. The particle oscillates hence around the point $(R, 0, 0)$. To make
sure that the motion is non-relativistic the maximum particle's speed must be much
smaller than the speed of light, i.e. we must have $\Phi R \omega_0 \ll 1$. Although being periodic, the motion (11.74) is not a *simple* harmonic one, see Eq. (11.49), and consequently a priori the particle emits radiation with all harmonic frequencies $N\omega_0$.
However, in the case at hand the characteristic time scale of the particle's acceleration coincides with its period T, and so the general estimate (11.48) applies with
$K \sim 1$. The particle is, hence, expected to predominantly emit radiation with the
first lowest harmonic frequencies. Below we perform explicitly the spectral analysis
of the emitted radiation, with a special care on these qualitative predictions.

We begin with the evaluation of the total mean power emitted during a cycle $\overline{W}$,
by means of Larmor's formula

$$\overline{W} = \frac{e^2}{6\pi} \overline{a^2} = \frac{e^2}{6\pi T} \int_0^T a^2(t)\, dt. \tag{11.75}$$

From the trajectory (11.74) we derive for the square of the instantaneous particle's
acceleration the expression

$$a^2(t) = \left(R\ddot{\varphi}\right)^2 + \left(R\dot{\varphi}^2\right)^2 = R^2 \omega_0^4 \left(\Phi^2 \sin^2(\omega_0 t) + \Phi^4 \cos^4(\omega_0 t)\right),$$

which, inserted in the integral (11.75), yields the total mean power

$$\overline{W} = \frac{e^2 R^2 \omega_0^4}{6\pi} \left(\frac{1}{2} \Phi^2 + \frac{3}{8} \Phi^4\right). \tag{11.76}$$

To determine, instead, the fraction of the power emitted with the frequency $N\omega_0$ we
must evaluate the Fourier coefficients of the electric dipole moment $\mathbf{D}(t) = e\mathbf{y}(t)$

$$\mathbf{D}_N = \frac{1}{T} \int_0^T e^{-iN\omega_0 t} \mathbf{D}(t)\, dt = \frac{eR}{T} \int_0^T e^{-iN\omega_0 t} \left(\cos\varphi(t), \sin\varphi(t), 0\right) dt.$$

Introducing the Bessel functions (11.63), and using the identities (11.64), the evaluation of the above integrals leads to

$$\mathbf{D}_N = \frac{eR}{2} \left(J_N(\Phi) + J_N(-\Phi), -i(J_N(\Phi) - J_N(-\Phi)), 0\right)$$

$$= eR J_N(\Phi) \cdot \begin{cases} (1, 0, 0), & \text{for } N \text{ even,} \\ (0, -i, 0), & \text{for } N \text{ odd.} \end{cases} \tag{11.77}$$

Once these coefficients are known, we can compute the Fourier coefficients of the
electric field (11.40), and hence the polarized spectral weights (11.21). It turns out
that the radiation is linearly polarized in all directions and for all frequencies $N\omega_0$.

However, for a given emission direction **n**, the even and odd harmonics have different, in general non orthogonal, polarization directions, see Problem 11.4. Disregarding the polarization, as well as the angular distribution, by inserting the Fourier coefficients (11.77) in the formula for the discrete spectral weights (11.42) we derive for the total power emitted with the frequency $N\omega_0$ the compact expression

$$\mathcal{W}_N = \frac{e^2 R^2 (N\omega_0)^4}{3\pi} J_N^2(\Phi). \tag{11.78}$$

According to the total mean-power formula (11.15), the discrete spectral weights (11.78) must sum up to give for all angular amplitudes Φ the expression (11.76)

$$\overline{W} = \sum_{N=1}^{\infty} \mathcal{W}_N. \tag{11.79}$$

To analyze the spectral distribution of the radiation in more detail, we now try to identify those weights $\mathcal{W}_N$ which account for most of the total power $\overline{W}$. From Eqs. (11.76) and (11.78) we obtain the fractional weights

$$\frac{\mathcal{W}_N}{\overline{W}} = \frac{N^4 J_N^2(\Phi)}{\frac{1}{4}\Phi^2 + \frac{3}{16}\Phi^4}. \tag{11.80}$$

Using the asymptotic behavior of the Bessel functions (11.69), and recalling Stirling's approximation for $N \to \infty$

$$N! \sim \sqrt{2\pi N} \left(\frac{N}{e}\right)^N, \tag{11.81}$$

we derive that the fraction (11.80) for large N entails the leading behavior

$$\frac{\mathcal{W}_N}{\overline{W}} \sim \frac{1}{N^{2N}}.$$

The higher harmonics are hence strongly suppressed. To quantify the degree of suppression we separately analyze the cases of small angular amplitudes, $\Phi \ll 1$, and of amplitudes of order unity, $\Phi \sim 1$. For small amplitudes the trajectory (11.74), modulo terms of order Φ^2, reduces to

$$\mathbf{y}(t) = (R, R\Phi \sin(\omega t), 0),$$

which is a *simple* harmonic motion, see Eq. (11.49). In this case the power of the fundamental harmonic practically exhausts the total mean power, $\mathcal{W}_1 \approx \overline{W}$. In fact, Eq. (11.80) for $N = 1$ confirms this prediction in that

$$\frac{\mathcal{W}_1}{\mathcal{W}} = \frac{J_1^2(\Phi)}{\frac{1}{4}\Phi^2 + \frac{3}{16}\Phi^4} = 1 - \Phi^2 + O(\Phi^4).$$

We have used the Taylor expansion of the Bessel function, easily derivable from the definition (11.63),

$$J_1(x) = \frac{x}{2} - \frac{x^3}{16} + O(x^5). \tag{11.82}$$

However, also for large angular amplitudes the situation remains qualitatively unchanged. For $\Phi = 1$, for instance, which corresponds to an amplitude of about 60°, Eq. (11.80) provides the relative weights

$$\frac{\mathcal{W}_N}{\mathcal{W}} = \frac{16}{7} N^4 J_N^2(1).$$

Using for $J_N(1)$ the numerical values listed in the handbooks one finds the ratios

$$\frac{\mathcal{W}_1}{\mathcal{W}} = 0.43, \qquad \frac{\mathcal{W}_1 + \mathcal{W}_2}{\mathcal{W}} = 0.91, \qquad \frac{\mathcal{W}_1 + \mathcal{W}_2 + \mathcal{W}_3}{\mathcal{W}} = 0.98.$$

We see thus that in practice almost all energy is carried away by the first three harmonics, as predicted by the general estimate (11.48) with $K \sim 1$.

11.4 Relativistic Spectral Analysis

In this section we perform the spectral analysis of the radiation emitted by a generic relativistic charged particle, with a focus on ultrarelativistic motions. First, in Sect. 11.4.1 we derive for the discrete and continuous spectral weights the exact integral representations (11.91) and (11.95), respectively. We then use these representations in Sect. 11.4.2 to determine the characteristic frequencies radiated by ultrarelativistic particles, and in Chap. 12 to investigate the spectral distribution of the synchrotron radiation.

11.4.1 Spectral Distribution of a Particle in Arbitrary Motion

A particle following a generic trajectory $\mathbf{y}(t)$ generates an electric field which in the wave zone assumes the form (7.59)

$$\mathbf{E}(t, \mathbf{x}) = \frac{e}{4\pi r} \frac{\mathbf{n} \times ((\mathbf{n} - \mathbf{v}) \times \mathbf{a})}{(1 - \mathbf{n} \cdot \mathbf{v})^3}. \tag{11.83}$$

To determine the spectral weights of the radiation emitted by the particle we must explicitly evaluate the continuous and discrete Fourier coefficients (11.4) and (11.7) of the field (11.83), and insert them in the general formulas (11.12) and (11.14), respectively. We perform these calculations separately for periodic and aperiodic trajectories.

Periodic trajectory. For a periodic motion we must compute for a fixed $\mathbf{x}$ the discrete Fourier coefficients

$$\mathbf{E}_N = \frac{1}{T} \int_0^T e^{-iN\omega_0 t}\, \mathbf{E}(t, \mathbf{x})\, dt. \tag{11.84}$$

Before proceeding with the calculation we recall that the kinematical variables $\mathbf{v}$ and $\mathbf{a}$ appearing in the field (11.83) are evaluated not at the time t, but at the time $t'(t, \mathbf{x})$ satisfying the delay equation (7.62)

$$t = t' + r - \mathbf{n} \cdot \mathbf{y}(t'). \tag{11.85}$$

For this reason, in the integral (11.84) it is convenient to change integration variable from t to t'. As $\mathbf{x}$ is fixed, the measure changes according to Eq. (10.4)

$$dt = (1 - \mathbf{n} \cdot \mathbf{v}(t'))\, dt'. \tag{11.86}$$

Inserting the field (11.83) in the integral (11.84) we so obtain the Fourier coefficients

$$\mathbf{E}_N = \frac{e}{4\pi r}\, e^{-iN\omega_0 r} \frac{1}{T} \int_0^T e^{-iN\omega_0 (t' - \mathbf{n} \cdot \mathbf{y}(t'))}\, \frac{\mathbf{n} \times \big((\mathbf{n} - \mathbf{v}(t')) \times \mathbf{a}(t')\big)}{(1 - \mathbf{n} \cdot \mathbf{v}(t'))^2}\, dt'. \tag{11.87}$$

Since the trajectory $\mathbf{y}(t')$ is periodic with period T, Eq. (11.85) implies that as t varies over a period, also t' varies over a period. The integral over t' in (11.87) is hence again between 0 and T. Henceforth, we denote the integration variable t' again by t. The integral (11.87) can be further simplified by means of an integration by parts, using the identities

$$\frac{d}{dt}\left(\frac{\mathbf{n} \times (\mathbf{n} \times \mathbf{v})}{1 - \mathbf{n} \cdot \mathbf{v}}\right) = \frac{\mathbf{n} \times \big((\mathbf{n} - \mathbf{v}) \times \mathbf{a}\big)}{(1 - \mathbf{n} \cdot \mathbf{v})^2}, \tag{11.88}$$

$$\frac{d}{dt}\, e^{-iN\omega_0 (t - \mathbf{n} \cdot \mathbf{y})} = -iN\omega_0 (1 - \mathbf{n} \cdot \mathbf{v})\, e^{-iN\omega_0 (t - \mathbf{n} \cdot \mathbf{y})}. \tag{11.89}$$

In this way the Fourier coefficients (11.87) take the simpler form

$$\mathbf{E}_N = \frac{ieN\omega_0}{4\pi r}\, e^{-iN\omega_0 r}\, \mathbf{n} \times \left(\mathbf{n} \times \frac{1}{T} \int_0^T e^{-iN\omega_0 (t - \mathbf{n} \cdot \mathbf{y})}\, \mathbf{v}\, dt\right). \tag{11.90}$$

The discrete spectral-weight formula (11.14) then yields for the angular distribution of the power radiated with frequency $N\omega_0$ the compact expression

$$\frac{dW_N}{d\Omega} = \frac{e^2(N\omega_0)^2}{8\pi^2} \left| \mathbf{n} \times \frac{1}{T} \int_0^T e^{-iN\omega_0(t-\mathbf{n}\cdot\mathbf{y})} \, \mathbf{v} \, dt \right|^2, \tag{11.91}$$

where we used the general identity $|\mathbf{a} \times (\mathbf{a} \times \mathbf{b})| = |\mathbf{a}| \cdot |\mathbf{a} \times \mathbf{b}|$. A particularly advantageous feature of Eq. (11.91) is that it expresses the spectral weights in terms of a line integral along the trajectory $\mathbf{y}(t)$.

Non-relativistic limit. It is easy to verify that in the non-relativistic limit the spectral weights (11.91), valid for arbitrary velocities, reduce to the weights (11.41). In fact, in this limit the microscopic delay $\mathbf{n} \cdot \mathbf{y}$ in the exponent of the integrand in (11.91) is negligible, so that an integration by parts leads to ($\mathbf{v} = d\mathbf{y}/dt$)

$$\frac{dW_N}{d\Omega} \approx \frac{e^2(N\omega_0)^4}{8\pi^2} \left| \mathbf{n} \times \frac{1}{T} \int_0^T e^{-iN\omega_0 t} \, \mathbf{y} \, dt \right|^2.$$

This expression coincides with (11.41), as the Fourier coefficients of the dipole moment $\mathbf{D}(t) = e\mathbf{y}(t)$ of a particle are given by

$$\mathbf{D}_N = \frac{1}{T} \int_0^T e^{-iN\omega_0 t} \, \mathbf{D}(t) \, dt = \frac{e}{T} \int_0^T e^{-iN\omega_0 t} \, \mathbf{y} \, dt.$$

Aperiodic trajectory. To compute the spectral weights of the radiation of a particle in aperiodic motion we proceed in the same way as above. Inserting the electric field (11.83) in Eq. (11.4), this time we find the Fourier coefficients

$$\mathbf{E}(\omega) = \frac{e}{4\pi r} e^{-i\omega r} \frac{1}{\sqrt{2\pi}} \int_{-\infty}^{\infty} e^{-i\omega(t-\mathbf{n}\cdot\mathbf{y})} \frac{\mathbf{n} \times ((\mathbf{n}-\mathbf{v}) \times \mathbf{a})}{(1-\mathbf{n}\cdot\mathbf{v})^2} \, dt \tag{11.92}$$

$$= \frac{ie\omega}{4\pi r} e^{-i\omega r} \, \mathbf{n} \times \left(\mathbf{n} \times \frac{1}{\sqrt{2\pi}} \int_{-\infty}^{\infty} e^{-i\omega(t-\mathbf{n}\cdot\mathbf{y})} \, \mathbf{v} \, dt \right). \tag{11.93}$$

The continuous spectral-distribution formula (11.12) then yields the spectral weights

$$\frac{d^2\varepsilon}{d\omega d\Omega} = \frac{e^2}{8\pi^2} \left| \frac{1}{\sqrt{2\pi}} \int_{-\infty}^{\infty} e^{-i\omega(t-\mathbf{n}\cdot\mathbf{y})} \frac{\mathbf{n} \times ((\mathbf{n}-\mathbf{v}) \times \mathbf{a})}{(1-\mathbf{n}\cdot\mathbf{v})^2} \, dt \right|^2 \tag{11.94}$$

$$= \frac{e^2\omega^2}{8\pi^2} \left| \mathbf{n} \times \frac{1}{\sqrt{2\pi}} \int_{-\infty}^{\infty} e^{-i\omega(t-\mathbf{n}\cdot\mathbf{y})} \, \mathbf{v} \, dt \right|^2. \tag{11.95}$$

As above, in the non-relativistic limit the spectral weights (11.94) reduce to the weights (11.35).

Improper integrals and distribution theory. Unfortunately, the line integrals appearing in the Fourier coefficients (11.93) and in the spectral weights (11.95), although simpler than the corresponding integrals appearing in (11.92) and (11.94), are, actually, improper *divergent* integrals. This problem originates from the fact that the integration by parts based on the identities (11.88) and (11.89) – which led from the *finite* integral (11.92) to the divergent one (11.93) – cannot be performed *naively*. The reason is that, contrary to what happens in the periodic case, the boundary term of the integration by parts is located at infinity, and, due to the oscillating factors $e^{-i\omega(t-\mathbf{n}\cdot\mathbf{y}(t))}$, the integrand does not admit finite limits for $t \to \pm\infty$. To overcome this difficulty, before performing in formula (11.92) the integration by parts, we regularize the integral via the replacement

$$\int_{-\infty}^{\infty} dt \quad \rightarrow \quad \int_{-L}^{L} dt,$$

where L is a cut-off. Of course, the truncated integral tends to the original one in the limit for $L \to \infty$. The crucial point is, however, that this limit exists in the *pointwise* sense, as well as in the *distributional* sense, if we consider the integral as a function of ω. For a finite L we can now perform the integration by parts in the usual way. Obviously, if we let L go to infinity pointwise, that is to say, for a fixed ω, the boundary terms involving the factors $e^{-i\omega(\pm L-\mathbf{n}\cdot\mathbf{y}(\pm L))}$ still oscillate and diverge, as does the integral remaining after the integration by parts. Clearly, by construction, the divergences of the boundary term and of the remaining integral cancel each other. However, we are also allowed to perform the limit for $L \to \infty$ of the whole expression in the distributional sense, and in this case it is easily seen that the boundary term admits a well-defined limit, which is actually zero. It follows that the expressions (11.93) and (11.95) are well-defined, provided that the improper integrals appearing therein are considered as limits in the *sense of distributions*, i.e. provided that one understands the replacement (see the definition (2.63)–(2.64))

$$\int_{-\infty}^{\infty} dt \quad \rightarrow \quad \mathcal{S}' - \lim_{L\to\infty} \int_{-L}^{L} dt. \tag{11.96}$$

We illustrate the procedure by verifying that for a particle in uniform linear motion the integral (11.93), if interpreted in the above sense, yields indeed the Fourier coefficients $\mathbf{E}(\omega) = 0$, in agreement with the fact that a particle not subject to accelerations does not emit radiation. Notice that for such a motion the expressions (11.92) and (11.94), which are anyhow well-defined, give the correct answer, as in this case $\mathbf{a}(t)$ vanishes identically. Considering, conversely, the formal expression (11.93) of the Fourier coefficients, according to the above recipe we must replace the integral appearing therein with

$$\int_{-\infty}^{\infty} e^{-i\omega(t-\mathbf{n}\cdot\mathbf{y})} \mathbf{v}\, dt = \mathbf{v} \int_{-\infty}^{\infty} e^{-i\omega t(1-\mathbf{n}\cdot\mathbf{v})}\, dt \tag{11.97}$$

$$\rightarrow \quad \mathbf{v} \left(\mathcal{S}' - \lim_{L\to\infty} \int_{-L}^{L} e^{-i\omega t(1-\mathbf{n}\cdot\mathbf{v})}\, dt \right),$$

where we have substituted the trajectory of a uniform linear motion $\mathbf{y}(t) = \mathbf{v}t$. In virtue of the representation of the δ-function (2.99), which in the one-dimensional case reads

$$\mathcal{S}' - \lim_{L \to \infty} \int_{-L}^{L} e^{-ikx} dk = 2\pi \delta(x),$$

by *definition* the formal integral (11.97) is thus equal to the distribution

$$\int_{-\infty}^{\infty} e^{-i\omega(t-\mathbf{n}\cdot\mathbf{y})} \mathbf{v} \, dt = 2\pi\mathbf{v} \, \delta\big(\omega(1 - \mathbf{n} \cdot \mathbf{v})\big) = \frac{2\pi\mathbf{v} \, \delta(\omega)}{1 - \mathbf{n} \cdot \mathbf{v}}. \tag{11.98}$$

In the Fourier coefficients (11.93) the integral (11.98) is multiplied by ω, and so, due to the identity $\omega \, \delta(\omega) = 0$, we find again $\mathbf{E}(\omega) = 0$.

Systems of charged particles. Formulas (11.94) and (11.95) admit simple generalizations to radiation processes involving more than a single charged particle, as for instance a collision between charged particles, or the decay of neutral particles in charged ones. The four-current of a system of particles with charges e_r and world lines y_r^μ is given by the familiar additive expression $j^\mu(x) = \sum_r e_r \int \delta^4(x - y_r) \, dy_r^\mu$. Therefore, thanks to the superposition principle, the total asymptotic electric field is given by a sum of Liénard-Wiechert fields of the kind (11.83), and hence its Fourier transform $\mathbf{E}(\omega)$ results in a sum of terms of the type (11.92). For a system of charged particles the formula (11.12) gives, therefore, the spectral weights

$$\frac{d^2\varepsilon}{d\omega d\Omega} = \frac{1}{8\pi^2} \left| \sum_r \frac{e_r}{\sqrt{2\pi}} \int_{-\infty}^{\infty} e^{-i\omega(t-\mathbf{n}\cdot\mathbf{y}_r)} \frac{\mathbf{n} \times ((\mathbf{n} - \mathbf{v}_r) \times \mathbf{a}_r)}{(1 - \mathbf{n} \cdot \mathbf{v}_r)^2} \, dt \right|^2. \tag{11.99}$$

We will see an interesting application of this formula in Sect. 11.4.3.

11.4.2 Characteristic Frequencies in the Ultrarelativistic Limit

We now perform a qualitative analysis of the spectral distribution of a generic *ultrarelativistic* particle in aperiodic motion, having hence speed $v \approx 1$. In particular, we want to find the range of frequencies over which the dominant part of the radiation is spread. In this regard we recall from Sect. 11.3 that the characteristic frequencies radiated by a non-relativistic particle lie in the range

$$\omega \lesssim \frac{1}{\mathcal{T}}, \tag{11.100}$$

where $\mathcal{T}$ is the characteristic time scale of the force acting on the particle. Let us now suppose that an ultrarelativistic particle crosses a zone where an external electromagnetic field $\{\mathbf{E}, \mathbf{B}\}$ is present, and that this field is significantly different from zero in a region of linear spatial dimensions L. Since the particle has a high speed, its trajectory will deviate only little from a linear trajectory. In particular, the *scattering angle* χ, i.e. the angle between the initial and final velocities of the particle, is then very small, as is the angular aperture α of the cone inside which the particle emits most part of its radiation. In fact, from Sect. 10.3 we know that this angle is related to v by

$$\alpha \sim \sqrt{1 - v^2}. \tag{11.101}$$

Scattering angle. To estimate the scattering angle we start from the Lorentz equation and from the work-energy theorem

$$\frac{d\mathbf{p}}{dt} = e(\mathbf{E} + \mathbf{v} \times \mathbf{B}), \qquad \frac{d\varepsilon}{dt} = e\,\mathbf{v} \cdot \mathbf{E}. \tag{11.102}$$

As the speed of the particle is close to the speed of light, the particle experiences the external fields only for a time of the order $\mathcal{T} \sim L/v \approx L$. The changes of its momentum and energy between the initial and final states are then given by

$$|\Delta\mathbf{p}| = e\left|\int_{-\infty}^{\infty} (\mathbf{E} + \mathbf{v} \times \mathbf{B})\,dt\right| \sim eC\mathcal{T}, \qquad \Delta\varepsilon = e\int_{-\infty}^{\infty} \mathbf{v} \cdot \mathbf{E}\,dt \sim eC\mathcal{T}, \tag{11.103}$$

where we denote by C the characteristic values of the magnitudes of the fields $\mathbf{E}$ and $\mathbf{B}$. Since the particle is deflected only little by the fields, the scattering angle equals the magnitude of the difference between the versors of its final and initial velocities $\mathbf{v}_f$ and $\mathbf{v}_i$

$$\chi = \left|\Delta\left(\frac{\mathbf{v}}{v}\right)\right| = \left|\frac{\mathbf{v}_f}{v_f} - \frac{\mathbf{v}_i}{v_i}\right|.$$

Using the relation $\mathbf{v} = \mathbf{p}/\varepsilon$ we can rewrite the scattering angle also as

$$\chi = \left|\Delta\left(\frac{\mathbf{p}}{|\mathbf{p}|}\right)\right|. \tag{11.104}$$

Furthermore, the kinematic relation $|\mathbf{p}|^2 = \varepsilon^2 - m^2$ implies that $|\mathbf{p}|\,\Delta|\mathbf{p}| = \varepsilon\Delta\varepsilon$, or, equivalently, $\Delta|\mathbf{p}| = \varepsilon\Delta\varepsilon/|\mathbf{p}| \approx \Delta\varepsilon$. In this way we obtain

$$\Delta\left(\frac{\mathbf{p}}{|\mathbf{p}|}\right) \sim \frac{\Delta\mathbf{p}}{|\mathbf{p}|} - \frac{\mathbf{p}\Delta\varepsilon}{|\mathbf{p}|^2} \sim \frac{1}{|\mathbf{p}|}(\Delta\mathbf{p} - \mathbf{v}\Delta\varepsilon) = \frac{\sqrt{1-v^2}}{m}(\Delta\mathbf{p} - \mathbf{v}\Delta\varepsilon). \tag{11.105}$$

The relations (11.103)–(11.105) then yield for the scattering angel the estimate

$$\chi \sim \sqrt{1 - v^2}\,\frac{eC\mathcal{T}}{m}. \tag{11.106}$$

The same result can be obtained also by integrating the Lorentz equation in (11.102), rewritten as in Eq. (2.38), over the time interval $[0, T]$. For the ratio between the scattering angle and the angle α (11.101) we then obtain the value

$$\frac{\chi}{\alpha} \sim \frac{eCT}{m}, \tag{11.107}$$

which is independent of the particle's velocity. The dimensionless expression (11.107) can be interpreted as the ratio between the non-relativistic "cyclotron frequency" eC/m, see Eq. (10.33), and the characteristic frequency $1/T$ of the radiation emitted by a non-relativistic particle subject to the same fields

$$\frac{\chi}{\alpha} \sim \frac{\left(\dfrac{eC}{m}\right)}{\left(\dfrac{1}{T}\right)}. \tag{11.108}$$

Since for ultrarelativistic particles both angles α and χ are very small, the spectral analysis of the emitted radiation requires to distinguish two different geometrical regimes: $\chi \ll \alpha$ and $\alpha \ll \chi$.

Characteristic frequencies for $\chi \ll \alpha$. If the scattering angle is much smaller than α, most of the particle's radiation is emitted inside a cone of angular aperture α with an axis parallel to $\mathbf{v}_i \approx \mathbf{v}_f = \mathbf{v}$, whose direction remains practically unchanged during the whole radiation process. It is thus sufficient to analyze the radiation emitted in the directions $\mathbf{n}$ close to $\mathbf{v}/v$, and so in the formula (11.94) for the spectral weights we can set

$$\mathbf{n} \approx \frac{\mathbf{v}}{v}, \qquad \mathbf{n} - \mathbf{v} \approx (1 - v)\, \mathbf{n}, \qquad \mathbf{y}(t) \approx \mathbf{v}t.$$

In this way Eq. (11.94) can be approximated by

$$\frac{d^2\varepsilon}{d\omega d\Omega} \approx \frac{e^2}{8\pi^2} \left| \frac{\mathbf{n}}{1 - v} \times \frac{1}{\sqrt{2\pi}} \int_{-\infty}^{\infty} e^{-i\omega t(1-v)}\, \mathbf{a}(t)\, dt \right|^2$$

$$= \frac{e^2}{8\pi^2(1 - v)^2} \left| \mathbf{n} \times \mathbf{a}((1 - v)\omega) \right|^2, \tag{11.109}$$

where $\mathbf{a}(\omega)$ denotes the Fourier transform of $\mathbf{a}(t)$. Considering that the particle is acted on by the external fields during the characteristic time T, the function $\mathbf{a}(t)$ varies appreciably on the same time scale T. Consequently, according to the Fourier-transform uncertainty inequality, the function $\mathbf{a}((1 - v)\omega)$ is appreciably different from zero for values of ω such that

$$(1 - v)\omega = \frac{1 - v^2}{1 + v}\, \omega \approx \frac{1}{2}\, (1 - v^2)\, \omega \lesssim \frac{1}{T}.$$

Expressing this relation in terms of the particle's energy ε, we find thus that the dominant part of the emitted radiation has frequencies lying in the range

$$\omega \lesssim \frac{1}{\mathcal{T}} \frac{1}{1 - v^2} = \frac{1}{\mathcal{T}} \left(\frac{\varepsilon}{m}\right)^2. \tag{11.110}$$

With respect to the non-relativistic characteristic frequencies $\omega \lesssim 1/\mathcal{T}$, the spectrum of an ultrarelativistic particle is, hence, violently shifted towards *high* frequencies. At the quantum level this means that an ultrarelativistic particle emits photons which are much *harder* than those emitted by a non-relativistic one.[3]

Characteristic frequencies for $\alpha \ll \chi$. If the scattering angle is much larger than α, the direction **n** in which at each instant the dominant part of the radiation is emitted changes during the motion. In this case the radiation emitted in a certain direction **n** originates only from a small arc γ of the trajectory, during which the particle's velocity forms with **n** an angle smaller than $\alpha \sim \sqrt{1 - v^2}$. Calling Δ the length of γ, and considering that during the whole radiation process the direction of the trajectory of length L changes by an angle χ, we find that along γ the direction of the velocity changes by the angle

$$\frac{\Delta}{L} \chi.$$

Equating this angle to α, and recalling that $\alpha \ll \chi$, for the length of γ we derive therefore the bound

$$\Delta = \frac{\alpha}{\chi} L \ll L.$$

Since Δ is much smaller than L, along the arc γ the external fields can be considered as constant. Moreover, since this arc is short, it can be approximated with a circular arc and, furthermore, since $v \approx 1$, along γ the motion can be considered as *uniform*. For this kinematic configuration we can then anticipate the result (12.21) of Chap. 12, that provides the characteristic frequencies radiated by an ultrarelativistic particle in uniform circular motion, caused, in that case, by a uniform constant magnetic field B. Via the substitution $B \to C$, for the present case Eq. (12.21) predicts therefore the characteristic frequencies

$$\omega \lesssim \frac{eC}{m} \left(\frac{\varepsilon}{m}\right)^2. \tag{11.111}$$

Notice that these frequencies exhibit the same energy-dependence of the frequencies (11.110), although the proportionality coefficient is different. In light of the relation (11.108) we can summarize the results (11.110) and (11.111) of this section, referring to the cases $\chi \ll \alpha$ and $\alpha \ll \chi$, respectively, by saying that an ultrarelativistic charged particle emits radiation with characteristic frequencies

---

[3]If the particle has a generic speed v, the quantities $\mathcal{T}$ and L are related by $1/\mathcal{T} \sim v/L$. This implies that, for a fixed spatial scale L, the non-relativistic frequencies (11.100) carry with respect to the ultrarelativistic ones (11.110) a further suppression factor v/c.

$$\omega \lesssim \widehat{\omega} \left(\frac{\varepsilon}{m}\right)^2, \tag{11.112}$$

where $\widehat{\omega}$ is the larger of the two "fundamental" frequencies eC/m and $1/T$.

11.4.3   Low Frequencies

In general the integrals appearing in the spectral weights (11.94) and (11.95) cannot be performed analytically. However, for *low* frequencies, that is to say, for frequencies which are much smaller than the characteristic ones (11.112), it is still possible to derive compact, as well as meaningful, expressions for the spectral weights. As prototypical case we consider a charged particle performing an unbounded motion, with initial and final asymptotic velocities $\mathbf{v}_i$ and $\mathbf{v}_f$. For low frequencies, in Eq. (11.94) the exponential $e^{-i\omega(t - \mathbf{n} \cdot \mathbf{y})}$ is approximately equal to unity, and so, with the aid of the identity (11.88), in the limit for $\omega \to 0$ we find the spectral weights

$$\frac{d^2\varepsilon}{d\omega d\Omega} \approx \frac{e^2}{16\pi^3} \left| \int_{-\infty}^{\infty} \frac{\mathbf{n} \times ((\mathbf{n} - \mathbf{v}) \times \mathbf{a})}{(1 - \mathbf{n} \cdot \mathbf{v})^2} \, dt \right|^2 \tag{11.113}$$

$$= \frac{e^2}{16\pi^3} \left| \int_{-\infty}^{\infty} \frac{d}{dt} \left(\frac{\mathbf{n} \times (\mathbf{n} \times \mathbf{v})}{1 - \mathbf{n} \cdot \mathbf{v}} \right) dt \right|^2 \tag{11.114}$$

$$= \frac{e^2}{16\pi^3} \left| \mathbf{n} \times \left(\frac{\mathbf{v}_f}{1 - \mathbf{n} \cdot \mathbf{v}_f} - \frac{\mathbf{v}_i}{1 - \mathbf{n} \cdot \mathbf{v}_i} \right) \right|^2. \tag{11.115}$$

This expression links, once more, the emitted radiation directly to its cause, i.e. the change of the velocity. In particular, in the ultrarelativistic limit $v_i \approx v_f \approx 1$, due to the factors $1/(1 - \mathbf{n} \cdot \mathbf{v})$, the low-frequency radiation is emitted mostly in the vicinity of the initial and final flight directions. Actually, if ϑ denotes the angle between $\mathbf{n}$ and $\mathbf{v}$, for $\vartheta = 0$ the vector $\mathbf{n} \times \mathbf{v}/(1 - \mathbf{n} \cdot \mathbf{v})$ is zero. However, if $v \approx 1$, for a very small angle of the order $\vartheta \sim \sqrt{1 - v^2}$ the magnitude of this vector is of the large order

$$\left| \frac{\mathbf{n} \times \mathbf{v}}{1 - \mathbf{n} \cdot \mathbf{v}} \right| \sim \frac{1}{\sqrt{1 - v^2}}, \tag{11.116}$$

while for intermediate angles this magnitude is of order unity, see Problem 11.2.

Infrared catastrophe. As the expression (11.115) is independent of the frequencies, the general spectral weights $d^2\varepsilon/d\omega d\Omega$ (11.94) admit a finite limit for $\omega \to 0$ and, consequently, also in the general case the number of emitted soft photons is infinite. More precisely, Eq. (11.115) yields for the angular distribution of the number N of emitted soft photons the expression

$$\frac{d^2N}{d\omega d\Omega} = \frac{e^2}{16\pi^3 \hbar \omega} \left| \mathbf{n} \times \left(\frac{\mathbf{v}_f}{1 - \mathbf{n} \cdot \mathbf{v}_f} - \frac{\mathbf{v}_i}{1 - \mathbf{n} \cdot \mathbf{v}_i} \right) \right|^2 = \frac{f(\mathbf{n})}{\omega}, \tag{11.117}$$

where $f(\mathbf{n})$ is a frequency-independent positive function. The number of photons emitted per unit solid angle with frequencies below a certain frequency ω_1

$$\int_0^{\omega_1} \frac{d^2 N}{d\omega d\Omega} \, d\omega = f(\mathbf{n}) \int_0^{\omega_1} \frac{d\omega}{\omega}$$

diverges hence again logarithmically for any $\omega_1 > 0$. Therefore, in any however small cone of directions an infinite number of soft photons is emitted. Therefore, as one might have expected, the infrared catastrophe persists also at the relativistic level, see Sect. 11.3.1. In particular, considering Eq. (11.117) in the non-relativistic limit, where $v_i \ll 1$ and $v_f \ll 1$, and integrating it over the solid angle, one retrieves the low-frequency spectral distribution (11.62).

Bremsstrahlung in the beta decay. As an interesting application of the low-frequency formula (11.115) we investigate the radiation that inevitably accompanies whatever production process of charged particles. The so generated radiation is sometimes called *innere bremsstrahlung* – internal braking radiation – to distinguish it from the radiation due to an acceleration caused by external forces. As prototypical example we consider the beta decay of neutrons. In this process a neutron of a neutral atom A with atomic number Z decays weakly in a *proton*, an *electron*, and an electron-*antineutrino*, giving rise to a positively charged ion B of atomic number $Z + 1$. Schematically, we thus have

$$A \rightarrow B + e + \bar{\nu}_e.$$

The energy of the emitted electron *varies*, from $\varepsilon = m_e = 0.51\mathrm{MeV}$ up to about $\varepsilon \sim 10\,\mathrm{MeV}$, because it must share the energy released by the nucleus with the antineutrino. There are two charged particles involved in this process, the electron and the ion B, so that the formula (11.115) must be generalized to a system of charged particles. Starting from the spectral weights (11.99) for a system of particles, and proceeding in the same way as in Eqs. (11.113)–(11.115), taking into account that the electron and the ion have opposite charges we obtain the low-frequency spectral weights

$$\frac{d^2 \varepsilon}{d\omega d\Omega} = \frac{e^2}{16\pi^3} \left| \mathbf{n} \times \left(\frac{\mathbf{v}_f^B}{1 - \mathbf{n} \cdot \mathbf{v}_f^B} - \frac{\mathbf{v}_i^B}{1 - \mathbf{n} \cdot \mathbf{v}_i^B} - \frac{\mathbf{v}_f^e}{1 - \mathbf{n} \cdot \mathbf{v}_f^e} + \frac{\mathbf{v}_i^e}{1 - \mathbf{n} \cdot \mathbf{v}_i^e} \right) \right|^2 , \tag{11.118}$$

where the symbols are self-explanatory. We schematize the decay process considering the electron and the ion as particles which initially are essentially at rest, $\mathbf{v}_i^e \sim \mathbf{v}_i^B \sim 0$, and successively experience a quasi-instantaneous acceleration.[4]

---

[4]As shown in Sect. 11.3.1, a *strictly* instantaneous acceleration would give rise to an infinite total radiated energy, see Eq. (11.57). Nevertheless, as implied by the general Eqs. (11.61) and (11.118), the low-frequency spectral weights are always independent of the duration of the acceleration, so that at low frequencies the assumption of an instantaneous acceleration has no physical consequences at all. Similarly, for generic frequencies the spectral weights admit sensible limits as

Since the ion is much heavier than the electron, its velocity and acceleration are negligible with respect to those of the latter, in particular $v_f^B \ll v_f^e$. This means that the dominant radiation is the one emitted by the electron. Considering, therefore, in Eq. (11.118) only the third term, setting $\mathbf{v}_f^e = \mathbf{v}$ and denoting the angle between $\mathbf{n}$ and $\mathbf{v}$ by ϑ, we obtain

$$\frac{d^2\varepsilon}{d\omega d\Omega} = \frac{e^2}{16\pi^3} \left| \frac{\mathbf{n} \times \mathbf{v}}{1 - \mathbf{n} \cdot \mathbf{v}} \right|^2 = \frac{e^2 v^2 \sin^2 \vartheta}{16\pi^3 (1 - v\cos\vartheta)^2}. \tag{11.119}$$

Integrating this expression over the angles we obtain the energy radiated per unit frequency interval by an electron produced with speed v

$$\frac{d\varepsilon}{d\omega} = \frac{e^2 v^2}{16\pi^3} \int_0^{2\pi} d\varphi \int_0^{\pi} \frac{\sin^2 \vartheta}{(1 - v\cos\vartheta)^2} \sin\vartheta \, d\vartheta = \frac{e^2}{4\pi^2} \left(\frac{1}{v} \ln\frac{1+v}{1-v} - 2 \right). \tag{11.120}$$

By construction, this result is valid for low frequencies. For higher frequencies the expression (11.120) represents an upper bound. In fact, in this case the phase factor $e^{-i\omega(t - \mathbf{n}\cdot\mathbf{y})}$, which we have neglected in the derivation of (11.120), in general generates in the integral (11.94) a destructive interference. On top of this, our *classical* picture breaks down for frequencies which penetrate the quantum regime, i.e. for frequencies of the order $\hbar\omega \approx \varepsilon = m_e/\sqrt{1 - v^2}$. Taking, thus, as maximum frequency the value $\omega_M = \varepsilon/\hbar$ – the electron can anyway not emit photons with energies exceeding its own energy – Eq. (11.120) yields for the total energy radiated by the electron the crude upper bound

$$\Delta\varepsilon \lesssim \int_0^{\omega_M} \frac{d\varepsilon}{d\omega} \, d\omega = \frac{e^2 \varepsilon}{4\pi^2 \hbar} \left(\frac{1}{v} \ln\frac{1+v}{1-v} - 2 \right).$$

The fraction of energy lost by the electron during its production process is therefore at most

$$\frac{\Delta\varepsilon}{\varepsilon} \lesssim \frac{\alpha}{\pi} \left(\frac{c}{v} \ln\frac{1 + v/c}{1 - v/c} - 2 \right), \tag{11.121}$$

where we have restored the speed of light and introduced the *fine structure constant*

$$\alpha = \frac{e^2}{4\pi\hbar c} \approx \frac{1}{137}. \tag{11.122}$$

In the non-relativistic limit, $v \ll c$, Eq. (11.121) yields the negligible energy loss

$$\frac{\Delta\varepsilon}{\varepsilon} \lesssim \frac{2\alpha}{3\pi} \left(\frac{v}{c} \right)^2.$$

---

the duration of the acceleration process tends to zero, which are, in turn, frequency-independent, see, for instance, Eq. (11.59). Of course, it would make no sense to integrate the resulting spectral weights over all frequencies.

On the other hand, in the ultrarelativistic limit, where $\varepsilon = m_e c^2/\sqrt{1 - v^2/c^2} \gg m_e c^2$, we find

$$\frac{\Delta\varepsilon}{\varepsilon} \lesssim \frac{2\alpha}{\pi} \left(\ln\left(\frac{2\varepsilon}{m_e c^2}\right) - 1 \right). \tag{11.123}$$

Since in the beta decay the electrons are produced with a maximum energy of the order of $\varepsilon \sim 10\,\mathrm{MeV} \approx 20 m_e c^2$, the relation (11.123) yields the estimate $\Delta\varepsilon/\varepsilon \lesssim 1\%$. This means that even in the ultrarelativistic regime the effect of the radiative energy loss is relatively small. Nevertheless, the *innere bremsstrahlung* emitted by the electrons of the beta decay has been observed experimentally – also in the decay of *free* neutrons – and it provides important insights into the phenomenology of the weak as well as the strong interactions, see, for instance, Ref. [8].

11.5   Spectral Distribution of a Generic Current

In this section we determine the spectral weights of the radiation of a charge distribution represented by a generic conserved four-current $j^\mu(x)$, not necessarily composed of point-like charges. We will see that these weights admit simple expressions in terms of the Fourier coefficients of the four-current $J^\mu(k^0, \mathbf{k})$, relative to *all* space-time variables, see Eqs. (11.130) and (11.141) below. As usual, we distinguish periodic charge distributions and aperiodic ones.

11.5.1   Periodic Charge Distribution

The current of a periodic charge distribution, with period $T = 2\pi/\omega_0$, admits an expansion in a Fourier series in the time variable, and a Fourier-transform representation in the spatial variables, of the form

$$j^\mu(x) = \frac{1}{(2\pi)^{3/2}} \sum_{N=-\infty}^{\infty} \int d^3 k\, e^{i(N\omega_0 t - \mathbf{k}\cdot\mathbf{x})} J_N^\mu(\mathbf{k}). \tag{11.124}$$

Performing the discrete and continuous inverse Fourier transforms with respect to the variables t and $\mathbf{x}$, respectively, we find that the Fourier coefficients of the current are given by

$$J_N^\mu(\mathbf{k}) = \frac{1}{(2\pi)^{3/2}T} \int_0^T dt \int d^3 x\, e^{-i(N\omega_0 t - \mathbf{k}\cdot\mathbf{x})} j^\mu(x). \tag{11.125}$$

To compute the spectral weights (11.14) we need the Fourier coefficients $\mathbf{E}_N$ (11.7) of the electric field in the wave zone (8.14), and to compute the latter we need, in turn, the potential in the wave zone (8.10). Inserting the Fourier representation (11.124) in Eq. (8.10) we obtain

$$\mathbf{A} = \frac{1}{(2\pi)^{3/2}4\pi r} \sum_{N=-\infty}^{\infty} \int d^3k \int d^3y\, e^{iN\omega_0(t-r+\mathbf{n}\cdot\mathbf{y})}\, e^{-i\mathbf{k}\cdot\mathbf{y}}\, \mathbf{J}_N(\mathbf{k})$$

$$= \frac{1}{(2\pi)^{3/2}4\pi r} \sum_{N=-\infty}^{\infty} \int d^3k \int d^3y\, e^{iN\omega_0(t-r)}\, e^{-i(\mathbf{k}-N\omega_0\mathbf{n})\cdot\mathbf{y}}\, \mathbf{J}_N(\mathbf{k}).$$

$$(11.126)$$

According to the identity (2.98) the integral over $\mathbf{y}$ produces the δ-function

$$\int d^3y\, e^{-i(\mathbf{k}-N\omega_0\mathbf{n})\cdot\mathbf{y}} = (2\pi)^3\, \delta^3(\mathbf{k}-N\omega_0\mathbf{n}),$$

which permits, in turn, to perform in Eq. (11.126) the integral over $\mathbf{k}$

$$\mathbf{A} = \frac{\sqrt{2\pi}}{2r} \sum_{N=-\infty}^{\infty} \int d^3k\, e^{iN\omega_0(t-r)}\, \delta^3(\mathbf{k}-N\omega_0\mathbf{n})\, \mathbf{J}_N(\mathbf{k})$$

$$= \frac{\sqrt{2\pi}}{2r} \sum_{N=-\infty}^{\infty} e^{iN\omega_0(t-r)}\, \boldsymbol{\mathcal{J}}_N.$$

$$(11.127)$$

In the series (11.127) we have introduced the four-vectors

$$\mathcal{J}_N^\mu = J_N^\mu(N\omega_0\mathbf{n}),$$

$$(11.128)$$

whose dependence on the emission direction $\mathbf{n}$ is understood. Given the potential (11.127), Eq. (8.14) yields the electric field in the wave zone

$$\mathbf{E}(t) = \frac{i\sqrt{2\pi}}{2r}\, \mathbf{n} \times \left(\mathbf{n} \times \sum_{N=-\infty}^{\infty} N\omega_0\, e^{iN\omega_0(t-r)}\, \boldsymbol{\mathcal{J}}_N \right).$$

Comparing this expression with the general expansion (11.6) we recover the Fourier coefficients

$$\mathbf{E}_N = \frac{i\sqrt{2\pi}}{2r}\, N\omega_0\, e^{-iN\omega_0 r}\, \mathbf{n} \times (\mathbf{n} \times \boldsymbol{\mathcal{J}}_N).$$

Inserting them in the general formula (11.14) we finally find the spectral weights

$$\frac{d\mathcal{W}_N}{d\Omega} = \pi(N\omega_0)^2 \big|\mathbf{n} \times \boldsymbol{\mathcal{J}}_N\big|^2.$$

$$(11.129)$$

These weights can be rewritten in a more transparent form by using the continuity equation $\partial_\mu j^\mu = 0$. Taking the four-divergence of Eq. (11.124), and swapping the derivatives with the summation sign and with the integral, we obtain the relation

$$\partial_\mu j^\mu(x) = \frac{i}{(2\pi)^{3/2}} \sum_{N=-\infty}^{\infty} \int d^3k \left(N\omega_0 J_N^0(\mathbf{k}) - \mathbf{k} \cdot \mathbf{J}_N(\mathbf{k}) \right) e^{i(N\omega_0 t - \mathbf{k} \cdot \mathbf{x})} = 0,$$

which, via inverse Fourier transform, yields the relations between the Fourier coefficients of the current

$$N\omega_0 J_N^0(\mathbf{k}) - \mathbf{k} \cdot \mathbf{J}_N(\mathbf{k}) = 0.$$

Evaluating these equations for the wave vectors $\mathbf{k} = N\omega_0 \mathbf{n}$ we find the further relations (see the definition (11.128))

$$\mathcal{J}_N^0 = \mathbf{n} \cdot \boldsymbol{\mathcal{J}}_N.$$

Using the latter, the modulus squared appearing in the spectral weights (11.129) can finally be recast as the four-dimensional scalar product

$$\left| \mathbf{n} \times \boldsymbol{\mathcal{J}}_N \right|^2 = \left| \boldsymbol{\mathcal{J}}_N \right|^2 - \left| \mathbf{n} \cdot \boldsymbol{\mathcal{J}}_N \right|^2 = -\mathcal{J}_N^{\mu*} \mathcal{J}_{N\mu},$$

so that these weights can be written in the equivalent form

$$\frac{dW_N}{d\Omega} = -\pi (N\omega_0)^2 \mathcal{J}_N^{\mu*} \mathcal{J}_{N\mu}. \tag{11.130}$$

Linear antenna. To exemplify the use of Eq. (11.129) we provide an alternative derivation of the formula (8.34) for the angular distribution of the radiation emitted by an antenna. According to Eqs. (11.125), (11.128) and (11.129) we must compute the Fourier coefficients $\mathbf{J}_N(\mathbf{k})$ of the spatial current density (see Eq. (8.29) with ω replaced by ω_0)

$$\mathbf{j}(t, \mathbf{x}) = I\, \delta(x)\, \delta(y) \sin\left(\omega_0 \left(\frac{L}{2} - |z| \right) \right) \cos(\omega_0 t)\, \mathbf{u}. \tag{11.131}$$

This current is periodic, with period $T = 2\pi/\omega_0$ and fundamental frequency ω_0, and, in particular, *monochromatic*. This implies that all Fourier coefficients $\mathbf{J}_N(\mathbf{k})$ in (11.125) vanish, apart from those relative to $N = \pm 1$. The antenna emits thus only radiation with the fundamental frequency ω_0. Equation (11.129) then yields for the unique non-vanishing spectral weight

$$\frac{dW_1}{d\Omega} = \pi \omega_0^2 \left| \mathbf{n} \times \boldsymbol{\mathcal{J}}_1 \right|^2 = \frac{d\overline{W}}{d\Omega}. \tag{11.132}$$

To evaluate the coefficient $\mathcal{J}_1$ we must insert the current density (11.131) in Eq. (11.125), setting $N = 1$ and $\mathbf{k} = \omega_0 \mathbf{n}$. We find

$$\mathcal{J}_1 = \frac{I\mathbf{u}}{T(2\pi)^{3/2}} \int_0^T dt\, e^{-i\omega_0 t} \cos(\omega_0 t) \int d^3x\, e^{i\omega_0 \mathbf{n}\cdot\mathbf{x}}\, \delta(x)\delta(y) \sin\left(\omega_0\left(\frac{L}{2} - |z|\right)\right)$$

$$= \frac{I\mathbf{u}}{2(2\pi)^{3/2}} \int_{-L/2}^{L/2} dz\, e^{i\omega_0 z \cos\vartheta} \sin\left(\omega_0\left(\frac{L}{2} - |z|\right)\right)$$

$$= \frac{I\mathbf{u}}{(2\pi)^{3/2}} \int_0^{L/2} dz \cos\left(\omega_0 z \cos\vartheta\right) \sin\left(\omega_0\left(\frac{L}{2} - z\right)\right),$$

where ϑ is the angle between the z axis and $\mathbf{n}$. The remaining integral over z is elementary and leads to

$$\mathcal{J}_1 = \frac{I\mathbf{u}}{(2\pi)^{3/2}\, \omega_0 \sin^2\vartheta} \left(\cos\left(\frac{\omega_0 L}{2} \cos\vartheta\right) - \cos\frac{\omega_0 L}{2}\right).$$

Inserting this expression in Eq. (11.132), recalling the position (8.30), and using that $|\mathbf{n}\times\mathbf{u}| = \sin\vartheta$, we retrieve the angular distribution (8.34).

Point-like particles. For the case of a point-like particle Eq. (11.129) should reproduce the spectral weights (11.91). To verify it we must compute the Fourier coefficients (11.128) relative to the spatial current

$$\mathbf{j}(x) = e\mathbf{v}(t)\, \delta^3(\mathbf{x} - \mathbf{y}(t)). \tag{11.133}$$

Inserting this current in Eq. (11.125), setting $\mathbf{k} = N\omega_0 \mathbf{n}$, and performing the integral over $\mathbf{x}$, we obtain

$$\mathcal{J}_N = \frac{e}{T} \int_0^T \frac{dt}{(2\pi)^{3/2}} \int d^3x\, e^{-i(N\omega_0 t - N\omega_0 \mathbf{n}\cdot\mathbf{x})}\, \delta^3(\mathbf{x} - \mathbf{y}(t))\, \mathbf{v}(t)$$

$$= \frac{e}{(2\pi)^{3/2}\, T} \int_0^T dt\, e^{-iN\omega_0(t - \mathbf{n}\cdot\mathbf{y})}\, \mathbf{v}.$$

Substituting these expressions in Eq. (11.129) we retrieve the spectral weights (11.91).

11.5.2   Aperiodic Charge Distribution

The four-current associated with an aperiodic charge distribution admits the four-dimensional continuous Fourier representation

$$j^{\mu}(x) = \frac{1}{(2\pi)^2} \int d\omega \int d^3k\, e^{i(\omega t - \mathbf{k}\cdot\mathbf{x})} J^{\mu}(\omega, \mathbf{k}), \qquad (11.134)$$

with inverse Fourier transform

$$J^{\mu}(\omega, \mathbf{k}) = \frac{1}{(2\pi)^2} \int d^4x\, e^{-i(\omega t - \mathbf{k}\cdot\mathbf{x})} j^{\mu}(x). \qquad (11.135)$$

We proceed as above, inserting the current (11.134) in the potential in the wave zone (8.10). In this case the integral over $\mathbf{y}$ produces the δ-function $(2\pi)^3\delta^3(\mathbf{k} - \omega\mathbf{n})$, which allows, in turn, to perform the integral over $\mathbf{k}$:

$$\begin{aligned}
\mathbf{A} &= \frac{1}{(2\pi)^2 4\pi r} \int d^3y \int d\omega \int \delta^3 k\, e^{i\omega(t-r+\mathbf{n}\cdot\mathbf{y})}\, e^{-i\mathbf{k}\cdot\mathbf{y}}\, \mathbf{J}(\omega, \mathbf{k}) \\
&= \frac{1}{2r} \int d\omega \int d^3k\, e^{i\omega(t-r)} \delta^3(\mathbf{k} - \omega\mathbf{n})\, \mathbf{J}(\omega, \mathbf{k}) \\
&= \frac{1}{2r} \int d\omega\, e^{i\omega(t-r)} \boldsymbol{J}_\omega.
\end{aligned} \qquad (11.136)$$

This time in the integral (11.136) we have introduced the ω-dependent four-vectors

$$\boldsymbol{J}_\omega^{\mu} = J^{\mu}(\omega, \omega\mathbf{n}), \qquad (11.137)$$

whose $\mathbf{n}$-dependence is again understood. With the potential (11.136) the electric field in the wave zone (8.14) becomes

$$\mathbf{E}(t) = \frac{i}{2r}\,\mathbf{n} \times \left(\mathbf{n} \times \int d\omega\, \omega\, e^{i\omega(t-r)} \boldsymbol{J}_\omega \right). \qquad (11.138)$$

Equating the representations (11.3) and (11.138) we find for the Fourier coefficients of the electric field the expressions, matching with the general formula (11.2),

$$\mathbf{E}(\omega) = \frac{i\sqrt{2\pi}}{2r}\, \omega e^{-i\omega r}\,\mathbf{n} \times (\mathbf{n} \times \boldsymbol{J}_\omega). \qquad (11.139)$$

Substituting them in Eq. (11.12) we obtain the spectral weights

$$\frac{d^2\varepsilon}{d\omega d\Omega} = \pi\omega^2|\mathbf{n} \times \boldsymbol{J}_\omega|^2 = \pi\omega^2 \left(|\boldsymbol{J}_\omega|^2 - |\mathbf{n} \cdot \boldsymbol{J}_\omega|^2 \right). \qquad (11.140)$$

Using, once more, the continuity equation $\partial_\mu j^{\mu} = 0$, from the Fourier representation of the current (11.134) we derive the relation $\omega J^0(\omega, \mathbf{k}) = \mathbf{k} \cdot \mathbf{J}(\omega, \mathbf{k})$, which, when evaluated for $\mathbf{k} = \omega\mathbf{n}$, reduces to

$$\boldsymbol{J}_\omega^0 = \mathbf{n} \cdot \boldsymbol{J}_\omega.$$

The spectral weights (11.140) can hence be recast in the equivalent form, resembling their discrete counterpart (11.130),

$$\frac{d^2\varepsilon}{d\omega d\Omega} = -\pi\omega^2 \mathcal{J}_\omega^{\mu*} \mathcal{J}_{\omega\mu}. \tag{11.141}$$

It is easy to see that in the case of a charged particle the formula (11.140) reproduces the spectral weights (11.95). In fact, inserting the current density (11.133) in the equation for the Fourier coefficients (11.135), and proceeding as above, one obtains

$$\mathcal{J}_\omega = \frac{e}{(2\pi)^2} \int_{-\infty}^{\infty} e^{-i\omega(t-\mathbf{n}\cdot\mathbf{y})} \mathbf{v}\, dt. \tag{11.142}$$

With these expression for $\mathcal{J}_\omega$ Eq. (11.140) then reduces to (11.95).

As noticed in Sect. 11.4.1, the integral (11.142) in general diverges. In the present context the origin of this divergence is self-evident: the quantity $\mathcal{J}_\omega = \mathbf{J}(\omega, \omega\mathbf{n})$ involves the Fourier transform $\mathbf{J}(\omega, \mathbf{k})$ of the *distribution* $\mathbf{j}(t, \mathbf{x})$, and as such it must be performed in the *sense of distributions*. As shown in Sect. 11.4.1, this mismatch can be settled a posteriori by restricting the integration domain in Eq. (11.142) to the finite interval $[-L, L]$, and by taking then the distributional limit of the resulting integral for $L \to \infty$.

11.6 Problems

11.1 Consider a charged particle passing at a large distance from a nucleus with an approximately constant velocity $\mathbf{v}$, as in Problems 8.5 and 11.1.

(a) Determine the characteristic frequencies radiated by the particle in the non-relativistic limit $v \ll 1$.

   Hint: The characteristic time $\mathcal{T}$ of the process can be determined from Newton's second law $m\mathbf{a} = e\mathbf{E}$, by considering the approximated trajectory (8.136).

(b) Determine the characteristic frequencies radiated by the particle in a given direction $\mathbf{n}$, if its velocity $\mathbf{v}$ is arbitrary.

   Hint: Starting from the general formula (11.94), taking into account that $\mathbf{v}$ is practically constant derive for the spectral distribution an approximated expression analogous to (11.109). Notice that, although in this case one must resort to the relativistic form (2.38) of Newton's second law, one can still use the approximated trajectory (8.136).

11.2 Consider the low-frequency angular distribution (11.119) of the radiation emitted by the electron in the beta decay, as a function of the angle ϑ between $\mathbf{n}$ and the flight direction, i.e.

$$f(\vartheta) = \frac{e^2 v^2 \sin^2 \vartheta}{16\pi^3 (1 - v \cos \vartheta)^2}.$$

(a) Find the minima and maxima of the function $f(\vartheta)$ for $\vartheta \in [0, \pi]$, and draw its plot.
(b) Use these results to confirm the estimate (11.116) for the maximum magnitude of the vector $\mathbf{n} \times \mathbf{v}/(1 - \mathbf{n} \cdot \mathbf{v})$ in the ultrarelativistic limit.

11.3 Consider a non-relativistic charged particle crossing a region where a constant electric field is present, as in Sect. 11.3.1.

(a) Show that the energy of the radiation emitted per unit solid angle in a direction $\mathbf{n}$ with polarization parallel to $\mathbf{e}_p$ is given by (see Sect. 11.3.1 for the notation)

$$\frac{d^2 \varepsilon^p}{d\Omega} = \left(\frac{e^2 a^2 T \sin^2 \beta}{8\pi^2} \right) \cos^2 \varphi_p. \tag{11.143}$$

Hint: Integrate the polarized spectral weights (11.55) over all frequencies, using the integral (11.56).
(b) Considering that the radiation is linearly polarized in all directions $\mathbf{n}$, verify that for $\varphi_p = 0$ the integral over the angles of Eq. (11.143) matches with the total radiated energy (11.57).
Hint: It is convenient to rewrite the product $a^2 \sin^2 \beta$ in the form

$$a^2 \sin^2 \beta = a^2 - (\mathbf{a} \cdot \mathbf{n})^2,$$

and to use the invariant integrals of Problem 2.6.

11.4 Consider a non-relativistic charged particle performing the anharmonic periodic motion (11.74).

(a) Show that for a *generic* non-relativistic periodic charge distribution the polarized spectral weights (11.21) can be expressed as

$$\frac{dW_N^p}{d\Omega} = \frac{(N\omega_0)^4}{8\pi^2} \left| \mathbf{e}_p \cdot \mathbf{D}_N \right|^2. \tag{11.144}$$

Hint: Insert the Fourier coefficients of the electric field (11.40) in Eq. (11.21), taking into account that the vectors $\mathbf{n}$ and $\mathbf{e}_p$ are orthogonal.
(b) Evaluate the polarized spectral weights (11.144) for the anharmonic motion (11.74), showing that the radiation is linearly polarized in all directions for all frequencies $N\omega_0$. For a given emission direction $\mathbf{n}$, determine the polarization directions for N even and for N odd.
Hint: Proceed as in Eqs. (11.54) and (11.55), taking into account that the Fourier coefficients $\mathbf{D}_N$ (11.77) have different expressions for even and odd N.

11.5 Consider a system of non-relativistic charged particles performing the simple harmonic motions (11.49).

(a) Show that, in general, the radiation emitted in a generic direction $\mathbf{n}$ is elliptically polarized.

(b) Find the conditions the vectors $\mathbf{b}_r$ and $\mathbf{c}_r$ must satisfy, in order that the radiation emitted in the direction $\mathbf{n} = (0, 0, 1)$ is circularly polarized.

   Hint: Use Eq. (11.44) and impose the conditions (11.26) or, equivalently, (11.29). Consider that the system emits only radiation with the fundamental frequency ω_0.

References

1. F. Bloch, A. Nordsieck, Note on the radiation field of the electron. Phys. Rev. **52**, 54 (1937)
2. T.D. Lee, M. Nauenberg, Degenerate systems and mass singularities. Phys. Rev. **133**, B1549 (1964)
3. M. Lavelle, D. McMullan, Collinearity, convergence and cancelling infrared divergences. JHEP **0603**, 026 (2006)
4. S. Weinberg, *The Quantum Theory of Fields I* (Cambridge University Press, Cambridge, 2005)
5. S. Weinberg, Infrared photons and gravitons. Phys. Rev. **140**, B516 (1965)
6. I.S. Gradshteyn, I.M. Rhyzik, *Table of Integrals, Series, and Products*, ed. by A. Jeffrey, D. Zwillinger (Academic Press, San Diego, 2007)
7. M. Abramowitz, I.A. Stegun, *Handbook of Mathematical Functions with Formulas, Graphs, and Mathematical Tables* (U.S. Government Printing Office, Washington DC, 1964)
8. J.S. Nico et al., Observation of the radiative decay mode of the free neutron. Nature **444**, 1059 (2006)

Chapter 12
Synchrotron Radiation

The radiation emitted by a charged particle in uniform circular motion is called *synchrotron radiation*. Radiation of this kind is generated, in particular, whenever a charged particle moves under the influence of a static magnetic field varying only little in space. Synchrotron radiation is hence emitted by circular high-energy particle accelerators, whose strong magnetic fields are generated by powerful superconducting electromagnets, but also by charged particles moving in the much weaker natural magnetic fields that wrap, for instance, the Earth and the planet Jupiter. However, some of the most spectacular ultrarelativistic sources of this type of radiation are extraterrestrial, as for instance the famous *Crab Nebula*, although their magnetic fields are considerably weaker than that of the Earth. In this chapter we investigate the main characteristics of the synchrotron radiation, especially its frequency and angular distributions and its polarization. We will see that, notably in the ultrarelativistic limit, this radiation possesses peculiar physical properties, which allow to undoubtedly establish its presence experimentally.

The trajectory of a particle in uniform circular motion with angular velocity ω_0 is periodic with period $T = 2\pi/\omega_0$. The particle emits hence radiation with the discrete frequency spectrum

$$\omega_N = N\omega_0, \qquad N = 1, 2, 3, \ldots,$$

were ω_0 is the *fundamental* harmonic frequency. We know, furthermore, that the total radiated power is given by the formula (10.34)

$$\mathcal{W} = \frac{e^2 v^2 \omega_0^2}{6\pi(1 - v^2)^2}, \tag{12.1}$$

where $v = R\omega_0$ is the particle's speed and R the radius of the orbit. If the circular motion is caused by a magnetic field of magnitude B, the fundamental frequency equals the *cyclotron frequency* (10.33)

© Springer International Publishing AG, part of Springer Nature 2018
K. Lechner, *Classical Electrodynamics*, UNITEXT for Physics,
https://doi.org/10.1007/978-3-319-91809-9_12

$$\omega_0 = \frac{eB}{\varepsilon},$$

(12.2)

where $\varepsilon = m/\sqrt{1 - v^2}$ is the energy of the particle. Below we consider, however, a generic uniform circular motion, independently of the force which causes it.

12.1 Non-Relativistic Synchrotron Radiation

Before facing the analysis of the radiation of a charged particle with arbitrary velocity, for comparison we summarize the basic properties of the synchrotron radiation in the non-relativistic limit $v \ll 1$. As the motion is a *simple harmonic* one, see Eq. (11.49), from Sect. 11.3 we know that in this limit the particle emits only radiation of frequency ω_0, corresponding to the harmonic of order $N = 1$. From the result (8.129) of Problem 8.1 we then deduce that the spectral weights (11.91) are given by

$$\frac{dW_N}{d\Omega} = \begin{cases} \dfrac{e^2 v^2 \omega_0^2}{32\pi^2} (1 + \cos^2 \vartheta) = \dfrac{dW}{d\Omega}, & \text{for } N = 1, \\ 0, & \text{for } N > 1, \end{cases}$$

(12.3)

where ϑ is the angle between the emission direction n and the axis of the orbit. For simplicity throughout this chapter we omit the bar denoting the time average, writing hence $dW/d\Omega$ in place of $d\overline{W}/d\Omega$. From Eqs. (12.3) we can determine the ratio between the intensities of the radiation in the plane of the orbit, at $\vartheta = \pi/2$, and the radiation emitted along its axis, at $\vartheta = 0$,

$$\frac{\left.\dfrac{dW}{d\Omega}\right|_{\text{plane}}}{\left.\dfrac{dW}{d\Omega}\right|_{\text{axis}}} = \frac{1}{2}.$$

(12.4)

This means that in the non-relativistic limit the angular distribution of the radiation has a quite *regular* shape, as there are no particulary privileged directions of emission. Integrating the first equation in (12.3) over the solid angle we obtain the total power

$$W = \frac{e^2 v^2 \omega_0^2}{6\pi},$$

(12.5)

to which the exact formula (12.1) reduces for $v \ll 1$. Concerning the polarization, the radiation emitted with polarization parallel to a generic direction $\mathbf{e}_p$ entails the angular distribution, see Problem 12.1,

$$\frac{dW^p}{d\Omega} = \cos^2 \varphi_p \frac{e^2 v^2 \omega_0^2}{32\pi^2} + \sin^2 \varphi_p \frac{e^2 v^2 \omega_0^2 \cos^2 \vartheta}{32\pi^2},$$

(12.6)

where φ_p is the angle $\mathbf{e}_p$ forms with the polarization direction $\mathbf{e}_\parallel$ lying in the plane of the orbit. From Eq. (12.6), and the characterizations provided in Sect. 11.2, we see that for a generic direction $0 < \vartheta < \pi/2$ the radiation is elliptically polarized. Conversely, along the axis of the orbit, for $\vartheta = 0$, it is circularly polarized, and in the plane of the orbit, for $\vartheta = \pi/2$, it is linearly polarized, with polarization parallel to the plane of orbit.

12.2 Spectral Analysis

Henceforth, we consider a particle with arbitrary speed $v < 1$. Starting point of the spectral analysis of the radiation is the explicit evaluation of the discrete spectral weights (11.91)

$$\frac{d\mathcal{W}_N}{d\Omega} = \frac{e^2 (N\omega_0)^2}{8\pi^2} \left| \mathbf{n} \times \frac{1}{T} \int_0^T e^{-iN\omega_0 (t - \mathbf{n} \cdot \mathbf{y})} \, \mathbf{v} \, dt \right|^2. \tag{12.7}$$

Taking as axis of the orbit the z axis, the particle's trajectory, velocity, and acceleration are then given by

$$\mathbf{y}(t) = R\,(\cos \varphi(t), \sin \varphi(t), 0), \tag{12.8}$$

$$\mathbf{v}(t) = v\,(-\sin \varphi(t), \cos \varphi(t), 0), \tag{12.9}$$

$$\mathbf{a}(t) = -\omega_0^2\, \mathbf{y}(t), \tag{12.10}$$

respectively, where the angular position at time t is

$$\varphi(t) = \omega_0 t. \tag{12.11}$$

Thanks to rotation invariance around the z axis, without loss of generality we can choose an emission direction $\mathbf{n}$ lying in the yz plane

$$\mathbf{n} = (0, \sin \vartheta, \cos \vartheta), \tag{12.12}$$

were, again, ϑ is the angle between $\mathbf{n}$ and the z axis. Using the identity

$$\omega_0\, \mathbf{n} \cdot \mathbf{y} = v \sin \vartheta \sin \varphi,$$

we can then rewrite the t integral appearing in Eq. (12.7) as the integral over φ

$$\frac{1}{T} \int_0^T e^{-iN\omega_0 (t - \mathbf{n} \cdot \mathbf{y})} \, \mathbf{v} \, dt = \frac{v}{2\pi} \int_0^{2\pi} e^{-iN(\varphi - v \sin \vartheta \sin \varphi)} (-\sin \varphi, \cos \varphi, 0) \, d\varphi.$$

Using the properties (11.65) and (11.66) of the Bessel functions $J_N(x)$ of integer order N, we then find

$$\frac{1}{T}\int_0^T e^{-iN\omega_0(t-\mathbf{n}\cdot\mathbf{y})}\,\mathbf{v}\,dt = v\left(iJ_N'(vN\sin\vartheta),\frac{1}{v\sin\vartheta}J_N(vN\sin\vartheta),0\right).$$
(12.13)

Considering the exterior product of the vectors (12.12) and (12.13), and calculating its magnitude, eventually the spectral weights (12.7) take the form

$$\frac{dW_N}{d\Omega} = \frac{e^2(N\omega_0)^2}{8\pi^2}\left(\cot^2\vartheta\,J_N^2(vN\sin\vartheta) + v^2 J_N'^2(vN\sin\vartheta)\right).$$
(12.14)

Non-relativistic limit. In the non-relativistic limit $v \ll 1$ we can resort to the expansions of the Bessel functions at fixed N (11.68), which yield for the spectral weights (12.14) the leading small-velocity behaviors

$$\frac{dW_N}{d\Omega} \approx \frac{e^2(N\omega_0)^2 v^{2N}}{8\pi^2}\left(\frac{N^N}{2^N N!}\right)^2 (\sin^2\vartheta)^{N-1}(1+\cos^2\vartheta).$$
(12.15)

The higher harmonics, say, up to order $N \sim 10$, are thus suppressed with respect to the fundamental harmonic $N = 1$ by a factor v^{2N-2}. Actually, this suppression persists for any N. In fact, in the limit for $N \to \infty$ the prefactor in (12.15) behaves as (use Stirling's formula (11.81) and consider that $ev/2 \ll 1$)

$$\left(\frac{v^N N^{N+1}}{2^N N!}\right)^2 \to \frac{N\left(\frac{ev}{2}\right)^{2N}}{2\pi} \to 0,$$

so that also for large N the spectral weights $dW_N/d\Omega$ are strongly suppressed with respect to $dW_1/d\Omega$. We confirm, thus, that in the non-relativistic limit the particle emits mainly radiation of the fundamental frequency ω_0. Notice that for $N = 1$ the approximated spectral weight (12.15) matches with the dipole formula (12.3).

Radiation along the axis of the orbit. The spectral weights of the radiation emitted in the direction of the axis of the orbit are obtained setting in (12.14) $\vartheta = 0$. Resorting again to the expansions (11.68) we find

$$\left.\frac{dW_1}{d\Omega}\right|_{\text{axis}} = \frac{e^2 v^2 \omega_0^2}{16\pi^2}, \qquad \left.\frac{dW_N}{d\Omega}\right|_{\text{axis}} = 0, \quad \text{for } N > 1. \qquad (12.16)$$

This means that along the axis for any velocity v the particle emits only radiation of the fundamental frequency. This features becomes self-evident if we write out the asymptotic electric field (11.83) for $\mathbf{n} = (0,0,\pm 1)$. In the case at hand, using the kinematical relations (12.8)–(12.10) and the delay equation (11.85), it takes the simple form

$$\mathbf{E}(t,r) = -\frac{ea(t')}{4\pi r} = -\frac{ea(t-r)}{4\pi r} = \frac{ev\omega_0}{4\pi r}\left(\cos(\omega_0(t-r)), \sin(\omega_0(t-r)), 0\right).$$

$$(12.17)$$

Along the axis the radiation is thus composed of a *monochromatic circularly polarized* wave of frequency ω_0, see the relations (5.91) and (5.93). As the electric field (12.17) can also be written in the form

$$\mathbf{E}(t,r) = \frac{e\omega_0^2}{4\pi r}\,\mathbf{y}(t-r),$$

the polarization is *right* circular in the direction $\mathbf{n} = (1,0,0)$, and *left* circular in the opposite direction $\mathbf{n} = (0,0,-1)$.

12.2.1 Dominant Frequencies in the Ultrarelativistic limit

As seen in Sect. 10.3, the radiation emitted by an ultrarelativistic charged particle forms an extremely collimated beam, being significantly different from zero only in a narrow cone around its flight direction. To analyze the spectrum emitted by such a particle it is then convenient to consider the total radiation, i.e. the radiation summed over all directions. The corresponding spectral weights are obtained by integrating the expressions (12.14) over the solid angle[1]

$$\mathcal{W}_N = \int \frac{d\mathcal{W}_N}{d\Omega}\,d\Omega = \frac{e^2 N\omega_0^2}{4\pi v}\left(2v^2 J_{2N}'(2Nv) - (1-v^2)\int_0^{2Nv} J_{2N}(y)\,dy\right).$$

$$(12.18)$$

A qualitative argument. Before presenting the quantitative analysis of the total spectral weights (12.18), we give a qualitative argument to establish the order of magnitude of the characteristic frequencies emitted by an ultrarelativistic particle. We recall from Sect. 10.3 that a particle with speed $v \approx 1$ emits mainly radiation in a cone, centered at the flight direction, having the small angular aperture $\alpha \sim \sqrt{1-v^2}$. Consequently, as the particle performs a circular motion with period $T = 2\pi/\omega_0$, the radiation emitted in a given direction comes only from a small arc of the orbit traversed by the particle in the time interval

$$\Delta t' \sim \frac{\alpha}{2\pi}\,T \sim \frac{\sqrt{1-v^2}}{\omega_0}.$$

A typical emission frequency would hence be

$$\omega' = \frac{1}{\Delta t'} \sim \frac{\omega_0}{\sqrt{1-v^2}}.$$

---

[1]For the details of the calculation we refer the reader to Ref. [1].

On the other hand, according to Eq. (11.86) the *emission* time $\Delta t'$ is associated with the *observation* time

$$\Delta t = (1 - \mathbf{n} \cdot \mathbf{v})\Delta t' \sim (1 - v)\Delta t' \sim (1 - v^2)\Delta t',$$

which corresponds to the observed frequency

$$\omega = \frac{1}{\Delta t} \sim \frac{\omega'}{1 - v^2}.$$

We expect, therefore, that an ultrarelativistic particle emits mainly radiation with frequencies

$$\omega \lesssim \frac{\omega_0}{(1 - v^2)^{3/2}}. \tag{12.19}$$

Actually, in the case at hand the spectrum is discrete, the allowed frequencies being given by $\omega_N = N\omega_0$. From (12.19) it then follows that the radiation contains harmonics up to the very high order

$$N \sim \frac{1}{(1 - v^2)^{3/2}}. \tag{12.20}$$

For the particular case of a circular motion caused by a magnetic field the fundamental frequency is given by the cyclotron formula (12.2), so that the dominant frequencies (12.19) take the form

$$\omega \lesssim \frac{eB}{m}\left(\frac{\varepsilon}{m}\right)^2. \tag{12.21}$$

These frequencies grow thus with the square of the energy.

Quantitative analysis. Going back to the quantitative expression (12.18), according to the above argument we expect that for velocities close to the velocity of light the dominant spectral weights $\mathcal{W}_N$ are those with N very large. Actually, it can be seen that the behavior for $N \gg 1$ of the sequence (12.18) depends significantly on the value of the, likewise large, number $1/\sqrt{1 - v^2}$. Through an asymptotic analysis of the Bessel functions appearing in Eq. (12.18) it is found that for $v \approx 1$ apart from numerical coefficients the spectral weights $\mathcal{W}_N$ entail the distinct behaviors[2]

$$\mathcal{W}_N \approx \begin{cases} e^2\omega_0^2 N^{1/3}, & \text{if } 1 \ll N \ll \dfrac{1}{(1 - v^2)^{3/2}}, \\[2mm] e^2\omega_0^2 \sqrt{N}\,(1 - v^2)^{1/4}\, e^{-\frac{2}{3}N(1-v^2)^{3/2}}, & \text{if } N \gg \dfrac{1}{(1 - v^2)^{3/2}}. \end{cases} \tag{12.22}$$

---

[2]For the derivation see, for instance, Refs. [1, 2].

For values of N which are large, but smaller than $1/(1-v^2)^{3/2}$, the spectral weights grow hence as $N^{1/3}$, while for N much larger than $1/(1-v^2)^{3/2}$ they are exponentially suppressed. The dominant harmonics of the ultrarelativistic synchrotron radiation are hence those corresponding to the orders

$$N \lesssim \frac{1}{(1-v^2)^{3/2}} = \left(\frac{\varepsilon}{m}\right)^3, \tag{12.23}$$

in agreement with the estimate (12.20).

Characteristic wavelengths of high-energy accelerators. The fundamental frequency of a circular high-energy accelerator, with accumulation ring of radius R, is given by $\omega_0 = v/R \approx 1/R$. From the relation (12.23) it then follows that the accelerator emits synchrotron radiation with characteristic wavelengths

$$\lambda = \frac{2\pi}{N\omega_0} \gtrsim 2\pi R \left(\frac{m}{\varepsilon}\right)^3.$$

Inserting the values reported in Sect. 10.2.2 one finds that the radiation emitted by the Cornell Synchrotron contained wavelengths down to 0.1 nm, corresponding to X-rays, while LEP reached much shorter wavelengths of the order 10^{-3} nm, corresponding to gamma rays. On the other hand, the radiation emitted by the LHC is peaked at the much longer wavelengths of the order 100 nm, which are not far away from the ultraviolet visible spectrum.

12.3 Angular Distribution

Rather than analyzing the angular distribution $dW_N/d\Omega$ of the radiation of a fixed frequency, see Eq. (12.14), we analyze the angular distribution of the total, i.e. summed over all frequencies, radiation. For this purpose we would have to sum the series

$$\frac{dW}{d\Omega} = \sum_{N=1}^{\infty} \frac{dW_N}{d\Omega},$$

an operation that is difficult to carry out analytically. Fortunately, in this case there exists an indirect method to perform this sum, based on the fundamental formulas (11.13) and (11.14), which consists in the evaluation of the time average

$$\frac{dW}{d\Omega} = \frac{r^2}{T} \int_0^T |\mathbf{E}|^2 dt, \tag{12.24}$$

in which $\mathbf{E}$ is the asymptotic electric field (11.83). Inserting the kinematical relations (12.8)–(12.10), and Eq. (12.12), a simple calculation gives for the square of this field

$$|\mathbf{E}|^2 = \frac{e^2 v^2 \omega_0^2}{16\pi^2 r^2} \cdot \frac{(1-v^2)\cos^2\vartheta + (v - \sin\vartheta\cos\varphi)^2}{(1 - v\sin\vartheta\cos\varphi)^6}, \qquad \varphi = \omega_0 t'(t, \mathbf{x}).$$
(12.25)

Although this expression looks rather complicated, the integral (12.24) can be performed analytically, and we postpone its calculation to Sect. 12.4.2. Here we anticipate the resulting angular distribution, given by the sum of the contributions (12.43) and (12.45),

$$\frac{dW}{d\Omega} = \frac{e^2 v^2 \omega_0^2}{32\pi^2} \cdot \frac{1 + \cos^2\vartheta - \frac{v^2}{4}(1+3v^2)\sin^4\vartheta}{(1 - v^2\sin^2\vartheta)^{7/2}}.$$
(12.26)

For velocities $v \ll 1$ we reobtain the angular distribution (12.3), which has a maximum at $\vartheta = 0$ and a minimum at $\vartheta = \pi/2$. Conversely, for $v \approx 1$ from the form of the denominator in Eq. (12.26) we see that $dW/d\Omega$ has a pronounced maximum in the vicinity of $\vartheta = \pi/2$, i.e. for all directions of the plane of the orbit. To determine the directions close to the plane of the orbit in which most of the radiation is emitted we resort to a standard argument: for these directions the factor $1/(1 - v^2\sin^2\vartheta)^{7/2}$ in Eq. (12.26) must remain of the same order of magnitude of its maximum $1/(1 - v^2)^{7/2}$. The corresponding angles ϑ must hence be such that

$$1 - v^2\sin^2\vartheta \sim 1 - v^2.$$

Setting $\beta = \pi/2 - \vartheta$, this conditions translates into

$$1 - v^2\cos^2\beta \approx 1 - v^2\left(1 - \frac{\beta^2}{2}\right) \approx 1 - v^2 + \frac{\beta^2}{2} \sim 1 - v^2 \quad \Rightarrow \quad \beta \sim \sqrt{1 - v^2}.$$

The dominant part of the radiation is thus emitted along directions which form with the plane of the orbit angles smaller than

$$\beta \sim \sqrt{1 - v^2} = \frac{m}{\varepsilon},$$

a conclusion that matches with the general qualitative prediction (10.42) of Sect. 10.3. Finally, we compute the ratio of the intensity of the radiation emitted in the plane of the orbit, at $\vartheta = \pi/2$, and the corresponding intensity emitted along its axis, at $\vartheta = 0$. From the angular distribution (12.26) we find

$$\frac{\left.\frac{dW}{d\Omega}\right|_{\text{plane}}}{\left.\frac{dW}{d\Omega}\right|_{\text{axis}}} = \frac{1}{8}\frac{4 + 3v^2}{(1 - v^2)^{5/2}}.$$

In the non-relativistic limit we retrieve Eq. (12.4), while for $v \approx 1$ we obtain the large ratio

$$\frac{\left.\dfrac{dW}{d\Omega}\right|_{\text{plane}}}{\left.\dfrac{dW}{d\Omega}\right|_{\text{axis}}} \approx \frac{7}{8}\left(\frac{\varepsilon}{m}\right)^5.$$

12.4 Polarization

To investigate the polarization of the radiation emitted in a given direction $\mathbf{n}$ it is convenient to introduce two appropriate directions orthogonal to each other and, of course, to $\mathbf{n}$. Given the geometry of a circular motion, for a generic emission direction $\mathbf{n} = (0, \sin\vartheta, \cos\vartheta)$ we choose the versor $\mathbf{e}_\parallel$ parallel to the plane of the orbit, and the versor $\mathbf{e}_\perp$ orthogonal to the former, more precisely,

$$\mathbf{e}_\parallel = (1,0,0), \quad \mathbf{e}_\perp = (0,\cos\vartheta, -\sin\vartheta), \quad \mathbf{e}_\parallel \cdot \mathbf{n} = \mathbf{e}_\perp \cdot \mathbf{n} = \mathbf{e}_\perp \cdot \mathbf{e}_\parallel = 0. \tag{12.27}$$

Our aim is to determine the intensity of the radiation with polarization parallel to $\mathbf{e}_\parallel$ and $\mathbf{e}_\perp$, respectively, or, more in general, to a generic direction

$$\mathbf{e}_p = \cos\varphi_p\,\mathbf{e}_\parallel + \sin\varphi_p\,\mathbf{e}_\perp, \tag{12.28}$$

where φ_p is the angle formed by the vectors $\mathbf{e}_p$ and $\mathbf{e}_\parallel$.

12.4.1 Polarization at Fixed Frequency

For a periodic system, the discrete spectral weights $dW_N^p/d\Omega$ of the radiation with frequency $N\omega_0$ and polarization parallel to a unit vector $\mathbf{e}_p$ are given by the general expression (11.21) where, in the case of a charged particle, the Fourier coefficients $\mathbf{E}_N$ of the electric field are given by Eq. (11.90). As $\mathbf{e}_p$ is orthogonal to $\mathbf{n}$ we have the identity, holding for an arbitrary vector $\mathbf{V}$,

$$\mathbf{e}_p \cdot (\mathbf{n} \times (\mathbf{n} \times \mathbf{V})) = -\mathbf{e}_p \cdot \mathbf{V}. \tag{12.29}$$

We obtain thus the polarized spectral weights

$$\frac{dW_N^p}{d\Omega} = 2r^2 \left|\mathbf{e}_p \cdot \mathbf{E}_N\right|^2 = \frac{e^2 N^2 \omega_0^2}{8\pi^2} \left|\mathbf{e}_p \cdot \frac{1}{T}\int_0^T e^{-iN\omega_0(t-\mathbf{n}\cdot\mathbf{y})}\,\mathbf{v}\,dt\right|^2,$$

where the integral appearing on the right hand side has been evaluated in Eq. (12.13). Inserting the vector (12.28), the spectral weights with generic polarization $\mathbf{e}_p$ then result in the diagonal expression

$$\frac{dW_N^p}{d\Omega} = \cos^2 \varphi_p \, \frac{dW_N^\perp}{d\Omega} + \sin^2 \varphi \, \frac{dW_N^\parallel}{d\Omega}, \qquad (12.30)$$

where the spectral weighs with polarization parallel to $e_\perp$ and $e_\parallel$ are given by

$$\frac{dW_N^\perp}{d\Omega} = \frac{e^2 (N\omega_0)^2}{8\pi^2} \cot^2 \vartheta \, J_N^2 (vN \sin \vartheta), \qquad (12.31)$$

$$\frac{dW_N^\parallel}{d\Omega} = \frac{e^2 (N\omega_0)^2}{8\pi^2} \, v^2 J_N'^2 (vN \sin \vartheta), \qquad (12.32)$$

respectively. The two addends of the total spectral weights (12.14) represent, hence, just the contributions of the polarizations along to the basis vectors $e_\parallel$ and $e_\perp$. From Eqs. (12.30)–(12.32), and from the general characterization (11.31) of Sect. 11.2, we see that for a generic direction $0 < \vartheta < \pi/2$ the radiation is *elliptically* polarized. In particular, as follows from the diagonal form (12.30), the principal axes of the ellipse described by the electric field are parallel to $e_\parallel$ and $e_\perp$, respectively. On the other hand, along the axis of the orbit, for $\vartheta = 0$, from the expansions (11.68) we see that the spectral weights (12.31) and (12.32) for $N > 1$ vanish, while those for $N = 1$ are equal to

$$\left. \frac{dW_1^\perp}{d\Omega} \right|_{\text{axis}} = \left. \frac{dW_1^\parallel}{d\Omega} \right|_{\text{axis}} = \frac{e^2 v^2 \omega_0^2}{32\pi^2}. \qquad (12.33)$$

These relations are in agreement with the total angular distributions (12.16), as well as with the general conditions for circular polarization (11.29). In fact, as we have previously deduced from Eq. (12.17), the radiation emitted along the axis of the orbit is *circularly* polarized. Conversely, for the radiation emitted in the plane of the orbit, for $\vartheta = \pi/2$, Eqs. (12.31) and (12.32) give

$$\left. \frac{dW_N^\parallel}{d\Omega} \right|_{\text{plane}} = \frac{e^2 (N\omega_0)^2}{8\pi^2} \, v^2 J_N'^2 (vN), \qquad \left. \frac{dW_N^\perp}{d\Omega} \right|_{\text{plane}} = 0. \qquad (12.34)$$

Therefore, the radiation emitted in the plane of the orbit is *linearly* polarized, parallel to $e_\parallel$, for all frequencies.

Non-relativistic limit. As formulas (12.30)–(12.32) have the same structure as their non-relativistic counterpart (12.6), they share with the latter the qualitative polarization features discussed above. In particular, in the non-relativistic limit the polarized spectral weights (12.31) and (12.32) reduce to (again the dominant weights are those of the fundamental harmonic)

$$\frac{dW^\perp}{d\Omega} \approx \frac{dW_1^\perp}{d\Omega} \approx \frac{e^2 v^2 \omega_0^2}{32\pi^2} \cos^2 \vartheta, \qquad \frac{dW^\parallel}{d\Omega} \approx \frac{dW_1^\parallel}{d\Omega} \approx \frac{e^2 v^2 \omega_0^2}{32\pi^2}, \qquad (12.35)$$

in agreement with (12.6). Integrating equations (12.35) over the solid angle, we find that the total weights of the radiation with orthogonal and parallel polarization are given by

$$W^\perp = \frac{1}{4}W, \qquad W^\parallel = \frac{3}{4}W,$$

respectively, where we have introduced the total non-relativistic power $W = e^2v^2\omega_0^2/6\pi$, see Eq. (12.5). The total polarized weights satisfy the sum rule $W = W^\perp + W^\parallel$, whereas their ratio amounts to

$$\frac{W^\parallel}{W^\perp} = 3. \tag{12.36}$$

In the non-relativistic limit the polarization of the total radiation is hence unbalanced in favor of $e_\parallel$. Below we will see that in the ultrarelativistic limit this *polarization effect* is further enhanced.

12.4.2 Overall Polarization

To examine the overall polarization of the radiation, independently of the frequency, we would have to sum the spectral weights (12.30) over all N. Alternatively, as done in Eq. (12.24), we could perform the time average of the polarized angular distribution (11.19)

$$\frac{dW^p}{d\Omega} = \frac{r^2}{T}\int_0^T (\mathbf{e}_p \cdot \mathbf{E})^2 dt. \tag{12.37}$$

We apply the, much simpler, second method, evaluating separately the intensities with orthogonal and parallel polarization, which are obtained setting in Eq. (12.37) $\mathbf{e}_p = \mathbf{e}_\perp$ and $\mathbf{e}_p = \mathbf{e}_\parallel$, respectively.

Orthogonal polarization. To determine the intensity of the radiation with orthogonal polarization we first must compute the scalar product between the electric field (11.83) and the vector $\mathbf{e}_\perp = (0, \cos\vartheta, -\sin\vartheta)$. Inserting therein the relations (12.8)–(12.10) and (12.12) we find

$$\mathbf{e}_\perp \cdot \mathbf{E} = \frac{ev\omega_0}{4\pi r} \cdot \frac{\cos\vartheta\sin\varphi}{(1 - v\sin\vartheta\cos\varphi)^3}, \tag{12.38}$$

where one must keep in mind that the angle $\varphi = \omega_0 t'(t, \mathbf{x})$ is an implicit function of t. For this reason it is convenient to change in the integral (12.37) the integration variable from t to φ, using the retarded-time relations (11.85) and (11.86). In particular one has

$$dt = (1 - \mathbf{n} \cdot \mathbf{v}(t')) \, dt' = \frac{1 - v \sin \vartheta \cos \varphi}{\omega_0} \, d\varphi. \tag{12.39}$$

Recalling that $\omega_0 = 2\pi/T$, from Eqs. (12.37)–(12.39) we then find for the angular distribution of the radiation with polarization parallel to $\mathbf{e}_\perp$

$$\frac{dW^\perp}{d\Omega} = \frac{r^2}{T} \int_0^T (\mathbf{e}_\perp \cdot \mathbf{E})^2 \, dt = \frac{e^2 v^2 \omega_0^2 \cos^2 \vartheta}{32 \pi^3} \, I(\beta), \tag{12.40}$$

where we have introduced the function

$$I(\beta) = \beta^5 \int_0^{2\pi} \frac{\sin^2 \varphi \, d\varphi}{(\beta - \cos \varphi)^5}, \qquad \beta = \frac{1}{v \sin \vartheta}.$$

To determine $I(\beta)$ we first perform an integration by parts

$$I(\beta) = -\frac{\beta^5}{4} \int_0^{2\pi} \frac{d}{d\varphi} \left(\frac{1}{(\beta - \cos \varphi)^4} \right) \sin \varphi \, d\varphi = \frac{\beta^5}{4} \int_0^{2\pi} \frac{\cos \varphi \, d\varphi}{(\beta - \cos \varphi)^4}.$$

The new integral, in turn, can be recast in the form

$$I(\beta) = \frac{\beta^5}{4} \int_0^{2\pi} \left(\frac{\beta}{(\beta - \cos \varphi)^4} - \frac{1}{(\beta - \cos \varphi)^3} \right) d\varphi$$

$$= -\frac{\beta^5}{4} \left(\frac{\beta}{6} \frac{d^3}{d\beta^3} + \frac{1}{2} \frac{d^2}{d\beta^2} \right) \int_0^{2\pi} \frac{d\varphi}{\beta - \cos \varphi}. \tag{12.41}$$

We have thus reduced the calculation of $I(\beta)$ to the evaluation of some derivatives, and to the determination of an elementary integral. Applying, for instance, the *residue theorem* one finds

$$\int_0^{2\pi} \frac{d\varphi}{\beta - \cos \varphi} = \frac{2\pi}{\sqrt{\beta^2 - 1}}. \tag{12.42}$$

Evaluating the remaining derivatives in (12.41) we eventually obtain

$$I(\beta) = \pi \frac{\beta^5 \left(\beta^2 + \frac{1}{4} \right)}{(\beta^2 - 1)^{7/2}} = \pi \frac{1 + \frac{1}{4} v^2 \sin^2 \vartheta}{(1 - v^2 \sin^2 \vartheta)^{7/2}}.$$

In this way Eq. (12.40) yields for the radiation emitted with polarization parallel to $\mathbf{e}_\perp$ the angular distribution

$$\frac{dW^\perp}{d\Omega} = \frac{e^2 v^2 \omega_0^2}{32 \pi^2} \cdot \frac{\left(1 + \frac{1}{4} v^2 \sin^2 \vartheta \right) \cos^2 \vartheta}{(1 - v^2 \sin^2 \vartheta)^{7/2}}. \tag{12.43}$$

Parallel polarization. Proceeding in the same way as above, for the radiation with polarization parallel to $\mathbf{e}_\parallel = (1, 0, 0)$ we find the angular distribution

$$\frac{dW^\parallel}{d\Omega} = \frac{r^2}{T} \int_0^T (\mathbf{e}_\parallel \cdot \mathbf{E})^2 \, dt = \frac{e^2 v^2 \omega_0^2}{32\pi^3} \, H(\beta), \qquad (12.44)$$

where we have set

$$H(\beta) = \beta^3 \int_0^{2\pi} \frac{(\beta \cos\varphi - 1)^2}{(\beta - \cos\varphi)^5} \, d\varphi, \qquad \beta = \frac{1}{v \sin\vartheta}.$$

The function $H(\beta)$ can be computed in a similar way as $I(\beta)$ by performing an appropriate decomposition of the integrand

$$H(\beta) = \beta^3 \int_0^{2\pi} \left(\frac{(\beta^2 - 1)^2}{(\beta - \cos\varphi)^5} - \frac{2\beta(\beta^2 - 1)}{(\beta - \cos\varphi)^4} + \frac{\beta^2}{(\beta - \cos\varphi)^3} \right) d\varphi$$

$$= \beta^3 \left(\frac{1}{24} (\beta^2 - 1)^2 \frac{d^4}{d\beta^4} + \frac{\beta}{3} (\beta^2 - 1) \frac{d^3}{d\beta^3} + \frac{\beta^2}{2} \frac{d^2}{d\beta^2} \right) \int_0^{2\pi} \frac{d\varphi}{\beta - \cos\varphi}.$$

Using again the integral (12.42), and evaluating the remaining derivatives, we find

$$H(\beta) = \pi \, \frac{\beta^3 \left(\beta^2 + \frac{3}{4}\right)}{(\beta^2 - 1)^{5/2}} = \pi \, \frac{1 + \frac{3}{4} v^2 \sin^2 \vartheta}{(1 - v^2 \sin^2 \vartheta)^{5/2}}.$$

Equation (12.44) then yields for the radiation emitted with polarization parallel to $\mathbf{e}_\parallel$ the angular distribution

$$\frac{dW^\parallel}{d\Omega} = \frac{e^2 v^2 \omega_0^2}{32\pi^2} \cdot \frac{1 + \frac{3}{4} v^2 \sin^2 \vartheta}{(1 - v^2 \sin^2 \vartheta)^{5/2}}. \qquad (12.45)$$

Summing the expressions (12.43) and (12.45) we obtain the formula (12.26) for the angular distribution of the total power, anticipated in Sect. 12.3. Evaluating them, instead, along the axis and in the plane of the orbit we obtain the values

$$\begin{cases} \dfrac{dW^\perp}{d\Omega} \bigg|_{\text{axis}} = \dfrac{e^2 v^2 \omega_0^2}{32\pi^2}, \\[3mm] \dfrac{dW^\perp}{d\Omega} \bigg|_{\text{plane}} = 0, \end{cases} \qquad \begin{cases} \dfrac{dW^\parallel}{d\Omega} \bigg|_{\text{axis}} = \dfrac{e^2 v^2 \omega_0^2}{32\pi^2}, \\[3mm] \dfrac{dW^\parallel}{d\Omega} \bigg|_{\text{plane}} = \dfrac{e^2 v^2 \omega_0^2}{32\pi^2} \dfrac{1 + \frac{3}{4} v^2}{(1 - v^2)^{5/2}}. \end{cases}$$

We retrieve thus in particular the results (12.33) and (12.34) derived for the individual frequencies: along the axis of the orbit the radiation is circularly polarized, and in the plane of the orbit it is linearly polarized parallel to $\mathbf{e}_\parallel$. However, from the

expression of $dW^\parallel/d\Omega\big|_{\text{plane}}$ we read off a feature that in the spectral weights of
the individual frequencies (12.34) is hidden: once we sum over all frequencies, in
the ultrarelativistic limit the radiation becomes very intense in the plane of the orbit.

Polarization of the total radiation. In the ultrarelativistic limit the dominant part of
the radiation is emitted in the vicinity of the plane of the orbit. As in this plane the
radiation is linearly polarized parallel to $e_\parallel$, we expect hence that in this limit the
total radiation is predominantly polarized along $e_\parallel$. To verify this prediction, and to
quantify this polarization effect of the total radiation, we must integrate the polar-
ized angular distributions (12.43) and (12.45) over the solid angle $d\Omega = \sin\vartheta d\vartheta d\varphi$.
Setting $y = \cos\vartheta$, from Eq. (12.43) we find for the total power of the radiation with
orthogonal polarization

$$W^\perp = \int \frac{dW^\perp}{d\Omega}\, d\Omega = \frac{e^2 v^2 \omega_0^2}{32\pi} \int_0^{\pi/2} \frac{(4 + v^2 \sin^2\vartheta)\cos^2\vartheta\sin\vartheta}{(1 - v^2\sin^2\vartheta)^{7/2}}\, d\vartheta$$

$$= \frac{e^2 v^2 \omega_0^2}{32\pi} \int_0^1 \frac{(4 + v^2)y^2 - v^2 y^4}{(v^2 y^2 + 1 - v^2)^{7/2}}\, dy.$$

The last integral can be evaluated via the change of variables $y = \frac{\sqrt{1-v^2}}{v}\tan x$, the
result being

$$W^\perp = \frac{e^2 v^2 \omega_0^2 (2 - v^2)}{48\pi(1 - v^2)^2} = \frac{2 - v^2}{8}\, W, \qquad (12.46)$$

where we have introduced the total emitted power W of Eq. (12.1). Similarly, for
the total power of the radiation emitted with parallel polarization we obtain

$$W^\parallel = \int \frac{dW^\parallel}{d\Omega}\, d\Omega = \frac{e^2 v^2 \omega_0^2}{32\pi} \int_0^{\pi/2} \frac{(4 + 3v^2\sin^2\vartheta)\sin\vartheta}{(1 - v^2\sin^2\vartheta)^{5/2}}\, d\vartheta$$

$$= \frac{e^2 v^2 \omega_0^2}{32\pi} \int_0^1 \frac{4 + 3v^2 - 3v^2 y^2}{(v^2 y^2 + 1 - v^2)^{5/2}}\, dy$$

$$= \frac{e^2 v^2 \omega_0^2 (6 + v^2)}{48\pi(1 - v^2)^2} = \frac{6 + v^2}{8}\, W. \qquad (12.47)$$

The partial powers (12.46) and (12.47), obviously, satisfy the sum rule $W^\perp + W^\parallel = W$, whereas their ratio is equal to

$$\frac{W^\parallel}{W^\perp} = \frac{6 + v^2}{2 - v^2}. \qquad (12.48)$$

In the non-relativistic limit this ratio reduces to 3, as in Eq. (12.36), while in the
ultrarelativistic limit we find the even larger value

$$\frac{\mathcal{W}^{\parallel}}{\mathcal{W}^{\perp}} = 7. \tag{12.49}$$

An ultrarelativistic particle emits, hence, radiation which is predominantly polarized in the plane of the orbit. From an experimental point of view this characteristics makes it quite easy to identify an incoming radiation as an *ultrarelativistic synchrotron radiation*, i.e. as a radiation emitted by a highly relativistic charged particle performing a circular motion, typically under the influence of a magnetic field.

12.5 Synchrotron Light

The radiation emitted by ultrarelativistic *synchrotrons* – the most powerful circular high-energy accelerators in use today – is also called *synchrotron light*. It has been observed for the first time in an electron synchrotron at the General Electric Company of Schenectady, New York, in 1947. Since then, the quantitative predictions (12.14) and (12.26) have been tested experimentally in several synchrotrons, and the measured angular and frequency distributions are in excellent agreement with these formulas. Whereas in high-energy particle accelerators synchrotron light represents a *dissipative* effect, nowadays this radiation is produced in specialized storage rings, and used for scientific as well as technical purposes in several research areas, as for instance condensed matter physics, biology and medicine, whenever very energetic photon beams are required. One of the advantages of this radiation is that in general it entails a wide range of frequencies, see (12.21), covering the visible, ultraviolet, and X-ray spectrum. With the aid of particular experimental insertion devices, *wigglers* or *undulators*, it is then possible to select from the spectrum the narrow frequency band required for a specific research activity.

Extraterrestrial sources. Synchrotron light is also produced by extraterrestrial sources, as for example the planet *Jupiter*, and the *Crab Nebula*, a supernova remnant of the Milky Way lying at a distance of about 6500 light years from Earth. Jupiter is surrounded by a rather intense magnetic field of the order $B \sim 10\,\text{gauss}$ – by comparison, the mean magnetic field of the Earth is about $0.4\,\text{gauss}$ – and it emits synchrotron light generated by electrons with energies ranging up to about $50\,\text{MeV}$, see Ref. [3], reaching hence the ultrarelativistic velocities $v \sim 0.9999$. For a typical energy value of $\varepsilon = 5\,\text{MeV}$ the relation (12.21) predicts the characteristic frequencies $\omega \sim 10\,\text{GHz}$, corresponding to *microwaves*. The relation (12.20) implies, furthermore, that the radiation includes harmonics up to the order $N \sim (5\,\text{MeV}/0.5\,\text{MeV})^3 = 1000$. Finally, the curvature radius of the orbits of these electrons is of the order $R = v/\omega_0 \approx 1/\omega_0 = \varepsilon/eB \sim 20\,\text{m}$. All these predictions are in good qualitative agreement with the astronomical observations. The synchrotron radiation emitted by the Crab Nebula originates, instead, from extremely relativistic electrons, reaching energies of the order $\varepsilon \sim 10^6\,\text{GeV}$, see Ref. [4, 5], although they move under the influence of a much weaker magnetic field, estimated to be on average $B \sim 3 \cdot 10^{-4}$ gauss. Correspondingly, the curvature radius of the orbits of the electrons is much larger, of the order $R = \varepsilon/eB \sim 10^{11}$ km. According to the general prediction (12.21) these electrons typically emit *gamma rays* with characteristic frequencies $\omega \sim 10^{13}$ GHz, corresponding to large photon energies of the order $\hbar\omega \sim 10$ MeV. In this case the radiation contains harmonics up to the extremely high order $N \sim (10^6 \, \text{GeV}/0.5 \, \text{MeV})^3 \approx 10^{28}$. However, from an observational point of view, the crucial element in the identification of the radiation emitted by the Crab Nebula as *synchrotron radiation*, accomplished by M.A. Vashakidze and V.A. Dombrovskij in the early fifties of the last century, was its high degree of *polarization*, see Eqs. (12.48) and (12.49).

12.6 Problems

12.1 Consider the synchrotron radiation emitted by a charged particle in the non-relativistic limit $v \ll 1$.

(a) Show that the unique non-vanishing Fourier coefficient of the electric field in the wave zone is given by

$$\mathbf{E}_1 = \frac{ev\omega_0 \, e^{-i\omega_0 r}}{8\pi r} \, \mathbf{n} \times \big(\mathbf{n} \times (-1, i, 0)\big). \tag{12.50}$$

Hint: Use the Eq. (11.44) for the Fourier coefficients of the electric field in the dipole approximation, and recall the particle's acceleration (12.10).

(b) Show that the angular distribution of the radiation emitted with polarization parallel to a generic direction $\mathbf{e}_p$ (12.28) has the expression (12.6).
Hint: Insert the Fourier coefficient (12.50) in the formula for the polarized spectral weights (12.29). Use the identity (12.29) and the relations (12.27) and (12.28).

References

1. J. Schwinger, L.L. DeRaad, K.A. Milton, W. Tsai, *Classical Electrodynamics* (Perseus Books Publishing, Reading MA, 1998)
2. L.D. Landau, E.M. Lifshitz, *The Classical Theory of Fields* (Addison-Wesley, Boston, 1951)
3. S.J. Bolton et al., Ultra-relativistic electrons in Jupiter's radiation belts. Nature **415**, 987 (2002)
4. A.M. Atoyan, F.A. Aharonian, On the mechanisms of gamma radiation in the crab nebula. Mon. Not. R. Astron. Soc. **278**, 525 (1996)
5. A.M. Hillas et al., The spectrum of TeV gamma rays from the crab nebula. Astrophys. J. **503**, 744 (1998)

Chapter 13
The Čerenkov Effect

In 1934 P.A. Čerenkov performed a series of experiments on the *luminescence* emitted by certain liquid solutions, if irradiated with gamma rays deriving from radioactive sources. During the experiments, which continued until 1938, Čerenkov realized that the gamma rays cause a very weak electromagnetic radiation, even in *pure* solvents such as water and benzene, resulting in a dim blue light. However, an in-depth analysis of the physical properties of the emitted light revealed that the observed effect could not be a luminescence phenomenon. In first place, the *Čerenkov radiation* was emitted along a *cone* of directions forming a well-defined angle with the direction of the incident gamma rays and, in second place, the radiation was *linearly polarized*, and, in fact, none of these properties are inherent in luminescence. In addition, the observed radiation had *universal character*, in that these distinctive features were independent of the specific properties of the solutions used, such as their temperature and chemical composition.

In 1937 I.E. Frank and I.M. Tamm [1] provided the theoretical explanation of the *Čerenkov effect*, relying on the hypothesis that the observed radiation was not caused directly by the gamma rays, but rather by *ultrarelativistic electrons* produced by the gamma rays via *Compton scattering*. They realized, in fact, that the Čerenkov radiation is generated by electrons traveling in a dielectric medium with a *constant* velocity exceeding the *velocity of light in the medium*. We recall that in a dielectric medium with *index of refraction n* the phase velocity of light is c/n. For $n > 1$ this velocity is thus smaller than c.

In this chapter we present the theory underlying the Čerenkov effect – providing in particular the theoretical explanation of the properties of the emitted radiation described above – by analyzing in detail the electromagnetic field generated by a charged particle in uniform linear motion in a dielectric medium. We will see that particles with velocities smaller and larger than the velocity of light in the medium, respectively, generate fields with radically different characteristics. In particular,

© Springer International Publishing AG, part of Springer Nature 2018
K. Lechner, *Classical Electrodynamics*, UNITEXT for Physics,
https://doi.org/10.1007/978-3-319-91809-9_13

whereas in the first case only a *Coulomb field* is present, in the second case, in addition, a *radiation field* is generated.

Macroscopic and microscopic aspects. The explanation of the Čerenkov effect provided by Frank and Tamm is based on *Maxwell's equations in matter*, which describe the dynamics of the *macroscopic* electromagnetic field in a medium. As is well known, these equations represent a simple way to take into account the *polarization charges* which are induced in a dielectric medium by the *free charges*, see, for instance, the standard textbook [2]. In fact, the macroscopic electromagnetic field is a superposition of the field produced by the free charges in vacuum, and of the field produced by the polarization charges. Therefore, as a charged particle in uniform linear motion in vacuum does not generate a radiation field, at the microscopic level the Čerenkov radiation must originate from the polarization charges. In fact, a charged particle moving in a dielectric medium deforms the atoms, or molecules, present along its path, which acquire so a non-vanishing electric dipole moment, that disappears immediately after the passage of the particle. The charges forming these dipole moments are hence subjected to a quasi-instantaneous acceleration, and so they become impulsive sources of electromagnetic waves, which are observed as Čerenkov radiation. However, it would be a difficult task to determine the resulting electromagnetic field by explicitly evaluating the coherent superposition of the plethora of these microscopic elementary waves. Vice versa, Maxwell's equations in matter represent a very efficient tool to evaluate the total electromagnetic field produced by the particle *and* by the polarization charges induced by the passage of the former. Nevertheless, by a slight abuse of language commonly adopted in this context, and for the sake of simplicity, we will refer to the total field as the *field produced by the particle*, as we will refer to the radiated energy as the *energy radiated by the particle*.

Based on Maxwell's equations in matter, in Sects. 13.1–13.3 we analyze the electromagnetic field produced by a charged particle in uniform linear motion with velocity $\mathbf{v}$ in a *non-dispersive* medium, that is to say, in a medium whose index of refraction n is independent of the frequency, distinguishing the cases $v < c/n$ and $v > c/n$. The analysis of these fields will allow us to establish the absence of radiation in the first case, and the presence of radiation in the second. However, the *idealized* simplifying assumption of a non-dispersive medium gives rise to an electromagnetic field which diverges on a conical surface, resembling a *Mach cone*, and, correspondingly, to an *infinite* emitted energy. In order to get rid of these unphysical *ultraviolet* divergences, in Sect. 13.4 we analyze the realistic, more complicated, case of a charged particle in uniform linear motion in a *dispersive* medium, finding that, in turn, the resulting electromagnetic field is *regular* throughout space. Based on this field, in Sect. 13.5 we finally derive the famous *Frank-Tamm formula* for the energy and the number of photons radiated by the particle per unit of traveled distance.

13.1 Maxwell's Equations in a Non-dispersive Medium

We consider a homogeneous medium with relative magnetic permeability equal to its vacuum value, $\mu_r = 1$, and with a *real* relative dielectric permittivity, also called *dielectric constant*, bigger than one, $\varepsilon_r > 1$. In this way we neglect, in particular, the absorption of the electromagnetic radiation, a hypothesis that is justified for frequencies far away from the resonance frequencies of the medium. In this section we, furthermore, assume that the dielectric medium is *non-dispersive*, meaning that the dielectric constant is independent of the frequencies. It follows that also the index of refraction

$$n = \sqrt{\varepsilon_r}$$

is the same for all frequencies. In a dielectric medium with these characteristics *Maxwell's equations in matter* take the form

$$-\frac{n^2}{c}\frac{\partial \mathbf{E}}{\partial t} + \nabla \times \mathbf{B} = \frac{\mathbf{j}}{c}, \tag{13.1}$$

$$\frac{1}{c}\frac{\partial \mathbf{B}}{\partial t} + \nabla \times \mathbf{E} = 0, \tag{13.2}$$

$$\nabla \cdot \mathbf{E} = \frac{\rho}{n^2}, \tag{13.3}$$

$$\nabla \cdot \mathbf{B} = 0, \tag{13.4}$$

where ρ and $\mathbf{j}$ represent the charge and current densities of the *free charges*, respectively. Equations (13.1)–(13.4) can be derived from the microscopic Maxwell equations (2.24) and (2.25), once one has reintroduced the velocity of light, through the replacements

$$\mathbf{E} \to n\mathbf{E}, \qquad \mathbf{B} \to \mathbf{B}, \qquad c \to \frac{c}{n}, \qquad \rho \to \frac{\rho}{n}, \qquad \mathbf{j} \to \frac{\mathbf{j}}{n}. \tag{13.5}$$

As the Bianchi identities (13.2) and (13.4) are the same of the microscopic equations, they admit the usual general solution

$$\mathbf{E} = -\nabla A^0 - \frac{1}{c}\frac{\partial \mathbf{A}}{\partial t}, \qquad \mathbf{B} = \nabla \times \mathbf{A}, \tag{13.6}$$

where the potentials A^0 and $\mathbf{A}$ are still defined modulo the gauge transformations $A^\mu \to A^\mu + \partial^\mu \Lambda$. However, in the presence of a medium it is more convenient to impose the *adapted* Lorenz gauge

$$\frac{n^2}{c}\frac{\partial A^0}{\partial t} + \nabla \cdot \mathbf{A} = 0, \tag{13.7}$$

because then Eqs. (13.1) and (13.3) assume the familiar form (exercise)

$$\Box_n A^\mu = \left(\frac{\rho}{n^2}, \frac{\mathbf{j}}{c}\right), \qquad \Box_n = \frac{n^2}{c^2}\frac{\partial^2}{\partial t^2} - \nabla^2. \tag{13.8}$$

As we see from the expression of the modified d'Alembertian $\Box_n$, in the absence of free charges, i.e. for $\rho = \mathbf{j} = 0$, in a dielectric medium the electromagnetic waves propagate at the speed c/n.

13.1.1   Charged Particle in Uniform Linear Motion

A charged particle traveling in a medium with a constant spatial velocity $\mathbf{v}$ has the constant four-velocity $u^\mu = (1, \mathbf{v}/c)/\sqrt{1 - v^2/c^2}$, so that according to Eq. (6.87) the charge and current densities are given by

$$\rho = e\,u^0 \int \delta^4(x - us)\,ds = e\,\delta^3(\mathbf{x} - \mathbf{v}t), \qquad \mathbf{j} = \rho\mathbf{v}. \tag{13.9}$$

Since $\mathbf{j}$ is proportional to ρ, it is sufficient to solve Eqs. (13.8) for $\mu = 0$

$$\Box_n A^0 = \frac{\rho}{n^2}. \tag{13.10}$$

In fact, according to Eqs. (13.8) and (13.9) the spatial potential is given in terms of A^0 by

$$\mathbf{A} = \frac{n^2 A^0}{c}\,\mathbf{v}. \tag{13.11}$$

Henceforth, for simplicity we choose as z axis the orbit of the particle, in which case the trajectory takes the form

$$\mathbf{y}(t) = (0, 0, vt).$$

Furthermore, according to the *cylindrical* symmetry of the problem, it is convenient to resort to the cylindrical coordinates $\{r, \varphi, z\}$, where r and φ are the polar coordinates in the xy plane. These coordinates are associated with the set of mutually orthogonal unit vectors $\{\mathbf{u}_r, \mathbf{u}_\varphi, \mathbf{u}_z\}$. In Sects. 13.2 and 13.3 we solve Maxwell's equation in matter (13.10) for the cases $v < c/n$ and $v > c/n$, respectively, and perform a detailed analysis of the properties of the resulting electromagnetic fields.

13.2   The Field for $v < c/n$

If the speed v of the particle is less than the speed of light in the medium c/n, Eq. (13.10) can be solved in exactly the same way as the analogous equation in free space, see Sect. 6.3.1,

$$\Box A^0 = \rho. \tag{13.12}$$

In that case we had $n = 1$ and $v < c$, and consequently $v < c/n$. Comparing Eqs. (13.10) and (13.12) we then see that the solution of the former can be obtained from the solution (6.93) of the latter through the replacements $e \to e/n^2$ and $c \to c/n$. As the time component of the four-potential (6.93) has the form

$$A^0 = \frac{e}{4\pi} \frac{u^0}{\sqrt{(ux)^2 - x^2}} = \frac{e}{4\pi} \frac{1}{\sqrt{(z - vt)^2 + \left(1 - \frac{v^2}{c^2}\right) r^2}},$$

in this way we derive the solution of equation (13.10)

$$A^0 = \frac{e}{4\pi n^2} \frac{1}{\sqrt{(z - vt)^2 + \left(1 - \frac{v^2 n^2}{c^2}\right) r^2}}. \tag{13.13}$$

Obviously, by construction the fields (13.13) and (13.11) satisfy the adapted Lorenz gauge (13.7). Formulas (13.6), (13.11) and (13.13) then yield the electromagnetic fields

$$\mathbf{E} = \frac{e}{4\pi n^2} \frac{\left(1 - \frac{v^2 n^2}{c^2}\right)(\mathbf{x} - \mathbf{v}t)}{\left((z - vt)^2 + \left(1 - \frac{v^2 n^2}{c^2}\right) r^2\right)^{3/2}}, \tag{13.14}$$

$$\mathbf{B} = \frac{e}{4\pi c} \frac{\left(1 - \frac{v^2 n^2}{c^2}\right) vr\, \mathbf{u}_\varphi}{\left((z - vt)^2 + \left(1 - \frac{v^2 n^2}{c^2}\right) r^2\right)^{3/2}} = \frac{n^2 \mathbf{v}}{c} \times \mathbf{E}, \tag{13.15}$$

and the associated Poynting vector

$$\mathbf{S} = c\,\mathbf{E} \times \mathbf{B} = \left(\frac{e}{4\pi n}\right)^2 \frac{\left(1 - \frac{v^2 n^2}{c^2}\right)^2 vr\left(r\mathbf{u}_z + (vt - z)\mathbf{u}_r\right)}{\left((z - vt)^2 + \left(1 - \frac{v^2 n^2}{c^2}\right) r^2\right)^3} = S_z \mathbf{u}_z + S_r \mathbf{u}_r. \tag{13.16}$$

In particular, the radial component of this vector is given by

$$S_r = \left(\frac{e}{4\pi n}\right)^2 \frac{\left(1 - \frac{v^2 n^2}{c^2}\right)^2 vr\,(vt - z)}{\left((z - vt)^2 + \left(1 - \frac{v^2 n^2}{c^2}\right) r^2\right)^3}. \tag{13.17}$$

As long as $v < c/n$, the fields (13.13)–(13.17) have the same qualitative properties of the corresponding fields produced by Maxwell's equations in free space. In particular, the electromagnetic fields (13.14) and (13.15) entail no singularities, apart from those on the trajectory. In fact, for $v < c/n$ their denominators vanish only for $\mathbf{x} = \mathbf{v}t$, i.e. for $z = vt$ and $r = 0$. From the expression (13.17) of the radial component of the Poynting vector it follows, furthermore, that the *net* radial energy flux is zero: behind the particle, for $z < vt$, the energy flux is outgoing ($S_r > 0$), while in front of the particle, for $z > vt$, it is incoming ($S_r < 0$), and, for symmetry reasons, these fluxes compensate each other. More in detail, the total energy emitted across a cylindrical surface of radius r, concentric with the orbit and with bases located at z_1 and z_2, turns out to be ($\Delta z = z_2 - z_1$, $u = z - vt$)

$$\Delta\varepsilon = 2\pi r \int_{z_1}^{z_2} dz \int_{-\infty}^{\infty} S_r\, dt = \frac{2\pi r \Delta z}{v} \int_{-\infty}^{\infty} S_r\, du = 0, \qquad (13.18)$$

where the conclusion follows from the antisymmetry of S_r as a function of the variable u. In conclusion, a charged particle moving in a medium with speed $v < c/n$ does not generate radiation.

13.2.1 Frequency Analysis and Evanescent Waves

In light of the comparison with the fields produced by a particle with speed $v > c/n$ it is useful to perform a *spectral analysis* of the potential (13.13). In general, such an analysis is meaningful in the presence of *radiation fields*. In contrast, in the case at hand the fields (13.14) and (13.15) at large distances decrease as $1/r^2$, and they are hence *Coulomb fields*, rather than radiation fields. Nevertheless, as we will see later on, it is instructive to investigate the properties of the temporal Fourier transform $A^0(\omega, \mathbf{x}) \equiv A^0(\omega)$ of the potential (13.13). This Fourier transform reads

$$A^0(\omega) = \frac{1}{\sqrt{2\pi}} \int_{-\infty}^{\infty} e^{-i\omega t} A^0(t, \mathbf{x})\, dt$$

$$= \frac{e}{\sqrt{2\pi}\, 4\pi n^2} \int_{-\infty}^{\infty} \frac{e^{-i\omega t}\, dt}{\sqrt{(z - vt)^2 + \left(1 - \frac{v^2 n^2}{c^2}\right) r^2}}. \qquad (13.19)$$

Performing the change of variables

$$t \rightarrow t(p) = \frac{z}{v} - \frac{rp}{v}\sqrt{1 - \frac{v^2 n^2}{c^2}},$$

the expression (13.19) can be recast in the form

$$A^0(\omega) = \frac{e}{(2\pi)^{3/2} n^2 v}\, e^{-i\omega z/v}\, K\!\left(\frac{\omega r}{v}\sqrt{1 - \frac{v^2 n^2}{c^2}}\right), \qquad (13.20)$$

where we have introduced the real special function[1]

$$K(x) = \frac{1}{2} \int_{-\infty}^{\infty} \frac{e^{ixp}}{\sqrt{p^2 + 1}} \, dp = \int_{0}^{\infty} \cos\left(x \sinh \beta\right) d\beta. \tag{13.21}$$

The second representation is obtained through the change of variables $p(\beta) = \sinh \beta$. Some important properties of this function are derived in Sect. 13.2.2.

$A^0(\omega)$ *as an evanescent wave*. The potential $A^0(\omega)$ (13.20) depends on the variable z only through the plane-wave phase $e^{-ik_z z}$, relative to the wave vector $k_z = \omega/v$. This term describes thus a wave propagating in the z direction with the velocity $v_z = \omega/k_z = v$, i.e. with the same velocity of the particle. On the other hand, the asymptotic behavior (13.27) of $K(x)$ implies that at large distances from the trajectory, i.e. for large r, $A^0(\omega)$ behaves as

$$A^0(\omega) \approx \frac{\beta}{\sqrt{r}} \, e^{-ik_z z} e^{-\gamma r}, \qquad \gamma = \frac{|\omega|}{v} \sqrt{1 - \frac{v^2 n^2}{c^2}}, \tag{13.22}$$

where β is a position-independent constant. The factor $1/\sqrt{r}$ in formula (13.22) represents, actually, a characteristic property of *cylindrical waves*. In fact, as we shall see in Sect. 13.5, in the same way as energy conservation implies for a spherical wave the presence of a factor $1/r$, for a cylindrical waves it requires a factor $1/\sqrt{r}$. However, in the potential (13.22) the factor $1/\sqrt{r}$ is superseded by the exponential decrease of the term $e^{-\gamma r}$. This means that asymptotically the potential $A^0(\omega)$ (13.20), actually, goes over to what is called an *evanescent* wave. To become a *true* wave, the real exponential $e^{-\gamma r}$ should be replaced by an oscillating exponential of the type $e^{-ik_r r}$. In conclusion, the spectral analysis confirms that a particle traveling in a medium with constant speed less than c/n does not generate electromagnetic waves.

13.2.2   The Special Function $K(x)$

Alternative representations of $K(x)$. For $x > 0$ the function $K(x)$ in (13.21) is linked to the analytically-continued Bessel and Neumann functions introduced in Sect. 11.3.2 through the relations (see, for instance, the handbook [3])

$$K(x) = \frac{i\pi}{2}\big(J_0(ix) + iY_0(ix)\big) = -\frac{i\pi}{2}\big(J_0(-ix) - iY_0(-ix)\big). \tag{13.23}$$

We can derive a further representation of $K(x)$, which will allow us to derive more easily some properties of this function that will turn out to be useful later on, by rely-

---

[1] $K(x)$ corresponds to the *modified Bessel function of the second type* of order zero. Usually, in handbooks, it is denoted by the symbol $K_0(x)$.

ing on complex analysis. Since, as follows from the Definition (13.21), this function is even

$$K(-x) = K(x),$$ (13.24)

it is not restrictive to consider only positive x. For a fixed $x > 0$ we then introduce the function of a complex variable

$$f(z) = \frac{e^{ixz}}{\sqrt{z^2 + 1}},$$

which is analytic in the upper half-plane, apart from a branch cut that we can direct along the semi-axis of equation $z(u) = iu$, with $u \in [1, \infty]$. As the line integral of $f(z)$ along a closed curve of the complex plane vanishes, as long as the interior of the curve does not contain poles of $f(z)$, we have

$$\oint_{\Gamma_{R,\varepsilon}} f(z)\, dz = 0,$$ (13.25)

where $\Gamma_{R,\varepsilon}$, with $\varepsilon > 0$ and $R > 1$, is the closed counter-clockwise oriented curve composed of the following paths:

- the interval $[-R, R]$ of the real axis;
- two quarters of a circle of radius R centered at the origin lying in the upper half-plane, with angular apertures $0 < \varphi < \pi/2$ and $\pi/2 < \varphi < \pi$, respectively;
- two half-lines parallel to the imaginary axis, of equations $z_\pm(u) = \pm\varepsilon + iu$, respectively, with $u \in [1, R]$;
- a semi-circle of radius ε centered at the point $z = i$.

Considering the limits of the identity (13.25) for $R \to \infty$ and for $\varepsilon \to 0$, the integrals along the three arcs of a circle tend to zero, and only the integrals along the real axis and along the two half-lines parallel to the imaginary axis survive. Since $\varepsilon u > 0$, according to the choice of our branch cut the limit for $\varepsilon \to 0$ of the denominator of the function $f(z)$ along the half-lines $z_\pm(u)$ becomes

$$\lim_{\varepsilon \to 0} \sqrt{z_\pm^2(u) + 1} = \lim_{\varepsilon \to 0} \sqrt{-(u^2 - 1) \pm 2i\varepsilon u + \varepsilon^2} = \pm i\sqrt{u^2 - 1}.$$

Taking into account that the half-lines have opposite orientations, in this way equation (13.25) yields the identity

$$\int_{-\infty}^{\infty} \frac{e^{ixp}}{\sqrt{p^2 + 1}}\, dp - 2\int_{1}^{\infty} \frac{e^{-xu}}{\sqrt{u^2 - 1}}\, du = 0.$$

It follows that, as long as $x > 0$, the function $K(x)$ admits the additional alternatives representations

$$K(x) = \int_1^\infty \frac{e^{-xu}}{\sqrt{u^2 - 1}} \, du = \int_0^\infty e^{-x \cosh \beta} \, d\beta, \tag{13.26}$$

where we have set $u(\beta) = \cosh \beta$.

Behavior of $K(x)$ ***for*** $x \to \infty$. The representations (13.26) are particularly useful to determine the behaviors of $K(x)$ for large and small arguments. For large x we apply the *saddle point method* to the second representation in (13.26). In fact, for $x \to \infty$ the values of β which dominate the integral are those corresponding to the minimum of $\cosh \beta$, i.e. those close to zero. Correspondingly, using the expansion

$$\cosh \beta = 1 + \frac{\beta^2}{2} + O(\beta^4),$$

we find

$$K(x) = e^{-x} \int_0^\infty e^{-\left(x\beta^2/2 + O(x\beta^4)\right)} \, d\beta.$$

Performing the rescaling $\beta \to \beta/\sqrt{x}$, and eventually considering also negative values of x, we so obtain the asymptotic behavior

$$K(x) = \frac{e^{-x}}{\sqrt{x}} \int_0^\infty e^{-\left(\beta^2/2 + O(\beta^4/x)\right)} \, d\beta = \sqrt{\frac{\pi}{2|x|}} \, e^{-|x|} \left(1 + O\left(\frac{1}{x}\right)\right). \tag{13.27}$$

Behavior of $K(x)$ ***for*** $x \to 0$. From any of the representations given above we see that in the limit for $x \to 0$ the function $K(x)$ diverges. To determine the nature of the divergence we consider the first representation in (13.26). To isolate the divergent part of the integral we first rescale the integration variable by $u \to u/x$, obtaining

$$K(x) = \int_x^\infty \frac{e^{-u}}{\sqrt{u^2 - x^2}} \, du = \int_x^1 \frac{e^{-u}}{\sqrt{u^2 - x^2}} \, du + \int_1^\infty \frac{e^{-u}}{\sqrt{u^2 - x^2}} \, du.$$

As in the limit for $x \to 0$ the second integral converges, it is sufficient to analyze the first integral

$$\int_x^1 \frac{e^{-u}}{\sqrt{u^2 - x^2}} \, du = \int_x^1 \frac{1}{\sqrt{u^2 - x^2}} \, du + \int_x^1 \frac{e^{-u} - 1}{\sqrt{u^2 - x^2}} \, du.$$

Since the function $(e^{-u} - 1)/u$ is bounded in the closed interval $[0, 1]$, for $x \to 0$ the second integral converges, and we remain with

$$\int_x^1 \frac{du}{\sqrt{u^2 - x^2}} = \operatorname{arccosh}\left(\frac{1}{x}\right) = -\ln\left(\frac{x}{2}\right) + O(x).$$

In the limit for $x \to 0$ the function $K(x)$ diverges hence logarithmically. More precisely, taking also negative values of the argument into account, it entails the expansion

$$K(x) = -\ln|x| + a + O(x),\tag{13.28}$$

where a is a constant.

Differential equation. Frequently, the special functions are defined through the differential equations they obey. The defining equation for $K(x)$ is

$$K'' + \frac{1}{x}K' - K = 0.\tag{13.29}$$

To verify that $K(x)$ satisfies this equation we use again the first representation in (13.26), calculating the quantity

$$\begin{aligned}
K'' + \frac{1}{x}K' &= \int_1^\infty \left(\frac{u^2}{\sqrt{u^2-1}} - \frac{1}{x}\frac{u}{\sqrt{u^2-1}}\right) e^{-ux}\, du \\
&= \int_1^\infty \left(\frac{u^2}{\sqrt{u^2-1}} - \frac{1}{x}\frac{d\sqrt{u^2-1}}{du}\right) e^{-ux}\, du.
\end{aligned}$$

An integration by parts then reduces this expression to (13.26).

The special function $\widetilde{K}(x)$. The second-order linear differential equation (13.29) has two linearly independent solutions. One solution is $K(x)$, and the other is the special function[2]

$$\widetilde{K}(x) = \int_{-1}^1 \frac{e^{xu}}{\sqrt{1-u^2}}\, du = \int_0^\pi e^{x\cos\vartheta}\, d\vartheta.\tag{13.30}$$

Also this function is even, $\widetilde{K}(-x) = \widetilde{K}(x)$, and it entails the asymptotic behaviors

$$\widetilde{K}(x) = \sqrt{\frac{\pi}{2|x|}}\, e^{|x|}\left(1 + O\left(\frac{1}{x}\right)\right), \qquad \widetilde{K}(x) = \pi + O(x),\tag{13.31}$$

to be compared with the behaviors (13.27) and (13.28) of $K(x)$. In particular, contrary to K, the function $\widetilde{K}$ does *not* define a distribution as it increases exponentially for $x \to \pm\infty$.

---

[2] $\widetilde{K}(x)$ is proportional to the *modified Bessel function of the first type* of order zero $I_0(x) = J_0(ix)$, namely $\widetilde{K}(x) = \pi I_0(x)$.

13.3 The Field for $v > c/n$

If the speed v of the particle exceeds the speed of light in the medium c/n, the solution of Maxwell's equation in matter (13.10) can no longer be derived from its solution in free space (13.13) by means of substitutions. In particular, for $v > c/n$ in large regions of space the radicand appearing in the potential (13.13) would be negative and, furthermore, on the conical surface $z = vt \pm r\sqrt{v^2 n^2/c^2 - 1}$ the potential would diverge. Nonetheless, we can still resort to the Green function method. Henceforth, we set $c = 1$, so that we have $vn > 1$. By definition, the *Green function in the medium* $G_n(x)$ associated with Eq. (13.8) must satisfy the kernel equation

$$\Box_n G_n(x) = \left(n^2 \frac{\partial^2}{\partial t^2} - \nabla^2 \right) G_n(x) = \delta(t)\, \delta^3(x). \tag{13.32}$$

As in the case of the kernel equation in free space (6.43), the solution of equation (13.32) is not unique. However, we again search for the *retarded* solution, subject to the condition $G_n(t, \mathbf{x}) = 0$ for $t < 0$. Moreover, although $\Box_n$ is no longer invariant under Lorentz transformations, it is still invariant under spatial rotations. Consequently, since only the isotropy subgroup of the Lorentz group was used in Sect. 6.2.1 to derive the retarded Green function in free space (6.59)

$$G(t, \mathbf{x}) = \frac{1}{2\pi} H(t)\, \delta(t^2 - |\mathbf{x}|^2), \tag{13.33}$$

to find the retarded solution of equation (13.32) we can take advantage from the fact that the Green function (13.33) satisfies the kernel equation (6.43), namely

$$\left(\frac{\partial^2}{\partial t^2} - \nabla^2 \right) G(t, \mathbf{x}) = \delta(t)\, \delta^3(x).$$

In fact, performing in the latter the replacement $t \to t/n$, and using the identity $\delta(t/n) = n\, \delta(t)$, we obtain

$$\left(n^2 \frac{\partial^2}{\partial t^2} - \nabla^2 \right) G\left(\frac{t}{n}, \mathbf{x} \right) = n\, \delta(t)\, \delta^3(x).$$

Comparing this identity with Eq. (13.32) we see that the kernel we search for is given by

$$G_n(x) = \frac{1}{n} G\left(\frac{t}{n}, \mathbf{x} \right) = \frac{1}{2\pi n} H(t)\, \delta(x_n^2), \qquad x_n^2 = \frac{t^2}{n^2} - |\mathbf{x}|^2. \tag{13.34}$$

Here, and in the following, we use for the modified scalar product of two four-vectors a^μ and b^μ the notation

$$(ab)_n = \frac{a^0 b^0}{n^2} - \mathbf{a} \cdot \mathbf{b}, \qquad a_n^2 = \frac{(a^0)^2}{n^2} - |\mathbf{a}|^2.$$

13.3.1 The Mach Cone

Once the Green function G_n is known, *formally* the retarded solution of equation (13.10) can be written as the convolution

$$A^0(x) = \frac{1}{n^2} (G_n * \rho)(x) = \frac{1}{n^2} \int G_n(x-y)\, \rho(y)\, d^4 y. \tag{13.35}$$

From Sect. 6.2.4 we, actually, know that *a priori* it is not guaranteed that the Green function method provides a solution of Maxwell's equations, especially if it involves singularities, as in the case at hand. In particular, in free space we have seen that, while the method yields a solution if $v < 1$, it fails if $v = 1$. In the present case we have even $v > 1/n$, where $1/n$ is the velocity of light in the medium, and so we have no *a priori* control over the Green function method. Accordingly, for the moment we will proceed formally, starting from the convolution (13.35), and then we shall verify *a posteriori* that the potential $A^0(x)$ so obtained satisfies Maxwell's equations, in particular that it is a *distribution*. We substitute, thus, in Eq. (13.35) for the charge density ρ the first expression in (13.9), and for the kernel $G_n(x)$ the expression (13.34). If for a given space-time point x^μ the quadratic form

$$f(s) = (x - us)_n^2 = x_n^2 - 2s(ux)_n + u_n^2 s^2$$

possesses two real zeros $s_\pm$, applying the usual procedure we then find the potential

$$\begin{aligned} A^0(x) &= \frac{eu^0}{2\pi n^3} \int H(t - u^0 s)\, \delta(f(s))\, ds \\ &= \frac{eu^0}{2\pi n^3} \left(\frac{H(t - u^0 s_+)}{|f'(s_+)|} + \frac{H(t - u^0 s_-)}{|f'(s_-)|} \right). \end{aligned} \tag{13.36}$$

Vice versa, at points x^μ for which $f(s)$ has no real zeros, the potential $A^0(x)$ vanishes. The value of the integral (13.36) depends hence, (i) on the existence of reals zeros of the function $f(s)$ and, (ii) on the signs of the numbers $t - u^0 s_\pm$. The zeros of $f(s)$ are given by

$$s_\pm = \frac{(ux)_n \pm \sqrt{(ux)_n^2 - u_n^2 x_n^2}}{u_n^2}, \qquad u_n^2 = \frac{1 - v^2 n^2}{(1 - v^2) n^2} < 0, \tag{13.37}$$

and we have

$$|f'(s_\pm)| = 2\sqrt{(ux)_n^2 - u_n^2 x_n^2} = \frac{2u^0}{n}\sqrt{(z-vt)^2 - (v^2n^2 - 1)r^2}. \quad (13.38)$$

Therefore, real zeros exist only in the space-time region

$$|z - vt| > r\sqrt{v^2n^2 - 1} \qquad \leftrightarrow \qquad \frac{r^2}{(z-vt)^2 + r^2} < \frac{1}{v^2n^2}. \quad (13.39)$$

This region corresponds to the interior of a *double cone* centered at the position of the particle and with axis the particle's trajectory, with angular aperture α given by

$$\sin\alpha = \frac{1}{vn}.$$

Out of this cone the electromagnetic field is, thus, anyway zero. For an x^μ staying in the interior of the cone we must now analyze the signs of the numbers $t - u^0 s_\pm$. Using the expressions (13.37) of the zeros $s_\pm$, a simple calculation gives

$$t - u^0 s_\pm = \frac{n}{v^2n^2 - 1}\left(vn(vt - z) \pm \sqrt{(z-vt)^2 - (v^2n^2 - 1)r^2}\right). \quad (13.40)$$

Since $vn > 1$, for $z > vt$ both numbers $t - u^0 s_\pm$ are hence *negative*. It follows that the potential (13.36), and so the electromagnetic field, vanish also in the interior of the *right* half of the double cone. Conversely, in the region

$$z < vt \quad (13.41)$$

both numbers $t - u^0 s_\pm$ are *positive*, so that both terms in (13.36) give a non-vanishing contribution. In conclusion, from the inequalities (13.39) and (13.41) we read off that at the time t the electromagnetic field produced by the particle is different from zero only in the region

$$z - vt < -r\sqrt{\frac{v^2n^2}{c^2} - 1},$$

where we have reintroduced the velocity of light. This region corresponds to the interior of a *simple* cone centered at the position of the particle, coaxial with the trajectory and oriented opposite to its velocity, with angular aperture given by

$$\alpha = \arcsin\left(\frac{c}{vn}\right). \quad (13.42)$$

Given the similarity between this cone and the acoustic wavefront created by an airplane flying at a supersonic velocity, we will refer to this surface as the *Mach cone*. For a fixed time t the parametric equation of this cone is

$$z(r) = vt - r\sqrt{\frac{v^2 n^2}{c^2} - 1}. \tag{13.43}$$

Finally, taking the inequalities (13.39) and (13.41) into account, formulas (13.36) and (13.38) yield the scalar potential

$$A^0(x) = \frac{e}{2\pi n^2} \frac{H\left(-z + vt - r\sqrt{\frac{v^2 n^2}{c^2} - 1}\right)}{\sqrt{(z - vt)^2 - \left(\frac{v^2 n^2}{c^2} - 1\right) r^2}}, \tag{13.44}$$

where H is the Heaviside function. Finally, since $A^0(x)$ depends on z and t only through the combination $z - vt$, it is immediately seen that the four-potential given by formulas (13.44) and (13.11) satisfies the adapted Lorenz gauge (13.7). Comparing the potential (13.44) with the corresponding potential (13.13) generated by a particle with speed $v < c/n$, we see that the two expressions *formally* exhibit the same functional dependencies on $\mathbf{x}$ and t, differing only by an overall factor of 2. However, despite this formal similarity, there are two crucial differences between the two expressions: whereas the potential (13.13) is non-vanishing throughout space and diverges only on the trajectory of the particle, the potential (13.44) diverges on the whole Mach cone (13.43) and it vanishes outside this cone.

From a mathematical point of view, the crucial property of the potential (13.44) is that it defines a *distribution* in $S'(\mathbb{R}^4)$. In fact, it is easily seen that, although singular, the function (13.44) is *locally integrable* on the Mach cone and, furthermore, for $x^\mu \to \infty$ it decreases as $1/|x^\mu|$. Consequently, also the electromagnetic fields associated via (13.6) with the four-potential (13.44) and (13.11) are distributions. It remains to prove that the fields so obtained solve indeed Maxwell's equations, which amounts to verify that the potentials (13.44) and (13.11) satisfy equations (13.7) and (13.10). We will provide an indirect proof of this in Sect. 13.4.2.

13.3.2   The Electromagnetic Field and its Ultraviolet Singularities

Although the four-potential A^μ given by Eqs. (13.44) and (13.11) is a distribution, the calculation of the associated field strength $F^{\mu\nu}$ requires some care. In fact, since both the denominator and the argument of the Heaviside function of the expression (13.44) vanish on the Mach cone, the derivatives appearing in (13.6) cannot be computed *naively*. A convenient way to bypass the problem is to introduce a *regularized* four-potential A_ε^μ, for instance

$$A_\varepsilon^0 = \frac{e}{2\pi n^2} \frac{H\left(-z + vt - \sqrt{\frac{v^2 n^2}{c^2} - 1}\,(r + \varepsilon)\right)}{\sqrt{(z - vt)^2 - \left(\frac{v^2 n^2}{c^2} - 1\right) r^2}}, \qquad \mathbf{A}_\varepsilon = \frac{n^2 \mathbf{v}}{c} A_\varepsilon^0, \quad (13.45)$$

where ε is a positive parameter. Since this four-potential entails the distributional limit

$$\mathcal{S}' - \lim_{\varepsilon \to 0} A_\varepsilon^\mu = A^\mu,$$

the field strength of the four-potential (13.11), (13.44) can be computed as the limit

$$F^{\mu\nu} = \mathcal{S}' - \lim_{\varepsilon \to 0} (\partial^\mu A_\varepsilon^\nu - \partial^\nu A_\varepsilon^\mu). \tag{13.46}$$

As the modified Heaviside function in (13.45) vanishes on the Mach cone, the advantage of the regularized potential A_ε^μ is that its derivatives can now be computed in the ordinary sense, see Problem 13.5 for an elementary illustration of the procedure. According to (13.46) the resulting electric and magnetic fields take the form

$$\mathbf{E} = \frac{e}{2\pi n^2} \left(1 - \frac{v^2 n^2}{c^2}\right) (\mathbf{x} - \mathbf{v}t) \cdot$$

$$\mathcal{S}' - \lim_{\varepsilon \to 0} \left(\frac{H\left(-z + vt - \sqrt{\frac{v^2 n^2}{c^2} - 1}\,(r + \varepsilon)\right)}{\left((z - vt)^2 - \left(\frac{v^2 n^2}{c^2} - 1\right) r^2\right)^{3/2}} - \frac{\delta\left(z - vt + r\sqrt{\frac{v^2 n^2}{c^2} - 1}\right)}{\sqrt{2\varepsilon}\left(\frac{v^2 n^2}{c^2} - 1\right) r^{3/2}} \right),$$

$$\mathbf{B} = \frac{n^2 \mathbf{v}}{c} \times \mathbf{E}.$$

$$\tag{13.47}$$

In the term containing the δ-function we simplified ε-dependencies that under the distributional limit for $\varepsilon \to 0$ drop out. Of course, the fields $\mathbf{E}$ and $\mathbf{B}$ are *ultraviolet* divergent on the Mach cone, see Sect. 11.3.1 for the terminology, as is the four-potential A^μ. Nevertheless, by construction these fields are well-defined, although *non-regular*, distributions. In fact, the result of the distributional limit in (13.47) cannot be written as a combination of ordinary functions and of δ-functions. In particular, none of the two terms between parentheses in (13.47) admit a distributional limit for $\varepsilon \to 0$: the first, because the resulting function would have non-integrable singularities on the Mach cone, and the second, because it is proportional to a δ-function supported on the Mach multiplied by $1/\sqrt{\varepsilon}$. However, by construction their sum converges to a distribution, as can be explicitly checked by applying both terms to a test function $\varphi(x) \in \mathcal{S}(\mathbb{R}^4)$ and by analyzing the limit for $\varepsilon \to 0$ of the resulting expressions, see Problem 13.7.

Comparing the fields $\mathbf{E}$ and $\mathbf{B}$ of formulas (13.47) with the corresponding expressions (13.14) and (13.15) of the fields generated by a particle with speed $v < c/n$, we see that the "regular" term of $\mathbf{E}$ in (13.47) has formally the same functional dependencies on $\mathbf{x}$ and t as the electric field (13.14). There is, however, a dramatic difference: whereas the field lines of the latter are *outgoing* from the particle, since $v > c/n$ the fields lines of the former are *incoming*. This apparent mismatch with the Coulomb law $\nabla \cdot \mathbf{E} = \rho/n^2$, see Eq. (13.3), which requires a *positive* electric flux coming out from the particle, is resolved by the δ-function contribution in (13.47). In fact, the latter entails an electric field supported on the Mach cone which is tangential to the cone and *outgoing* from the particle. Actually, both fluxes are divergent, but their sum is finite and positive and equal to e/n^2, see Problem 13.8.

Singular energy emission. As the fields (13.47) are ultraviolet divergent on the Mach cone, so is the associated Poynting vector $\mathbf{S} = c\,\mathbf{E} \times \mathbf{B}$. Nevertheless, in the complement of this cone $\mathbf{S}$ is regular, and it is easy to see that in its neighborhood it becomes *orthogonal* to it, see Problem 13.4. In particular, from formulas (13.47) we derive that this vector entails the decomposition $\mathbf{S} = S_z \mathbf{u}_z + S_r \mathbf{u}_r$, where in the complement of the Mach cone the radial component is given by

$$
S_r = \left(\frac{e}{2\pi n}\right)^2 \frac{\left(1 - \frac{v^2 n^2}{c^2}\right)^2 vr\,(vt - z)}{\left((z - vt)^2 - \left(\frac{v^2 n^2}{c^2} - 1\right)r^2\right)^3} \, H\left(-z + vt - r\sqrt{\frac{v^2 n^2}{c^2} - 1}\right).
$$

$$(13.48)$$

In contrast to the radial component (13.17) of the Poynting vector for a motion with speed $v < c/n$, the radial component (13.48) is non-vanishing only inside the Mach cone, where it is *positive*. Consequently, a particle moving with speed $v > c/n$ generates a net non-vanishing radially outgoing electromagnetic energy flux. Actually, since S_r diverges on the Mach cone, considering the same cylindrical surface as in Eq. (13.18) we find that, unfortunately, the energy radiated per unit of traveled distance is *infinite*, see Problem 13.3,

$$
\frac{\Delta\varepsilon}{\Delta z} = 2\pi r \int_{-\infty}^{\infty} S_r \, dt = 2\pi r \int_{\frac{1}{v}\left(z + r\sqrt{\frac{v^2 n^2}{c^2} - 1}\right)}^{\infty} S_r \, dt = \infty. \qquad (13.49)
$$

However, as we shall find out in Sect. 13.4, eventually the ultraviolet singularities on the Mach cone we came across in this section turn out to be *unphysical*: in fact, as we will see, they are a mere artefact of our poorly realistic idealization of a charged particle moving in a *non-dispersive* dielectric medium.

13.3.3   Frequency Analysis and Čerenkov Angle

The electromagnetic fields (13.47) – singular on the Mach if considered in the time domain – appear more regular if analyzed in the frequency domain. Consequently, the spectral analysis of these fields allows to investigate several basic features of the emitted radiation, notwithstanding these singularities. As in the case of a particle with speed $v < c/n$, it is more convenient to base the spectral analysis of the radiation on the temporal Fourier transform $A^0(\omega, \mathbf{x}) = A^0(\omega)$ of the potential (13.44). Taking the Heaviside function into account we obtain

$$A^0(\omega) = \frac{1}{\sqrt{2\pi}} \int_{-\infty}^{\infty} e^{-i\omega t} A^0(t, \mathbf{x}) \, dt \qquad (13.50)$$

$$= \frac{e}{(2\pi)^{3/2} n^2} \int_{\frac{1}{v}\left(z + r\sqrt{\frac{v^2 n^2}{c^2} - 1}\right)}^{\infty} \frac{e^{-i\omega t} \, dt}{\sqrt{(z - vt)^2 - \left(\frac{v^2 n^2}{c^2} - 1\right) r^2}}.$$

Performing the change of variable

$$t \to t(u) = \frac{z}{v} + \frac{ru}{v} \sqrt{\frac{v^2 n^2}{c^2} - 1}$$

this time we find

$$A^0(\omega) = \frac{e}{(2\pi)^{3/2} n^2 v} e^{-i\omega z/v} L\left(\frac{\omega r}{v} \sqrt{\frac{v^2 n^2}{c^2} - 1}\right), \qquad (13.51)$$

where we have introduced the complex special function

$$L(x) = \int_1^{\infty} \frac{e^{-ixu}}{\sqrt{u^2 - 1}} \, du = \int_0^{\infty} e^{-ix \cosh \beta} \, d\beta, \qquad (13.52)$$

which replaces the function $K(x)$ we have introduced for $v < c/n$. Actually, the representations (13.26) and (13.52) imply the formal relation

$$L(x) = K(ix), \qquad (13.53)$$

meaning that the functions K and L are analytic continuations of each other from the real to the imaginary axis, as can be seen also from formulas (13.23) and (13.54). This implies that, in a sense, also the potentials (13.20) and (13.51) constitute analytic continuations of each other: from the region $v < c/n$ to the region $v > c/n$.

The special function $L(x)$. For $x > 0$ the function $L(x)$ can be seen to be a linear combination of the Bessel and Neumann functions of order zero[3]

$$L(x) = -\frac{i\pi}{2} \left(J_0(x) - iY_0(x)\right). \qquad (13.54)$$

This implies that the main properties of this function can be derived from the corresponding properties of the Bessel and Neumann functions, see Sect. 11.3.2. For-

---

[3]The function $H_0^{(2)}(x) = J_0(x) - iY_0(x)$ is called *Hankel function* of order zero.

mally, resorting to the relation (13.53), these properties could also be inferred from
those of $K(x)$. First of all, the definition (13.52) implies the parity relation[4]

$$L^*(x) = L(-x), \tag{13.55}$$

in agreement with the fact that $A^0(t, \mathbf{x})$ is a real field. It follows that it is sufficient
to analyze the properties of $L(x)$ for $x > 0$. From the properties (11.67), (11.68),
(11.72) and (11.73) we derive the expansions for large and small arguments, to be
compared with the expansions (13.27) and (13.28) of the function $K(x)$,

$$L(x) = \sqrt{\frac{\pi}{2|x|}}\, e^{-i(x+\pi/4)} \left(1 + O\left(\frac{1}{x}\right)\right), \quad L(x) = -\ln\frac{|x|}{2} - \gamma - \frac{i\pi}{2}\,\varepsilon(x) + O(x), \tag{13.56}$$

where $\varepsilon(x)$ is the sign function and γ is the *Euler-Mascheroni constant*. In addition,
since $L(x)$ is a linear combination of the functions $J_0(x)$ and $Y_0(x)$, and the latter
both satisfy the differential equation (11.70) with $N = 0$, also $L(x)$ satisfies the
same equation

$$L'' + \frac{1}{x}\,L' + L = 0, \tag{13.57}$$

to be compared with the differential equation (13.29) satisfied by $K(x)$. Finally, the
functions $L(x)$ and $L^*(x)$ constitute a complete set of solutions of equation (13.57).

The electromagnetic field in the wave zone. From the large-x expansion in (13.56)
we derive that in the *wave zone*, i.e. for large r, modulo terms of order $O(1/r^{3/2})$
the potential (13.51) assumes the expression

$$\begin{aligned}
A^0(\omega) &= \frac{e}{4\pi n^2} \cdot \frac{e^{-i\pi/4}}{\left(\frac{v^2 n^2}{c^2} - 1\right)^{1/4}\sqrt{|\omega|vr}}\, e^{-i\frac{\omega}{v}\left(z + r\sqrt{\frac{v^2 n^2}{c^2}-1}\right)} \\
&= \frac{\beta}{\sqrt{r}}\, e^{-i(k_z z + k_r r)} = \frac{\beta}{\sqrt{r}}\, e^{-i\mathbf{k}\cdot\mathbf{x}},
\end{aligned} \tag{13.58}$$

where β is a position-independent complex constant. This implies that in the wave
zone the scalar potential

$$A^0(t, \mathbf{x}) = \frac{1}{\sqrt{2\pi}} \int_{-\infty}^{\infty} e^{i\omega t} A^0(\omega)\, d\omega$$

results in a continuous superposition of *plane waves* of the form

$$\frac{\beta}{\sqrt{r}}\, e^{i(\omega t - \mathbf{k}\cdot\mathbf{x})} + c.c. \tag{13.59}$$

---

[4]Notice that, as the relation (13.54) holds only for $x > 0$, formulas (13.54) and (13.55) do *not*
imply that $Y_0(x)$ is even and that $J_0(x)$ is odd. In fact, $J_0(x)$ is *even*, while $Y_0(x)$ has a branch cut
on the negative semi-axis, see the behavior (11.73).

In contrast to the wave-zone potential (13.22) generated by a particle with speed $v <$ c/n, the potential (13.58) entails thus an electromagnetic wave-zone field which is a superposition of *true* waves with wave vector ($\omega > 0$)

$$\mathbf{k} = \left(k_z = \frac{\omega}{v}, \; k_r = \frac{\omega}{v} \sqrt{\frac{v^2 n^2}{c^2} - 1}, \; k_\varphi = 0 \right), \quad |\mathbf{k}| = \sqrt{k_r^2 + k_z^2} = \frac{n\omega}{c}.$$
(13.60)

The phase velocity of these waves is given by

$$\frac{\omega}{|\mathbf{k}|} = \frac{c}{n}$$

and coincides hence with the velocity of light in the medium. The propagation directions of the waves (13.59) are identified by the angle Θ_C that $\mathbf{k}$ forms with the z axis, known as *Čerenkov angle*, which is determined by the equation

$$\cos \Theta_C = \frac{k_z}{|\mathbf{k}|} = \frac{c}{nv}.$$
(13.61)

Notice that this angle is defined as long as the condition $v > c/n$ holds. Since $k_z > 0$, the propagation directions of the waves (13.59) lie hence on a *forward* cone, coaxial with the trajectory of the particle and with aperture Θ_C, called *Čerenkov cone*. In water, for instance, which in the visible spectrum has an index of refraction of about $n = 4/3$, these waves are generated by charged particles traversing it with speeds $v > 3c/4$, and as v varies between $3c/4$ and c, their propagation directions vary between $\Theta_C = 0$ and $\Theta_C = \arccos(3/4) \approx 41.4°$. The Čerenkov angle is related to the angle α (13.42) of the Mach cone by the complementarity relation

$$\Theta_C + \alpha = \frac{\pi}{2}.$$

The Mach and Čerenkov cones are thus orthogonal to each other. Finally, a simple calculation reveals that the wave vector $\mathbf{k}$ (13.60) is parallel to the Poynting vector $\mathbf{S} = c\,\mathbf{E} \times \mathbf{B}$ of the fields (13.47) evaluated close to the Mach cone, and has its same orientation, see Problem 13.4.

In summary, a charged particle moving in a non-dispersive medium with index of refraction n with a speed $v > c/n$ creates an electromagnetic field that in the wave zone results in a continuous superposition of cylindrical waves of all frequencies. These waves propagate with the velocity c/n parallel to the directions of the Čerenkov cone, forming with the trajectory the angle $\Theta_C = \arccos(c/nv)$. On the other hand, the total electromagnetic field is different from zero only inside the Mach cone and it diverges on its surface. In fact, as we will see in more detail in Sect. 13.5.2, in the frequency domain the ultraviolet divergences encountered in Sect. 13.3.2 arise from the presence in the radiation of arbitrarily high frequencies, see the comment after Eq. (13.103). We postpone the determination of the emitted energy and the analysis of the polarization properties of the radiation to Sect. 13.5, where we consider the more realistic case of a dispersive medium.

13.4   Dispersive Medium

As seen in Sect. 13.3, the electromagnetic field generated by a particle moving with a speed greater than c/n in a non-dispersive medium diverges on the Mach cone, and also the radiated energy is infinite. In this section we show that, actually, these *anomalies* are consequences of the unrealistic hypothesis that the index of refraction n of the medium is the same for all frequencies. In fact, although many dielectric media have an index of refraction which in the visible spectrum varies only little, in general this index is a rather complicated function $n(\omega)$ of the frequency: One says that the medium is *dispersive*. In a dispersive medium, waves of different frequencies ω propagate hence with the different phase velocities $c/n(\omega)$. The precise form of the function $n(\omega)$ depends noticeable on the molecular properties of the material and, in particular, on the presence of resonance frequencies. Nevertheless, this function entails the general properties

$$\begin{cases} n(\omega) < 1, & \text{for } \omega > \omega_m, \\ \lim_{\omega \to \infty} n(\omega) = 1, \end{cases} \tag{13.62}$$

where ω_m is a limit-frequency very close to the highest resonance frequency. In particular, for large values of ω the function $n(\omega)$ entails the universal asymptotic behavior $n(\omega) \approx 1 - \omega_p^2/2\omega^2$, where ω_p is the *plasma frequency* of the medium. From the first property in (13.62) we draw the important conclusion that the frequency band for which $n(\omega) > 1$ is *bounded*. As we shall see, this feature ensures (i) that the electromagnetic field generated by the particle is everywhere regular, and (ii) that the radiated energy is finite. In this section we shall, moreover, assume that $n(\omega)$ is *real*, meaning that we still neglect the effects of *absorption*. Finally, we will also suppose that the material is homogeneous and isotropic, and *linear*, so that $n(\omega)$ depends neither on the position $\mathbf{x}$, nor on the electromagnetic field itself.

13.4.1   Maxwell's Equations in a Dispersive Medium

In a dispersive medium the dynamics of the electromagnetic field is no longer governed by the local Maxwell equations in matter (13.1)–(13.4). In fact, in this case its dynamics must be based on the temporal Fourier transforms of these equations amended by the replacement $n \to n(\omega)$, namely

$$-\frac{n^2(\omega)}{c} i\omega \mathbf{E}(\omega) + \nabla \times \mathbf{B}(\omega) = \frac{\mathbf{j}(\omega)}{c}, \tag{13.63}$$

$$\frac{i\omega}{c} \mathbf{B}(\omega) + \nabla \times \mathbf{E}(\omega) = 0, \tag{13.64}$$

$$\nabla \cdot \mathbf{E}(\omega) = \frac{\rho(\omega)}{n^2(\omega)}, \tag{13.65}$$

$$\nabla \cdot \mathbf{B}(\omega) = 0. \tag{13.66}$$

As usual, with the symbol $f(\omega) = f(\omega, \mathbf{x})$ we denote the temporal Fourier transform of the function of four variables $f(t, \mathbf{x})$. To take into account the formal presence of negative frequencies we set

$$n(-\omega) = n(\omega).$$

The space-time fields $\mathbf{E}(t, \mathbf{x})$ and $\mathbf{B}(t, \mathbf{x})$ are then *defined* as the inverse temporal Fourier transforms of the fields $\mathbf{E}(\omega)$ and $\mathbf{B}(\omega)$, subjected to Maxwell's equations in matter (13.63)–(13.66). As usual, the Bianchi identities (13.64) and (13.66) can be solved in terms of a four-potential $A^\mu(\omega) = A^\mu(\omega, \mathbf{x})$ via the relations

$$\mathbf{E}(\omega) = -\nabla A^0(\omega) - \frac{i\omega}{c} \mathbf{A}(\omega), \tag{13.67}$$

$$\mathbf{B}(\omega) = \nabla \times \mathbf{A}(\omega), \tag{13.68}$$

and the fields (13.67) and (13.68) are then invariant under the gauge transformations

$$A^0(\omega) \to A^0(\omega) + \frac{i\omega}{c} \Lambda(\omega), \qquad \mathbf{A}(\omega) \to \mathbf{A}(\omega) - \nabla \Lambda(\omega). \tag{13.69}$$

In this case it is convenient to impose the adapted Lorenz gauge (see Eq. (13.7))

$$\frac{i\omega}{c} n^2(\omega) A^0(\omega) + \nabla \cdot \mathbf{A}(\omega) = 0. \tag{13.70}$$

With this gauge-fixing the remaining Maxwell equations in matter (13.63) and (13.65) reduce to the equations for the functions $A^\mu(\omega)$, see Problem 13.2,

$$-\left(\frac{n^2(\omega)}{c^2} \omega^2 + \nabla^2\right) A^\mu(\omega) = \left(\frac{\rho(\omega)}{n^2(\omega)}, \frac{\mathbf{j}(\omega)}{c}\right), \tag{13.71}$$

to be compared with the corresponding equations (13.8) in a non-dispersive medium. Finally, we *define* the space-time four-potential as the inverse Fourier transform

$$A^\mu(t, \mathbf{x}) = \frac{1}{\sqrt{2\pi}} \int_{-\infty}^{\infty} e^{i\omega t} A^\mu(\omega) \, dt. \tag{13.72}$$

In basis of the relations (13.67) and (13.68) and of the definition (13.72), the space-time fields $\mathbf{E}(t, \mathbf{x})$ and $\mathbf{B}(t, \mathbf{x})$ are expressed in terms of $A^\mu(t, \mathbf{x})$ by the usual relations (13.6), so that the Maxwell tensor is still given by $F^{\mu\nu} = \partial^\mu A^\nu - \partial^\nu A^\mu$. However, if $n(\omega)$ is not constant, the fields $F^{\mu\nu}$ and A^μ do no longer satisfy local differential equations such as (13.1)–(13.4) and (13.8).

In conclusion, to determine the electromagnetic field produced by a current $j^\mu(t, \mathbf{x})$ in a dispersive medium one must first determine its temporal Fourier transform $j^\mu(\omega)$, then solve the differential equations (13.70) and (13.71), and finally determine the electromagnetic fields using formulas (13.72) and (13.6).

13.4.2   Charged Particle in Uniform Linear Motion

For a charged particle in uniform linear motion formulas (13.9) imply the linear relation between the charge and current densities $\mathbf{j}(\omega) = \rho(\omega)\mathbf{v}$. From the structure of Eqs. (13.71) we then derive that the spatial and temporal components of the four-potential are related by

$$\mathbf{A}(\omega) = \frac{n^2(\omega)A^0(\omega)}{c}\,\mathbf{v}. \tag{13.73}$$

Therefore, it is again sufficient to solve Eq. (13.71) for the component $A^0(\omega)$. For this purpose we must first determine the temporal Fourier transform of the charge density in (13.9). We find

$$\rho(\omega) = \frac{e}{\sqrt{2\pi}} \int_{-\infty}^{\infty} e^{-i\omega t}\, \delta(z - vt)\, \delta^2(\mathbf{r})\, dt = \frac{e}{\sqrt{2\pi}\,v}\, e^{-i\omega z/v}\, \delta^2(\mathbf{r}),$$

where we have introduced the two-dimensional δ-function $\delta^2(\mathbf{r}) = \delta(x)\delta(y)$. The time component of equation (13.71) takes thus the form

$$\left(\frac{n^2(\omega)}{c^2}\, \omega^2 + \nabla^2 \right) A^0(\omega) = -\frac{e}{\sqrt{2\pi}\,vn^2(\omega)}\, e^{-i\omega z/v}\, \delta^2(\mathbf{r}). \tag{13.74}$$

To solve this equation it is convenient to switch to cylindrical coordinates and to decompose the three-dimensional Laplacian according to

$$\nabla^2 = \partial_z^2 + \widehat{\nabla}^2, \qquad \widehat{\nabla}^2 = \partial_x^2 + \partial_y^2 = \partial_r^2 + \frac{1}{r}\, \partial_r + \frac{1}{r^2}\, \partial_\varphi^2, \tag{13.75}$$

where $r = \sqrt{x^2 + y^2}$. In writing the two-dimensional Laplacian $\widehat{\nabla}^2$ in terms of polar coordinates as in (13.75) we disregarded possible singularities of $A^0(\omega)$ at the origin of the xy plane, i.e. for $r = 0$. We shall address this problem below, see equation (13.80). According to the cylindrical symmetry inherent in the physical process at hand we search for a factorized solution of equation (13.74) of the form (rotation invariance requires $A^0(\omega) = A^0(\omega, z, r, \varphi)$ to be independent of φ)

$$A^0(\omega) = \frac{e}{(2\pi)^{3/2} n^2(\omega) v} e^{-i\omega z/v} I(\omega, r).$$ (13.76)

Equation (13.74) goes then over to the differential equation for the function $I(\omega, r)$

$$\left(\widehat{\nabla}^2 + \frac{\omega^2}{v^2} \left(\frac{v^2 n^2(\omega)}{c^2} - 1 \right) \right) I(\omega, r) = -2\pi \delta^2(\mathbf{r}),$$ (13.77)

which for $r \neq 0$ reduces to

$$\left(\partial_r^2 + \frac{1}{r} \partial_r + \frac{\omega^2}{v^2} \left(\frac{v^2 n^2(\omega)}{c^2} - 1 \right) \right) I(\omega, r) = 0.$$ (13.78)

First we search for an appropriate "particular" solution of equation (13.78). Recalling that the special functions K and L satisfy the differential equations (13.29) and (13.57), respectively, we recognize that Eq. (13.78) is satisfied by the expression

$$\begin{aligned}
I(\omega, r) = {} & H\left(\frac{c}{n(\omega)} - v \right) K\left(\frac{\omega r}{v} \sqrt{1 - \frac{v^2 n^2(\omega)}{c^2}} \right) \\
& + H\left(v - \frac{c}{n(\omega)} \right) L\left(\frac{\omega r}{v} \sqrt{\frac{v^2 n^2(\omega)}{c^2} - 1} \right),
\end{aligned}$$ (13.79)

where H is the Heaviside function. The function $I(\omega, r)$ has thus different determinations according to whether, for a given frequency ω, one has $v < c/n(\omega)$ or $v > c/n(\omega)$.

To prove that the function (13.79) satisfies the complete equation (13.77) in the distributional sense – thereby checking also the correct overall normalization of $I(\omega, r)$ – we must analyze the singularities of the functions $L(u)$ and $K(u)$ at $u = 0$. As both these functions entail the same logarithmic singularity $-\ln|u|$, see the expansions (13.28) and (13.56), the function (13.79) entails the expansion around $r = 0$

$$I(\omega, r) = -\ln r + a + O(r).$$

Since the Green function of the two-dimensional Laplacian $\widehat{\nabla}^2$ is the logarithm, more precisely, see Problem 6.3,

$$\widehat{\nabla}^2 \ln r = 2\pi \delta^2(\mathbf{r}),$$

the contribution singular at $r = 0$ of the quantity $\widehat{\nabla}^2 I(\omega, r)$ is given by

$$\left(\widehat{\nabla}^2 I(\omega, r) \right)_{\text{sing}} = \widehat{\nabla}^2(-\ln r) = -2\pi \delta^2(\mathbf{r}).$$ (13.80)

It follows that the scalar potential (13.76), with $I(\omega, r)$ defined in (13.79), satisfies Maxwell's equation in matter (13.74) in the distributional sense. Moreover, as the velocity of the particle is given by $\mathbf{v} = (0, 0, v)$, it is easily seen that the four-potential $A^\mu(\omega)$ given by Eqs. (13.73) and (13.76) satisfies the adapted Lorenz gauge (13.70). In the case of a non-dispersive medium the potential (13.76) reduces to the solutions (13.20) for $v < c/n$, and (13.51) for $v > c/n$, respectively. Since the potential (13.51) for $v > c/n$ is the Fourier transform of the potential $A^0(t, \mathbf{x})$ (13.44), the analysis above thus provides also an indirect proof that the latter is, actually, a solution of Maxwell's equations (13.10) and (13.7), as anticipated at the end of Sect. 13.3.1.

Uniqueness and causality. We analyze briefly the issues of uniqueness and causality regarding the solution (13.76) of Maxwell's equations in matter, showing in particular that the differential equation (13.77) does not admit *physical* solutions other than (13.79). Let us first consider the set of frequencies for which $v < c/n(\omega)$, in which case the function (13.79) reduces to $I = K$. In this case the homogeneous equation (13.78) is of the type (13.29), which actually possesses the two linearly independent solutions K (13.26) and $\tilde{K}$ (13.30). Therefore, since $\tilde{K}$ is regular at $r = 0$, see (13.31), the general solution of the nonhomogeneous equation (13.77) is given by the combination

$$I = K + a\tilde{K},$$

where a is a real constant. However, in the limit for $r \to \infty$ the function $\tilde{K}$ diverges exponentially, see (13.31), and so it is *not* a distribution. This means that we are forced to choose $a = 0$. We are hence left with the set of frequencies for which $v > c/n(\omega)$, in which case the function (13.79) reduces to $I = L$. In this case the homogeneous equation (13.78) is of the type (13.57), a real equation, which possesses the two linearly independent solutions L and L^*. From the behavior of L for $r \to 0$ in (13.56) it follows that the imaginary part of L is regular at $r = 0$, and therefore the general solution of the nonhomogeneous equation (13.77) is given by

$$I = (1 - b)L + bL^*,$$

where b is a real constant. However, the behavior of $L(x)$ for large x in (13.56) shows that for large r the function L appearing in (13.79) reduces to the radially *outgoing* wave e^{-irk_r}, where according to (13.60) we have $k_r > 0$ for $\omega > 0$, whereas L^* reduces to the wave e^{irk_r} radially *incoming* from infinity. If we want to preserve *causality*, which requires that the radiation is always emitted by charged particles and never absorbed, we are hence forced to choose $b = 0$. Notice that in the absence of dispersion this is equivalent to choose the *retarded* Green function (13.34), in place of the advanced one.

13.4.3 Coulomb Fields and Radiation Fields

In basis of equations (13.72), (13.73) and (13.76) a particle in uniform linear motion in a dispersive medium generates the four-potential (henceforth we set $c = 1$)

$$
A^0(t, \mathbf{x}) = \frac{e}{(2\pi)^2 v} \int_{-\infty}^{\infty} e^{-i\frac{\omega}{v}(z-vt)} \frac{I(\omega, r)}{n^2(\omega)}\, d\omega,
$$
$$
\mathbf{A}(t, \mathbf{x}) = \frac{e\,\mathbf{u}_z}{(2\pi)^2} \int_{-\infty}^{\infty} e^{-i\frac{\omega}{v}(z-vt)} I(\omega, r)\, d\omega,
$$
(13.81)

where the function $I(\omega, r)$ is given in (13.79). From the expansions (13.27) and (13.56) it follows that for large r, or large $|\omega|$, modulo terms of order $O(1/r^{3/2})$ this function behaves as

$$
I(\omega, r) \approx
\begin{cases}
\left(\dfrac{\pi v}{2|\omega|r\sqrt{1 - v^2 n^2(\omega)}} \right)^{1/2} e^{-\frac{|\omega|r}{v}\sqrt{1-v^2 n^2(\omega)}}, & n(\omega) < 1/v, \\[4mm]
\left(\dfrac{\pi v}{2|\omega|r\sqrt{v^2 n^2(\omega) - 1}} \right)^{1/2} e^{-i\pi/4}\, e^{-i\frac{\omega r}{v}\sqrt{v^2 n^2(\omega)-1}}, & n(\omega) > 1/v.
\end{cases}
$$
(13.82)

Absence of ultraviolet singularities and disappearance of the Mach cone. In a dispersive medium the four-potential $A^\mu(t, \mathbf{x})$ (13.81) is regular throughout space, except for the position of the particle and, in particular, there is no longer a Mach cone entailing a singular wavefront. To see it we evaluate, say, the scalar potential in (13.81) on a *pseudo* Mach cone (13.43) relative to an arbitrary index of refraction n_a such that $v > 1/n_a$, corresponding to a Mach cone with a generic angular aperture $\alpha = \arcsin(1/v n_a)$,

$$
A^0\Big|_{\text{Mach}} = \frac{e}{(2\pi)^2 v} \int_{-\infty}^{\infty} e^{i\frac{\omega r}{v}\sqrt{v^2 n_a^2 - 1}} \frac{I(\omega, r)}{n^2(\omega)}\, d\omega.
$$
(13.83)

The integral over ω appearing in this formula is convergent. In fact, in virtue of the general properties (13.62), for $\omega > \omega_m$ we have $n(\omega) < 1$, and so for sufficiently large ω the index of refraction satisfies the inequality $n(\omega) < 1/v$. From the behaviors (13.82) it then follows that for large frequencies the function $I(\omega, r)$ decreases exponentially, and consequently the integral (13.83) is finite for any n_a. As the potential (13.81) is everywhere regular, in the presence of dispersion the concept of *Mach cone* itself looses thus its meaning. Notice, however, that in the absence of dispersion the integral (13.83) would reproduce the known ultraviolet divergences. In fact, setting $n(\omega) = n_a$ for all ω, and choosing a speed v greater than $1/n_a$, for large ω the function $I(\omega, r)$ would have the oscillatory behavior of the second expression in (13.82), which would precisely compensate the factor $e^{i\frac{\omega r}{v}\sqrt{v^2 n_a^2 - 1}}$ of the integrand in (13.83). The integral over ω would thus diverge for large frequencies, and so the potential $A^0\big|_{\text{Mach}}$ would again be infinite. Of course, this in line with the general paradigm that in the frequency domain the ultraviolet divergences arise for $\omega \to \infty$.

Coulomb fields and radiation fields. Let us finally analyze the behavior of the fields $A^\mu(t, \mathbf{x})$ and $F^{\mu\nu}(t, \mathbf{x})$ at large distances r from the particle's trajectory. In basis of the decomposition (13.79) the four-potential (13.81) can be written as the superposition

$$A^\mu = A_K^\mu + A_L^\mu, \tag{13.84}$$

where A_K^μ represents a *Coulomb field*, and A_L^μ a *radiation field*. This terminology is motivated by the large-r behaviors

$$A_K^\mu \sim \frac{1}{r}, \qquad F_K^{\mu\nu} \sim \frac{1}{r^2}, \tag{13.85}$$

$$A_L^\mu \sim \frac{1}{\sqrt{r}}, \qquad F_L^{\mu\nu} \sim \frac{1}{\sqrt{r}}. \tag{13.86}$$

We first derive the behavior of A_K^μ. This potential collects the frequencies for which $v < 1/n(\omega)$, which in the case of a non-dispersive medium were associated with evanescent waves, see Sect. 13.2.1. For these frequencies the function I equals K and so, according to the first behavior in (13.82), it is essentially an exponentially decreasing function of the product ωr. In the integrals (13.81) we can then perform the rescaling of the integration variable $\omega \to \omega/r$, entailing the change of the measure $d\omega \to d\omega/r$. Thanks to the exponential decrease of the function I the resulting integral over ω converges, and consequently the behavior of A_K^μ for large r is determined by the $1/r$ prefactor coming from the rescaling of the measure. On the other hand, in the computation of the field strength associated with A_K^μ the partial derivatives ∂_μ produce a further factor of ω, which causes $F_K^{\mu\nu}$ to decrease as $1/r^2$, like a Coulomb field.[5] Vice versa, the potential A_L^μ involves the *bounded* set of frequencies for which $v > 1/n(\omega)$. For these frequencies the function I equals L, and its large-r behavior is thus the oscillatory one in the second line of (13.82), entailing a prefactor $1/\sqrt{r}$. Consequently, A_L^μ decreases as $1/\sqrt{r}$ and so does the associated electromagnetic field $F_L^{\mu\nu}$. Finally, as anticipated at the end of Sect. 13.2.1, in the presence of a cylindrical symmetry an oscillating field that at large distances decreases as $1/\sqrt{r}$, like $F_L^{\mu\nu}$, represents a *true* radiation field. In fact, the associated Poynting vector decreases as $S \sim (F_L^{\mu\nu})^2 \sim C/r$, and so the energy flux through a cylindrical surface of radius r and length L in the limit for $r \to \infty$ becomes constant and non-vanishing: $d\varepsilon/dt \sim S \cdot 2\pi r L \sim CL$.

---

[5]Clearly, depending on the precise form of the function $n(\omega)$, the fields A_K^μ and $F_K^{\mu\nu}$ could decrease even faster as indicated in (13.85).

13.5 The Frank-Tamm Formula

In this section we determine the energy radiated by a particle traveling in a dispersive medium with speed v, assuming that the set of frequencies for which $v > 1/n(\omega)$ is non-empty. Due to the stationary character of the phenomenon the physically meaningful quantity is the energy radiated per unit frequency interval per unit of traveled distance

$$\frac{d^2\varepsilon}{dzd\omega}.$$

In Sect. 13.5.2 we shall evaluate this quantity from *first principles*, i.e. by relying on the electromagnetic fields determined in Sect. 13.4, obtaining the *Frank-Tamm formula* (13.102). In the next section we present, instead, an interesting heuristic derivation of this formula.

13.5.1 Heuristic Argument

We start from the general formula (11.95) for the spectral weights of the radiation emitted by a particle performing a generic aperiodic motion

$$\frac{d^2\varepsilon}{d\omega d\Omega} = \frac{e^2\omega^2}{16\pi^3} \left| \mathbf{n} \times \int_{-\infty}^{\infty} e^{-i\omega(t-\mathbf{n}\cdot\mathbf{y})} \mathbf{v}\, dt \right|^2. \tag{13.87}$$

We recall that this formula holds in free space, having index of refraction $n = 1$, and, obviously, the velocity of the particle must satisfy $v < 1$. Of course, if the particle is not accelerated the emitted energy $d^2\varepsilon/d\omega d\Omega$ is zero. The expression (13.87) yields the energy emitted per unit frequency interval during the whole trajectory. If the motion is unbounded and, in addition, the particle's acceleration lasts for an infinite time, in general the quantity $d^2\varepsilon/d\omega d\Omega$ is *infinite*. However, in such a case the energy emitted during a finite time interval, say, between the instants $-T$ and T, will nevertheless be finite. To compute this energy we must restrict the integration interval in Eq. (13.87) between $-T$ and T. If finally we want to compute the average energy radiated per unit time per unit frequency interval we must divide the expression so obtained by $2T$, and then take its limit for $T \to \infty$

$$\frac{d^3\varepsilon}{dt d\omega d\Omega} = \frac{e^2\omega^2}{16\pi^3} \lim_{T\to\infty} \frac{1}{2T} \left| \mathbf{n} \times \int_{-T}^{T} e^{-i\omega(t-\mathbf{n}\cdot\mathbf{y})} \mathbf{v}\, dt \right|^2. \tag{13.88}$$

Let us now consider a particle in uniform linear motion with trajectory $\mathbf{y} = \mathbf{v}t$. In this case the integral appearing in (13.88) can be computed analytically

$$\left| \mathbf{n} \times \int_{-T}^{T} e^{-i\omega(t-\mathbf{n}\cdot\mathbf{y})} \mathbf{v}\, dt \right|^2 = 4\left(v^2 - (\mathbf{n}\cdot\mathbf{v})^2\right) \frac{\sin^2\left((1-\mathbf{n}\cdot\mathbf{v})\,\omega T\right)}{(1-\mathbf{n}\cdot\mathbf{v})^2\omega^2}.$$

(13.89)

To compute the limit for $T \to \infty$ appearing in Eq. (13.88) we resort to the distributional limit, see Problem 13.6,

$$\mathcal{S}' - \lim_{T\to\infty} \frac{\sin^2(Tx)}{Tx^2} = \pi\delta(x).$$

(13.90)

Setting in this formula $x = (1-\mathbf{n}\cdot\mathbf{v})\,\omega$, and using the result (13.89), for the limit appearing in (13.88) we so obtain

$$\lim_{T\to\infty} \frac{1}{2T} \left| \mathbf{n} \times \int_{-T}^{T} e^{-i\omega(t-\mathbf{n}\cdot\mathbf{y})} \mathbf{v}\, dt \right|^2 = 2\pi\left(v^2 - (\mathbf{n}\cdot\mathbf{v})^2\right)\delta((1-\mathbf{n}\cdot\mathbf{v})\omega)$$

$$= \frac{2\pi}{\omega}\left(v^2-1\right)\delta(1-\mathbf{n}\cdot\mathbf{v})$$

$$= \frac{2\pi}{\omega}\left(v^2-1\right)\delta(1-v\cos\vartheta),$$

where ϑ is the angle between $\mathbf{v}$ and the emission direction $\mathbf{n}$. Dividing equation (13.88) by v, and reintroducing the velocity of light c, we finally obtain a formula for the angular distribution of the energy radiated per unit frequency interval per unit of traveled distance dz

$$\frac{d^3\varepsilon}{dz\,d\omega\,d\Omega} = \frac{e^2\omega}{8\pi^2 vc}\left(\frac{v^2}{c^2}-1\right)\delta\left(1-\frac{v}{c}\cos\vartheta\right).$$

(13.91)

As expected, since the velocity v of the particle is always smaller than c, the argument of the δ-function in (13.91) is different from zero for all ϑ and, therefore, no radiation arises in whatever direction.

Analytic continuation. Formula (13.91), which in vacuum is actually *empty*, admits a natural "analytic continuation" to the case of a particle traveling with constant speed v in a dielectric dispersive medium. In fact, according to the general recipe (13.5) we must replace c with c/n and e with e/n, and then, since our medium is dispersive, we must replace n with $n(\omega)$. In this way the *formal* equation (13.91) goes over into the meaningful equation

$$\frac{d^3\varepsilon}{dz\,d\omega\,d\Omega} = \frac{e^2\omega}{8\pi^2 vn(\omega)c}\left(\frac{v^2 n^2(\omega)}{c^2}-1\right)\delta\left(1-\frac{vn(\omega)}{c}\cos\vartheta\right).$$

(13.92)

In fact, formula (13.92) now predicts the presence of radiation for all frequencies ω which satisfy the inequality $n(\omega) > c/v$. Furthermore, it predicts that the radiation with frequency ω is emitted along a *cone* of directions forming with the particle's trajectory the frequency-dependent angle ϑ determined by the equation

$$\cos\vartheta = \frac{c}{vn(\omega)},$$

which for a non-dispersive medium reduces to the Čerenkov angle (13.61). We can sum the emitted energy (13.92) over all directions, by using the integral

$$\int \delta\left(1 - \frac{vn(\omega)}{c}\cos\vartheta\right) d\omega = 2\pi \int_{-1}^{1} \delta\left(1 - \frac{vn(\omega)}{c}\cos\vartheta\right) d\cos\vartheta$$

$$= \frac{2\pi c}{vn(\omega)} H\left(v - \frac{c}{n(\omega)}\right),$$

where H is the Heaviside function. Integrating formula (13.92) over all angles we so derive that the energy radiated by the particle per unit frequency interval per unit of traveled distance is given by

$$\frac{d^2\varepsilon}{dzd\omega} = \begin{cases} \dfrac{e^2\omega}{4\pi c^2}\left(1 - \dfrac{c^2}{v^2 n^2(\omega)}\right), & \text{if } n(\omega) > \dfrac{c}{v}, \\[2ex] 0, & \text{if } n(\omega) < \dfrac{c}{v}. \end{cases} \tag{13.93}$$

This formula has been first derived by Frank and Tamm in 1937 in their explanation of the Čerenkov effect [1]. We will investigate its physical implications in Sects. 13.5.2 and 13.6.

13.5.2   Derivation of the Frank-Tamm Formula

The above derivation is interesting for the mathematical as well as physical tools involved. However, being heuristic, it may appear more or less convincing. In this section we derive formula (13.93) from first principles, i.e. from the fields $\mathbf{E}$ and $\mathbf{B}$ produced by the particle via the four-potential (13.81). We start from the energy $\Delta\varepsilon$ radiated by the particle during the whole trajectory across a fixed cylindrical surface coaxial with the trajectory, of radius r and of length $\Delta z = z_2 - z_1$. Given the four-potential (13.81), the fields $\mathbf{E}$ and $\mathbf{B}$ depend on z and t only through the combination $z - vt$, and so the integral over z becomes trivial. We then obtain

$$\Delta\varepsilon = (2\pi r\Delta z)\int_{-\infty}^{\infty} (\mathbf{E}\times\mathbf{B})\cdot\mathbf{u}_r\,dt = (2\pi r\Delta z)\int_{-\infty}^{\infty} (\mathbf{E}^*(\omega)\times\mathbf{B}(\omega))\cdot\mathbf{u}_r\,d\omega, \tag{13.94}$$

where we used *Plancherel's theorem*. Actually, for a non-dispersive medium with constant index of refraction n such that $v > c/n$, both integrals appearing in (13.94) would be divergent. For the first integral this has been shown in (13.49), and for the second integral it follows from the Frank-Tamm formula (13.102), see the comment after Eq. (13.103). Conversely, for a dispersive medium Eq. (13.94) is well defined and it implies that the energy emitted per unit frequency interval per unit of traveled distance is equal to

$$\frac{d^2\varepsilon}{dzd\omega} = 2\pi r \left(\mathbf{E}^*(\omega) \times \mathbf{B}(\omega)\right) \cdot \mathbf{u}_r + c.c. \tag{13.95}$$

where at second member the limit for $r \to \infty$ is understood. The result will hence be different from zero only if for large r the fields $\mathbf{E}(\omega)$ and $\mathbf{B}(\omega)$ decrease as $1/\sqrt{r}$. To express the latter in terms of only the scalar potential $A^0(\omega)$, given by formulas (13.76) and (13.79), we substitute the spatial potential (13.73) in the general relations (13.67) and (13.68)

$$\mathbf{E}(\omega) = -\left(\boldsymbol{\nabla} + i\omega n^2(\omega)\,\mathbf{v}\right) A^0(\omega), \tag{13.96}$$

$$\mathbf{B}(\omega) = -n^2(\omega)\,\mathbf{v} \times \boldsymbol{\nabla} A^0(\omega). \tag{13.97}$$

The evaluation of the emitted energy (13.95) is facilitated by the results of Sect. 13.4. For the frequencies for which $v < 1/n(\omega)$ formulas (13.76) and (13.82) imply that for large r the potential $A^0(\omega)$ decreases exponentially, and so the emitted energy (13.95) is zero. Therefore, there is no radiation with these frequencies. Vice versa, for the frequencies for which $v > 1/n(\omega)$ the potential $A^0(\omega)$ decreases as $1/\sqrt{r}$. In fact, for these frequencies formulas (13.76) and (13.82) yield the asymptotic expansion, holding modulo terms of order $1/r^{3/2}$,

$$A^0(\omega) = \frac{e}{4\pi n^2(\omega)} \cdot \frac{e^{-i\pi/4}}{(v^2 n^2(\omega) - 1)^{1/4}\sqrt{|\omega|vr}}\, e^{-i\frac{\omega}{v}\left(z + r\sqrt{v^2 n^2(\omega) - 1}\right)}. \tag{13.98}$$

Notice that for a non-dispersive medium this expansion reduces to (13.58). Since for the evaluation of the emitted energy (13.95) it is sufficient to retain in the fields (13.96) and (13.97) the terms of order $1/\sqrt{r}$, in the calculation of the gradient $\boldsymbol{\nabla} A^0(\omega)$ it is sufficient to consider the derivatives of the exponential in (13.98). Modulo terms of order $1/r^{3/2}$ from Eqs. (13.96), (13.97) and (13.98) we so obtain

$$\boldsymbol{\nabla} A^0(\omega) = -\frac{i\omega}{v}\left(\mathbf{u}_z + \sqrt{v^2 n^2(\omega) - 1}\,\mathbf{u}_r\right) A^0(\omega), \tag{13.99}$$

$$\mathbf{E}(\omega) = \frac{i\omega}{v}\sqrt{v^2 n^2(\omega) - 1}\left(\mathbf{u}_r - \sqrt{v^2 n^2(\omega) - 1}\,\mathbf{u}_z\right) A^0(\omega), \tag{13.100}$$

$$\mathbf{B}(\omega) = i\omega n^2(\omega)\sqrt{v^2 n^2(\omega) - 1}\, A^0(\omega)\,\mathbf{u}_\varphi. \tag{13.101}$$

Linear polarization. Formulas (13.60), (13.100) and (13.101) imply the obvious relations $\mathbf{k} \perp \mathbf{E}(\omega)$, $\mathbf{k} \perp \mathbf{B}(\omega)$ and $\mathbf{E}(\omega) \perp \mathbf{B}(\omega)$. Moreover, from the above equations we see that the vector $\mathbf{E}(\omega)$ can be rewritten in the factorized form $\mathbf{E}(\omega) = \boldsymbol{\mathcal{E}}(\omega, \mathbf{x})\, e^{-i\mathbf{k}\cdot\mathbf{x}}$, see formula (13.58), and consequently the *electric field in the wave zone* can be written as the superposition of elementary waves

$$\mathbf{E}(t, \mathbf{x}) = \frac{1}{\sqrt{2\pi}} \int_> e^{i\omega t} \mathbf{E}(\omega) \, d\omega = \frac{1}{\sqrt{2\pi}} \int_> e^{i k \cdot x} \boldsymbol{\mathcal{E}}(\omega, \mathbf{x}) \, d\omega,$$

where the symbol $>$ indicates that the integral is only over the frequencies for which $v > 1/n(\omega)$. Furthermore, from Eqs. (13.98) and (13.100) we infer that the polarization vector $\boldsymbol{\mathcal{E}}(\omega, \mathbf{x})$ is *real* for all frequencies, apart from the constant phase $ie^{-i\pi/4} = e^{i\pi/4}$. From the polarization criterion (5.92) it then follows that the Čerenkov radiation is *linearly polarized* for all frequencies. More precisely, from formula (13.100) we see that the electric field lies in the plane formed by the vectors $\mathbf{k}$ and $\mathbf{v}$, in agreement with the observations made by Čerenkov.

Inserting the fields (13.100) and (13.101) in the equation (13.95) for the emitted energy we obtain the intermediate result

$$\frac{d^2\varepsilon}{dz d\omega} = \frac{1}{v} \, 4\pi r n^2(\omega) \, \omega^2 \big(v^2 n^2(\omega) - 1\big)^{3/2} \big|A^0(\omega)\big|^2,$$

and from the expansion (13.98) we find that the scalar potential has the modulus squared, considering $\omega > 0$,

$$\big|A^0(\omega)\big|^2 = \left(\frac{e}{4\pi n^2(\omega)}\right)^2 \frac{1}{v\omega r \sqrt{v^2 n^2(\omega) - 1}}.$$

From these relations we finally derive the *Frank-Tamm formula* for the radiated energy, valid for the frequencies for which $v > c/n(\omega)$,

$$\frac{d^2\varepsilon}{dz d\omega} = \frac{e^2 \omega}{4\pi c^2} \left(1 - \frac{c^2}{v^2 n^2(\omega)}\right), \tag{13.102}$$

where we have reinserted the velocity of light. As this formula involves only the index of refraction of the medium, and none of its other physical properties, it entails *universal* character, as the radiation observed by Čerenkov. Finally, integrating equation (13.102) over the frequencies we obtain the total energy radiated per unit of traveled distance

$$\frac{d\varepsilon}{dz} = \frac{e^2}{4\pi c^2} \int_> \omega \left(1 - \frac{c^2}{v^2 n^2(\omega)}\right) d\omega, \tag{13.103}$$

where the symbol $>$ denotes again that for a given speed v the integral extends only over the frequencies for which $v > c/n(\omega)$. As the set of these frequencies is *bounded*, the emitted energy is hence always *finite*. Conversely, in a non-dispersive medium with index of refraction $n(\omega) = n_0$ for all ω, for a speed v greater than c/n_0 the integral (13.103) would extend over all frequencies, giving thus an infinite emitted energy, in accordance with the likewise ultraviolet-divergent integral (13.49). Formula (13.103) tells us, in particular, that the divergence of the total energy emitted by a particle in a non-dispersive medium is caused by the radiation with frequencies tending to infinity.

Photon emission. Dividing Eq. (13.102) by the energy $\hbar\omega$ of a photon we obtain a simple formula for the number N of photons emitted by the particle per unit frequency interval per unit of traveled distance

$$\frac{d^2N}{dz\,d\omega} = \frac{e^2}{4\pi c^2 \hbar}\left(1 - \frac{c^2}{v^2 n^2(\omega)}\right). \tag{13.104}$$

The total number of photons emitted per unit of traveled distance is thus equal to

$$\frac{dN}{dz} = \frac{\alpha}{c}\int_{>}\left(1 - \frac{c^2}{v^2 n^2(\omega)}\right)d\omega, \tag{13.105}$$

where α is the fine structure constant (11.122). As the integral (13.105) extends over a finite frequency band, the number of emitted photons is always finite. As an example we estimate the number of photons emitted in the visible spectrum by an electron traveling in pure water at an ultrarelativistic speed $v \approx c$. In the visible spectrum water has the practically constant index of refraction $n(\omega) \approx 4/3$, and so we have

$$1 - \frac{c^2}{v^2 n^2(\omega)} = 1 - \frac{9}{16} = \frac{7}{16}.$$

Introducing the boundary wavelengths of the visible spectrum $\lambda_1 = 800\,\mathrm{nm}$ and $\lambda_2 = 400\,\mathrm{nm}$, and recalling that $\omega = 2\pi c/\lambda$, Eq. (13.105) then gives for the number of emitted photons

$$\frac{dN}{dz} = \frac{\alpha}{c}\int_{\omega_1}^{\omega_2}\left(1 - \frac{c^2}{v^2 n^2(\omega)}\right)d\omega = \frac{7\pi\alpha}{8}\left(\frac{1}{\lambda_1} - \frac{1}{\lambda_1}\right) \approx 250/\mathrm{cm}. \tag{13.106}$$

This means that while the electron covers in water a distance of one centimeter, it emits about 250 photons with frequencies of the visible spectrum. Actually, Eq. (13.106) provides the general estimate

$$\frac{dN}{dz} \approx \frac{\alpha}{\lambda} = \frac{1}{137\lambda},$$

meaning that while the particle travels a distance of 137 times a considered wavelength, it emits about one photon of that wavelength.

13.6 Čerenkov Detectors

An experimental device that employs the Čerenkov effect to reveal elementary particles is called a *Čerenkov detector*. It is usually made of a container filled with a transparent material – the so-called *radiator*, for instance, pure water or quartz – which acts as a dielectric medium. The Čerenkov radiation, generally of low intensity, caused by the passage of a charged particle in the radiator is collected and analyzed by a series of photodetectors. Since the radiation is emitted on concentric cones, it is easy to reconstruct the flight direction of the charged particle. Measuring the number of emitted photons, and knowing the emission angle of the radiation, Eqs. (13.61) and (13.104) then allow to determine the velocity of the particle. Although the potential of the Čerenkov effect as basis for a detector was clear from its discovery, it was only the advent of *photomultipliers*, able to reveal with high efficiency and rapid response even low-intensity radiations, that in 1951 allowed J.V. Jelley to develop the first device suitable for experiments.

Kamiokande and the physics of neutrinos. Neutrinos interact only *weakly* with matter. Nevertheless, there is a non-vanishing probability that a high-energy neutrino, for instance, of astrophysical origin, interacts with an atom thereby transferring most of its energy to a charged particle, typically an electron or a muon. The latter, in turn, when passing through a radiator produces Čerenkov light and signals so the transit of the neutrino. Čerenkov detectors have been used in the neutrino physics experiments carried out by the observatories *Kamiokande* and *Super-Kamiokande* in the Kamioka Mines in Japan. Super-Kamiokande used a cylindrical steel tank, with a height and diameter of about 40 m, containing as radiator 50000 tons of ultra-pure water. Its free surface is covered by about 11000 photomultipliers. The experiments conducted in Kamioka have led to fundamental breakthroughs in the physics of elementary particles. In 1987 Kamiokande revealed for the first time a neutrino flux from the explosion of a supernova in the *Large Magellanic Cloud*, and in 1988 it realized the first direct observation of the solar neutrino flux. In 1998 the experiments of Super-Kamiokande provided the first experimental evidence of *neutrino oscillations*, a fundamental discovery which implies that neutrinos necessarily have a non-vanishing mass. Čerenkov detectors played a likewise fundamental role in the discovery of the *antiproton* in the Bevatron at the Lawrence Berkeley National Laboratory in 1955, and in the discovery of the *charm* quark at the Brookhaven National Laboratory in 1974.

13.7 Problems

13.1 *Fresnel's drag coefficient*. Assuming the existence of the *ether*, in 1818 A.-J. Fresnel proposed that if a liquid with index of refraction n flows with respect to the laboratory with velocity v, then the velocity of a light ray propagating in the liquid in the same direction has with respect to the laboratory the approximated value

$$c^* \approx \frac{c}{n} + \left(1 - \frac{1}{n^2}\right) v, \tag{13.107}$$

a prediction that was verified experimentally by H. Fizeau in 1851. The factor $1 - 1/n^2$ is known as *Fresnel's drag coefficient*. According to the theories prevailing at the time, the velocity of light in the laboratory was to be the simple sum of the velocity of the medium and of the velocity of light in the medium, i.e. $v + c/n$. The unexpected term $-v/n^2$ in Eq. (13.107), representing a braking effect with respect to $v + c/n$, was interpreted as due to the ether contained in the liquid, partially hampering the propagation of light.

(a) Derive the *relativistic* expression of c^* and compare it with Fresnel's formula (13.107).
   Hint: Determine the wave vector k^μ with respect to the medium by solving the homogeneous d'Alembert equation $\Box_n A^\mu = 0$, see formulas (13.8), and perform an appropriate Lorentz transformation to compute $k^{*\mu}$ in the laboratory. Alternatively, apply the relativistic transformation laws for the velocities. The result is

$$c^* = \frac{\dfrac{c}{n} + v}{1 + \dfrac{v}{nc}}.$$

(b) Verify that $c^* \le c$, as long as $n \ge 1$.

13.2 Verify that the fields (13.67) and (13.68) are invariant under the gauge transformations (13.69), where $\Lambda(\omega)$ is an arbitrary complex function of ω and $\mathbf{x}$. Show that with the gauge-fixing (13.70) Maxwell's equations in matter (13.63) and (13.65) take the form (13.71).

13.3 Verify that the integral (13.49), giving the total energy radiated per unit of traveled distance by a particle propagating with speed $v > c/n$ in a non-dispersive medium, is divergent.
Hint: Perform the change of variables

$$t(w) = \frac{1}{v}\left(z + rw\sqrt{\frac{v^2 n^2}{c^2} - 1}\right).$$

13.4 Consider the Poynting vector $\mathbf{S} = c\mathbf{E} \times \mathbf{B}$ associated with the fields (13.47) in the complement of the Mach cone.

(a) Write out explicitly its longitudinal and radial components S_z and S_r.
   Hint: As the fields (13.47) formally have the same functional dependencies as the fields (13.14) and (13.15) for $v < c/n$, the answer can be read off from formula (13.16).
(b) Evaluate $\mathbf{S}$ near the Mach cone and verify that there it has the same direction of the wave vector $\mathbf{k}$ (13.60).

13.5 Consider the function of a real variable $f(x) = H(x) \ln x$, where H is the Heaviside function.

(a) Prove that f defines a distribution in $\mathcal{S}'(\mathbb{R})$.
    Hint: f is locally integrable and polynomially bounded.
(b) Determine the distributional derivative df/dx of f, being aware that neither $H(x)/x$ nor $\delta(x) \ln x$ are distributions.
    Hint: Introduce the regularized distributions $f_\varepsilon = H(x) \ln(x + \varepsilon)$, with $\varepsilon > 0$, entailing the distributional limit

$$\mathcal{S}' - \lim_{\varepsilon \to 0} f_\varepsilon = f,$$

and use the relation

$$\mathcal{S}' - \lim_{\varepsilon \to 0} \frac{df_\varepsilon}{dx} = \frac{df}{dx}.$$

(c) Show that df/dx when applied to a test function $\varphi \in \mathcal{S}(\mathbb{R})$ gives

$$\frac{df}{dx}(\varphi) = \int_1^\infty \frac{\varphi(x)}{x}\, dx + \int_0^1 \frac{\varphi(x) - \varphi(0)}{x}\, dx.$$

13.6 Prove the distributional limit (13.90).
Hint: By definition one must show that the ordinary limits

$$\lim_{T \to \infty} \frac{1}{T} \int_{-\infty}^\infty \frac{\sin^2(Tx)}{x^2}\, \varphi(x)\, dx = \pi \varphi(0)$$

hold for every $\varphi \in \mathcal{S}(\mathbb{R})$. To prove these limits rescale x by x/T, move the limit under the integral sign, and use the integral (11.56).

13.7 *Electromagnetic fields as non-regular distributions.* Verify explicitly that the limit defining the electric field in (13.47) as a non-regular distribution exists.
Solution: One must show that the sequence of numbers

$$\int \frac{H(-z + vt - a(r + \varepsilon))\, \varphi(t, x, y, z)\, d^4x}{((z - vt)^2 - a^2 r^2)^{3/2}} - \int \frac{\varphi(t, x, y, vt - ar)}{\sqrt{2\varepsilon}\, a^2 r^{3/2}}\, dx\, dy\, dt$$
$$\equiv F_\varepsilon^1(\varphi) + F_\varepsilon^2(\varphi)$$

admits finite limits for $\varepsilon \to 0$ for all $\varphi \in \mathcal{S}(\mathbb{R}^4)$, where $a = \sqrt{v^2 n^2/c^2 - 1}$. The first integral can be rewritten as (replace z by $z(u) = -u + vt - ar$)

$$
\begin{aligned}
F_\varepsilon^1(\varphi) &= \int dx\,dy\,dt \int_{\varepsilon a}^\infty \frac{\varphi(t,x,y,-u+vt-ar)}{(2aur+u^2)^{3/2}}\,du \\
&= \int dx\,dy\,dt \int_1^\infty \frac{\varphi(t,x,y,-u+vt-ar)}{(2aur+u^2)^{3/2}}\,du \\
&\quad + \int dx\,dy\,dt \int_{\varepsilon a}^1 \frac{\varphi(t,x,y,-u+vt-ar)-\varphi(t,x,y,vt-ar)}{(2aur+u^2)^{3/2}}\,du \\
&\quad + \int dx\,dy\,dt\,\varphi(t,x,y,vt-ar)\int_{\varepsilon a}^1 \frac{du}{(2aur+u^2)^{3/2}}.
\end{aligned}
$$

The first term of $F_\varepsilon^1(\varphi)$ is ε-independent, and the second term admits a finite limit for $\varepsilon \to 0$. The divergence has thus been confined to the third term. In fact, since

$$
\int_{\varepsilon a}^1 \frac{du}{(2aur+u^2)^{3/2}} = \frac{1}{\sqrt\varepsilon}\int_a^{1/\varepsilon}\frac{du}{(2aur+\varepsilon u^2)^{3/2}} = \frac{1}{\sqrt{2\varepsilon}\,a^2 r^{3/2}}+O(1),
$$

this divergence is canceled by $F_\varepsilon^2(\varphi)$.

13.8 *The flux of the electric field confined inside the Mach cone*. Verify explicitly that, in agreement with the Coulomb law (13.3), the flux $\Phi(\mathbf{E})$ of the electric field in (13.47) across a sphere of radius R centered at the particle's position is equal to e/n^2.

Solution: To perform the surface integral over the sphere we set $z = vt - R\cos\vartheta$, $r = R\sin\vartheta$, and we write the surface element as $d\Sigma = R^2\sin\vartheta\,d\vartheta\,d\varphi\,\mathbf{n}$. Since we have $(\mathbf{x}-vt)\cdot\mathbf{n} = R$, the flux across the sphere of the electric field $\mathbf{E}$ in (13.47) can then be written as the ordinary limit ($a = \sqrt{v^2n^2/c^2-1}$)

$$
\begin{aligned}
\Phi(\mathbf{E}) &= \int_{|\mathbf{x}-vt|=R} \mathbf{E}\cdot d\Sigma \\
&= \frac{e}{n^2}\lim_{\varepsilon\to0}\left(-\int_0^{\vartheta_0}\frac{a^2\sin\vartheta\,d\vartheta}{(\cos^2\vartheta-a^2\sin^2\vartheta)^{3/2}} + \sqrt{\frac{R}{2\varepsilon}}\int\frac{\delta(\cos\vartheta-a\sin\vartheta)}{\sqrt{\sin\vartheta}}\,d\vartheta\right),
\end{aligned}
$$

where the upper extremum ϑ_0 is determined via the Heaviside function in (13.47) by the equation

$$
R(\cos\vartheta_0 - a\sin\vartheta_0) = a\varepsilon. \tag{13.108}
$$

In the limit for $\varepsilon \to 0$ we have $\vartheta_0 \to \arcsin(c/vn)$, which is the angular aperture of the Mach cone (13.42). Outside this cone the electric field is, in fact, zero. Notice that in the limit for $\varepsilon \to 0$ the first "regular" term of $\Phi(\mathbf{E})$ gives a negative divergent flux, while the second term, supported on the Mach cone, gives a positive divergent flux. For $\varepsilon > 0$ the first integral can be computed analytically and gives

$$-\int_0^{\vartheta_0} \frac{a^2 \sin\vartheta\, d\vartheta}{(\cos^2\vartheta - a^2\sin^2\vartheta)^{3/2}} = -\frac{\cos\vartheta}{\sqrt{\cos^2\vartheta - a^2\sin^2\vartheta}}\bigg|_0^{\vartheta_0}$$

$$= 1 - \frac{1}{\sqrt{\varepsilon}}\frac{R\sin\vartheta_0 + \varepsilon}{\sqrt{2R\sin\vartheta_0 + \varepsilon}} = 1 - \sqrt{\frac{R}{2\varepsilon}}\frac{1}{(1+a^2)^{1/4}} + O(\sqrt{\varepsilon}), \qquad (13.109)$$

where we used the Eq. (13.108) for ϑ_0. The second integral can be computed applying the rule (2.76)

$$\sqrt{\frac{R}{2\varepsilon}}\int \frac{\delta(\cos\vartheta - a\sin\vartheta)}{\sqrt{\sin\vartheta}}\, d\vartheta = \sqrt{\frac{R}{2\varepsilon}}\frac{1}{(1+a^2)^{1/4}}. \qquad (13.110)$$

Summing the contributions (13.109) and (13.110) we see that the divergent parts cancel out, and taking the limit for $\varepsilon \to 0$ we obtain $\Phi(\mathbf{E}) = e/n^2$. Notice that eventually the positive flux e/n^2 comes from the regular term, although for $\varepsilon \to 0$ this term – as a whole – tends to minus infinity.

References

1. I.M. Frank, I.E. Tamm, Coherent visible radiation of fast electrons passing through matter. Compt. Rend. Acad. Sci. USSR **14**, 109 (1937)
2. J.D. Jackson, *Classical Electrodynamics* (Wiley, New York, 1998)
3. M. Abramowitz, I.A. Stegun, *Handbook of Mathematical Functions with Formulas, Graphs, and Mathematical Tables* (U.S. Government Printing Office, Washington, 1964)

Chapter 14
Radiation from an Infinite Wire

In this chapter we perform a case study of the electromagnetic radiation generated by a straight infinite conducting wire, carrying a generic homogeneous time-
dependent current $I(t)$. Being an infinitely extended physical system, a straight wire
represents a prototypical *unbounded* system of electric charges, entailing a four-
current which has at each time a *non-compact* spatial support. The physical properties of the electromagnetic field produced by an infinite wire are, therefore, expected
to differ substantially from the general properties of the fields created by compact
charge distributions – typically a single charged particle, or a system of charged particles confined to a bounded region – analyzed in detail in Chaps. 6–8. On the other
hand, a current flowing in a straight wire possesses *cylindrical symmetry*, as does
the charged particle in linear motion originating the Čerenkov effect. This peculiar
geometrical feature allows for a close comparison of the physical and mathematical aspects of the electromagnetic fields generated by the two systems. Below we
anticipate the main similarities and differences.

The fields of an infinite wire, of a charged particle, and of the Čerenkov effect. The
main distinctive feature of the electromagnetic field created by an infinite wire with
respect to the field created by a particle is its behavior at large distances from the
source, where it is expected to become a *radiation field*. In fact, while the radiation
fields of the particle in (7.48) decrease as $1/r$, the radiation field of the wire (14.23)
decreases as $1/\sqrt{r}$, where in the latter case r is the distance from the wire. Since
the origin of this stronger asymptotic behavior is the cylindrical symmetry of the
underlying geometry, it is also shared by the radiation field of the Čerenkov effect,
see formulas (13.81)–(13.86). Since a homogenous current flowing in a wire entails
a vanishing charge density, it creates the *velocity field* (14.14) – actually a magnetic
field – which is of the relativistic order $1/c$, while a particle generates a velocity
field – actually an electric field – which is of order zero. Concerning the *radiation
fields*, *alias* acceleration fields, the radiation field of the wire (14.23) is a relativistic
effect of order $1/c^{3/2}$, in contrast to the radiation fields of the particle in (7.48)
which are of the smaller order $1/c^2$. An important consequence is that, whereas the
energy radiated by the particle is of order $1/c^3$, the energy radiated by the wire is of

© Springer International Publishing AG, part of Springer Nature 2018
K. Lechner, *Classical Electrodynamics*, UNITEXT for Physics,
https://doi.org/10.1007/978-3-319-91809-9_14

the larger order $1/c^2$. These relativistic features of the radiation of the wire parallel those of the Čerenkov effect, if the velocity v of the charged particle moving in the medium is close to the velocity of light c. For instance, the total energy $d\varepsilon/dz$ radiated by the particle per unit of distance traveled in the medium is of order $1/c^2$, see Eq. (13.103).

A further fundamental difference between the radiation of a wire and that of a charged particle concerns the finiteness properties of the *total* radiated energy. Let us consider a wire whose current $I(t)$ rises continuously from an initial constant value I_a to a final constant value I_b. The analogous dynamics of a charged particle would be a trajectory whose velocity changes continuously from an initial value $\mathbf{v}_i$ to a final value $\mathbf{v}_f$. The basic difference between the two phenomena is that, while in the case of the particle the total energy $\Delta\varepsilon$ radiated during the deflection process is *finite* – see, for instance, Eq. (11.57) for the total energy radiated in the non-relativistic limit – the total energy $d\varepsilon/dz$ radiated by the wire per unit length is *infrared divergent*, unless the initial and final currents coincide, i.e. $I_b = I_a$. As we will see, this *universal* infrared divergence is caused by the too slow decrease in time of the radiation field of the wire. In fact, unless $I_a = I_b$, in the limit for $t \to \infty$ this field decreases only as fast as $1/\sqrt{t}$, see formulas (14.29). Conversely, as we have seen in Chap. 13, no such infrared divergences arise in the Čerenkov effect. In particular, the total energy $d\varepsilon/dz$ (13.103) radiated per unit of traveled distance converges for small frequencies.[1] The physical reason of this difference is that, while the geometry of the wire entails a *true* cylindrical symmetry – at fixed time its electromagnetic field is translation invariant in the direction of the wire – the electromagnetic field of the particle of the Čerenkov effect represents a *transient* phenomenon, whose intensity, at fixed time, decreases along the trajectory of the particle, as one moves away from it. For a fixed position $\mathbf{x}$, this implies, in turn, that the radiation field of the Čerenkov effect decreases in time faster than the field of the wire, namely (at least) as $1/t^{3/2}$. This behavior can be derived, for instance, from the exact formulas for the fields of the Čerenkov effect in a non-dispersive medium (13.47), by performing first their large-r expansion, as in (14.26).

In contrast to the infrared divergences, a wire and a charged particle entail the *same* kind of ultraviolet divergences, if the current $I(t)$ and the particle's velocity $\mathbf{v}(t)$, respectively, change *discontinuously* in time. In fact, in the same way as the total energy $\Delta\varepsilon$ (11.57) radiated by the particle becomes infinite if its velocity jumps discontinuously from $\mathbf{v}_i$ to $\mathbf{v}_f$ (because its acceleration during the transition is infinite), if the current of the wire changes discontinuously from I_a to I_b, subjecting thus a large number of charged particles to an infinite instantaneous acceleration, the total energy $d\varepsilon/dz$ radiated by the wire per unit length acquires an ultraviolet divergence, in addition to its *intrinsic* infrared divergence. However, in a sense, the ultraviolet divergence of the field (14.12) generated by the *prototypical* discontinuous current (14.3) resembles more the ultraviolet divergence on the Mach cone of the field (13.47) of the Čerenkov effect in a non-dispersive

---

[1] We make here the far-from-obvious assumption that the index of refraction $n(\omega)$ maintains its physical meaning also for $\omega \to 0$.

medium, rather than the ultraviolet divergence of the field generated by a particle having the discontinuous velocity $\mathbf{v}(t) = H(-t)\,\mathbf{v}_i + H(t)\,\mathbf{v}_f$. In fact, while for a fixed t the fields (13.47) and (14.12) exhibit inverse-power spatial singularities, the non-relativistic field of the particle (7.65), being proportional to the acceleration $\mathbf{a}(t-r) = (\mathbf{v}_f - \mathbf{v}_i)\,\delta(t-r)$, exhibits δ-like spatial singularities.

14.1   Kinematics of the Current

Choosing as z axis the line of the wire the conserved four-current takes the form

$$j^\mu(t,\mathbf{x}) = (0,0,0, I(t)\,\delta^2(x,y)). \tag{14.1}$$

In the first four sections of this chapter we focus on *aperiodic* currents $I(t)$, whereas Sect. 14.5 is devoted to *periodic* ones. We require our aperiodic currents to be (i) *continuous*, and (ii) to approach for large times sufficiently fast constant values, in the past as well as in the future,

$$\lim_{t\to-\infty} I(t) = I_a, \qquad \lim_{t\to\infty} I(t) = I_b. \tag{14.2}$$

The simplest current of this type is the idealized *step-function* current

$$I_{\mathrm{sf}}(t) = I_a H(-t) + I_b H(t) = I_a + (I_b - I_a)H(t), \tag{14.3}$$

which at $t = 0$ jumps discontinuously from I_a to I_b. As usual, $H(t)$ denotes the Heaviside function. Despite its "unphysical" discontinuity at $t = 0$, the current (14.3) has the advantage of being tractable analytically and, most importantly, it will allow us to establish the general physical properties of the electromagnetic field generated by a *generic* current. In fact, a generic current $I(t)$ satisfying the asymptotic conditions (14.2) can always be decomposed in the sum of the step-function current and of a *fluctuation* $\Delta I(t)$,

$$I(t) = I_{\mathrm{sf}}(t) + \Delta I(t), \tag{14.4}$$

where the latter obeys the limits[2]

$$\lim_{t\to\pm\infty} \Delta I(t) = 0. \tag{14.5}$$

As the current $\Delta I(t)$ decreases rapidly for large times, thanks to the *superposition principle* the decomposition (14.4) will allow us to characterize the asymptotic

---

[2]As we supposed $I(t)$ to be continuous, the current $\Delta I(t)$ at $t = 0$ entails the finite discontinuity $I_a - I_b$. However, we will see that the latter is easier to handle than the overall variation $I_b - I_a$ of the current $I(t)$ between $t = -\infty$ and $t = \infty$.

behavior of the electromagnetic field generated by a generic current $I(t)$ in a *universal* manner.

Green function method. In the first instance, to compute the electromagnetic field generated by the four-current (14.1) we shall resort to the standard Green function method. However, in doing so, at the very beginning we have to face an unexpected problem: for the usual textbook-wire carrying a *constant* current I the Green function method fails, producing, actually, an *infinite* four-potential. Of course, this is not a too great surprise as we know that a priori this method is not guaranteed to yield a well-defined solution of Maxwell's equations, a fortiori if the four-current at hand (14.1) is of non-compact spatial support. Nonetheless, we will see that, by means of a *minimal* adaptation of the method, also in the case of an infinite wire the Green function method can be applied successfully.

14.2 Potentials and Fields

As only the z component of the current (14.1) is non vanishing, only the z component of the four-potential will be different from zero

$$A^\mu(t, \mathbf{x}) = (0, 0, 0, A(t, \mathbf{x})). \tag{14.6}$$

For simplicity we set $A(t, \mathbf{x}) = A^z(t, \mathbf{x})$. Inserting the four-current (14.1) in the general solution (6.65) of Maxwell's equations we would obtain the potential

$$A(t, \mathbf{x}) = \frac{1}{4\pi} \int \frac{I(t - |\mathbf{x} - \mathbf{y}|)}{|\mathbf{x} - \mathbf{y}|}\, \delta(y^1)\, \delta(y^2)\, d^3y = \frac{1}{4\pi} \int \frac{I\big(t - \sqrt{p^2 + r^2}\,\big)}{\sqrt{p^2 + r^2}}\, dp,$$

where $r = \sqrt{(x^1)^2 + (x^2)^2}$, and we performed the change of integration variable $y^3(p) = p + x^3$. Making the further change of variable $p(u) = r\sqrt{u^2 - 1}$, we would end up with the integral

$$A(t, \mathbf{x}) = \frac{1}{2\pi} \int_1^\infty \frac{I(t - ru)}{\sqrt{u^2 - 1}}\, du. \tag{14.7}$$

As $A(t, \mathbf{x})$ represents a *retarded* potential, for a fixed t the convergence properties of the integral (14.7) for $u \to \infty$ are controlled by the behavior of the current $I(t)$ for large negative times, for which it tends to the constant value I_a. This implies that, as long as $I_a \neq 0$, the integral (14.7) diverges logarithmically for all t and for all r, since the integrand for large u behaves as I_a/u. Vice versa, for $I_a = 0$ the integral would converge. To overcome this problem we write the current as the sum

$$I(t) = \big(I(t) - I_a\big) + I_a.$$

The term between brackets vanishes now for $t \to -\infty$ and generates thus, via Eq. (14.7), a *convergent* potential. On the other hand, the constant current I_a generates the well-known static vector potential $\mathbf{A} = -(I_a \ln r / 2\pi) \mathbf{u}_z$. Therefore, in the general case a well-defined solution of Maxwell's equations, replacing the ill-defined expression (14.7), is given by the potential

$$A(t, \mathbf{x}) = \frac{1}{2\pi} \int_1^\infty \frac{I(t - ru) - I_a}{\sqrt{u^2 - 1}} \, du - \frac{I_a \ln r}{2\pi} \tag{14.8}$$

$$A(t, \mathbf{x}) = \frac{1}{2\pi} \int_0^\infty \frac{I(t - r - \beta) - I_a}{\sqrt{\beta^2 + 2r\beta}} \, d\beta - \frac{I_a \ln r}{2\pi}, \tag{14.9}$$

where the variant (14.9) has been obtained by setting $u = \beta / r - 1$. As the four-potential is of the form (14.6), the unique non-vanishing components of the electromagnetic field are given by

$$E^3 = -\dot{A}, \qquad B^1 = \partial_2 A, \qquad B^2 = -\partial_1 A. \tag{14.10}$$

The electric field is thus always parallel to the wire, and, since the potential (14.8) depends only on t and on the radial coordinate r, the field lines of $\mathbf{B}$ are circles concentric with the z axis:

$$\mathbf{E} = E^3 \, \mathbf{u}_z, \qquad \mathbf{B} = B_\varphi \mathbf{u}_\varphi.$$

As in the case of the Čerenkov effect, we use a set of cylindrical coordinates $\{r, \varphi, z\}$, with the associated set of mutually orthogonal unit vectors $\{\mathbf{u}_r, \mathbf{u}_\varphi, \mathbf{u}_z\}$.

14.2.1 The Electromagnetic Field of the Step-Function Current

Before proceeding with the analysis of the electromagnetic field generated by a generic current $I(t)$ via the potential (14.8), we perform a case study of the field generated by the step-function current $I_{\text{sf}}(t)$ (14.3), which entails the advantage that the integral (14.8) can be computed analytically. In fact, according to the above prescription we must insert the current

$$I_{\text{sf}}(t - ru) - I_a = (I_b - I_a) H(t - ru)$$

in the integral (14.8), leading to

$$A_{\text{sf}}(t, \mathbf{x}) = \frac{I_b - I_a}{2\pi} \int_1^\infty \frac{H(t - ru)}{\sqrt{u^2 - 1}} \, du - \frac{I_a \ln r}{2\pi}$$

$$= \frac{(I_b - I_a)H(t - r)}{2\pi} \ln\left(\frac{t}{r} + \sqrt{\frac{t^2}{r^2} - 1}\right) - \frac{I_a \ln r}{2\pi}, \qquad (14.11)$$

where we set $u = \cosh s$ and used the identity

$$\text{arccosh} \, u = \ln\left(u + \sqrt{u^2 - 1}\right).$$

The relations (14.10) then yield the fields

$$\mathbf{E}_{\text{sf}} = -\frac{(I_b - I_a)H(t - r)}{2\pi\sqrt{t^2 - r^2}} \mathbf{u}_z, \qquad \mathbf{B}_{\text{sf}} = \frac{t}{r} \mathbf{u}_r \times \mathbf{E}_{\text{sf}} + \frac{I_a \mathbf{u}_\varphi}{2\pi r}. \qquad (14.12)$$

The derivative of the Heaviside function $\delta(t - r)$ does not contribute to the fields, since the argument of the logarithm in (14.11) for $t = r$ is equal to unity. For $t < 0$ the electromagnetic field is composed only of the static magnetic field $\mathbf{B}_a = I_a \mathbf{u}_\varphi/2\pi r$, created by the current I_a, whereas for $t > 0$ it acquires the "radiation" fields $\mathbf{E}_{\text{sf}}$ and $\mathbf{B}_{\text{rad}} = (t/r)\mathbf{u}_r \times \mathbf{E}_{\text{sf}}$, see Sect. 14.3, which are non-vanishing outside a cylinder of radius $r = t$. The associated Poynting vector $\mathbf{S}_{\text{sf}} = \mathbf{E}_{\text{sf}} \times \mathbf{B}_{\text{rad}} = (t/r)E_{\text{sf}}^2 \mathbf{u}_r$ points indeed outwards in the radial direction for all $r < t$. At the *wavefront* $r = t$, which propagates at the velocity of light, due to the discontinuity of the step-function current at $t = 0$ these fields are divergent.

If we consider the fields (14.12) at a fixed radial position r and let the time $t > r$ increase indefinitely, the electric field $\mathbf{E}_{\text{sf}}$ tends to zero as $1/t$, while the magnetic field $\mathbf{B}_{\text{sf}}$, thanks to the prefactor t/r, tends to $\mathbf{B}_b = I_b \mathbf{u}_\varphi/2\pi r$, which is the static field created by the current I_b. This is in line with the step-function current (14.3), which for $t \to \infty$ tends indeed to I_b. In contrast, the potential (14.11) for $t \to \infty$ fails to reach the limit potential $A_b = -I_b \ln r/2\pi$, in that for large t it behaves as

$$A_{\text{sf}}(t, \mathbf{x}) \to \frac{I_b - I_a}{2\pi} \ln\frac{t}{r} - \frac{I_a \ln r}{2\pi} = -\frac{I_b \ln r}{2\pi} + \frac{I_b - I_a}{2\pi} \ln t. \qquad (14.13)$$

In a similar way, the potential $A_{\text{sf}}^\gamma(t, \mathbf{x}) = A_{\text{sf}}(t - \gamma, \mathbf{x})$, generated by the step-function current $I_\gamma(t) = I_a H(\gamma - t) + I_b H(t - \gamma)$ jumping at an *earlier* time $\gamma < 0$, for $\gamma \to -\infty$ fails to reach the limit $-I_b \ln r/2\pi$. None of these occurrences represents, however, a physical or mathematical inconsistency. In fact, concerning the non-existence of the limit (14.13) we recall that, contrary to the fields, the potentials are non-observable physical quantities, and the potential (14.13) gives indeed rise to the correct limiting fields. Similarly, although the current $I_\gamma(t)$ which generates the potential $A_{\text{sf}}^\gamma(t, \mathbf{x})$ converges for $\gamma \to -\infty$ to the constant current I_b, a *priori* the Green function method does not guarantee the existence of the (distributional) limit for $\gamma \to -\infty$ of the associated potential. Rather the opposite is true: *if* the potential $A_{\mathrm{sf}}^{\gamma}(t, \mathbf{x})$ would converge for $\gamma \to -\infty$ in the distributional sense, then the limiting potential would automatically satisfy Maxwell's equations.

14.2.2   General Properties of the Electromagnetic Field

We return now to the expression of the potential (14.9) generated by a generic current $I(t)$. Its second term generates the known static *velocity fields*, decreasing as $1/r$,

$$\mathbf{E}_v = 0, \qquad \mathbf{B}_v = \frac{I_a \mathbf{u}_{\varphi}}{2\pi r c}. \tag{14.14}$$

We restored the velocity of light to underline that the electromagnetic field created by a wire represents a *genuine* relativistic effect, in that it is produced exclusively by moving charges. The first term in (14.8) generates instead a *radiation field*, decreasing as $1/\sqrt{r}$, that we will analyze in detail in Sect. 14.3.

The fields close to the wire. Close to the wire, for small radial distances r at fixed t, the integral of the potential (14.9) exhibits a logarithmic divergence. To see it we decompose the integral in the sum $\int_0^1 d\beta + \int_1^{\infty} d\beta$, where the second integral admits a finite limit for $r \to 0$. We rewrite the first integral in the form

$$\int_0^1 \frac{I(t - r - \beta) - I_a}{\sqrt{\beta^2 + 2r\beta}}\, d\beta = \int_0^1 \frac{I(t - r - \beta) - I(t - r)}{\sqrt{\beta^2 + 2r\beta}}\, d\beta \tag{14.15}$$

$$+ \left(I(t - r) - I_a\right) \int_0^1 \frac{1}{\sqrt{\beta^2 + 2r\beta}}\, d\beta. \tag{14.16}$$

The integral in (14.15) admits a finite limit for $r \to 0$, whereas the integral in (14.16) diverges as $-\ln r + O(1)$. From formulas (14.9), (14.10) and (14.16) we then derive that near the wire the potential and the fields take the expected form

$$A(t, \mathbf{x}) \approx -\frac{I(t) \ln r}{2\pi}, \qquad \mathbf{E}(t, \mathbf{x}) \approx \frac{\dot{I}(t) \ln r}{2\pi}\, \mathbf{u}_z, \qquad \mathbf{B}(t, \mathbf{x}) \approx \frac{I(t) \mathbf{u}_{\varphi}}{2\pi r}.$$

The fields for large positive times. To analyze the behavior of the fields in the limit for $t \to \infty$ at a fixed radial distance r, we resort to the decomposition of the current (14.4). The step-function current generates the potential $A_{\mathrm{sf}}(t, \mathbf{x})$ (14.11) and the fields (14.12). It remains, therefore, to analyze the potential $\Delta A(t, \mathbf{x})$ generated by the fluctuation $\Delta I(t)$ according to Eq. (14.9). We have thus

$$A(t, \mathbf{x}) = A_{\mathrm{sf}}(t, \mathbf{x}) + \Delta A(t, \mathbf{x}), \qquad \Delta A(t, \mathbf{x}) = \frac{1}{2\pi} \int_0^{\infty} \frac{\Delta I(t - r - \beta)}{\sqrt{\beta^2 + 2r\beta}}\, d\beta,$$

$$\tag{14.17}$$

where we used the condition (14.5) for $t \to -\infty$. In order to analyze the behavior of the potential $\Delta A(t, \mathbf{x})$ for large "retarded" times $t' = t - r$, we divide the integral in the regions $\int_0^\infty d\beta = \int_0^a d\beta + \int_a^{t'/2} d\beta + \int_{t'/2}^\infty d\beta$, where a is a fixed positive constant smaller than $t'/2$. The first integral for large t' goes to zero as fast as does $\Delta I(t')$

$$\int_0^a \frac{\Delta I(t' - \beta)}{\sqrt{\beta^2 + 2r\beta}} \, d\beta \quad \to \quad \Delta I(t') \int_0^a \frac{1}{\sqrt{\beta^2 + 2r\beta}} \, d\beta.$$

The second integral can be bound by (performing the shift $\beta \to \beta + t'$)

$$\left| \int_a^{t'/2} \frac{\Delta I(t' - \beta)}{\sqrt{\beta^2 + 2r\beta}} \, d\beta \right| \leq \int_{a-t'}^{-t'/2} \frac{|\Delta I(-\beta)|}{\sqrt{a^2 + 2ra}} \, d\beta. \tag{14.18}$$

Supposing that $\Delta I(t)$ decreases monotonously for large negative times, according to the *mean value theorem* the second integral in (14.18) goes to zero for $t' \to \infty$ at least as $t' \Delta I(-t'/2)$. Finally, in the third region the integral (14.17) can be recast in the form

$$\int_{t'/2}^\infty \frac{\Delta I(t' - \beta)}{\sqrt{\beta^2 + 2r\beta}} \, d\beta = \frac{1}{t'} \int_{-t'/2}^\infty \frac{\Delta I(-\beta)}{\sqrt{\left(1 + \frac{\beta}{t'}\right)^2 + \frac{2r}{t'} \left(1 + \frac{\beta}{t'}\right)}} \, d\beta$$

$$\to \frac{1}{t'} \int_{-\infty}^\infty \Delta I(-\beta) \, d\beta,$$

where, to take the limit for $t' \to \infty$ under the integral sign, we applied the corollary of *Lebesgue's dominated convergence theorem* reported in Sect. 16.3.1. Therefore, the potential $\Delta A(t, \mathbf{x})$ in (14.17) decays as $1/t$

$$\Delta A(t, \mathbf{x}) \approx \frac{1}{2\pi t} \int_{-\infty}^\infty \Delta I(-\beta) \, d\beta. \tag{14.19}$$

As this behavior is *subleading* with respect to the $\ln t$-divergence (14.13) of the step-function-current potential (14.11), we conclude that the large-t behaviors of the potential and the fields generated by a generic current $I(t)$ coincide with the large t-behaviors of the expressions (14.11) and (14.12) relative to the step-function current, namely

$$A(t, \mathbf{x}) \approx \frac{(I_b - I_a) \ln t}{2\pi}, \quad \mathbf{E}(t, \mathbf{x}) \approx -\frac{(I_b - I_a)}{2\pi t} \, \mathbf{u}_z, \quad \mathbf{B}(t, \mathbf{x}) \approx \frac{I_b \mathbf{u}_\varphi}{2\pi r}. \tag{14.20}$$

These large-t behaviors are, therefore, *universal*.

14.3   Potentials and Fields in the Wave Zone

An analysis of the radiation requires to evaluate the electromagnetic fields in the *wave zone*, i.e. at large distances from their sources, which in the case at hand corresponds to large radial distances from the wire. For this purpose it is convenient to use for the potential again the integral representation (14.9). Keeping the *phase* $t - r$ fixed, for large r this integral behaves as

$$A(t, \mathbf{x}) = \frac{1}{2\pi\sqrt{2r}} \int_0^\infty \frac{I(t - r - \beta) - I_a}{\sqrt{\beta}} \, d\beta - \frac{I_a \ln r}{2\pi} + O\left(\frac{1}{r^{3/2}}\right). \quad (14.21)$$

The first, time-dependent term, in this formula decreases as $1/\sqrt{r}$ and represents the *potential in the wave zone*, see the corresponding behavior (13.86) of the Čerenkov potential (13.81). In particular, since this term is of the form $f(t - r)/\sqrt{r}$, the potential (14.21) – modulo terms of order $1/r$ – satisfies the *cylindrical wave relations*

$$\boldsymbol{\nabla} A = -\dot{A}\,\mathbf{u}_r.$$

Formulas (14.10) then yield the *fields in the wave zone*

$$\mathbf{E} = -\dot{A}\,\mathbf{u}_z, \qquad \mathbf{B} = \mathbf{u}_r \times \mathbf{E}, \qquad \mathbf{S} = \mathbf{E} \times \mathbf{B} = E^2 \mathbf{u}_r. \quad (14.22)$$

In particular, the electric wave-zone field admits the integral representation

$$\mathbf{E}(t, \mathbf{x}) = -\frac{\mathbf{u}_z}{2\pi\,c^{3/2}\sqrt{2r}} \int_0^\infty \frac{\dot{I}\left(t - \frac{r}{c} - \beta\right)}{\sqrt{\beta}} \, d\beta, \quad (14.23)$$

where we have reintroduced the velocity of light to emphasize its peculiar relativistic behavior. In fact, the field (14.23) is of order $1/c^{3/2}$, in contrast to the electric wave-zone field of a charged particle in (7.48), which is of the smaller order $1/c^2$. Correspondingly, from the radiation formula $\Delta\varepsilon \propto cE^2$ it follows that the energy radiated by an infinite wire is of order $1/c^2$, as is the energy (13.102) radiated in the Čerenkov effect, which likewise entails a cylindrical symmetry. This is in contrast to the $1/c^3$-dependence of the energy radiated by a single charged particle. From Eqs. (14.22) and (14.23) we infer that the wave-zone fields decrease as

$$\mathbf{E} \sim \frac{1}{\sqrt{r}}, \qquad \mathbf{B} \sim \frac{1}{\sqrt{r}},$$

exactly as the Čerenkov fields (13.86): Like the latter they constitute, in fact, *cylindrical radiation fields*. Since the equation (14.23) for the electric wave-zone field entails the proportionality relation $\mathbf{E} \propto \dot{I}$, the occurrence of radiation requires currents that vary with time, namely, once more, *accelerated charges*. Although the fields (14.22) share with the Čerenkov fields the asymptotic behavior $1/\sqrt{r}$, a straight wire emits cylindrical electromagnetic waves which propagate in all *radial* directions $\mathbf{u}_r$, whereas the Čerenkov waves propagate in directions which form with the z axis the Čerenkov angle (13.61). Finally, from Eq. (14.22) we also see that for a fixed emission direction $\mathbf{u}_r$ the fields $\mathbf{E}$ and $\mathbf{B}$ have constant directions. This means that the emitted radiation is *linearly* polarized.

Radiation fields of the step-function current. For the step-function current (14.3) formula (14.21) yields the wave-zone potential

$$A_{\mathrm{sf}}(t, \mathbf{x}) = \frac{I_b - I_a}{2\pi\sqrt{2r}} \int_0^\infty \frac{H(t - r - \beta)}{\sqrt{\beta}} \, d\beta = \frac{(I_b - I_a)H(t - r)}{\pi\sqrt{2r}} \sqrt{t - r}.$$
(14.24)

The radiation fields (14.22) then take the form

$$\mathbf{E}_{\mathrm{sf}}(t, \mathbf{x}) = -\frac{(I_b - I_a)H(t - r)}{2\pi\sqrt{2r(t - r)}} \, \mathbf{u}_z, \qquad \mathbf{B}_{\mathrm{sf}}(t, \mathbf{x}) = \frac{(I_b - I_a)H(t - r)}{2\pi\sqrt{2r(t - r)}} \, \mathbf{u}_\varphi.$$
(14.25)

These expressions match with the large-r expansions of the exact formulas (14.11) and (14.12), by keeping the phase $t - r$ fixed. In Eq. (14.12), for instance, this amounts to the replacement

$$\frac{1}{\sqrt{t^2 - r^2}} = \frac{1}{\sqrt{(t - r)(2r + (t - r))}} \to \frac{1}{\sqrt{2r}\sqrt{t - r}}.$$
(14.26)

Both radiation fields in formulas (14.25) entail the large-t behavior $1/\sqrt{t}$, which seems, however, to conflict with the general behaviors (14.20). There is, of course, no contradiction: the limits for $t \to \infty$ and for $r \to \infty$ simply do not commute.

General large-t behavior of the radiation fields. As we will see in Sect. 14.3.1, for physical purposes the large-t behavior of the radiation fields (14.22) is, actually, of even greater importance than that of the exact fields (14.20). We can analyze it resorting to the same techniques of Sect. 14.2.2, by using the decomposition (14.4) to write the wave-zone potential in (14.21) in the form

$$A(t, \mathbf{x}) = A_{\mathrm{sf}}(t, \mathbf{x}) + \Delta A(t, \mathbf{x}), \qquad \Delta A(t, \mathbf{x}) = \frac{1}{2\pi\sqrt{2r}} \int_0^\infty \frac{\Delta I(t' - \beta)}{\sqrt{\beta}} \, d\beta.$$
(14.27)

The wave-zone potential $A_{\mathrm{sf}}(t, \mathbf{x})$ of the step-function current is now given by (14.24), and we have introduced again the retarded time $t' = t - r$. In decomposing the integral in (14.27) as $\int_0^\infty d\beta = \int_0^a d\beta + \int_a^{t'/2} d\beta + \int_{t'/2}^\infty d\beta$, it is again the last term which gives the leading contribution for $t' \to \infty$

$$\Delta A(t,\mathbf{x}) \approx \frac{1}{2\pi\sqrt{2r}} \int_{t'/2}^{\infty} \frac{\Delta I(t'-\beta)}{\sqrt{\beta}}\, d\beta = \frac{1}{2\pi\sqrt{2rt'}} \int_{-t'/2}^{\infty} \frac{\Delta I(-\beta)}{\sqrt{1+\frac{\beta}{t'}}}\, d\beta$$

$$\rightarrow \frac{1}{2\pi\sqrt{2rt}} \int_{-\infty}^{\infty} \Delta I(-\beta)\, d\beta.$$

(14.28)

Since for $t \rightarrow \infty$ the step-function-current potential (14.24) grows as $\sqrt{t}$, the potential $\Delta A(t,\mathbf{x})$, decreasing as $1/\sqrt{t}$, represents thus a subleading contribution. It would create radiation fields that decrease as $1/t^{3/2}$. It follows that a generic current $I(t)$, satisfying the asymptotic conditions (14.2), generates radiation fields which for large positive times decay in a *universal* manner as the radiation fields (14.25) of the step-function current, namely as $1/\sqrt{t}$. Formulas (14.24) and (14.25) imply so the universal large-t behaviors in the wave zone

$$A(t,\mathbf{x}) \approx \frac{(I_b - I_a)\sqrt{t}}{\pi\sqrt{2r}}, \quad \mathbf{E}(t,\mathbf{x}) \approx -\frac{I_b - I_a}{2\pi\sqrt{2rt}}\,\mathbf{u}_z, \quad \mathbf{B}(t,\mathbf{x}) \approx \frac{I_b - I_a}{2\pi\sqrt{2rt}}\,\mathbf{u}_\varphi.$$

(14.29)

14.3.1 Radiated Energy: Ultraviolet and Infrared Singularities

Thanks to the cylindrical geometry of the physical system, the energy radiated per unit time per unit length of the wire, i.e. the power radiated per unit length, can be read off from the corresponding equation of the Čerenkov effect (13.94)

$$\frac{d^2\varepsilon}{dzdt} = 2\pi r\,(\mathbf{E} \times \mathbf{B}) \cdot \mathbf{u}_r = 2\pi r E^2 = 2\pi r \dot{A}^2,$$

(14.30)

where we have used the wave-zone relations (14.22). Let us first investigate the properties of this expression for the radiation produced by the step-function current (14.3). In basis of the equation (14.25) for the corresponding wave-zone fields, the emitted energy (14.30) turns out to be proportional to the square of the overall variation of the current

$$\frac{d^2\varepsilon}{dzdt} = \frac{(I_b - I_a)^2 H(t - r)}{4\pi(t - r)}.$$

(14.31)

However, the total energy radiated by the wire per unit length during the whole process

$$\frac{d\varepsilon}{dz} = \int_{-\infty}^{\infty} \frac{d^2\varepsilon}{dzdt}\, dt = \frac{(I_b - I_a)^2}{4\pi} \int_{r}^{\infty} \frac{1}{t - r}\, dt$$

(14.32)

is *infinite*. Actually, there are two sources of divergences. There is an *ultraviolet* divergence in the integral (14.32) localized on the wavefront $t = r$, which is due to the sudden jump of the step-function current (14.3) at $t = 0$ that forces the charged particles in the wire to undergo an infinite acceleration. This "unphysical" singularity can, thus, be cured by letting the current rise *continuously* from I_a to I_b, see below. The second divergence of the integral (14.32) arises from the $1/t$-behavior of the integrand for $t \to \infty$, and it is hence of the *infrared* type. It is caused, in turn, by the slow $1/\sqrt{t}$-decay of the fields (14.25). As the latter behavior is universal, see the relations (14.29), we conclude that a generic current $I(t)$ generates radiation whose total energy emitted per unit length is always infinite, unless the overall change of the current $I_b - I_a$ vanishes. Eventually, this *intrinsic* divergence of the emitted energy (14.32) originates from the cylindrical geometry of the problem, which requires a four-current (14.1) of non-compact support; see Problem 14.5 for an example of a compact wire. Finally, concerning the physical origin of the phenomenon, i.e. the external forces which furnish the power supply needed to accelerate the current, we must conclude that the energy employed by these forces per unit length to push the current from I_a to I_b is likewise infinite.

Continuous currents. We illustrate the general features outlined above in the example of a continuous *template* current $I_c(t)$, which during the finite time interval $2T$ increases linearly from I_a to I_b according to the linear law

$$I_c(t) = \begin{cases} I_a, & \text{for } t < -T, \\ \dfrac{I_b + I_a}{2} + \dfrac{I_b - I_a}{2T}\, t, & \text{for } -T < t < T, \\ I_b, & \text{for } t > T. \end{cases} \qquad (14.33)$$

The electric wave-zone field generated by this current can be calculated easily from the general formula (14.23), as the derivative of the current (14.33) reads simply

$$\dot{I}_c(t) = \frac{I_b - I_a}{2T}\left(H(t + T) - H(t - T)\right). \qquad (14.34)$$

The result is (see also Problem 14.4)

$$\mathbf{E}(t, \mathbf{x}) = -\frac{(I_b - I_a)\mathbf{u}_z}{2\pi T \sqrt{2r}}\left(H(t - r + T)\sqrt{t - r + T} - H(t - r - T)\sqrt{t - r - T}\right). \qquad (14.35)$$

This field is continuous for all values of t. In particular, for a fixed r, it vanishes for $t < r - T$, it increases in magnitude in the interval $r - T < t < r + T$, and for $t > r + T$ it decreases in magnitude with the large-t behavior (14.29). As the field (14.35) is bounded for all t, so is the radiated power (14.30) is finite for all t, attaining a maximum at $t = r + T$

$$\left.\frac{d^2\varepsilon}{dz\,dt}\right|_{\text{max}} = \frac{(I_b - I_a)^2}{2\pi T}.$$

However, as implied by the general formulas (14.29) and (14.30), in the limit for $t \to \infty$ the radiated power has the same universal infrared behavior as the power (14.31) radiated by the step-function current, namely

$$\frac{d^2\varepsilon}{dzdt} \approx \frac{(I_b - I_a)^2}{4\pi t}. \tag{14.36}$$

There is, thus, a fundamental difference between the radiation of a charged particle and that of an infinite wire: whereas the total energy (11.57) emitted by an accelerated particle undergoing a finite velocity change $\Delta\mathbf{v}$ is finite as long as the acceleration time $2T$ is finite, the total energy emitted by a wire per unit length is always infinite, even for a non-vanishing transition time $2T$, unless the overall current change $I_b - I_a$ is zero.

14.4 Spectral Analysis

Assuming, momentarily, that the total energy emitted per unit length by the wire is finite – choosing hence $I_a = I_b$ – we can integrate the radiation equation (14.30) over all times. Introducing the temporal Fourier transform $\widehat{A}(\omega)$ of the wave-zone potential $A(t, \mathbf{x})$, where in $\widehat{A}(\omega)$ we understand as usual the dependence on $\mathbf{x}$, i.e. on r, we then obtain

$$\frac{d\varepsilon}{dz} = 2\pi r \int_{-\infty}^{\infty} \dot{A}^2 \, dt = 4\pi r \int_{0}^{\infty} \omega^2 |\widehat{A}(\omega)|^2 d\omega.$$

The continuous spectral weights of the radiation are thus given by

$$\frac{d^2\varepsilon}{dzd\omega} = 4\pi r \, \omega^2 |\widehat{A}(\omega)|^2. \tag{14.37}$$

The Fourier transform $\widehat{A}(\omega)$ of the wave-zone potential (14.21) can be reduced to the Fourier transform $\widehat{I}(\omega)$ of the current $I(t)$. In particular, since in the spectral weights (14.37) the factor $|\widehat{A}(\omega)|^2$ is multiplied by ω^2, it is sufficient to compute the product (for $\omega \geq 0$)

$$\begin{aligned}
\omega\widehat{A}(\omega) &= \frac{\omega}{(2\pi)^{3/2}\sqrt{2r}} \int_{-\infty}^{\infty} e^{-i\omega t} \int_{0}^{\infty} \frac{I(t-\beta) - I_a}{\sqrt{\beta}} \, d\beta \, dt \\
&= \frac{1}{2\pi\sqrt{2r}} \int_{0}^{\infty} \frac{e^{-i\omega\beta}}{\sqrt{\beta}} \left(\omega\widehat{I}(\omega) - \sqrt{2\pi}\, I_a\omega\, \delta(\omega) \right) d\beta \\
&= \frac{\sqrt{\omega}\, \widehat{I}(\omega)}{\pi\sqrt{2r}} \int_{0}^{\infty} e^{-iu^2} du = \frac{e^{-i\pi/4}}{2\sqrt{2\pi r}} \sqrt{\omega}\, \widehat{I}(\omega),
\end{aligned} \tag{14.38}$$

where we have used the Fresnel integral

$$\int_0^\infty e^{ix^2}\,dx = \frac{\sqrt{\pi}}{2}\,e^{i\pi/4}. \tag{14.39}$$

The spectral weights (14.37) take thus the simple form

$$\frac{d^2\varepsilon}{dz\,d\omega} = \frac{1}{2}\left|\sqrt{\omega}\,\widehat{I}(\omega)\right|^2. \tag{14.40}$$

The same result can be obtained by expressing the Fourier transform of the exact potential $A(t,\mathbf{x})$ (14.8) in terms of the special function $L(x)$ (13.52), for the derivation see the calculation (14.81) of Problem 14.1,

$$\omega\widehat{A}(\omega) = \frac{\omega}{2\pi}\,\widehat{I}(\omega)L(r\omega), \tag{14.41}$$

and by using then the expansion (13.56) of $L(x)$ to evaluate in Eq. (14.37) the limit of the quantity $|\sqrt{r}\,\omega\widehat{A}(\omega)|$ for $r \to \infty$. Notice, however, that the simple formula (14.41) gives the Fourier transform $\omega\widehat{A}(\omega)$ of $-i$ times the time derivative $\dot{A}(t,\mathbf{x})$ of the potential. The determination of the Fourier transform $\widehat{A}(\omega)$ of $A(t,\mathbf{x})$ is much more delicate and, in particular, it cannot be deduced from Eq. (14.41) by simply dividing both sides by ω, see Problem 14.1.

Small-frequency analysis and step-function current. As first example we perform the spectral analysis of the step-function current (14.3). To evaluate its Fourier transform we need the Fourier transform $\widehat{H}(\omega)$ of the Heaviside function $H(t)$

$$\widehat{H}(\omega) = -\frac{i}{\sqrt{2\pi}}\,\mathcal{P}\frac{1}{\omega} + \sqrt{\frac{\pi}{2}}\,\delta(\omega), \tag{14.42}$$

where $\mathcal{P}$ indicates the principal-value prescription (2.85). Formula (14.42) can be derived by writing the Heaviside function in the form $H(t) = (\varepsilon(t)+1)/2$, and by using then the Fourier transforms of the constant (2.93) and of the sign function (2.95). The Fourier transform of the step-function current (14.3) is then

$$\widehat{I}_{\rm sf}(\omega) = -\frac{i(I_b - I_a)}{\sqrt{2\pi}}\,\mathcal{P}\frac{1}{\omega} + \sqrt{\frac{\pi}{2}}\,(I_a + I_b)\,\delta(\omega). \tag{14.43}$$

As the δ-function does not contribute to the product $\sqrt{\omega}\,\widehat{I}_{\rm sf}(\omega)$ the spectral weights (14.40) become simply

$$\frac{d^2\varepsilon}{dz\,d\omega} = \frac{(I_b - I_a)^2}{4\pi\omega}. \tag{14.44}$$

As happened in the formula (14.32) for the emitted energy in the time domain, by integrating equation (14.44) over all frequencies we find that the total energy

$d\varepsilon/dz$ radiated per unit length is infinite: again, we have an ultraviolet divergence for $\omega \to \infty$, and an infrared one for $\omega \to 0$. In fact, according to a standard argument, a singularity for $t \to \infty$ ($t \to 0$) in the time domain is reflected as a singularity for $\omega \to 0$ ($\omega \to \infty$) in the frequency domain.

The $1/\omega$-behavior for small frequencies of the spectral weights (14.44) is, actually, universal, in that it holds for a generic current $I(t)$ subject to the boundary conditions (14.2), as one might have expected from the corresponding universal large-t behavior of the emitted energy in the time domain, see Eq. (14.36). In fact, according to the decomposition (14.4), the Fourier transform of $I(t)$ can be written as

$$\widehat{I}(\omega) = \widehat{I}_{\mathrm{sf}}(\omega) + \widehat{\Delta I}(\omega),$$

where, by virtue of the *Riemann-Lebesgue lemma*, thanks to the fast-decay conditions (14.5) the function $\widehat{\Delta I}(\omega)$ is continuous for all $\omega \in \mathbb{R}$. Formulas (14.40) and (14.43) imply thus the general small-frequency behavior of the spectral weights

$$\frac{d^2\varepsilon}{dzd\omega} \approx \frac{(I_b - I_a)^2}{4\pi\omega}.$$

Continuous current. As next example we consider the continuous current $I_{\mathrm{c}}(t)$ (14.33), which increases continuously from I_a to I_b during the time interval $2T$. In the limit for $T \to 0$ this current reduces to the step-function current (14.3). To perform the spectral analysis of the emitted radiation we write the current in the form

$$I_{\mathrm{c}}(t) = I_a H\left(-t - T\right) + \left(\frac{I_a + I_b}{2} + \frac{I_b - I_a}{2T}\,t\right) \chi_{[-T,T]}(t) + I_b H\left(t - T\right),$$

$$(14.45)$$

where χ is the characteristic function (2.75). Knowing the Fourier transform of the Heaviside function (14.42), and using the rules (2.96) and (2.97), it is eyas to calculate the Fourier transforms of the first and third terms of $I_{\mathrm{c}}(t)$. As the Fourier transform of the second term amounts to a simple integral, the Fourier transform of the current (14.45) thus results in

$$\widehat{I}_{\mathrm{c}}(\omega) = -\frac{i(I_b - I_a)}{\sqrt{2\pi}\,T} \frac{\sin(\omega T)}{\omega^2} + \sqrt{\frac{\pi}{2}}\,(I_a + I_b)\,\delta(\omega). \qquad (14.46)$$

Notice that in the distributional limit for $T \to 0$ this expression reduces to the Fourier transform of the step-function current (14.43). The spectral weights (14.40) of the current (14.45) take thus the form

$$\frac{d^2\varepsilon}{dzd\omega} = \frac{(I_b - I_a)^2 \sin^2(\omega T)}{4\pi T^2 \omega^3}. \qquad (14.47)$$

As was to be expected – in contrast to the step-function current (14.3) the current (14.45) is continuous – for large frequencies the spectral weights (14.47) decrease faster than those of the step-function current (14.44), namely as $1/\omega^3$. Conversely, since for large times the current (14.45) entails the same total variation $I_b - I_a$ as the step-function current, the singular behavior for $\omega \to 0$ of the spectral weights (14.47) and (14.44) is exactly the same. In fact, as seen in Sect. 14.3.1 on general grounds, the total radiated energy $d\varepsilon/dz$ is always infrared divergent. To obtain a finite total radiated energy we must consider an *accelerated* current which, nonetheless, attains the same initial and finite values, $I_b = I_a$. The following example illustrates this possibility.

Impulsive current. As last example we consider the impulsive current

$$I(t) = I_a + I_M \, e^{-t^2/2T^2}, \tag{14.48}$$

where T is the characteristic time of the impulse, and I_M is its amplitude. This current varies appreciably around the background value I_a during a time of order T, and for $t \to \pm\infty$ it reaches the same values $I_a = I_b$. For the spectral analysis it is sufficient to recall the Fourier transform of a Gaussian function, leading to the Fourier transform of the current (14.48)

$$\widehat{I}(\omega) = \sqrt{2\pi}\, I_a \delta(\omega) + T I_M \, e^{-\omega^2 T^2/2}.$$

The spectral weights (14.40) take thus the form

$$\frac{d^2\varepsilon}{dz\,d\omega} = \frac{T^2 I_M^2 \, \omega}{2}\, e^{-\omega^2 T^2}, \tag{14.49}$$

implying that the wire emits radiation with the dominant frequencies $\omega \lesssim 1/T$. The spectral weights (14.49) are regular for all frequencies and, in particular, the total energy radiated per unit length is now finite:

$$\frac{d\varepsilon}{dz} = \frac{T^2 I_M^2}{2} \int_0^\infty \omega e^{-\omega^2 T^2}\, d\omega = \frac{I_M^2}{4}. \tag{14.50}$$

According to the decomposition (14.4), for an impulsive current of the form (14.48) only the current-fluctuation

$$\Delta I(t) = I_M \, e^{-t^2/2T^2}$$

generates radiation fields. In fact, since $I_b = I_a$, the leading term of the wave-zone potential $A(t, \mathbf{x})$ in (14.27), i.e. the step-function-current potential $A_{\mathrm{sf}}(t, \mathbf{x})$ (14.24), is identically zero. Therefore, in this case the potential $A(t, \mathbf{x})$ equals the fluctuation $\Delta A(t, \mathbf{x})$ (14.28), which decreases as $1/\sqrt{t}$. Correspondingly, the radiation fields $\mathbf{E}$

and **B** of formulas (14.22) decrease as $1/t^{3/2}$. This ensures that the integral giving according to Eq. (14.30) the total energy radiated per unit length

$$\frac{d\varepsilon}{dz} = 2\pi r \int_{-\infty}^{\infty} E^2 \, dt$$

is *convergent*. In fact, it equals the value (14.50). For a further example, see Problem 14.2.

14.5 Periodic Currents

In this section we consider a generic periodic current $I(t)$ of period T, entailing hence the fundamental frequency $\omega_0 = 2\pi/T$. For the moment we make the mild regularity assumption that $I(t)$ is an element of the space $L^2[0, T]$, see the definition (2.101), meaning that

$$\int_0^T |I(t)|^2 \, dt < \infty.$$

It then admits the Fourier expansion[3]

$$I(t) = \sum_{N=-\infty}^{\infty} e^{iN\omega_0 t} I_N, \quad I_N = \frac{1}{T} \int_0^T e^{-iN\omega_0 t} I(t) \, dt, \quad I_{-N} = I_N^*, \quad (14.51)$$

which, according to a historical result of *functional analysis* obtained by L. Carleson in 1966 [4], converges almost everywhere to $I(t)$. Being an element of $L^2[0, T]$ the current $I(t)$ is modulus integrable. In particular, its *fluctuating* part $I(t) - I_0$, where I_0 is the constant term in the expansion (14.51), admits then the primitive function

$$M(t) = \int_0^t \left(I(s) - I_0 \right) ds, \quad (14.52)$$

which is periodic with period T and *absolutely continuous*. This last property ensures the validity of the *fundamental theorem of calculus*, which we will apply below. The function $M(t)$ is much more regular than $I(t)$, in particular it is almost everywhere differentiable and obeys the relation

$$\dot{M}(t) = I(t) - I_0. \quad (14.53)$$

---

[3]For basic properties of the Fourier series, and for other notions of *functional analysis* used in this section, as for instance the concept of *absolute continuity*, we refer the reader to the Refs. [1–3].

14.5.1 Regularity Properties of Potentials and Fields

The retarded potential generated via Maxwell's equation by a periodic current is formally still given by Eq. (14.8), if we replace the background current I_a with the mean current I_0,

$$A(t,\mathbf{x}) = \frac{1}{2\pi} \int_1^\infty \frac{I(t-ru) - I_0}{\sqrt{u^2-1}}\, du - \frac{I_0 \ln r}{2\pi}. \tag{14.54}$$

However, is not obvious that this integral is finite, since for $u \to \infty$ the numerator $I(t-ru) - I_0$ does not admit a limit. Therefore, before proceeding any further, let us show that the integral representation (14.54) is well defined. Separating the contributions for small and large u we rewrite it in the form

$$2\pi A(t,\mathbf{x}) + I_0 \ln r = \int_1^2 \frac{I(t-ru) - I_0}{\sqrt{u^2-1}}\, du + \int_2^\infty \frac{I(t-ru) - I_0}{\sqrt{u^2-1}}\, du. \tag{14.55}$$

The first integral converges trivially. In the second integral we use the relation (14.53) and perform an integration by parts, which is allowed since $M(t)$ is absolutely continuous,

$$\int_2^\infty \frac{I(t-ru) - I_0}{\sqrt{u^2-1}}\, du = -\frac{1}{r} \int_2^\infty \frac{1}{\sqrt{u^2-1}} \frac{dM(t-ru)}{du}\, du$$

$$= -\frac{1}{r} \int_2^\infty \frac{uM(t-ur)}{(u^2-1)^{3/2}}\, du + \frac{M(t-2r)}{\sqrt{3}\, r}. \tag{14.56}$$

The integral in (14.56) is now convergent, because for large u the integrand behaves as $M(t-ur)/u^2$, where $M(t-ur)$ is bounded since $M(t)$ is a periodic continuous function. Moreover, since $M(t)$ and $I(t)$ both are functions belonging to $L^1[0,T]$, formulas (14.55) and (14.56) imply that the potential $A(t,\mathbf{x})$ is a *continuous* function of time. Furthermore, from either of the above expressions it follows that also $A(t,\mathbf{x})$ is periodic with period T.

Concerning the electromagnetic field, the defining relations (14.10) – involving distributional derivatives – guarantee that $\mathbf{E}(t,\mathbf{x})$ and $\mathbf{B}(t,\mathbf{x})$ are well-defined distributions, periodic in time with period T. However, it is less obvious which are the regularity properties of these fields as functions of t and r. As we will see below, and in Sect. 14.5.2 in the frequency domain, the presence, or absence, of singularities in the fields depends crucially on further regularity properties of the current $I(t)$.

Potentials and fields in the wave zone. In the wave zone, the potential (14.54) reduces to the potential (14.21) with I_a replaced by I_0

$$A(t,\mathbf{x}) = \frac{1}{2\pi\sqrt{2r}} \int_0^\infty \frac{I(t-r-\beta) - I_0}{\sqrt{\beta}}\, d\beta, \tag{14.57}$$

where we omitted the static term. The electromagnetic wave-zone fields are still given by formulas (14.22) and, in particular, the electric field admits the integral representation (14.23), that for convenience we rewrite here setting $c = 1$

$$\mathbf{E}(t, \mathbf{x}) = -\frac{\mathbf{u}_z}{2\pi\sqrt{2r}} \int_0^\infty \frac{\dot{I}(t - r - \beta)}{\sqrt{\beta}} \, d\beta. \tag{14.58}$$

As the fields $\mathbf{E}$ and $\mathbf{B}$ decay as $1/\sqrt{r}$ they represent, of course, radiation fields. However, in contrast to the fields generated by aperiodic currents, the fields (14.22) are now periodic in time and so they never entail *infrared* divergences, as the latter are associated with singular behaviors of the fields for $t \to \infty$. Similarly, in the frequency domain the absence of infrared divergences is ensured by the *discreteness* of the emission spectrum, see Sect. 14.5.2. In fact, as the lowest emitted frequency $\omega_0 = 2\pi/T$ is different from zero, no divergence for $\omega \to 0$ can ever arise in any physical quantity. On the other hand, the electromagnetic fields generated by periodic and aperiodic currents are subject to the same type of *ultraviolet* divergences, as the latter are caused by too violent, for instance, discontinuous, variations of the current $I(t)$ for finite values of t.

Wave-zone fields of the square-wave current. A prototypical example of a periodic current is the alternating *square-wave* current defined by

$$I(t) = \frac{I_m}{2} \left(H(t) - H(-t) \right), \quad \text{for } t \in [-T/2, T/2], \tag{14.59}$$

and continued periodically for all $t \in \mathbb{R}$. This function belongs to $L^2[0, T]$, but it entails finite discontinuities at the discrete times $t = nT/2$, with $n \in \mathbb{Z}$, where it jumps by I_m. Notice that in the case at hand the mean current I_0 vanishes, see formulas (14.51). In a sense, the square-wave current (14.59) is the periodic analog of the aperiodic step-function current (14.3). To evaluate the potential (14.57) for the current (14.59) we first rearrange the integral over β according to

$$\int_0^\infty (\cdots) \, d\beta \to \sum_{n=0}^\infty \int_{nT}^{nT+T} (\cdots) \, d\beta,$$

and then, exploiting the periodicity of the current, in the nth term of the series we perform the change of variable $\beta \to \beta - nT$. Moreover, since the potential (14.57) is invariant under the replacement $t - r \to t - r + T$, it is not restrictive to confine the variable $t - r$ to the interval $[-T/2, T/2]$. In this way a straightforward computation yields the wave-zone potential, for $t - r \in [-T/2, T/2]$,

$$A(t, \mathbf{x}) = \frac{I_{\mathrm{m}}}{2\pi\sqrt{2r}} \sum_{n=0}^{\infty} \Big\{ H(t-r)\Big[2\sqrt{t-r+nT} - 2\sqrt{t-r+nT+T/2}$$

$$+ \sqrt{nT+T} - \sqrt{nT}\,\Big]$$

$$+ H(r-t)\Big[2\sqrt{t-r+nT+T} - 2\sqrt{t-r+nT+T/2}$$

$$- \sqrt{nT+T} + \sqrt{nT}\,\Big] \Big\}.$$

$$(14.60)$$

Since the single terms of the two series multiplying the Heaviside functions for large n decrease as $1/n^{3/2}$, the series (14.60) converges for all $t-r \in [-T/2, T/2]$. Moreover, as anticipated above on general grounds, the potential (14.60) is a continuous function of $t-r$. In particular, its value at $t-r = -T/2$ coincides with its value at $t-r = T/2$. On the other hand, this potential is only piecewise differentiable with respect to time. In particular, in the interval $[-T/2, T/2]$ it is non-differentiable from the right at the points $t-r = -T/2$ and $t-r = 0$. Actually, the resulting distributional derivative is the electric field in the wave zone

$$E^3(t, \mathbf{x}) = -\dot{A}(t, \mathbf{x})$$

$$= -\frac{I_{\mathrm{m}}}{2\pi\sqrt{2r}} \Bigg(\frac{H(t-r)}{\sqrt{t-r}} + \sum_{n=1}^{\infty} \bigg(\frac{1}{\sqrt{t-r+nT}} - \frac{1}{\sqrt{t-r+nT-T/2}} \bigg) \Bigg).$$

$$(14.61)$$

Notice that, as $A(t, \mathbf{x})$ is continuous, the field $E^3(t, \mathbf{x})$ contains no δ-functions. The expression (14.61) could also be derived, more quickly, by inserting the derivative of the square-wave current (14.59) as a function for all $t \in \mathbb{R}$

$$\dot{I}(t) = I_{\mathrm{m}} \sum_{n=-\infty}^{\infty} \Big(\delta\big(t - nT\big) - \delta\big(t - nT - T/2\big) \Big) \qquad (14.62)$$

in the equation (14.58) for the electric field in the wave zone, see Problem 14.3.

One may notice that the series (14.60) contains time-independent terms which, consequently, do not contribute to the electric field (14.61). Nonetheless, in the potential (14.60) these terms are needed to make the series converge. Similarly, the series (14.61) in the interval $[-T/2, T/2]$ converges for $t-r \neq -T/2$, since the single terms decrease as $1/n^{3/2}$. Actually, considered as a function of the whole real axis, the electric field (14.61) has inverse-square-root divergences in the right vicinities of the points $t-r = nT/2$. At a fixed position r this field has hence a kind of *sawtooth profile* with *divergent* peaks of alternating sign, occurring at the times $t = r + nT/2$. Consequently, according to the radiation equation (14.30), the mean power emitted by the wire per unit length, i.e. the energy $d\varepsilon/dz$ emitted during a period per unit length divided the period, is *infinite*,

$$\frac{d\overline{W}}{dz} = \frac{1}{T}\frac{d\varepsilon}{dz} = \frac{2\pi r}{T} \int_0^T E^2(t, \mathbf{x})\, dt = \infty. \qquad (14.63)$$

The physical origin of this ultraviolet divergence is again the infinite acceleration of the charged particles of the wire, carrying the current (14.59), at the discontinuity instants $t = nT/2$. As we will see in the next sections, more regular (continuous) currents generate electromagnetic fields which are bounded over the whole period, leading thus to a finite mean power.

14.5.2 Discrete Spectral Weights

As seen above, as long as $I(t) \in L^2[0, T]$ the potential (14.54) is a continuous periodic function of time. Therefore, it admits the Fourier decomposition

$$A(t, \mathbf{x}) = \sum_{N=-\infty}^{\infty} e^{iN\omega_0 t} A_N, \quad A_N = \frac{1}{T} \int_0^T e^{-iN\omega_0 t} A(t, \mathbf{x})\, dt, \quad A_{-N} = A_N^*,$$

$$(14.64)$$

which converges almost everywhere to $A(t, \mathbf{x})$, see, for instance, Ref. [3]. To perform the spectral analysis of the radiation we compute from Eq. (14.30) the mean power emitted per unit length, supposing for the moment that it is finite,

$$\frac{d\overline{\mathcal{W}}}{dz} = \frac{1}{T} \int_0^T \frac{d^2\varepsilon}{dz\, dt}\, dt = \frac{2\pi r}{T} \int_0^T \dot{A}^2(t, \mathbf{x})\, dt = 4\pi r \sum_{N=1}^{\infty} (N\omega_0)^2 |A_N|^2.$$

$$(14.65)$$

In the last step we applied Parseval's identity to the function $\dot{A}(t, \mathbf{x})$, whose Fourier coefficients are equal to $iN\omega_0 A_N$. The mean power of the radiation with frequency $N\omega_0$ emitted per unit length of the wire is thus given by (the large-r limit of)

$$\frac{d\mathcal{W}_N}{dz} = 4\pi r (N\omega_0)^2 |A_N|^2.$$

$$(14.66)$$

To relate the Fourier coefficients A_N to those of the current I_N (14.51) we insert the exact potential $A(t, \mathbf{x})$ (14.54) in the definition (14.64)

$$
\begin{aligned}
A_N &= \frac{1}{2\pi T} \int_0^T e^{-iN\omega_0 t} \int_1^\infty \frac{I(t - ru) - I_0}{\sqrt{u^2 - 1}}\, du\, dt - \frac{I_0 \ln r}{2\pi} \delta_N^0 \\
&= \frac{1}{2\pi T} \int_0^T e^{-iN\omega_0 t} \left(I(t) - I_0 \right) dt \int_1^\infty \frac{e^{-iN\omega_0 ru}}{\sqrt{u^2 - 1}}\, du - \frac{I_0 \ln r}{2\pi} \delta_N^0 \\
&= \frac{1}{2\pi} \left(I_N - \delta_N^0 I_0 \right) L(N\omega_0 r) - \frac{I_0 \ln r}{2\pi} \delta_N^0,
\end{aligned}
$$

$$(14.67)$$

where we met again the special function $L(x)$ of Eq. (13.52). To evaluate the discrete spectral weights (14.66) we only need the form of the coefficients A_N in the limit for $r \to \infty$, for $N \geq 1$, which can be derived from the asymptotic expansion in (13.56)

$$A_N \to \frac{e^{-i\pi/4}\, e^{-iN\omega_0 r}}{\sqrt{8\pi r N \omega_0}}\, I_N. \qquad (14.68)$$

As the large-r behavior of the quantity $L(N\omega_0 r)$ coincides with its large-N behavior, from the asymptotic relation (14.68) we derive that the Fourier series of the potential entails better convergence properties than that of the current, in that $A_N \sim I_N/\sqrt{N}$. Inserting the Fourier coefficients (14.68) in formula (14.66) we find the spectral weights ($\omega_0 = 2\pi/T$)

$$\frac{dW_N}{dz} = \frac{\pi N}{T}\, |I_N|^2, \qquad (14.69)$$

which are the discrete counterparts of the continuous spectral weights (14.40). There is, hence, again a simple relation between the current and the energy emitted at a given frequency. For instance, the square-wave current (14.59) entails the Fourier coefficients

$$I_N = \begin{cases} 0, & \text{for } N \text{ even,} \\ -\dfrac{iI_m}{\pi N}, & \text{for } N \text{ odd,} \end{cases} \qquad (14.70)$$

leading thus to the spectral weights (for N odd)

$$\frac{dW_N}{dz} = \frac{I_m^2}{\pi N T}. \qquad (14.71)$$

14.5.3 Total Radiated Power

Summing up the spectral weights (14.69) we obtain for the total mean power emitted per unit length of the wire the series

$$\frac{dW}{dz} = \frac{\pi}{T} \sum_{N=1}^{\infty} N |I_N|^2. \qquad (14.72)$$

The finiteness of this physical quantity depends, thus, crucially on the fall-off properties of the Fourier coefficients I_N of the current, which, in turn, control the convergence properties of its Fourier series (14.51). If we only require, as we did until now, that $I(t) \in L^2[0, T]$, then Parseval's identity

$$\frac{1}{T} \int_0^T I^2(t)\, dt = \sum_{N=-\infty}^{\infty} |I_N|^2 < \infty \qquad (14.73)$$

is not enough to ensure that the total mean power (14.72) is finite. For instance, the Fourier coefficients of the square-wave current (14.70) decrease to slowly – just as

$1/N$ – to yield a convergent total mean power, as we have already established in the time domain in Eq. (14.63).

To obtain a finite total mean power we need a more regular current profile than the discontinuous square-wave current (14.59). However, in general a *generic* continuous current may still produce a divergent total mean power. For instance, even if a current $I(t)$ is *absolutely continuous*, it is only guaranteed that its Fourier coefficients are bounded by $|I_N| < a/|N|$, for some constant a. If, on the other hand, $I(t)$ is absolutely continuous and piecewise of class C^2, then it can be shown that its Fourier coefficients decay faster, namely $|I_N| < b/N^2$ for some constant b, so that the series of the total mean power (14.72) converges.[4] A simple example is the *triangle-wave* current

$$I(t) = C|t|, \quad \text{for } t \in [-T/2, T/2], \tag{14.74}$$

continued periodically for all $t \in \mathbb{R}$. The derivative of this current

$$\dot{I}(t) = C\big(H(t) - H(-t)\big), \quad \text{for } t \in [-T/2, T/2], \tag{14.75}$$

is, actually, a piecewise constant function. Correspondingly, its Fourier coefficients

$$I_N = \begin{cases} \dfrac{CT}{4}, & \text{for } N = 0, \\ 0, & \text{for } N \text{ even, } N \neq 0, \\ -\dfrac{CT}{\pi^2 N^2}, & \text{for } N \text{ odd,} \end{cases} \tag{14.76}$$

decrease as $1/N^2$. The triangle-wave current (14.74) emits hence the finite total mean power (14.72)

$$\frac{d\overline{W}}{dz} = \frac{TC^2}{\pi^3} \sum_{N=1}^{\infty} \frac{1}{(2N+1)^3}. \tag{14.77}$$

Analyzing the problem in the time domain, it is easily seen that the electric field $\mathbf{E}(t, \mathbf{x})$ generated by the triangle-wave current (14.74) is a continuous function for all t, and so it is obvious that the total mean power (14.72) – which can also be expressed as in Eq. (14.63) – is finite. To see it explicitly we insert in the general formula for the electric wave-zone field (14.58) the time derivative of the triangle-wave current (14.75). By inspection, the resulting electric field equals formally the potential generated by the square-wave current (14.59) via Eq. (14.57), if one sets

$$I_m = -2C.$$

---

[4]Actually, for the bound $|I_N| < b/N^2$ to hold, it is sufficient that the current $I(t)$ is absolutely continuous and admits a piecewise absolutely continuous first derivative $\dot{I}(t)$.

Consequently, as the potential (14.60) generated by the square-wave current is continuous, so is the electric field generated by the triangle-wave current.

More generally, if we consider a current $I(t)$ of class C^k, its kth derivative $I^k(t)$ is continuous and is thus an element of $L^2[0, T]$. Applying Parseval's identity (14.73) to the function $I^k(t)$ we then see that the series

$$\sum_{N=-\infty}^{\infty} N^{2k} |I_N|^2$$

is convergent. This implies that the Fourier coefficients of the current are subject to the bound $|I_N| < a/|N|^k$ for some constant a. Correspondingly, the spectral weights (14.69) are bounded by

$$\frac{dW_N}{dz} < \frac{\pi a^2}{T N^{2k-1}}.$$

Therefore, for more and more regular currents, which vary hence slower in time, the radiation with high harmonic frequencies $N\omega_0$ is more and more suppressed.

14.6   Problems

14.1 *Fourier transform of the step-function-current potential*. Determine the Fourier transform $\widehat{A}(\omega)$ of the potential (14.11)

$$A(t, \mathbf{x}) = \frac{(I_b - I_a)H(t - r)}{2\pi} \ln\left(\frac{t}{r} + \sqrt{\frac{t^2}{r^2} - 1}\right) - \frac{I_a \ln r}{2\pi} \tag{14.78}$$

generated by the step-function current $I_{\mathrm{sf}}(t)$ (14.3).

Solution: As any potential created by a current satisfying the boundary conditions (14.2), for large t the potential (14.78) diverges as $\ln t$. Consequently, its Fourier transform cannot be computed by means of an *integral*. In the following we present a strategy to bypass this problem for the potential generated by a generic current $I(t)$ subject to the conditions (14.2). The first step consists in the introduction of a regularized current $I_\lambda(t)$, with λ a positive parameter, entailing the limits

$$\lim_{\lambda \to 0} I_\lambda(t) = I(t), \qquad \lim_{t \to \pm\infty} I_\lambda(t) = I_a. \tag{14.79}$$

Since the difference

$$\mathcal{I}_\lambda(t) = I_\lambda(t) - I_a \tag{14.80}$$

has the same asymptotic behaviors (14.5) of the fluctuation $\Delta I(t)$, the regularized current $I_\lambda(t)$ creates via Eq. (14.8) a regularized potential $A_\lambda(t, \mathbf{x})$ which for large times decreases as $1/t$, see formulas (14.17) and (14.19). Although this behavior does not imply that the potential $A_\lambda(t, \mathbf{x})$ is integrable in modulus, it allows us, nevertheless, to compute its Fourier transform $\widehat{A}_\lambda(\omega)$ by means of an integral. This means that we can determine the latter from the integral representation (14.8) via the usual integral

$$
\begin{aligned}
\widehat{A}_\lambda(\omega) &= \frac{1}{(2\pi)^{3/2}} \int e^{-i\omega t} \int_1^\infty \frac{I_\lambda(t - ru) - I_a}{\sqrt{u^2 - 1}} \, du \, dt - \frac{I_a \ln r}{\sqrt{2\pi}} \delta(\omega) \\
&= \frac{1}{(2\pi)^{3/2}} \int e^{-i\omega t} \left(I_\lambda(t) - I_a \right) dt \int_1^\infty \frac{e^{-ir\omega u}}{\sqrt{u^2 - 1}} \, du - \frac{I_a \ln r}{\sqrt{2\pi}} \delta(\omega) \\
&= \frac{1}{2\pi} \widehat{\mathcal{I}}_\lambda(\omega) L(r\omega) - \frac{I_a \ln r}{\sqrt{2\pi}} \delta(\omega),
\end{aligned}
\tag{14.81}
$$

where $L(x)$ is the special function (13.52), and we have introduced the Fourier transform

$$
\widehat{\mathcal{I}}_\lambda(\omega) = \widehat{I}_\lambda(\omega) - \sqrt{2\pi} \, I_a \, \delta(\omega)
\tag{14.82}
$$

of the regularized current-fluctuation $\mathcal{I}_\lambda(t)$ (14.80). However, the crucial point is that if the distributional limit[5]

$$
\mathcal{S}' - \lim_{\lambda \to 0} A_\lambda(t, \mathbf{x}) = A(t, \mathbf{x})
\tag{14.83}
$$

holds, then – as the Fourier transform is a continuous operation in $\mathcal{S}'$ – the sought Fourier transform $\widehat{A}(\omega)$ can be computed via the distributional limit

$$
\mathcal{S}' - \lim_{\lambda \to 0} \widehat{A}_\lambda(\omega) = \widehat{A}(\omega),
\tag{14.84}
$$

where $\widehat{A}_\lambda(\omega)$ is the Fourier transform of $A_\lambda(t, \mathbf{x})$.

Let us now return to the step-function current $I(t) = I_a + (I_b - I_a) H(t)$ of Eq. (14.3). In this case, a convenient regularization satisfying the limits (14.79) is given by

$$
I_\lambda(t) = I_a + (I_b - I_a) H(t) \, e^{-\lambda t},
\tag{14.85}
$$

and the associated potential $A_\lambda(t, \mathbf{x})$ clearly obeys the distributional limit (14.83). The regularized current-fluctuation (14.80) then takes the simple form

$$
\mathcal{I}_\lambda(t) = (I_b - I_a) H(t) \, e^{-\lambda t},
\tag{14.86}
$$

---

[5]Since the spatial coordinate $\mathbf{x}$ is a mere spectator, it is sufficient to consider the one-dimensional distribution space $\mathcal{S}' = \mathcal{S}'(\mathbb{R})$ relative to the time coordinate.

with Fourier transform

$$\widehat{\mathcal{I}}_\lambda(\omega) = -\frac{i(I_b - I_a)}{\sqrt{2\pi}\,(\omega - i\lambda)}.\tag{14.87}$$

We are thus left with the determination of the distributional limit (14.84), where the potential $\widehat{A}_\lambda(\omega)$ is given by formula (14.81) in which one has to substitute the Fourier transform $\widehat{\mathcal{I}}_\lambda(\omega)$ (14.87). The distributional limit for $\lambda \to 0$ of the expression (14.87) equals the Fourier transform of the distributional limit for $\lambda \to 0$ of the regularized current-fluctuation (14.86), i.e. the unregularized current-fluctuation $\mathcal{I}(t) = I(t) - I_a = (I_b - I_a)H(t)$. Knowing the Fourier transform of the Heaviside function (14.42), we so obtain the well-defined distribution

$$\mathcal{S}' - \lim_{\lambda \to 0} \widehat{\mathcal{I}}_\lambda(\omega) = \widehat{\mathcal{I}}(\omega) = (I_b - I_a)\left(-\frac{i}{\sqrt{2\pi}}\,\mathcal{P}\frac{1}{\omega} + \sqrt{\frac{\pi}{2}}\,\delta(\omega)\right),\tag{14.88}$$

involving a δ-function supported in $\omega = 0$. However, as the special function $L(r\omega)$ has logarithmic and sign-function singularities at $\omega = 0$, see the expansions (13.56), this raises the problem that the distributional limit of the product appearing in (14.81) in general cannot be computed as a *product of limits*. Notice that the problem is not due to the ultraviolet singularity of the step-function current (14.3) at $t = 0$, but rather to its infrared behavior (14.2) for $t \to \pm\infty$. For instance, in the case of the continuous current $I_c(t)$ (14.45), the Fourier transform of the unregularized current-fluctuation has the expression (see Eq. (14.46))

$$\widehat{\mathcal{I}}_c(\omega) = \widehat{I}_c(\omega) - \sqrt{2\pi}\,I_a\,\delta(\omega) = (I_b - I_a)\left(-\frac{i}{\sqrt{2\pi}}\,\frac{\sin(\omega T)}{\omega^2 T} + \sqrt{\frac{\pi}{2}}\,\delta(\omega)\right),$$

which contains the same *universal* δ-function as the Fourier transform (14.88). Nevertheless, since the existence of the limit (14.84) is guaranteed by the existence of the limit (14.83), there remains only the technical problem to determine the former explicitly.

Distributional limit. To determine the limit (14.84) we must separate in the product of distributions appearing in (14.81) the terms which become singular at $\omega = 0$. In light of the small-argument expansion (13.56) of the function $L(x)$ we operate the splitting

$$\widehat{\mathcal{I}}_\lambda(\omega)L(r\omega) = \widehat{\mathcal{I}}_\lambda(\omega)\left(L(r\omega) + \ln|\omega| + \frac{i\pi}{2}\,\varepsilon(\omega)\right) - \widehat{\mathcal{I}}_\lambda(\omega)\left(\ln|\omega| + \frac{i\pi}{2}\,\varepsilon(\omega)\right),\tag{14.89}$$

where $\varepsilon(\omega)$ is the sign function. In fact, since the first bracket is regular for $\omega \to 0$, in the first term we now can use the limit of the current-fluctuation (14.88). The singularities have indeed been confined to the second term, see (14.87),

$$\widehat{\mathcal{I}}_\lambda(\omega) \left(\ln|\omega| + \frac{i\pi}{2}\,\varepsilon(\omega) \right) =$$

$$\frac{I_b - I_a}{\sqrt{2\pi}\,(\omega^2 + \lambda^2)} \left(\lambda \ln|\omega| + \frac{\pi}{2}\,|\omega| + i\left(\frac{\pi}{2}\,\lambda\,\varepsilon(\omega) - \omega \ln|\omega| \right) \right). \tag{14.90}$$

Under the distributional limit for $\lambda \to 0$ the third term goes to zero, and the fourth term admits a trivial limit, proportional to $\mathcal{P}(1/\omega) \ln\omega$. The first and second terms taken separately diverge for $\lambda \to 0$, while their sum admits a finite limit. To see it we apply them to a test function $\varphi(\omega) \in \mathcal{S}(\mathbb{R})$

$$\left(\frac{\lambda \ln|\omega| + \frac{\pi}{2}\,|\omega|}{\omega^2 + \lambda^2} \right)(\varphi) =$$

$$\int \frac{\ln|\omega| + \ln\lambda}{\omega^2 + 1}\,\varphi(\lambda\omega)\,d\omega + \frac{\pi}{2}\int_{|\omega|>M} \frac{|\omega|\,\varphi(\omega)}{\omega^2 + \lambda^2}\,d\omega + \frac{\pi}{2}\int_{-M}^{M} \frac{|\omega|\,\varphi(\omega)}{\omega^2 + \lambda^2}\,d\omega, \tag{14.91}$$

where we have introduced an arbitrary positive separation constant M. As λ goes to zero, in the first integral in (14.91) we can replace $\varphi(\lambda\omega)$ with $\varphi(0)$. The limit for $\lambda \to 0$ of the second integral is trivial. In the third integral, before taking the limit we must separate a term which becomes singular at $\omega = 0$

$$\int_{-M}^{M} \frac{|\omega|\,\varphi(\omega)}{\omega^2 + \lambda^2}\,d\omega = \int_{-M}^{M} \frac{|\omega|\,(\varphi(\omega) - \varphi(0))}{\omega^2 + \lambda^2}\,d\omega + \varphi(0)\int_{-M}^{M} \frac{|\omega|}{\omega^2 + \lambda^2}\,d\omega.$$

The first term admits a trivial limit for $\lambda \to 0$, and the integral of the second term is elementary. So we obtain

$$\int_{-M}^{M} \frac{|\omega|\,\varphi(\omega)}{\omega^2 + \lambda^2}\,d\omega = \int_{-M}^{M} \frac{\varphi(\omega) - \varphi(0)}{|\omega|}\,d\omega - 2\varphi(0)\ln\frac{\lambda}{M} + O(\lambda^2).$$

Thanks to the integral

$$\int \frac{d\omega}{\omega^2 + 1} = \pi$$

we then see that the divergences proportional to $\ln\lambda$ cancel between the first and the last terms in (14.91), and we obtain the finite limit

$$\lim_{\lambda \to 0} \left(\frac{\lambda \ln|\omega| + \frac{\pi}{2}\,|\omega|}{\omega^2 + \lambda^2} \right)(\varphi) = \frac{\pi}{2}\int_{|\omega|>M} \frac{\varphi(\omega)}{|\omega|}\,d\omega + \frac{\pi}{2}\int_{-M}^{M} \frac{\varphi(\omega) - \varphi(0)}{|\omega|}\,d\omega$$
$$+ \pi \ln M\varphi(0). \tag{14.92}$$

We used the vanishing of the integral ($\omega = e^y$)

$$\int_{-\infty}^{\infty} \frac{\ln|\omega|}{\omega^2 + 1}\,d\omega = 2\int_{0}^{\infty} \frac{\ln\omega}{\omega^2 + 1}\,d\omega = \int_{-\infty}^{\infty} \frac{y\,dy}{\cosh y} = 0.$$

Notice that the expression (14.92) is M-independent, as it must be by construction, but that the limit for $M \to 0$, or for $M \to \infty$, cannot be taken analytically. However, a convenient choice is $M = 1$.

Collecting the above results, eventually the distributional limit for $\lambda \to 0$ of the expression (14.81), i.e. the sought Fourier transform of the potential (14.78), can be written as the sum

$$\widehat{A}(\omega) = \widehat{A}_{\mathrm{r}}(\omega) + \widehat{A}_{\mathrm{nr}}(\omega) \tag{14.93}$$

of the regular distribution (where we include also the δ-functions)

$$\widehat{A}_{\mathrm{r}}(\omega) = \frac{I_b - I_a}{2\sqrt{2\pi}} \left(\frac{1}{2|\omega|} - \frac{iL(r\omega)}{\pi\omega} - \left(\gamma + \ln \frac{M}{2} \right) \delta(\omega) \right) - \frac{I_b + I_a}{2\sqrt{2\pi}} \ln r \, \delta(\omega),$$

and of the non-regular one

$$\widehat{A}_{\mathrm{nr}}(\varphi) = -\frac{I_b - I_a}{4\sqrt{2\pi}} \left(\int_{|\omega|>M} \frac{\varphi(\omega)}{|\omega|} \, d\omega + \int_{-M}^{M} \frac{\varphi(\omega) - \varphi(0)}{|\omega|} \, d\omega \right), \tag{14.94}$$

where γ is the Euler–Mascheroni constant coming from the expansion (13.56). As we see, the Fourier transform of the potential (14.78) cannot be written in the *ill-defined* simple form, see Eq. (14.88),

$$\widehat{A}(\omega) = \frac{1}{2\pi} \widehat{I}(\omega) L(r\omega) - \frac{I_a \ln r}{\sqrt{2\pi}} \delta(\omega),$$

suggested by the regularized potential (14.81). On the other hand, the product $\omega \widehat{A}(\omega)$ admits the much more simpler expression (14.41). In fact, it can be derived from Eqs. (14.81), (14.82) and (14.84) as the limit (the multiplication by a coordinate is a *continuous* map in $\mathcal{S}'$)

$$\omega \widehat{A}(\omega) = \mathcal{S}' - \lim_{\lambda \to 0} \left(\omega \widehat{A}_{\lambda}(\omega) \right) = \frac{1}{2\pi} \mathcal{S}' - \lim_{\lambda \to 0} \left(\omega \widehat{I}_{\lambda}(\omega) L(r\omega) \right)$$
$$\frac{1}{2\pi} \mathcal{S}' - \lim_{\lambda \to 0} \left(\omega \widehat{I}_{\lambda}(\omega) L(r\omega) \right) = \frac{\omega}{2\pi} \widehat{I}(\omega) L(r\omega), \tag{14.95}$$

where the last limit could be taken trivially, because the "joined" singularities of $L(r\omega)$ and $\widehat{I}(\omega)$ at $\omega = 0$ are disarmed by the factor ω. For the step-function current, for instance, it is straightforward to check that the Fourier transform (14.93) multiplied by ω agrees with formula (14.95). In fact, in this case, inserting for $\widehat{I}(\omega)$ the Fourier transform of the step-function current (14.43), formula (14.95) yields

$$\omega \widehat{A}(\omega) = -\frac{i(I_b - I_a)}{(2\pi)^{3/2}} L(r\omega).$$

The same result is obtained from formula (14.93), if one takes into account that the quantity $(\omega\widehat{A}_{\mathrm{nr}})(\varphi) = \widehat{A}_{\mathrm{nr}}(\omega\varphi)$ results from the expression (14.94) by performing everywhere the replacement $\varphi(\omega) \to \omega\varphi(\omega)$. This amounts, in particular, to set $\varphi(0) \to 0$.

14.2 *Radiation produced by a shock-wave current*. Consider a straight wire carrying a constant current I_a, whose section at the instant $t = 0$ is suddenly traversed by a charge Q. The resulting total current has hence the form $I(t) = I_a + Q\delta(t)$.

(a) Show that the potential in the wave zone in (14.21) is given by

$$A(t, \mathbf{x}) = \frac{QH(t - r)}{2\pi\sqrt{2r}\sqrt{t - r}}. \tag{14.96}$$

(b) Show that its Fourier transform is equal to

$$\widehat{A}(\omega) = \frac{Q\,e^{-i\pi/4}e^{-i\omega r}}{4\pi\sqrt{|\omega|r}}.$$

Hint: Use the Fresnel integral (14.39).

(c) Show that the spectral weights of the emitted radiation are given by

$$\frac{d^2\varepsilon}{dz\,d\omega} = \frac{Q^2\omega}{4\pi},$$

and compare the divergences arising for the total energy $d\varepsilon/dz$ radiated per unit length with those of the step-function current, see Eq. (14.44).

(d) Determine the electric field in the wave zone $\mathbf{E}(t, \mathbf{x})$ associated with the potential (14.96).

Hint: The distributional time derivative of the potential (14.96) cannot be computed *naively*. In particular, a term like $\delta(t - r)/\sqrt{t - r}$ would be meaningless. To bypass this difficulty introduce the regularized potential

$$A_\varepsilon(t, \mathbf{x}) = \frac{QH(t - r)}{2\pi\sqrt{2r}\sqrt{t - r + \varepsilon}},$$

where ε is a positive parameter, and then proceed as in Problem 13.5.

14.3 Consider an infinite wire carrying the square-wave current (14.59).

(a) Show that this current generates in the wave zone the electric field, viewed as a function for all $t \in \mathbb{R}$,

$$\mathbf{E}(t, \mathbf{x}) = -\frac{I_m\mathbf{u}_z}{2\pi\sqrt{2r}} \sum_{n=-\infty}^{\infty} \left(\frac{H(t - r - nT)}{\sqrt{t - r - nT}} - \frac{H(t - r - nT - T/2)}{\sqrt{t - r - nT - T/2}} \right). \tag{14.97}$$

Hint: Insert the derivative of the square-wave current (14.62) in the general equation (14.58).

(b) Show that the electric field (14.97) when restricted to the fundamental domain $t - r \in [-T/2, T/2]$ reproduces the expression (14.61).

14.4 Consider an infinite wire carrying the continuous current (14.33).

(a) Show that the exact potential generated by the current has the form

$$
A(t, \mathbf{x}) = \frac{I_b - I_a}{4\pi T} \left(H\big(t - r + T\big)\big((t + T)\,\text{arccosh}\,\frac{t + T}{r} - \sqrt{(t + T)^2 - r^2}\big) \right.
$$
$$
\left. - H\big(t - r - T\big)\big((t - T)\,\text{arccosh}\,\frac{t - T}{r} - \sqrt{(t - T)^2 - r^2}\big) \right) - \frac{I_a \ln r}{2\pi}.
$$
$$
\tag{14.98}
$$

Hint: Use the integral representation (14.8).

(b) Verify that the potential (14.98) tends for $T \to 0$ to the step-function-current potential (14.11), and that for $t \to \infty$ it entails the universal asymptotic behavior given in (14.20).

(c) Compute the electric field $\mathbf{E}(t, \mathbf{x})$ associated with the potential (14.98) and verify, (i) that it is a continuous function of time, and (ii) that for large times it has the universal asymptotic behavior given in (14.20).

(d) Evaluate the field $\mathbf{E}(t, \mathbf{x})$ in the wave zone, i.e. for large r, and check that it agrees with the expression (14.35).
Hint: The retarded time $t' = t - r$ must be kept fixed.

14.5 *Radiation from a closed wire.* Consider a closed conducting wire with parametric equation $\mathbf{y}(\lambda)$, which carries a generic current $I(t)$.

(a) Show that the four-current j^μ of the wire is given by

$$
j^0(t, \mathbf{x}) = 0, \qquad \mathbf{j}(t, \mathbf{x}) = I(t) \oint \delta^3(\mathbf{x} - \mathbf{y}(\lambda))\, \mathbf{m}(\lambda)\, d\lambda, \tag{14.99}
$$

where $\mathbf{m}(\lambda) = d\mathbf{y}(\lambda)/d\lambda$ is the vector tangent to the wire.

(b) Determine the vector potential in the wave zone (8.10)

$$
\mathbf{A}(t, \mathbf{x}) = \frac{1}{4\pi r} \oint I\big(t - r + \mathbf{n} \cdot \mathbf{y}(\lambda)\big)\, \mathbf{m}(\lambda)\, d\lambda, \tag{14.100}
$$

and verify that its time derivative satisfies the identity $\mathbf{n} \cdot \dot{\mathbf{A}}(t, \mathbf{x}) = 0$. Conclude that the electric wave-zone field is given by $\mathbf{E}(t, \mathbf{x}) = -\dot{\mathbf{A}}(t, \mathbf{x})$. Show that if the wire lies in a plane, no radiation is emitted in the direction orthogonal to the plane.

(c) Use the general equations (5.101), (11.4) and (11.12) to derive for the continuous spectral weights of the emitted radiation the formula

$$\frac{d^2\varepsilon}{d\omega d\Omega} = \frac{\left|\omega \widehat{I}(\omega)\right|^2}{8\pi^2} \left|\oint e^{i\omega \mathbf{n}\cdot\mathbf{y}(\lambda)} \, \mathbf{m}(\lambda) \, d\lambda\right|^2, \tag{14.101}$$

where $\widehat{I}(\omega)$ is the Fourier transform of an aperiodic current $I(t)$. Perform the same computation to determine their discrete counterparts (11.14) for a periodic current (14.51)

$$\frac{dW_N}{d\Omega} = \frac{(N\omega_0)^2 \left|I_N\right|^2}{8\pi^2} \left|\oint e^{iN\omega_0 \mathbf{n}\cdot\mathbf{y}(\lambda)} \, \mathbf{m}(\lambda) \, d\lambda\right|^2. \tag{14.102}$$

(d) Show that if an aperiodic current $I(t)$ is continuous and entails the initial and final conditions (14.2), the total energy $d\varepsilon/d\Omega$ radiated per unit solid angle is finite. This means that no infrared divergences arise, as implied by the compact geometry of the wire. Conclude that the electricity generator employs a finite amount of energy to make the current rise continuously from I_a to I_b.
Hint: Write the current in the form $I(t) = I_c(t) + \Delta I(t)$, where $I_c(t)$ is the continuous current (14.33). This implies that $\Delta I(t)$ is a bounded continuous current satisfying the fast fall-off boundary conditions (14.5). Use the Fourier transform (14.46) of $I_c(t)$ to show that the integral over all frequencies of the spectral weights (14.101) is convergent. Notice, in particular, that the modulus squared of the integral appearing in (14.101) is bounded by the square of the perimeter of the wire.

(e) *Ultraviolet divergences of the radiation of the step-function current.* Prove that the energy required to *instantaneously* change the current of the wire from a constant value I_a to a constant value I_b is *infinite*, in that the energy carried away by the radiation emitted during the process would be infinite.

Solution: From the Fourier transform (14.43) of the step-function current (14.3) we obtain for the first factor of the spectral weights (14.101) the ω-independent value

$$\frac{\left|\omega \widehat{I}_{sf}(\omega)\right|^2}{8\pi^2} = \frac{(I_b - I_a)^2}{16\pi^3}. \tag{14.103}$$

To evaluate the integral over ω of the second factor it is convenient to rewrite it in the form

$$\int_0^\infty \left|\oint e^{i\omega \mathbf{n}\cdot\mathbf{y}(\lambda)} \, \mathbf{m}(\lambda) \, d\lambda\right|^2 d\omega =$$

$$\frac{1}{2}\int_{-\infty}^\infty \oint d\lambda \oint d\mu \, e^{i\omega \mathbf{n}\cdot(\mathbf{y}(\lambda)-\mathbf{y}(\mu))} \, \mathbf{m}(\lambda) \cdot \mathbf{m}(\mu) \, d\omega =$$

$$\pi \oint d\lambda \oint d\mu \, \delta(\mathbf{n}\cdot(\mathbf{y}(\lambda) - \mathbf{y}(\mu))) \, \mathbf{m}(\lambda) \cdot \mathbf{m}(\mu) = \tag{14.104}$$

$$\pi \oint \left(\frac{1}{|\mathbf{n} \cdot \mathbf{m}(\lambda)|} + \sum_{i=1}^{N} \frac{\mathbf{m}(\lambda) \cdot \mathbf{m}(\mu_i)}{|\mathbf{n} \cdot \mathbf{m}(\mu_i)|} \right) d\lambda, \tag{14.105}$$

where we applied the properties (2.83) and (2.98) of the δ-function, using that $\mu = \lambda$ is a zero of the argument of the δ-function appearing in (14.104). The N parameters μ_i, different from λ, are such that the vectors $\mathbf{y}(\lambda) - \mathbf{y}(\mu_i)$ are orthogonal to $\mathbf{n}$, and in general they are complicated functions of λ and $\mathbf{n}$. For topological reasons there is at least one such μ_i. Regarding the first integral in (14.105), for generic directions $\mathbf{n}$ there is always some point λ_0 on the wire, depending on $\mathbf{n}$, such that $\mathbf{n} \cdot \mathbf{m}(\lambda_0) = 0$. In the vicinity of this point we have thus $\mathbf{n} \cdot \mathbf{m}(\lambda) \approx (\lambda - \lambda_0) \mathbf{n} \cdot \mathbf{m}'(\lambda_0)$, and so the first integral in (14.105) is logarithmically divergent. This is sufficient to conclude that the expression (14.105) is infinite.

(f) **Radiation from a circular wire.** Determine the spectral weights (14.101) for a circular wire of radius R lying in the xy plane. Specialize the analysis to the step-function current (14.3), analyzing the ultraviolet behavior of the spectral weights. Compare the results with the general analysis of issue (e) and with the ultraviolet divergences entailed by the spectral weights (14.44) of the radiation of the step-function current when flowing in an infinite wire.

Solution: For a circular wire we have $\mathbf{y}(\lambda) = R(\cos \lambda, \sin \lambda, 0)$ and $\mathbf{m}(\lambda) = R(-\sin \lambda, \cos \lambda, 0)$, where $\lambda \in [0, 2\pi]$. Due to the rotational symmetry of the problem it is not restrictive to choose $\mathbf{n} = (0, \sin \vartheta, \cos \vartheta)$. Using the definition of the Bessel function $J_1(x)$ in (11.63) we then obtain the integral

$$\oint e^{i\omega \mathbf{n} \cdot \mathbf{y}(\lambda)} \mathbf{m}(\lambda) \, d\lambda = -2\pi R \, J_1(\omega R \sin \vartheta)(i, 0, 0), \tag{14.106}$$

so that Eq. (14.101) yields the spectral weights

$$\frac{d^2\varepsilon}{d\omega d\Omega} = \frac{1}{2} \left| \omega R \widehat{I}(\omega) \right|^2 J_1^2(\omega R \sin \vartheta). \tag{14.107}$$

For the step-function current (14.3) the first factor is a constant, see Eq. (14.103). From the large-argument expansion of the Bessel functions (11.67) we then derive that the spectral weights (14.107) entail the large ω-behavior

$$\frac{d^2\varepsilon}{d\omega d\Omega} \sim \frac{1}{\omega} \sin^2 \left(\omega R \sin \vartheta - \frac{\pi}{4} \right). \tag{14.108}$$

It follows that the integral $\int_0^\infty (d^2\varepsilon/d\omega d\Omega) d\omega$ is ultraviolet divergent, in agreement with the general analysis of issue (e). The large-ω behavior (14.108) resembles the analogous behavior (14.44) of the spectral weights produced by the step-function current in an infinite wire, although the former are oscillating.

In fact, the occurrence of ultraviolet divergences is insensitive to the compactness of the wire, which represents instead an infrared property.

14.6 *Circular antenna.* Consider a circular antenna of radius R in which flows the periodic current $I(t) = I_0 \cos(\omega t)$.

(a) Apply the results of Problem 14.5 to show that the antenna emits only radiation of frequency ω, and that the angular distribution of the emitted power, reintroducing the velocity of light, results in

$$\frac{dW}{d\Omega} = \frac{1}{8c^3} \left(\omega R I_0\right)^2 J_1^2 \left(\frac{\omega R}{c} \sin\vartheta\right), \qquad (14.109)$$

where ϑ is the angle between the emission direction **n** and the axis of the antenna.
Hint: Use formula (14.102) for $N = 1$, and the integral (14.106).

(b) Analyze the form of the emission pattern (14.109) and compare it qualitatively with the emission pattern (8.34) of a linear antenna.
Solution: Since $J_1(0) = 0$, no radiation is emitted for $\vartheta = 0$, i.e. in the direction orthogonal to the antenna, as shown in general in issue (b) of Problem 14.5. Since the first non-trivial zero of the function $J_1(x)$ occurs at $x = 3.83$, for values of the dimensionless parameter $\omega R/c = 2\pi R/\lambda$ greater than or equal to 3.83 there is a finite number of additional directions $0 < \vartheta \le \pi/2$ for which $dW/d\Omega$ vanishes, with relative maxima in between. There are no such additional directions if $\omega R/c < 3.83$. The same qualitative features occur *formally* for the angular distribution (8.34) of the linear antenna; see, however, issue (c).

(c) *Small circular antenna.* Analyze the form of the emission pattern (14.109) for a *small circular* antenna, i.e. an antenna for which $R \ll \lambda$, and compare it to the emission pattern (8.62) of a *short linear* antenna, both situations that can be analyzed by means of the non-relativistic multipole expansion (see Sect. 8.3.1 and the condition (8.43)).
Solution: For a small circular antenna we have $\omega R/c = 2\pi R/\lambda \ll 1$, and so we can use the small-argument expansion (11.68) of $J_1(x)$, i.e. $J_1(x) = x/2 + O(x^3)$. Formula (14.109) takes thus the simple form

$$\frac{dW}{d\Omega} = \frac{I_0^2 (\omega R)^4}{32 c^5} \sin^2 \vartheta. \qquad (14.110)$$

This emission pattern looks identical to the pattern of the short linear antenna (8.62), in which ϑ is still the angle between **n** and the z axis: both entail a maximum emission in the xy plane, and they vanish along the z axis. However, despite this mathematical correspondence, these two patterns are physically in contrast with each other. In fact, there is a crucial *physical* difference between the two types of antennas: in the linear antenna the charged particles are accelerated along the z axis, whereas in the circular antenna their acceleration is always orthogonal to this axis. Recall, in this regard, that in the non-relativistic approximation according to Larmor's angular-distribution-formula (7.66) the dominant emission occurs always in the direction orthogonal to the acceleration. This puzzle is solved, once we realize that the four-current (14.1), entailing a vanishing charge density j^0, has vanishing dipole and quadrupole moments, see formulas (8.47) and (8.116). It follows that the radiation emitted by a circular antenna for $R \ll \lambda$ is dominated by the *magnetic* dipole radiation, whose emission pattern (14.110) – of order $1/c^5$ rather than of order $1/c^3$ as the electric dipole power (8.62) – in general has a different form of the universal angular distribution (8.52) of the electric dipole radiation. Actually, Eq. (14.110) could also be derived by evaluating the magnetic dipole moment (8.112) for the current (14.99), by inserting it in the equation (8.118) for the vector potential, and by finally resorting to the general formula (8.18) for the emitted power. Notice, in particular, that, since the electric dipole and quadrupole moments are zero, the vector potential (8.118) satisfies automatically the identity $\mathbf{n} \cdot \mathbf{A} = 0$, so that $\mathbf{E} = -\dot{\mathbf{A}}/c$, as shown in general in issue (b) of Problem 14.5.

(d) Analyze the polarization of the emitted radiation in the general case.

Solution: As in issue (f) of Problem 14.5 we choose the emission direction $\mathbf{n} = (0, \sin \vartheta, \cos \vartheta)$ parallel to the yz plane. Using the integral (14.106), we can thus evaluate explicitly the potential (14.100) for the current $I(t) = I_0 \cos(\omega t)$. For the electric field in the wave zone we then find

$$\mathbf{E}(t, \mathbf{x}) = -\frac{1}{c} \dot{\mathbf{A}}(t, \mathbf{x}) = -\frac{\omega R I_0}{2rc^2} \cos\left(\omega \left(t - \frac{r}{c}\right)\right) J_1\left(\frac{\omega R}{c} \sin \vartheta\right) (1, 0, 0).$$

As this vector is constantly parallel to the x axis, we conclude that for all emission directions $\mathbf{n}$ the radiation is *linearly* polarized, with polarization parallel to the plane of the antenna.

References

1. M. Reed, B. Simon, *Methods of Modern Mathematical Physics - I Functional Analysis* (Academic Press, New York, 1980)
2. W. Rudin, *Functional Analysis* (McGraw-Hill, New York, 1991)
3. K.R. Stromberg, *Introduction to Classical Real Analysis* (Wadsworth International Group, Belmont, 1981)
4. L. Carleson, On convergence and growth of partial sums of Fourier series. Acta Math. **116**, 135 (1966)

Part III
Selected Topics

Chapter 15
Radiation Reaction

A system of accelerated charged particles emits electromagnetic radiation, that carries four-moment. Consequently, if the total four-momentum is to be conserved, during the acceleration process the four-momentum of the particles cannot remain constant. The effects caused in this way, indirectly, by the radiation on the dynamics of the system of particles are generically summarized with the term *radiation reaction*. We begin the analysis of this fundamental phenomenon starting from the simplest system of electric charges, namely a single point-like particle, whose dynamics is governed by the system of coupled equations

$$\partial_\mu F^{\mu\nu} = e \int \delta^4(x - y)\, dy^\nu, \qquad \partial_{[\mu} F_{\nu\rho]} = 0, \tag{15.1}$$

$$\frac{dp^\mu}{ds} = e F^{\mu\nu}(y) u_\nu. \tag{15.2}$$

In previous chapters, to face the solution of this system we have determined, in the first place, the exact solution of Maxwell's equations (15.1) for a known particle's world line $y^\mu(s)$. The resulting electromagnetic field is the sum of the Liénard-Wiechert (7.38), which henceforth we will denote by the symbol $\mathcal{F}^{\mu\nu}$, and of a generic external field $F_{\text{in}}^{\mu\nu}$ satisfying the homogenous Maxwell equations:

$$F^{\mu\nu} = \mathcal{F}^{\mu\nu} + F_{\text{in}}^{\mu\nu}. \tag{15.3}$$

Substituting the field (15.3) in the Lorentz equation (15.2) the latter *formally* goes over into the closed equation for the unknown world line $y^\mu(s)$

$$\frac{dp^\mu}{ds} = e \mathcal{F}^{\mu\nu}(y) u_\nu + e F_{\text{in}}^{\mu\nu}(y) u_\nu. \tag{15.4}$$

© Springer International Publishing AG, part of Springer Nature 2018
K. Lechner, *Classical Electrodynamics*, UNITEXT for Physics,
https://doi.org/10.1007/978-3-319-91809-9_15

In general, this equation cannot be solved analytically and, actually, until now we have addressed its solution adopting tacitly a pragmatic *perturbative* procedure. Below we briefly recall its main steps.

Perturbative scheme. Assuming the external field $F_{\text{in}}^{\mu\nu}$ as known, in general we determined preliminarily the world line $y^\mu(s)$ of the particle considering in equation (15.4) only the term $eF_{\text{in}}^{\mu\nu}(y)u_\nu$. In doing so, we disregarded thus the *radiation reaction force*

$$e\mathcal{F}^{\mu\nu}(y)u_\nu. \tag{15.5}$$

This force is also called *self-force* as it is proportional to the *self-field* $\mathcal{F}^{\mu\nu}(y)$, which represents the action of the field $\mathcal{F}^{\mu\nu}(x)$, generated by the particle, on the particle itself. Successively, we have inserted the world line $y^\mu(s)$ so determined in the Liénard-Wiechert field (7.38), and eventually we have evaluated the latter at large distances from the trajectory to calculate the four-momentum radiated by the particle. Moreover, in a few cases, relying on total four-momentum conservation we were also able to quantify indirectly the effects of the radiation reaction. We recall, for instance, the forward boost experienced by the particle in the Thomson scattering, Sect. 8.4.2, the energy loss and consequent slowdown of a particle in an accelerator, Sect. 10.2, and the collapse of the classical hydrogen atom, Sect. 8.4.4. In the light of the Lorentz equation (15.4), the *direct* cause of this *back-reaction* of the radiation on the dynamics of the particle must necessarily be the radiation reaction force (15.5), as it is the only term that in our perturbative solution scheme has been disregarded.

Internal inconsistency of classical electrodynamics. At this point we run, however, into the basic obstacle that the self-field $\mathcal{F}^{\mu\nu}(y)$ – the field generated by the particle evaluated at the particle's position – is always *infinite*. In fact, from formulas (7.40)–(7.42) we see that in the vicinity of the trajectory the Liénard-Wiechert field is dominated by the velocity field and, correspondingly, if at a given time t we let $\mathbf{x}$ approach the particle's position $\mathbf{y}(t)$, the field diverges as[1]

$$\mathcal{F}^{\mu\nu}(t,\mathbf{x})\Big|_{\mathbf{x}\to\mathbf{y}(t)} \sim \frac{1}{|\mathbf{x}-\mathbf{y}(t)|^2}. \tag{15.6}$$

This means that the radiation reaction force (15.5)

$$e\mathcal{F}^{\mu\nu}(y)u_\nu = e\mathcal{F}^{\mu\nu}(t,\mathbf{y}(t))u_\nu \tag{15.7}$$

is always divergent, and so the Lorentz equation (15.4) – that was supposed to govern the dynamics of a charged particle – becomes meaningless. We must thus draw the dramatic conclusion that *classical electrodynamics in its original formulation, that is to say, as a theory based on the fundamental equations* (2.20)–(2.22), *is*

---

[1]As $\mathbf{x}$ approaches $\mathbf{y}(t)$, the retarded time as determined by the delay equation (7.19) behaves as $t'(t,\mathbf{x}) = t + O(|\mathbf{x}-\mathbf{y}(t)|)$. Consequently, in this limit the distance $R = |\mathbf{x}-\mathbf{y}(t')|$ appearing in the fields (7.40)–(7.42) becomes proportional to $|\mathbf{x}-\mathbf{y}(t)|$.

internally inconsistent. This inconsistency is the reason for why we postponed a systematic investigation of the radiation-reaction phenomenon to this section.

Apart from the conceptual obstacle represented by the infinite Lorentz force, it is obvious that the perturbation procedure adopted so far is only of limited validity. In fact, the dynamics of a charged particle is determined both by the external field and by the radiation reaction force, and these two forces have to be considered *at once*. It is, therefore, imperative to establish an equation of motion for the particle that takes into account both forces at the same time. From this point of view, the most dramatic outcome of this chapter is the replacement of the *divergent* Lorentz equation (15.4) with the *finite* Lorentz-Dirac equation (15.19). As we shall see, this replacement respects the *postulates of special relativity* and it is, furthermore, compatible with all conservation laws of electrodynamics in its original formulation, in particular with local four-momentum conservation. However, this *unavoidable* replacement, in turn, introduces a peculiar type of *causality violation* in classical electrodynamics: owing to the specific characteristics of the Lorentz-Dirac equation, the effects caused by the electromagnetic fields on the charges occur slightly *before* the fields themselves begin to act. This violation represents a *phenomenological*, as well as *logical*, inconsistency, that eventually can be cured only by the *paradigm shift* to *quantum electrodynamics*.

Point-like particles and ultraviolet divergences. The divergence of the self-field $\mathcal{F}^{\mu\nu}(y)$ leaves noticeable traces also in the total four-momentum of the electromagnetic field, as anticipated in Sect. 2.3.3 for the case of a static particle. In fact, again due to the behavior (15.6), in the vicinity of a charged particle the electromagnetic energy-momentum tensor (2.137) diverges as

$$T_{\text{em}}^{\mu\nu} \sim \frac{1}{|\mathbf{x} - \mathbf{y}(t)|^4}, \tag{15.8}$$

a behavior that represents a *non-integrable* singularity with respect to the spatial measure d^3x. As a consequence, the electromagnetic four-momentum integrals $P_{\text{em}}^\mu = \int T_{\text{em}}^{0\mu} \, d^3x$ are always *divergent*. We shall address this problem separately in Chap. 16. It is clear that the origin of both above problems – divergent radiation reaction force and infinite electromagnetic four-momentum – resides in the *point-like* nature of the charged particle: the bad short-distance behavior of the field (15.6) is, indeed, due to the fact that the charge e has been "compressed" to a point. The divergences of the radiation reaction force and of the total electromagnetic four-momentum are, hence, of *ultraviolet* nature. Vice versa, a point-like particle at the classical level does not entail any *infrared* divergences. In fact, since at large distances from the particle the Liénard-Wiechert fields (7.48) decrease as $1/|\mathbf{x}|^2$, for large $|\mathbf{x}|$ the four-momentum integrals $\int T_{\text{em}}^{0\mu} \, d^3x$ are indeed convergent.[2] Notice

---

[2]The delay equation (7.19) implies that for a fixed t for $|\mathbf{x}| \to \infty$ the retarded time $t'(t, \mathbf{x})$ tends to $-\infty$. If the acceleration $\mathbf{a}(t)$ for $t \to -\infty$ vanishes sufficiently fast, in particular if it lasts for a finite time, then, in basis of formulas (7.46)–(7.48), the field $\mathcal{F}^{\mu\nu}(t, \mathbf{x})$ for large $|\mathbf{x}|$ decreases as $1/|\mathbf{x}|^2$. In the opposite case, in formulas (7.48) the acceleration fields would dominate, and so

that the infrared divergences met in Sect. 11.3.1 were, instead, of *quantum mechanical* origin.

A particle with a more regular charge distribution, as, for instance, a surface charge distribution on a small rigid sphere, would create an electromagnetic field without singularities throughout space. However, such a distribution would conflict with the principles of relativity: the rigidity constraint would require internal *forces at a distance*, that would violate causality, and, in particular, the compensation of the electrostatic repulsion of the charge distribution would require the introduction of new binding forces of non-electromagnetic origin. In order to preserve the postulates of special relativity, together with the economy inherent in the *minimal* formulation of electrodynamics, which does not envisage forces other than those of electromagnetic nature, we prefer to keep the particles point-like and to replace, instead, the Lorentz equation with the Lorentz-Dirac equation.

15.1 Radiation Reaction Forces: Qualitative Analysis

We start our systematic analysis of the radiation reaction phenomenon with some general qualitative considerations about its relevance in the dynamics of a charged particle. There are, in fact, several situations where *locally* the radiation forces can be treated as a perturbation and, possibly, neglected. We call *locally negligible* a force that affects the *instantaneous* motion of a particle only marginally. Although locally negligible, radiation reaction forces can still have significant *cumulative* effects. Of course, it can also happen that the radiation reaction forces are compensated by suitable external forces, as for instance the radio-frequency resonant cavities in a synchrotron, or the alternating current generators which keep the electrons of an antenna in a permanent oscillatory motion.

Locally negligible radiation reaction forces. We implement the above definition by means of the following criterion: radiation reaction forces can be considered as locally negligible, if the energy $\Delta\varepsilon_{\mathrm{rad}}$ dissipated via radiation by a charged particle during a certain time interval is small with respect to the variation $\Delta\varepsilon$ of the particle's energy during the same interval. In the following we shall apply the criterion in the non-relativistic limit, so that $\Delta\varepsilon = \Delta(mv^2/2)$. Let us denote with $\mathcal{T}$ the characteristic time scale of the external force, with a the average magnitude of the acceleration during this time, and with v the initial velocity of the particle. During the time $\mathcal{T}$ its velocity changes thus by

$$\Delta v \sim a\mathcal{T}.$$

$\mathcal{F}^{\mu\nu}(t, \mathbf{x})$ would decrease only as $1/|\mathbf{x}|$. In this *unphysical* situation the particle would be subject to an *eternal* acceleration, and consequently its electromagnetic field would entail an *infrared divergent* four-momentum P_{em}^μ.

According to Larmor's formula (7.67), during the same time interval the particle radiates the energy

$$\Delta\varepsilon_{\text{rad}} \sim \frac{e^2 a^2}{6\pi c^3}\, \mathcal{T}.$$

To estimate the change in energy $\Delta\varepsilon$ we must distinguish the regimes of small velocity changes, $\Delta v \ll v$, and of large velocity changes, $\Delta v \sim v$ or also $\Delta v \gg v$. For small velocity changes we have

$$\Delta\varepsilon \approx mv\Delta v \sim mva\mathcal{T},$$

leading to the ratio

$$\frac{\Delta\varepsilon_{\text{rad}}}{\Delta\varepsilon} \sim \frac{e^2 a}{6\pi mvc^3} \sim \frac{e^2}{6\pi mc^3}\frac{1}{\mathcal{T}}\frac{\Delta v}{v}. \tag{15.9}$$

The quantity

$$\tau = \frac{e^2}{6\pi mc^3} \tag{15.10}$$

appearing in this estimate has the dimension of a time, being related to the classical radius r_0 of the particle by $\tau = 2r_0/3c$. It attains a maximum for the lightest charged particle – the electron – in which case it amounts to

$$\tau = 0.6 \cdot 10^{-23}\, s.$$

This very small time scale will play a fundamental role in this chapter. In particular, the ratio (15.9) can be recast in the form

$$\frac{\Delta\varepsilon_{\text{rad}}}{\Delta\varepsilon} \sim \frac{\tau}{\mathcal{T}}\frac{\Delta v}{v} \ll \frac{\tau}{\mathcal{T}}. \tag{15.11}$$

On the other hand, for large velocity changes Δv the kinetic energy of the particle changes by

$$\Delta\varepsilon \sim \frac{1}{2}m(\Delta v)^2 \sim ma^2\mathcal{T}^2,$$

leading to the ratio

$$\frac{\Delta\varepsilon_{\text{rad}}}{\Delta\varepsilon} \sim \frac{\tau}{\mathcal{T}}. \tag{15.12}$$

We see thus that, independently of the size of the relative velocity changes, locally the radiation reaction forces are negligible if the time scale $\mathcal{T}$ during which the external forces vary significantly is large with respect to the fundamental time scale τ, i.e. if $\mathcal{T} \gg \tau$. Conversely, in general the radiation reaction forces can no longer be neglected if the external forces change rapidly in time, i.e. if during the very short time τ they are subjected to an appreciable relative change. We will come back to forces of this type in Sect. 15.2.6 in connection with the *preacceleration* phenomenon.

Cumulative effects. The above analysis has local validity. In fact, even if the external forces vary on time scales $T \gg \tau$, the radiation reaction forces may induce appreciable *cumulative* effects. An electron in a non-relativistic synchrotron, for instance, in the absence of resonant cavities after a sufficiently long time stops circulating, having dissipated all its kinetic energy via radiation. Similarly, the collapse of the classical hydrogen atom is due to a cumulative effect, although the energy (8.110) radiated during a cycle is negligible, see Problem 15.8. On the other hand, as seen in Sect. 10.2.2, in an ultrarelativistic synchrotron the radiation reaction can entail important consequences also *locally*, causing the arrest of the particle in a tiny fraction of a cycle. In these cases the radiation reaction phenomenon certainly cannot be disregarded – not even locally – and in some situations the radiation reaction forces may even dominate over the external forces.

15.1.1   Heuristic Argument for the Lorentz-Dirac Equation

Although we must abandon the original Lorentz equation (15.4), irretrievably compromised by the divergent self-force (15.7), in the search for an alternative dynamics we insist on a relativistic equation of motion of the type

$$\frac{dp^\mu}{ds} = f^\mu, \tag{15.13}$$

where f^μ represents the total four-force acting on the particle. However, this four-vector cannot be arbitrary, as the first member of Eq. (15.13) satisfies the kinematical identity $u_\mu dp^\mu/ds = mu_\mu w^\mu = 0$. The four-force is, thus, subject to the algebraic constraint

$$u_\mu f^\mu = 0. \tag{15.14}$$

As seen in Sect. 2.2.3, the relation (15.14) ensures in particular that equation (15.13) amounts only to three functionally independent differential equations. Aim of this section is to provide a heuristic argument in favor of a new equation of motion, the Lorentz-Dirac equation (15.19), based on the constraint (15.14) and on covariance and conservation arguments. A more formal "derivation" will be given in Sect. 15.2. We start from the total four-momentum radiated by a charged particle per unit proper time in all directions, as given by the relativistic Larmor formula (10.7)

$$\frac{dP_{\text{rad}}^\mu}{ds} = -\frac{e^2}{6\pi} w^2 u^\mu. \tag{15.15}$$

If the total four-momentum is to be conserved, the particle must thus *lose* this four-momentum. Therefore, in the presence of an external field its equation of motion must be of the form

$$\frac{dp^\mu}{ds} = \frac{e^2}{6\pi} \, w^2 u^\mu + \cdots + eF_{\mathrm{in}}^{\mu\nu}(y)u_\nu, \tag{15.16}$$

where the ellipsis stands for possible additional terms. In fact, the Larmor term cannot be the unique term present at the second member, because then equation (15.16) would conflict with the constraint (15.14)

$$u_\mu \left(\frac{e^2}{6\pi} \, w^2 u^\mu + eF_{\mathrm{in}}^{\mu\nu}(y)u_\nu \right) = \frac{e^2}{6\pi} \, w^2 \neq 0. \tag{15.17}$$

We have actually anticipated the possibility of additional contributions to the emitted four-momentum – not reaching infinity – when in Sect. 10.1.2 we have established the precise meaning of Eq. (15.15). The algebraic inconsistency (15.17) not only shows that these contributions are necessarily present, but it also suggests their form. In fact, differentiating the identity $u_\mu w^\mu = 0$ with respect to s we find the further identity

$$w_\mu w^\mu + u_\mu \frac{dw^\mu}{ds} = 0 \quad \leftrightarrow \quad u_\mu \frac{dw^\mu}{ds} = -w^2. \tag{15.18}$$

As a consequence, a completion of Eq. (15.16) which is consistent with the constraint (15.14) is represented by the *Lorentz-Dirac equation*

$$\frac{dp^\mu}{ds} = \frac{e^2}{6\pi} \left(\frac{dw^\mu}{ds} + w^2 u^\mu \right) + eF_{\mathrm{in}}^{\mu\nu}(y)u_\nu. \tag{15.19}$$

H.A. Lorentz derived the non-relativistic version (15.116) of this equation in 1904, while P.A.M. Dirac obtained its relativistic completion (15.19) in 1938. Comparing equation (15.19) with the ill-defined Lorentz equation (15.4) we see that the term proportional to e^2 represents now a *finite* radiation reaction four-force. By dividing both members of (15.19) by the mass, the Lorentz-Dirac equation can be recast in the equivalent form

$$w^\mu = \tau \left(\frac{dw^\mu}{ds} + w^2 u^\mu \right) + \frac{e}{m} \, F_{\mathrm{in}}^{\mu\nu}(y)u_\nu, \tag{15.20}$$

where τ is the fundamental time scale introduced in (15.10). Apart from the *purely heuristic* character of the above argument, it is important to realize that there is no way to *truly derive* the Lorentz-Dirac equation (15.19) from the Lorentz equation (15.4), simply because the latter, being divergent, is meaningless. Ultimately, the Lorentz-Dirac equation must be *postulated*.

15.2 Lorentz-Dirac Equation

In this section, we first retrieve the Lorentz-Dirac equation from a *regularized* version of the original Lorentz equation. For simplicity, initially we consider again a single particle in the presence of an external field. The new dynamics implied by the Lorentz-Dirac equation for a system of charged particles will be presented at the end of Sect. 15.2.2. Sections 15.2.4–15.2.6 are devoted to a detailed analysis of the *new* physical implications of the Lorentz-Dirac equation, illustrated by several example, whose general conclusions will then be drawn in Sect. 15.3. In general, the equation of motion of a relativistic particle must meet the following *minimal* requirements:

(1) Lorentz invariance;
(2) absence of divergent terms;
(3) consistency with the algebraic identity $u_\mu dp^\mu/ds = 0$;
(4) compatibility with four-momentum conservation.

In this chapter we shall deal mainly with the first three requirements. The, no less important, fourth requirement will be treated separately in Chap. 16, as its implementation relies predominantly on distribution theory. In particular, as we shall explain at the end of Sect. 15.2.4, the Lorentz-Dirac equation cannot be derived from a variational principle, and so one can no longer resort to Noether's theorem to establish the presence of conservation laws.

15.2.1 Regularization and Renormalization

Let us return to the covariant expression of the Liénard-Wiechert field (7.38)

$$\mathcal{F}^{\mu\nu}(x) = \frac{e}{4\pi(uL)^3}\Big(L^\mu u^\nu + L^\mu\big((uL)w^\nu - (wL)u^\nu\big) - (\mu \leftrightarrow \nu)\Big), \quad (15.21)$$

where the vector field $L^\mu(x)$ is defined by

$$L^\mu(x) = x^\mu - y^\mu(s). \tag{15.22}$$

We recall that the kinematical variables $y(s)$, $u(s)$ and $w(s)$ are evaluated at the retarded proper time $s(x)$, determined by the delay relations

$$(x - y(s))^2 = 0, \qquad x^0 > y^0(s). \tag{15.23}$$

Our procedure to derive a finite equation of motion from the divergent one (15.4) resembles a method commonly used in *relativistic quantum field theory* to cure *quantum* ultraviolet divergences. It consists of two steps: a *regularization*, followed by a *renormalization*.

Regularization. A regularization requires first of all the introduction of a *regulator* ε, which in the case at hand will be a positive real number with the dimension of a length. Being an *ultraviolet* regulator, eventually it must tend to zero. With every $\varepsilon > 0$ we associate a *regularized* Liénard-Wiechert field,

$$\mathcal{F}^{\mu\nu}(x) \rightarrow \mathcal{F}_\varepsilon^{\mu\nu}(x), \tag{15.24}$$

subject to the *pointwise* limit in the complement of the world line

$$\lim_{\varepsilon \to 0} \mathcal{F}_\varepsilon^{\mu\nu}(x) = \mathcal{F}^{\mu\nu}(x), \qquad \forall\, x^\mu \neq y^\mu(s).$$

Most importantly, we require the field $\mathcal{F}_\varepsilon^{\mu\nu}(x)$ to be regular throughout space, included the world line, so that the regularized *self-field*

$$\mathcal{F}_\varepsilon^{\mu\nu}(y), \qquad y^\mu = y^\mu(s), \tag{15.25}$$

is *finite* for every $\varepsilon > 0$ and for every s. Our regularization procedure includes, furthermore, the replacement of the particle's mass m with an ε-dependent parameter m_ε, whose form will be specified later on. For the moment, we only anticipate that it will *not* entail the limit $\lim_{\varepsilon \to 0} m_\varepsilon = m$. A priori we could also introduce a regularized charge e_ε, but in the case at hand it is not needed.

Renormalization. We propose as new equation of motion for a charged particle

$$\lim_{\varepsilon \to 0} \left(m_\varepsilon \frac{du^\mu}{ds} - e\mathcal{F}_\varepsilon^{\mu\nu}(y)u_\nu - eF_{\text{in}}^{\mu\nu}(y)u_\nu \right) = 0, \tag{15.26}$$

provided that for an appropriate choice of the parameter m_ε the limit in (15.26) exists for every s. This condition is actually rather restrictive, in that m_ε multiplies a particular four-vector, i.e. the four-acceleration $w^\mu = du^\mu/ds$. It follows, in fact, that the only divergent terms of $e\mathcal{F}_\varepsilon^{\mu\nu}(y)u_\nu$ that can be "absorbed" by choosing appropriately m_ε are those proportional to w^μ. This final step is called *renormalization*, that in the present case amounts to a *mass renormalization*. If such an m_ε exists, the proposed equation (15.26) satisfies automatically the above requirements (2) and (3). Property (3) follows from the fact that in (15.26) the four-vector u_ν is always contracted with an antisymmetric tensor. On the other hand, the requirement (1) will be automatically satisfied if the regularization (15.24) preserves Lorentz invariance. With this we mean that for every $\varepsilon > 0$ under a Lorentz transformation the field $\mathcal{F}_\varepsilon^{\mu\nu}(x)$ transforms as a two-index *tensor field*.

Lorentz-invariant regularization. We implement the procedure outlined above resorting to a specific Lorentz-invariant regularization. It consists in choosing as regularized Liénard-Wiechert field $\mathcal{F}_\varepsilon^{\mu\nu}(x)$ the one obtained by replacing in the expression (15.21) the retarded time $s(x)$ (15.23) with the *regularized* retarded time $s_\varepsilon(x)$ defined by

$$(x - y(s_\varepsilon))^2 = \varepsilon^2, \qquad x^0 > y^0(s_\varepsilon). \tag{15.27}$$

We thus have

$$\mathcal{F}_\varepsilon^{\mu\nu}(x) = \mathcal{F}^{\mu\nu}(x)\Big|_{s(x)\to s_\varepsilon(x)}. \qquad (15.28)$$

In virtue of the Lorentz invariance of the *future light cone*, see Sect. 5.2.3, the function $s_\varepsilon(x)$ is a Lorentz scalar, on the same footing as $s(x)$, and consequently the field (15.28) is indeed a *tensor field*. The same conclusion can be reached by observing that the regularization based on formulas (15.27) and (15.28) is equivalent to replace the Green function $G(x) = H(x^0)\,\delta(x^2)/2\pi$ of the d'Alembertian $\square = \partial^\mu\partial_\mu$ with the *regularized* manifestly Lorentz-invariant Green function

$$G_\varepsilon(x) = \frac{1}{2\pi}\,H(x^0)\,\delta(x^2 - \varepsilon^2). \qquad (15.29)$$

In fact, introducing the regularized four-potential $\mathcal{A}_\varepsilon^\mu = G_\varepsilon * j^\mu$, and repeating the steps that led from Eq. 7.9 to the Liénard-Wiechert field (7.38), it is easily seen that the field (15.28) coincides with the field strength $\mathcal{F}_\varepsilon^{\mu\nu} = \partial^\mu\mathcal{A}_\varepsilon^\nu - \partial^\nu\mathcal{A}_\varepsilon^\mu$. It remains to verify that the regularized self-field (15.25) is finite for all ε and for all s. This follows from the fact that the factor $(u_\varepsilon L_\varepsilon)$ appearing in the denominator of this field, see formula (15.38), being the scalar product between two time-like vectors never vanishes. Below we illustrate this property explicitly for the field of a particle in uniform linear motion.

Regularized Liénard-Wiechert field of particle in uniform linear motion. As the world line of a particle in uniform linear motion has the simple form

$$y^\mu(s) = u^\mu s, \qquad w^\mu(s) = 0, \qquad u^2 = 1,$$

the associated regularized Liénard-Wiechert field can be evaluated analytically. For this purpose we first must determine the regularized retarded time $s_\varepsilon(x)$, by solving the conditions (15.27)

$$(x - us_\varepsilon)^2 = \varepsilon^2 \quad \Rightarrow \quad s_\varepsilon(x) = (ux) - \sqrt{(ux)^2 - x^2 + \varepsilon^2}.$$

The minus sign in front of the square root is implied by the inequality $x^0 > y^0(s_\varepsilon)$, i.e. $x^0 > u^0 s_\varepsilon$. Since, according to the recipe (15.28), in the field (15.21) we must operate the replacement $s(x) \to s_\varepsilon(x)$, i.e. $L^\mu \to L_\varepsilon^\mu = x^\mu - u^\mu s_\varepsilon(x)$, we need the scalar product

$$(uL_\varepsilon) = (ux) - s_\varepsilon(x) = \sqrt{(ux)^2 - x^2 + \varepsilon^2}.$$

Since the four-acceleration w^μ is identically zero, from the Liénard-Wiechert formula (15.21) we so derive the regularized field

$$\mathcal{F}_\varepsilon^{\mu\nu}(x) = \frac{e}{4\pi} \frac{x^\mu u^\nu - x^\nu u^\mu}{\left((ux)^2 - x^2 + \varepsilon^2\right)^{3/2}}. \tag{15.30}$$

This field is manifestly Lorentz covariant, and in the pointwise limit for $\varepsilon \to 0$ it reduces to the *bare* fiel (6.96) of a particle in uniform linear motion, that on its world line diverges. In contrast, the regularized field (15.30) is regular throughout space. In particular, on the world line, i.e. for $x^\mu = y^\mu(s) = u^\mu s$, its denominator reduces to the finite value $4\pi\varepsilon^3$, while its numerator vanishes. The regularized self-field is thus finite, more precisely,

$$\mathcal{F}_\varepsilon^{\mu\nu}(y(s)) = 0. \tag{15.31}$$

This means that a particle in uniform linear motion experiences no self-interaction, a conclusion that obviously is in line with the fact that a non-accelerated charged particle does emit electromagnetic radiation. Finally, since to have a particle in uniform linear motion the external field $F_{\text{in}}^{\mu\nu}$ must vanish, in this particular case equation (15.26) is trivially satisfied for any choice of the parameter m_ε.

15.2.2 Derivation of the Lorentz-Dirac Equation

We now evaluate explicitly the limit (15.26) for a particle in arbitrary motion. For this purpose we must preliminary determine the behavior of the regularized self-field $\mathcal{F}_\varepsilon^{\mu\nu}(y)$ for $\varepsilon \to 0$. Actually, for a generic world line $y^\mu(s)$ the limit $\lim_{\varepsilon \to 0} \mathcal{F}_\varepsilon^{\mu\nu}(y)$ does not exist. In fact, as shown in Sect. 15.2.3 below, the regularized self-field entails the expansion in a *Laurent series* around $\varepsilon = 0$

$$\mathcal{F}_\varepsilon^{\mu\nu}(y) = \frac{e}{8\pi\varepsilon}\left(u^\mu w^\nu - u^\nu w^\mu\right) - \frac{e}{6\pi}\left(u^\mu \frac{dw^\nu}{ds} - u^\nu \frac{dw^\mu}{ds}\right) + O(\varepsilon). \tag{15.32}$$

Notice that for a uniform linear motion this expansion consistently reduces to (15.31). Using the expansion (15.32), together with the kinematical identity (15.18), for the regularized radiation reaction force appearing in (15.26) we then obtain the expansion

$$e\mathcal{F}_\varepsilon^{\mu\nu}(y)u_\nu = \frac{e^2}{6\pi}\left(\frac{dw^\mu}{ds} + w^2 u^\mu\right) - \frac{e^2}{8\pi\varepsilon}w^\mu + O(\varepsilon).$$

As expected, in the limit for $\varepsilon \to 0$ this four-force diverges. However, the divergent term is proportional to the four-acceleration w^μ. Thanks to this circumstance the limit (15.26) takes the form

$$\lim_{\varepsilon \to 0}\left(\left(m_\varepsilon + \frac{e^2}{8\pi\varepsilon}\right)\frac{du^\mu}{ds} - \frac{e^2}{6\pi}\left(\frac{dw^\mu}{ds} + w^2 u^\mu\right) - eF_{\text{in}}^{\mu\nu}(y)u_\nu + O(\varepsilon)\right) = 0. \tag{15.33}$$

The divergent term can thus be absorbed by choosing for the parameter m_ε the expression tending to $-\infty$

$$m_\varepsilon = m - \frac{e^2}{8\pi\varepsilon},$$

where m identifies the physical mass of the particle. After this mass renormalization the limit (15.33) exists, and the resulting equation of motion is indeed the Lorentz-Dirac equation (15.19). Adopting a terminology in use in quantum field theory, we can summarize the above procedure by saying that *the ultraviolet divergent part of the radiation force has been removed by a renormalization of the particle's mass*.

Lorentz-Dirac equations for a system of charged particles. To generalize equation (15.19) to a system of N particles it is suffices to take into account in the equation of motion of each particle the Liénard-Wiechert fields $\mathcal{F}_s^{\mu\nu}(x)$ generated by the remaining $N-1$ particles. The resulting Lorentz-Dirac equation for the rth particle thus reads

$$\frac{dp_r^\mu}{ds_r} = \frac{e_r^2}{6\pi}\left(\frac{dw_r^\mu}{ds_r} + w_r^2 u_r^\mu\right) + e_r F_r^{\mu\nu}(y_r)u_{r\nu}, \qquad (r = 1,\ldots,N), \quad (15.34)$$

where the total "external" field $F_r^{\mu\nu}(x)$ acting on it is given by

$$F_r^{\mu\nu}(x) = F_{\text{in}}^{\mu\nu}(x) + \sum_{s\neq r} \mathcal{F}_s^{\mu\nu}(x). \qquad (15.35)$$

As this field is now regular at $x = y_r$, Eqs. (15.34) constitute a well-defined system of $4N$ coupled differential equations for the N unknown world lines $y_r^\mu(s_r)$. Only $3N$ out of these equations are functionally independent.

15.2.3 Small ε-Expansion of the Regularized Self-Field

To derive the Laurent expansion (15.32) of the regularized self-field $\mathcal{F}_\varepsilon^{\mu\nu}(y) = \mathcal{F}_\varepsilon^{\mu\nu}(y(s))$, according to the delay equations (15.27) we must preliminarily determine for every fixed s the retarded time s_ε, such that

$$(y(s) - y(s_\varepsilon))^2 = \varepsilon^2, \qquad (15.36)$$

$$y^0(s) > y^0(s_\varepsilon) \quad \Leftrightarrow \quad s_\varepsilon < s. \qquad (15.37)$$

In basis of formulas (15.21) and (15.28) the regularized self-field can then be written in the form

$$\mathcal{F}_\varepsilon^{\mu\nu}(y(s)) = \frac{e}{4\pi(u_\varepsilon L_\varepsilon)^3}\left(L_\varepsilon^\mu u_\varepsilon^\nu + L_\varepsilon^\mu L_{\varepsilon\gamma}\left(u_\varepsilon^\gamma w_\varepsilon^\nu - w_\varepsilon^\gamma u_\varepsilon^\nu\right) - (\mu \leftrightarrow \nu)\right),$$

$$(15.38)$$

where we have set

$$u_\varepsilon^\mu = u^\mu(s_\varepsilon), \qquad w_\varepsilon^\mu = w^\mu(s_\varepsilon), \qquad L_\varepsilon^\mu = y^\mu(s) - y^\mu(s_\varepsilon). \tag{15.39}$$

To perform the expansion of the field (15.38) around $\varepsilon = 0$ we need the expansion of s_ε around s in powers of ε, which can be derived from the delay equation (15.36). Since in the limit for $\varepsilon \to 0$ the retarded time s_ε tends to s, it is convenient to introduce a *positive* parameter Δ, depending on s and ε, according to

$$s = s_\varepsilon + \Delta, \qquad \lim_{\varepsilon \to 0} \Delta = 0. \tag{15.40}$$

To solve the delay equation (15.36) perturbatively we resort to the expansion in powers of Δ

$$y^\mu(s) - y^\mu(s_\varepsilon) = y^\mu(s) - y^\mu(s - \Delta) = \Delta u^\mu(s) - \frac{1}{2}\Delta^2 w^\mu(s) + O(\Delta^3),$$

so that the former takes the form

$$(y(s) - y(s_\varepsilon))^2 = \left(\Delta u(s) - \frac{1}{2}\Delta^2 w(s) + O(\Delta^3)\right)^2 = \Delta^2 + O(\Delta^4) = \varepsilon^2.$$

For a fixed s we obtain thus the relation

$$\Delta = \varepsilon(1 + O(\varepsilon^2)), \tag{15.41}$$

allowing us to trade the expansion in powers of ε for an expansion in powers of Δ.

Expansions in powers of Δ. Since in the denominator of the field (15.38) appears the factor $(u_\varepsilon L_\varepsilon)^3$, and L_ε^μ is of order Δ, it is necessary to expand the numerator of this field up to terms of order Δ^3. Setting $u^\mu = u^\mu(s)$ and $w^\mu = w^\mu(s)$, and using the elementary Taylor expansions

$$L_\varepsilon^\mu = \Delta u^\mu - \frac{1}{2}\Delta^2 w^\mu + \frac{1}{6}\Delta^3 \frac{dw^\mu}{ds} + O(\Delta^4),$$

$$u_\varepsilon^\mu = u^\mu - \Delta w^\mu + \frac{1}{2}\Delta^2 \frac{dw^\mu}{ds} + O(\Delta^3),$$

$$w_\varepsilon^\mu = w^\mu - \Delta \frac{dw^\mu}{ds} + O(\Delta^2),$$

for the various terms appearing in (15.38) we so obtain

$$(u_\varepsilon L_\varepsilon) = \Delta + O(\Delta^3),$$

$$L_\varepsilon^\mu u_\varepsilon^\nu - L_\varepsilon^\nu u_\varepsilon^\mu = -\frac{1}{2}\Delta^2(u^\mu w^\nu - u^\nu w^\mu) + \frac{1}{3}\Delta^3\left(u^\mu \frac{dw^\nu}{ds} - u^\nu \frac{dw^\mu}{ds}\right) + O(\Delta^4),$$

$$u_\varepsilon^\gamma w_\varepsilon^\nu - w_\varepsilon^\gamma u_\varepsilon^\nu = u^\gamma w^\nu - u^\nu w^\gamma - \Delta\left(u^\gamma \frac{dw^\nu}{ds} - u^\nu \frac{dw^\gamma}{ds}\right) + O(\Delta^2),$$

$$L_\varepsilon^\mu L_{\varepsilon\gamma}(u_\varepsilon^\gamma w_\varepsilon^\nu - w_\varepsilon^\gamma u_\varepsilon^\nu) - (\mu \leftrightarrow \nu) = \Delta^2 u^\mu w^\nu - \Delta^3 u^\mu \frac{dw^\nu}{ds} - (\mu \leftrightarrow \nu) + O(\Delta^4).$$

Inserting these expansions in the regularized self-field (15.38) we finally find

$$\mathcal{F}_\varepsilon^{\mu\nu}(y(s)) = \frac{e}{8\pi\Delta}(u^\mu w^\nu - u^\nu w^\mu) - \frac{e}{6\pi}\left(u^\mu \frac{dw^\nu}{ds} - u^\nu \frac{dw^\mu}{ds}\right) + O(\Delta).$$

$$(15.42)$$

Since in basis of the relation (15.41) Δ differs from ε by terms of order ε^3, in this expansion we can replace Δ with ε, so that the result is (15.32).

15.2.4   Basic Properties of the Lorentz-Dirac Equation

The role of the equation. We insist that there is no way to *logically infer* the Lorentz-Dirac equation from the fundamental equations of electrodynamics, i.e. from the Maxwell and Lorentz equations (15.1) and (15.2). Correspondingly, however convincing the "derivation" presented in Sect. 15.2.2 may appear, it is important to realize that this derivation – like many similar ones present in the literature – constitutes nothing else than an *argument* in its favor. Nonetheless, we promote the Lorentz-Dirac equation to a *new* fundamental equation of electrodynamics – replacing the original Lorentz equation – in that, as we shall show in Chap. 16, it is *dictated* by the basic paradigm of *local four-momentum conservation*. In fact, it was not without reason that Dirac based his relativistic derivation of the equation on conservation arguments. However, from the point of view of the *foundations* of the theory, there remains the fact that the Lorentz-Dirac equation must be considered as a new *postulate* of classical electrodynamics.

So far we have dealt with the dynamics of a system of charged particles considering in the first instance the external forces and, in case, the forces of mutual interaction, treating, thus, the radiation reaction forces as a perturbation. Now that we have available a system of finite equations of motion which take into account the radiation reaction, the analysis of the dynamics of the system must be based on the corresponding system of equations (15.34). In the forthcoming sections we will illustrate possible procedures to deal with this system, by means of several examples.

Radiation reaction four-force and Schott term. We write the Lorentz-Dirac equation for a single particle (15.19) in the form

$$\frac{dp^\mu}{ds} = \Gamma^\mu + eF_{\mathrm{in}}^{\mu\nu}(y)u_\nu,$$

$$(15.43)$$

where the four-vector

$$\Gamma^\mu = \frac{e^2}{6\pi} \left(\frac{dw^\mu}{ds} + w^2 u^\mu \right) \tag{15.44}$$

represents the *radiation reaction four-force*. It consists of two contributions. The second, called *Larmor term*, is intimately related to four-momentum conservation. In contrast, the first contribution, called *Schott term*, is necessary to make Γ^μ compatible with the constraint $u_\mu dp^\mu/ds = 0$. In fact, the kinematical relation (15.18) implies the identity

$$u_\mu \Gamma^\mu = 0. \tag{15.45}$$

The force Γ^μ represents a *relativistic* correction to the external Lorentz force $eF_{\text{in}}^{\mu\nu}(y)u_\nu$, starting with terms of order $1/c^3$. The simplest way to see it is to rewrite equation (15.43) in the equivalent form (15.20), and to notice that the time τ (15.10) is of order $1/c^3$. As one may expect, and as we will see more in detail in Sect. 15.4, the fact that the radiation reaction force starts with terms of order $1/c^3$ is strictly related with the fact that in the non-relativistic limit, i.e. in the dipole approximation, the energy radiated by a charged particle is of this same order.

Four-momentum balance. As the origin of the Schott term does not reside in four-momentum conservation, this term should not take part of the *overall* energy-momentum balance of a radiation process, as we have implicitly assumed in all analyses of these processes performed so far. To verify this expectation we choose an external field confined to a bounded region and consider a particle that in the initial state has not yet entered this region, while in the final state it has already left it. Integrating equation (15.43) between these two states we then find that when passing through this region the four-momentum of the particle changes by

$$\Delta p^\mu = e \int_i^f F_{\text{in}}^{\mu\nu}(y)u_\nu \, ds + \frac{e^2}{6\pi} \int_i^f w^2 u^\mu \, ds + \frac{e^2}{6\pi} \left(w_f^\mu - w_i^\mu \right). \tag{15.46}$$

In regions where the external field vanishes the particle performs a uniform linear motion,[3] and so we have $w_f^\mu = 0 = w_i^\mu$. Consequently, the Schott term drops out from Eq. (15.46). An identical conclusion is reached if the particle performs a quasi-periodic motion, since in that case we have $w_f^\mu \approx w_i^\mu$. The total change of the four-momentum (15.46) is thus determined by the Larmor term – which we know to represent, actually, the radiated four-momentum reaching infinity – and, obviously, by the external field $F_{\text{in}}^{\mu\nu}$ acting on the particle, which in the presence of more particles must be replaced by the field $F_r^{\mu\nu}$ (15.35).

Violation of determinism and supplementary conditions. Although the Schott term does not participate to the overall four-momentum exchange, it has far-reaching consequences on the *local* dynamics of the particle, as it contains the *third* derivative of the trajectory $\mathbf{y}(t)$. Indeed, this feature leads to the dramatic conclusion that, once

---

[3] At this stage of the discussion we are disregarding the *preacceleration* phenomenon, see Sects. 15.2.6 and 15.3.

one has fixed the usual initial data $\mathbf{y}(0)$ and $\mathbf{v}(0)$, the Lorentz-Dirac equation does not admit a unique solution, in apparent violation of the Newtonian determinism. In fact, once these initial conditions have been chosen, there exist infinite different trajectories $\mathbf{y}(t)$, associated with the infinite different values $\mathbf{a}(0)$ of the initial acceleration. This circumstance is in net contrast with the experimental observation that in nature the acceleration is, actually, determined by the *dynamics*, i.e. by the forces acting on the particle. From a theoretical point of view, the fundamental flaw of the Lorentz-Dirac equation remains, however, that, being a third-order differential equation, it has lost the *predictive power* inherent in Newton's second law $m\mathbf{a} = \mathbf{F}$ as a second-order differential equation. As eventually we do not want to renounce to the Newtonian determinism, a possible solution of this problem is represented by the hypothesis that not all solutions of the Lorentz-Dirac equation correspond to motions *realized in nature*. If this is the case, we must identify a criterion that selects the physically allowed motions, without compromising Lorentz invariance. Considering, for definiteness, a particle that performs an *unbounded motion*, see Sect. 7.1, and assuming that the external fields vanish sufficiently fast at spatial infinity, then there exists a natural set of such *supplementary conditions*. In fact, in this case the particle is subject to the external fields essentially for a limited time, and so it is natural to assume that for large times its acceleration tends to zero, whereas its velocity tends to a constant vector with a magnitude less than the speed of light. Correspondingly, we impose the Lorentz-invariant supplementary conditions[4]

$$\lim_{s \to +\infty} w^\mu(s) = 0, \qquad \lim_{s \to +\infty} u^\mu(s) = u_\infty^\mu. \tag{15.47}$$

We do not impose analogous conditions for $s \to -\infty$, for a reason that will be clarified below. Notice that from a three-dimensional point of view the conditions (15.47) amount to

$$\lim_{t \to +\infty} \mathbf{a}(t) = 0, \qquad \lim_{t \to +\infty} \mathbf{v}(t) = \mathbf{v}_\infty, \qquad |\mathbf{v}_\infty| < 1. \tag{15.48}$$

Of course, under certain regularity conditions, the requirement of the asymptotic vanishing of the acceleration implies the asymptotic constancy of the velocity. The second conditions in (15.47) and (15.48) may hence become redundant.

Third-order determinism. In principle there exists an alternative – more pragmatic although less ambitious – strategy to the imposition of the supplementary conditions (15.47). Let us suppose that we measure at the initial time $t = 0$ not only the position and the velocity of the particle, but also its acceleration. With these three initial data the Lorentz-Dirac equation admits a unique solution $\mathbf{y}(t)$, and so we could predict the position of the particle for any subsequent time $t > 0$. However, apart

---

[4]In reality, the conditions (15.47) and (15.48) hold also for motions confined to a bounded region. In fact, for physical reasons, a particle confined to a bounded region cannot be fed by an external field forever. Correspondingly, it radiates energy until eventually it reaches a constant or vanishing velocity.

from being in conflict with the Newtonian determinism, this *third-order determinism* fails, for experimental as well as stability reasons. Let us illustrate the problem in the elementary case of a *free* particle for which, in particular, the external field $F_{in}^{\mu\nu}$ vanishes. In order to ascertain whether the particle performs a uniform linear motion, or if its acceleration is zero, the observer would measure the velocity of the particle at different times and, given the inevitability of experimental errors, eventually he would find for the acceleration a non-vanishing value $w^\mu(0)$, although very small. On the other hand, from the general solution of the Lorentz-Dirac equation for a free particle, see Eqs. (15.61) and (15.59), one sees that for any non-vanishing initial value $w^\mu(0)$ the four-acceleration grows so violently that the velocity of the particle tends rapidly to the velocity of light. Repeating the measurement at a later time, and finding within the experimental errors the same finite velocity of the initial time, the observer would, therefore, conclude that theory and experiment are in *maximal* disagreement. Conversely, if the measured initial acceleration $w^\mu(0)$ would be *strictly* zero, then according to the general solution (15.61) the acceleration would remain zero for all times, in agreement with the observation. Therefore, we conclude that the only way to make the theory agree with experiments would be to carry out measurements with *vanishing* errors, furnishing for the acceleration the value zero, but this is experimentally impossible.

Explicit violation of time-reversal invariance and arrow of time. From Sect. 2.2.2 we know that the fundamental equations of electrodynamics (2.20)–(2.22) are invariant under time reversal. In the case of a single charged particle this invariance ensures that if the configuration

$$\Sigma = \{\mathbf{y}(t), \mathbf{E}(t, \mathbf{x}), \mathbf{B}(t, \mathbf{x})\} \tag{15.49}$$

satisfies these equations, so does the configuration

$$\Sigma^* = \{\mathbf{y}(-t), \mathbf{E}(-t, \mathbf{x}), -\mathbf{B}(-t, \mathbf{x})\}. \tag{15.50}$$

Here $\mathbf{E}$ and $\mathbf{B}$ denote the total fields, given by the sum of the Liénard-Wiechert fields and of the external field. Subsequently, in Sect. 6.2.3, we found out that this invariance is subject to a *spontaneous* violation, in the sense that only one of the two configurations Σ and Σ^* is realized in nature, namely the one entailing radiation which is *emitted* from the charged particle, rather than absorbed. Now that we have eliminated the Liénard-Wiechert field and replaced the Lorentz equation with the Lorentz-Dirac equation, the situation changes again drastically: as the latter contains terms that are linear in a derivative of *odd* order of $\mathbf{y}(t)$, i.e. a third derivative, the Lorentz-Dirac equation violates *explicitly* time-reversal invariance. In particular, while the quantities dp^μ/ds and $F_{in}^{\mu\nu}(y)u_\nu$ are *vectors* under time reversal, the radiation reaction four-force Γ^μ (15.44) is a *pseudovector* under this transformation, see Sect. 2.2.2. Therefore, if now we understand that in (15.49) and (15.50) the fields $\mathbf{E}$ and $\mathbf{B}$ are the components of the external field $F_{in}^{\mu\nu}$, then, if the configuration Σ is a solution of equation (15.43), the configuration Σ^* in general is *not*. Since it is Γ^μ to violate time-reversal invariance, this asymmetry holds thus also in the absence of external fields. In particular, as we will see in Sect. 15.2.5, an important consequence of the explicit violation of time-reversal invariance is that the asymptotic behaviors of the particle's velocity for $t \to +\infty$ and $t \to -\infty$ are radically different, even if $F_{in}^{\mu\nu} = 0$. This is the reason for why the supplementary conditions (15.47) and (15.48) must be imposed only for large *positive* times; for explicit examples see Sects. 15.2.5, 15.2.6 and 15.3.

Theoretically there exists, of course, a world in which the *arrow of time* is inverted. In fact, the Lorentz-Dirac equation has been derived from the regularized Lorentz equation (15.26), in which $\mathcal{F}^{\mu\nu}$ is the *retarded* Liénard-Wiechert field (15.21). If in place of this field we had used the *advanced* Liénard-Wiechert field associated with the four-potential A_{adv}^μ defined by formulas (6.42) and (6.60), which entails a radiation that, incoming from infinity, propagates *towards* the particle, then the resulting Lorentz-Dirac equation would have been Eq. (15.43) with $-\Gamma^\mu$ in place of Γ^μ. If subject to this equation, the particle would absorb four-momentum, rather than emit it, and the conditions (15.48) should be imposed for large *negative* times. It is thus clear that classical electrodynamics, as a theory founded on the Lorentz-Dirac equation, violates explicitly time-reversal invariance by choosing the *same* preferred arrow of time inherent in the radiation. In a sense, the existence of a preferred arrow of time in our world represents one of the great mysteries of nature, that becomes even more profound if one realizes that the *same* arrow of time is inherent in the *second law of thermodynamics*, which requires the entropy of an isolated system to increase as t increases, as well as in the *expansion of the Universe*.

Lorentz-Dirac equation and variational principle. Once we have replaced the Lorentz equations (2.20) – which, we recall, can be derived from the action (4.13) – with the system of Lorentz-Dirac equations (15.34), there arises the natural question whether also the latter can be derived from a variational principle. The question is of basic importance, as the existence of an action, thanks to Noether's theorem, would automatically guarantee the validity of the main conservation laws. Unfortunately, the appearance of the third derivatives of the coordinates $y_r^\mu(s_r)$ in the system (15.34) prevents the existence of such an action. To show it we consider a single particle in the presence of a generic external field $F_{in}^{\mu\nu} = \partial^\mu A_{in}^\nu - \partial^\nu A_{in}^\nu$, in which case the system (15.34) reduces to the Lorentz-Dirac equation (15.43). In this case we know already that for a vanishing radiation reaction force Γ^μ the action would be given by formula (4.13), with A^μ replaced by A_{in}^μ. To reproduce the force Γ^μ (15.44), which contains a term linear in the third derivative of $y^\mu(s)$, one must add to the action (4.13) terms which are quadratic in $y^\mu(s)$, with a total of three derivatives. Accordingly, enforcing the reparameterization invariance of the world line, in addition to Lorentz invariance, the total action should have the general form

$$I = -m \int ds - e \int A_{in}^\mu \, dy_\mu + e^2 \int \left(a \frac{dy^\mu}{ds} \frac{d^2 y_\mu}{ds^2} + b \, y^\mu \frac{d^3 y_\mu}{ds^3} \right) ds, \quad (15.51)$$

where a and b are dimensionless constants. The third term is equal to $e^2 a \int u^\mu w_\mu ds$ and is hence identically zero. Similarly, the forth term amounts to a total derivative which does not contribute to the equations of motion,

$$y^\mu \frac{d^3 y_\mu}{ds^3} = \frac{d}{ds}\left(y^\mu \frac{d^2 y_\mu}{ds^2}\right) - \frac{dy^\mu}{ds}\frac{d^2 y_\mu}{ds^2} = \frac{d}{ds}\left(y^\mu \frac{d^2 y_\mu}{ds^2}\right).$$

Therefore, regardless of the values of a and b, the action (15.51) gives rise to the equation of motion $dp^\mu/ds = eF_{\text{in}}^{\mu\nu} u_\nu$, rather than to (15.43). As anticipated above, a negative consequence of the fact that the Lorentz-Dirac equation cannot be derived from a variational principle is that the main conservation laws – in particular the form of the energy-momentum tensor – must be established essentially from scratch, see Chap. 16.

15.2.5 Free Charged Particle

In a few simple situations the Lorentz-Dirac equation can be solved exactly, an instructive example being a free charged particle. For this case in Eq. (15.43) we must set $F_{\text{in}}^{\mu\nu} = 0$, so that only the radiation reaction four-force survives,

$$\frac{dp^\mu}{ds} = \frac{e^2}{6\pi}\left(\frac{dw^\mu}{ds} + w^2 u^\mu\right). \tag{15.52}$$

Non-relativistic limit and runaway solutions. Before facing the general solution of equation (15.52) we solve it in the non-relativistic limit. To determine its form in this limit one must reintroduce the velocity of light and expand its second member in powers of $1/c$, keeping only the lowest-order terms. The resulting space-time components of equation (15.52) read

$$\frac{d\varepsilon}{dt} = \frac{e^2}{6\pi c^3}\,\mathbf{v}\cdot\frac{d\mathbf{a}}{dt} + O\!\left(\frac{1}{c^5}\right), \tag{15.53}$$

$$\frac{d\mathbf{p}}{dt} = \frac{e^2}{6\pi c^3}\frac{d\mathbf{a}}{dt} + O\!\left(\frac{1}{c^5}\right), \tag{15.54}$$

where for the moment we have kept the relativistic energy and momentum ε and $\mathbf{p}$. The second members of these equations – representing the radiation reaction – start with terms of order $1/c^3$ as anticipated in Sect. 15.2.4. Performing the explicit expansion of equation (15.52) it turns out that in Eq. (15.53) at order $1/c^3$ both the Larmor term and the Schott term contribute, while in Eq. (15.54) only the Schott term contributes. As the kinematical relation $\varepsilon^2 = c^2 \mathbf{p}^2 + m^2 c^4$ implies the known identity

$$\varepsilon \frac{d\varepsilon}{dt} = c^2 \mathbf{p} \cdot \frac{d\mathbf{p}}{dt} \quad \leftrightarrow \quad \frac{d\varepsilon}{dt} = \mathbf{v} \cdot \frac{d\mathbf{p}}{dt}, \tag{15.55}$$

the *work-energy theorem* (15.53) follows, actually, from *Newton's second law* (15.54). This is, obviously, a consequence of the fact that the Lorentz-Dirac equation has only three functionally independent components. Taking into account that in the non-relativistic limit we have $\mathbf{p} = m\mathbf{v}$, Eq. (15.54) eventually reduces to the simple form

$$\mathbf{a} = \tau \frac{d\mathbf{a}}{dt}, \tag{15.56}$$

where τ is the characteristic time (15.10). The equation obtained shows that the explicit violation of the time-reversal invariance of the Lorentz-Dirac equation survives also in the non-relativistic limit. In fact, under the transformation $t \to -t$ Eq. (15.56) goes over into $\mathbf{a} = -\tau \, d\mathbf{a}/dt$. Consequently, if a trajectory with velocity $\mathbf{v}(t)$ is a solution of the equation of motion (15.56), the trajectory obtained via time reversal, namely $\mathbf{v}^*(t) = -\mathbf{v}(-t)$, in general is not. More in detail, if we impose on the acceleration and on the velocity the initial conditions $\mathbf{a}(0)$ and $\mathbf{v}(0)$, respectively, Eq. (15.56) admits the unique solution

$$\mathbf{a}(t) = e^{t/\tau} \mathbf{a}(0) \quad \Rightarrow \quad \mathbf{v}(t) = \tau \big(e^{t/\tau} - 1 \big) \, \mathbf{a}(0) + \mathbf{v}(0). \tag{15.57}$$

Therefore, unless $\mathbf{a}(0)$ vanishes, the velocity of the transformed trajectory

$$\mathbf{v}^*(t) = -\mathbf{v}(-t) = -\tau \big(e^{-t/\tau} - 1 \big) \, \mathbf{a}(0) - \mathbf{v}(0)$$

is not of the form (15.57), and so does not solve (15.56). The solution (15.57) entails, however, a more serious problem of *phenomenological* nature: although the particle is not subject to external forces it accelerates *spontaneously*, and for $t \to +\infty$ its velocity increases exponentially unless the initial acceleration $\mathbf{a}(0)$ vanishes. For this reason the solutions (15.57) are called *runaway solutions*. Vice versa, for $t \to -\infty$ the velocity $\mathbf{v}(t)$ tends to the constant value $\mathbf{v}(0) - \tau \mathbf{a}(0)$. Due to this "unrealistic" behavior, not all solutions given in (15.57) are acceptable as *physical* trajectories. In fact, the role of the supplementary conditions established in Sect. 15.2.4 is precisely to remove the *unphysical* trajectories from the general solution (15.57): the conditions (15.48) oblige us, indeed, to choose $\mathbf{a}(0) = 0$. In this way, the velocity at time t becomes the velocity of a *uniform linear motion* $\mathbf{v}(t) = \mathbf{v}(0)$, as appropriate for a *free* particle.

Exact solution. The general solution (15.57) of the non-relativistic Lorentz-Dirac equation (15.53) entails a particle's speed that for $t \to +\infty$ tends to infinity, actually a behavior that invalidates the non-relativistic approximation itself. Nevertheless, our main conclusions carry over to the relativistic case. To show it we proceed to the exact solution of equation (15.52). For this purpose we rewrite it in the form

$$w^\mu = \tau \left(\frac{dw^\mu}{ds} + w^2 u^\mu \right), \tag{15.58}$$

and perform the change of variable

$$s \to \lambda(s) = e^{s/\tau}, \qquad \frac{d}{ds} = \frac{\lambda}{\tau} \frac{d}{d\lambda}, \tag{15.59}$$

where we have set again $c = 1$. Denoting the derivative $d/d\lambda$ with a *prime* we then have

$$w^\mu = \frac{\lambda}{\tau} u'^\mu, \qquad \frac{dw^\mu}{ds} = \frac{\lambda}{\tau^2} \left(\lambda u''^\mu + u'^\mu \right),$$

so that Eq. (15.58) reduces to

$$u''_\mu + (u'u')u_\mu = 0, \tag{15.60}$$

where we have introduced the notation (7.32). Thanks to the kinematical identity $(uu') = \tau(uw)/\lambda = 0$, contracting equation (15.60) with u'^μ we obtain

$$(u'u'') = \frac{1}{2} (u'u')' = 0 \quad \Rightarrow \quad (u'u') = -K^2,$$

where K is a positive or vanishing constant. In this way Eq. (15.60) reduces eventually to the linear differential equation $u''_\mu = K^2 u_\mu$, with general solution

$$u^\mu(s) = A^\mu e^{K\lambda} + B^\mu e^{-K\lambda}, \qquad w^\mu(s) = \frac{\lambda K}{\tau} \left(A^\mu e^{K\lambda} - B^\mu e^{-K\lambda} \right), \tag{15.61}$$

where A^μ and B^μ are constant vectors. As we must have $u^2(s) = 1$ for all s, these vectors are subject to the constraints

$$A^\mu A_\mu = 0 = B^\mu B_\mu, \qquad A^\mu B_\mu = \frac{1}{2}. \tag{15.62}$$

In particular, $A^0 = |\mathbf{A}|$ and $B^0 = |\mathbf{B}|$. The solutions (15.61) exhibit again a runaway behavior. In fact, in the limit for $s \to +\infty$ the parameter $\lambda(s)$ (15.59) increases exponentially, and so, if $K \neq 0$, for large positive s all components of the four-velocity and of the four-acceleration (15.61) diverge. Using that $u^0(s) = dt/ds$, from formulas (15.59) and (15.61) one derives, in particular, that for large positive proper times s the coordinate time t increases as

$$t \approx \frac{\tau |\mathbf{A}| e^{K\lambda}}{K\lambda} \approx \frac{\tau |\mathbf{A}| e^{Ke^{s/\tau}}}{Ke^{s/\tau}}.$$

Accordingly, the energy as a function of time diverges as

$$\varepsilon(t) = mu^0(s) \approx m|\mathbf{A}|\, e^{K\lambda} \approx \frac{mt}{\tau} \ln\left(\frac{t}{\tau}\right).$$

Notice that, although this behavior is essentially linear in time, since τ is a very small time scale the energy of the particle grows rapidly. Correspondingly, the particle's velocity tends to the velocity of light

$$\mathbf{v}_+ = \lim_{s\to+\infty} \mathbf{v} = \lim_{s\to+\infty} \frac{\mathbf{u}}{u^0} = \frac{\mathbf{A}}{|\mathbf{A}|}, \qquad |\mathbf{v}_+| = 1.$$

In contrast, in the limit for $s \to -\infty$ we have $\lambda(s) \to 0$, and consequently the four-velocity admits the finite limit

$$\lim_{s\to-\infty} u^\mu(s) = A^\mu + B^\mu.$$

The velocity tends thus to a constant vector with a magnitude less than the speed of light[5]

$$\mathbf{v}_- = \lim_{s\to-\infty} \frac{\mathbf{u}}{u^0} = \frac{\mathbf{A}+\mathbf{B}}{|\mathbf{A}|+|\mathbf{B}|}, \qquad |\mathbf{v}_-| < 1.$$

Finally, we must impose the supplementary conditions (15.47). Since $s \to +\infty$ amounts to $\lambda \to \infty$, from the general solution (15.61) we see that the four-acceleration $w^\mu(s)$ vanishes for $s \to +\infty$ only if $K = 0$, in which case the solution reduces to

$$w^\mu(s) = 0, \qquad u^\mu(s) = A^\mu + B^\mu.$$

We conclude, thus, that in the case of a free particle the unique solutions of the Lorentz-Dirac equation which are compatible with the supplementary conditions (15.47) are *uniform linear motions*, as expected. The situation encountered in this section is prototypical: (i) the general solution of the Lorentz-Dirac equation exhibits always a runaway behavior, and (ii) the role of the supplementary conditions is precisely to eliminate these unphysical solutions. However, in Sects. 15.2.6 and 15.3 we will see that – in contrast to the case of a free particle – in the presence of external forces the solutions of the Lorentz-Dirac equation which survive the conditions (15.47) in general violate *causality*, although they do not exhibit runaway behaviors.

15.2.6 One-Dimensional Motion: Preacceleration

In this section we investigate the motion of a particle in the presence of an external field $F_{\text{in}}^{\mu\nu}(x)$ consisting only of a static static electric field of constant direction

---

[5]The choices $\mathbf{A} = 0$ and/or $\mathbf{B} = 0$ are forbidden by the constraints (15.62).

$\mathbf{E}(z) = (0, 0, E(z))$. We shall consider only motions that occur in the same direction of the field. Even if it seems rather elementary, this example incorporates all problematic aspects inherent in the Lorentz-Dirac equation for a particle subject to a generic external field. For the moment we make no assumption about the functional dependence of the field $E(z)$ on z, apart from the asymptotic conditions

$$\lim_{|z| \to \infty} E(z) = 0.$$

Since the particle moves on the z axis, it is sufficient to solve the z component of the Lorenz-Dirac equation (15.20). Setting

$$u = u^3,$$

and denoting the derivative d/ds with a *prime*, we then have

$$u^\mu = (u^0, 0, 0, u), \qquad (u^0)^2 - u^2 = 1, \qquad w^\mu = (u^{0\prime}, 0, 0, u').$$

The components of the external four-force take thus the form

$$eF_{\text{in}}^{\mu\nu} u_\nu = \left(eEu, 0, 0, eEu^0\right). \tag{15.63}$$

Moreover, from the kinematical constraint $u^0 = \sqrt{1 + u^2}$ we derive the relation

$$u^{0\prime} = \frac{d}{ds} \sqrt{1 + u^2} = \frac{uu'}{\sqrt{1 + u^2}},$$

which, in turn, implies that

$$w^2 = (u^{0\prime})^2 - u'^2 = -\frac{u'^2}{1 + u^2}. \tag{15.64}$$

Considering as unknown the function $u(s)$, the z component of the Lorentz-Dirac equation (15.20) then becomes

$$u' = \tau \left(u'' - \frac{u'^2 u}{1 + u^2} \right) + \frac{F}{m} u^0, \qquad F = eE, \tag{15.65}$$

where F represents the external *force* in three-dimensional notation. Introducing the new unknown function $V(s)$ in place of $u(s)$, via the trigonometric relations

$$u = \sinh V, \qquad u^0 = \cosh V, \tag{15.66}$$

after some algebra equation (15.65) goes over into the simpler equation for V

$$V' = \tau V'' + \frac{F}{m}. \tag{15.67}$$

Below we shall also use the kinematical relations

$$V' = \frac{dV}{du} u' = \frac{u'}{u_0} = \frac{du}{dt}. \tag{15.68}$$

Asymptotic analysis. Let us consider a generic solution of the differential equation (15.67). Since at spatial infinity the external field vanishes, in the limits for $s \to \pm\infty$ this equation assumes the asymptotic form $V' \approx \tau V''$ entailing the general runaway solution[6]

$$V(s) \approx Ke^{s/\tau} + D, \tag{15.69}$$

where K and D are arbitrary constants. Notice that, although the external force at spatial infinity vanishes, for $s \to +\infty$ the four-velocity $u(s) = \sinh V(s)$ grows exponentially, unless K vanishes. In particular, if $E = 0$ throughout space, formula (15.69) yields an exact solution for the free particle, see Eqs. (15.61) and Problem 15.6. The supplementary conditions (15.47) enforce, hence, again the choice $K = 0$.

External field confined to an interval. To investigate the *local* characteristics of the solutions of the Lorentz-Dirac equation (15.67) satisfying the conditions (15.47), we consider the prototypical situation of a *uniform* electric field $E(z)$ vanishing, say, outside the interval $[z_a, z_b]$. Denoting the proper times at which the particle passes through z_a and z_b by a and b, respectively, the external force F in (15.65) then takes the form

$$F(s) = eE(z(s)) = eE_0\chi_{[a,b]}(s), \tag{15.70}$$

where $\chi_{[a,b]}(s)$ is the characteristic function of the proper-time interval $[a, b]$, and E_0 is the magnitude of the electric field in the interval $[z_a, z_b]$. For concreteness, we assume that the particle enters the interval $[z_a, z_b]$ from the left and leaves it from the right. For the force (15.70), the general solution of equation (15.67) – actually a *complete first integral* – is given by

$$V'(s) = \frac{F(s)}{m} + \frac{eE_0}{m}\left(H(a - s)\,e^{(s-a)/\tau} - H(b - s)\,e^{(s-b)/\tau}\right) + \frac{K}{\tau}\,e^{s/\tau}, \tag{15.71}$$

where K is an arbitrary constant, see Problem 15.3. Notice that, since for $s > b$ both above Heaviside functions vanish, for $s \to +\infty$ the solutions (15.71) exhibit the universal asymptotic runaway behavior (15.69). To select the physical solutions we must again impose the conditions (15.47), which in the case at hand reduce to

---

[6]We assume here that the particle is not in a *bound state*, but that it performs an unbounded motion. In this case for $s \to \pm\infty$ it tends to the asymptotic positions $z = \pm\infty$.

$$\lim_{s \to +\infty} u' = 0, \qquad \lim_{s \to +\infty} u = u_\infty. \tag{15.72}$$

Formulas (15.66) then imply, in particular, that the energy $\varepsilon = mu^0$ admits a finite limit for $s \to +\infty$. Since according to the relations (15.68) we have $V' = u'/u^0$, the conditions (15.72) imply that the function $V(s)$ must obey the limit

$$\lim_{s \to +\infty} V'(s) = 0. \tag{15.73}$$

It then follows that the *physical* solution of the Lorentz-Dirac equation is given by equation (15.71) with $K = 0$. Introducing the relativistic momentum $p = mu$, and recalling from formulas (15.68) the relation $V' = du/dt$, we can then recast this "solution" in the form

$$\frac{dp}{dt} = F(s) + F_{\text{rr}}(s), \tag{15.74}$$

where we have introduced the *spatial* radiation reaction force

$$F_{\text{rr}}(s) = eE_0 \left(H(a - s)\, e^{(s-a)/\tau} - H(b - s)\, e^{(s-b)/\tau} \right). \tag{15.75}$$

To show that this is indeed the correct interpretation of the quantity $F_{\text{rr}}(s)$ we write out the z component of the third-order Lorentz-Dirac equation (15.43), of which the second-order Eq. (15.74) represents an exact first integral. For the external field at hand, the z component of equation (15.43) reads, see formulas (15.63) and (15.70),

$$\frac{dp}{dt} = F(s) + \frac{\Gamma^3(s)}{u^0(s)}, \tag{15.76}$$

where $\Gamma^3(s)$ is the z component of the radiation reaction four-force (15.44). Comparing equations (15.74) and (15.76) we see, therefore, that $F_{\text{rr}}(s)$ equals precisely the z component of the spatial radiation reaction force[7]

$$F_{\text{rr}}(s) = \frac{\Gamma^3(s)}{u^0(s)}. \tag{15.77}$$

Preacceleration and violation of causality. The second-order differential equation (15.74) represents in all respects the relativistic generalization of Newton's second law $\mathbf{F} = m\mathbf{a}$, taking into account the radiation reaction. In addition to the external force $F(s)$ (15.70) there appears, in fact, the radiation reaction force $F_{\text{rr}}(s)$ (15.75). However, while the former acts only during the proper-time interval $a < s < b$, the latter is different from zero for *all* $s < b$, and, in particular, for $s < a$ it acts in the same direction as the external field. Therefore, we reach the dramatic conclusion that

---

[7]The four-force f^μ is defined by the equation $dp^\mu/ds = f^\mu$, while the spatial relativistic force $\mathbf{F}$ is defined by $d\mathbf{p}/dt = \mathbf{F}$. These quantities are thus related by $\mathbf{F} = \mathbf{f}/u^0$.

the particle undergoes a *preacceleration* for all times $s < a$, during which there is *no external force* acting on it. In fact, this behavior is in sharp conflict with causality, as the *effect*, i.e. the acceleration, precedes the *cause*, i.e. the external force. On the other hand, thanks to the exponentials present in (15.75), the radiation reaction force, responsible for the preacceleration, is significantly different from zero only in the small intervals $a - \tau \lesssim s \lesssim a$ and $b - \tau \lesssim s \lesssim b$. This means that this force causes an appreciable distortion of the profile of the external force (15.70) only if $b - a$ is of the same order of τ, that is to say, if the characteristic time scale $b - a$ of the external field is extremely small, i.e. of the order $\tau \sim 10^{-23}$ s.

In conclusion, the reduction of the Lorentz-Dirac equation to a second-order differential equation by means of the conditions (15.47) – a process that, we repeat, is unavoidable if we insist on the Newtonian determinism and if we want to get rid of the runaway solutions – has introduced a *causality violation* in the form of a preacceleration, which manifests itself on time scales of order τ. As we will see in Sect. 15.3, these conclusions have general validity. There remains, of course, the question whether the causality violation introduced by the Lorentz-Dirac equation can, actually, be observed in experiments. We postpone the answer to this fundamental question to Sect. 15.3.

Emission of radiation and hyperbolic motion. By multiplying the equation of motion (15.74) by the particle's velocity $v = u/u_0 = p/\varepsilon$, we obtain the work-energy theorem

$$\frac{d\varepsilon}{dt} = \mathcal{W}_{\mathrm{ext}}(s) + \mathcal{W}_{\mathrm{rad}}(s), \quad \mathcal{W}_{\mathrm{ext}}(s) = vF(s), \quad \mathcal{W}_{\mathrm{rad}}(s) = vF_{\mathrm{rr}}(s). \quad (15.78)$$

The power $\mathcal{W}_{\mathrm{ext}}(s)$ furnished by the external force $F(s)$ is different from zero only in the proper-time interval $[a, b]$. Conversely, the "radiated power" $\mathcal{W}_{\mathrm{rad}}(s)$ is appreciably different from zero in the small intervals $a - \tau \lesssim s \lesssim a$ and $b - \tau \lesssim s \lesssim b$. Enforcing energy conservation, we conclude hence that the particle emits an appreciable amount of radiation only in the vicinities of the regions where the external field *varies*, whereas in the regions where $E(z)$ is constant the radiation is negligible. In particular, as we see from the formula (15.75) for the radiation reaction force, for proper times s in the interval $[a, b]$ we have $\mathcal{W}_{\mathrm{rad}}(s) < 0$, whereas for $s < a$ we have $\mathcal{W}_{\mathrm{rad}}(s) > 0$. This means that, while the particle moves in the region where there is a non-vanishing external field, it *emits* radiation, as expected, while during the preacceleration period, i.e. for $s < a$, it is as if it would *absorb* radiation. This anomalous behavior occurs, however, while the particle is in the regime were causality is, anyhow, violated. Nevertheless, since the total radiated energy

$$\varepsilon_{\mathrm{rad}} = -\int_{-\infty}^{\infty} \mathcal{W}_{\mathrm{rad}}(s) \, dt = -\int_{-\infty}^{\infty} u(s) F_{\mathrm{rr}}(s) \, ds \qquad (15.79)$$

is also equal to the integral from $t = -\infty$ to $t = \infty$ of the relativistic Larmor power (10.9), which is always greater than or equal to zero, this energy is *positive*, for details see Problem 15.7.

Is is interesting to analyze the limiting case where the constant electric field E_0 extends over the whole space, which corresponds to send in the formulas above a to $-\infty$ and b to $+\infty$. In this limit the radiation reaction force $F_{\text{rr}}(s)$ (15.75) and, consequently, the radiated power $\mathcal{W}_{\text{rad}}(s)$ vanish, and so Eq. (15.74) reduces to the equation of a *hyperbolic motion* (see Sect. 7.1)

$$\frac{dp}{dt} = F(s) = eE_0, \tag{15.80}$$

whose solution is

$$\mathbf{y}(t) = \left(0, 0, \sqrt{t^2 + \left(\frac{m}{eE_0}\right)^2}\right). \tag{15.81}$$

The same result (15.80) is obtained if one inserts the force $F(s) = eE_0$ directly in the equation of motion (15.67), and enforces then the supplementary condition (15.73). The particle moves, thus, as if it were only under the influence of the electric field $\mathbf{E} = (0, 0, E_0)$, knowing nothing about radiation reaction. We conclude, therefore, that a particle moving in the presence of a constant uniform electric field in the same direction of the field, and which obeys the Lorentz-Dirac equation subject to the conditions (15.47), experiences *no self-interaction*. If, in addition, we enforce energy conservation, we conclude furthermore that the particle does not emit radiation. In particular, comparing equations (15.80) and (15.76) we see that for a hyperbolic motion the component Γ^3 of the radiation reaction four-force vanishes, and so, thanks to the identity $u_\mu \Gamma^\mu = 0$, it vanishes as a whole: $\Gamma^\mu = 0$. This can be verified also directly by inserting the trajectory (15.81) in the definition of Γ^μ (15.44). Stated differently, for a hyperbolic motion the *Schott term* compensates exactly the *Larmor term* in formula (15.44), so that – for this very particular case – the Lorentz-Dirac equation reduces to the original Lorentz equation $dp^\mu/ds = eF_{\text{in}}^{\mu\nu}(y)u_\nu$. There is, of course, no contradiction with the argument (15.46) according to which the emitted four-momentum is determined solely by the Larmor term. In fact, for a hyperbolic motion the four-acceleration $w^\mu(s)$ never vanishes and for $s \to \pm\infty$ it actually diverges, even though for $t \to \pm\infty$ the spatial acceleration $\mathbf{a}(t)$ associated with the trajectory (15.81) vanishes as $1/t^3$. However, eventually one must bear in mind that an infinitely extended static electric field does not exist in nature: outside a certain region it must necessarily vanish. But, as seen above, precisely in the region where the electric field transits from a vanishing to a non-vanishing value, or vice versa, the particle emits radiation and experiences a non-vanishing radiation reaction force. For an estimate of the emitted energy, see Problem 15.7.

15.3   Rohrlich's Integro-Differential Equation

Aim of this section is to analyze the general characteristics of a generic solution $y^\mu(s)$ of the Lorentz-Dirac equation (15.20), subject to the supplementary conditions (15.47). As exemplified in Sects. 15.2.5 and 15.2.6, if the external field is sufficiently regular, the conditions

$$y^\mu(0) = y_0^\mu, \qquad u^\mu(0) = u_0^\mu, \tag{15.82}$$

$$\lim_{s \to +\infty} w^\mu(s) = 0, \tag{15.83}$$

determine a unique solution of this equation. We begin by dealing with the supplementary condition (15.83). A standard method to impose on a differential equation of order n a boundary condition like (15.83) consists in transforming the equation in an integro-differential equation of order $n - 1$, that automatically takes into account the boundary condition. In general, there exist several ways to perform such a reduction. Here we follow the method of F. Rohrlich [1], which has, in particular, the advantage of preserving manifest Lorentz invariance. Accordingly, we recast the Lorentz-Dirac equation in the form

$$m\left(w^\mu - \tau \frac{dw^\mu}{ds}\right) = \frac{e^2}{6\pi}\, w^2 u^\mu + e F_{\text{in}}^{\mu\nu} u_\nu \equiv \mathcal{F}^\mu. \tag{15.84}$$

As the Larmor term $e^2 w^2 u^\mu / 6\pi$ represents the four-momentum, emitted per unit proper time, that reaches infinity, and $e F_{\text{in}}^{\mu\nu} u_\nu$ is the external four-force, we interpret the vector $\mathcal{F}^\mu(s)$ as the total *effective* four-force acting on the particle at proper time s.

A qualitative argument for preacceleration. Before proceeding any further, we give a qualitative argument for the unavoidable presence of preacceleration effects in a generic solution of equation (15.84). Since τ is a very small time scale, neglecting terms of order τ^2 we can, in fact, rewrite equation (15.84) in the form

$$m w^\mu(s - \tau) \approx \mathcal{F}^\mu(s),$$

which, in turn, is equivalent to

$$m w^\mu(s) \approx \mathcal{F}^\mu(s + \tau). \tag{15.85}$$

The particle's acceleration at time s is thus determined not by the value of the effective force at the same time s, but rather by its value at the *advanced* time $s' \sim s + \tau$. The particle experiences, hence, a preacceleration. Obviously, the effect is observable only if the effective force varies appreciably during a time of order τ.

We now proceed to the reduction of the third-order differential equation (15.84) to a second-order differential equation, which incorporates the asymptotic condition

(15.83). We first multiply equation (15.84) by $e^{-s/\tau}$, rewriting it in the form

$$-m\tau \frac{d}{ds}\left(e^{-s/\tau}w^{\mu}(s)\right) = e^{-s/\tau}\mathcal{F}^{\mu}(s).$$

We then integrate this equation between a generic initial instant s and a final instant b, obtaining the – still equivalent – equation

$$m\left(e^{-s/\tau}w^{\mu}(s) - e^{-b/\tau}w^{\mu}(b)\right) = \frac{1}{\tau}\int_{s}^{b} e^{-q/\tau}\mathcal{F}^{\mu}(q)\,dq. \tag{15.86}$$

Now we are in a position to impose the boundary condition (15.83) requiring that

$$\lim_{b\to+\infty} w^{\mu}(b) = 0.$$

Taking the limit for $b \to +\infty$ of both members of equation (15.86), we then find

$$me^{-s/\tau}w^{\mu}(s) = \frac{1}{\tau}\int_{s}^{\infty} e^{-q/\tau}\mathcal{F}^{\mu}(q)\,dq. \tag{15.87}$$

The non-trivial point of this procedure is that, despite the presence of the damping factor $e^{-b/\tau}$, for a *generic* solution of the Lorentz-Dirac equation the product $e^{-b/\tau}w^{\mu}(b)$ appearing in (15.86) *diverges* in the limit for $b \to +\infty$. In fact, as follows, for instance, from the solution of the free particle (15.61), or from the formulas (15.66) and (15.69) for the one-dimensional motion, for large proper times b the four-acceleration in general increases faster than an exponential, namely as

$$w^{\mu}(b) \sim Ke^{b/\tau}e^{Ke^{b/\tau}}.$$

Performing finally the change of variable $q(k) = \tau k + s$, Eq. (15.87) goes over into *Rohrlich's integro-differential equation*

$$mw^{\mu}(s) = \int_{0}^{\infty} e^{-k}\mathcal{F}^{\mu}(s + \tau k)\,dk. \tag{15.88}$$

This equation is now of *second order* in the derivatives of y^{μ}, although highly non-linear in that in general $\mathcal{F}^{\mu}$ is a complicated function of y^{μ}, u^{μ} and of the four-acceleration w^{μ} itself. Correspondingly, if the external field $F_{in}^{\mu\nu}$ is sufficiently regular, it admits a unique solution $y^{\mu}(s)$ once the *standard* initial conditions (15.83) have been fixed, as required by the Newtonian determinism. As anticipated above, one of the advantages of Rohrlich's equation is its manifest Lorentz invariance. On the other hand, one of its main drawbacks is that, due to its involved non-linearity, hardly it is useful to analyze the effects of the radiation reaction concretely. For instance, even in the simplest situation of a free particle, where the effective force (15.84) reduces to $\mathcal{F}^{\mu} = e^2 w^2 u^{\mu}/6\pi$, it is not straightforward to solve equation

(15.88) analytically. Conversely, in this case it is easy to check that the general solution (15.61) of the Lorentz-Dirac equation does not fulfill Rohrlich's equation (15.88), unless the constant K is zero. In particular, for $K \neq 0$ the integral in (15.88) would not even converge.

15.3.1   Preacceleration and Causality Violation

As, ultimately, Rohrlich's equation (15.88) represents the *physical* equation of motion of a classical relativistic charged particle – playing the role of Newton's second law – we now investigate the general properties of its solutions. First of all, the acceleration $w^\mu(s)$ at time s appearing on its left hand side depends not only on the value of the effective force $\mathcal{F}^\mu$ at time s, but also on its values at all subsequent times $s' = s + \tau k$, a fact that signals again a causality violation in the form of a preacceleration. However, thanks to the presence of the damping factor e^{-k}, which suppresses the contributions to the integral coming from values $k \gg 1$, the times that contribute predominantly to the acceleration at time s are those of the order $s' \sim s + \tau$, in agreement with our qualitative argument (15.85). To quantify the effect of the causality violation we recast equation (15.88) in the equivalent form

$$mw^\mu(s) = \mathcal{F}^\mu(s) + \Delta\mathcal{F}^\mu(s), \tag{15.89}$$

$$\Delta\mathcal{F}^\mu(s) \equiv \int_0^\infty e^{-k}\big(\mathcal{F}^\mu(s + \tau k) - \mathcal{F}^\mu(s)\big)\, dk. \tag{15.90}$$

We have hence divided the total force appearing at the right hand side of Eq. (15.88) in two contributions. The first, $\mathcal{F}^\mu(s)$, is *local* in that it depends only on the time s. The second, $\Delta\mathcal{F}^\mu(s)$, codifies instead the causality violation. Comparing equation (15.89) with the Lorentz-Dirac equation (15.84) we deduce that the latter equals, actually, the Schott term (see formula (15.44))

$$\Delta\mathcal{F}^\mu(s) = m\tau \frac{dw^\mu}{ds} = \frac{e^2}{6\pi} \frac{dw^\mu}{ds}.$$

From the form (15.89) of Rohrlich's equation we infer that the causality violation is experimentally detectable, if the force $\Delta\mathcal{F}^\mu$ – responsible for the preacceleration – can be distinguished from the local force $\mathcal{F}^\mu$, i.e. if it is of the same order of magnitude of the latter. To provide an estimate for $\Delta\mathcal{F}^\mu$ we use that the values of k which dominate the integral (15.90) are those of order unity. Setting thus $\mathcal{F}^\mu(s + \tau k) \sim \mathcal{F}^\mu(s + \tau)$, we obtain the estimate

$$\Delta\mathcal{F}^\mu(s) \sim \int_0^\infty e^{-k}\big(\mathcal{F}^\mu(s + \tau) - \mathcal{F}^\mu(s)\big)\, dk = \mathcal{F}^\mu(s + \tau) - \mathcal{F}^\mu(s). \tag{15.91}$$

This relation tells us that $\Delta\mathcal{F}^\mu$ equals the variation of $\mathcal{F}^\mu$ on a time scale of order τ. If this variation is of the same order of $\mathcal{F}^\mu$, the causality violation is thus detectable. Given the form (15.84) of the effective force $\mathcal{F}^\mu$, we thus reach the following general conclusion: *the preacceleration phenomenon becomes experimentally observable if the external field $F_{\text{in}}^{\mu\nu}$ experienced by the particle along its trajectory varies appreciably during the small time τ.* Notice that this condition allows also for *static* external fields, as for instance the electric field of the one-dimensional motion of Sect. 15.2.6. For comparison, during the time τ light covers the extremely small distance $c\tau = 2r_0/3$, where r_0 is the classical radius of the particle (8.81).

Causality violation and quantum mechanics. We now investigate the effects caused *microscopically*, i.e. at the quantum level, by external fields that vary so rapidly that *macroscopically*, i.e. in a classical world, they would entail an observable violation of causality, as discussed above. As the following qualitative arguments are based on the laws of *non-relativistic* quantum mechanics, strictly speaking they apply in the instantaneous rest frame of the particle, where s reduces to the coordinate time t. We begin by recalling that the energy scale $\Delta\varepsilon$ at which the quantum-production process of particle-antiparticle pairs out of the vacuum induces observable effects is of the order $\Delta\varepsilon \sim 2m$. On the other hand, for fields varying on a generic characteristic time scale $\mathcal{T}$ Heisenberg's uncertainty principle predicts an indeterminacy in energy of the order $\Delta\varepsilon \sim \hbar/\mathcal{T}$. To reach the threshold $2m$ the fields must thus vary significantly on a time scale of the order

$$\mathcal{T} \sim \frac{\hbar}{2m} \sim \frac{4\pi\hbar}{e^2}\frac{e^2}{6\pi m} = \frac{\tau}{\alpha} \approx 137\,\tau,$$

where we have introduced the fine structure constant (11.122). But, as seen above, in order to observe in classical electrodynamics a violation of causality we need fields varying with a characteristic time scale of order τ, a time scale that is hence by a factor 137 *smaller* than $\mathcal{T}$. This means that such fields would operate already in a deep quantum regime, giving rise in particular to the pair-production phenomenon. We conclude, therefore, that *the classical violation of causality is shielded by quantum effects.* In fact, in the regime where this violation would come to light, classical electrodynamics is no longer valid being superseded by quantum mechanics, and so eventually the causality violation it would entail remains *unobservable*.

A similar argument can be given for a particle moving under the influence of an external force, that makes it move with a characteristic frequency ω. Correspondingly, in the non-relativistic limit, the particle emits mainly radiation of frequency ω, see Sect. 11.3. As can be inferred from the defining equation (15.84), in this case also the effective force $\mathcal{F}^\mu$ varies with frequency ω. From the relation (15.91) we then derive the estimate

$$\Delta\mathcal{F}^\mu(s) \sim \int_s^{s+\tau} \frac{d\mathcal{F}^\mu(p)}{dp}\,dp \sim \omega \int_s^{s+\tau} \mathcal{F}^\mu(p)\,dp \sim \omega\tau\mathcal{F}_0^\mu,$$

where $\mathcal{F}_0^\mu$ denotes an average value of $\mathcal{F}^\mu$. The causality-violating force $\Delta\mathcal{F}^\mu$ is hence of the same order of $\mathcal{F}^\mu$ if the frequency is very large, i.e. $\omega \sim 1/\tau$, implying that the particle emits radiation of very small wavelengths, i.e. of the order of the classical radius of the particle, $\lambda = 2\pi/\omega \sim \tau \sim r_0$. On the other hand, classical electrodynamics maintains its validity only as long as the wavelengths do not drop under the *Compton wavelength*

$$\lambda_C = \frac{\hbar}{m} \sim \frac{r_0}{\alpha} \approx 137\, r_0,$$

in that for $\lambda \lesssim \lambda_C$ the quantum nature of the electromagnetic field, that is to say, its composition in terms of photons, comes to light. But, as seen above, in order to observe a causality violation experimentally, we need wavelengths of the order $\lambda \sim r_0$, which are by a factor 137 smaller than λ_C. Therefore, for such short wavelengths the electromagnetic field is already in a strong quantum regime, and, consequently, possible *acausal* effects are again unobservable.

Summarizing, the Lorentz-Dirac equation, in Rohrlich's integro-differential form, results in a violation of *temporal causality* – the particle experiences the effects of a force *before* the force acts on it – in light of which classical electrodynamics turns into a *logically inconsistent* theory. However, from a phenomenological point of view, this violation occurs on distance-, energy-, and time-scales, where classical electrodynamics is no longer valid, being superseded by *quantum electrodynamics*. Therefore, no experiment can ever detect this violation.

15.4 Relativistic Two-Body Problem

In Newtonian physics, a pair of charged particles subject to their mutual electro-magnetic interaction forms an isolated two-body system. Conversely, in a relativis-tic framework, a two-particle system can no longer be considered as *isolated*, in that the Newtonian *action at a distance* is replaced by the electromagnetic field. This has two important consequences: Newton's third law – the principle of *action and reac-tion* – ceases to be valid, and, technically speaking, the mutual-interaction forces are no longer *conservative*. In a sense, at the relativistic level the two-body system turns into an isolated three-body system, the third *body* being represented by the electro-magnetic field, which acquires its own independent life. In this section we analyze the dynamics of this system with an emphasis on its distinctive *relativistic* features, and with a particular focus on the four-momentum balance. We shall concentrate mainly on a *scattering process*, where the particles come in from infinity, deflect each other, and then go out again to infinity. In the non-relativistic limit, their orbits would thus be hyperbolas.

Four-momentum conservation and asymptotic properties. In the presence of two charged particles the conservation of the total four-momentum is expressed by the equation, see formulas (2.150),

$$\frac{d}{dt}\left(p_1^\mu + p_2^\mu + P_{\rm em}^\mu\right) = 0,\tag{15.92}$$

where p_1^μ and p_2^μ are the four-momenta of the particles, and $P_{\rm em}^\mu = \int T_{\rm em}^{0\mu}\, d^3x$ is the total four-momentum of the electromagnetic field. As we will briefly review in Sect. 15.4.1, in the non-relativistic limit the momentum $\mathbf{P}_{\rm em}$ of the field is zero, and its energy $\varepsilon_{\rm em} = P_{\rm em}^0$ reduces to the *potential energy* of the particles, namely $\varepsilon_{\rm em} = e_1 e_2/4\pi|\mathbf{y}_1 - \mathbf{y}_2|$. Conversely, at the relativistic level, due to the generation of radiation, in the four-momentum balance of the system the electromagnetic field plays a fundamental role. Unfortunately, however, in general it is not possible to evaluate the integrals $\int T_{\rm em}^{0\mu}\, d^3x$ analytically, and so it is difficult to establish explicit expressions for the electromagnetic four-momentum $P_{\rm em}^\mu$. In this section we shall, therefore, adopt a different strategy: we will enforce the Lorentz-Dirac equations of the particles, *virtually* in Rohrlich's form (15.88), together with the conservation equation (15.92), to determine the form of $P_{\rm em}^\mu$. Furthermore, as the Lorentz-Dirac equations are highly non-linear, we will resort to a non-relativistic expansion in powers of $1/c$. In this way, we will also be in a position to compare the results of this section with the non-relativistic analysis of the emitted radiation of Sect. 8.3, based on the multipole expansion.

At this stage of our presentation, the concrete usefulness of Eq. (15.92) is, however, undermined by the fact that, as recalled at the beginning of the chapter, the integrals $\int T_{\rm em}^{0\mu}\, d^3x$ are, actually, *divergent*. Nevertheless, in Chap. 16 we shall show that it is possible to "renormalize" the field-energy-momentum tensor $T_{\rm em}^{\mu\nu}$, in such a way that it gives rise to a four-momentum $P_{\rm em}^\mu$ which (i) is finite, (ii) satisfies equation (15.92), and (iii) vanishes at the infinite past

$$\lim_{t\to-\infty} P_{\rm em}^\mu = 0.\tag{15.93}$$

To clarify the meaning of the asymptotic property (15.93), holding specifically for the case at hand, we anticipate from Chap. 16 the intuitive result that the (renormalized) four-momentum $P_{\rm em}^\mu$ of the field produced by a *free* charged particle is zero. This information is useful, because in the limit for $t \to -\infty$ the two particles stay at an infinite distance from each other, and so, as we keep only the solutions of the Lorenz-Dirac equations which obey the supplementary conditions (15.47), their accelerations tend to zero. Therefore, the fields of both particles approach asymptotically the field of a free particle and, consequently, the four-momentum of each field tends to zero. In addition, also the four-momentum due to the interference between the two fields vanishes asymptotically, again because in the limit for $t \to -\infty$ the two particles move infinitely apart from each other. This implies the limit (15.93). Concerning, instead, the behavior of $P_{\rm em}^\mu$ for large positive times, due to the production of bremsstrahlung during the scattering process, for $t \to +\infty$ the four-momentum $P_{\rm em}^\mu$ reaches a finite *non-vanishing* limit. In particular, if we denote the variation between the asymptotic instants $t = +\infty$ and $t = -\infty$ with the symbol Δ, Eq. (15.92) yields the overall four-momentum-balance equation

$$\Delta(p_1^\mu + p_2^\mu) = -\Delta P_{\text{em}}^\mu. \tag{15.94}$$

We will make use of Eqs. (15.92)–(15.94) in Sects. 15.4.1–15.4.4, in particular for the purpose of constraining the explicit form of P_{em}^μ.

15.4.1   Relativistic and Non-relativistic Scattering Processes

Non-relativistic scattering. Before proceeding to the investigation of the relativistic scattering process, we briefly rederive its known characteristics in the non-relativistic limit, which amounts to an analysis at zero order in $1/c$, thereby illustrating the use of Eqs. (15.92)–(15.94). Denoting the trajectories of the particles, as usual, by $\mathbf{y}_1(t)$ and $\mathbf{y}_1(t)$, their "non-relativistic Lorentz-Dirac equations" amount to the Newtonian equations of motion

$$\frac{d\mathbf{p}_1}{dt} = \frac{e_1 e_2 (\mathbf{y}_1 - \mathbf{y}_2)}{4\pi |\mathbf{y}_1 - \mathbf{y}_2|^3}, \qquad \frac{d\mathbf{p}_2}{dt} = -\frac{e_1 e_2 (\mathbf{y}_1 - \mathbf{y}_2)}{4\pi |\mathbf{y}_1 - \mathbf{y}_2|^3}. \tag{15.95}$$

Summing them, thanks to Newton's principle of *action and reaction* holding in non-relativistic physics, one finds that the total momentum of the particles is conserved

$$\frac{d}{dt}(\mathbf{p}_1 + \mathbf{p}_2) = 0.$$

From Eq. (15.92) it then follows that the momentum of the field $\mathbf{P}_{\text{em}}$ is separately conserved, and so the knowledge of the boundary value (15.93) finally allows to conclude that $\mathbf{P}_{\text{em}} = 0$. Concerning the kinetic energies $\varepsilon_i = m_i v_i^2/2$ of the particles, the equations of motion (15.95) imply for their sum the work-energy theorem

$$\frac{d(\varepsilon_1 + \varepsilon_2)}{dt} = \frac{e_1 e_2}{4\pi} \frac{(\mathbf{v}_1 - \mathbf{v}_2) \cdot (\mathbf{y}_1 - \mathbf{y}_2)}{|\mathbf{y}_1 - \mathbf{y}_2|^3} = -\frac{d\varepsilon_{\text{pt}}}{dt},$$

where

$$\varepsilon_{\text{pt}} = \frac{e_1 e_2}{4\pi |\mathbf{y}_1 - \mathbf{y}_2|} \tag{15.96}$$

is the potential energy of the particles. Therefore, in the non-relativistic approximation the total conserved energy is given by the *mechanical energy* $\varepsilon_1 + \varepsilon_2 + \varepsilon_{\text{pt}}$. Formulas (15.92) and (15.93), for $\mu = 0$, then imply that the energy of the electromagnetic field equals the potential energy: $\varepsilon_{\text{em}} = \varepsilon_{\text{pt}}$. Moreover, since for large relative distances ε_{pt} vanishes, the total kinetic energy of the particles $\varepsilon_1 + \varepsilon_2$ for $t \to -\infty$ and for $t \to +\infty$ is the same. This implies that in the non-relativistic limit equation (15.94) reduces to

$$\Delta(p_1^\mu + p_2^\mu) = 0 = \Delta P_{\text{em}}^\mu. \tag{15.97}$$

We retrieve so the well known fact, that at the non-relativistic level the electromagnetic field plays no role in the overall four-momentum balance of a Coulomb scattering between two particles.

Relativistic scattering. For a relativistic scattering process equations (15.95) must be replaced by the exact Lorentz-Dirac equations (15.34)

$$\frac{dp_1^\mu}{ds_1} = \frac{e_1^2}{6\pi}\left(\frac{dw_1^\mu}{ds_1} + w_1^2 u_1^\mu\right) + e_1 \mathcal{F}_2^{\mu\nu}(y_1)u_{1\nu},$$

$$\frac{dp_2^\mu}{ds_2} = \frac{e_2^2}{6\pi}\left(\frac{dw_2^\mu}{ds_2} + w_2^2 u_2^\mu\right) + e_2 \mathcal{F}_1^{\mu\nu}(y_2)u_{2\nu}, \tag{15.98}$$

where $\mathcal{F}_1^{\mu\nu}(x)$ and $\mathcal{F}_2^{\mu\nu}(x)$ are the Liénard-Wiechert fields of the two particles, respectively. As we are interested only in the *physical* solutions of these equations, which fulfill, in addition, Rohrlich's equation (15.88), we enforce the supplementary conditions (15.47). Consequently, both world lines fulfill the asymptotic relations

$$\lim_{s\to\pm\infty} w^\mu(s) = 0, \qquad \lim_{s\to\pm\infty} u^\mu(s) = u_\pm^\mu, \tag{15.99}$$

where we have also included the limits for $s \to -\infty$, which play the role of *initial conditions*. For a relativistic scattering process, Eqs. (15.97) are violated because of the four-momentum carried away by the radiation, and, moreover, in this case the particles follow no longer hyperbolic orbits. Nevertheless, if one integrates Eqs. (15.98) between $s_{1,2} = -\infty$ and $s_{1,2} = +\infty$, thanks to the asymptotic conditions (15.99) the Schott terms drop out, and so for the total variations of the four-momenta of the particles one obtains the expressions

$$\Delta p_1^\mu = \frac{e_1^2}{6\pi}\int w_1^2\, u_1^\mu\, ds_1 + e_1 \int \mathcal{F}_2^{\mu\nu}(y_1)\, u_{1\nu}\, ds_1,$$

$$\Delta p_2^\mu = \frac{e_2^2}{6\pi}\int w_2^2\, u_2^\mu\, ds_2 + e_2 \int \mathcal{F}_1^{\mu\nu}(y_2)\, u_{2\nu}\, ds_2.$$

Summing them up, and using for the Liénard-Wiechert four-potentials the intermediate expression (7.11), eventually one obtains the relativistic generalization of the four-momentum-balance equation (15.97)

$$\Delta(p_1^\mu + p_2^\mu) = -\Delta P_{\text{em}}^\mu = \frac{e_1^2}{6\pi}\int w_1^2\, u_1^\mu\, ds_1 + \frac{e_2^2}{6\pi}\int w_2^2\, u_2^\mu\, ds_2 +$$

$$\frac{e_1 e_2}{\pi}\int ds_1 \int ds_2\, (y_1^\mu - y_2^\mu)\,(u_2 u_1)\,\big(H(y_1^0 - y_2^0) - H(y_2^0 - y_1^0)\big)\,\delta'\big((y_1 - y_2)^2\big). \tag{15.100}$$

The first line contains the Larmor terms, which represent the radiation emitted by each particle *individually*. The term in the second line, proportional to $e_1 e_2$, represents, instead, the four-momentum due to the *interference* of these radiations and looks, however, rather complicated. Therefore, to analyze concretely the various relativistic corrections contained in formula (15.100), it is more convenient to go back to Eqs. (15.98), perform their non-relativistic expansions, sum them up, and finally integrate them between $t = -\infty$ and $t = +\infty$. In the sections to follow we adopt this alternative procedure.

15.4.2   Non-relativistic Expansion of the Lorentz-Dirac Equations

Given that the radiation reaction forces start with terms of order $1/c^3$, to obtain the first non-trivial relativistic corrections we must expand the Lorentz-Dirac equations (15.98) up to terms of this order. From the expansion of the Lorentz-Dirac equation for the free particle (15.54) we derive that, up to terms of order $1/c^3$, the spatial components of equations (15.98) take the form

$$\frac{d\mathbf{p}_1}{dt} = \frac{e_1^2}{6\pi c^3}\frac{d\mathbf{a}_1}{dt} + \mathbf{F}_{21}, \qquad \frac{d\mathbf{p}_2}{dt} = \frac{e_2^2}{6\pi c^3}\frac{d\mathbf{a}_2}{dt} + \mathbf{F}_{12}, \qquad (15.101)$$

where the, not yet expanded, mutual-interaction forces are given by

$$\mathbf{F}_{21} = e_1\left(\mathbf{E}_2(y_1) + \frac{\mathbf{v}_1}{c} \times \mathbf{B}_2(y_1)\right), \quad \mathbf{F}_{12} = e_2\left(\mathbf{E}_1(y_2) + \frac{\mathbf{v}_2}{c} \times \mathbf{B}_1(y_2)\right). \qquad (15.102)$$

Here the pairs $\{\mathbf{E}_1, \mathbf{B}_1\}$ and $\{\mathbf{E}_2, \mathbf{B}_2\}$ denote the Liénard-Wiechert fields of the two particles, respectively. Similarly, since we have $e_1\mathbf{v}_1 \cdot \mathbf{E}_2(y_1) = \mathbf{v}_1 \cdot \mathbf{F}_{21}$ and $e_2\mathbf{v}_2 \cdot \mathbf{E}_1(y_2) = \mathbf{v}_2 \cdot \mathbf{F}_{12}$, in basis of the work-energy theorem for a free particle (15.53), the time components of equations (15.98), expanded up terms of order $1/c^3$, take the form

$$\frac{d\varepsilon_1}{dt} = \frac{e_1^2}{6\pi c^3}\mathbf{v}_1 \cdot \frac{d\mathbf{a}_1}{dt} + \mathbf{v}_1 \cdot \mathbf{F}_{21}, \qquad \frac{d\varepsilon_2}{dt} = \frac{e_2^2}{6\pi c^3}\mathbf{v}_2 \cdot \frac{d\mathbf{a}_2}{dt} + \mathbf{v}_2 \cdot \mathbf{F}_{12}. \qquad (15.103)$$

As usual, the work-energy theorems (15.103) are consequences of Newton's second laws (15.101), see the kinematical relations (15.55).

Non-relativistic expansion of the mutual-interaction forces. From formulas (15.101)–(15.103) we see that we must expand the electric fields up to terms of order $1/c^3$, while it is sufficient to expand the magnetic fields up to terms of order $1/c^2$. For this purpose, we can resort to the general non-relativistic expansions of the Liénard-Wiechert fields (7.78)

$$\mathbf{E} = \frac{e}{4\pi}\left(\frac{\mathbf{R}}{R^3} - \frac{1}{2c^2 R}\left(\mathbf{a} + (\widehat{\mathbf{R}}\cdot\mathbf{a})\widehat{\mathbf{R}} + \frac{(3(\widehat{\mathbf{R}}\cdot\mathbf{v})^2 - v^2)\widehat{\mathbf{R}}}{R}\right)\right.$$
$$\left. + \frac{2}{3c^3}\frac{d\mathbf{a}}{dt}\right) + O\left(\frac{1}{c^4}\right),$$

$$\mathbf{B} = \frac{e}{4\pi c}\mathbf{v}\times\frac{\mathbf{R}}{R^3} + O\left(\frac{1}{c^3}\right),$$

(15.104)

where $\mathbf{R} = \mathbf{x} - \mathbf{y}$ and $\widehat{\mathbf{R}} = \mathbf{R}/R$. We recall that in the expansions (15.104) the quantities $\mathbf{y}$, $\mathbf{v}$ and $\mathbf{a}$ denote the kinematical variables of the particle, creating the fields, evaluated at time t. For instance, if we want to calculate $\mathbf{E}_1(y_2)$ and $\mathbf{B}_1(y_2)$, in formulas (15.104) we must set $e = e_1$, $\mathbf{x} = \mathbf{y}_2$, $\mathbf{y} = \mathbf{y}_1$, $\mathbf{v} = \mathbf{v}_1$ and $\mathbf{a} = \mathbf{a}_1$. Inserting the expressions so obtained in formulas (15.102), we obtain the expansions of the mutual-interaction forces up to terms of order $1/c^3$

$$\mathbf{F}_{12} = \frac{e_1 e_2}{4\pi}\left(\frac{\mathbf{r}}{r^3} + \frac{1}{2c^2 r^2}\left(\left[v_1^2 - 2(\mathbf{v}_1\cdot\mathbf{v}_2) - 3(\widehat{\mathbf{r}}\cdot\mathbf{v}_1)^2\right]\widehat{\mathbf{r}} + 2(\mathbf{v}_2\cdot\widehat{\mathbf{r}})\mathbf{v}_1\right)\right.$$
$$\left. - \frac{1}{2c^2 r}(\mathbf{a}_1 + (\widehat{\mathbf{r}}\cdot\mathbf{a}_1)\widehat{\mathbf{r}}) + \frac{2}{3c^3}\frac{d\mathbf{a}_1}{dt}\right),$$

$$\mathbf{F}_{21} = \frac{e_1 e_2}{4\pi}\left(-\frac{\mathbf{r}}{r^3} - \frac{1}{2c^2 r^2}\left(\left[v_2^2 - 2(\mathbf{v}_1\cdot\mathbf{v}_2) - 3(\widehat{\mathbf{r}}\cdot\mathbf{v}_2)^2\right]\widehat{\mathbf{r}} + 2(\mathbf{v}_1\cdot\widehat{\mathbf{r}})\mathbf{v}_2\right)\right.$$
$$\left. - \frac{1}{2c^2 r}(\mathbf{a}_2 + (\widehat{\mathbf{r}}\cdot\mathbf{a}_2)\widehat{\mathbf{r}}) + \frac{2}{3c^3}\frac{d\mathbf{a}_2}{dt}\right),$$

(15.105)

where we have introduced the relative position vector $\mathbf{r}$ at time t, and the associated versor $\widehat{\mathbf{r}}$,

$$\mathbf{r} = \mathbf{y}_2 - \mathbf{y}_1, \qquad \widehat{\mathbf{r}} = \frac{\mathbf{r}}{r}.$$

Inserting the forces (15.105) in Eqs. (15.101) and (15.103), we obtain the relativistic generalizations of Eqs. (15.95), valid up to terms of order $1/c^3$ included.

15.4.3 Relativistic Momentum Balance

Summing the equations of motion (15.101) we find

$$\frac{d}{dt}(\mathbf{p}_1 + \mathbf{p}_2) = \frac{1}{6\pi c^3}\frac{d}{dt}\left(e_1^2\,\mathbf{a}_1 + e_2^2\,\mathbf{a}_2\right) + \mathbf{F}_{12} + \mathbf{F}_{21}.$$

(15.106)

In the sum $\mathbf{F}_{12} + \mathbf{F}_{21}$ the term of order zero in $1/c$, that is to say, the Coulomb field, drops out, but the terms of order $1/c^2$ and $1/c^3$ survive. However, a simple calculation shows that this sum can be written as the total derivative

$$\mathbf{F}_{12} + \mathbf{F}_{21} = \frac{e_1 e_2}{4\pi} \frac{d}{dt} \left(-\frac{1}{2rc^2} \left(\mathbf{v}_1 + \mathbf{v}_2 + \left(\widehat{\mathbf{r}} \cdot (\mathbf{v}_1 + \mathbf{v}_2) \right) \widehat{\mathbf{r}} \right) + \frac{2}{3c^3} \left(\mathbf{a}_1 + \mathbf{a}_2 \right) \right).$$

Substituting this expression in equation (15.106), and comparing the resulting formula with Eq. (15.92), i.e. the conservation law

$$\frac{d}{dt} \left(\mathbf{p}_1 + \mathbf{p}_2 + \mathbf{P}_{\text{em}} \right) = 0, \tag{15.107}$$

we find that the momentum of the electromagnetic field at time t has the expression

$$\mathbf{P}_{\text{em}} = \frac{e_1 e_2}{8\pi c^2 r} \left(\mathbf{v}_1 + \mathbf{v}_2 + \left(\widehat{\mathbf{r}} \cdot (\mathbf{v}_1 + \mathbf{v}_2) \right) \widehat{\mathbf{r}} \right) - \frac{e_1 + e_2}{6\pi c^3} \left(e_1 \mathbf{a}_1 + e_2 \mathbf{a}_2 \right). \tag{15.108}$$

Actually, Eq. (15.107) determines the momentum $\mathbf{P}_{\text{em}}$ only modulo a constant vector. To obtain (15.108) we have, indeed, enforced the boundary condition (15.93), which ensures that for $t \to -\infty$ the momentum $\mathbf{P}_{\text{em}}$ vanishes. Notice, in this regard, that for $t \to \pm\infty$ the quantities $\mathbf{a}_1$, $\mathbf{a}_2$ and $1/r$ vanish. This implies that $\mathbf{P}_{\text{em}}$ vanishes also in the limit for $t \to +\infty$. The fact that for a generic time t the vector $\mathbf{P}_{\text{em}}$ is non-vanishing, signals that during the scattering process the particles continuously exchange momentum with the electromagnetic field.

Let us now analyze the various contributions of the electromagnetic momentum (15.108). $\mathbf{P}_{\text{em}}$ does not contain terms of order $1/c$. This is, actually, a direct consequence of the definition $\mathbf{P}_{\text{em}} = \frac{1}{c} \int (\mathbf{E} \times \mathbf{B}) \, d^3 x$, and of the fact that the magnetic field $\mathbf{B}$ is intrinsically of order $1/c$. The terms of order $1/c^2$ in (15.108), involving the velocities of the particles, represent *kinetic* contributions to the momentum, see below. The terms of order $1/c^3$ represent, instead, the *virtual* instantaneous radiation, which disappears when the particles cease to be accelerated. Finally, thanks to Eq. (15.107) and to the fact that $\mathbf{P}_{\text{em}}$ vanishes for both limits $t \to \pm\infty$, according to the overall momentum-balance equation (15.94) the total momentum of the particles in the initial and final states is the same (as happens in the non-relativistic case)

$$\Delta(\mathbf{p}_1 + \mathbf{p}_2) = -\Delta\mathbf{P}_{\text{em}} = - \left(\mathbf{P}_{\text{em}}^{\infty} - \mathbf{P}_{\text{em}}^{-\infty} \right) = 0. \tag{15.109}$$

This is, actually, an expected result. In fact, from Sects. 8.4 and 10.1.1, see formulas (8.54) and (10.10), we know that the total momentum radiated by a system of charged particles is, indeed, a quantity of order $1/c^5$, i.e. $\Delta\mathbf{P}_{\text{em}} = O(1/c^5)$. The overall momentum-balance-equation (15.94) then implies that $\Delta(\mathbf{p}_1 + \mathbf{p}_2) = O(1/c^5)$. In contrast, the result (15.109) would only allow to infer that $\Delta(\mathbf{p}_1 + \mathbf{p}_2) = O(1/c^4)$. It follows, in particular, that the second member of the spatial components of equation (15.100), of which the relation (15.109) represents the expansion up to terms of order $1/c^3$, is a quantity of order $1/c^5$.

15.4.4 Relativistic Energy Balance

Adding the two equations in (15.103) we find

$$\frac{d}{dt}\left(\varepsilon_1 + \varepsilon_2\right) = \frac{1}{6\pi c^3}\left(e_1^2\, \mathbf{v}_1 \cdot \frac{d\mathbf{a}_1}{dt} + e_2^2\, \mathbf{v}_2 \cdot \frac{d\mathbf{a}_2}{dt}\right) + \mathbf{v}_1 \cdot \mathbf{F}_{21} + \mathbf{v}_2 \cdot \mathbf{F}_{12}.$$
(15.110)

Starting from formulas (15.105), a simple calculation shows that the sum of the "relative powers" appearing in equation (15.110) can be recast in the form, see Problem 15.4,

$$\mathbf{v}_1 \cdot \mathbf{F}_{21} + \mathbf{v}_2 \cdot \mathbf{F}_{12} = -\frac{d\varepsilon_{\text{pt}}}{dt} + \frac{e_1 e_2}{6\pi c^3}\left(\mathbf{v}_1 \cdot \frac{d\mathbf{a}_2}{dt} + \mathbf{v}_2 \cdot \frac{d\mathbf{a}_1}{dt}\right),$$
(15.111)

where we have introduced the *relativistic* potential energy

$$\varepsilon_{\text{pt}} = \frac{e_1 e_2}{4\pi r}\left(1 + \frac{1}{2c^2}\left(\mathbf{v}_1 \cdot \mathbf{v}_2 + (\hat{\mathbf{r}} \cdot \mathbf{v}_1)(\hat{\mathbf{r}} \cdot \mathbf{v}_2)\right)\right).$$
(15.112)

Inserting the expression (15.111) in formula (15.110) we obtain the energy-balance equation

$$\frac{d}{dt}\left(\varepsilon_1 + \varepsilon_2 + \varepsilon_{\text{pt}}\right) = \frac{1}{6\pi c^3}\left(e_1\mathbf{v}_1 + e_2\mathbf{v}_2\right) \cdot \frac{d}{dt}\left(e_1\mathbf{a}_1 + e_2\mathbf{a}_2\right)$$
$$= \frac{1}{6\pi c^3}\frac{d}{dt}\left(\left(e_1\mathbf{v}_1 + e_2\mathbf{v}_2\right) \cdot \left(e_1\mathbf{a}_1 + e_2\mathbf{a}_2\right)\right) - \frac{1}{6\pi c^3}\left|e_1\mathbf{a}_1 + e_2\mathbf{a}_2\right|^2.$$

Comparing this equation with the time component of equation (15.92)

$$\frac{d}{dt}\left(\varepsilon_1 + \varepsilon_2 + \varepsilon_{\text{em}}\right) = 0,$$
(15.113)

we derive for the electromagnetic energy at time t the expression

$$\varepsilon_{\text{em}} = \varepsilon_{\text{pt}} - \frac{1}{6\pi c^3}\left(e_1\mathbf{v}_1 + e_2\mathbf{v}_2\right) \cdot \left(e_1\mathbf{a}_1 + e_2\mathbf{a}_2\right)$$
$$+ \frac{1}{6\pi c^3}\int_{-\infty}^{t}\left|e_1\mathbf{a}_1 + e_2\mathbf{a}_2\right|^2 dt.$$
(15.114)

As in the case of the electromagnetic momentum $\mathbf{P}_{\text{em}}$, Eq. (15.113) determines the energy ε_{em} only modulo a constant, that we have fixed with the aid of the asymptotic condition (15.93), i.e. $\lim_{t \to -\infty} \varepsilon_{\text{em}} = 0$. In Eq. (15.114) the first term is the relativistic potential energy (15.112), to be compared with the Newtonian expression (15.96), and the second represents the energy of the *virtual* radiation, that fades out together with the acceleration. Both these terms vanish for $t \to \pm\infty$. Conversely, the third term – involving the entire history of the particle between $t = -\infty$ and the considered instant t – represents the *real* radiation. This term comprises the energies of the radiations emitted by each particle individually, proportional to e_1^2 and e_2^2, respectively, and the energy associated to the interference between these radiations, proportional to the product $e_1 e_2$. In the limit for $t \to -\infty$ this term vanishes, in agreement with the fact that initially, before the particles deflect each other, there is no radiation. Finally, if we evaluate equation (15.114) at $t = -\infty$ and $t = +\infty$, in basis of the energy-balance equation (15.94), for the decrease of the particles' energy during the whole scattering process we find the expression

$$\Delta(\varepsilon_1 + \varepsilon_2) = -\Delta\varepsilon_{\text{em}} = -(\varepsilon_{\text{em}}^{\infty} - \varepsilon_{\text{em}}^{-\infty}) = -\frac{1}{6\pi c^3} \int_{-\infty}^{\infty} \left| e_1 \mathbf{a}_1 + e_2 \mathbf{a}_2 \right|^2 dt,$$

(15.115)

which is, of course, in agreement with Larmor's formula (8.57) for the power radiated by a system of charged particles

$$W = \frac{1}{6\pi c^3} \left| e_1 \mathbf{a}_1 + e_2 \mathbf{a}_2 \right|^2.$$

However, one must bear in mind that it would not be correct to determine the energy ε_{em} of the electromagnetic field at time t, by enforcing the identification $W = d\varepsilon_{\text{em}}/dt$, which would lead to the wrong expression $\varepsilon_{\text{em}} = \int_{-\infty}^{t} |e_1 \mathbf{a}_1 + e_2 \mathbf{a}_2|^2 dt/6\pi c^3$. In fact, the latter represents only the energy of the electromagnetic field that eventually manages to *escape to infinity*, whereas formula (15.114) represents its total instantaneous energy, which is such that $\varepsilon_1 + \varepsilon_2 + \varepsilon_{\text{em}}$ is constant for all t. Finally, one can compare equation (15.115) with the time component of the four-momentum-balance equation (15.100). One sees that the diagonal Larmors terms, proportional to e_1^2 and e_2^2, respectively, are reproduced correctly in the non-relativistic limit. Furthermore, one can infer that the $\mu = 0$ component of the interference term in the second line of formula (15.100) at leading order in $1/c$ is equal to

$$-\frac{e_1 e_2}{3\pi c^3} \int_{-\infty}^{\infty} (\mathbf{a}_1 \cdot \mathbf{a}_2) \, dt.$$

15.4.5   The Lagrangian at Order $1/c^2$

Up to terms of order $1/c^3$ the dynamics of the particles is governed by the equations of motion (15.101), where the forces $\mathbf{F}_{12}$ and $\mathbf{F}_{21}$ are those in (15.105). As these equations have been derived by *eliminating* the Liénard-Wiechert fields – those of mutual interaction as well as the self-interaction fields – from the original Lorentz equations of the particles, one may ask whether the resulting *effective* equations of motion (15.101) can be derived from a Lagrangian or not. The answer is affirmative only relative to the relativistic corrections of order $1/c^2$. More precisely, a direct calculation shows that, neglecting the terms of order $1/c^3$, Eqs. (15.101) can be derived from the Lagrangian

$$L_{(2)} = \frac{1}{2} m_1 v_1^2 + \frac{1}{2} m_2 v_2^2 + \frac{1}{8} m_1 \frac{v_1^4}{c^2} + \frac{1}{8} m_2 \frac{v_2^4}{c^2}$$
$$- \frac{e_1 e_2}{4\pi r} \left(1 - \frac{1}{2c^2} (\mathbf{v}_1 \cdot \mathbf{v}_2 + (\widehat{\mathbf{r}} \cdot \mathbf{v}_1)(\widehat{\mathbf{r}} \cdot \mathbf{v}_2)) \right).$$

Without entering into the details of the calculation, we limit ourselves to the observation that the terms of $L_{(2)}$ of the type v^4/c^2 come from the expansion of the action of the free particle $-mc \int ds$, up to terms of order $1/c^2$,

$$-mc^2 \int \sqrt{1 - \frac{v^2}{c^2}}\, dt = \int \left(-mc^2 + \frac{1}{2} mv^2 + \frac{1}{8} m \frac{v^4}{c^2} \right) dt + O\left(\frac{1}{c^4} \right).$$

Correspondingly, the terms of $L_{(2)}$ of the type v^4/c^2 reproduce in equations (15.101) the terms coming from the expansion in powers of $1/c$ of the relativistic momenta appearing at their left hand sides,

$$\mathbf{p} = \frac{m\mathbf{v}}{\sqrt{1 - \frac{v^2}{c^2}}} = m\mathbf{v} + \frac{mv^2}{2c^2} \mathbf{v} + O\left(\frac{1}{c^4} \right).$$

Furthermore, the terms of $L_{(2)}$ proportional to $e_1 e_2$ reproduce the Coulomb terms $\pm e_1 e_2 \mathbf{r}/4\pi r^3$ of the forces $\mathbf{F}_{12}$ and $\mathbf{F}_{21}$, and all their terms of order $1/c^2$.

In contrast, the terms of order $1/c^3$ in Eqs. (15.101) cannot be derived from a Lagrangian, as these terms are linear in the third derivatives of the position vectors $\mathbf{y}_1(t)$ and $\mathbf{y}_2(t)$. The reason is essentially the same as the one that prevented an action for the exact Lorentz-Dirac equation, see Sect. 15.2.4. In fact, focusing on one of the two particles, Eqs. (15.101) have the structure

$$m\mathbf{a} = \frac{e^2}{6\pi c^3} \frac{d\mathbf{a}}{dt} + \cdots.$$

The term at second member, being linear in $d\mathbf{a}/dt = d^3\mathbf{y}/dt^3$, can descend only from a Lagrangian which is quadratic in $\mathbf{y}$ and involves a total of three derivatives. This Lagrangian should thus have the form

$$L_{(3)} = \frac{1}{2} mv^2 + \frac{e^2}{6\pi c^3} \left(k_1 \mathbf{v} \cdot \mathbf{a} + k_2 \mathbf{y} \cdot \frac{d\mathbf{a}}{dt} \right),$$

where k_1 and k_2 are dimensionless constants. However, the terms between parentheses amount to a total derivative

$$k_1 \, \mathbf{v} \cdot \mathbf{a} + k_2 \, \mathbf{y} \cdot \frac{d\mathbf{a}}{dt} = \frac{d}{dt} \left(k_2 \, \mathbf{y} \cdot \mathbf{a} + \frac{k_1 - k_2}{2} \, v^2 \right),$$

and so the Lagrangian $L_{(3)}$ produces the equation of motion of a free particle. It follows that also the non-relativistic Lorentz-Dirac equation cannot be derived from a variational principle.

15.5   Problems

15.1 *Non-relativistic Lorentz-Dirac equation*. Consider a charged particle moving in the presence of a generic external field $F_{in}^{\mu\nu} = \{\mathbf{E}, \mathbf{B}\}$.

(a) Show that in the non-relativistic limit, i.e. for speeds $v \ll c$, the Lorentz-Dirac equation (15.19), and its integro-differential version (15.88), reduce to

$$m\mathbf{a} = m\tau \frac{d\mathbf{a}}{dt} + \mathbf{f}, \tag{15.116}$$

$$m\mathbf{a}(t) = \int_0^\infty e^{-k} \, \mathbf{f}(t + \tau k) \, dk, \tag{15.117}$$

respectively, where the external force is given by $\mathbf{f} = e(\mathbf{E} + \mathbf{v} \times \mathbf{B}/c)$.
Hint: Use the expansion of the Lorentz-Dirac equation for a free particle (15.54).

(b) Verify that Eq. (15.117) represents a first integral of equation (15.116).

(c) Relying on Eq. (15.117), establish the conditions under which in the non-relativistic limit the radiation reaction is locally negligible. Compare the answer with the general criterion $\tau \ll T$, following from the estimates (15.11) and (15.12) of Sect. 15.1.

(d) Suppose that $\mathbf{B}$ is zero, and that $\mathbf{E}$ is different from zero only in a bounded region of space, where it is constant and uniform. Determine explicitly the second member of Eq. (15.117), and compare the resulting equation with formula (15.71) for $K = 0$. Discuss the causality violation entailed by Eq. (15.117) in this case.
Hint: Formula (15.71) represents the general solution of the relativistic equation of motion (15.67), which formally, in turn, is identical to the non-relativistic equation (15.116).

(e) Assume, henceforth, that the external fields vary only little on the time scale τ. Show that in this case equation (15.117) can be approximated by

$$m\mathbf{a} = \mathbf{f} + \tau \frac{d\mathbf{f}}{dt} = \mathbf{f}_{\text{eff}}. \tag{15.118}$$

Hint: For slowly varying external forces you can set $\mathbf{f}(t + \tau k) \approx \mathbf{f}(t) + \tau k \, \dot{\mathbf{f}}(t)$.

(f) Show that up to terms of order $1/c^4$ the effective force defined in equation (15.118) can be written as

$$\mathbf{f}_{\text{eff}} = e\left(\mathbf{E} + \frac{\mathbf{v}}{c} \times \mathbf{B}\right) + e\tau\left(\dot{\mathbf{E}} + \frac{\mathbf{v}}{c} \times \dot{\mathbf{B}} + \frac{e}{mc}\mathbf{E} \times \mathbf{B}\right). \quad (15.119)$$

Hint: At lowest order in v/c and τ the acceleration $d\mathbf{v}/dt$ can be replaced with $e\mathbf{E}/m$.

15.2 *Radiation reaction in the Thomson scattering.* Apply the non-relativistic Eqs. (15.118) and (15.119) – valid if the external fields vary only little on the time scale τ – to the *Thomson scattering*, see Sect. 8.4.2. In this case the external fields are the plane-wave fields $\mathcal{E}$ and $\mathcal{B}$ of period $T = 2\pi/\omega$ given in (8.68). The above approximation is hence valid for periods $T \gg \tau$, or for frequencies $\omega \ll 1/\tau$. Consider as zero-order solution of equation (15.118) the trajectory (8.72), entailing the velocity

$$\mathbf{v}(t) = \left(\frac{e\mathcal{E}_0}{m\omega}\sin(\omega t), 0, 0\right). \quad (15.120)$$

Recall that the requirement $v \ll c$ amounts to $e\mathcal{E}_0/m\omega c \ll 1$, see (8.72).

(a) Substituting the zero-order solution, and the fields (8.68), in the effective force (15.119), show that the radiation reaction turns the velocity (15.120) into

$$\mathbf{v}(t) = \left(\frac{e\mathcal{E}_0}{m\omega}\sin(\omega t) + \frac{e\mathcal{E}_0\tau}{m}\cos(\omega t), 0, 0\right). \quad (15.121)$$

(b) Using (15.121) show that the effective force (15.119) at first order in τ becomes

$$\mathbf{f}_{\text{eff}} = e\mathcal{E}_0\big(\cos(\omega t) - \omega\tau\sin(\omega t)\big)\mathbf{u}_x + \frac{e^2\mathcal{E}_0^2\tau}{2mc}\big(3\cos(2\omega t) + 1\big)\mathbf{u}_z.$$

(c) Evaluate the average of this force over a period T

$$\langle\mathbf{f}_{\text{eff}}\rangle = \frac{e^2\mathcal{E}_0^2\tau}{2mc}\mathbf{u}_z = \frac{e^4\mathcal{E}_0^2}{12\pi m^2 c^4}\mathbf{u}_z. \quad (15.122)$$

This result can also be derived directly from Eq. (15.118), by observing that the average of the force $\tau(d\mathbf{f}/dt)$ over a period is zero. From the velocity (15.121) and the fields (8.68) one then finds

$$\langle\mathbf{f}_{\text{eff}}\rangle = \langle\mathbf{f}\rangle = e\left\langle\mathcal{E} + \frac{\mathbf{v}}{c} \times \mathcal{B}\right\rangle = \frac{e}{c}\langle\mathbf{v} \times \mathcal{B}\rangle = \frac{e}{c}\langle v_x\mathcal{B}_y\rangle\mathbf{u}_z = \frac{e^2\mathcal{E}_0^2\tau}{2mc}\mathbf{u}_z.$$

Formula (15.122) provides a proof of the prediction (8.84) of Sect. 8.4.2, based on conservation arguments, according to which in the Thomson scattering the particle is subject to a *continuous* momentum transfer in the same direction of the incident wave, of the order $1/c^4$.

15.3 Verify that for an external force $F(s)$ of the form (15.70), Eq. (15.71) constitutes a first integral of equation (15.67).

Hint: Take into account that the derivative of the characteristic function is given by $d\chi_{[a,b]}(x)/dx = \delta(x - a) - \delta(x - b)$, and check that eventually all δ-functions cancel out.

15.4 Prove via a direct calculation that the external forces (15.105) satisfy the identity (15.111). Alternatively, use the identity,

$$\mathbf{v}_1 \cdot \mathbf{F}_{21} + \mathbf{v}_2 \cdot \mathbf{F}_{12} = e_1 \mathbf{v}_1 \cdot \mathbf{E}_2(y_1) + e_2 \mathbf{v}_2 \cdot \mathbf{E}_1(y_2),$$

together with the intermediate formula for the electric field (7.77).

15.5 *Energy conservation in the presence of an external field*. Consider a charged particle moving in the presence of an *electrostatic* external field, where $\mathbf{E}_{in}(t, \mathbf{x}) = -\nabla\varphi(\mathbf{x})$ and $\mathbf{B}_{in}(t, \mathbf{x}) = 0$, generated by a static charge distribution as, for instance, the electric field between the plates of a capacitor.

(a) Show that in this case the external field $F_{in}^{\mu\nu}$ cannot be considered strictly as a *free* field, in that the four-vector $K^\mu = \partial_\nu F_{in}^{\nu\mu}$ is non-vanishing. What is the physical interpretation of the non-vanishing components of K^μ?

(b) Considering that the total electromagnetic field is given by formula (15.3), verify that in this case the total energy-momentum tensor $T^{\mu\nu} = T_{em}^{\mu\nu} + T_p^{\mu\nu}$ violates the continuity equation, in that

$$\partial_\mu T^{\mu\nu} = K_\mu F^{\mu\nu} \neq 0.$$

Hint: Repeat the calculations (2.140) and (2.141) taking into account that now $\partial_\mu F^{\mu\alpha} \neq j^\alpha$, and use the Lorentz equation (15.4) disregarding ultraviolet divergences.

(c) Show that, nevertheless, the total energy of the system $\varepsilon = \int T^{00} d^3x$ is conserved, being given by

$$\varepsilon = \varepsilon_{em} + \varepsilon_p + e\varphi(\mathbf{y}),$$

where ε_{em} is the energy of the Liénard-Wiechert field of the particle, and $\varepsilon_p = m/\sqrt{1 - v^2}$.

15.6 Consider a free charged particle moving on a straight line, subject to the Lorentz-Dirac equation, and the corresponding exact solution (15.69) for the variable $V(s)$, defined by formulas (15.66),

$$V(s) = Ke^{s/\tau} + D. \tag{15.123}$$

Verify that (15.123) constitutes a particular exact solution (15.61) for a free particle moving in three-dimensional space, for suitable four-vectors A^μ and B^μ satisfying the constraints (15.62).

15.7 Consider a charged particle moving on the z axis in the presence of a constant electric field confined to a bounded region, as in Sect. 15.2.6. In this case, once one has enforced the supplementary conditions (15.47), the Lorentz-Dirac equation goes over into Eq. (15.71) with $K = 0$, i.e.

$$V'(s) = \frac{F(s)}{m} + \frac{eE_0}{m}\left(H(a-s)\,e^{(s-a)/\tau} - H(b-s)\,e^{(s-b)/\tau}\right). \qquad (15.124)$$

Let v_0 be the velocity of the particle for $s \to -\infty$, corresponding to $z \to -\infty$.

(a) Show that with this initial condition Eq. (15.124) admits the unique solution

$$V(s) = \mathcal{A}H(b-s)\left(s - b + \tau\left(1 - e^{(s-b)/\tau}\right)\right)$$
$$-\mathcal{A}H(a-s)\left(s - a + \tau\left(1 - e^{(s-a)/\tau}\right)\right) + V_0 + \mathcal{A}(b-a),$$
$$(15.125)$$

where $\mathcal{A} = eE_0/m$ is the "non-relativistic" acceleration of the particle, and $\sinh V_0 = v_0/\sqrt{1 - v_0^2}$, see formulas (15.66).

(b) Show that, while the particle passes from $z = -\infty$ to $z = +\infty$, its energy increases by

$$\Delta\varepsilon = m\left(\cosh(V_0 + \mathcal{A}(b-a)) - \cosh V_0\right). \qquad (15.126)$$

Hint: The energy of the particle at proper time s is given by $\varepsilon(s) = mu^0(s) = m\cosh V(s)$.

(c) Denote by $\Delta z = z_b - z_a$ the width of the zone where the electric field is non-vanishing. Show that at first order in τ one has

$$\Delta z = \frac{1}{\mathcal{A}}\left(\cosh(V_0 + \mathcal{A}(b-a)) - \cosh V_0\right) + \tau\left(\sinh(V_0 + \mathcal{A}(b-a)) - \sinh V_0\right).$$
$$(15.127)$$

Hint: Use the relation $\Delta z = \int_a^b u(s)\,ds = \int_a^b \sinh V(s)\,ds$, and neglect in formula (15.125) the exponentials.

(d) According to formula (15.79) the total energy radiated by the particle is given by

$$\varepsilon_{\text{rad}} = -\int_{-\infty}^{\infty} u(s)F_{\text{rr}}(s)\,ds, \qquad (15.128)$$

where $F_{\text{rr}}(s)$ is the radiation reaction force (15.75). Using the same approximation as in issue (c), and recalling that $\tau = e^2/6\pi m$, show that

$$\varepsilon_{\text{rad}} = \frac{e^2\mathcal{A}}{6\pi}\left(\sinh(V_0 + \mathcal{A}(b-a)) - \sinh V_0\right). \qquad (15.129)$$

Hint: Integrating the first equation in (15.78) between $t = -\infty$ and $t = +\infty$ one obtains for the total variation of the particle's energy the exact relation

$$\Delta\varepsilon = eE_0\Delta z - \varepsilon_{\text{rad}}.$$

Determine ε_{rad} from this relation, using the results (15.126) and (15.127). Alternatively, insert in Eq. (15.128) the radiation reaction force $F_{\text{rr}}(s)$ (15.75), and use that $u(s) = \sinh V(s)$.

(e) The spatial radiation reaction force (15.75) is related to the radiation reaction four-force Γ^μ by Eq. (15.77). From formulas (15.44) and (15.45) one then infers the alternative expression

$$F_{\text{rr}}(s) = \frac{\Gamma^3(s)}{u^0(s)} = \frac{\Gamma^0(s)}{u(s)} = \frac{e^2}{6\pi u(s)}\left(\frac{dw^0(s)}{ds} + w^2(s)u^0(s)\right).$$

Equation (15.128) then entails for the radiated energy the exact expression

$$\varepsilon_{\text{rad}} = -\frac{e^2}{6\pi}\left(\int_{-\infty}^{\infty} w^2u^0\,ds + w^0(\infty) - w^0(-\infty)\right) = -\frac{e^2}{6\pi}\int_{-\infty}^{\infty} w^2u^0\,ds,$$
$$\tag{15.130}$$

in agreement with the relativistic Larmor formula (10.9). Evaluate the integral (15.130) using the same approximation as in issue (c), and compare the result with Eq. (15.129).

Hint: Formulas (15.64) and (15.68) imply the relation $w^2 = -V'^2$, and furthermore $u^0(s) = \cosh V(s)$.

(f) Perform the non-relativistic limit of the radiated energy (15.129), and interpret the result in the light of Larmor's formula (7.67).

(g) Evaluate the radiated energy (15.129) in the ultrarelativistic limit, and compare the result with the analysis of Sect. 10.2.1, in particular with the formula for the emitted power (10.28).

Hint: In the ultrarelativistic limit one has $v_0 \approx 1$ and $V_0 \to \infty$, so that

$$\cosh(V_0 + x) \approx \sinh(V_0 + x) \approx \frac{1}{2}\,e^{V_0+x}.$$

From Eqs. (15.127) and (15.129) it then follows that at first order in τ one has $\varepsilon_{\text{rad}} \approx e^2\mathcal{A}^2\Delta z/6\pi$. Moreover, in the ultrarelativistic limit $\Delta z \approx \Delta t$.

15.8 Verify that the expression (8.110) for the relative energy decrease of the classical hydrogen atom during a cycle, according to which *locally* the radiation reaction force is negligible, respects the general estimate (15.12).

Hint: Use the kinematical relations (8.107).

Reference

1. F. Rohrlich, *Classical Charged Particles* (Addison-Wesley, Reading, 1965)

Chapter 16
Renormalization of the Electromagnetic Energy-Momentum Tensor

16.1 Singularities of the Electromagnetic Energy-Momentum Tensor

Close to the position $\mathbf{y}(t)$ of a charged particle, the electromagnetic field it creates diverges as $F^{\mu\nu} \sim 1/r^2$, where $r = |\mathbf{x} - \mathbf{y}(t)|$ is the distance from the particle. Correspondingly, the electromagnetic energy-momentum tensor

$$T_{em}^{\mu\nu} = F^{\mu\alpha} F_{\alpha}{}^{\nu} + \frac{1}{4} \eta^{\mu\nu} F^{\alpha\beta} F_{\alpha\beta} \tag{16.1}$$

diverges as

$$T_{em}^{\mu\nu} \sim \frac{1}{r^4}. \tag{16.2}$$

For instance, a static particle at the origin, with position vector $\mathbf{y}(t) = (0, 0, 0)$, generates the fields

$$\mathbf{E} = \frac{e\mathbf{x}}{4\pi r^3}, \qquad \mathbf{B} = 0, \tag{16.3}$$

and formulas (2.144)–(2.146) then yield the components

$$T_{em}^{00} = \frac{1}{2} E^2 = \frac{1}{2} \left(\frac{e}{4\pi}\right)^2 \frac{1}{r^4}, \tag{16.4}$$

$$T_{em}^{0i} = 0, \tag{16.5}$$

$$T_{em}^{ij} = \frac{1}{2} E^2 \delta^{ij} - E^i E^j = \frac{1}{2} \left(\frac{e}{4\pi}\right)^2 \frac{1}{r^4} \left(\delta^{ij} - 2 \frac{x^i x^j}{r^2}\right). \tag{16.6}$$

The pronounced $1/r^4$-singularity of the electromagnetic energy-momentum tensor near the particle's position gives rise to two interrelated conceptual difficulties, which eventually *spoil the four-momentum-conservation paradigm* inherent in the

© Springer International Publishing AG, part of Springer Nature 2018
K. Lechner, *Classical Electrodynamics*, UNITEXT for Physics,
https://doi.org/10.1007/978-3-319-91809-9_16

original formulation of classical electrodynamics. Below we explain these difficulties in detail.

Infinite total four-moment. The first difficulty has been mentioned several times before: due to the singular behavior (16.2) of the tensor $T_{\text{em}}^{\mu\nu}$ in the vicinities of the charged particles, in general the total electromagnetic four-momentum integrals $P_{\text{em}}^{\mu} = \int T_{\text{em}}^{0\mu} d^3x$ are divergent. In the case of a static particle, for instance, from the formulas above we obtain

$$\varepsilon_{\text{em}} = \int T_{\text{em}}^{00} d^3x = \frac{1}{2}\left(\frac{e}{4\pi}\right)^2 \int \frac{1}{r^4} d^3x = \infty, \quad P_{\text{em}}^{i} = \int T_{\text{em}}^{0i} d^3x = 0,$$
$$(16.7)$$

and so only the total energy diverges, while for a particle in arbitrary motion also the total momentum $\mathbf{P}_{\text{em}}$ would be infinite, see e.g. formula (16.46) in the limit for $\varepsilon \to 0$. More in general, since $T_{\text{em}}^{\mu\nu}$ is singular only at the positions of the particles, the four-momentum $P_{\text{em}}^{\mu}(V) = \int_V T_{\text{em}}^{0\mu} d^3x$ contained in a generic finite volume V *enclosing at least one particle* is divergent, while the four-momentum contained in a volume which does not enclose particles is always finite. In the various investigations, performed so far, of the energy content of the radiation emitted by a system of charged particles – see, for instance, the systematic analysis of Chap. 8 – the problem of the infinite four-momentum of the electromagnetic field has never emerged directly, because the radiated power (8.18) involves only the electromagnetic field at *large distances* from the particles. In addition, the radiated power refers to *differences* of energy values, and, in a sense, in the differences the infinities cancel out. However, if we want to give a precise meaning to the statement "the total four-momentum is conserved", it is necessary that at each instant the four-momentum is a *finite* physical quantity.

The energy-momentum tensor is not a distribution. The second difficulty caused by the singularity (16.2) consists in the fact that the ten component fields of the tensor $T_{\text{em}}^{\mu\nu}$ are not distributions, i.e. elements of the space $\mathcal{S}' = \mathcal{S}'(\mathbb{R}^4)$. In fact, the behavior $1/|\mathbf{x}|^4$ constitutes a singularity which for fixed t is locally *non-integrable* in $\mathbb{R}^3$ around the point $\mathbf{x} = 0$. The mathematical origin of this bad behavior is that the tensor $T_{\text{em}}^{\mu\nu}$ (16.1) is a sum of *products* of the distributions $F^{\mu\nu}$, and, in fact, products of distributions in general are not distributions. A negative consequence of $T_{\text{em}}^{\mu\nu}$ not being a distribution is that this tensor does not admit partial derivatives, and so, in particular, the four-divergence $\partial_\mu T_{\text{em}}^{\mu\nu}$ is ill-defined. It follows, in turn, that the *proof* of the continuity equation for the total energy-momentum tensor $T^{\mu\nu} = T_{\text{em}}^{\mu\nu} + T_{\text{p}}^{\mu\nu}$ given before in Sect. 2.4.3 can be considered, at most, as a *heuristic argument*. In fact, the problem is far from being a *mathematical subtlety*. Indeed, the expression derived in (2.140) for the four-divergence of the tensor $T_{\text{em}}^{\mu\nu}$, namely

$$\partial_\nu T_{\text{em}}^{\nu\mu} = -\sum_r e_r \int F^{\mu\nu}(y_r)\, u_{r\nu}\, \delta^4(x - y_r)\, ds_r, \qquad (16.8)$$

is meaningless, in that all N terms of its second member are *divergent*. More precisely, since the total electromagnetic field

$$F^{\mu\nu}(x) = F_{\text{in}}^{\mu\nu}(x) + \sum_s \mathcal{F}_s^{\mu\nu}(x) \tag{16.9}$$

appearing in Eq. (16.8) comprises the Liénard-Wiechert field $\mathcal{F}_r^{\mu\nu}(x)$ of the rth particle, the term $F^{\mu\nu}(y_r)$ comprises the *infinite* self-field $\mathcal{F}_r^{\mu\nu}(y_r)$, see the behavior (15.6). In addition, the proof of the continuity equation for $T^{\mu\nu}$ of Sect. 2.4.3 relied crucially on the original Lorentz equations (2.20), see formula (2.142), but now we know that these equations involve the *infinite* self-forces of the particles. Aim of the present chapter is the construction of a *new* total energy-momentum tensor $T^{\mu\nu}$ which:

(1) transforms as a *tensor* under Poincaré transformations;
(2) constitutes a distribution;
(3) admits finite four-momentum integrals;
(4) satisfies the continuity equation $\partial_\mu T^{\mu\nu} = 0$.

Even though the conceptual strategy underlying our approach for the construction of an energy-momentum tensor with these properties is rather simple, see Sect. 16.2, the proof that the tensor $T^{\mu\nu}$ so obtained actually fulfills the conditions (1)–(4) is technically slightly involved, see Refs. [1, 2] for details. For this reason, in Sect. 16.2.1 we present a *heuristic* implementation of the approach for a particle performing an arbitrary motion, whereas in Sect. 16.3 we implement the approach *explicitly* for the simpler case of a particle in uniform linear motion. Its generalization to a generic system of charged particles will then be presented, without proofs, in Sect. 16.4. The key point of our approach is the construction of the *renormalized electromagnetic energy-momentum tensor* $\mathbb{T}_{\text{em}}^{\mu\nu}$ of Eq. (16.49), which turns out to be *uniquely* determined. This tensor has been first derived by E.G.P. Rowe in his seminal paper [3], applying a technique in which $\mathbb{T}_{\text{em}}^{\mu\nu}$ appears as a sum of multiple derivatives of distributions, rather than as a distributional limit like Eq. (16.49).

The main conclusion of this chapter, however, is that the new total energy-momentum tensor $T^{\mu\nu} = \mathbb{T}_{\text{em}}^{\mu\nu} + T_{\text{p}}^{\mu\nu}$, actually, satisfies the conditions (1)–(3) for arbitrary trajectories of the particles, while condition (4), namely the requirement of local four-momentum conservation, forces the trajectories of the particles to obey the *Lorentz-Dirac equations* (15.34).

16.2 Regularization and Renormalization

Consider a single charged particle following a *generic* world line $y^\mu(s)$ in the presence of an external field $F_{\text{in}}^{\mu\nu}$, which we suppose to be regular throughout spacetime. The total electromagnetic field is then given by $F^{\mu\nu} = F_{\text{in}}^{\mu\nu} + \mathcal{F}^{\mu\nu}$, where $\mathcal{F}^{\mu\nu}$ is the Liénard-Wiechert field (15.21). To construct an energy-momentum tensor $T^{\mu\nu}$ entailing the above properties (1)–(4) we follow a procedure which parallels the one adopted in Chap. 15, consisting of a regularization followed by a renormalization.

Since the energy-momentum tensor of the particle $T_{\mathrm{p}}^{\mu\nu}$ in (16.15) satisfies already the properties (1)–(3), we will be mainly concerned with the tensor $T_{\mathrm{em}}^{\mu\nu}$.

Regularization. As the field $F_{\mathrm{in}}^{\mu\nu}$ is regular, the singularities of the tensor $T_{\mathrm{em}}^{\mu\nu}$ (16.1) are localized along the world line $y^{\mu}(s)$. This suggests to adopt again the Lorentz-invariant regularization introduced in Sect. 15.2.1. Resorting hence to the regularized Liénard-Wiechert $\mathcal{F}_{\varepsilon}^{\mu\nu}$ (15.28), we introduce the regularized total field

$$F_{\varepsilon}^{\mu\nu} = F_{\mathrm{in}}^{\mu\nu} + \mathcal{F}_{\varepsilon}^{\mu\nu}, \tag{16.10}$$

which is regular for all $x^{\mu} \in \mathbb{R}^4$ and, in particular, for $x^{\mu} = y^{\mu}(s)$, i.e. along the world line. More precisely, for every $\varepsilon > 0$ the six component fields of the tensor $F_{\varepsilon}^{\mu\nu}(x)$ are *bounded functions of class* C^{∞} on $\mathbb{R}^4$. We illustrate theses properties for the fields (16.3) produced by a static particle, setting $F_{\mathrm{in}}^{\mu\nu} = 0$. In this case the field $F_{\varepsilon}^{\mu\nu}$ can be derived by choosing in the regularized field of a generic uniform linear motion (15.30) the four-velocity $u^{\mu} = (1, 0, 0, 0)$. In this way one obtains the regularized fields

$$\mathbf{E}_{\varepsilon} = \frac{e\mathbf{x}}{4\pi(r^2 + \varepsilon^2)^{3/2}}, \qquad \mathbf{B}_{\varepsilon} = 0, \tag{16.11}$$

to be compared with the singular fields (16.3). The fields (16.11) are indeed bounded and of class C^{∞}, and as such they are regular throughout space, included the point $\mathbf{x} = 0$ where the particle sits. Correspondingly, the regularized total electromagnetic energy $\varepsilon_{\mathrm{em}\,\varepsilon}$ is finite. In fact, using the elementary integral

$$\int \frac{r^2 d^3 x}{(r^2 + 1)^3} = \frac{3\pi^2}{4},$$

in place of (16.7) we now find the finite result

$$\varepsilon_{\mathrm{em}\,\varepsilon} = \frac{1}{2} \int E_{\varepsilon}^2 \, d^3 x = \frac{1}{2} \left(\frac{e}{4\pi}\right)^2 \int \frac{r^2 d^3 x}{(r^2 + \varepsilon^2)^3} = \frac{3e^2}{128\varepsilon}. \tag{16.12}$$

Notice that in the limit for $\varepsilon \to 0$ the energy $\varepsilon_{\mathrm{em}\,\varepsilon}$ diverges as $1/\varepsilon$.

Returning to an arbitrary world line $y^{\mu}(s)$, we now take advantage from the fact that products of bounded functions of class C^{∞} are again bounded functions of class C^{∞}. Consequently, unlike the *bare* energy-momentum tensor $T_{\mathrm{em}}^{\mu\nu}$ (16.1), the *regularized* energy-momentum tensor defined by

$$T_{\varepsilon}^{\mu\nu} = F_{\varepsilon}^{\mu\alpha} F_{\varepsilon\,\alpha}{}^{\nu} + \frac{1}{4} \eta^{\mu\nu} F_{\varepsilon}^{\alpha\beta} F_{\varepsilon\,\alpha\beta} \tag{16.13}$$

constitutes a distribution for all $\varepsilon > 0$. Obviously, for every $x^{\mu} \neq y^{\mu}(s)$ we have the *pointwise* limit

$$\lim_{\varepsilon \to 0} T_{\varepsilon}^{\mu\nu}(x) = T_{\mathrm{em}}^{\mu\nu}(x).$$

However, due to the singularities present at $x^\mu = y^\mu(s)$, this limit does *not* exist if performed in the topology of $\mathcal{S}'$. In fact, the tensor $T_{\text{em}}^{\mu\nu}$ is not an element of $\mathcal{S}'$. We illustrate these features again for the fields of a static particle. In this case the regularized energy density has the expression

$$T_\varepsilon^{00}(x) = \frac{1}{2} E_\varepsilon^2 = \frac{1}{2}\left(\frac{e}{4\pi}\right)^2 \frac{r^2}{(r^2 + \varepsilon^2)^3},$$

which for every $\mathbf{x} \neq 0$ admits the pointwise limit, see Eq. (16.4),

$$\lim_{\varepsilon \to 0} T_\varepsilon^{00}(x) = \frac{1}{2}\left(\frac{e}{4\pi}\right)^2 \frac{1}{r^4} = T_{\text{em}}^{00}(x).$$

But, indeed, $T_{\text{em}}^{00}(x)$ is not a distribution.

Renormalization. Before we can perform the limit for $\varepsilon \to 0$ in the topology of $\mathcal{S}'$ of the tensor $T_\varepsilon^{\mu\nu}$ (16.13), we first must identify its "divergent part" – that we denote with the symbol $\widehat{T}_\varepsilon^{\mu\nu}$ – and then subtract it. The renormalization process consists in this subtraction procedure, and the quantity $\widehat{T}_\varepsilon^{\mu\nu}$ is called *counterterm*. We require the counterterm to fulfill the following properties:

(a) it must be a *tensor* under Poincaré transformations;
(b) its *support* must be the world line of the particle,

$$\widehat{T}_\varepsilon^{\mu\nu}(x) = 0, \quad \text{for } x^\mu \neq y^\mu(s);$$

(c) it must be *symmetric* and *traceless*,

$$\widehat{T}_\varepsilon^{\mu\nu} = \widehat{T}_\varepsilon^{\nu\mu}, \qquad \widehat{T}_\varepsilon^{\mu\nu}\eta_{\mu\nu} = 0;$$

(d) it must be such that the limiting tensor defined by

$$\mathbb{T}_{\text{em}}^{\mu\nu} = \mathcal{S}' - \lim_{\varepsilon \to 0}\left(T_\varepsilon^{\mu\nu} - \widehat{T}_\varepsilon^{\mu\nu}\right) \tag{16.14}$$

exists, where the symbol "$\mathcal{S}' - \lim$" denotes the limit in the sense of distributions;
(e) it must be such that, if the particle's world line is subject to an appropriate equation of motion, then the new total energy-momentum tensor $T^{\mu\nu}$ defined by

$$T^{\mu\nu} = \mathbb{T}_{\text{em}}^{\mu\nu} + T_{\text{p}}^{\mu\nu}, \qquad T_{\text{p}}^{\mu\nu} = m\int u^\mu u^\nu \delta^4(x - y)\, ds, \tag{16.15}$$

satisfies the continuity equation

$$\partial_\mu T^{\mu\nu} = 0, \tag{16.16}$$

where it is understood that the four-divergence is computed in the distributional
sense.

The tensor $\mathbb{T}^{\mu\nu}_{\mathrm{em}}$ defined in Eq. (16.14) represents the *renormalized electromagnetic
energy-momentum tensor*, and it replaces *for all purposes* the original ill-defined
energy-momentum tensor $T^{\mu\nu}_{\mathrm{em}}$ (16.1). Let us now explain the meaning of the above
requirements on $\widehat{T}^{\mu\nu}_{\varepsilon}$. Since $T^{\mu\nu}_{\varepsilon}$, by construction, is a tensor, property (a) ensures
that also the object $\mathbb{T}^{\mu\nu}_{\mathrm{em}}$ defined by the limit (16.14) is a *tensor*. Property (b) is
motivated by the following two facts. First, in the complement of the world line the
original energy-momentum tensor $T^{\mu\nu}_{\mathrm{em}}$ is regular, as well a conserved, and so there
is no reason to modify its values for $x^{\mu} \neq y^{\mu}(s)$. Second, the form of $T^{\mu\nu}_{\mathrm{em}}$ in the
complement of the world line has a firm phenomenological basis, as we saw, for
instance, from its mixed components $T^{0i}_{\mathrm{em}} = (\mathbf{E} \times \mathbf{B})^{i}$, which quantify the energy-
balance of all radiation phenomena. Therefore, since we do not want to spoil the
agreement between theory and experiment, the renormalization process should not
modify the values of $T^{\mu\nu}_{\mathrm{em}}$ in the complement of the trajectory. Consequently, the
counterterm $\widehat{T}^{\mu\nu}_{\varepsilon}$ can be non-vanishing at most on the particle's world line. The
requirements (c) follow from the properties of the tensor $T^{\mu\nu}_{\varepsilon}$ to be symmetric and
traceless, implying that also its divergent part must be so. Property (d) ensures that
$\mathbb{T}^{\mu\nu}_{\mathrm{em}}$ is a distribution, guaranteeing thus, in particular, that the distributional derivatives $\partial_{\mu}\mathbb{T}^{\mu\nu}_{\mathrm{em}}$ are well defined. The meaning of requirement (e) is self-evident: as
anticipated, its realization will force the particle to satisfy the Lorentz-Dirac equa-
tion (15.19). Finally, it can be shown that the requirements (a)–(e) fix the form of the
counterterm $\widehat{T}^{\mu\nu}_{\varepsilon}$ uniquely, modulo terms which are of order ε in the topology of $\mathcal{S}'$
(for an indirect proof see Ref. [3]). Since in Eq. (16.14) we take the limit for $\varepsilon \to 0$,
this then implies that the renormalized energy-momentum tensor $\mathbb{T}^{\mu\nu}_{\mathrm{em}}$ is *uniquely*
determined.

16.2.1   Heuristic Renormalization of $T^{\mu\nu}_{\mathrm{em}}$

We now determine the form of the counterterm $\widehat{T}^{\mu\nu}_{\varepsilon}$ for a particle in *arbitrary*
motion, by enforcing the requirements (a)–(e) via heuristic arguments. According
to property (b), this tensor must be proportional to the distribution $\delta^{3}(\mathbf{x} - \mathbf{y}(t))$ or,
in conformity with the covariance requirement (a), to the Lorentz-invariant quantity

$$\int \delta^{4}(x - y) \, ds = \sqrt{1 - v^{2}(t)} \, \delta^{3}(\mathbf{x} - \mathbf{y}(t)).$$

Next, in basis of requirement (d), the counterterm must cancel the divergent part of
$T^{\mu\nu}_{\varepsilon}$, and so in the limit for $\varepsilon \to 0$ it must necessarily diverge. As the regularized
energy (16.12) diverges as $1/\varepsilon$, the tensor $\widehat{T}^{\mu\nu}_{\varepsilon}$ should hence diverge in the same
way. Since, in addition, the energy (16.12) is also proportional to e^{2}, the counterterm
should eventually have the form

$$\widehat{T}_{\varepsilon}^{\mu\nu} = \frac{1}{\varepsilon} \left(\frac{e}{4\pi}\right)^2 \int H^{\mu\nu} \delta^4(x - y) \, ds, \tag{16.17}$$

where, in basis of properties (a) and (c), $H^{\mu\nu}$ must be a *symmetric traceless tensor*. Being defined intrinsically on the world line $y^\mu = y^\mu(s)$, $H^{\mu\nu}$ must depend on the kinematical variables y^μ, u^μ, w^μ and, possibly, their higher derivatives. Moreover, considering that $\widehat{T}_{\varepsilon}^{\mu\nu}$ must have the dimension of an energy density, and that ε has the dimension of a length, $H^{\mu\nu}$ must be dimensionless. Since the unique dimensionless kinematical variable is u^μ, $H^{\mu\nu}$ must thus be of the form $H^{\mu\nu} = au^\mu u^\nu + b\eta^{\mu\nu}$, where a and b are dimensionless constants. Finally, the constraint $H^{\mu\nu}\eta_{\mu\nu} = 0$ imposes that the coefficients a and b are related by $b = -a/4$, so that finally the counterterm (16.17) takes the form

$$\widehat{T}_{\varepsilon}^{\mu\nu} = \frac{a}{\varepsilon} \left(\frac{e}{4\pi}\right)^2 \int \left(u^\mu u^\nu - \frac{1}{4}\eta^{\mu\nu}\right) \delta^4(x - y) \, ds. \tag{16.18}$$

Notice that contributions to $H^{\mu\nu}$ of the type $y^\mu w^\nu + y^\nu w^\mu - \frac{1}{2}\eta^{\mu\nu}y^\rho w_\rho$, or $y^\mu \partial^\nu + y^\nu \partial^\mu - \frac{1}{2}\eta^{\mu\nu}y^\rho \partial_\rho$, which are likewise dimensionless, symmetric and traceless, are excluded because they are not invariant under the space-time translation $y^\mu \to y^\mu + a^\mu$, see requirement (a). With the counterterm (16.18) eventually the renormalized electromagnetic energy-momentum tensor (16.14) takes the form

$$\mathbb{T}_{em}^{\mu\nu} = \mathcal{S}' - \lim_{\varepsilon \to 0} \left(T_{\varepsilon}^{\mu\nu} - \frac{a}{\varepsilon}\left(\frac{e}{4\pi}\right)^2 \int \left(u^\mu u^\nu - \frac{1}{4}\eta^{\mu\nu}\right)\delta^4(x - y)\,ds\right), \tag{16.19}$$

where the unique undetermined element is the number a. The latter should be determined by imposing the requirements (d) and (e), i.e. by proving that for an appropriate choice of a the limit (16.19) exists and that, for the same choice, the resulting total energy-momentum tensor is conserved. In fact, it can be shown that the limit (16.19) exists only for the choice

$$a = \frac{\pi^2}{2},$$

and that the resulting total energy-momentum tensor (16.15) is conserved *if* the particle satisfies the Lorentz-Dirac equation (15.19), see Ref. [1]. In Sect. 16.3 we present the proofs of these results for a particle in uniform linear motion.

16.3 Explicit Renormalization of $T_{em}^{\mu\nu}$ for a Free Particle

In the absence of external fields, $F_{in}^{\mu\nu} = 0$, a particle is *free* and performs hence a uniform linear motion. In this case, if we are able to prove that the expression (16.19) satisfies properties (d) and (e) in a given inertial frame, then, thanks to the manifest Lorentz invariance of our approach, these properties hold automatically in every inertial frame. Therefore, for a free particle, it is sufficient to prove these properties, for instance, in the *rest frame* of the particle. In practice it suffices thus (i) to prove the existence of the limit (16.19) in the relatively simple case of a *static* particle and (ii), as the energy-momentum tensor of a free particle is identically conserved,

$$\partial_\mu T_p^{\mu\nu} = \int \frac{dp^\nu}{ds} \, \delta^4(x-y) \, ds = 0, \tag{16.20}$$

to verify that the resulting energy-momentum tensor (16.19) satisfies separately the continuity equation $\partial_\mu \mathbb{T}_{em}^{\mu\nu} = 0$. In the absence of external fields, the regularized total field (16.10) reduces to $F_\varepsilon^{\mu\nu} = \mathcal{F}_\varepsilon^{\mu\nu}$, and for a static particle the regularized Liénard-Wiechert field $\mathcal{F}_\varepsilon^{\mu\nu}$, in turn, is given by formulas (16.11). To establish the components of the regularized electromagnetic energy-momentum tensor $T_\varepsilon^{\mu\nu}$ (16.13) it is then sufficient to perform in the expressions (16.4)–(16.6) the replacement $\mathbf{E} \to \mathbf{E}_\varepsilon$

$$T_\varepsilon^{00} = \frac{1}{2} \left(\frac{e}{4\pi}\right)^2 \frac{r^2}{(r^2 + \varepsilon^2)^3},$$

$$T_\varepsilon^{0i} = 0, \tag{16.21}$$

$$T_\varepsilon^{ij} = \frac{1}{2} \left(\frac{e}{4\pi}\right)^2 \frac{\delta^{ij} r^2 - 2x^i x^j}{(r^2 + \varepsilon^2)^3}.$$

Notice that this tensor satisfies still the tracelessness condition $\eta_{\mu\nu} T_\varepsilon^{\mu\nu} = 0$. It is also straightforward to write down the components of the counterterm in (16.19), since in the case at hand we have $u^\mu = (1,0,0,0)$ and $\int \delta^4(x-y(s)) \, ds = \delta^3(\mathbf{x})$. By inserting the components (16.21) in formula (16.19), we then see that to prove property (d) we must establish the existence of the distributional limits

$$\mathbb{T}_{em}^{00} = \frac{1}{2} \left(\frac{e}{4\pi}\right)^2 S' - \lim_{\varepsilon \to 0} \left(\frac{r^2}{(r^2 + \varepsilon^2)^3} - \frac{3a}{2\varepsilon} \, \delta^3(\mathbf{x})\right), \tag{16.22}$$

$$\mathbb{T}_{em}^{0i} = 0, \tag{16.23}$$

$$\mathbb{T}_{em}^{ij} = \frac{1}{2} \left(\frac{e}{4\pi}\right)^2 S' - \lim_{\varepsilon \to 0} \left(\frac{\delta^{ij} r^2 - 2x^i x^j}{(r^2 + \varepsilon^2)^3} - \frac{a}{2\varepsilon} \, \delta^{ij} \, \delta^3(\mathbf{x})\right), \tag{16.24}$$

for an appropriate choice of the constant a.

16.3.1   Existence of $\mathbb{T}_{em}^{\mu\nu}$ as a Distribution

To evaluate the limits (16.22)–(16.24) we frequently will need to move the limit for $\varepsilon \to 0$ under the integral sign, an operation that is not always allowed. In this regard, it is useful to recall the following theorem concerning certain sequences of absolutely integrable functions, i.e. functions belonging to the space $L^1 = L^1(\mathbb{R}^D)$, see the definition (2.101).

Lebesgue's dominated convergence theorem. Let $\{f_n\} \in L^1$ be a sequence of functions such that:

(a) there exists the pointwise limit (almost everywhere with respect to Lebesgue measure)

$$\lim_{n\to\infty} f_n(x) = f(x); \tag{16.25}$$

(b) there exists a positive function $g \in L^1$ such that (almost everywhere with respect to Lebesgue measure)

$$\left|f_n(x)\right| \leq g(x), \quad \forall n. \tag{16.26}$$

Then $f \in L^1$, and the sequence $\{f_n\}$ converges to f in the topology of L^1,

$$L^1 - \lim_{n\to\infty} f_n = f.$$

Corollary. Under the hypotheses of Lebesgue's dominated convergence theorem one can interchange the limit with the integral sign, i.e.

$$\lim_{n\to\infty} \int f_n \, d^D x = \int \lim_{n\to\infty} f_n \, d^D x.$$

Proof We have the bound

$$\left| \int f_n \, d^D x - \int f \, d^D x \right| = \left| \int (f_n - f) \, d^D x \right| \leq \int \left| f_n - f \right| d^D x = \|f_n - f\|_{L^1}.$$

Thanks to Lebesgue's dominated convergence theorem, the sequence $\{f_n\}$ converges to f in the topology of L^1. Therefore, in the limit for $n \to \infty$ the last term of the above bound converges to a zero and, consequently, so does the first. In basis of the pointwise limit (16.25) we then obtain

$$0 = \lim_{n\to\infty} \int f_n \, d^D x - \int f \, d^D x = \lim_{n\to\infty} \int f_n \, d^D x - \int \lim_{n\to\infty} f_n \, d^D x.$$

$\square$

In the following, the role of the discrete index $n \to \infty$ will be played by the *continuous* index $\varepsilon \to 0$, corresponding to the identification $f_n(x) \leftrightarrow f_\varepsilon(x)$. In the situations we will meet, the pointwise limit (16.25) will always exist trivially, and so, to move the limit under the integral sign, it suffices to find a positive ε-independent function $g \in L^1$ satisfying the uniform bound (16.26), i.e. $\left|f_\varepsilon(x)\right| \leq g(x)$.

Existence of $\mathbb{T}_{em}^{00}$. We begin the proof of property (d) by proving the existence of the limit (16.22), which regards the electromagnetic *energy density*. According to the definition of the limit in the sense of distributions (2.64), we must show that, for an appropriate constant a, the ordinary limits[1]

$$\mathbb{T}^{00}_{\text{em}}(\varphi) = \frac{1}{2} \left(\frac{e}{4\pi} \right)^2 \lim_{\varepsilon \to 0} \left(\int \frac{r^2 \varphi(\mathbf{x})}{(r^2 + \varepsilon^2)^3} \, d^3x - \frac{3a}{2\varepsilon} \varphi(0) \right) \tag{16.27}$$

exist for any test function $\varphi(\mathbf{x}) \in \mathcal{S} = \mathcal{S}(\mathbb{R}^3)$. We begin this proof by adding and subtracting in the numerator of the integrand in (16.27) the constant $\varphi(0)$. Using the integral

$$\int \frac{r^2 d^3x}{(r^2 + \varepsilon^2)^3} = \frac{3\pi^2}{4\varepsilon}, \tag{16.28}$$

the limit (16.27) can then be recast in the equivalent form

$$\mathbb{T}^{00}_{\text{em}}(\varphi) = \frac{1}{2} \left(\frac{e}{4\pi} \right)^2 \lim_{\varepsilon \to 0} \left(\int \frac{r^2 (\varphi(\mathbf{x}) - \varphi(0))}{(r^2 + \varepsilon^2)^3} \, d^3x + \frac{3}{2\varepsilon} \left(\frac{\pi^2}{2} - a \right) \varphi(0) \right). \tag{16.29}$$

The limit for $\varepsilon \to 0$ of the integral in (16.29) now exists for any $\varphi(\mathbf{x})$. To show it we separate in the integral the small values of r from the large ones, by setting

$$\int \frac{r^2 (\varphi(\mathbf{x}) - \varphi(0))}{(r^2 + \varepsilon^2)^3} \, d^3x = \int f_\varepsilon(\mathbf{x}) \, d^3x, \tag{16.30}$$

where the functions $f_\varepsilon(\mathbf{x})$ have different expressions according to whether $r < 1$ or $r > 1$, namely

$$f_\varepsilon(\mathbf{x}) = \frac{r^2 (\varphi(\mathbf{x}) - \varphi(0) - x^i \partial_i \varphi(0))}{(r^2 + \varepsilon^2)^3} H(1 - r) + \frac{r^2 (\varphi(\mathbf{x}) - \varphi(0))}{(r^2 + \varepsilon^2)^3} H(r - 1). \tag{16.31}$$

For $r < 1$ we have subtracted a term proportional to $x^i \partial_i \varphi(0)$ which does not contribute to the integral (16.30), because the integral over the solid angle $\int x^i d\Omega = r \int n^i d\Omega$ is zero. In the integral (16.30) we can now move the limit for $\varepsilon \to 0$ under the integral sign, by applying the corollary of the dominated convergence theorem. In fact, the sequence of functions $f_\varepsilon(\mathbf{x})$ can be bounded uniformly in ε by a function $g(\mathbf{x}) \in L^1(\mathbb{R}^3)$, see the condition (16.26),

$$|f_\varepsilon(\mathbf{x})| \leq \frac{|\varphi(\mathbf{x}) - \varphi(0) - x^i \partial_i \varphi(0)|}{r^4} H(1 - r) + \frac{2\|\varphi\|}{r^4} H(r - 1) = g(\mathbf{x}), \tag{16.32}$$

where $\|\varphi\|$ denotes the maximum of $|\varphi(\mathbf{x})|$ in $\mathbb{R}^3$. The positive function $g(\mathbf{x})$ belongs indeed to $L^1(\mathbb{R}^3)$ because, on the one hand, for $r \to 0$ the function $\varphi(\mathbf{x}) - \varphi(0) - x^i \partial_i \varphi(0)$ vanishes as r^2, and so $g(\mathbf{x})$ increases just as $1/r^2$, and, on the other hand, for $r \to \infty$ the function $g(\mathbf{x})$ vanishes as $1/r^4$. Consequently, $g(\mathbf{x})$ is

---

[1] The prescribed test function space would be $\mathcal{S}(\mathbb{R}^4)$. However, since in the static case the dependence on the time variable t is trivial, it is not restrictive to consider the *spatial* test function space $\mathcal{S}(\mathbb{R}^3)$.

absolutely integrable in $\mathbb{R}^3$. Taking, thus, in Eq. (16.30) the limit for $\varepsilon \to 0$ under the integral sign, and evaluating the trivial pointwise limit for $\varepsilon \to 0$ of the sequence $f_\varepsilon(\mathbf{x})$ (16.31), we obtain for any $\varphi(\mathbf{x})$ the finite result

$$
\lim_{\varepsilon \to 0} \int \frac{r^2(\varphi(\mathbf{x}) - \varphi(0))}{(r^2 + \varepsilon^2)^3}\, d^3x = \int_{r<1} \frac{\varphi(\mathbf{x}) - \varphi(0) - x^i \partial_i \varphi(0)}{r^4}\, d^3x
$$
$$
+ \int_{r>1} \frac{\varphi(\mathbf{x}) - \varphi(0)}{r^4}\, d^3x. \tag{16.33}
$$

It follows that the limit (16.29) exists, if and only if $a = \pi^2/2$. The limit (16.33) can be recast in a simpler form. In fact, if in the integral for $r < 1$ we adopt the convention that the integration over the angles precedes the integration over r, the term $x^i \partial_i \varphi(0)$ drops out, and we can write the sum of the two integrals in (16.33) as a single integral over $\mathbb{R}^3$. If we understand, therefore, that the integration over the angles precedes the integration over r – it is said that the integral is *conditionally convergent* – eventually the renormalized electromagnetic energy density (16.29), as a distribution, takes the simple form

$$
\mathbb{T}_{\mathrm{em}}^{00}(\varphi) = \frac{1}{2}\left(\frac{e}{4\pi}\right)^2 \int \frac{\varphi(\mathbf{x}) - \varphi(0)}{r^4}\, d^3x. \tag{16.34}
$$

Existence of $\mathbb{T}_{\mathrm{em}}^{ij}$. In the same way it can be shown that – for the same value of a – the limit (16.24) for the renormalized Maxwell's stress tensor exists, see Problem 16.1. It turns out that applied to a test function it is given by

$$
\mathbb{T}_{\mathrm{em}}^{ij}(\varphi) = \frac{1}{2}\left(\frac{e}{4\pi}\right)^2 \int \frac{\varphi(\mathbf{x}) - \varphi(0)}{r^4}\left(\delta^{ij} - 2\frac{x^i x^j}{r^2}\right) d^3x, \tag{16.35}
$$

where the integral is again conditionally convergent. In conclusion, in the case of a free particle the energy-momentum tensor (16.19) for the chosen value of a defines a distribution, as required by condition (d). Henceforth, we set thus $a = \pi^2/2$.

16.3.2   Continuity Equation for $\mathbb{T}_{\mathrm{em}}^{\mu\nu}$

According to our last requirement (e), the total energy-momentum tensor (16.15) should be conserved if the particle satisfies an appropriate equation of motion. In the case of a free particle, in light of the identity (16.20), we need thus to show that the distribution-valued tensor $\mathbb{T}_{\mathrm{em}}^{\mu\nu}$ defined by formulas (16.5), (16.34) and (16.35) satisfies the continuity equation

$$
\partial_\mu \mathbb{T}_{\mathrm{em}}^{\mu\nu} = 0. \tag{16.36}
$$

The $\nu = 0$ component of this equation is identically satisfied, because $\mathbb{T}^{i0}_{em}$ is zero and $\mathbb{T}^{00}_{em}$ is time-independent. It remains, therefore, to verify the $\nu = j$ component of Eq. (16.36), which reduces to the non-trivial condition for Maxwell's stress tensor

$$\partial_i \mathbb{T}^{ij}_{em} = 0. \tag{16.37}$$

To prove that the tensor $\mathbb{T}^{ij}_{em}$ (16.35) satisfies this condition, it is more convenient to go back to its representation (16.24), and to use that the partial derivatives ∂_μ constitute *continuous* operations in $\mathcal{S}'$, a property that allows to interchange distributional limits with distributional derivatives. Taking thus the spatial divergence of the tensor (16.24), we obtain

$$\partial_i \mathbb{T}^{ij}_{em} = \frac{1}{2} \left(\frac{e}{4\pi} \right)^2 \mathcal{S}' - \lim_{\varepsilon \to 0} \left(\partial_i \left(\frac{\delta^{ij} r^2 - 2x^i x^j}{(r^2 + \varepsilon^2)^3} \right) - \frac{\pi^2}{4\varepsilon} \partial_j \delta^3(\mathbf{x}) \right). \tag{16.38}$$

The first term is a C^∞ function, and therefore its derivatives can be computed in the ordinary sense

$$\partial_i \left(\frac{\delta^{ij} r^2 - 2x^i x^j}{(r^2 + \varepsilon^2)^3} \right) = -\frac{6\varepsilon^2 x^j}{(r^2 + \varepsilon^2)^4} = \partial_j \left(\frac{\varepsilon^2}{(r^2 + \varepsilon^2)^3} \right).$$

Equation (16.38) can therefore be recast in the alternative form

$$\partial_i \mathbb{T}^{ij}_{em} = \frac{1}{2} \left(\frac{e}{4\pi} \right)^2 \partial_j \left(\mathcal{S}' - \lim_{\varepsilon \to 0} \left(\frac{\varepsilon^2}{(r^2 + \varepsilon^2)^3} - \frac{\pi^2}{4\varepsilon} \delta^3(\mathbf{x}) \right) \right),$$

where, in turn, we have interchanged the derivative ∂_j with the limit for $\varepsilon \to 0$. This last operation is allowed, if the limit of the distribution between parenthesis exists. In fact, this limit exists and, in particular, it vanishes. To prove it we must show that, for any $\varphi(\mathbf{x}) \in \mathcal{S}$, in the limit for $\varepsilon \to 0$ the sequence of numbers

$$\int \frac{\varepsilon^2 \varphi(\mathbf{x})}{(r^2 + \varepsilon^2)^3} d^3 x - \frac{\pi^2}{4\varepsilon} \varphi(0) = \int \frac{\varepsilon^2 (\varphi(\mathbf{x}) - \varphi(0))}{(r^2 + \varepsilon^2)^3} d^3 x = \int h_\varepsilon(\mathbf{x}) d^3 x \tag{16.39}$$

converges to zero. We have used the integral

$$\int \frac{d^3 x}{(r^2 + \varepsilon^2)^3} = \frac{\pi^2}{4\varepsilon^3},$$

and we have introduced the sequence of functions

$$h_\varepsilon(\mathbf{x}) = \frac{\varphi(\varepsilon \mathbf{x}) - \varphi(0)}{\varepsilon (r^2 + 1)^3}.$$

In the last term of Eq. (16.39) we can now move the limit for $\varepsilon \to 0$ under the integral sign, by resorting again to the corollary of the dominated convergence theorem. In this case the sequence $h_\varepsilon(\mathbf{x})$ can be bounded uniformly in ε by writing

$$\varphi(\varepsilon\mathbf{x}) - \varphi(0) = \varepsilon \int_0^1 \mathbf{x} \cdot \nabla\varphi(|\,\varepsilon\mathbf{x})\,d\,| \quad \Rightarrow \quad |\varphi(\varepsilon\mathbf{x}) - \varphi(0)| \leq \varepsilon r\,\|\nabla\varphi\|,$$

where $\|\nabla\varphi\|$ denotes the maximum of $|\nabla\varphi(\mathbf{x})|$ in $\mathbb{R}^3$. We obtain thus the bound

$$|h_\varepsilon(\mathbf{x})| \leq \frac{r\|\nabla\varphi\|}{(r^2+1)^3} = g(\mathbf{x}) \in L^1(\mathbb{R}^3).$$

On the other hand, the sequence $h_\varepsilon(\mathbf{x})$ admits the pointwise limit

$$\lim_{\varepsilon\to 0} h_\varepsilon(\mathbf{x}) = \frac{x^i \partial_i \varphi(0)}{(r^2+1)^3}.$$

Therefore, taking in Eq. (16.39) the limit for $\varepsilon \to 0$ under the integral sign, we obtain

$$\lim_{\varepsilon\to 0}\left(\int \frac{\varepsilon^2 \varphi(\mathbf{x})}{(r^2+\varepsilon^2)^3}\,d^3x - \frac{\pi^2}{4\varepsilon}\,\varphi(0)\right) = \int \lim_{\varepsilon\to 0} h_\varepsilon(\mathbf{x})\,d^3x = \int \frac{x^i \partial_i \varphi(0)}{(r^2+1)^3}\,d^3x = 0,$$

where the conclusion follows from the vanishing of the angle integral $\int x^i d\Omega = r \int n^i d\Omega$. This concludes the proof of the identity (16.37).

16.3.3 Finite Energy of the Electromagnetic Field

The proof of the existence of $T_{\text{em}}^{\mu\nu}$ as a distribution, given in Sect. 16.3.1, has furnished, in particular, a *constructive definition* of the renormalized electromagnetic energy density T_{em}^{00}, namely formula (16.34). Concerning the electromagnetic energy itself, this formula allows in particular to express the electromagnetic energy contained in a volume V as[2]

$$\varepsilon_{\text{em}}(V) = \int_V T_{\text{em}}^{00}\,d^3x = \int T_{\text{em}}^{00}\,\chi_V\,d^3x = T_{\text{em}}^{00}(\chi_V)$$

$$= \frac{1}{2}\left(\frac{e}{4\pi}\right)^2 \int \frac{\chi_V(\mathbf{x}) - \chi_V(0)}{r^4}\,d^3x, \tag{16.40}$$

---

[2] As the characteristic function $\chi_V(\mathbf{x})$ is not continuous in $\mathbb{R}^3$, it is actually not an element of $\mathcal{S}(\mathbb{R}^3)$. Consequently, a priori the quantities $T_{\text{em}}^{00}(\chi_V)$ are ill-defined. This problem can be remedied in a rigorous way, by "approximating" $\chi_V(\mathbf{x})$ with a sequence of functions $\chi_V^n(\mathbf{x}) \in \mathcal{S}(\mathbb{R}^3)$ such that almost everywhere $\lim_{n\to\infty} \chi_V^n(\mathbf{x}) = \chi_V(\mathbf{x})$, and by *defining* then $T_{\text{em}}^{00}(\chi_V)$ as the limit $\lim_{n\to\infty} T_{\text{em}}^{00}(\chi_V^n)$.

where $\chi_V(\mathbf{x})$ denotes the characteristic function of the volume V. Indeed, the expression (16.40) entails the correct physical properties necessary to identify it as the *energy contained in the volume* V. We list them below.

- The energy (16.40) is *finite* for any volume V whose *boundary* does not contain the particle, i.e. the origin. In fact, the function $\chi_V(\mathbf{x}) - \chi_V(0)$ vanishes in a neighborhood of $\mathbf{x} = 0$, rendering thus the singularity of the factor $1/r^4$ at $\mathbf{x} = 0$ innocuous. Notice that the same restriction on the position of the particle applies to the volume integrals of the energy-momentum tensor of the particle $T_p^{\mu\nu}$ in (16.15), which are likewise ill-defined if the particle sits at the boundary of V.
- If V does not contain the particle the number $\chi_V(0)$ is zero, and accordingly formula (16.40) reduces to

$$\varepsilon_{\text{em}}(V) = \frac{1}{2}\left(\frac{e}{4\pi}\right)^2 \int_V \frac{1}{r^4}\, d^3x = \int_V T_{\text{em}}^{00}\, d^3x,$$

where T_{em}^{00} is the original energy density (16.4). Therefore, restricted to volumes not containing the particle, the renormalized energy density $\mathbb{T}_{\text{em}}^{00}$ is equivalent to T_{em}^{00}.
- If V_R denotes the volume of a sphere of radius R centered at the origin we have

$$\chi_{V_R}(\mathbf{x}) - \chi_{V_R}(0) = H(R - r) - 1 = -H(r - R).$$

For the energy contained in the sphere formula (16.40) then yields the finite value

$$\varepsilon_{\text{em}}(V_R) = -\frac{1}{2}\left(\frac{e}{4\pi}\right)^2 \int_{r>R} \frac{1}{r^4}\, d^3x = -\frac{e^2}{8\pi R}. \tag{16.41}$$

Notice that, although negative, this energy is an *increasing* function of R.
- Taking the limit of Eq. (16.41) for $R \to \infty$ we derive that the total energy of the electromagnetic field is zero,

$$\varepsilon_{\text{em}}(\mathbb{R}^3) = 0. \tag{16.42}$$

Notice the net difference between the result (16.42) and the prediction (16.7) of the original energy-momentum tensor. Since the components $\mathbb{T}_{\text{em}}^{0i}$ of the renormalized energy-momentum tensor are identically zero, see (16.23), the result (16.42) implies that the total four-momentum of the electromagnetic field of a static particle is zero,

$$P_{\text{em}}^\mu = \int \mathbb{T}_{\text{em}}^{0\mu}\, d^3x = 0. \tag{16.43}$$

- In non-relativistic physics, usually the infinite electrostatic self-energy (16.7) is subtracted *by hand*, to make the *potential energy* of a system of charged particles finite, see Problem 2.8. It is easy to realize that Eq. (16.42) reproduces precisely this subtraction.

Electromagnetic four-momentum of a particle in uniform linear motion. Thanks to the manifest Lorentz invariance of our approach, the results derived above for a static particle extend automatically to a particle performing a generic uniform linear motion, i.e. to a free particle. In particular, the energy-momentum tensor (16.19) defines a distribution also for the electromagnetic field generated by a free particle, and Eq. (16.36) ensures that this tensor is conserved,

$$\partial_\mu \mathbb{T}_{\text{em}}^{\mu\nu} = 0. \tag{16.44}$$

Moreover, since $\mathbb{T}_{\text{em}}^{\mu\nu}$ is distribution-valued, also in the case of a free particle the spatial integrals yielding the four-momentum of the electromagnetic field contained in a volume V

$$P_{\text{em}}^\mu(V) = \int_V \mathbb{T}_{\text{em}}^{0\mu} \, d^3x = \int \mathbb{T}_{\text{em}}^{0\mu} \chi_V \, d^3x = \mathbb{T}_{\text{em}}^{0\mu}(\chi_V)$$

are finite for any V. In particular, if at a given instant the particle is outside V, then $P_{\text{em}}^\mu(V)$ coincides with the four-momentum of the original energy-momentum tensor,

$$P_{\text{em}}^\mu(V) = \int_V T_{\text{em}}^{0\mu} \, d^3x.$$

Finally, again thanks to Lorentz invariance, Eq. (16.43) implies that also the total four-momentum of the electromagnetic field of a particle in uniform linear motion is zero,

$$P_{\text{em}}^\mu = \int \mathbb{T}_{\text{em}}^{0\mu} \, d^3x = 0, \tag{16.45}$$

a result that we have used in Sect. 15.4 to constrain the form of P_{em}^μ for generic motions, see Eq. (15.93).

Renormalization, relativistic invariance, and conservation. From the above results we infer a deep relationship between the *covariance* and *finiteness* properties of the total four-momentum, a relationship that, actually, we have already established on general grounds in Sects. 2.4.1 and 2.4.2. Consider, in fact, the total four-momentum $P_{\text{em}\,\varepsilon}^\mu$ entailed by the regularized energy-momentum tensor $T_\varepsilon^{\mu\nu}$ (16.13) in the case of a particle in uniform linear motion with velocity $\mathbf{v}$. This four-momentum can be determined explicitly by inserting the regularized Liénard-Wiechert field (15.30) in the components $T_\varepsilon^{0\mu}$ of the tensor (16.13), and by calculating then the four spatial integrals

$$P_{\text{em}\,\varepsilon}^\mu = \int T_\varepsilon^{0\mu} \, d^3x = \frac{e^2}{32\varepsilon\sqrt{1-v^2}} \left(\frac{3+v^2}{4}, \mathbf{v} \right). \tag{16.46}$$

Notice that for $\mathbf{v} = 0$ this result reproduces the regularized electrostatic energy $\varepsilon_{\text{em}\,\varepsilon}$ of Eq. (16.12). However, the unpleasant feature of Eq. (16.46) is that, although $T_\varepsilon^{\mu\nu}$ is a tensor, the four quantities $P_{\text{em}\,\varepsilon}^\mu$ do not form a four-vector. Similarly, also the four-momentum integrals associated with the covariant counterterm $\widehat{T}_\varepsilon^{\mu\nu}$ (16.18)

$$\widehat{P}^\mu_{\text{em}\,\varepsilon} = \int \widehat{T}_\varepsilon^{0\mu}\, d^3x = \frac{e^2}{32\varepsilon}\left(u^\mu - \frac{1}{4}\sqrt{1-v^2}\,\eta^{\mu 0}\right)$$

do not form a four-vector. On the other hand, it is immediately seen that for every $\varepsilon > 0$ we have the equality $P^\mu_{\text{em}\,\varepsilon} - \widehat{P}^\mu_{\text{em}\,\varepsilon} = 0$, a relation that, in basis of the definition (16.14), in the limit for $\varepsilon \to 0$ reproduces the covariant result (16.45). The explanation of this apparent mismatch is that, although both $T_\varepsilon^{\mu\nu}$ and $\widehat{T}_\varepsilon^{\mu\nu}$ are tensors, they are not conserved, $\partial_\mu T_\varepsilon^{\mu\nu} \neq 0 \neq \partial_\mu \widehat{T}_\varepsilon^{\mu\nu}$. In contrast, $\mathbb{T}_{\text{em}}^{\mu\nu}$ is a tensor, it satisfies the continuity equation (16.44), and the associated total four-momentum P^μ_{em} (16.45) is a four-vector. From Sect. 2.4.2 we know, in fact, that in order for the spatial integrals $\int T^{0\mu}\, d^3x$ to form a four-vector, in general it is not sufficient that $T^{\mu\nu}$ is a tensor: it must also be conserved. For the case at hand this implies that a *regularization* may very well preserve Lorentz invariance, but without a *renormalization*, i.e. the subsequent subtraction of the divergent counterterms, leading eventually to a conserved energy-momentum tensor, the regularization *alone* would spoil the Lorentz invariance of the total four-momentum: It is *as if* the divergent terms of the energy-momentum tensor, once integrated over whole space, would violate Lorentz invariance.

16.4   General Renormalization of $T_{\text{em}}^{\mu\nu}$ and Four-Momentum Conservation

The renormalization procedure carried out in Sect. 16.3 for the case of a static particle can be applied conceptually in the same way to a generic system of charged particles, the outcome being a total energy-momentum tensor $T^{\mu\nu}$ having the properties (1)–(4) envisaged in Sect. 16.1. However, this time the intermediate computations are less trivial, in particular, we now have $\partial_\mu \mathbb{T}_{\text{em}}^{\mu\nu} \neq 0$, and so for the derivations of the main results we refer the reader to Ref. [1]. We consider the usual system of N charged particles in the presence of an external field $F_{\text{in}}^{\mu\nu}$ satisfying the homogenous Maxwell equations. Since each particle generates a Liénard-Wiechert $\mathcal{F}_r^{\mu\nu}$, $r = 1, \ldots, N$, the total electromagnetic field is hence given by

$$F^{\mu\nu} = F_{\text{in}}^{\mu\nu} + \sum_r \mathcal{F}_r^{\mu\nu}. \tag{16.47}$$

We begin by regularizing each Liénard-Wiechert field according to the prescription (15.28), and we denote the corresponding regularized field by $\mathcal{F}_{r\,\varepsilon}^{\mu\nu}$. Accordingly, we introduce the total regularized electromagnetic field

$$F_\varepsilon^{\mu\nu} = F_{\text{in}}^{\mu\nu} + \sum_r \mathcal{F}_{r\,\varepsilon}^{\mu\nu},$$

and the corresponding regularized electromagnetic energy-momentum tensor

$$T_\varepsilon^{\mu\nu} = F_\varepsilon^{\mu\alpha} F_{\varepsilon\,\alpha}{}^\nu + \frac{1}{4}\,\eta^{\mu\nu} F_\varepsilon^{\alpha\beta} F_{\varepsilon\,\alpha\beta}. \tag{16.48}$$

The renormalized electromagnetic energy-momentum tensor $\mathbb{T}_{\text{em}}^{\mu\nu}$ then arises from a natural generalization of the prescription (16.19), namely

$$\mathbb{T}_{\text{em}}^{\mu\nu} = \mathcal{S}' - \lim_{\varepsilon \to 0}\left(T_\varepsilon^{\mu\nu} - \frac{\pi^2}{2\varepsilon}\sum_r \left(\frac{e_r}{4\pi}\right)^2 \int \left(u_r^\mu u_r^\nu - \frac{1}{4}\,\eta^{\mu\nu}\right)\delta^4(x - y_r)\,ds_r \right). \tag{16.49}$$

The total counterterm is thus simply the sum of the counterterms of each particle. This is a consequence of the fact that in the regularized energy-momentum tensor (16.48) the products of the Liénard-Wiechert fields of *different* particles in the limit for $\varepsilon \to 0$ give rise to singularities of the type $1/r^2$, which are well defined in the sense of distributions. One can then prove the following theorems.

Theorem I. The distributional limit (16.49) exists for arbitrary non-intersecting world lines y_r^μ of the particles.

Theorem II. The four-momentum of the electromagnetic field contained in a generic finite volume V

$$P_{\text{em}}^\mu(V) = \int_V \mathbb{T}_{\text{em}}^{0\mu}\,d^3x = \mathbb{T}_{\text{em}}^{0\mu}(\chi_V) \tag{16.50}$$

is finite. The total four-momentum of the field

$$P_{\text{em}}^\mu = \int \mathbb{T}_{\text{em}}^{0\mu}\,d^3x = \mathbb{T}_{\text{em}}^{0\mu}(1) \tag{16.51}$$

is likewise finite, if in the limit for $t \to -\infty$ the accelerations of the particles vanish sufficiently fast. For a rigorous definition of the quantities $\mathbb{T}_{\text{em}}^{0\mu}(\chi_V)$ and $\mathbb{T}_{\text{em}}^{0\mu}(1)$, see footnote 2 in Sect. 16.3.3, and for the meaning of the asymptotic conditions on the accelerations of the particles, see footnote 2 in the introductory section of Chap. 15.

Theorem III. For arbitrary world lines y_r^μ of the particles the four-divergence of the tensor (16.49) is equal to

$$\partial_\nu \mathbb{T}_{\text{em}}^{\nu\mu} = -\sum_r \int \left(\frac{e_r^2}{6\pi}\left(\frac{dw_r^\mu}{ds_r} + w_r^2 u_r^\mu \right) + e_r F_r^{\mu\nu}(y_r)\,u_{r\nu} \right)\delta^4(x - y_r)\,ds_r, \tag{16.52}$$

where $F_r^{\mu\nu}$ denotes the total external field experienced by the rth particle,

$$F_r^{\mu\nu} = F_{\text{in}}^{\mu\nu} + \sum_{s\neq r} \mathcal{F}_s^{\mu\nu}. \tag{16.53}$$

The fundamental identity (16.52) is the *well-defined* counterpart of the *formal* relation (16.8). From the comparison one sees that it is *as if* in formula (16.8) the divergent self-force $e_r \mathcal{F}_r^{\mu\nu}(y_r) u_{r\nu}$ had been replaced by the finite radiation reaction force

$$\frac{e_r^2}{6\pi} \left(\frac{dw_r^\mu}{ds_r} + w_r^2 u_r^\mu \right).$$

The identity (16.52) generalizes the identity (16.44) holding for a single particle in uniform linear motion, and reduces to it for $N = 1$, $w_r^\mu = 0$ and $F_{in}^{\mu\nu} = 0$.

Conservation of the total energy-momentum tensor. The energy-momentum tensor of the particles (2.138)

$$T_p^{\mu\nu} = \sum_r m_r \int u_r^\mu u_r^\nu \, \delta^4(x - y_r) \, ds_r \tag{16.54}$$

satisfies the identity (2.141)

$$\partial_\nu T_p^{\nu\mu} = \sum_r \int \frac{dp_r^\mu}{ds_r} \delta^4(x - y_r) \, ds_r.$$

Theorem III then implies that, for arbitrary world lines of the particles, the total energy-momentum of the system

$$T^{\mu\nu} = \mathbb{T}_{em}^{\mu\nu} + T_p^{\mu\nu} \tag{16.55}$$

satisfies the identity

$$\partial_\nu T^{\nu\mu} = \sum_r \int \left(\frac{dp_r^\mu}{ds_r} - \frac{e_r^2}{6\pi} \left(\frac{dw_r^\mu}{ds_r} + w_r^2 u_r^\mu \right) - e_r F_r^{\mu\nu}(y_r) u_{r\nu} \right) \delta^4(x - y_r) \, ds_r. \tag{16.56}$$

If we want the total four-momentum of the system to be locally conserved, namely that $\partial_\nu T^{\nu\mu} = 0$, we then conclude that each particle must satisfy the *Lorentz-Dirac equation* (15.34)

$$\frac{dp_r^\mu}{ds_r} = \frac{e_r^2}{6\pi} \left(\frac{dw_r^\mu}{ds_r} + w_r^2 u_r^\mu \right) + e_r F_r^{\mu\nu}(y_r) u_{r\nu}. \tag{16.57}$$

Finally, from Eqs. (16.51), (16.54) and (16.55) it follows that the total *conserved* and *finite* four-moment P^μ of the system has the expected form

$$P^\mu = \int T^{0\mu} \, d^3x = \int \left(\mathbb{T}_{em}^{0\mu} + T_p^{0\mu} \right) d^3x = P_{em}^\mu + \sum_r p_r^\mu,$$

a result that we have used in Sect. 15.4 to constrain the explicit form of P_{em}^μ, see Eq. (15.92). Ultimately, it is thus the paradigm of *local four-momentum conserva-*

tion which irrevocably subjects each charged particle to the Lorentz-Dirac equation, even though, in the absence of an action, Neother's theorem does not apply. This conservation law must, therefore, be considered as the *ultimate* cause of all the problematic physical consequences the Lorentz-Dirac equation entails, see Chap. 15.

Uniqueness and finite counterterms. Once a physical quantity has found to be divergent and its divergences have been removed by means of a renormalization, the finite quantity obtained in this way is defined *intrinsically* only modulo *finite counterterms*, i.e. counterterms which have the same structure as the divergent ones, but are finite. In the case at the hand, given the form of the divergent counterterms of the electromagnetic energy-momentum tensor $\mathbb{T}_{\text{em}}^{\mu\nu}$ of Eq. (16.49), the finite counterterms must (i) be supported on the particles' world lines, (ii) have the same dimensions as the counterterm (16.18), (iii) be symmetric in μ and ν and traceless, (iv) be Lorentz invariant. Considering, for simplicity, a single charged particle, these restrictions allow just for two linearly independent finite counterterms. They correspond to replace the total energy-momentum tensor $T^{\mu\nu}$ (16.55) with

$$T'^{\mu\nu} = T^{\mu\nu} - \frac{e^2}{6\pi} \int \left(c_1(u^\mu w^\nu + u^\nu w^\mu) + c_2(u^\nu \partial^\mu + u^\mu \partial^\nu) \right) \delta^4(x - y)\, ds,$$

where c_1 and c_2 are dimensionless constants. From the identity (16.56) we then derive that the modified energy-momentum tensor satisfies the identity, see the definition (16.53),

$$\partial_\nu T'^{\nu\mu} = \int \left(\frac{dp^\mu}{ds} - \frac{e^2}{6\pi} \left((1 + c_1) \frac{dw^\mu}{ds} + u^\mu(w^2 + c_1 w^\nu \partial_\nu + c_2 \Box) \right) \right.$$
$$\left. - e F_{\text{in}}^{\mu\nu}(y)\, u_\nu \right) \delta^4(x - y)\, ds.$$

However, due to the derivatives ∂_ν and $\Box = \partial^\mu \partial_\mu$ acting on the δ-function, which cannot be removed by an integration by parts, there exists no equation of motion of the particle – of whatever form – such that $\partial_\nu T'^{\nu\mu} = 0$. Therefore, if we insist on four-momentum conservation, we must set $c_1 = c_2 = 0$. This proves that the total energy-momentum tensor $T^{\mu\nu}$ defined by Eqs. (16.49), (16.54) and (16.55), and the resulting Lorentz-Dirac equations (16.57), are *uniquely* determined.

16.5   Problems

16.1 Show that for $a = \pi^2/2$ the distributional limit (16.24) exists and corresponds to the expression (16.35), by proceeding as follows.

(a) Show that when applying formula (16.24) to a test function $\varphi(\mathbf{x})$ one obtains

$$\mathbb{T}_{em}^{ij}(\varphi) = \frac{1}{2}\left(\frac{e}{4\pi}\right)^2 \lim_{\varepsilon \to 0} \left(\int \frac{\left(\delta^{ij}r^2 - 2x^i x^j\right)\left(\varphi(\mathbf{x}) - \varphi(0)\right)}{(r^2 + \varepsilon^2)^3}\, d^3 x \right.$$
$$\left. + \frac{1}{2\varepsilon}\left(\frac{\pi^2}{2} - a\right)\delta^{ij}\varphi(0)\right). \tag{16.58}$$

Hint: Write the product $x^i x^j$ in the form $x^i x^j = n^i n^j r^2$, and use the angle integral $\int n^i n^j\, d\Omega = 4\pi\delta^{ij}/3$, together with the integral (16.28).

(b) In Eq. (16.58) move the limit for $\varepsilon \to 0$ under the integral sign, by proceeding in the same way as after Eq. (16.30), i.e. by applying the corollary of the dominated convergence theorem of Sect. 16.3.1.

References

1. K. Lechner, P.A. Marchetti, Variational principle and energy-momentum tensor for relativistic electrodynamics of point charges. Ann. Phys. **322**, 1162 (2007)
2. K. Lechner, Radiation reaction and four-momentum conservation for point-like dyons. J. Phys. A **39**, 11647 (2006)
3. E.G.P. Rowe, Structure of the energy tensor in the classical electrodynamics of point particles. Phys. Rev. D **18**, 3639 (1978)

Chapter 17
The Electromagnetic Field of a Massless Particle

Massless charged particles represent an interesting challenge of elementary particle physics: From an experimental point of view, no such particles have ever been observed in nature, and, from a theoretical point of view, the existence of a consistent quantum field theory describing their dynamics is still an open problem; see the seminal Ref. [1] for a non-perturbative argument against the existence of a quantum field theory of *unconfined* massless charges, and Ref. [2] for the problematic aspects involved in the cancelation of their perturbative infrared divergences. If a consistent quantum theory exists, an appropriate semiclassical limit plausibly would then give rise to a consistent classical theory as well. In this sense, a direct construction of the classical electrodynamics of massless charged particles can, in turn, shed new light on the possible existence of such particles in nature. In this chapter we present the two fundamental steps of this construction. The first step consists in the determination of the exact electromagnetic field generated by a generic massless particle. This amounts to solve Maxwell's equations for a particle performing an arbitrary *light-like* motion, i.e. a motion with speed $v = 1$. The basic results of this chapter, in this respect, are Eq. (17.39), giving the field for a bounded light-like trajectory, and Eq. (17.51), giving the field for an unbounded like-light trajectory. For comparison, in Sect. 17.5 we analyze the known field of a hyperbolic motion.

The second step regards the determination of the equation of motion of a massless charged particle, taking into account the radiation reaction. As we have seen in Chap. 15 for the case of a massive particle, due to the singularities of the electromagnetic field – which for a massless particle have an even more involved structure – the derivation of the equation of motion in compatibility with *four-momentum conservation* relies on the renormalization of the electromagnetic energy-momentum tensor, and on the associated continuity equation. The renormalization of this tensor, and the resulting Lorentz-Dirac equation of motion, will lead us to the, may be surprising, conclusion that a classical massless charged particle does *not emit radiation* and, consequently, experiences *no radiation reaction*. The topics of this chapter will be presented mainly without derivations; for the proofs of the results of Sects. 17.1–17.3, see Refs. [3, 4], for the proofs regarding Sect. 17.4, see Ref. [5].

© Springer International Publishing AG, part of Springer Nature 2018
K. Lechner, *Classical Electrodynamics*, UNITEXT for Physics,
https://doi.org/10.1007/978-3-319-91809-9_17

17.1   Limiting Procedures

A massless particle moves wit h the speed of light, and so we must parameterize its world line $y^\mu(\lambda)$ with a generic parameter λ instead of the proper time s. Correspondingly, in this chapter we introduce the parameterization-dependent four-velocity and four-acceleration $u^\mu(\lambda) = dy^\mu(\lambda)/d\lambda$ and $w^\mu(\lambda) = du^\mu(\lambda)/d\lambda$, respectively, where the former is subject to be light-like

$$u^2 = \left(1 - v^2\right)\left(u^0\right)^2 = 0.$$

Occasionally, we will also resort to the four-velocity "vector"

$$v^\mu = \frac{u^\mu}{u^0} = (1, \mathbf{v}). \tag{17.1}$$

Correspondingly, the four-current of the particle takes the reparameterization-invariant form

$$j^\mu(x) = e \int u^\mu(\lambda)\, \delta^4\big(x - y(\lambda)\big)\, d\lambda. \tag{17.2}$$

The equations we want to solve are the usual Maxwell equations

$$\partial_{[\mu} F_{\nu\rho]} = 0, \qquad \partial_\mu F^{\mu\nu} = j^\nu. \tag{17.3}$$

Returning momentarily to a *time-like* world line $y^\mu(\lambda)$, parameterized with a generic parameter λ, the Liénard-Wiechert field (7.38) is still composed of a velocity field $V^{\mu\nu}$ and of an acceleration field $A^{\mu\nu}$

$$\mathcal{F}^{\mu\nu} = V^{\mu\nu} + A^{\mu\nu}. \tag{17.4}$$

The latter, being reparameterization invariant, maintains the same form as in (7.38), while the former acquires a factor u^2

$$V^{\mu\nu} = \frac{e\, u^2}{4\pi(uL)^3}\left(L^\mu u^\nu - L^\nu u^\mu\right), \tag{17.5}$$

$$A^{\mu\nu} = \frac{e}{4\pi(uL)^3}\, L^\mu\big((uL)w^\nu - (wL)u^\nu\big) - (\mu \leftrightarrow \nu). \tag{17.6}$$

In formulas (17.5) and (17.6) the kinematical variables are evaluated at the parameter $\lambda(x)$, determined by the usual retarded-time relations

$$L^2 = 0, \qquad L^0 > 0, \qquad L^\mu = x^\mu - y^\mu(\lambda). \tag{17.7}$$

Since for a light-like trajectory we have $u^2 = 0$, a naive *extrapolation* would lead to the expectation that a massless particle produces (i) no velocity field, and (ii) a non-vanishing acceleration field given by formula (17.6). However, as established previously, see Sects. 6.2.4 and 7.2.2, for particles on light-like trajectories the Green function method, that led us to formulas (17.4)–(17.6), in general fails to provide a solution to Maxwell's equations. Therefore, the above expectations cannot be trusted. Actually, as we will see, the failure of this method manifests itself in different ways, according to whether the trajectory of the particle is *unbounded*, tending in the infinite past to a straight line, or *bounded*, being confined to a finite spatial region for all times. Nevertheless, it can be seen that also for light-like trajectories the Liénard-Wiechert acceleration field $A^{\mu\nu}$ (17.6) defines a *distribution*, and hence this field is expected to play a basic role also in the solution of Maxwell's equations for massless particles.

17.1.1 The Singularity String

The main difference between time-like and light-like trajectories is the singularity locus of the corresponding electromagnetic fields. Since, as anticipated, the field $A^{\mu\nu}$ (17.6) is present for both kinds of trajectories, this locus is the set of space-time points where the factor $(uL) = u^\mu L_\mu$ in the denominator of formula (17.6) vanishes. For time-like trajectories, at a time t this factor vanishes only on the particle's position, i.e. for $\mathbf{x} = \mathbf{y}(t)$. This follows from the fact that the scalar product between a time-like vector and a light-like vector only vanishes when the latter is identically zero, see also footnote 1 in Chap. 15. Conversely, for a light-like world line $y^\mu(\lambda)$ we have $u^2 = (u^0)^2 - |\mathbf{u}|^2 = 0$, and therefore the scalar product $(uL) = u^0 L^0 - \mathbf{u} \cdot \mathbf{L} = |\mathbf{u}| \, |\mathbf{L}| - \mathbf{u} \cdot \mathbf{L}$ vanishes whenever the spatial vectors $\mathbf{L} = \mathbf{x} - \mathbf{y}(\lambda)$ and $\mathbf{u}$ are aligned, that is to say, whenever the equality

$$\frac{\mathbf{x} - \mathbf{y}(\lambda)}{|\mathbf{x} - \mathbf{y}(\lambda)|} = \frac{\mathbf{u}(\lambda)}{|\mathbf{u}(\lambda)|} = \mathbf{v}(\lambda) \qquad (17.8)$$

holds, where $\mathbf{v}(\lambda)$ is the particle's velocity. If we parameterize the world line with the time, in which case we have $y^\mu(\lambda) = (\lambda, \mathbf{y}(\lambda))$, the delay relations (17.7) become

$$|\mathbf{x} - \mathbf{y}(\lambda)| = t - \lambda,$$

and so the condition for the singularity locus (17.8) translates into

$$\mathbf{x} = \mathbf{y}(\lambda) + (t - \lambda)\mathbf{v}(\lambda), \qquad t - \lambda \geq 0.$$

Introducing the parameter $b = t - \lambda$ in place of λ, we then obtain the *singularity string* at time t

$$\boldsymbol{\Gamma}(t, b) = \mathbf{y}(t - b) + b\mathbf{v}(t - b), \qquad b \geq 0. \qquad (17.9)$$

At time t, the electromagnetic field generated by a massless charged particle is thus expected to be singular at all points of the curve $\boldsymbol{\Gamma}(t, b)$. Due to the condition $b \geq 0$, this string has one endpoint at the particle's position, i.e. $\boldsymbol{\Gamma}(t, 0) = \mathbf{y}(t)$, and hence it appears "attached" to the particle. As the particle moves, the string (17.9) sweeps out a two-dimensional *singularity surface* Γ, which admits the Lorentz-invariant parameterization

$$\Gamma^\mu(\lambda, b) = y^\mu(\lambda) + b u^\mu(\lambda), \qquad b \geq 0. \tag{17.10}$$

The expression (17.10) is also invariant under a reparameterization of the world line $\lambda \to \lambda'(\lambda)$, if we accompany it with the transformation $b \to b' = (d\lambda'(\lambda)/d\lambda)\, b$. The surface Γ has a boundary at $b = 0$, which is the particle's world line $y^\mu(\lambda)$. There can be an additional boundary for $b \to \infty$, see below. In general, given a generic two-dimensional surface $\Gamma^\mu(\lambda, b)$, *Poincaré duality*[1] associates with it a *canonical* antisymmetric distribution-valued tensor – essentially the δ-function on Γ – given by

$$Q^{\mu\nu}(x) = e \int \frac{\partial \Gamma^\mu(\lambda, b)}{\partial b} \frac{\partial \Gamma^\nu(\lambda, b)}{\partial \lambda} \, \delta^4(x - \Gamma(\lambda, b)) \, db \, d\lambda - (\mu \leftrightarrow \nu)$$

$$= e \int_{-\infty}^{\infty} d\lambda \int_{0}^{\infty} b \left(u^\mu(\lambda) w^\nu(\lambda) - u^\nu(\lambda) w^\mu(\lambda) \right) \delta^4(x - \Gamma(\lambda, b)) db, \tag{17.11}$$

where the charge e has been introduced for later convenience. The tensor $Q^{\mu\nu}$ is reparameterization invariant, and it will play a major role regarding the electromagnetic field produced by the particle. In particular, a quantity that will become relevant later on is the four-divergence of this tensor. Applying $Q^{\mu\nu}$ to the derivative ∂_μ of a test function $\varphi(x) \in \mathcal{S}(\mathbb{R}^4)$, and using the explicit form of $\Gamma^\mu(\lambda, b)$ (17.10), a straightforward calculation gives

$$(\partial_\mu Q^{\mu\nu})(\varphi) = -Q^{\mu\nu}(\partial_\mu \varphi) = -e \int_{-\infty}^{\infty} d\lambda \, u^\nu(\lambda) \int_{0}^{\infty} \frac{\partial}{\partial b} \varphi\big(y(\lambda) + b u(\lambda)\big) db$$

$$= -e \int_{-\infty}^{\infty} dt \, v^\nu(t - b) \int_{0}^{\infty} \frac{\partial}{\partial b} \varphi\big(t, \boldsymbol{\Gamma}(t, b)\big) db, \tag{17.12}$$

where $\boldsymbol{\Gamma}(t, b)$ is the singularity string (17.9). To obtain the second line we have parameterized the world line with the time, $y^0(\lambda) = \lambda = t$, and we have shifted the integration variable t in $t - b$. We will further elaborate on the expressions (17.12) below.

Bounded and unbounded trajectories. As anticipated above, the structure of the electromagnetic field generated by massless charges entails basically different features according to whether the trajectories are bounded or unbounded. For a bounded

---

[1] See Sect. 21.4.2 for details.

trajectory, as, for instance, a cyclotron orbit, by definition the position vector $\mathbf{y}(t)$ satisfies the limitation

$$|\mathbf{y}(t)| < l, \quad \text{for all } t.$$

In this case, for a fixed time t the singularity string $\mathbf{\Gamma}(t, b)$ (17.9) for $b \to \infty$ reaches infinity, and correspondingly the singularity surface Γ (17.10) has as only boundary the particle's world line $y^{\mu}(\lambda)$, corresponding to $b = 0$. Consequently, since for $b \to \infty$ we have $|\mathbf{\Gamma}(t, b)| \to \infty$, the total derivative with respect to b in the integral (17.12) receives only a contribution from $b = 0$, and the result is

$$\partial_{\mu} Q^{\mu\nu} = j^{\nu}, \tag{17.13}$$

where j^{μ} is the current (17.2).

Conversely, for an unbounded trajectory the distinctive feature is its behavior at the infinite past. This asymmetry between past and future is intimately related to the fact that we search for a *retarded* solution of Maxwell's equations, which at each instant is sensitive to the whole past history of the particle up to this instant. To be definite, we assume that for $t \to -\infty$ the trajectory approaches *sufficiently fast* a *linear* trajectory, with asymptotic velocity $\mathbf{v}_{\infty}$,

$$\mathbf{y}(t) \approx \mathbf{v}_{\infty} t, \qquad |\mathbf{v}_{\infty}| = 1. \tag{17.14}$$

In the following we also use the notation

$$v_{\infty}^{\mu} = (1, \mathbf{v}_{\infty}). \tag{17.15}$$

The main new feature of an unbounded trajectory with respect to a bounded one is that for the former the singularity string $\mathbf{\Gamma}(t, b)$ (17.9) for each t has a *finite* extension. In fact, at time t its end points for $b \to 0$ and for $b \to \infty$ are

$$\mathbf{\Gamma}(t, 0) = \mathbf{y}(t), \qquad \mathbf{\Gamma}(t, \infty) = \mathbf{v}_{\infty} t. \tag{17.16}$$

The endpoints for $b \to \infty$ form thus the trajectory $\mathbf{y}_{\mathcal{L}}(t) = \mathbf{v}_{\infty} t$ of a *virtual* linear light-like motion with velocity $\mathbf{v}_{\infty}$. The world lines corresponding to the two trajectories in (17.16) are the real world line of the particle $y^{\mu}(\lambda)$, and the virtual world line

$$y_{\mathcal{L}}^{\mu}(\lambda) = \lambda v_{\infty}^{\mu}. \tag{17.17}$$

The boundary of the singularity surface (17.10) is, hence, composed by the world lines $y^{\mu}(\lambda)$ and $y_{\mathcal{L}}^{\mu}(\lambda)$. As a consequence, the four-divergence (17.12) of the field $Q^{\mu\nu}$ receives now a non-vanishing contribution also from $b \to \infty$. In fact, for an unbounded motion, in place of Eq. (17.13) $Q^{\mu\nu}$ satisfies the modified equation

$$\partial_{\mu} Q^{\mu\nu} = j^{\nu} - j_{\mathcal{L}}^{\nu}, \tag{17.18}$$

where $j_{\mathcal{L}}^{\mu}$ is the current associated with the virtual world line (17.17)

$$j_{\mathcal{L}}^{\mu}(x) = e \int v_{\infty}^{\mu}\, \delta^4(x - \lambda v_{\infty})\, d\lambda. \tag{17.19}$$

For the geometrical interpretation of formula (17.18), see Sect. 21.4.2 on Poincaré duality.

17.1.2　Regularizations

Since for massless particles the Green function method is not reliable, another efficient method to solve Maxwell's equations in this case is to resort to dedicated limiting procedures in the space of distributions, as illustrated in Sect. 6.3.2. These procedures rely on appropriate regularizations, leading to well-defined potentials or field strength which then produce, as limiting cases, solutions of Maxwell's equations. For the case at hand there are two regularizations suitable for these purposes.

ε-regularization. Consider a *bounded* light-like world line $y^{\mu}(\lambda)$. In this case we can introduce a potential A^{μ} inherited, actually, from the Green function method. In fact, for such a world line the retarded-time conditions (17.7) still allow for a unique solution $\lambda(x)$ for all space-time points x^{μ}, and we can write down, at least formally, the Liénard-Wiechert potential (7.16)

$$\mathcal{A}^{\mu}(x) = \frac{eu^{\mu}}{4\pi(uL)}\bigg|_{\lambda=\lambda(x)} = \frac{eu^{\mu}(\lambda)}{4\pi(x - y(\lambda))^{\nu} u_{\nu}(\lambda)}\bigg|_{\lambda=\lambda(x)}. \tag{17.20}$$

In the case of a bounded motion this expression defines, indeed, a distribution. To see it, we apply it to a test function $\varphi(x) \in \mathcal{S}(\mathbb{R}^4)$. The related computation has been performed in Sect. 7.2.2, and the result was formula (7.27)

$$\mathcal{A}^{\mu}(\varphi) = \frac{e}{4\pi} \int \frac{v^{\mu}(t)}{r}\, \varphi(t + r, \mathbf{x} + \mathbf{y}(t))\, d^4x, \qquad r = |\mathbf{x}|. \tag{17.21}$$

As shown in Sect. 7.2.2, these four functionals *are* distributions – independently of the velocity of the particle – in that the unique property used in the proof was that the trajectory $\mathbf{y}(t)$ is of compact support. However, there remain two major challenges to be faced: the first is the determination of the field strength $F^{\mu\nu}$ associated with $\mathcal{A}^{\mu}$, and the second is the proof that this field strength is actually a solution of Maxwell's equations. Both these tasks are made difficult by the intricate singularities present at the surface Γ (17.10) – the denominator in formula (17.20) contains, in fact, the factor (uL) vanishing precisely on Γ – and to overcome them it is convenient to resort to the ε-regularization, proven powerful in Sects. 15.2.1 and 16.2. Correspondingly, we replace the condition $L^2 = 0$ in (17.7) with the condition

$$L^2 = (x - y(\lambda_\varepsilon))^2 = \varepsilon^2, \tag{17.22}$$

which determines a regularized retarded parameter $\lambda_\varepsilon(x)$, and then we define the regularized potential

$$\mathcal{A}_\varepsilon^\mu(x) = \left. \frac{eu^\mu}{4\pi(uL)} \right|_{\lambda=\lambda_\varepsilon(x)}, \tag{17.23}$$

in place of the potential (17.20). The denominator (uL) in Eq. (17.23) is now the scalar product between the time-like vector L^μ and the non-vanishing light-like vector u^μ, and therefore it never vanishes. It follows that the four components of the potential (17.23) are C^∞-functions, and so, in particular, their partial derivatives appearing in the field strength (17.27) can be computed in the ordinary sense. The distributional form of the potential (17.23), to be compared with formula (17.21), is

$$\mathcal{A}_\varepsilon^\mu(\varphi) = \frac{e}{4\pi} \int \frac{v^\mu(t)}{r_\varepsilon} \varphi\big(t + r_\varepsilon, \mathbf{x} + \mathbf{y}(t)\big) \, d^4x, \qquad r_\varepsilon = \sqrt{r^2 + \varepsilon^2}. \tag{17.24}$$

Since, formally, the four-potential (17.23) can also be written as the convolution $\mathcal{A}_\varepsilon^\mu = G_\varepsilon * j^\mu$, where G_ε is the regularized retarded Green function (15.29), it is not a surprise that it still satisfies the Lorenz gauge

$$\partial_\mu \mathcal{A}_\varepsilon^\mu = 0, \tag{17.25}$$

as can be verified via an explicit computation. From formulas (17.21) and (17.24) it follows immediately that the potential $\mathcal{A}_\varepsilon^\mu$ converges to $\mathcal{A}^\mu$ in the distributional sense

$$\mathcal{S}' - \lim_{\varepsilon \to 0} \mathcal{A}_\varepsilon^\mu = \mathcal{A}^\mu. \tag{17.26}$$

Thanks to the limit (17.26) the regularized electromagnetic field

$$F_\varepsilon^{\mu\nu} = \partial^\mu \mathcal{A}_\varepsilon^\nu - \partial^\nu \mathcal{A}_\varepsilon^\mu \tag{17.27}$$

then entails the limit

$$\mathcal{S}' - \lim_{\varepsilon \to 0} F_\varepsilon^{\mu\nu} = \partial^\mu \mathcal{A}^\nu - \partial^\nu \mathcal{A}^\mu = F^{\mu\nu}. \tag{17.28}$$

The ε-regularization has the advantage to preserve the manifest Lorentz invariance. In Sect. 17.2 we will use it, (i) to determine explicitly the field $F^{\mu\nu}$ – by means of the relation (17.28) – and, (ii) to show that this field satisfies Maxwell's equations (17.3) for a generic bounded light-like trajectory.

β-**regularization.** For *unbounded* light-like motions the ε-regularization unfortunately fails, as both integrals (17.21) and (17.24) are divergent. The reason is that for $t \to -\infty$ the trajectory $\mathbf{y}(t)$ becomes linear, see (17.14), and, as shown after Eq. (7.27), for motions that in the infinite past become linear and light-like, the integral (17.21) diverges when $\mathbf{x}$ goes to infinity along the direction of $\mathbf{v}_\infty r$, and when at the same time t goes to minus infinity as $-r$. The same happens with the integral (17.24), in that for large r we have $r_\varepsilon \approx r$. Therefore, for unbounded trajectories we must resort to a different regularization. We introduce a *speed* $\beta < 1$ and associate with this parameter the regularized *time-like* world line

$$y_\beta^\mu(\lambda) = \left(\frac{y^0(\lambda)}{\beta}, \mathbf{y}(\lambda) \right). \tag{17.29}$$

This world line entails the regularized trajectory $\mathbf{y}_\beta(t) = \mathbf{y}(\beta t)$, differing from the original one simply by a rescaling of the time. Its velocity is hence given by

$$\mathbf{v}_\beta(t) = \frac{d\mathbf{y}_\beta(t)}{dt} = \beta \mathbf{v}(\beta t), \tag{17.30}$$

so that the regularized speed $|\mathbf{v}_\beta(t)| = \beta$ is constant and smaller than the speed of light for all t. Therefore, the regularized world line $y_\beta^\mu(\lambda)$ (17.29) creates a *standard* well-defined Liénard-Wiechert potential of the form (17.20), that we denote by $\mathcal{A}_\beta^\mu$, and the associated well-defined Liénard-Wiechert field $F_\beta^{\mu\nu}$ can be decomposed as in Eqs. (17.4)–(17.6)

$$F_\beta^{\mu\nu} = \partial^\mu \mathcal{A}_\beta^\nu - \partial^\nu \mathcal{A}_\beta^\mu = V_\beta^{\mu\nu} + A_\beta^{\mu\nu}. \tag{17.31}$$

Obviously, theses fields satisfy Maxwell's equations

$$\partial^{[\mu} F_\beta^{\nu\rho]} = 0, \qquad \partial_\mu F_\beta^{\mu\nu} = j_\beta^\nu, \tag{17.32}$$

where j_β^μ, the current associated with the regularized world line (17.29), satisfies the trivial distributional limit

$$\mathcal{S}' - \lim_{\beta \to 1} j_\beta^\mu = j^\mu. \tag{17.33}$$

As explained in Sect. 6.3.2 – distributional limits and distributional derivatives commute – from Eqs. (17.32) and from the limit (17.33) it then follows that, *if* the distributional limit

$$\mathcal{S}' - \lim_{\beta \to 1} F_\beta^{\mu\nu} \equiv F^{\mu\nu} \tag{17.34}$$

exists, then the resulting field $F^{\mu\nu}$ satisfies automatically Maxwell's equations (17.3). Notice that, contrary to the ε-regularization, this procedure does not imply, or require, that the potential $\mathcal{A}_\beta^\mu$ admits a distributional limit for $\beta \to 1$ and, in fact, it will not. Furthermore, the β-regularization breaks manifest Lorentz invariance, which is, however, recovered in the limit for $\beta \to 1$. Finally, it is obvious that this regularization can also be applied to bounded motions, but in that case the ε-regularization turns out to be more efficient.

17.2   The Electromagnetic Field for a Bounded Trajectory

To solve Maxwell's equations for a bounded light-like motion we resort to the ε-regularization, formulas (17.22)–(17.28). From Eqs. (17.25) and (17.27) we derive the relation

$$\partial_\nu F_\varepsilon^{\nu\mu} = \Box \mathcal{A}_\varepsilon^\mu \equiv j_\varepsilon^\mu, \tag{17.35}$$

which serves as a *definition* of the current j_ε^μ. A simple calculation, based on the explicit expression of the regularized potential $\mathcal{A}_\varepsilon^\mu$ (17.23), and on the relations analogous to (7.33)–(7.37) of the massive case – except that now $L^2 = \varepsilon^2$ – yields for the d'Alembertian of the potential $\mathcal{A}_\varepsilon^\mu$ the explicit expression

$$j_\varepsilon^\mu = \frac{\varepsilon^2 e}{4\pi}\left(\frac{g^\mu(uL) - u^\mu(gL)}{(uL)^4} + \frac{3(wL)}{(uL)^5}\big((wL)u^\mu - (uL)w^\mu\big) \right), \quad g^\mu = \frac{dw^\mu}{d\lambda}. \tag{17.36}$$

A crucial point is now the determination of the distributional limit for $\varepsilon \to 0$ of this current. First of all, this limit exists, thanks to the definition (17.35) and to the existence of the limit (17.26). Next, since j_ε^μ is proportional to ε^2, this limit can have as support at most the set of space-time points where, in the limit for $\varepsilon \to 0$, the scalar product (uL) appearing in the denominators of (17.36) vanishes, i.e. precisely the singular surface Γ (17.10). Last, according to the definition (17.35) the current j_ε^μ is conserved and Lorentz covariant, and so must be its limit. This information essentially forces the limit of j_ε^μ to be proportional to the particle's current (17.2). An explicit calculation gives indeed

$$S' - \lim_{\varepsilon \to 0} j_\varepsilon^\mu = j^\mu.$$

Since the field $F^{\mu\nu}$ (17.28) automatically satisfies the Bianchi identity, by taking the limit for $\varepsilon \to 0$ of the first and third members of equation (17.35) we then conclude that $F^{\mu\nu}$ is, actually, a solution of Maxwell's equations (17.3). It then only remains to determine this field explicitly.

17.2.1   Derivation of the Field

To determine the field $F^{\mu\nu}$ we start from the regularized field $F_\varepsilon^{\mu\nu}$ (17.27), which is the field strength of the potential (17.23). Since the derivatives involved in this relation can be computed in the ordinary sense, the result is formally nothing else than the field given by formulas (17.4)–(17.6), modulo the replacement $\lambda(x) \to \lambda_\varepsilon(x)$. However – and this is one of the advantages of the ε-regularization – since for a light-like trajectory u^2 vanishes, the regularized velocity field $V_\varepsilon^{\mu\nu}$ following from Eq. (17.5) is identically zero. We remain, therefore, only with the regularized acceleration field

$$F_\varepsilon^{\mu\nu} = A_\varepsilon^{\mu\nu} = A^{\mu\nu}\big|_{\lambda=\lambda_\varepsilon(x)}. \tag{17.37}$$

Since the unregularized acceleration field $A^{\mu\nu}$ (17.6) *is* a distribution, and the *point-wise* limit for $\varepsilon \to 0$ of $A_\varepsilon^{\mu\nu}$ obviously equals $A^{\mu\nu}$, the *distributional* limit for $\varepsilon \to 0$ of $A_\varepsilon^{\mu\nu}$ certainly contains the field $A^{\mu\nu}$. However, a careful calculation of the four-divergence of the field (17.6) leads to the surprising result

$$\partial_\mu A^{\mu\nu} = \frac{1}{2} j^\nu. \tag{17.38}$$

This means that the electric field A^{i0} associated with $A^{\mu\nu}$ entails the flux $e/2$ across a closed surface containing the particle, i.e. only half of the flux required by Maxwell's equations. This implies that $A^{\mu\nu}$ cannot be the unique term arising from the distributional limit of $A_\varepsilon^{\mu\nu}$ for $\varepsilon \to 0$. Indeed, due to the singularities arising in the expression (17.6) in the limit for $\varepsilon \to 0$ from the prefactor $1/(uL)^3$ at space-time points x^μ belonging to the singularity surface Γ (17.10), an explicit calculation, requiring, as usual, to apply the expression (17.37) to a test function $\varphi(x) \in \mathcal{S}(\mathbb{R}^4)$, and only then to perform the limit for $\varepsilon \to 0$, leads to the result

$$F^{\mu\nu} = \mathcal{S}' - \lim_{\varepsilon \to 0} F_\varepsilon^{\mu\nu} = \mathcal{S}' - \lim_{\varepsilon \to 0} A_\varepsilon^{\mu\nu} = A^{\mu\nu} + \frac{1}{2} Q^{\mu\nu}, \tag{17.39}$$

where $Q^{\mu\nu}$ is the δ-function (17.11) supported on Γ. Thanks to the identity (17.13), holding for bounded trajectories, the total field (17.39) satisfies now indeed the Maxwell equation $\partial_\mu F^{\mu\nu} = j^\nu$, as it must by construction. As we see, half of the flux of the electric field $E^i = F^{i0}$ across a closed surface comes from the naive acceleration field A^{i0}, and the other half comes from the δ-function Q^{i0} supported on the singularity string $\Gamma(t, b)$ (17.9). The surface integral of Q^{i0} equals, indeed, the number of intersections, with sign, of the closed surface with the string $\Gamma(t, b)$, multiplied by the charge e. Since one endpoint of the string is the particle's position, if the particle is inside the closed surface the flux integral of Q^{i0} is, therefore, always equal to e. Notice also that, whereas the fields $Q^{\mu\nu}$ and $A^{\mu\nu}$ satisfy the individual flux equations (17.13) and (17.38), we have $\partial_{[\mu} A_{\nu\rho]} \neq 0$ and $\partial_{[\mu} Q_{\nu\rho]} \neq 0$. Nevertheless, by construction, the total field (17.39) satisfies the Bianchi identity $\partial_{[\mu} A_{\nu\rho]} + \frac{1}{2} \partial_{[\mu} Q_{\nu\rho]} = 0$.

For a bounded trajectory, one could also apply the β-regularization $F_\beta^{\mu\nu} = V_\beta^{\mu\nu} + A_\beta^{\mu\nu}$ of Eq. (17.31). In that case, the regularized velocity field $V_\beta^{\mu\nu}$, obtained from formula (17.5) via the replacement $y^\mu(\lambda) \to y_\beta^\mu(\lambda)$, would be non-vanishing. However, due to the prefactor $u^2 = (1 - \beta^2)(u^0)^2$ in formula (17.5), in the case of a bounded trajectory the distributional limit for $\beta \to 1$ of the field $V_\beta^{\mu\nu}$ can be seen to vanish. On the other hand, for uniqueness reasons, the distributional limit for $\beta \to 1$ of the regularized acceleration field $A_\beta^{\mu\nu}$ gives again the field (17.39).

In a sense, the string $\Gamma(t, b)$ (17.9) resembles the *Dirac string* attached to a magnetic monopole, see Sect. 21.3.1. However, contrary to the latter, the former is *physically observable* in that the electromagnetic field (17.39) entails a δ-function-singularity supported on this string. In particular, from Eq. (17.9) one derives that the velocity at time t of a point of the string with coordinate b – equal to the component of the vector $\partial \Gamma(t, b)/\partial t = \mathbf{v}(t - b) + b\mathbf{a}(t - b)$ orthogonal to the string – is equal to $\mathbf{v}(t - b)$, and so its speed equals the speed of light. The dynamics of the singularity string respects, hence, *causality*.

17.3 The Electromagnetic Field for an Unbounded Trajectory

Since for an unbounded trajectory the ε-regularization yields the vector potential (17.24) which is not distribution-valued, in this case we resort to the β-regularization. As seen in Sect. 17.1.2, if the regularized field (17.31)

$$F_\beta^{\mu\nu} = \partial^\mu \mathcal{A}_\beta^\nu - \partial^\nu \mathcal{A}_\beta^\mu = V_\beta^{\mu\nu} + A_\beta^{\mu\nu} \tag{17.40}$$

admits the distributional limit (17.34), then the resulting field strength $F^{\mu\nu}$ satisfies automatically Maxwell's equations. We recall that $\mathcal{A}_\beta^\mu$ denotes the standard Liénard-Wiechert potential (17.20), or (17.21), generated by the world line $y_\beta^\mu(\lambda)$ (17.29).

Construction of a vector potential. The most efficient way to prove the existence of the limit (17.34) is via the construction of a potential A^μ for the envisaged limiting field strength $F^{\mu\nu}$. However, this potential cannot be the distributional limit

$$\mathcal{S}' - \lim_{\beta \to 1} \mathcal{A}_\beta^\mu,$$

as this limit does not exist. In particular, the pointwise limit for $\beta \to 1$ of $\mathcal{A}_\beta^\mu$ would amount to the potential (17.21) relative to the world line $y^\mu(\lambda)$, which is not a distribution. To construct a potential we take advantage from the fact that the world line $y^\mu(\lambda)$ entails for $t \to -\infty$ the same "bad" behavior as the linear world line $y_\mathcal{L}^\mu(\lambda)$ (17.17), i.e. $\mathbf{y}(t) \approx \mathbf{v}_\infty t$, and that we know how to cure the bad behavior of the *naive* Liénard-Wiechert potential (6.114) generated by the latter, namely by means of the gauge transformation (6.115), (6.116). Since the world line $y_\beta^\mu(\lambda)$ entails for $t \to -\infty$ the asymptotic behavior $\mathbf{v}_\beta(t) \approx \beta\mathbf{v}_\infty$, see Eq. (17.30), in the case at hand we introduce the regularized time-like four-velocity

$$U_\beta^\mu = \frac{(1, \beta\mathbf{v}_\infty)}{\sqrt{1 - \beta^2}}, \qquad U_\beta^2 = 1,$$

and the corresponding gauge function (6.115),

$$\Lambda_\beta = \frac{e}{4\pi} \ln \left| (U_\beta x) - \sqrt{(U_\beta x)^2 - x^2} \right|,$$

$$\partial^\mu \Lambda_\beta = -\frac{e U_\beta^\mu}{4\pi \sqrt{(U_\beta x)^2 - x^2}} + \frac{e}{4\pi} \left(1 + \frac{(U_\beta x)}{\sqrt{(U_\beta x)^2 - x^2}} \right) \frac{x^\mu}{x^2}. \tag{17.41}$$

We then consider the distributional limit

$$A^\mu = \mathcal{S}' - \lim_{\beta \to 1} \left(\mathcal{A}_\beta^\mu + \partial^\mu \Lambda_\beta \right) \tag{17.42}$$

$$= \mathcal{S}' - \lim_{\beta \to 1} \left(\mathcal{A}_\beta^\mu - \frac{e U_\beta^\mu}{4\pi \sqrt{(U_\beta x)^2 - x^2}} \right) + \frac{e x^\mu}{2\pi x^2} H(v_\infty x), \tag{17.43}$$

where the last term is a distribution, see the potential (6.117). The existence of the remaining limit in (17.43) can be established by applying it to a test function φ, and by interchanging then the limit with the integral sign. Recalling the general form of the distributional Liénard-Wiechert potential (17.21), and recognizing that the second term in (17.43) is the Liénard-Wiechert potential of a uniform linear motion (6.113), the result can be seen to be

$$A^\mu(\varphi) = \frac{e}{4\pi} \int \frac{1}{r} \left(v^\mu(t)\, \varphi\big(t+r, \mathbf{x}+\mathbf{y}(t)\big) - v_\infty^\mu\, \varphi\big(t+r, \mathbf{x}+\mathbf{v}_\infty t\big) \right) d^4 x$$

$$+ \frac{e}{2\pi} \int \frac{x^\mu}{x^2} H(v_\infty x)\, \varphi(x)\, d^4 x. \tag{17.44}$$

The crucial point of the proof that the limit in (17.43) exists, and equals the functional (17.44), is that the two integrands in the first line of (17.44) have exactly the *same* asymptotic behavior for $t \to -\infty$. Therefore, in their difference the divergences that prevented the potential (17.21) from becoming a distribution cancel out.

Derivation of the field. The importance of the construction of a potential consists in the fact that the existence of the limit (17.42) ensures the existence of the limit (17.34), as the gauge term $\partial^\mu \Lambda_\beta$ does not contribute to the field strength $F_\beta^{\mu\nu}$ (17.40). It follows, in particular, that the limiting field strength is given by $F^{\mu\nu} = \partial^\mu A^\nu - \partial^\nu A^\mu$. We remain hence only with the technical problem to determine $F^{\mu\nu}$ explicitly. Since, like in the bounded case, the pointwise limit $A^{\mu\nu}$ (17.6) of the regularized acceleration field $A_\beta^{\mu\nu}$ *is* a distribution, it will again be a piece of $F^{\mu\nu}$. However, in this case an explicit calculation shows that the four-divergence of this field does no longer satisfy Eq. (17.38), but rather the equation

$$\partial_\mu A^{\mu\nu} = \frac{1}{2} \left(j^\nu - j_\mathcal{L}^\nu \right), \tag{17.45}$$

where $j_\mathcal{L}^\mu$ is the current (17.19) of the virtual world line $y_\mathcal{L}^\mu(\lambda)$. Notice the consistency of equation (17.45) with the identity $\partial_\nu \partial_\mu A^{\mu\nu} = 0$, implied by the commutativity of distributional derivatives. Another distinctive feature of the field $A^{\mu\nu}$ of an unbounded motion, with respect to a bounded one, is that for space-time points x^μ such that $\mathbf{x} \cdot \mathbf{v}_\infty > t$, the retarded-time conditions (17.7) admit no solution, whereas for $\mathbf{x} \cdot \mathbf{v}_\infty < t$ they always admit a unique solution. This means that at a time t the field $A^{\mu\nu}$ vanishes *beyond* the plane orthogonal to the virtual trajectory $\mathbf{y}_\mathcal{L}(t^*) = \mathbf{v}_\infty t^*$, passing through the position $\mathbf{v}_\infty t$, i.e.

$$A^{\mu\nu}(t, \mathbf{x}) = 0, \quad \text{for } \mathbf{x} \cdot \mathbf{v}_\infty > t. \tag{17.46}$$

The explicit determination of the distributional limit of the acceleration field $A^{\mu\nu}_\beta$, obtained from formula (17.6) via the replacement $y^\mu(\lambda) \to y^\mu_\beta(\lambda)$, leads to

$$\mathcal{S}' - \lim_{\beta \to 1} A^{\mu\nu}_\beta = A^{\mu\nu} + \frac{1}{2} Q^{\mu\nu}, \tag{17.47}$$

exactly as in the case of a bounded trajectory, see the limit (17.39). However, from the relation (17.18) for an unbounded motion, and from Eq. (17.45), this time we obtain

$$\partial_\mu \left(A^{\mu\nu} + \frac{1}{2} Q^{\mu\nu} \right) = j^\nu - j^\nu_\mathcal{L}, \tag{17.48}$$

meaning that the limiting field (17.47) does not satisfy Maxwell's equations. The solution of the puzzle arises from a careful calculation of the distributional limit for $\beta \to 1$ of the regularized velocity field $V^{\mu\nu}_\beta$ in Eq. (17.40). In fact, contrary to a bounded motion, and despite the presence of the factor $u^2 = (1 - \beta^2)(u^0)^2$ in the field (17.5) with $y^\mu(\lambda)$ replaced by $y^\mu_\beta(\lambda)$, for an unbounded motion this field entails now the non-vanishing limit of a *shock wave*, see formula (6.119),

$$\mathcal{S}' - \lim_{\beta \to 1} V^{\mu\nu}_\beta = \frac{e\,(v^\mu_\infty x^\nu - v^\nu_\infty x^\mu)}{2\pi x^2} \, \delta(v_\infty x) \equiv F^{\mu\nu}_{\text{sw}}, \tag{17.49}$$

where v^μ_∞ is the asymptotic velocity vector (17.15). As we know from the Sect. 6.3.2 on the electromagnetic field produced by a charged massless particle in linear motion, the shock-wave field (17.49) satisfies the Maxwell equation

$$\partial_\mu F^{\mu\nu}_{\text{sw}} = j^\nu_\mathcal{L}. \tag{17.50}$$

From Eqs. (17.48) and (17.50) it then follows that the distributional limit for $\beta \to 1$ of the field (17.40)

$$F^{\mu\nu} = A^{\mu\nu} + \frac{1}{2} Q^{\mu\nu} + F^{\mu\nu}_{\text{sw}} \tag{17.51}$$

satisfies Maxwell's equations (17.3), as it must by construction. The necessary presence of the shock wave in the solution (17.51) can also be inferred from the limiting situation of an unbounded *linear* trajectory, in which case both the acceleration field

$A^{\mu\nu}$ (17.6) and the δ-function $Q^{\mu\nu}$ (17.11) vanish. In this limiting case, formula (17.51) thus correctly reduces to the pure shock wave (6.119). Notice also that the position of the shock-wave plane in (17.49), i.e. $(v_\infty x) = t - \mathbf{x} \cdot \mathbf{v}_\infty = 0$, coincides precisely with the plane (17.46) beyond which the acceleration field $A^{\mu\nu}$ – the only *regular* contribution of the field (17.51) – vanishes. In a sense, as the motion of the charged particle "starts" for $t \to -\infty$ essentially as a *linear light-like* trajectory, the first component of the electromagnetic field that is created is the shock wave $F_{\rm sw}^{\mu\nu}$, propagating at the speed of light, and so the other components of the field, $A^{\mu\nu}$ and $Q^{\mu\nu}$, must necessarily remain confined behind the shock wave, as these fields are created by the *real* particle, whose motion is curved and, therefore, always remains behind the *virtual* particle creating the shock wave.

While the peculiar composition of the solutions (17.39) and (17.51), and, in particular, the special singular behavior of the acceleration field $A^{\mu\nu}$ near the singularity surface $\Gamma^\mu(\lambda, b)$, and near the real and virtual world lines $y^\mu(\lambda)$ and $y_{\mathcal{L}}^\mu(\lambda)$, implied by Eq. (17.45), may deserve further investigation, the fundamental conclusion of this chapter remains, however, that *Maxwell's equations admit well-defined solutions also for light-like trajectories*, a property that, given the intricate singularities involved, was *a priori* far from obvious.

17.4   Four-Momentum Conservation and Absence of Radiation

As seen in Sect. 16.4, the equation of motion of a charged particle, taking into account the radiation reaction, is eventually determined by the requirement of local four-momentum conservation. The implementation of this paradigm relies again on the construction of a distribution-valued energy-momentum tensor for the electromagnetic field. The main difference with respect to the case of a massive charged particle is that for a massless one the divergent part of the electromagnetic energy-momentum tensor $T_{\rm em}^{\mu\nu}$ (2.137), i.e. the *counterterm* called $\widehat{T}_\varepsilon^{\mu\nu}$ in Sect. 16.2, see Eq. (16.18), is now supported on the singularity surface $\Gamma^\mu(\lambda, b)$ (17.10), rather than on the world line $y^\mu(\lambda)$. In fact, the singularities of the tensor $T_{\rm em}^{\mu\nu}$ are inherited from those of the field $F^{\mu\nu}$. Moreover, as in the case of a massive particle, see Sect. 16.4, the divergent part of the electromagnetic energy-momentum tensor is determined only modulo *finite counterterms*, which for a massless particle are supported, in turn, on $\Gamma^\mu(\lambda, b)$. This ambiguity must be carefully taken into account.

For definiteness, we consider a particle on a *bounded* trajectory, in which case we can resort to the ε-regularization (17.22) and to the corresponding regularized electromagnetic field $F_\varepsilon^{\mu\nu}$ (17.37), containing only the acceleration field $A_\varepsilon^{\mu\nu}$. Our starting point is then the resulting regularized energy-momentum tensor

$$T_\varepsilon^{\mu\nu} = A_\varepsilon^{\mu\alpha} A_{\varepsilon\,\alpha}{}^\nu + \frac{1}{4}\,\eta^{\mu\nu} A_\varepsilon^{\alpha\beta} A_{\varepsilon\,\alpha\beta}, \tag{17.52}$$

as in Eq. (16.13). For simplicity, we omit here the external field $F_{\text{in}}^{\mu\nu}$. The determination of the divergent part $\widehat{T}_\varepsilon^{\mu\nu}$ of the tensor (17.52) – under the distributional limit for $\varepsilon \to 0$ – is based on similar techniques as the ones adopted in Sects. 16.2.1 and 16.3 for massive particles, and leads now to the more complicated expression, supported on the singularity surface,[2]

$$
\widehat{T}_\varepsilon^{\mu\nu} = \frac{e^2}{16\pi} \int b^2 \left\{ -\frac{8b^2 w^2}{3\varepsilon^4} u^\mu u^\nu + \frac{1}{\varepsilon^2} \left(4w^2 \eta^{\mu\nu} - 4bw^2 u^{(\mu} \partial^{\nu)} + \right.\right.
$$
$$
\left. \frac{5b^2}{6} w^2 u^\mu u^\nu \Box - \frac{b^2}{3} G^{\mu\alpha} G^{\nu\beta} \partial_\alpha \partial_\beta \right)
$$
$$
\left. + w^2 \ln \varepsilon^2 \left(\partial^\mu \partial^\nu + \frac{\eta^{\mu\nu}}{2} \Box - bu^{(\mu} \partial^{\nu)} \Box + \frac{b^2}{8} u^\mu u^\nu \Box^2 \right) \right\} \delta^4(x - \Gamma(\lambda, b)) \, db \, d\lambda,
$$

$$(17.53)$$

where the tensor $G^{\mu\nu} = u^\mu w^\nu - u^\nu w^\mu$ transforms covariantly under a world line reparameterization $\lambda \to \lambda'(\lambda)$, i.e. by a factor $(d\lambda/d\lambda')^6$. Taking into account the transformation law of b given after Eq. (17.10), it is then immediately seen that the tensor (17.53) is symmetric, traceless and reparamaterization invariant. Formula (17.53) represents the massless counterpart of the counterterm (16.18) for a massive charged particle. By construction, the distributional limit

$$
\mathbb{T}_{\text{em}}^{*\mu\nu} = \mathcal{S}' - \lim_{\varepsilon \to 0} \left(T_\varepsilon^{\mu\nu} - \widehat{T}_\varepsilon^{\mu\nu} \right), \tag{17.54}
$$

representing the *provisional* renormalized electromagnetic energy-momentum tensor, exists and defines a distribution. An explicit calculation reveals that its distributional four-divergence is equal to the expression, still supported on the singularity surface,

$$
\partial_\mu \mathbb{T}_{\text{em}}^{*\mu\nu} = \frac{e^2}{96\pi} \int b^2 w^2 (3\partial^\nu - bu^\nu \Box) \Box \, \delta^4(x - \Gamma(\lambda, b)) \, db \, d\lambda. \tag{17.55}
$$

Would this be the whole story, there would exist no particle's equation of motion at all guaranteeing four-momentum conservation. In fact, a massless particle entails the four-momentum $p^\mu(\lambda) = g(\lambda) u^\mu(\lambda)$ – where $g(\lambda)$ is a scalar *einbein* ensuring the reparameterization invariance of $p^\mu(\lambda)$ – and the reparameterization-invariant energy-momentum tensor

---

[2]The term multiplying the logarithmic divergence $\ln \varepsilon^2$ in formula (17.53), for dimensional reasons would require an additional, *finite* term, amounting to the replacement $\ln \varepsilon^2 \to \ln(\varepsilon^2/l^2)$, where l is a length scale. However, the tensor proportional to $\ln \varepsilon^2$ in Eq. (17.53) has vanishing four-divergence. Actually, it can be seen to be of the form $\ln \varepsilon^2 \partial_\alpha K^{\alpha\mu\nu}$, where $K^{\alpha\mu\nu} = -K^{\mu\alpha\nu}$, and therefore the term $\ln l^2 \partial_\alpha K^{\alpha\mu\nu}$ represents an irrelevant contribution to the energy-momentum tensor, see Eqs. (3.81) and (3.82).

$$T_{\mathrm{p}}^{\mu\nu} = \int g\, u^\mu u^\nu \delta^4(x - y(\lambda))\, d\lambda, \qquad \partial_\mu T_{\mathrm{p}}^{\mu\nu} = \int \frac{dp^\nu}{d\lambda}\, \delta^4(x - y(\lambda))\, d\lambda.$$

$$(17.56)$$

Consequently, since the four-divergence $\partial_\mu \mathbb{T}_{\mathrm{em}}^{*\mu\nu}$ in (17.55) is supported on the singularity surface, and $\partial_\mu T_{\mathrm{p}}^{\mu\nu}$ is supported on the world line, there exists no equation of motion of the form $dp^\mu/d\lambda = f^\mu$ such that the sum of these two four-divergences cancels. However, as anticipated at the beginning of the section, the divergent counterterm (17.53) is *intrinsically* determined only modulo *finite* terms $\mathcal{D}^{\mu\nu}$, supported on the singularity surface, which, for dimensional and covariance reasons, must be combinations of the terms multiplying $\ln \varepsilon^2$ in Eq. (17.53). There arises, therefore, the question whether there exists a tensor $\mathcal{D}^{\mu\nu}$ such that the four-divergence of the modified electromagnetic energy-momentum tensor

$$\mathbb{T}_{\mathrm{em}}^{\mu\nu} = \mathbb{T}_{\mathrm{em}}^{*\mu\nu} + \mathcal{D}^{\mu\nu} \qquad (17.57)$$

is *supported on the world line*. Such a tensor can indeed be found, and it is given by the symmetric, traceless and reparameterization-invariant combination

$$\mathcal{D}^{\mu\nu} = \frac{e^2}{16\pi} \int b^2 \left(- w^2 \partial^\mu \partial^\nu + \frac{w^2}{2}\, \eta^{\mu\nu} \Box \right.$$

$$\left. + \frac{b^2 w^2}{24}\, u^\mu u^\nu \Box^2 - \frac{b^2}{12}\, G^{\mu\alpha} G^{\nu\beta} \partial_\alpha \partial_\beta \right) \delta^4(x - \Gamma(\lambda, b))\, db\, d\lambda.$$

$$(17.58)$$

Actually, a simple calculation using Eq. (17.55) reveals that

$$\partial_\mu \mathbb{T}_{\mathrm{em}}^{\mu\nu} = \partial_\mu \left(\mathbb{T}_{\mathrm{em}}^{*\mu\nu} + \mathcal{D}^{\mu\nu} \right) = 0, \qquad (17.59)$$

an identity that has to be compared with the corresponding identity (16.52) for a massive particle (for $N = 1$ and $F_{\mathrm{in}}^{\mu\nu} = 0$). Equation (17.59) states that the *renormalized energy-momentum tensor of the electromagnetic field generated by a massless particle is conserved*, a feature that implies that no exchange of energy and momentum occurs between the particle and the field. In fact, owing to Eqs. (17.56) and (17.59), the total four-momentum conservation condition $\partial_\mu T^{\mu\nu} = \partial_\mu(\mathbb{T}_{\mathrm{em}}^{\mu\nu} + T_{\mathrm{p}}^{\mu\nu}) = 0$ implies the equation of motion $dp^\mu/d\lambda = 0$, meaning that the particle experiences *no radiation reaction*. In other words, based on the above construction, a *massless charged particle does not emit radiation*. A straightforward generalization of the construction of the renormalized energy-momentum tensor (17.57) shows that in the presence of an external field $F_{\mathrm{in}}^{\mu\nu}$ the continuity equation $\partial_\mu T^{\mu\nu} = 0$ implies the *standard* Lorentz equation $dp^\mu/d\lambda = eF_{\mathrm{in}}^{\mu\nu} u_\nu$, meaning that the particle is subjected only to the external force, and radiates no four-momentum. These conclusions, that have been envisaged for the first time in Ref. [6], signal that the emission of radiation, i.e. of photons, by a massless charged particle must be a genuinely quantum effect, provided that a consistent quantum field theory for such particles exists, see, for instance, Ref. [1]. Finally, in light of the fact that a particle in hyperbolic motion does likewise not emit radiation, see the last paragraph of Sect. 15.2.6, the absence of radiation from a particle in light-like motion should not be too surprising. In fact, the disappearance of the radiation of a hyperbolic motion traces back to the fact that the trajectory for $t \rightarrow \pm\infty$ becomes light-like.

Uniqueness. There remains, of course, the question about the uniqueness of the tensor $\mathbb{T}_{em}^{\mu\nu}$ constructed in (17.57). A priori, we should also accept tensors that differ from $\mathbb{T}_{em}^{\mu\nu}$ by the addition of further finite counterterms $\Delta^{\mu\nu}$ supported on the singularity surface, of the form (17.58), where the integrand is a *polynomial, symmetric, traceless, Lorentz-invariant and reparameterization-covariant* combination of u^μ and of its derivatives, and of b. Of course, in addition, the four-divergence of $\Delta^{\mu\nu}$ must be supported on the world line,

$$\partial_\mu \Delta^{\mu\nu} = - \int f^\nu \, \delta^4(x - y(\lambda)) \, d\lambda, \qquad (17.60)$$

where $f^\mu = f^\mu(\lambda)$ is the putative self-force. In fact, only in this case the equation of motion $dp^\mu/d\lambda = f^\mu$ would guarantee the validity of the conservation law $\partial_\mu T^{\mu\nu} = \partial_\mu(\mathbb{T}_{em}^{\mu\nu} + \Delta^{\mu\nu} + T_p^{\mu\nu}) = 0$. According to the above requirements, the vector f^μ defined in Eq. (17.60) must be a polynomial of u^μ and of its derivatives, and under a reparameterization it must transform as $f^\mu \rightarrow f'^\mu = (d\lambda/d\lambda')f^\mu$. Furthermore, apart from a factor of e^2, it must have the dimension of an inverse length squared. All these requirements constrain f^μ to be of the form $f^\mu = Ke^2u^\mu\square$, where K is a dimensionless constant. But clearly a derivative operator, like the d'Alembertian, cannot appear in an equation of motion defined on a world line. This implies that f^μ must necessarily vanish, and consequently we have $\partial_\mu \Delta^{\mu\nu} = 0$. On the other hand, a modification of $\mathbb{T}_{em}^{\mu\nu}$ with identically vanishing four-divergence is physically irrelevant.[3] Therefore, the renormalized electromagnetic energy-momentum tensor defined by Eqs. (17.57), (17.54) and (17.58) is uniquely determined.

17.5 The Electromagnetic Field for a Hyperbolic Trajectory

The world line of a hyperbolic motion along the z axis, parameterized with the proper time s, is given by the expression (7.1)

$$y^\mu(s) = (k\sinh(s/k), 0, 0, k\cosh(s/k)), \quad k > 0, \qquad (17.61)$$

and entails the trajectory (7.2)

---

[3]Since the equation $\partial_\mu \Delta^{\mu\nu} = 0$ must hold as an *algebraic* identity, presumably all its solutions are of the general form $\Delta^{\mu\nu} = \partial_\alpha K^{\alpha\mu\nu}$, where $K^{\alpha\mu\nu} = \int k^{\alpha\mu\nu} \delta^4(x - \Gamma(\lambda, b)) \, db \, d\lambda$ and $k^{\alpha\mu\nu} = -k^{\mu\alpha\nu}$.

$$\mathbf{y}(t) = \left(0, 0, \sqrt{t^2 + k^2}\right), \qquad \mathbf{v}(t) = \left(0, 0, \frac{t}{\sqrt{t^2 + k^2}}\right). \qquad (17.62)$$

The particle comes in from infinity from the right, reaches the minimum distance from the origin k at the time $t = 0$, and then goes again out to infinity to the right. This motion is time-like for all $|t| < \infty$, although for large times the speed tends to the speed of light, $\lim_{t \to \pm\infty} v(t) = 1$. In this section we analyze the electromagnetic field generated by the trajectory (17.62). As we will see, in a sense, this field interpolates between the field (17.51) generated by an unbounded curved light-like motion, and the shock-wave field (6.119) generated by a linear light-like motion.

Since for large negative times a hyperbolic motion becomes light-like, according to the general analysis of Sect. 7.2.2 also in this case the Green function method fails to provide a solution of Maxwell's equations. In fact, for the world line (17.61) the Liénard-Wiechert potential (17.20) is not a distribution, see also below. Nevertheless, it turns out that the velocity field $V^{\mu\nu}$ (17.5) and the acceleration field $A^{\mu\nu}$ (17.6) produced by the world line (17.61) *are* distributions, and so these fields certainly play a role in the solution of Maxwell's equations. However, as seen in Problem 7.2, for the world line (17.61) the retarded-time conditions (17.7) admit no solution for the space-time points $(t, \mathbf{x})$ for which $t + z < 0$. This means that within this region the above fields vanish:

$$V^{\mu\nu}(t, \mathbf{x}) = A^{\mu\nu}(t, \mathbf{x}) = 0, \quad \text{for} \ \ v_\infty^\mu x_\mu = t + z < 0. \qquad (17.63)$$

We have introduced the vector $v_\infty^\mu = (1, 0, 0, -1)$, coinciding with the four-velocity of the particle for $t \to -\infty$. Notice the similarity between the relations (17.63) and the vanishing-condition (17.46) for the acceleration field of an unbounded light-like motion, where in the present case $\mathbf{v}_\infty = (0, 0, -1)$. Conversely, in the region $t + z > 0$ the retarded-time conditions (17.7) – equivalent to Eq. (7.19) – can be solved analytically for the retarded coordinate time $t' = y^0(s)$, the solution being

$$t'(t, \mathbf{x}) = \frac{t(k^2 - t^2 + |\mathbf{x}|^2) - z\sqrt{(k^2 - t^2 + |\mathbf{x}|^2)^2 - 4k^2(z^2 - t^2)}}{2(z^2 - t^2)}. \qquad (17.64)$$

However, as first realized by H. Bondi and T. Gold in Ref. [7], the putative Liénard-Wiechert field $V^{\mu\nu} + A^{\mu\nu}$ fails to satisfy Maxwell's equations in the region $t + z = 0$, i.e. on the *shock-wave* plane $z = -t$, orthogonal to the trajectory, propagating along the z axis at the speed of light. In particular, it turned out that, whereas the field $V^{\mu\nu} + A^{\mu\nu}$ vanishes for $z < -t$, it assumes finite *non-vanishing* values if one approaches the plane $z = -t$ from the right; for explicit expressions of these fields, see, for instance, Ref. [8]. More precisely, from this Heaviside-function discontinuity located at $z + t = 0$ one derives the flux equation

$$\partial_\mu(V^{\mu\nu} + A^{\mu\nu}) = j^\nu - \frac{ek^2}{\pi(x^2 + y^2 + k^2)^2}\, v_\infty^\nu \delta(t + z). \qquad (17.65)$$

The second term on the right hand side, corresponding to a fictitious charge density spread over the plane $z = -t$, quantifies the failure of $V^{\mu\nu} + A^{\mu\nu}$ to satisfy Maxwell's equations. As found by Bondi and Gold via a limiting procedure, based on a regularized world line which is strictly time-like for all $-\infty \le t \le \infty$, the *regularized* velocity field (17.5) gives indeed rise also to the shock-wave field

$$F^{\mu\nu}_{\mathrm{sw}(k)} = \frac{e\,(v^{\mu}_{\infty}x^{\nu} - v^{\nu}_{\infty}x^{\mu})}{2\pi(x^{\alpha}x_{\alpha} - k^2)}\,\delta(v^{\beta}_{\infty}x_{\beta}) = -\frac{e\,(v^{\mu}_{\infty}x^{\nu} - v^{\nu}_{\infty}x^{\mu})}{2\pi(x^2 + y^2 + k^2)}\,\delta(t + z). \quad (17.66)$$

The mechanism producing this field is thus analogous to the limit (17.49) generating – in the case of an unbounded light-like trajectory – the shock-wave field out of the velocity field, except for the parameter k appearing in Eq. (17.66). In conclusion, a charged particle in hyperbolic motion generates the electromagnetic field

$$F^{\mu\nu} = V^{\mu\nu} + A^{\mu\nu} + F^{\mu\nu}_{\mathrm{sw}(k)}. \quad (17.67)$$

Equations (17.65)–(17.67) imply, in particular, that this field satisfies, indeed, the Maxwell equation $\partial_{\mu}F^{\mu\nu} = j^{\nu}$.

The fields of hyperbolic and light-like motions. It is worthwhile to compare the expression (17.67) with the field (17.51) generated by an unbounded light-like trajectory. First of all, in both cases there appears a shock-wave field whose origin lies in the infinite past, i.e. in the region $t \to -\infty$, where the trajectory becomes *linear* and *light-like*. This field then leads its own independent life, although for finite times the motion of the particle is no longer linear and light-like. The only difference between the shock waves of equations (17.51) and (17.67) is that the δ-profile of the latter decays slower along the plane, like $1/(x^2 + y^2 + k^2)$, while the former decays as $1/(x^2 + y^2)$. This is due to the fact that a hyperbolic motion becomes light-like only for $t \to \pm\infty$, unless $k = 0$. In particular, under the distributional limit for $k \to 0$ the additional current at the right hand side of equation (17.65) reduces to minus the current $j^{\mu}_{\mathcal{L}}$ (17.19). In the same limit, the deformed shock wave (17.66) reduces to the standard shock wave (17.49). In this sense, Eq. (17.65) resembles hence Eq. (17.48). Finally, the field (17.51) contains the δ-like contribution $Q^{\mu\nu}$, supported on the singularity surface, which is absent in the field (17.67), since for a hyperbolic motion the fields $V^{\mu\nu}$ and $A^{\mu\nu}$ are everywhere regular, except for the world line and for the shock-wave plane. On the other hand, the field (17.51) of an unbounded light-like motion contains no velocity field $V^{\mu\nu}$, as the latter is proportional to $(1 - v^2)$.

Construction of a vector potential. Although the Liénard-Wiechert potential (17.20) is not a distribution, in the complement of the shock-wave plane $t + z = 0$ its ordinary derivatives $\partial^{\mu}\mathcal{A}^{\nu} - \partial^{\nu}\mathcal{A}^{\mu}$ equal indeed the field $V^{\mu\nu} + A^{\mu\nu}$. For this reason, it is worthwhile to analyze it a little bit closer. For the world line (17.61) we have $y^{\mu}(s)u_{\mu}(s) = 0$, and consequently the expression (17.20) simplifies to

$$\mathcal{A}^\mu(t, \mathbf{x}) = \frac{ev^\mu(t')}{4\pi\big(t - v_z(t')z\big)} H(t + z), \tag{17.68}$$

where $v_z(t) = t/\sqrt{t^2 + k^2}$, and the retarded time t' is given in Eq. (17.64). We have explicitly introduced the Heaviside function $H(t + z)$, to stress that the potential vanishes beyond the shock-wave plane. As the potential (17.68) is expected to become singular *at* the shock-wave plane, we analyze it close to this plane. From formula (17.64) one recovers that for $t + z \to 0^+$ the retarded time tends to minus infinity as

$$t'(t, \mathbf{x}) \to -\frac{x^2 + y^2 + k^2}{2(t + z)}. \tag{17.69}$$

This behavior signals that the shock wave is, actually, generated by the charged particle when – in the infinite past – it stays at $z = \infty$. From the asymptotic relation (17.69) we derive that for $t + z \to 0^+$ the Liénard-Wiechert potential (17.68) entails the Laurent expansion

$$\mathcal{A}^\mu(t, \mathbf{x}) = \left(\frac{ev_\infty^\mu}{4\pi}\left(\frac{1}{t + z} + \frac{2k^2 z}{(x^2 + y^2 + k^2)^2}\right) + O(t + z)\right) H(t + z). \tag{17.70}$$

As we see, the singular part of this potential

$$\mathcal{A}_{\text{sing}}^\mu(t, \mathbf{x}) = \frac{ev_\infty^\mu}{4\pi(t + z)} H(t + z)$$

is *not* a distribution.[4] On the other hand, if we *formally* compute the field strength $\partial^\mu \mathcal{A}_{\text{sing}}^\nu - \partial^\nu \mathcal{A}_{\text{sing}}^\mu$ we obtain zero, because $\partial^\mu(t + z) = v_\infty^\mu$. This is the reason for why the field $V^{\mu\nu} + A^{\mu\nu}$ has finite limits when x^μ approaches the shock-wave plane from the right, as anticipated above. Furthermore, if we ignore the term $\mathcal{A}_{\text{sing}}^\mu$, the derivative of the Heaviside function in (17.70) does not contribute to the field strength of $\mathcal{A}^\mu(t, \mathbf{x})$, since $\partial^\mu H(t + z) = v_\infty^\mu \delta(t + z)$. This implies that, if we subtract from the potential (17.68) the term $\mathcal{A}_{\text{sing}}^\mu(t, \mathbf{x})$, the remaining potential (i) is a distribution, and (ii) its field strength equals the field $V^{\mu\nu} + A^{\mu\nu}$. Then it only remains to find a potential for the shock wave (17.66). Such a potential can easily be read off from the shock-wave potentials of a linear light-like motion given in Sect. 6.3.2, see formulas (6.117) and (6.124), modulo the replacement $x^\alpha x_\alpha \to x^\alpha x_\alpha - k^2$. In this way, we conclude that a vector potential for the field strength (17.67) is given by

$$A^\mu(t, \mathbf{x}) = \frac{e}{4\pi}\left(\frac{v^\mu(t')}{t - v_z(t')z} - \frac{v_\infty^\mu}{t + z} + \frac{2x^\mu}{x^\alpha x_\alpha - k^2}\right) H(t + z). \tag{17.71}$$

---

[4]In fact, the function $1/(t + z)$ is non-integrable, and, in addition, the presence of the Heaviside function $H(t + z)$ prevents the principal-value "regularization" $\mathcal{P}\big(\frac{1}{t+z}\big)$.

References

1. G. Morchio, F. Strocchi, Confinement of massless charged particles in QED_4 and of charged particles in QED_3. Ann. Phys. **172**, 267 (1986)
2. M. Lavelle, D. McMullan, Collinearity, convergence and cancelling infrared divergences. JHEP **0603**, 026 (2006)
3. F. Azzurli, K. Lechner, The Liénard–Wiechert field of accelerated massless charges. Phys. Lett. A **377**, 1025 (2013)
4. F. Azzurli, K. Lechner, Electromagnetic fields and potentials generated by massless charged particles. Ann. Phys. **349**, 1 (2014)
5. K. Lechner, Electrodynamics of massless charged particles. J. Math. Phys. **56**, 022901 (2015)
6. B.P. Kosyakov, Massless interacting particles. J. Phys. A **41**, 465401 (2008)
7. H. Bondi, T. Gold, The field of a uniformly accelerated charge, with special reference to the problem of gravitational acceleration. Proc. R. Soc. London A **229**, 416 (1955)
8. J. Franklin, D. Griffiths, The fields of a charged particle in hyperbolic motion. Am. J. Phys. **82**, 755 (2014); Erratum ibid **83**, 278 (2015)

Chapter 18
Massive Vector Fields

According to Maxwell's equations in the form (6.30) and to the linearized version of Einstein's equations (9.14), formally identical, a generic system of accelerated charged bodies emits electromagnetic as well as gravitational waves, both propagating at the velocity of light. In basis of the classical equations (6.30) and (9.14), at the quantum level the electromagnetic radiation is composed of photons and the gravitational one of gravitons, both particles of zero mass. Similarly, the dynamics of the eight gluon vector potentials, mediating the strong interactions, at the linearized level is described by a Lagrangian analogous to the one of the electromagnetic field, see Eq. (3.42), and, correspondingly, also gluons are massless particles propagating at the velocity of light. However, according to the *color-confinement* paradigm inherent in Quantum Chromodynamics, the theory describing the strong interactions at the quantum level, the mediators of these interactions cannot propagate freely, and so in nature there exist no *gluon waves*.

Weak interactions. In the physics of the fundamental interactions the weak interactions play a special role, in that they are mediated by *massive* vector bosons, namely by the particles $W^{\pm}$ and Z^0 with masses $m_W = 80.385 \pm 0.015\,\text{GeV}$ and $m_Z = 91.1876 \pm 0.0021\,\text{GeV}$, respectively. Aim of this chapter is to investigate a few important properties of these vector bosons, which at the Lagrangian level are described by *massive vector fields*, analyzing them in the framework of classical field theory. Obviously, a classical analysis cannot reveal all fundamental characteristics of these particles. In fact, a self-consistent and phenomenologically viable theory of the weak interactions can be formulated only within the framework of *quantum field theory*. In particular, the classical theory cannot account for the finite *mean lifetime* of these particles, which is of the order of 10^{-25} s. Moreover, our simplified treatment will not take into account the *mutual* interactions of these vector bosons, a fundamental characteristic they share with the gluons and the gravitons, but not with the photons.

© Springer International Publishing AG, part of Springer Nature 2018
K. Lechner, *Classical Electrodynamics*, UNITEXT for Physics,
https://doi.org/10.1007/978-3-319-91809-9_18

In the following, as a brief selective summary of this chapter, we collect the distinctive features of a fundamental interaction mediated by a massive vector boson, of mass m, with respect to an interaction mediated by a massless one, as the electromagnetic interaction.

- If an interaction is mediated by a massive vector boson, a static particle of charge Q creates the *exponentially* decaying scalar potential

$$A^0(r) = \frac{Q \, e^{-mr/\hbar}}{4\pi r},$$

  in place of the electrostatic potential $A^0(r) = Q/4\pi r$. The resulting interaction is thus of short range, the *range* being equal to the Compton wavelength of the vector boson $\hbar/m$.
- A massive vector boson entails *three* physical degrees of freedom, instead of the two transverse degrees of freedom of the photon. In fact, there appears a *longitudinal* physical state of helicity *zero*.
- Whereas in the case of massive vector fields the four-current j^μ (2.16) continues to be conserved, the *local* gauge invariance (2.44) is violated. This circumstance does not conflict with Noether's theorem, since conservation laws are associated with *global* invariances.[1]
- In the case of massive vector fields an accelerated charged particle emits radiation with *minimum* frequency $\omega_{\min} = m/\hbar$. Since a wave of frequency ω is composed of vector bosons with energy $\hbar\omega$, due to energy conservation, for very large m no radiation can thus be produced – even if the vector bosons were stable – except in high-energy accelerators.
- Contrary to photons, massive vector bosons are not subject to the *infrared catastrophe*. In fact, since the emitted energy is always finite, and since the minimum energy of an emitted vector boson is m, the number of emitted particles is always *finite*.
- As opposed to photons, the massive vector bosons of the weak interactions, having a finite mean lifetime, are *unstable* particles and, hence, they cannot give rise to a true *radiation*.

Regardless of the weak interactions, the results of the present chapter are interesting in themselves in that they illustrate the drastic conceptual changes classical electrodynamics would undergo, had the photon only a very small non-vanishing mass. In fact, we will consider a *universal* model, in which a generic massive vector potential A^μ is minimally coupled to a conserved current j^μ, namely via a term in the Lagrangian of the form $A^\mu j_\mu$.

---

[1] The gauge transformations (2.44) are called *local*, because the parameter $\Lambda(x)$ depends on the space-time coordinate x^μ. In fact, the current of point-like charges (2.16) is *identically* conserved, and this conservation law is not related to any global invariance principle. However, when also the charged particles are described by *fields*, then the existence of a conserved current j^μ is associated with a *global* gauge transformation with Λ constant, see Problem 3.10'and Sect. 4.4.

18.1   Lagrangian and Dynamics

A *massive* vector field is a vector field A^μ whose dynamics, in the presence of an external current j^μ, is described by the Lagrangian

$$\mathcal{L} = -\frac{1}{4} F^{\mu\nu} F_{\mu\nu} + \frac{1}{2} M^2 A^\mu A_\mu - j^\mu A_\mu, \qquad (18.1)$$

where the Maxwell tensor is defined, as in the massless case, by $F^{\mu\nu} = \partial^\mu A^\nu - \partial^\nu A^\mu$. With respect to the Lagrangian $\mathcal{L}$ of equation (3.38), giving rise to Maxwell's equation (2.22), the Lagrangian (18.1) contains a term quadratic in A^μ where – for dimensional reasons – M must be a parameter with the dimension of an inverse length. In the following we will also use the parameter with the dimension of a length

$$L = \frac{1}{M}. \qquad (18.2)$$

We anticipate that M is related to the mass m of the vector boson by the relation (compare also with the mass terms for the mediator fields of the weak interactions in Eq. (3.44))

$$m = \hbar M. \qquad (18.3)$$

As usual, we require the current to be independent of A^μ and to satisfy the continuity equation

$$\partial_\mu j^\mu = 0. \qquad (18.4)$$

As anticipated, due to the term proportional to M^2 the Lagrangian (18.1) is no longer invariant under the local gauge transformations $A^\mu \rightarrow A^\mu + \partial^\mu \Lambda$.

18.1.1   Equations of Motion and Degrees of Freedom

We begin the analysis of the dynamics of A^μ by deriving the Euler-Lagrange equations associated with the Lagrangian $\mathcal{L}$ (18.1). Since formula (3.39) is still valid, we derive the relations

$$\Pi^{\mu\nu} = \frac{\partial \mathcal{L}}{\partial(\partial_\mu A_\nu)} = -F^{\mu\nu}, \qquad \frac{\partial \mathcal{L}}{\partial A^\nu} = M^2 A^\nu - j^\nu, \qquad (18.5)$$

leading to the equality

$$\partial_\mu \frac{\partial \mathcal{L}}{\partial(\partial_\mu A_\nu)} - \frac{\partial \mathcal{L}}{\partial A^\nu} = -\partial_\mu F^{\mu\nu} - M^2 A^\nu + j^\nu.$$

The Euler-Lagrange equations (3.8) then yield for A^μ the equation of motion

$$\partial_\mu F^{\mu\nu} = j^\nu - M^2 A^\nu. \tag{18.6}$$

Applying to both members of this equation the operator ∂_ν, and taking into account the continuity equation (18.4), we derive the scalar equation $M^2 \partial_\nu A^\nu = 0$. Since $M \neq 0$, the four components of the vector field A^μ are thus subject to the constraint

$$\partial_\mu A^\mu = 0. \tag{18.7}$$

Although this constraint *formally* coincides with the Lorenz gauge-fixing (5.11), it is important to realize that in the case at hand it arises *dynamically*, i.e. as consequence of the equations of motion, rather than as consequence of a symmetry principle. With the constraint (18.7) the first member of equation (18.6) simplifies to

$$\partial_\mu F^{\mu\nu} = \Box A^\nu - \partial^\nu (\partial_\mu A^\mu) = \Box A^\nu,$$

and, consequently, this equation of motion is equivalent to the system of equations

$$(\Box + M^2) A^\mu = j^\mu, \tag{18.8}$$
$$\partial_\mu A^\mu = 0. \tag{18.9}$$

Equation (18.8) would confer to each of the four components of A^μ *one* degree of freedom. However, the constraint (18.9) determines one of these components in terms of the three remaining ones. We conclude, therefore, that a vector field A^μ whose dynamics is governed by the Lagrangian (18.1) entails *three* physical degrees of freedom.

18.1.2   Energy-Momentum Tensor

To investigate the four-momentum balance of a physical system in a consistent way, we need the system to be *isolated*. For this purpose we consider a *free* massive vector field and set $j^\mu = 0$. In this case the energy-momentum tensor can be derived from the general formulas of Sect. 3.3. The prescription (3.71), together with the conjugate momenta (18.5), yields the *canonical* energy-momentum tensor

$$\tilde{T}^{\mu\nu} = \Pi^{\mu\alpha} \partial^\nu A_\alpha - \eta^{\mu\nu} \mathcal{L} = -F^{\mu\alpha} \partial^\nu A_\alpha + \eta^{\mu\nu} \left(\frac{1}{4} F^{\alpha\beta} F_{\alpha\beta} - \frac{1}{2} M^2 A^\alpha A_\alpha \right).$$

To derive the *symmetric* energy-momentum tensor we resort to the symmetrization procedure of Sect. 3.4. Since the derivative terms of the Lagrangians (3.77) and (18.1) are the same, the tensors $V^{\rho\mu\nu}$ and $\phi^{\rho\mu\nu}$ defined in formulas (3.83) and

(3.87), respectively, remain the same as for the free electromagnetic field. Equation (3.95) is hence still valid

$$\phi^{\rho\mu\nu} = -F^{\rho\mu}A^{\nu}.$$

However, using the equation of motion (18.6) with $j^{\mu} = 0$, for the four-divergence of the tensor $\phi^{\rho\mu\nu}$ we now obtain the different expression

$$\partial_{\rho}\phi^{\rho\mu\nu} = -\partial_{\rho}F^{\rho\mu}A^{\nu} - F^{\rho\mu}\partial_{\rho}A^{\nu} = M^2 A^{\mu}A^{\nu} + F^{\mu\alpha}\partial_{\alpha}A^{\nu}.$$

Consequently, the symmetric energy-momentum tensor $T^{\mu\nu} = \widetilde{T}^{\mu\nu} + \partial_{\rho}\phi^{\rho\mu\nu}$ takes the form

$$T^{\mu\nu} = F^{\mu\alpha}F_{\alpha}{}^{\nu} + \frac{1}{4}\eta^{\mu\nu}F^{\alpha\beta}F_{\alpha\beta} + M^2\left(A^{\mu}A^{\nu} - \frac{1}{2}\eta^{\mu\nu}A^{\alpha}A_{\alpha}\right), \quad (18.10)$$

differing thus from the electromagnetic energy-momentum tensor $T^{\mu\nu}_{\text{em}}$ (2.137) by terms proportional to M^2.

18.2 Plane Wave Solutions

We now proceed to the general solution of the system of equations (18.8) and (18.9) in empty space, in which case it simplifies to

$$\left(\Box + M^2\right)A^{\mu} = 0, \qquad \partial_{\mu}A^{\mu} = 0. \quad (18.11)$$

Once more it is convenient to resort to the Fourier transform, by writing the equivalent system

$$(k^2 - M^2)\widehat{A}^{\mu} = 0, \qquad k_{\mu}\widehat{A}^{\mu} = 0, \quad (18.12)$$

in which, as usual, $\widehat{A}^{\mu}(k)$ denotes the four-dimensional Fourier transform of the field $A^{\mu}(x)$. The first equation in (18.12) admits a general solution analogous to (5.69)

$$\widehat{A}^{\mu}(k) = \delta\left(k^2 - M^2\right)f^{\mu}(k), \quad (18.13)$$

where the $f^{\mu}(k)$ are arbitrary functions of k^{μ}, subject to the reality constraint $f^{\mu*}(k) = f^{\mu}(-k)$. Since we have $k^2 = (k^0)^2 - |\mathbf{k}|^2$, we obtain

$$\delta(k^2 - M^2) = \delta\left((k^0)^2 - \omega^2\right) = \frac{1}{2\omega}\left(\delta(k^0 - \omega) + \delta(k^0 + \omega)\right), \quad (18.14)$$

where we have introduced the *frequency*

$$\omega(\mathbf{k}) = \sqrt{|\mathbf{k}|^2 + M^2}. \quad (18.15)$$

The solution (18.13) takes thus the form

$$\widehat{A}^\mu(k) = \frac{1}{2\omega} \left(\delta(k^0 - \omega)\, \varepsilon^\mu(\mathbf{k}) + \delta(k^0 + \omega)\, \varepsilon^{\mu*}(-\mathbf{k}) \right), \tag{18.16}$$

where the *polarization vector* is defined by

$$\varepsilon^\mu(\mathbf{k}) = f^\mu(\omega, \mathbf{k}). \tag{18.17}$$

The expression (18.16) formally coincides with the solution for the free electromagnetic vector potential (5.69), apart from the fundamental difference that the frequency is here given by formula (18.15). Finally, the second equation in (18.12) imposes on the polarization vector the constraint

$$k_\mu \varepsilon^\mu = 0, \qquad k^0 = \omega. \tag{18.18}$$

Performing the inverse Fourier transform of the potential (18.16) we then obtain the general solution of the system (18.11)

$$A^\mu(x) = \frac{1}{(2\pi)^2} \int \frac{d^3k}{2\omega} \left(e^{ik\cdot x} \varepsilon^\mu(\mathbf{k}) + c.c. \right), \tag{18.19}$$

to be compared with the solution for the electromagnetic vector potential (5.73). Despite the formal similarity of these solutions there is a basic difference: according to the definition (18.15) the frequency $\omega = \omega(\mathbf{k})$ appearing in the potential (18.19) is always greater than the positive constant M.

18.2.1 Elementary Waves and Wave Packets

As in the case of the electromagnetic field, the solution (18.19) is a continuous superposition of the *elementary waves*

$$A^\mu_{\text{el}}(x) = \varepsilon^\mu e^{ik\cdot x} + c.c., \qquad k_\mu \varepsilon^\mu = 0, \qquad k^0 = \omega, \qquad k^2 = M^2. \tag{18.20}$$

Nevertheless, the waves associated with a massive vector field exhibit fundamental differences with respect to the electromagnetic ones. Below we analyze them in detail.

Dispersion and group velocity. From the form of the phase $k \cdot x = \omega t - \mathbf{k} \cdot \mathbf{x}$ it follows that for a fixed wave vector $\mathbf{k}$ the elementary waves (18.20) are *plane* and *monochromatic*, and that they move in the direction of $\mathbf{k}$ with the *phase velocity* $\omega/|\mathbf{k}|$. However, in view of the formula for the frequency (18.15), the latter exceeds the velocity of light. Moreover, the frequency and the wave number are no longer linked by a linear relation, $\omega \propto |\mathbf{k}|$, as happens for electromagnetic waves: the waves

(18.20) are, in fact, *dispersive*. This implies that the velocity which is physically meaningful is not the phase velocity, but rather the *group velocity* $\mathbf{V} = \partial\omega/\partial\mathbf{k}$, which is the propagation velocity of a *wave packet* peaked around the wave vector $\mathbf{k}$. From Eq. (18.15) we obtain the explicit expressions

$$\mathbf{V} = \frac{\partial\omega}{\partial\mathbf{k}} = \frac{\mathbf{k}}{\sqrt{|\mathbf{k}|^2 + M^2}} = \frac{\mathbf{k}}{\omega}, \qquad V = \sqrt{1 - \frac{M^2}{\omega^2}}. \tag{18.21}$$

The speed $V = |\mathbf{k}|/\omega$ depends thus on the frequency, but it is always smaller than the speed of light.

Degrees of freedom, helicity, spin. For a given wave vector $\mathbf{k}$ the polarization vector $\varepsilon^\mu = (\varepsilon^0, \varepsilon)$ is constrained by Eq. (18.18)

$$k_\mu \varepsilon^\mu = \omega\varepsilon^0 - \mathbf{k}\cdot\varepsilon = 0, \quad \text{i.e.} \quad \varepsilon^0 = \mathbf{V}\cdot\varepsilon. \tag{18.22}$$

It follows that only the spatial components ε can be chosen freely, and hence there are only *three* independent physical polarization states. For a fixed $\mathbf{k}$, the three-dimensional vector ε can be further decomposed in a *longitudinal* component $\varepsilon_\parallel$ parallel to $\mathbf{k}$, and in a *transverse* component $\varepsilon_\perp$ orthogonal to $\mathbf{k}$,

$$\varepsilon = \varepsilon_\parallel + \varepsilon_\perp. \tag{18.23}$$

The vector $\varepsilon_\perp$ can be further decomposed in two mutually orthogonal vectors perpendicular to $\mathbf{k}$. Under a spatial rotation by an angle φ around $\mathbf{k}$, the component $\varepsilon_\perp$ rotates as a *vector*, while the component $\varepsilon_\parallel$ remains invariant. Recalling the concept of *helicity* introduced in Sect. 5.3.3, it then follows that the two irreducible components of $\varepsilon_\perp$ have helicity ± 1, respectively, while the longitudinal component $\varepsilon_\parallel$ has helicity 0. The fundamental new feature with respect to the electromagnetic waves is, hence, the appearance of a *physical* longitudinal polarization state of helicity *zero*. The physical *reality* of this new state becomes evident if we write out the electric and magnetic fields $E^i = \partial^i A_{\text{el}}^0 - \partial^0 A_{\text{el}}^i$ and $B^i = -\varepsilon^{ijk}\partial^j A_{\text{el}}^k$, associated with the plane wave (18.20), taking into account the relations (18.21) and (18.22),

$$\mathbf{E} = -i\omega\left(\varepsilon_\perp + \frac{M^2}{\omega^2}\,\varepsilon_\parallel\right) e^{ik\cdot x} + c.c., \qquad \mathbf{B} = \frac{\mathbf{k}}{\omega}\times\mathbf{E}. \tag{18.24}$$

With respect to the electromagnetic case the magnetic field remains hence *transverse*, $\mathbf{B}\perp\mathbf{k}$, and we have still $\mathbf{B}\perp\mathbf{E}$. In contrast, the electric field acquires a component parallel to the propagation direction, proportional to M^2, $\mathbf{E}\cdot\mathbf{k}\propto M^2$, and, moreover, the magnitudes of these fields satisfy now the inequality $B < E$.

Spin. Given the general relation existing between helicity and *spin*, see Sect. 5.3.3, we conclude that the spin of the massive particles associated at the quantum level with the waves (18.20) – the intermediate vector bosons – takes the values $-\hbar, 0, +\hbar$. Usually one says that a massive vector boson lives in a *triplet state*.

Four-momentum. To analyze the energy and momentum content of the elementary wave A^{μ}_{el} (18.20) we insert it in the energy-momentum tensor (18.10). For this purpose it is convenient to resort to the *wave relations* (henceforth, for simplicity we write A^{μ} instead of A^{μ}_{el})

$$\partial_{\mu} A^{\nu} = k_{\mu} \widehat{A}^{\nu}, \qquad k_{\mu} \widehat{A}^{\mu} = 0, \qquad k^2 = M^2, \tag{18.25}$$

where we have set

$$\widehat{A}^{\mu} = i \varepsilon^{\mu} e^{ik \cdot x} + c.c. \tag{18.26}$$

Equation (18.10) then yields the energy-momentum tensor, see Problem 18.1,

$$T^{\mu\nu} = -k^{\mu} k^{\nu} \widehat{A}^2 + M^2 \left(A^{\mu} A^{\nu} - \widehat{A}^{\mu} \widehat{A}^{\nu} \right) + \frac{1}{2} M^2 \eta^{\mu\nu} \left(\widehat{A}^2 - A^2 \right). \tag{18.27}$$

Considering the average $\langle \cdot \rangle$ of this expression over a volume large with respect to the wavelength $\lambda = 2\pi/|\mathbf{k}|$, and noting that $\langle A^{\mu} A^{\nu} \rangle = \langle \widehat{A}^{\mu} \widehat{A}^{\nu} \rangle = \varepsilon^{\mu*} \varepsilon^{\nu} + \varepsilon^{\nu*} \varepsilon^{\mu}$, we find the simple result

$$\langle T^{\mu\nu} \rangle = -2 k^{\mu} k^{\nu} \left(\varepsilon^{\alpha*} \varepsilon_{\alpha} \right), \tag{18.28}$$

as all terms proportional to M^2 drop out. The four-momentum contained in a volume $\mathcal{V}$ is hence given by[2]

$$P^{\mu} = \langle T^{\mu 0} \rangle \mathcal{V} = -2 \omega k^{\mu} \mathcal{V} \left(\varepsilon^{\alpha*} \varepsilon_{\alpha} \right). \tag{18.29}$$

From this formula we derive the relation $\mathbf{P}/P^0 = \mathbf{k}/\omega = \mathbf{V}$, meaning that the *packet* contained in the volume $\mathcal{V}$ travels, actually, with the group velocity (18.21).

Mass and Compton wavelength. From the last statement we infer that, at the quantum level, an elementary wave with wave vector $\mathbf{k}$ is composed of vector bosons propagating with velocity $\mathbf{V} = \mathbf{k}/\omega$. To determine the mass m of these particles we resort to the *de Broglie relations*, stating that a wave of wave vector k^{μ} is associated with particles of four-momentum $p^{\mu} = \hbar k^{\mu}$. The relation $k^2 = M^2$ then yields the mass

$$m^2 = p^2 = \hbar^2 M^2, \qquad \text{i.e. } m = \hbar M. \tag{18.30}$$

We are now in a position to give a physical interpretation of the length $L = 1/M$, that we were obliged to introduce in the Lagrangian (18.1) for dimensional reasons. In fact, this length equals the *Compton wavelength* of the particle

$$L = \frac{\hbar}{m}.$$

---

[2]Below we will see that the scalar product $\varepsilon^{\alpha*} \varepsilon_{\alpha}$ is negative definite, so that the energy P^0 in Eq. (18.29) is always positive.

Finally, the formula of the group velocity (18.21) is also consistent with the de Broglie relations $p^\mu = \hbar k^\mu$, which imply indeed $\mathbf{p}/p^0 = \mathbf{k}/\omega = \mathbf{V}$.

Decoupling of the longitudinal state. Using the constraint on the polarization vector (18.22) we can recast the energy-momentum tensor (18.28) in the form

$$\langle T^{\mu\nu} \rangle = 2k^\mu k^\nu \left(|\varepsilon|^2 - |\mathbf{V} \cdot \varepsilon|^2 \right). \tag{18.31}$$

The decomposition (18.23) then allows to separate out the contributions to the average four-momentum of the transverse and longitudinal polarization states

$$\langle T^{\mu\nu} \rangle = 2k^\mu k^\nu \left(|\varepsilon_\perp|^2 + \left(1 - V^2 \right) |\varepsilon_\parallel|^2 \right).$$

As we see, for ultrarelativistic speeds $V \approx 1$ the contribution of the longitudinal state is suppressed by the factor $1 - V^2 = M^2/\omega^2$, see the relations (18.21). Said differently, for increasing frequencies the longitudinal component of the wave *decouples* from the dynamics, in that it contributes less and less to the four-momentum if compared to the transverse component. Correspondingly, for $\omega \gg M$ the electric field $\mathbf{E}$ of the wave in (18.24) becomes *transverse*, and the magnitude of the magnetic field $\mathbf{B}$ reduces to that of $\mathbf{E}$. Looking at this situation from the vector-boson point of view we recall that, in general, for ultrarelativistic speeds $v \approx 1$ the mass m of a particle becomes negligible in that its energy $m/\sqrt{1 - v^2}$ is much greater than m. It follows that an ultrarelativistic vector boson behaves essentially as a *massless* one, and so it excites mainly its two transverse degrees of freedom of spin $\pm\hbar$. The corpuscular point of view matches, thus, with the classical analysis.

18.3 Generation of Massive Vector Fields

The vector potential A^μ generated by a generic source j^μ is determined by the system of nonhomogeneous differential equations (18.8) and (18.9)

$$(\Box + M^2) A^\mu = j^\mu, \qquad \partial_\mu A^\mu = 0. \tag{18.32}$$

As in the case of Maxwell's equations, we address the solution of this system by resorting to the Green function method. We write thus the solution in the form

$$A^\mu(x) = \int G_M(x - y) j^\mu(y) \, d^4y, \tag{18.33}$$

and insert it in the first equation of (18.32)

$$(\Box + M^2) A^\mu(x) = \int (\Box + M^2) G_M(x - y) j^\mu(y) \, d^4y = j^\mu(x).$$

The Green function $G_M(x)$ of a massive vector field must hence satisfy the kernel equation

$$(\Box + M^2)G_M(x) = \delta^4(x). \tag{18.34}$$

Since we want to preserve Lorentz invariance and causality, we search for a *retarded* Green function satisfying the set of conditions, analogous to (6.43)–(6.45),

$$(\Box + M^2)G_M(x) = \delta^4(x), \tag{18.35}$$
$$G_M(\Lambda x) = G_M(x), \quad \forall \Lambda \in SO(1,3)_c, \tag{18.36}$$
$$G_M(x) = 0, \quad \forall t < 0. \tag{18.37}$$

Once more, we have reduced the determination of the potential $A^\mu(x)$ to the search for a suitable Green function. We recall that the conditions (18.36) and (18.37) imply that $G_M(x)$ vanishes outside the light cone, i.e. for $x^2 = t^2 - |\mathbf{x}|^2 < 0$, as we have shown in full generality in Sect. 6.2.1. Therefore, $G_M(x)$ can be different from zero only in the space-time region $t \geq |\mathbf{x}|$. Notice also that, thanks to the continuity equation (18.4), the *ansatz* (18.33) satisfies automatically the second equation in (18.32), i.e. the constraint $\partial_\mu A^\mu = 0$, independently of the form of the Green function $G_M(x)$.

18.3.1 Static Sources and Yukawa Potential

In Sect. 18.4 we show that the system (18.35)–(18.37) admits a unique solution, giving several equivalent representations of the resulting Green function $G_M(x)$, see formulas (18.48)–(18.56). We anticipate here the representation (18.56)

$$G_M(x) = \frac{1}{2(2\pi)^2 r} \int_{-\infty}^{\infty} e^{i\omega t} \Big(H(\omega^2 - M^2) \, e^{-i\omega r \sqrt{1 - M^2/\omega^2}} \\ + H(M^2 - \omega^2) \, e^{-r\sqrt{M^2 - \omega^2}} \Big) \, d\omega, \tag{18.38}$$

where $r = |\mathbf{x}|$, and H denotes the Heaviside function. We now apply this formula to investigate the properties of the potential (18.33) generated by a generic *static* source j^μ of compact support:

$$j^0(t, \mathbf{x}) = \rho(\mathbf{x}), \qquad \rho(\mathbf{x}) = 0, \text{ for } |\mathbf{x}| > l, \qquad \mathbf{j}(t, \mathbf{x}) = 0.$$

For a static source it is convenient to perform in the integral (18.33) the change of variable $y^0 \to T = x^0 - y^0$, leading to the static four-potential

$$A^0(\mathbf{x}) = \int G_M(T, \mathbf{x} - \mathbf{y}) \, \rho(\mathbf{y}) \, dT \, d^3y, \qquad \mathbf{A}(\mathbf{x}) = 0. \tag{18.39}$$

In the expression of $A^0(\mathbf{x})$ the Green function (18.38) appears integrated over the time variable. Using the relations $\int e^{i\omega T} dT = 2\pi\delta(\omega)$, $H(-M^2) = 0$, and $H(M^2) = 1$, we find

$$\int G_M(T, \mathbf{x})\, dT = \frac{2\pi}{2(2\pi)^2 r} \int_{-\infty}^{\infty} \delta(\omega)\, e^{i\omega t} \Big(H(\omega^2 - M^2) e^{-i\omega r\sqrt{1 - M^2/\omega^2}}$$
$$+ H(M^2 - \omega^2) e^{-r\sqrt{M^2 - \omega^2}} \Big)\, d\omega = \frac{e^{-Mr}}{4\pi r}.$$

Equation (18.39) then yields the scalar potential

$$A^0(\mathbf{x}) = \frac{1}{4\pi} \int \frac{e^{-M|\mathbf{x}-\mathbf{y}|}}{|\mathbf{x}-\mathbf{y}|}\, \rho(\mathbf{y})\, d^3 y, \tag{18.40}$$

which generalizes the electrostatic potential (6.15) to the case of a massive vector field. For a static particle sitting at the origin, with charge density $\rho(\mathbf{x}) = Q\delta^3(\mathbf{x})$, formula (18.40) reduces to the *Yukawa potential*

$$A^0(\mathbf{x}) = \frac{Q e^{-Mr}}{4\pi r}, \tag{18.41}$$

generalizing the *Coulomb potential* $A^0(\mathbf{x}) = Q/4\pi r$. Finally, to analyze the large-distance behavior of the potential $A^0(\mathbf{x})$ (18.40) generated by a generic static source of compact support, we resort to the usual identifications

$$|\mathbf{x}-\mathbf{y}| \approx r - \mathbf{n}\cdot\mathbf{y}, \qquad \frac{1}{|\mathbf{x}-\mathbf{y}|} \approx \frac{1}{r}, \qquad \mathbf{n} = \frac{\mathbf{x}}{r}.$$

Formula (18.40) then entails the expansion

$$A^0(\mathbf{x}) = \frac{e^{-Mr}}{4\pi r}\, f(\mathbf{n}) + O\left(\frac{e^{-Mr}}{r^2}\right), \quad \text{where } f(\mathbf{n}) = \int e^{M(\mathbf{n}\cdot\mathbf{y})} \rho(\mathbf{y})\, d^3 y.$$

At large distances $A^0(\mathbf{x})$ behaves, thus, in the same way as the Yukawa potential (18.41), apart from the modulation factor $f(\mathbf{n})$ depending on the directions.

Finite range of the interaction and electroweak vector bosons. The potential (18.41) contains, apart from the Coulomb term $Q/4\pi r$, the exponential damping factor $e^{-Mr} = e^{-r/L}$. We deduce hence that the interaction mediated by a massive vector field is of *finite range* in that the potential, and hence the field strength $F^{\mu\nu}$ and the force, are strongly suppressed for distances r which are much larger than the *range of the interaction* $L = 1/M$. In contrast, the electromagnetic interaction represents an interaction of *infinite range*. From Sect. 18.2.1 we know furthermore that L equals the Compton wavelength of the vector boson mediating the interaction at the quantum level, $L = \hbar/m$. For the weak interactions, mediated by the particles $W^\pm$ and Z^0, we obtain thus an extremely small range, of the order

$$L_{\text{weak}} = \frac{\hbar}{m_W} \approx \frac{\hbar}{m_Z} \approx 10^{-16} \,\text{cm}.$$

Nuclear interactions and pions. Apart from the *elementary* massive *vector* mediators of the weak interactions, in nuclear physics there is class of *composed* massive mediators of spin *zero*, the *pions* $\pi^\pm$ and π^0. These particles, composed of quarks, constitute the *effective* intermediate *scalar* bosons of the nuclear interactions between neutrons and protons, which are a *residual* form of the *strong* interactions. At the Lagrangian level the pions are thus described by massive scalar fields, see Sects. 5.2 and 5.3.3 for their free propagation, and Problems 3.1 and 3.2. The pions, more precisely, are *pseudoscalar* particles, and correspondingly they actually are described by pseudoscalar fields, i.e. scalar fields which under a *parity* transformation change their sign, see Sect. 1.4.3. Concerning the large-distance behavior of the static potential, the vectorial nature of a massive mediator field is inessential, and so also a scalar massive mediator field entails a Yukawa-type potential of the form (18.41). Since the masses of the pions are of the order $m_{\pi^\pm} \approx 140\,\text{MeV}$ and $m_{\pi^0} \approx 135\,\text{MeV}$, we therefore derive that the range of the nuclear interactions

$$L_{\text{nucl}} = \frac{\hbar}{m_\pi} \approx 1.4 \cdot 10^{-13} \,\text{cm} \tag{18.42}$$

is by about a factor of thousand larger than the range of the weak interactions. In fact, in 1935 H. Yukawa predicted the existence of intermediate bosons for the nuclear interactions with a mass of the order of $100\,\text{MeV}$, knowing that the dimensions of the nuclei are of the order of $10^{-13}\,\text{cm}$ [1].

Attractive and repulsive interactions. A more detailed analysis shows that the nucleons, actually, generate an effective potential of the form (18.41) in which, however, due to the *scalar* nature of the intermediate bosons, i.e. the pions, the charge Q must be replaced by $-Q$. The resulting potential energy $U(r)$ for the interaction between two particles of *equal* charge, i.e. $U(r) = QA^0 = -Q^2 e^{-Mr}/4\pi r$, is thus an *increasing* function of their distance r. This implies that the Yukawa potential generates a force between the nucleons $\mathbf{F} = -\nabla U$ which is always *attractive*. It is clear that this force, counteracting the electrostatic repulsion of the protons, plays a fundamental role for the stability of the atomic nuclei. The attractive character of the force between the nucleons, as mediated by a scalar potential, has the same origin of the attractive character of the gravitational interaction mediated by a two-index tensor potential, i.e. the *metric*, see Sect. 9.2. In fact, it can be seen that a tensor potential with an even (odd) number of indices entails an intermediate boson of even (odd) spin, and that it gives rise to an *attractive* (*repulsive*) force between particles of equal charge. Ultimately, it is this general feature to imply that masses attract each other, while electric charges of the same sign repel each other. Indeed, the electromagnetic vector potential A^μ has just *one* index.

18.4 Retarded Green Function

18.4.1 Uniqueness

Before addressing the solution of the system of equations (18.35)–(18.37) we show that the solution, if it exists, is uniquely determined. This is equivalent to prove that the associated homogeneous system

$$(\Box + M^2)F(x) = 0, \quad F(\Lambda x) = F(x), \ \forall \Lambda \in SO(1,3)_c, \quad F(x) = 0, \ \forall t < 0, \tag{18.43}$$

admits no solution. For this purpose it is convenient to apply the Fourier transform to the homogeneous system, sending the unknown distribution $F(x)$ in $\widehat{F}(k)$. We begin by analyzing the first equation in (18.43), which becomes

$$(k^2 - M^2)\widehat{F}(k) = 0. \tag{18.44}$$

The distribution $\widehat{F}(k)$ must therefore be of the form

$$\widehat{F}(k) = \delta(k^2 - M^2)f(k),$$

where, since we want $F(x)$ to be real, $f(k)$ is a complex function subject to the constraint $f^*(k) = f(-k)$. Moreover, since, according to the second condition in (18.43), $\widehat{F}(k)$ must be Lorentz invariant under $SO(1,3)_c$, $f(k)$ must be a Lorentz-invariant function defined on the hyperboloid $k^2 = M^2$. It is then necessarily of the form $f(k) = a + ib\,\varepsilon(k^0)$, where a and b are real constants. We recall that the *sign* $\varepsilon(k^0)$ of the time component of a time-like vector is a Lorentz-invariant quantity, see Theorem II of Sect. 5.2.3. We obtain so the two linearly independent solutions

$$\widehat{F}_1(k) = \delta(k^2 - M^2), \qquad \widehat{F}_2(k) = i\varepsilon(k^0)\,\delta(k^2 - M^2), \tag{18.45}$$

to be compared with the analogous solutions (6.49) for $M = 0$. Notice that for $M \neq 0$ there exists no solution of equation (18.44) generalizing the solution $\widehat{F}_3(k) = \delta^4(k)$, see formula (6.50). Performing the inverse Fourier transform of the distributions (18.45), using for the integral over k^0 formula (18.14), we obtain

$$F_1(x) = \frac{1}{(2\pi)^2} \int \frac{d^3k}{\omega}\, \cos(\omega t - \mathbf{k} \cdot \mathbf{x}), \quad F_2(x) = -\frac{1}{(2\pi)^2} \int \frac{d^3k}{\omega}\, \sin(\omega t - \mathbf{k} \cdot \mathbf{x}), \tag{18.46}$$

where $\omega = \sqrt{|\mathbf{k}|^2 + M^2}$. Since in the limit for $M \to 0$ these distributions reduce to the expressions given in (6.51), we conclude that none of them vanishes for $t < 0$. The Green function $G_M(x)$ is, therefore, unique.

18.4.2   Representations of the Green Function

Contrary to the Green function associated with a massless vector field, i.e. the Green function of the d'Alembertian (6.59)

$$G(x) = \frac{1}{2\pi} H(t)\, \delta(x^2) = \frac{\delta(t-r)}{4\pi r}, \tag{18.47}$$

the Green function associated with a massive vector field cannot be expressed in terms of elementary functions and/or distributions. Nevertheless, it admits several equivalent integral representations. Below we present some of them which turn out to be useful in different circumstances.

$$G_M(x) = -\frac{1}{(2\pi)^4}\, \mathcal{S}' - \lim_{\gamma \to 0^+} \int e^{ik\cdot x} \frac{1}{k^2 - M^2 - i\gamma\varepsilon(k^0)}\, d^4k \tag{18.48}$$

$$= -\frac{1}{(2\pi)^4} \int e^{ik\cdot x} \left(\mathcal{P}\frac{1}{k^2 - M^2} + i\pi\varepsilon(k^0)\,\delta\!\left(k^2 - M^2\right) \right) d^4k \tag{18.49}$$

$$= -\frac{iH(t)}{(2\pi)^3} \int e^{ik\cdot x}\, \varepsilon(k^0)\,\delta\!\left(k^2 - M^2\right) d^4k \tag{18.50}$$

$$= -\frac{2H(t)}{(2\pi)^4} \int e^{ik\cdot x}\, \mathcal{P}\frac{1}{k^2 - M^2}\, d^4k \tag{18.51}$$

$$= \frac{H(t)}{(2\pi)^3} \int \frac{d^3k}{\omega}\, \sin(\omega t - \mathbf{k}\cdot\mathbf{x}) \tag{18.52}$$

$$= \frac{2H(t)}{(2\pi)^2 r} \int_M^\infty \sin\left(r\sqrt{\omega^2 - M^2}\right) \sin(\omega t)\, d\omega \tag{18.53}$$

$$= \frac{H(t)}{2\pi} \left(\delta(x^2) - \frac{MH(x^2)}{2\sqrt{x^2}}\, J_1\left(M\sqrt{x^2}\right) \right) \tag{18.54}$$

$$= \frac{1}{4\pi} \left(\frac{1}{r}\,\delta(t-r) - \frac{MH(t-r)}{\sqrt{t^2 - r^2}}\, J_1\left(M\sqrt{t^2 - r^2}\right) \right) \tag{18.55}$$

$$= \frac{1}{2(2\pi)^2 r} \int_{-\infty}^\infty e^{i\omega t} \left(H(\omega^2 - M^2)\, e^{-i\omega r\sqrt{1 - M^2/\omega^2}} \right.$$
$$\left. + H(M^2 - \omega^2)\, e^{-r\sqrt{M^2 - \omega^2}} \right) d\omega. \tag{18.56}$$

We recall that $\mathcal{S}' - \lim$ denotes the limit in the sense of distributions, H the Heaviside function, ε the sign function, and J_1 the Bessel function of order $N = 1$, see Sect. 11.3.2. We have set $r = |\mathbf{x}|$, and we have introduced the composite principal value, see formula (2.86),

$$\mathcal{P}\frac{1}{k^2 - M^2} = \mathcal{P}\frac{1}{(k^0)^2 - \omega^2} = \frac{1}{2\omega} \left(\mathcal{P}\frac{1}{k^0 - \omega} - \mathcal{P}\frac{1}{k^0 + \omega} \right), \tag{18.57}$$

where $\omega = \sqrt{|\mathbf{k}|^2 + M^2}$. Most of the integrals appearing in formulas (18.48)–(18.56) are formally divergent. In fact, they make sense only if performed in the sense of distributions, as explained in Sect. 11.4.1. In the limit for $M \to 0$ all above formulas reduce to representations of the electromagnetic Green function $G(x)$ (18.47). Each representation highlights certain properties of the distribution $G_M(x)$, and hides others. Formulas (18.49)–(18.51), for instance, are manifestly Lorentz invariant, while from formulas (18.54) and (18.55) it is evident that in the limit for $M \to 0$ we have $G_M(x) \to G(x)$, a property that is less clear in the other representations. From formulas (18.54) and (18.55) we see, in particular, that $G_M(x)$ is different from zero only for $t \geq r$, as required by causality. The main difference between the kernels $G(x)$ and $G_M(x)$ is that the support of the former is the *boundary* of the future light cone, while the support of the latter includes, in addition, the *interior* of this cone. This difference is intimately related with the fact that the photons, mediating the electromagnetic interaction, travel at the speed of light, while the speed of the vector bosons associated with a massive vector field is always less than the speed of light. Finally, the *Fourier transform* of the kernel $G_M(x)$ can be read off from formula (18.49),

$$\widehat{G}_M(k) = -\frac{1}{(2\pi)^2} \left(\mathcal{P} \frac{1}{k^2 - M^2} + i\pi\varepsilon(k^0)\,\delta(k^2 - M^2) \right), \qquad (18.58)$$

an expression that in the limit for $M \to 0$ reduces to the Fourier transform of the Green function of the d'Alembertian (6.78).

18.4.3 Derivation of the Representations

In this section we provide some technical details involved in the derivation of formulas (18.48)–(18.56), relying in particular on the uniqueness property proven in Sect. 18.4.1. This implies that to show that a certain expression equals $G_M(x)$ it suffices to verify that it satisfies the conditions (18.35)–(18.37).

Derivation of formulas (18.48) **and** (18.49). In basis of the known distributional limits

$$\mathcal{S}' - \lim_{\gamma \to 0^\pm} \frac{1}{x \pm i\gamma} = \mathcal{P}\frac{1}{x} \mp i\pi\delta(x),$$

the expressions (18.48) and (18.49) represent the same distribution. To verify that they satisfy equation (18.35), we notice that the Fourier transform of the latter amounts to

$$(-k^2 + M^2)\,\widehat{G}_M(k) = \frac{1}{(2\pi)^2}.$$

This equation is trivially satisfied by the expression (18.58), which, in turn, is equivalent to formula (18.49). The property (18.36) is obvious, since the expression

(18.49) is manifestly Lorentz invariant. Finally, to show that the expression (18.49) satisfies the condition (18.37), we perform in both terms appearing in (18.49) the integral over k^0. The integral of the second term can be read off directly from the Fourier-formulas (18.45) and (18.46)

$$-\frac{i\pi}{(2\pi)^4} \int e^{ik\cdot x}\varepsilon(k^0)\,\delta\big(k^2 - M^2\big)\,d^4k = \frac{1}{(2\pi)^3}\int \frac{d^3k}{2\omega}\,\sin(\omega t - \mathbf{k}\cdot\mathbf{x}),$$

(18.59)

where $\omega = \sqrt{|\mathbf{k}|^2 + M^2}$. To perform the integral over k^0 in the first term of (18.49) we use the composite principal value (18.57), together with the Fourier transform of the principal value (2.95)

$$\int e^{ik^0 t}\,\mathcal{P}\frac{1}{k^2 - M^2}\,dk^0 = \frac{1}{2\omega}\int e^{ik^0 t}\left(\mathcal{P}\frac{1}{k^0 - \omega} - \mathcal{P}\frac{1}{k^0 + \omega}\right)dk^0$$

$$= \frac{1}{2\omega}\int e^{ik^0 t}\left(e^{i\omega t}\,\mathcal{P}\frac{1}{k^0} - e^{-i\omega t}\,\mathcal{P}\frac{1}{k^0}\right)dk^0 = \frac{i\pi}{2\omega}\,\varepsilon(t)\left(e^{i\omega t} - e^{-i\omega t}\right).$$

Multiplying this result by $e^{-i\mathbf{k}\cdot\mathbf{x}}$, and performing the integral over $\mathbf{k}$, we obtain

$$-\frac{1}{(2\pi)^4}\int e^{ik\cdot x}\,\mathcal{P}\frac{1}{k^2 - M^2}\,d^4k = \frac{\varepsilon(t)}{(2\pi)^3}\int \frac{d^3k}{2\omega}\,\sin(\omega t - \mathbf{k}\cdot\mathbf{x}). \quad (18.60)$$

Summing equations (18.59) and (18.60) we find for formula (18.49) the alternative expression

$$G_M(x) = \frac{H(t)}{(2\pi)^3}\int \frac{d^3k}{\omega}\,\sin(\omega t - \mathbf{k}\cdot\mathbf{x}), \quad (18.61)$$

which satisfies the condition (18.37) in a manifest manner.

Derivation of formulas (18.50)–(18.52). The derivation of these formulas is now straightforward. Formula (18.52) coincides with Eq. (18.61). Formulas (18.50) and (18.51) are obtained by multiplying the identities (18.59) and (18.60) with $2H(t)$, respectively, and by using then the representation (18.52).

Derivation of formula (18.53). This representation can be derived by performing in formula (18.52) the integral over the solid angle associated with the vector $\mathbf{k}$. We have $d^3k = k^2dk\,d\varphi\,\sin\vartheta d\vartheta$, and we can resort to rotation invariance to set $\mathbf{x} = (0, 0, r)$. In this way we have $\mathbf{k}\cdot\mathbf{x} = kr\cos\vartheta$. Equation (18.52) then becomes

$$G_M(x) = \frac{H(t)}{(2\pi)^3}\int_0^\infty \frac{k^2dk}{\omega}\int_0^{2\pi}d\varphi\int_0^\pi \sin\vartheta d\vartheta\,\sin(\omega t - kr\cos\vartheta)$$

$$= \frac{H(t)}{(2\pi)^2 r}\int_0^\infty \frac{kdk}{\omega}\big(\cos(\omega t - kr) - \cos(\omega t + kr)\big).$$

Replacing the integration variable k with $\omega = \sqrt{k^2 + M^2}$ we obtain the representation (18.53).

Derivation of formulas (18.54) *and* (18.55). These formulas represent the same object, written in different ways. Formula (18.54) satisfies in a manifest way properties (18.36) and (18.37). Hence, it is sufficient to show that the expression (18.55) satisfies the kernel equation (18.35). In what follows we address the solution of this equation using a technique similar to the one adopted in Sect. 6.2.1 to determine the kernel of the d'Alembertian G (18.47). Since the latter satisfies the equation $\Box G = \delta^4(x)$, and since in the limit for $M \to 0$ the kernel G_M reduces to G, it is convenient to set

$$G_M = G + B, \tag{18.62}$$

and to consider the distribution B as the new unknown. Substituting the position (18.62) in Eq. (18.35) we find that B must satisfy the equation

$$(\Box + M^2)B = -M^2 G. \tag{18.63}$$

Let us preliminarily restrict to the region $t \neq r$. Since G is proportional to $\delta(t - r)$, in this region B must hence solve the equation

$$(\Box + M^2)B = 0. \tag{18.64}$$

Moreover, since G_M is Lorentz invariant, also B must be so. This implies that B can depend on x^μ only through the combination $x^\mu x_\mu = t^2 - r^2$. Without loss of generality we can therefore set

$$B = \frac{1}{u} f(u), \qquad u \equiv M\sqrt{t^2 - r^2}. \tag{18.65}$$

In fact, we are searching for a G_M which is different from zero only for $t \geq r$. Since B depends on **x** only through its magnitude r, in Eq. (18.64) we can replace the d'Alembertian with

$$\Box \to \frac{\partial^2}{\partial t^2} - \frac{1}{r}\frac{\partial^2}{\partial r^2} r. \tag{18.66}$$

Substituting the *ansatz* for B in (18.65) in Eq. (18.64), and working out the derivatives, one finds that $f(u)$ must satisfy the second-order differential equation

$$u^2 f'' + u f' + (u^2 - 1)f = 0, \tag{18.67}$$

where the *prime* denotes the derivative with respect to u. Recalling the properties of the special functions of Sect. 11.3.2, more precisely equation (11.70), we recognize that $f(u)$ must be a linear combination of the Bessel and Neumann functions of order $N = 1$, i.e. of $J_1(u)$ and $Y_1(u)$. Given that B can be different from zero only for $t \geq r$, we then conclude that it is of the form

$$B = H(t - r)\frac{1}{u}\left(aJ_1(u) + bY_1(u)\right) \equiv B_J + B_Y. \tag{18.68}$$

Finally, the real coefficients a and b must be determined such that the expression (18.68) satisfies equation (18.63) in the sense of distributions. For this purpose we recall the behaviors of the Bessel and Neumann functions for small arguments (11.68) and (11.73), respectively, yielding the expansions

$$J_1(u) = \frac{u}{2} + O(u^3), \qquad Y_1(u) = -\frac{2}{\pi u} + O(u). \tag{18.69}$$

Consequently, in the limit for $u \to 0$, i.e. for $t \to r$, the function B_Y behaves as

$$B_Y \to -\frac{2bH(t-r)}{\pi M^2(t^2 - r^2)},$$

which corresponds to a non-integrable singularity in $\mathbb{R}^4$. This implies that B_Y is *not* a distribution, and so in formula (18.68) we must set $b = 0$. Vice versa, the function B_J entails the expansion

$$B_J = \frac{a}{2} H(t-r)\big(1 + O(t^2 - r^2)\big), \tag{18.70}$$

and defines hence a distribution. Inserting now the expression for B (18.68), with $b = 0$, in the first member of equation (18.63) we find

$$(\Box + M^2)B = a(\Box + M^2)\left(H(t-r)\frac{J_1}{u}\right) = aH(t-r)(\Box + M^2)\left(\frac{J_1}{u}\right)$$
$$+ 2a\,\partial_\mu H(t-r)\,\partial^\mu\left(\frac{J_1}{u}\right) + \frac{aJ_1}{u}\,\Box H(t-r). \tag{18.71}$$

The first term of the last member of this equation vanishes by construction, see (18.64) and (18.68), since according to the expansions (18.69) $J_1(u)/u$ is regular at $t = r$. To evaluate the second term we calculate the derivatives

$$\partial_\mu H(t-r) = \left(1, -\frac{\mathbf{x}}{r}\right)\delta(t-r) = \frac{x_\mu}{r}\,\delta(t-r),$$
$$\partial^\mu\left(\frac{J_1}{u}\right) = \frac{Mx^\mu}{\sqrt{x^2}}\frac{d}{du}\left(\frac{J_1}{u}\right),$$

leading to

$$\partial_\mu H(t-r)\,\partial^\mu\left(\frac{J_1}{u}\right)^{\!\bullet} = \frac{M}{r}\sqrt{t^2 - r^2}\,\delta(t-r)\frac{d}{du}\left(\frac{J_1}{u}\right) = 0.$$

We have applied the rule (2.72), and used that the function $(d/du)(J_1/u)$ – being of order $O(u) = O(\sqrt{t^2 - r^2})$ – is regular at $t = r$. To evaluate the third term in (18.71) we resort again to the replacement (18.66)

$$\Box H(t-r) = \left(\frac{\partial^2}{\partial t^2} - \frac{\partial^2}{\partial r^2} - \frac{2}{r}\frac{\partial}{\partial r}\right)H(t-r) = -\frac{2}{r}\frac{\partial}{\partial r}H(t-r) = \frac{2}{r}\delta(t-r).$$

Since for $u = 0$ the function $J_1(u)/u$ equals $1/2$, formula (18.71) eventually reduces to

$$(\Box + M^2)B = \frac{a}{r}\,\delta(t-r) = 4\pi a G,$$

where we have introduced the kernel G (18.47). Therefore, since we want B to satisfy equation (18.63), we must choose for a the value $a = -M^2/4\pi$. Formula (18.68) then yields

$$B = -\frac{MH(t-r)}{4\pi\sqrt{t^2-r^2}}\,J_1\big(M\sqrt{t^2-r^2}\big),$$

so that Eq. (18.62) reduces to formula (18.55).

Derivation of formula (18.56). To derive formula (18.56) it is convenient to introduce the *temporal* Fourier transform $\widetilde{G}_M(\omega, \mathbf{x})$ of the Green function,

$$G_M(t, \mathbf{x}) = \frac{1}{\sqrt{2\pi}}\int e^{i\omega t}\,\widetilde{G}_M(\omega, \mathbf{x})\,d\omega, \tag{18.72}$$

and to consider, correspondingly, the temporal Fourier transform of the kernel equation (18.35)

$$\left(\nabla^2 + \omega^2 - M^2\right)\widetilde{G}_M(\omega, \mathbf{x}) = -\frac{1}{\sqrt{2\pi}}\,\delta^3(\mathbf{x}). \tag{18.73}$$

We solve this equation preliminarily in the region $r \neq 0$. Since $\widetilde{G}_M(\omega, \mathbf{x})$ depends on $\mathbf{x}$ only through r, we can replace the Laplacian with

$$\nabla^2 \to \frac{1}{r}\frac{\partial^2}{\partial r^2}\,r.$$

We thus see that Eq. (18.73) has oscillatory or exponential solutions according to whether $|\omega| > M$ or $|\omega| < M$:

$$\widetilde{G}_M(\omega, \mathbf{x}) = \frac{1}{r}\,H\big(\omega^2 - M^2\big)\left(a_1\,e^{-i\omega r\sqrt{1-M^2/\omega^2}} + a_2\,e^{i\omega r\sqrt{1-M^2/\omega^2}}\right)$$
$$+\frac{1}{r}\,H\big(M^2 - \omega^2\big)\left(b_1\,e^{-r\sqrt{M^2-\omega^2}} + b_2\,e^{r\sqrt{M^2-\omega^2}}\right). \tag{18.74}$$

Notice that the exponents in the first line are $\omega\sqrt{1 - M^2/\omega^2}$, and not $\sqrt{\omega^2 - M^2}$, since we want $G_M(t, \mathbf{x})$ to be *real*. In fact, for this purpose we must have $\widetilde{G}_M^*(\omega, \mathbf{x}) = \widetilde{G}_M(-\omega, \mathbf{x})$. The coefficients a_i and b_i must now be determined such that Eq. (18.73) holds in the distributional sense. The coefficient b_2 must anyhow vanish, because $e^{r\sqrt{M^2-\omega^2}}$ is not a distribution. Inserting the resulting expression (18.74) in the first member of the kernel equation (18.73), and recalling the distributional identity (2.119), we obtain

$$\left(\nabla^2 + \omega^2 - M^2\right) \widetilde{G}_M(\omega, \mathbf{x}) =$$
$$- 4\pi\left((a_1 + a_2)H(\omega^2 - M^2) + b_1 H(M^2 - \omega^2)\right)\delta^3(\mathbf{x}).$$

The comparison with Eq. (18.73) then yields the constraints

$$a_1 + a_2 = \frac{1}{2(2\pi)^{3/2}}, \qquad b_1 = \frac{1}{2(2\pi)^{3/2}}.$$

Finally, imposing that $G_M(t, \mathbf{x})$ vanishes for $t < 0$, one finds that a_2 must be zero. The simplest way to see it is to use that in the limit for $M \to 0$ the expression (18.74) must reduce to the temporal Fourier transform $\widetilde{G}(\omega, \mathbf{x})$ of the kernel $G(t, \mathbf{x})$ (18.47), which indeed vanishes for $t < 0$. One has, in fact,

$$\widetilde{G}(\omega, \mathbf{x}) = \frac{1}{\sqrt{2\pi}} \int e^{-i\omega t} G(t, \mathbf{x})\, dt = \frac{e^{-i\omega r}}{2(2\pi)^{3/2} r}.$$

With $a_1 = b_1 = 1/2(2\pi)^{3/2}$, and $a_2 = b_2 = 0$, the Fourier transform (18.74) eventually reduces to

$$\widetilde{G}_M(\omega, \mathbf{x}) = \frac{1}{2(2\pi)^{3/2} r} \left(H(\omega^2 - M^2)\, e^{-i\omega r \sqrt{1 - M^2/\omega^2}} \right.$$
$$\left. + H(M^2 - \omega^2)\, e^{-r\sqrt{M^2 - \omega^2}} \right)..$$
(18.75)

Inserting this expression in Eq. (18.72) we obtain formula (18.56).

18.5 Radiation

In this section we analyze the radiation generated by a generic source $j^\mu(x)$, starting from the integral representation of the resulting vector potential $A^\mu(x)$ (18.33). As in the electromagnetic case, we take the source to be of compact spatial support, and we focus on the large-distance behavior of the produced field.

18.5.1 The Vector Field in the Wave Zone

In the case of a massive vector field it is more convenient to investigate the properties of the radiation "frequency by frequency", i.e. to perform a *spectral analysis*. Correspondingly, we decompose the current into its fixed-frequency contributions

$$j^{\mu}(t, \mathbf{x}) = \frac{1}{\sqrt{2\pi}} \int e^{i\omega t} \widetilde{j}^{\mu}(\omega, \mathbf{x}) \, d\omega. \tag{18.76}$$

Regarding the Green function $G_M(t, \mathbf{x})$, it is convenient to use the representation (18.56), equivalent to formula (18.72) with $\widetilde{G}_M(\omega, \mathbf{x})$ given in (18.75). The solution (18.33) then takes the form

$$
\begin{aligned}
A^{\mu}(x) &= \int dy^0 \, d^3y \, G_M(t - y^0, \mathbf{x} - \mathbf{y}) \, j^{\mu}(y^0, \mathbf{y}) \\
&= \frac{1}{2\pi} \int dy^0 \, d^3y \int e^{i\omega(t - y^0)} \, \widetilde{G}_M(\omega, \mathbf{x} - \mathbf{y}) \, d\omega \int e^{i\omega' y^0} \, \widetilde{j}^{\mu}(\omega', \mathbf{y}) \, d\omega' \\
&= \int d\omega \, e^{i\omega t} \int d^3y \, \widetilde{G}_M(\omega, \mathbf{x} - \mathbf{y}) \, \widetilde{j}^{\mu}(\omega, \mathbf{y}),
\end{aligned}
\tag{18.77}
$$

where the integrals over y^0 and ω' have been performed using the identity (2.98)

$$\int e^{i(\omega' - \omega)y^0} dy^0 = 2\pi\delta(\omega' - \omega).$$

Inserting the temporal Fourier transform of the Green function (18.75) in Eq. (18.77) we find the exact potential

$$
\begin{aligned}
A^{\mu}(x) = \frac{1}{2(2\pi)^{3/2}} \int d\omega \, e^{i\omega t} \int \frac{d^3y}{|\mathbf{x} - \mathbf{y}|} \Big(& H(\omega^2 - M^2) \, e^{-i\omega|\mathbf{x} - \mathbf{y}|\sqrt{1 - M^2/\omega^2}} \\
& + H(M^2 - \omega^2) \, e^{-|\mathbf{x} - \mathbf{y}|\sqrt{M^2 - \omega^2}} \Big) \widetilde{j}^{\mu}(\omega, \mathbf{y}).
\end{aligned}
$$

To analyze the behavior of $A^{\mu}(x)$ in the wave zone, i.e. for large values of $r = |\mathbf{x}|$, we use the standard expansions

$$|\mathbf{x} - \mathbf{y}| \to r - \mathbf{n} \cdot \mathbf{y}, \qquad \frac{1}{|\mathbf{x} - \mathbf{y}|} \to \frac{1}{r}, \qquad \mathbf{n} = \frac{\mathbf{x}}{r}.$$

In this way we arrive at the wave-zone potential

$$
\begin{aligned}
A^{\mu}(x) = \frac{1}{2(2\pi)^{3/2}r} \int d\omega \, e^{i\omega t} \int d^3y \Big(& H(\omega^2 - M^2) \, e^{-i\omega(r - \mathbf{n} \cdot \mathbf{y})\sqrt{1 - M^2/\omega^2}} \\
& + H(M^2 - \omega^2) \, e^{\mp(r - \mathbf{n} \cdot \mathbf{y})\sqrt{M^2 - \omega^2}} \Big) \widetilde{j}^{\mu}(\omega, \mathbf{y}).
\end{aligned}
\tag{18.78}
$$

Until now we proceeded in the same way as for the electromagnetic field, and, in particular, we have retrieved the familiar large-distance behavior $1/r$. However, comparing formula (18.78) with the analogous expression of the electromagnetic potential (8.145), we observe the presence of a new term – the one in the second line of equation (18.78) – which in the limit for $M \to 0$ actually vanishes, thanks to the Heaviside function $H(M^2 - \omega^2)$. Nevertheless, also for $M \neq 0$, in the limit for large r this term vanishes exponentially, and so it does not contribute to the potential in the wave zone. The wave-zone potential of a massive vector field is thus finally given by

$$A^\mu(x) = \frac{1}{2(2\pi)^{3/2}r} \int_{|\omega| \geq M} d\omega \, e^{i\omega\left(t - r\sqrt{1-M^2/\omega^2}\right)} \int d^3y \, e^{i\omega\mathbf{n}\cdot\mathbf{y}\sqrt{1-M^2/\omega^2}} \, \widetilde{j}^\mu(\omega, \mathbf{y}).$$
(18.79)

Of course, in the limit for $M \to 0$ this formula reduces still to (8.145).

Decomposition in elementary waves. In the spatial region where the current j^μ vanishes the exact potential satisfies the *free* equation $(\Box + M^2)A^\mu = 0$, and so we expect that at large distances from the source the potential reduces to a *radiation field*, i.e. to a superposition of the elementary waves (18.20). To check this hypothesis we express the potential in terms of its temporal Fourier transform $\widetilde{A}^\mu(\omega, \mathbf{x})$

$$A^\mu(x) = \frac{1}{\sqrt{2\pi}} \int e^{i\omega t} \widetilde{A}^\mu(\omega, \mathbf{x}) \, d\omega.$$
(18.80)

Comparing formulas (18.79) and (18.80) we see that in the wave zone the potential is indeed a frequency-superposition of waves of the type (18.20), given by

$$e^{i\omega t} \widetilde{A}^\mu(\omega, \mathbf{x}) = \begin{cases} \varepsilon^\mu e^{ik\cdot x}, & \text{for } |\omega| \geq M, \\ \\ 0, & \text{for } |\omega| < M, \end{cases}$$
(18.81)

where the wave vector k^μ and the polarization vector ε^μ are expressed in terms of the propagation velocity $\mathbf{V}$ and of the frequency ω by (see Problem 18.2 and the relations (18.21))

$$k^\mu = (\omega, \omega\mathbf{V}), \qquad \mathbf{V} = \mathbf{n}\sqrt{1 - \frac{M^2}{\omega^2}} = \mathbf{n}V, \qquad k^2 = M^2, \qquad (18.82)$$

$$\varepsilon^\mu = \frac{1}{4\pi r} \int e^{i\omega\mathbf{V}\cdot\mathbf{y}} \, \widetilde{j}^\mu(\omega, \mathbf{y}) \, d^3y, \qquad k_\mu\varepsilon^\mu = 0. \qquad (18.83)$$

Below we shall also make use of the algebraic relation

$$k_\mu \widetilde{A}^\mu(\omega, \mathbf{x}) = 0, \qquad (18.84)$$

implied by the second relation in (18.83) or, alternatively, by the constraint $\partial_\mu A^\mu = 0$. From (18.81) we infer the absence of radiation with frequencies $|\omega| < M$, which clearly reflects the fact that the frequencies of the elementary waves associated with a free massive vector field are intrinsically greater than M, see the dispersion relation (18.15). From formulas (18.81) and (18.83) we see that a current $\widetilde{j}^\mu(\omega, \mathbf{y})$ of frequency ω generates an elementary wave of the same frequency, exactly as happens in electrodynamics. However, in contrast to electrodynamic, a current $j^\mu(x)$ – coupled to a massive vector field – containing only components $\tilde{j}^\mu(\omega, \mathbf{y})$ with frequencies $|\omega| < M$, emits *no radiation* at all. Moreover, if the *characteristic* frequencies of the current $j^\mu(x)$ are well below M, the current emits a negligible radiation. At the quantum level, in basis of the de Broglie relation $p^\mu = \hbar k^\mu$, the condition $|\omega| \geq M$ amounts to the obvious fact that the vector bosons forming the radiation have necessarily energies $\hbar|\omega|$ greater than their mass m, see formulas (18.30).

18.6 Spectral Analysis

To perform the spectral analysis of the emitted radiation we must return to the fundamental radiation equations of Sect. 7.4, in that formulas (11.12) and (11.14) have been derived specifically for the electromagnetic radiation. Obviously, in the limit for $M \to 0$ we must recover those formulas. We start again from the general expression of the emitted four-momentum (7.55)

$$\frac{d^2 P^\mu}{dt\, d\Omega} = r^2 \big(T^{\mu i} n^i \big), \qquad (18.85)$$

where the limit for $r \to \infty$ is understood, and the energy-momentum tensor $T^{\mu\nu}$ is now given by Eq. (18.10). For definiteness we suppose that the current is *aperiodic*, so that we can integrate the emitted four-momentum (18.85) over all times

$$\frac{dP^\mu}{d\Omega} = r^2 n^i \int_{-\infty}^{\infty} T^{\mu i}\, dt. \qquad (18.86)$$

Thanks to the quadratic form of the tensor $T^{\mu\nu}$ (18.10), and to the Fourier representation of the potential A^μ (18.80), applying Plancherel's theorem the integral (18.86) can be transformed in an integral over frequencies. For this purpose we note that formulas (18.80)–(18.83) imply the relations, valid modulo terms of order $1/r^2$,

$$\partial^\mu A^\nu = \frac{i}{\sqrt{2\pi}} \int e^{i\omega t} k^\mu \tilde{A}^\nu \, d\omega, \qquad (18.87)$$

where henceforth we denote the potential $\tilde{A}^\mu(\omega, \mathbf{x})$ by $\tilde{A}^\mu$. For the integral of a product of two factors of the type (18.87) Plancherel's theorem gives

$$\int_{-\infty}^{\infty} \partial^\mu A^\nu \, \partial^\alpha A^\beta \, dt = \int_{-\infty}^{\infty} k^\mu k^\alpha \tilde{A}^{\nu*} \tilde{A}^\beta \, d\omega. \qquad (18.88)$$

Thanks to the constraint (18.84), and to the relation $k^2 = M^2$, the integral over all times of the energy-momentum tensor (18.10) then reduces to (see Problem 18.3)

$$\int_{-\infty}^{\infty} T^{\mu\nu}\, dt = -\int_{-\infty}^{\infty} k^\mu k^\nu \widetilde{A}^*_\alpha \widetilde{A}^\alpha\, d\omega. \tag{18.89}$$

Consequently, the radiation formula (18.86) takes the form

$$\frac{dP^\mu}{d\Omega} = -r^2 n^i \int_{-\infty}^{\infty} k^i k^\mu \widetilde{A}^*_\alpha \widetilde{A}^\alpha\, d\omega.$$

Using the relations (18.81)–(18.83) we then obtain for the four-momentum emitted per unit frequency interval and per unit solid angle the expression

$$\frac{d^2 P^\mu}{d\omega d\Omega} = -2r^2 n^i k^i k^\mu \widetilde{A}^*_\alpha \widetilde{A}^\alpha = -2r^2 V \omega k^\mu \varepsilon^*_\alpha \varepsilon^\alpha. \tag{18.90}$$

In this formula the frequencies are considered as positive, and larger than M, and the frequency-dependent speed V of the wave is given in (18.82) As in the case of the elementary waves, see the relations (18.28) and (18.29), it is sufficient to analyze the emitted energy $\Delta\varepsilon = \Delta P^0$. In fact, according to Eqs. (18.82) and (18.90) the emitted momentum is related to the emitted energy by $\Delta\mathbf{P} = \Delta\varepsilon\mathbf{V}$.

Formula (18.90) can be represented in a more transparent way, if we recognize that ε^μ is related in a simple way to the four-dimensional Fourier transform $J^\mu(k^0, \mathbf{k})$ of the current $j^\mu(x)$. In fact, from the integral representation (18.76) we find

$$J^\mu(k^0, \mathbf{k}) = \frac{1}{(2\pi)^2} \int e^{-ik\cdot x} j^\mu(x)\, d^4x = \frac{1}{(2\pi)^{3/2}} \int e^{i\mathbf{k}\cdot\mathbf{x}} \widetilde{j}^\mu(k^0, \mathbf{x})\, d^3x. \tag{18.91}$$

Comparing this equation with the polarization vector (18.83), and using the relations (18.82), we see that we can recast the latter in the form

$$\varepsilon^\mu = \frac{1}{r}\sqrt{\frac{\pi}{2}}\, \mathcal{J}^\mu_\omega, \qquad \mathcal{J}^\mu_\omega \equiv J^\mu(\omega, \omega\mathbf{V}). \tag{18.92}$$

The $\mu = 0$ component of equation (18.90) then yields for the spectral weights of the emitted radiation the compact expression

$$\frac{d^2\varepsilon}{d\omega d\Omega} = -\pi\omega^2 V \mathcal{J}^{\mu*}_\omega \mathcal{J}_{\omega\mu}. \tag{18.93}$$

Using the continuity equation $\partial_\mu j^\mu = 0$, equivalent to the algebraic equation $k_\mu J^\mu = 0$, i.e. $J^0 = \mathbf{V}\cdot\mathbf{J}$, the spectral weights finally can be rewritten as

$$\frac{d^2\varepsilon}{d\omega d\Omega} = \pi\omega^2 V \left(\left|\mathcal{J}_\omega\right|^2 - \left|\mathbf{V}\cdot\mathcal{J}_\omega\right|^2 \right). \tag{18.94}$$

It is understood that

$$\frac{d^2\varepsilon}{d\omega d\Omega} = 0, \quad \text{for} \quad \omega < M.$$

Massive and massless vector fields. Equation (18.94) represents the generalization of the analogous formula for the electromagnetic radiation (11.140), to the case of a massive vector field. Apart from the *formal* similarities, we stress the following basic differences. In Eq. (18.94) the Fourier coefficient $\mathcal{J}_\omega = \mathbf{J}(\omega, \omega\mathbf{V})$ is evaluated for a wave vector $k^\mu = (\omega, \omega\mathbf{V})$ satisfying the constraint $k^2 = M^2$, rather than $k^2 = 0$. Next, whereas the electromagnetic radiation contains arbitrarily low frequencies, for a massive vector field the spectrum is restricted to the frequencies $\omega \geq M$. Another difference between Eqs. (11.140) and (18.94) is the speed V appearing as prefactor in the latter. This implies that for a massive vector field there is no radiation emitted at *threshold*, i.e. radiation with frequency $\omega = M$, since for $\omega = M$ we have $V = 0$, see (18.21). A vector boson with frequency ω slightly above M, corresponding to an energy $\hbar\omega$ slightly above its mass m, is thus produced practically at rest, $V \approx 0$. Vice versa, from the kinematical relations (18.82) we see that for high frequencies, $\omega \gg M$, the wave vector k^μ and the velocity $\mathbf{V}$ of the wave reduce to their respective expressions of the electromagnetic field

$$k^\mu \to (\omega, \omega\mathbf{n}), \qquad \mathbf{V} \to \mathbf{n}, \qquad V \to 1.$$

In this limit the spectral weights (18.94) approach thus the spectral weights (11.140) of the electromagnetic radiation. From a quantum point of view this conclusion is not unexpected since, as observed previously, a vector boson with an energy $\hbar\omega$ considerably larger than its rest mass m behaves as a vector boson with negligible mass, despite the presence of a spin-zero mode.

Transverse and longitudinal polarization states. Since according to the relation (18.92) the spatial vectors $\mathcal{J}_\omega$ and ε are parallel, it is easy to extract from formula (18.94) the spectral weights with transverse and longitudinal polarizations, i.e. the intensities of the radiation whose polarization vector ε is orthogonal and parallel to the propagation direction $\mathbf{n} = \mathbf{V}/V$, respectively. Introducing the parallel and orthogonal components of $\mathcal{J}_\omega$, i.e. $\mathcal{J}_\omega^{\|} = \mathbf{n} \cdot \mathcal{J}_\omega$ and $\mathcal{J}_\omega^{\perp} = \mathcal{J}_\omega - \mathcal{J}_\omega^{\|}\mathbf{n}$, and using the identity

$$\left|\mathcal{J}_\omega\right|^2 - \left|\mathbf{V} \cdot \mathcal{J}_\omega\right|^2 = \left|\mathcal{J}_\omega^\perp\right|^2 + \left(1 - V^2\right)\left|\mathcal{J}_\omega^\|\right|^2,$$

the total spectral weights (18.94) decompose into the orthogonal and parallel contributions

$$\begin{aligned} \frac{d^2\varepsilon^\perp}{d\omega\,d\Omega} &= \pi\omega^2 V\left|\mathcal{J}_\omega^\perp\right|^2, \\ \frac{d^2\varepsilon^\|}{d\omega\,d\Omega} &= \pi\omega^2 V\left(1 - V^2\right)\left|\mathcal{J}_\omega^\|\right|^2 = \pi M^2 V\left|\mathcal{J}_\omega^\|\right|^2. \end{aligned} \qquad (18.95)$$

For low frequencies, $\omega \approx M$, corresponding to speeds $V \ll 1$, the transverse and longitudinal radiations exhibit hence comparable spectral weights. Conversely, for high frequencies, $\omega \gg M$, corresponding to ultrarelativistic speeds, $V \approx 1$, the transverse radiation dominates over the longitudinal one, as anticipated in Sect. 18.2.1. Notice also that in the electromagnetic limit $M \to 0$ the longitudinal radiation disappears for any ω.

Periodic systems. For a periodic current of period T and fundamental frequency $\omega_0 = 2\pi/T$ one can introduce the discrete Fourier coefficients

$$J_N^\mu(\mathbf{k}) = \frac{1}{T} \int_0^T dt \, \frac{1}{(2\pi)^{3/2}} \int d^3 x \, e^{-i(N\omega_0 t - \mathbf{k}\cdot\mathbf{x})} j^\mu(x), \qquad (18.96)$$

where N is an integer. Proceeding as above, one then finds that the angular distribution of the power of the radiation emitted with (positive) frequency $\omega_N = N\omega_0$ is given by (see Problem 18.5)

$$\frac{dW_N}{d\Omega} = \pi(N\omega_0)^2 V_N \left(|\boldsymbol{\mathcal{J}}_N|^2 - |\mathbf{V}_N \cdot \boldsymbol{\mathcal{J}}_N|^2 \right), \qquad (18.97)$$

where we have set

$$\boldsymbol{\mathcal{J}}_N = \mathbf{J}_N(N\omega_0 \mathbf{V}_N), \qquad \mathbf{V}_N = \mathbf{n} \sqrt{1 - \frac{M^2}{(N\omega_0)^2}}.$$

For a periodic current the radiation can contain only frequencies $N\omega_0$ which are greater than M. The harmonic of lowest order present in the radiation corresponds, therefore, to the order

$$N^* = \left[\frac{M}{\omega_0} \right] + 1,$$

where the symbol $[\,\cdot\,]$ stands for the *integer part* of a real number. Correspondingly, the energies of the emitted vector bosons have the quantized values $\varepsilon_N = \hbar N\omega_0$, where $N \geq N^*$. In particular, if the only non-vanishing Fourier coefficients of the current $J_N^\mu(\mathbf{k})$ (18.96) are those with $|N| < N^*$, then no radiation is emitted at all.

18.6.1 Spectrum of a Single Particle

The current density of a charged particle on a trajectory $\mathbf{y}(t)$ has the standard form

$$\mathbf{j}(t, \mathbf{x}) = e\mathbf{v}(t)\delta^3(\mathbf{x} - \mathbf{y}(t)). \qquad (18.98)$$

For the sake of concreteness, in the following we restrict our attention to aperiodic trajectories. To evaluate the continuous spectral weights (18.93) we then need the Fourier coefficients $\boldsymbol{\mathcal{J}}_\omega$, which, according to the definition in (18.92), equal the four-dimensional Fourier transform $\mathbf{J}(k^0, \mathbf{k})$ (18.91) of the current density (18.98), evaluated at $(k^0, \mathbf{k}) = (\omega, \omega\mathbf{V})$. The result reads

$$\mathcal{J}_\omega = \frac{e}{(2\pi)^2} \int_{-\infty}^{\infty} e^{-i\omega(t-\mathbf{V}\cdot\mathbf{y}(t))}\, \mathbf{v}(t)\, dt, \tag{18.99}$$

an expression differing from its electromagnetic counterpart (11.142) only by the replacement $\mathbf{n} \to \mathbf{V} = \mathbf{n}\sqrt{1-M^2/\omega^2}$. To improve the convergence properties of the integral (18.99), which is meaningful only if understood in the sense of distributions, we perform an integration by parts based on the identity

$$e^{-i\omega(t-\mathbf{V}\cdot\mathbf{y}(t))} = \frac{i}{\omega(1-\mathbf{V}\cdot\mathbf{v}(t))}\,\frac{d}{dt}\, e^{-i\omega(t-\mathbf{V}\cdot\mathbf{y}(t))}.$$

This yields

$$\mathcal{J}_\omega = \frac{-ie}{(2\pi)^2\omega}\int_{-\infty}^{\infty} e^{-i\omega(t-\mathbf{V}\cdot\mathbf{y})}\,\frac{(1-\mathbf{V}\cdot\mathbf{v})\mathbf{a}+(\mathbf{V}\cdot\mathbf{a})\mathbf{v}}{(1-\mathbf{V}\cdot\mathbf{v})^2}\, dt, \tag{18.100}$$

where it is understood that the kinematical variables $\mathbf{y}$, $\mathbf{v}$ and $\mathbf{a}$ are evaluated at the time t. Inserting the Fourier coefficients (18.100), or alternatively (18.99), in Eq. (18.94) we obtain formulas that generalize the electromagnetic spectral weights (11.94) and (11.95), respectively.

Non-relativistic limit. For non-relativistic charged particles, $v \ll 1$, it is convenient to rewrite the exponent in the integral (18.100) in the form

$$t - \mathbf{V}\cdot\mathbf{y}(t) = -\mathbf{V}\cdot\mathbf{y}(0) + \int_0^t \left(1 - \mathbf{V}\cdot\mathbf{v}(t')\right) dt'. \tag{18.101}$$

The term $\mathbf{V}\cdot\mathbf{y}(0)$ contributes only to the phase of $\mathcal{J}_\omega$, which in the spectral weights (18.94) is irrelevant. Moreover, since $V < 1$, for non-relativistic particles we can set

$$1 - \mathbf{V}\cdot\mathbf{v} \approx 1.$$

The Fourier coefficients (18.100) then simplify to

$$\mathcal{J}_\omega \approx \frac{-ie\,e^{i\omega\mathbf{V}\cdot\mathbf{y}(0)}}{(2\pi)^2\omega}\int_{-\infty}^{\infty} e^{-i\omega t}\,\mathbf{a}(t)\, dt = -\frac{ie\,e^{i\omega\mathbf{V}\cdot\mathbf{y}(0)}\,\mathbf{a}(\omega)}{(2\pi)^{3/2}\omega}, \tag{18.102}$$

where $\mathbf{a}(\omega)$ is the Fourier transform of the acceleration $\mathbf{a}(t)$. Equation (18.94) then yields for the spectral weights of a non-relativistic particle

$$\frac{d^2\varepsilon}{d\omega d\Omega} = \frac{e^2 V}{8\pi^2}\left(\left|\mathbf{a}(\omega)\right|^2 - V^2\left|\mathbf{n}\cdot\mathbf{a}(\omega)\right|^2\right). \tag{18.103}$$

If we want to analyze these spectral weights in terms of the polarization, from the form of the Fourier coefficients (18.102) and from the decomposition (18.95), we see that for a given frequency ω the transverse radiation is maximum in the plane orthogonal to the acceleration, $\mathbf{n} \perp \mathbf{a}(\omega)$, while the longitudinal one is maximum in the direction $\mathbf{n} \parallel \mathbf{a}(\omega)$. Nevertheless, the total radiation (18.103) has a maximum in the plane orthogonal to the acceleration, as happens for a charged particle in electrodynamics. An exception to this rule occurs for the low frequencies $\omega \approx M$, corresponding to emission speeds $V \ll 1$, which actually form an *isotropic* radiation, although of low intensity since for $\omega \to M$ we have $V \to 0$. Finally, integrating the spectral weights (18.103) over the solid angle we obtain for the total energy emitted per unit frequency interval (see Problem 18.4)

$$\frac{d\varepsilon}{d\omega} = V\left(3 - V^2\right) \frac{e^2 |\mathbf{a}(\omega)|^2}{6\pi} = \sqrt{1 - \frac{M^2}{\omega^2}} \left(1 + \frac{M^2}{2\omega^2}\right) \frac{e^2 |\mathbf{a}(\omega)|^2}{3\pi}, \quad \omega \geq M, \tag{18.104}$$

an equation that generalizes *Larmor's formula* (11.37) for an interaction mediated by massive vector bosons, and reduces to it in the limit for $M \to 0$. More in detail, inserting the Fourier coefficients (18.102) in the polarized spectral weights (18.95), and integrating them over the solid angle, one finds that the total spectral weights (18.104) split up in the transverse and longitudinal contributions

$$\frac{d\varepsilon^\perp}{d\omega} = \sqrt{1 - \frac{M^2}{\omega^2}} \frac{e^2 |\mathbf{a}(\omega)|^2}{3\pi}, \qquad \frac{d\varepsilon^\parallel}{d\omega} = \sqrt{1 - \frac{M^2}{\omega^2}} \frac{M^2}{\omega^2} \frac{e^2 |\mathbf{a}(\omega)|^2}{6\pi},$$

respectively. For low frequencies $\omega \approx M$ both radiations vanish. For $M = 0$ the longitudinal radiation vanishes, while the transverse one reduces to the electromagnetic one, and the same happens also in the high frequency regime $\omega \gg M$.

Let us now suppose that the charged particle generating the radiation is subject to a force with typical time scale T, in which case the function $\mathbf{a}(\omega)$ is appreciably different from zero only for frequencies ω ranging up to about $1/T$. As in the electromagnetic case, we then see that in the non-relativistic limit the particle emits predominantly radiation with frequencies $\omega \lesssim 1/T$. However, since the minimum frequency is M, if the time T is so large that $1/T \ll M$, the intensity of the total radiation is strongly suppressed. Viceversa, if $1/T \gg M$, the dominant frequencies are much larger than the minimum one, $\omega \gg M$, and for these frequencies equation (18.104) reduces to

$$\frac{d\varepsilon}{d\omega} \approx \frac{e^2 |\mathbf{a}(\omega)|^2}{3\pi},$$

which coincides with the electromagnetic intensity (11.37).

Absence of infrared divergences. Integrating equation (18.104) over all frequencies we find for the total emitted energy the finite value

$$\varepsilon_T = \frac{e^2}{3\pi} \int_M^\infty \sqrt{1 - \frac{M^2}{\omega^2}} \left(1 + \frac{M^2}{2\omega^2}\right) |\mathbf{a}(\omega)|^2 \, d\omega. \tag{18.105}$$

In fact, at the lower end $\omega = M$ the integrand vanishes, whereas for large frequencies it behaves as $|\mathbf{a}(\omega)|^2$, and the integral $\int |\mathbf{a}(\omega)|^2\, d\omega = \int |\mathbf{a}(t)|^2\, dt$ is finite. At the quantum level, since radiation of frequency ω is composed of vector bosons of mass $m = \hbar M$ with energy $\hbar\omega$, and since the minimum frequency is M, denoting the total number of emitted particles by n_T it follows that

$$\varepsilon_T \geq n_T m.$$

We obtain thus the crude bound $n_T \leq \varepsilon_T/m$, meaning that the total number of emitted particles is always *finite*. The physical origin of this property is indeed very simple: as the emitted energy is always finite, and the emitted vector bosons have as minimum energy their mass, their number is necessarily finite. In conclusion, for a fundamental interaction mediated by one, ore more, massive vector fields, the so-called *infrared catastrophe* never occurs, a phenomenon that, conversely, inevitably accompanies any electromagnetic radiation process, see Sect. 11.3.1.

18.6.2 Pion Bremsstrahlung

The vector bosons of the weak interactions $W^\pm$ and Z^0 are too heavy and too short-lived, see below, to give rise to a *true* radiation even in high-energy accelerators as the LHC, although their production rate in this pp machine is considerably high. In contrast, the mass of the pions is only about one tenth of the mass of the nucleons and, correspondingly, in ultrarelativistic collisions of heavy ions occurring in high energy accelerators these particles are, actually, produced copiously, giving rise to a bremsstrahlung of *pions* [2, 3]. More recently, the emission of a *neutral* pion bremsstrahlung, i.e. a bremsstrahlung consisting of π^0 particles, from ultrarelativistic protons accelerated by supernova remnants, has been inferred experimentally from the observation of the gamma rays produced by the π^0-decays into two photons [4]. These phenomena represent the nuclear counterpart of the electromagnetic bremsstrahlung caused, in this case, by the acceleration or deceleration of nuclei and nucleons. Actually, a realistic analysis of these phenomena requires the use of quantum theory. In particular, the expression of the spectral weights (18.94) must be adapted to the nuclear interactions, as this formula is valid for *vector* bosons – particles of spin one – while pions are intermediate *scalar* bosons – particles of spin zero. Moreover, nuclei in general have a non-vanishing spin, a feature that cannot be neglected. Nevertheless, it can be seen that the quantum spectral weights of the pion bremsstrahlung have an expression similar to formula (18.94), involving in particular the Fourier coefficients $\mathcal{J}_\omega$ (18.100), see references [5, 6]. However, apart from these qualitative considerations, one should bear in mind that the intermediate bosons of the weak and nuclear interactions, eventually, cannot generate a *true* radiation because, unlike the photons, they are *unstable* particles. In fact, their mean lifetime T is

$$
\mathcal{T} \approx \begin{cases} 3 \cdot 10^{-25}\,\text{s}, & \text{for} \quad W^{\pm} \text{ and } Z^0, \\ 2.6 \cdot 10^{-8}\,\text{s}, & \text{for} \quad \pi^{\pm}, \\ 8 \cdot 10^{-17}\,\text{s}, & \text{for} \quad \pi^0. \end{cases} \tag{18.106}
$$

18.7   Problems

18.1 Derive equation (18.27) by inserting the wave relations (18.25) in formula (18.10).

18.2 Using the continuity equation $\partial_\mu j^\mu = 0$ show that the polarization vector ε^μ defined in (18.83) satisfies the constraint $k_\mu \varepsilon^\mu = 0$.

18.3 Prove equation (18.89) using the relations (18.82), (18.84) and (18.88).

18.4 Derive formula (18.104) for the total energy emitted per unit frequency interval by integrating the spectral weights (18.103) over the angles, using the invariant integrals of Problem 2.6 and the expression for the speed V in (18.21).

18.5 Noticing that, given the Fourier coefficients (18.96), a periodic current admits the Fourier series

$$
j^\mu(x) = \frac{1}{(2\pi)^{3/2}} \sum_{N=-\infty}^{\infty} \int e^{i(N\omega_0 t - \mathbf{k}\cdot\mathbf{x})} J_N^\mu(\mathbf{k})\, d^3 k,
$$

elaborate the details that lead from the general radiation formula (18.85) to the discrete spectral weights (18.97).

References

1. H. Yukawa, On the interaction of elementary particles. Proc. Phys. Math. Soc. Jpn. **17**, 48 (1935)
2. W. Heisenberg, Über die Entstehung von Mesonen in Vielfachprozessen. Z. Physik **126**, 569 (1949)
3. T. Stahl, M. Uhlig, B. Müller, W. Greiner, D. Vasak, Pion and photon bremsstrahlung in a heavy ion reaction model with friction. Z. Phys. A **327**, 311 (1987)
4. M. Ackermann et. al., Detection of the characteristic pion-decay signature in supernova remnants. Science **339**, 807 (2013)
5. D. Vasak, H. Stöcker, B. Müller, W. Greiner, Pion bremsstrahlung and critical phenomena in relativistic nuclear collisions. Phys. Lett. B **93**, 243 (1980)
6. D. Vasak, B. Müller, W. Greiner, Pion radiation from fast heavy ions. Phys. Scripta **22**, 25 (1980)

Chapter 19
Electrodynamics of p-Branes

From a theoretical point of view, classical electrodynamics admits a series of conceptually consistent generalizations. We have already analyzed an important one in Chap. 18 dedicated to massive vector fields, corresponding – at the quantum level – to the possibility of replacing the intermediate vector boson of the interaction, i.e. the photon, with a massive particle. Other generalizations of physical interest are:

- the Bianchi identity can be modified to take into account the presence of *magnetic* charges;
- the charged particles can be replaced with *extended* charges occupying a p-dimensional volume, so-called *p-branes*;
- Maxwell's equations can be formulated in a space-time of *arbitrary* dimensions;
- the electromagnetic field can be subject to a *self-interaction*, a famous example being represented by the *Born-Infeld Lagrangian*, see Ref. [1].

The first above generalization regards the fundamental question of the compatibility of the basic principles of classical and quantum electrodynamics with the existence of magnetic monopoles in nature, and it is thus of prominent theoretical as well as experimental relevance. It will be treated in detail in Chaps. 20 and 21. The other above generalizations – of more speculative character – play a fundamental role in *superstring theory* and in *M-theory*. These theories, living in ten and eleven space-time dimensions, respectively, are candidates to unify the four fundamental interactions of nature, and, in particular, to provide a consistent formulation of *quantum gravity*. Aim of the present chapter is to provide an elementary introduction to the classical electrodynamics of charged p-branes, propagating in a generic D-dimensional Minkowski space-time. According to the terminology of superstring theory, a 0-brane is nothing else than a point-like *particle*, a 1-brane corresponds to a *string*, a 2-brane to a *membrane*, and so.

The first three generalizations mentioned above emerge in a natural way if Maxwell's equations are translated in the language of *differential forms*. This

© Springer International Publishing AG, part of Springer Nature 2018
K. Lechner, *Classical Electrodynamics*, UNITEXT for Physics,
https://doi.org/10.1007/978-3-319-91809-9_19

mathematical formalism, borrowed from *differential geometry*, has a variety of applications in mathematics as well as in theoretical physics. In *algebraic topology*, for instance, it is at the basis of the so-called *de Rham cohomology*, a fundamental instrument for the classification of abstract topological spaces, see e.g. Ref. [2]. In physics, the formalism of differential forms allows to represent many fundamental equations in an *intrinsic* notation avoiding the explicit reference to *indices*, and it provides, in particular, a geometric interpretation for a variety of global as well as local conservation laws. Accordingly, in Sect. 19.1 we provide a *pragmatic* introduction to the formalism of differential forms, appropriate for the above-mentioned generalizations, without entering into the deeper meaning inherent in this formalism in the framework of pure mathematics. As a first application, in Sect. 19.1.2 we rewrite Maxwell's equations in this alternative formalism. Taking inspiration from the *universal* form that these equations assume in the language of differential forms, in Sects. 19.2 and 19.3 we formulate the generalized Maxwell and Lorentz equations governing the electrodynamics of charged p-branes moving in a flat space-time of arbitrary dimensions. The guiding principles for the construction of a consistent dynamics will be again, (i) invariance requirements, and (ii) conservation laws, and the most efficient technical tool to implement them will be – once more – the *variational method*.

19.1   Differential Forms: A Practical Introduction

In this section we present synthetically the basics of the language of differential forms, being interested primarily, as anticipated, in its *practical* properties, rather than in its deeper mathematical meaning.[1] For concreteness, we will present the formalism in a D-dimensional space-time endowed with a Minkowski metric $\eta^{\mu\nu}$, and inverse $\eta_{\mu\nu}$, with signature

$$\mathrm{diag}\,(\eta^{\mu\nu}) = (1, -1, \dots, -1) = \mathrm{diag}\,(\eta_{\mu\nu}). \tag{19.1}$$

The Greek indices μ, ν, ρ etc. take now the values $0, 1, \dots, D-1$. Our space-time has thus one time dimension and $D-1$ spatial dimensions. We will still use the notation $x^\mu = (x^0, x^i)$, where the Latin indices i, j, k etc. take the values $1, \dots, D-1$, and we will again denote the spatial coordinates collectively by the symbol $\mathbf{x} = \{x^i\}$.

Components of a differential form. A differential p-form field, or more simply, a p-form, corresponds to a *completely antisymmetric* tensor field of rank p. Accordingly, its *components* are identified by a tensor field $\Phi_{\mu_1\cdots\mu_p}(x)$ such that

$$\Phi_{\mu_1\mu_2\cdots\mu_p}(x) = -\Phi_{\mu_2\mu_1\cdots\mu_p}(x), \text{ etc.} \tag{19.2}$$

---

[1] For a textbook of *differential geometry* paying particular attention to the language of differential forms in connection to its applications to physics, see, for instance, Ref. [3].

For the moment we will assume that these components are functions on $\mathbb{R}^D$ of class C^∞. The integer p is called the *degree* of the form. A 0-form corresponds thus to a scalar field $\Phi(x)$, a 1-form to a vector field $A^\mu(x)$, a 2-form to a two-index antisymmetric tensor field $B^{\mu\nu}(x)$, and so on. Henceforth, we will omit to indicate explicitly the dependence of the component fields on the space-time coordinate x. Since, due to the antisymmetry property (19.2), the components $\Phi_{\mu_1\cdots\mu_p}$ vanish if two of their indices are equal, in a D-dimensional space-time the highest degree of a form is D. The degree p of a differential form is thus subject to the limitations

$$0 \leq p \leq D.$$

Property (19.2) implies, furthermore, that the number of algebraically *independent* components of a p-form is given by the binomial coefficient

$$\binom{D}{p} = \frac{D!}{p!\,(D-p)!}. \tag{19.3}$$

Correspondingly, for a fixed x the p-forms form a vector space of dimension $\binom{D}{p}$.

Canonical basis. Being a vector space, in the space of p-forms one can introduce a *canonical* basis whose elements are denoted by

$$\{dx^{\mu_p} \wedge dx^{\mu_{p-1}} \wedge \cdots \wedge dx^{\mu_1}\}, \tag{19.4}$$

and which by definition are subject to the algebraic identifications

$$dx^{\mu_p} \wedge dx^{\mu_{p-1}} \wedge \cdots \wedge dx^{\mu_1} = -dx^{\mu_{p-1}} \wedge dx^{\mu_p} \wedge \cdots \wedge dx^{\mu_1}, \ \text{ etc.} \tag{19.5}$$

Notice that thanks to these identifications the number of linearly independent elements of the basis (19.4) is $\binom{D}{p}$, precisely as many as the independent components of the tensor $\Phi_{\mu_1\cdots\mu_p}$. With an appropriate convention on the overall normalization, in *intrinsic* notation a p-form takes thus the form

$$\Phi_p = \frac{1}{p!}\, dx^{\mu_p} \wedge \cdots \wedge dx^{\mu_1}\, \Phi_{\mu_1\cdots\mu_p}. \tag{19.6}$$

Exterior product of differential forms. The *exterior product*, also called *wedge product*, or simply the product, $A_p \wedge B_q$ between a p-form A_p and a q-form B_q is defined as the $(p+q)$-form

$$A_p \wedge B_q = \left(\frac{1}{p!}\, dx^{\mu_p} \wedge \cdots \wedge dx^{\mu_1}\, A_{\mu_1\cdots\mu_p}\right) \wedge \left(\frac{1}{q!}\, dx^{\nu_q} \wedge \cdots \wedge dx^{\nu_1}\, B_{\nu_1\cdots\nu_q}\right)$$

$$\equiv \frac{1}{p!}\frac{1}{q!}\, dx^{\mu_p} \wedge \cdots \wedge dx^{\mu_1} \wedge dx^{\nu_q} \wedge \cdots \wedge dx^{\nu_1}\, A_{[\mu_1\cdots\mu_p}B_{\nu_1\cdots\nu_q]}. \tag{19.7}$$

If $p + q > D$, we set $A_p \wedge B_q = 0$. In terms of components the exterior product between differential forms amounts, thus, to the *completely antisymmetrized product* of the corresponding tensor fields. From the definition (19.7) it follows that this product is *associative*. Moreover, the antisymmetry property (19.5) implies the *graded* commutation relation

$$A_p \wedge B_q = (-)^{pq} B_q \wedge A_p, \qquad (19.8)$$

where with the symbol $(-)^n$, with $n \in \mathbb{N}$, we understand the sign-factor $(-1)^n$. It follows that the square of a generic differential form of *odd* degree Φ_{2p+1} vanishes identically

$$\Phi_{2p+1} \wedge \Phi_{2p+1} = 0.$$

The wedge product $\wedge$ between differential forms is reminiscent of the cross product $\times$ between vectors in three-dimensional space. In fact, the wedge product of the 1-forms $A = dx^i a_i$ and $B = dx^i b_i$, with $i = 1, 2, 3$, amounts to the cross product $\mathbf{a} \times \mathbf{b}$ of the associated vectors:

$$A \wedge B = dx^i \wedge dx^j \, a_{[i} b_{j]} = \frac{1}{2} \, dx^i \wedge dx^j (a_i b_j - a_j b_i) = \frac{1}{2} \, dx^i \wedge dx^j \varepsilon_{ijk} (\mathbf{a} \times \mathbf{b})^k.$$

Once we have introduced the exterior product of differential forms, the elements of the basis (19.4) can be interpreted as multiple exterior products of the elements of the basis $\{dx^\mu\}$ for 1-forms. In fact, the identifications (19.5) descend from the anticommutation relations

$$dx^\mu \wedge dx^\nu = -dx^\nu \wedge dx^\mu, \qquad (19.9)$$

which can be viewed as a particular case of the general commutation relations (19.8), corresponding to $A_1 = dx^\mu$ and $B_1 = dx^\nu$.

Hodge duality. The binomial identity

$$\binom{D}{p} = \binom{D}{D-p}$$

implies that in a D-dimensional space-time the linear vector spaces of p-forms and $(D-p)$-forms have the same dimensions, see formula (19.3). The *Hodge dual*, denoted by the symbol $*$, is a map that realizes an isomorphism between these two vector spaces. By definition, it associates with a generic p-form Φ (19.6) the *dual* $(D-p)$-form

$$* \, \Phi = \frac{1}{(D-p)!} \, dx^{\mu_{D-p}} \wedge \cdots \wedge dx^{\mu_1} \, \widetilde{\Phi}_{\mu_1 \cdots \mu_{D-p}}, \qquad (19.10)$$

whose components are given by

$$\widetilde{\Phi}_{\mu_1\cdots\mu_{D-p}} = \frac{1}{p!}\,\varepsilon_{\mu_1\cdots\mu_{D-p}}{}^{\nu_1\cdots\nu_p}\,\Phi_{\nu_1\cdots\nu_p}. \tag{19.11}$$

In Eq. (19.11) we have introduced the Levi-Civita tensor in a generic D-dimensional space-time:

$$\varepsilon^{\mu_1\cdots\mu_D} = \begin{cases} 1, & \text{if } \mu_1\cdots\mu_D \text{ is an even permutation of } 0,1,\ldots,D-1, \\ -1, & \text{if } \mu_1\cdots\mu_D \text{ is an odd permutation of } 0,1,\ldots,D-1, \\ 0, & \text{if at least two indices are equal.} \end{cases}$$

$$\tag{19.12}$$

The Hodge-duality operation can be iterated and, given the trivial identity $D - (D-p) = p$, if applied twice, it associates to a p-form again a p-form. Thanks to Lorentz invariance we then have necessarily

$$*^2\Phi \propto \Phi,$$

meaning that the square of the operator $*$ is proportional to the identity operator 1. With the normalization chosen for the Hodge dual in Eq. (19.11), it turns out that, actually, the operator $*^2$ reduces to ± 1 depending on the values of p and D. More precisely, if $*^2$ operates on a p-form in a D-dimensional space-time endowed with the Minkowski metric (19.1) it is equal to

$$*^2 = (-1)^{(D+1)(p+1)}. \tag{19.13}$$

To prove this identity one needs the formula for the multiple contractions between two Levi-Civita tensors

$$\varepsilon_{\mu_1\cdots\mu_p\alpha_1\cdots\alpha_{D-p}}\,\varepsilon^{\nu_1\cdots\nu_p\alpha_1\cdots\alpha_{D-p}} = (-)^{D+1}p!\,(D-p)!\,\delta^{\nu_1}_{[\mu_1}\cdots\delta^{\nu_p}_{\mu_p]}, \tag{19.14}$$

generalizing the identities (1.40) and (1.41), see Problem 19.3. Since $*^2 = \pm 1$, the operator $*$ admits always an *inverse*, equal to $\pm*$.

Self-dual forms. In a space-time of *even* dimensions $D = 2N$ the Hodge dual of an N-form is again an N-form. Correspondingly, it makes sense to ask whether there exist N-forms Φ_N which are *(anti-)self-dual*, that is to say, N-forms which satisfy the relation

$$*\,\Phi_N = \pm\Phi_N. \tag{19.15}$$

A necessary condition for the existence of N-forms with this property can be obtained by applying the operator $*$ to Eq. (19.15), and then by enforcing this equation again,

$$*^2\,\Phi_N = \pm*\,\Phi_N = \Phi_N. \tag{19.16}$$

Furthermore, for $D = 2N$ and $p = N$ the identity (19.13) reduces to

$$*^2 = (-)^{N+1}. \tag{19.17}$$

Comparing equations (19.16) and (19.17) we see that a necessary condition for the existence of (anti-)self-dual N-forms in a $2N$-dimensional Minkowski space-time is that N is *odd*. Finally, it is not difficult to show that this condition is also *sufficient*. We therefore conclude that (anti-)self-dual differential forms exist only in space-times of dimensions

$$D = 2, 6, 10, 14, \ldots \tag{19.18}$$

In particular, no such forms exist for $D = 4$. As we will see in Sect. 20.2.1, only in space-times which allow for (anti-)self-dual differential forms, electric and magnetic charges can be identified. It follows that this identification can never occur in our four-dimensional Minkowski space-time, where these charges remain thus necessarily distinct *quantum numbers*.

19.1.1   Exterior Derivative and Poincaré Lemma

The *exterior derivative* d, sometimes also called the *differential*, is defined as the operator that maps a generic p-form Φ (19.6) in the $(p + 1)$-form $d\Phi$ defined by

$$d\Phi = d\left(\frac{1}{p!} \, dx^{\mu_p} \wedge \cdots \wedge dx^{\mu_1} \, \Phi_{\mu_1 \cdots \mu_p} \right) = \frac{1}{p!} \, dx^{\mu_p} \wedge \cdots \wedge dx^{\mu_1} \wedge dx^{\mu} \partial_{[\mu} \Phi_{\mu_1 \cdots \mu_p]}. \tag{19.19}$$

Formally, we have thus $d = dx^{\mu} \partial_{\mu}$. The exterior derivative satisfies the *graded* Leibniz's rule

$$d\left(A_p \wedge B_q \right) = A_p \wedge dB_q + (-)^q dA_p \wedge B_q, \tag{19.20}$$

the proof being left as exercise. A fundamental property of the exterior derivative is that it is a *nilpotent operator* of degree two, that is to say it satisfies the identity

$$d^2 = 0.$$

In fact, from the definition (19.19) we obtain

$$dd\Phi = \frac{1}{p!} \, dx^{\mu_p} \wedge \cdots \wedge dx^{\mu_1} \wedge dx^{\mu} \wedge dx^{\nu} \, \partial_{[\nu} \partial_{\mu} \Phi_{\mu_1 \cdots \mu_p]} = 0, \tag{19.21}$$

where the conclusion stems from the fact that the partial derivatives commute when they operate on functions of class C^∞.

Closed forms and exact forms. A p-form Φ_p is called *closed* if

$$d\Phi_p = 0, \tag{19.22}$$

and it is called *exact* if there exists a $(p-1)$-form Φ_{p-1} such that

$$\Phi_p = d\Phi_{p-1}. \tag{19.23}$$

If a form Φ_p is exact, the form Φ_{p-1}, in turn, is defined modulo the addition of an arbitrary closed $(p-1)$-form. Thanks to the fact that $d^2 = 0$, *every exact form is closed*, but not all closed forms are exact, as signaled by the fundamental *Poincaré lemma*. The statement of this lemma requires the notion of a *contractible set*.

Contractible sets. A subset C of a generic topological space $\mathcal{M}$ is said to be *contractible* to a point $y \in C$, if there exists a *continuous* map F from the direct product space $[0,1] \times C$ in C,

$$F : (\lambda, x) \to F(\lambda, x) \in C, \quad \text{with } 0 \le \lambda \le 1, \ x \in C,$$

such that

$$F(0, x) = x, \quad \forall\, x \in C,$$
$$F(1, x) = y, \quad \forall\, x \in C.$$

The *initial* map $F(0, \cdot)$ is hence the identity map on C, whereas the *final* map $F(1, \cdot)$ is the constant map on C sending all its points x in y. Notice that F must be continuous with respect to both spaces of the direct product $[0,1] \times C$. Qualitatively, a subset C of $\mathcal{M}$ is contractible if it can be deformed continuously, remaining always inside C, until it shrinks to one of its points y without encountering obstacles. For instance, in $\mathcal{M} = \mathbb{R}^2$ a circle, or an annulus, are non-contractible sets (to none of their points), whereas a disk is contractible to its center or to any of its points. Similarly, in $\mathcal{M} = \mathbb{R}^3$ a sphere S^2 is non-contractible, whereas a three-dimensional ball is contractible. Another example of a non-contractible subset of $\mathbb{R}^3$ is the set $C = \mathbb{R}^3 \setminus P$, with P a generic point of $\mathbb{R}^3$, an example that we will resume in Sect. 21.3. In general, all spaces without "topological complications" and without "defects" turn out to be contractible.

Poincaré lemma. *A differential form which is closed in a contractible open set C of $\mathbb{R}^D$ is exact in C.*

In particular, if a form is closed in whole $\mathbb{R}^D$ then it is exact in $\mathbb{R}^D$, since the latter is a contractible space. Frequently, the Poincaré lemma is enounced in the compact form "every closed form is locally exact", meaning that if we restrict the zone where the form is defined to a region so small that it becomes contractible, then within this region the form turns into an exact one. The systematic classification of the closed but not exact forms in a given non-contractible space is the principal purpose of the *de Rham cohomology*, mentioned in the introduction to the chapter. The examples of non-contractible spaces given above suggest that the analysis of the closed but not exact forms in a given space is intimately related with the topological properties of the space itself. In fact, closed, but non-exact, differential forms exist only in spaces endowed with a non-trivial "topology". Actually, in *differential geometry* this important link is realized concretely via a one-to-one map called *Poincaré duality*, see Sect. 21.4.2.

Finally, we mention that the Poincaré lemma represents a generalization of a known theorem of elementary *Mathematical Analysis*, stating that the *form*

$$\Phi_1 = dx f(x, y) + dy\, g(x, y) \tag{19.24}$$

is closed in a simply connected open subset of $\mathbb{R}^2$, if and only if there it is exact, see Problem 19.8. Notice, however, that, whereas in $\mathbb{R}^2$ open sets are simply connected if and only if they are contractible, in a space $\mathbb{R}^D$ with $D > 2$ not all simply connected open sets are contractible. For instance, the set $\mathbb{R}^3 \setminus P$ and the sphere S^2 are simply connected, but non-contractible. The hypothesis of the Poincaré lemma requires, thus, more than the simple connectedness of the set C.

Distribution-valued differential forms. The linear vector space of the *distribution-valued differential p-forms* – sometimes also called *p-currents* – represents an enlargement of the space of differential p-forms. By definition, all $\binom{D}{p}$ component fields $\Phi_{\mu_1\cdots\mu_p}$ of a distribution-valued p-form are elements of $\mathcal{S}'(\mathbb{R}^D)$. Nevertheless, also for distribution-valued p-forms we continue to use the *symbolic* notation

$$\Phi_p = \frac{1}{p!}\, dx^{\mu_p} \wedge \cdots \wedge dx^{\mu_1}\, \Phi_{\mu_1\cdots\mu_p}(x).$$

In this larger space, the exterior product $A_p \wedge B_q$ between two generic distribution-valued differential forms is no longer defined. On the other hand, all *linear* algebraic properties of p-forms introduced so far hold true also for distribution-valued differential forms. In particular, the exterior derivative d introduced in Eq. (19.19) remains well defined, because all distributions of $\mathcal{S}'(\mathbb{R}^D)$ admit partial derivatives belonging again to $\mathcal{S}'(\mathbb{R}^D)$. Moreover, this operator is still nilpotent,

$$d^2 = 0, \tag{19.25}$$

since the distributional partial derivatives ∂_μ and ∂_ν always *commute*, see Eq. (19.21). Closed and exact forms are defined again as in the case of *regular* forms, see the relations (19.22) and (19.23), and, thanks to the nilpotency of the exterior derivative (19.25), every exact form is also closed. By their very nature as distributions, distribution-valued p-forms are defined in whole $\mathbb{R}^D$ – a contractible space – and consequently it is not too surprising that the following fundamental theorem holds, which *identifies* closed forms with exact forms.

Poincaré lemma for distribution-valued differential forms. A *distribution-valued differential p-form is closed, if and only if it is exact.*

The strong point of this theorem is that distribution-valued differential forms in general are *non-differentiable* in the ordinary sense: actually, they can be rather *patho-*

logical objects. For instance, they can exhibit finite or infinite discontinuities, and δ-like behaviors on points, curves, surfaces, or on any higher-dimensional submanifold of $\mathbb{R}^D$.

19.1.2 Maxwell's Equations in the Language of Differential Forms

In this section we rewrite Maxwell's equations in the language of differential forms. More precisely, according to the *intrinsic* distributional nature of these equations, see Sect. 2.3, we will translate them in the space of distribution-valued differential forms. In this new framework we shall reanalyze, in particular, the general solution of the Bianchi identity, and the continuity equation of the four-current. In the language of differential forms, we associate with the antisymmetric electromagnetic tensor $F^{\mu\nu}$ the 2-form

$$F = \frac{1}{2}\, dx^\nu \wedge dx^\mu F_{\mu\nu}, \tag{19.26}$$

and with the four-current j^μ the 1-form

$$j - dx^\mu j_\mu. \tag{19.27}$$

We will now see that in terms of the differential forms F and j, Eqs. (2.21) and (2.22) can be reexpressed as the equations between 3-forms

$$dF = 0, \tag{19.28}$$
$$d * F = *j. \tag{19.29}$$

Bianchi identity and gauge invariance. First of all, we verify that Eq. (19.28) is equivalent to the Bianchi identity (2.21). For this purpose we recall that the latter can be rewritten, in turn, in the alternative form $\partial_{[\rho} F_{\mu\nu]} = 0$, see Eqs. (2.39)–(2.41). To verify the equivalence it is then sufficient to write out explicitly the 3-form

$$dF = d\left(\frac{1}{2}\, dx^\nu \wedge dx^\mu F_{\mu\nu}\right) = \frac{1}{2}\, dx^\nu \wedge dx^\mu \wedge dx^\rho\, \partial_{[\rho} F_{\mu\nu]}.$$

We derive thus the double implication

$$dF = 0 \quad \Leftrightarrow \quad \partial_{[\rho} F_{\mu\nu]} = 0,$$

as we wanted to prove. Next, we address the problem of the general solution of equation (19.28), which requires F to be a *closed* 2-form. Thanks to the Poincaré lemma for distribution-valued differential forms, the two-form F must thus be *exact*. Therefore, there exists a 1-form $A = dx^\nu A_\nu$ – a vector potential – such that

$$F = dA. \tag{19.30}$$

Writing out the differential of A, from the definition (19.26) we find

$$F = \frac{1}{2} \, dx^\nu \wedge dx^\mu F_{\mu\nu} = d \left(dx^\nu A_\nu \right) = dx^\nu \wedge dx^\mu \, \partial_{[\mu} A_{\nu]}.$$

Equating the component fields we then find

$$F_{\mu\nu} = 2 \, \partial_{[\mu} A_{\nu]} = \partial_\mu A_\nu - \partial_\nu A_\mu, \tag{19.31}$$

in agreement with the general solution (2.42). On the other hand, it is easy to realize that there exist infinitely many different 1-forms A such $F = dA$. In fact, given two arbitrary vector potentials A and A' we have

$$F = dA = dA' \quad \Rightarrow \quad d \left(A' - A \right) = 0.$$

The 1-form $A' - A$ must therefore be closed, and so – again thanks to the Poincaré lemma – it must be exact. This means that there exists a 0-form, i.e. a scalar field Λ, such that $A' - A = d\Lambda$, or

$$A' = A + d\Lambda. \tag{19.32}$$

Writing out the differential we find

$$dx^\mu A'_\mu = dx^\mu A_\mu + dx^\mu \, \partial_\mu \Lambda \quad \Leftrightarrow \quad A'_\mu = A_\mu + \partial_\mu \Lambda.$$

We see thus that two solutions A and A' of the Bianchi identity (19.28) differ by the *gauge transformation* (19.32). Vice versa, it is obvious that if A satisfies equation (19.30), then this equation is satisfied also by the 1-form $A + d\Lambda$. We have, thus, retrieved the results of Sect. 2.2.4. In particular, in the language of differential forms the relations (2.46) simply read

$$dF = 0 \quad \Leftrightarrow \quad F = dA, \quad \text{with } A \approx A + d\Lambda. \tag{19.33}$$

Maxwell's equation. We now prove the equivalence between Eqs. (2.22) and (19.29). We begin by noticing that the latter is an equation between 3-forms. In fact, the form $*j$, being the Hodge dual of a 1-form, is a 3-form. Similarly, since $*F$ – the Hodge dual of a 2-form – is again a 2-form, also $d * F$ is a 3-form. Consequently, we can take advantage from the properties of Hodge duality – an invertible map – to rewrite equation (19.29) as an equation between 1-forms. Applying the $*$ operator to both members of this equation, and noticing that the identity (19.13) for $D = 4$ and $p = 1$ yields $*^2 = 1$, in this way we obtain the equivalent equation between 1-forms

$$* \, d * F = *^2 j = j. \tag{19.34}$$

We now write out explicitly the first member of this equation, applying the rules derived in the previous sections,

$$
\begin{aligned}
* d * F &= * d \left(\frac{1}{2} dx^\nu \wedge dx^\mu \widetilde{F}_{\mu\nu} \right) = * \left(\frac{1}{2} dx^\nu \wedge dx^\mu \wedge dx^\rho \partial_{[\rho} \widetilde{F}_{\mu\nu]} \right) \\
&= * \left(\frac{1}{3!} dx^\nu \wedge dx^\mu \wedge dx^\rho \left(3 \partial_{[\rho} \widetilde{F}_{\mu\nu]} \right) \right) = dx^\mu \left(\frac{1}{3!} \varepsilon_{\mu\nu_1\nu_2\nu_3} \left(3 \partial^{\nu_1} \widetilde{F}^{\nu_2\nu_3} \right) \right) \\
&= dx^\mu \left(\frac{1}{3!} \varepsilon_{\mu\nu_1\nu_2\nu_3} \, 3 \partial^{\nu_1} \frac{1}{2!} \varepsilon^{\nu_2\nu_3\alpha_1\alpha_2} F_{\alpha_1\alpha_2} \right) \\
&= dx^\mu \left(\frac{1}{4} \varepsilon_{\mu\nu_1\nu_2\nu_3} \varepsilon^{\nu_2\nu_3\alpha_1\alpha_2} \partial^{\nu_1} F_{\alpha_1\alpha_2} \right) \\
&= dx^\mu \left(\frac{1}{4} (-) 2! \, 2! \, \delta_\mu^{\alpha_1} \delta_{\nu_1}^{\alpha_2} \partial^{\nu_1} F_{\alpha_1\alpha_2} \right) = dx^\mu (\partial^\nu F_{\nu\mu}).
\end{aligned}
$$
$$(19.35)$$

In the second member of the second line we have used the identity (1.37), and in the fourth line we have applied the identity for the Levi-Civita tensor (19.14) with $D = 4$ and $p = 2$. Equation (19.34) takes thus the form

$$
dx^\mu (\partial^\nu F_{\nu\mu}) = dx^\mu j_\mu,
$$

which amounts to Maxwell's equation (2.22).

Current conservation. The first member of equation (19.29) is a closed form. Therefore, for consistency, also its second member must be so:

$$
d * j = 0. \tag{19.36}
$$

To investigate the content of this constraint, which imposes the vanishing of a 4-form, it is again more convenient to consider its Hodge dual $* d * j = 0$. The latter amounts indeed to the vanishing of a 0-form, i.e. of a scalar field. With steps similar to those performed above to derive equation (19.35), this time we find

$$
\begin{aligned}
* d * j &= * d \left(\frac{1}{3!} dx^\rho \wedge dx^\nu \wedge dx^\mu \widetilde{j}_{\mu\nu\rho} \right) = * \left(\frac{1}{3!} dx^\rho \wedge dx^\nu \wedge dx^\mu \wedge dx^\sigma \partial_{[\sigma} \widetilde{j}_{\mu\nu\rho]} \right) \\
&= * \left(\frac{1}{4!} dx^\rho \wedge dx^\nu \wedge dx^\mu \wedge dx^\sigma \left(4 \partial_{[\sigma} \widetilde{j}_{\mu\nu\rho]} \right) \right) = \frac{1}{4!} \varepsilon_{\nu_1\nu_2\nu_3\nu_4} \left(4 \partial^{\nu_1} \widetilde{j}^{\nu_2\nu_3\nu_4} \right) \\
&= \frac{1}{4!} \varepsilon_{\nu_1\nu_2\nu_3\nu_4} \, 4 \partial^{\nu_1} \varepsilon^{\nu_2\nu_3\nu_4\alpha} j_\alpha = \frac{1}{3!} \varepsilon_{\nu_1\nu_2\nu_3\nu_4} \varepsilon^{\nu_2\nu_3\nu_4\alpha} \partial^{\nu_1} j_\alpha \\
&= \frac{1}{3!} 1! \, 3! \, \delta_{\nu_1}^\alpha \partial^{\nu_1} j_\alpha = \partial_\alpha j^\alpha.
\end{aligned}
$$
$$(19.37)$$

Equation (19.36) is thus equivalent to the continuity equation for the four-current

$$
d * j = 0 \quad \Leftrightarrow \quad \partial_\mu j^\mu = 0.
$$

Notice also that if we apply to the identity (19.37) again the $*$ operator, and use that the relation (19.13) for $D = 4$ and $p = 4$ gives $*^2 = -1$, we find the further identity between 4-forms

$$d * j = -\frac{1}{4!} \, dx^{\mu_1} \wedge dx^{\mu_2} \wedge dx^{\mu_3} \wedge dx^{\mu_4} \, \varepsilon_{\mu_4 \mu_3 \mu_2 \mu_1} \partial_\mu j^\mu.$$

From the examples worked out above we see that, on the one hand, the formalism of differential forms is *advantageous*, as it allows to state tensorial equations without writing out explicitly the indices, and, on the other, if applied in the space of distributions, it allows to define the components of the involved p-forms *globally*, i.e. in whole $\mathbb{R}^4$, even in the presence of potential singularities. In addition, in this wider framework, the Poincaré lemma abolishes the distinction between closed and exact forms, a property that will significantly simplify the analysis of electromagnetic fields with singularities extending over curves or surfaces, see Sect. 21.4. Conversely, the main *drawback* of the formalism of differential forms is that it does not apply to tensors which are not completely antisymmetric, as for instance the *metric* tensor $g_{\mu\nu}(x)$ of General Relativity.

19.2   Maxwell's Equations for Charged p-Branes

The simplest non-trivial extended charge is represented by a threadlike distribution of charge – a charged *string* – corresponding to a 1-brane. Whereas a particle during its time evolution describes a boundaryless world line, a string moving throughout space describes a two-dimensional surface. If the string is *closed* on itself, forming thus a closed curve, and if we consider the time interval $-\infty < t < \infty$, also the surface swept out by the string is boundaryless. Similarly, a 2-brane is a two-dimensional surface which during its time evolution describes a three-dimensional world volume, and so on. We will limit our analysis to *closed* p-branes, which will then draw $(p + 1)$-dimensional *boundaryless* world volumes, leaving them move in a Minkowski space-time of arbitrary dimension D. In this way, the integer p can also assume values greater than two, being subject only to the limitations $0 \le p \le D - 1$. Before addressing the dynamics of a p-brane, as intermediate step we generalize the electrodynamics of charged particles to a space-time of arbitrary dimension D.

19.2.1   *Electrodynamics in D Dimensions*

In analogy to the four-dimensional case we endow the space-time $\mathbb{R}^D$ with the Minkowski metric $\eta^{\mu\nu}$, for the notations see Sect. 19.1. We then postulate that the laws of physics are invariant under the D-dimensional Lorentz group

$$O(1, D-1) = \{\text{real } D \times D \text{ matrices } \Lambda \, / \, \Lambda^T \eta \Lambda = \eta\}.$$

The Lorentz matrices $\Lambda^\mu{}_\nu$ obey thus still the constraint $\Lambda^\alpha{}_\mu \Lambda^\beta{}_\nu \eta_{\alpha\beta} = \eta_{\mu\nu}$. Correspondingly, we assume that under a Poincaré transformation the coordinates transform according to the law $x^\mu \to x'^\mu = \Lambda^\mu{}_\nu x^\nu + a^\mu$, and that the space-time tensors transform as in the four-dimensional case, see the law (1.28). In such a space-time the electromagnetic field is still represented by a two-index antisymmetric tensor $F^{\mu\nu}$, and Maxwell's equations are still given by

$$\partial_{[\mu} F_{\nu\rho]} = 0, \qquad \partial_\mu F^{\mu\nu} = j^\nu. \tag{19.38}$$

In a space-time of dimension $D > 4$ the electric field $\mathbf{E}$ is still a spatial *vector*, $E^i = F^{i0}$, while the magnetic field is represented by the spatial antisymmetric tensor $B^{ij} = F^{ij}$. It is thus no longer equivalent to a spatial vector $\mathbf{B}$. A further, rather obvious, difference with respect to a four-dimensional space-time appears in the definition of the D-current j^μ, appearing in Eqs. (19.38). In fact, denoting the world line of the particle by $y^\mu(\lambda) = (y^0(\lambda), \mathbf{y}(\lambda))$, and its velocity by $\mathbf{v}(t) = d\mathbf{y}(t)/dt$, Lorentz invariance and world-line reparameterization invariance enforce the choice

$$j^\mu(x) = e \int \frac{dy^\mu}{d\lambda}\, \delta^D\big(x - y(\lambda)\big)\, d\lambda = e\,(1, \mathbf{v}(t))\, \delta^{D-1}\big(\mathbf{x} - \mathbf{y}(t)\big). \tag{19.39}$$

This expression still satisfies identically the conservation law $\partial_\mu j^\mu = 0$. Introducing the 2-form and the 1-form

$$F = \frac{1}{2}\, dx^\nu \wedge dx^\mu F_{\mu\nu}, \qquad j = dx^\mu j_\mu,$$

with a calculation analogous to (19.35) one finds that in the language of differential forms equation (19.38) are still of the form (19.28), (19.29), apart from a minus sign when D is odd

$$dF = 0, \qquad d * F = (-)^D * j. \tag{19.40}$$

The general solution of the Bianchi identity $dF = 0$ is still $F = dA$, or $F_{\mu\nu} = \partial_\mu A_\nu - \partial_\nu A_\mu$. Finally, the Lorentz equation reads again

$$\frac{dp^\mu}{ds} = eF^{\mu\nu} u_\nu, \qquad u^\mu = \frac{dy^\mu}{ds}, \qquad p^\mu = mu^\mu, \tag{19.41}$$

where the D-dimensional proper time is given by

$$ds = \sqrt{\frac{dy^\mu}{d\lambda} \frac{dy^\nu}{d\lambda}\, \eta_{\mu\nu}}\, d\lambda = \sqrt{1 - v^2}\, dt.$$

Writing out explicitly the space and time components of the Lorentz equation in (19.41), we now find the component equations ($\varepsilon = p^0$)

$$\frac{dp^i}{dt} = e\big(E^i - B^{ij}v^j\big), \qquad \frac{d\varepsilon}{dt} = e\mathbf{v} \cdot \mathbf{E}.$$

19.2.2   World Volume and Reparameterization Invariance

As for the case of a relativistic particle, starting point for the setup of the relativistic dynamics of a p-brane is its relativistic *kinematics*. As in the case of the particle, the invariances to be implemented are *Lorentz invariance* – an easy task if we resort to the tensor formalism – and *reparameterization invariance*. A point-like particle has no spatial extension and so in a D-dimensional space-time its *configuration* at a given instant is described by its $D - 1$ spatial position coordinates $y^i = (y^1, \ldots, y^{D-1})$. Similarly, the spatial *profile* of a p-extended object at a given instant is described by the $D - 1$ spatial coordinate functions

$$y^i(\boldsymbol{\lambda}) = (y^1(\boldsymbol{\lambda}), \ldots, y^{D-1}(\boldsymbol{\lambda})) \tag{19.42}$$

of the *p parameters* $(\lambda^1, \ldots, \lambda^p)$, that we denote collectively with the symbol $\boldsymbol{\lambda}$.

The sphere as a 2-brane. By way of example, we construct the profile of a two-dimensional sphere with unit radius embedded in a four-dimensional space-time, which constitutes a 2-brane. In this case we can consider as parameters the polar angles $\lambda^1 = \vartheta$ and $\lambda^2 = \varphi$, so that the parametric equation (19.42) takes the known form

$$y^1(\boldsymbol{\lambda}) = \sin \lambda^1 \cos \lambda^2, \tag{19.43}$$
$$y^2(\boldsymbol{\lambda}) = \sin \lambda^1 \sin \lambda^2, \tag{19.44}$$
$$y^3(\boldsymbol{\lambda}) = \cos \lambda^1. \tag{19.45}$$

Obviously, the profile of a sphere can be represented in infinitely many different ways, according to the choice of the parameters. For instance, using as parameters the Cartesian coordinates of the xy plane, namely $\lambda'^1 = x$ and $\lambda'^2 = y$, one obtains the alternative representation (for the upper half-sphere)

$$y'^1(\boldsymbol{\lambda}') = \lambda'^1, \tag{19.46}$$
$$y'^2(\boldsymbol{\lambda}') = \lambda'^2, \tag{19.47}$$
$$y'^3(\boldsymbol{\lambda}') = \sqrt{1 - (\lambda'^1)^2 - (\lambda'^2)^2}. \tag{19.48}$$

Since we must have $y'^i(\boldsymbol{\lambda}') = y^i(\boldsymbol{\lambda})$, the two above representations of the sphere are related by a locally invertible map between the two sets of parameters, namely by the *reparameterization*

$$\lambda'^1(\boldsymbol{\lambda}) = \sin \lambda^1 \cos \lambda^2, \qquad \lambda'^2(\boldsymbol{\lambda}) = \sin \lambda^1 \sin \lambda^2. \tag{19.49}$$

Returning to a generic p-brane, two representations $y^i(\boldsymbol{\lambda})$ and $y'^i(\boldsymbol{\lambda}')$ describe the same profile, if there exist p invertible functions $\boldsymbol{\lambda}'(\boldsymbol{\lambda})$ of the p parameters $\boldsymbol{\lambda}$ such that

$$y'^i(\boldsymbol{\lambda}') = y'^i(\boldsymbol{\lambda}'(\boldsymbol{\lambda})) = y^i(\boldsymbol{\lambda}). \tag{19.50}$$

We refer to the identification (19.50) by saying that the spatial profile $y^i(\boldsymbol{\lambda})$ of a brane is *reparameterization invariant*. Since all physical *observables* referring to a p-brane must be insensitive to the particular way we parameterize its profile, they must, hence, be invariant under an arbitrary reparameterization. Reparameterization invariance will, thus, be one of our guiding principles for the generalization of Maxwell's equations to p-branes.

Dynamical branes and world volume. Once we have specified the configuration of a p-brane at a given instant, we can now address its dynamics. At every instant t the brane will have a different profile, and its motion will thus be described by the $D - 1$ functions of $p + 1$ variables

$$y^i(t, \boldsymbol{\lambda}). \tag{19.51}$$

These functions generalize the time-dependent position $y^i(t)$ of a point-particle. As in that case, see Sect. 2.1, to give the time-dependent profile (19.51) a Lorentz-invariant form we add the time coordinate t, ending up with the space-time profile

$$y^\mu(t, \boldsymbol{\lambda}) = \big(t, y^i(t, \boldsymbol{\lambda})\big). \tag{19.52}$$

Finally, again in analogy to the particle, to recast the profile (19.52) in a *manifestly* Lorentz-invariant form we describe its time evolution not by means of the coordinate time t, but rather by means of an arbitrary parameter $\lambda^0(t)$. More in general, since we can choose a different function $\lambda^0(t)$ for any point $\boldsymbol{\lambda}$ of the brane, it is convenient to introduce an evolution parameter of the form

$$\lambda^0(t, \boldsymbol{\lambda}) \quad \leftrightarrow \quad t(\lambda^0, \boldsymbol{\lambda}) \equiv y^0(\lambda^0, \boldsymbol{\lambda}). \tag{19.53}$$

In this way, the time evolution of the spatial profile (19.51) can be recast in the equivalent form $y^i(\lambda^0, \boldsymbol{\lambda})$, so that the profile (19.52) turns into the Lorentz-covariant space-time profile of the brane

$$y^\mu(\lambda) = (y^0(\lambda), y^i(\lambda)), \qquad \lambda \equiv (\lambda^0, \boldsymbol{\lambda}). \tag{19.54}$$

In conclusion, the time evolution of a p-brane is represented in a manifestly Lorentz invariant way by the D functions $y^\mu(\lambda)$ of the $p + 1$ parameters λ. These parameters describe the $(p + 1)$-dimensional *world volume* swept out by the brane during its motion, generalizing the world line swept out by a moving point-particle. Henceforth, we assume that the functions $y^\mu(\lambda)$ are sufficiently regular and, in particular, twice differentiable with respect to the parameters λ. As anticipated in

(19.54), we denote the $p+1$ parameters $\boldsymbol{\lambda}$ and λ^0 collectively with the symbol λ. Furthermore, we shall use Latin indices a, b, c etc., assuming the values $0, 1, \ldots, p$, to characterize the single parameters $\{\lambda^a\} = \lambda$, and we will denote the *dimension* of the world volume by the integer

$$n = p + 1.$$

Despite the index notation, in the n-dimensional *internal* space of the brane we do not introduce a Minkowski metric, although we keep our convention that the sum over repeated indices is always understood. In particular, the indices a, b, c etc. will never be raised or lowered.

Reparameterization invariance. By construction, the world volume described by the D functions $y^\mu(\lambda)$ in (19.54) is subject to the reparameterization invariance of *all* n parameters λ. In fact, two world volumes $y^\mu(\lambda)$ and $y'^\mu(\lambda')$ descending one from the other through an invertible and regular transformation of the n parameters $\lambda \to \lambda'(\lambda)$, namely such that

$$y'^\mu(\lambda') = y^\mu(\lambda), \tag{19.55}$$

are physically indistinguishable. Finally, it is easy to reconstruct the time evolution of the spatial profile (19.51) from the covariant parameterization of the world volume (19.54). In fact, it is sufficient to invert the temporal component $y^0(\lambda)$ of the latter to express λ^0 as a function of t and $\boldsymbol{\lambda}$,

$$y^0(\lambda^0, \boldsymbol{\lambda}) = t \quad \to \quad \lambda^0(t, \boldsymbol{\lambda}), \tag{19.56}$$

and to insert it in the spatial components of the world volume (19.54)

$$y^i(\lambda^0, \boldsymbol{\lambda}) \quad \to \quad y^i(\lambda^0(t, \boldsymbol{\lambda}), \boldsymbol{\lambda}) \equiv y^i(t, \boldsymbol{\lambda}). \tag{19.57}$$

19.2.3 Electric Current

To couple a p-brane to an electromagnetic field, in the first place we must identify a conserved electric *current* associated with the brane, generalizing the current of a point-particle

$$j^\mu(x) = e \int \frac{dy^\mu}{d\lambda} \, \delta^D(x - y(\lambda)) \, d\lambda. \tag{19.58}$$

Unfortunately, this expression does not admit a *natural* generalization, since the role of the covariant velocity $dy^\mu/d\lambda$ is now played by a set of generalized velocities, namely by the *tangent vectors*

$$U_a^\mu = \frac{\partial y^\mu}{\partial \lambda^a}. \tag{19.59}$$

The problem is that the current associated with a brane must be *invariant* under reparameterizations, and so it cannot carry any *internal* index a. To determine this current we impose the following conditions, of obvious meaning:

(1) the current must be a *tensor* field $j^{\mu\nu\cdots}(x)$;

(2) $j^{\mu\nu\cdots}(x)$ can be different from zero only on the world volume of the brane;

(3) $j^{\mu\nu\cdots}(x)$ must be reparameterization invariant;

(4) the current must satisfy identically the continuity equation

$$\partial_\mu j^{\mu\nu\cdots}(x) = 0. \tag{19.60}$$

Property (2) – together with the Pincaré-invariance requirement (1) – implies that the current must involve the Dirac δ-function $\delta^D(x - y(\lambda))$. Correspondingly, the linear measure $d\lambda$ appearing in the current (19.58) must be replaced with the n-dimensional measure on the world volume $d^n\lambda$. From the form of the current (19.58) we infer, furthermore, that the integrand should involve the generalized velocities U_a^μ. Condition (3), i.e. reparameterization invariance, then forces these velocities to appear in the integrand via a *homogeneous* function $\mathcal{P}(U)$ of degree n, as we will see in a moment. In summary, the current must be of the form

$$\int \mathcal{P}(U)\, \delta^D(x - y(\lambda))\, d^n\lambda. \tag{19.61}$$

To show that $\mathcal{P}(U)$ must be a homogeneous function of degree n we impose the invariance of the integral (19.61) under the particular reparameterization that rescales all parameters by a constant k: $\lambda^a \to \lambda'^a = \lambda^a/k$. For this case we have

$$d^n\lambda' = \frac{1}{k^n}\, d^n\lambda, \qquad U_a'^\mu = \frac{\partial y^\mu}{\partial \lambda'^a} = k\frac{\partial y^\mu}{\partial \lambda^a} = kU_a^\mu.$$

Since under a generic reparameterization we have $y'(\lambda') = y(\lambda)$, see (19.55), from the invariance condition

$$\int \mathcal{P}(U)\, \delta^D(x - y(\lambda))\, d^n\lambda = \int \mathcal{P}(U')\, \delta^D(x - y'(\lambda'))\, d^n\lambda'$$

$$= \frac{1}{k^n} \int \mathcal{P}(kU)\, \delta^D(x - y(\lambda))\, d^n\lambda$$

we derive the sought homogeneity relation

$$\mathcal{P}(kU) = k^n\, \mathcal{P}(U). \tag{19.62}$$

However, for a generic homogeneous function $\mathcal{P}(U)$ of degree n the integral (19.61) is not invariant under an *arbitrary* reparameterization. On the other hand, from the form of the point-particle current (19.58) we can argue that $\mathcal{P}(U)$ must be a *polynomial* in U_a^μ, in which case the relation (19.62) requires this polynomial to be of degree n. Actually, below we show that the unique polynomial of degree n which ensures full reparameterization invariance of the integral (19.61) is the one corresponding to the completely *antisymmetric tensorial current* of rank n

$$j^{\mu_1\cdots\mu_n}(x) = e \int U_{a_1}^{\mu_1} \cdots U_{a_n}^{\mu_n} \varepsilon^{a_1\cdots a_n} \delta^D(x - y(\lambda))\, d^n\lambda \tag{19.63}$$

$$= n!\, e \int U_0^{[\mu_1} \cdots U_p^{\mu_n]} \delta^D(x - y(\lambda))\, d^n\lambda. \tag{19.64}$$

In formula (19.63) $\varepsilon^{a_1\cdots a_n}$ is the Levita-Civita tensor, see the definition (19.12), and the variant (19.64) can be derived from (19.63) using the identity

$$U_{a_1}^{\mu_1} \cdots U_{a_n}^{\mu_n} \varepsilon^{a_1\cdots a_n} = U_{a_1}^{[\mu_1} \cdots U_{a_n}^{\mu_n]} \varepsilon^{a_1\cdots a_n} = n!\, U_0^{[\mu_1} \cdots U_p^{\mu_n]}.$$

For $n = p + 1 = 1$ the current (19.64) reduces to the point-particle current (19.58). However, for $n > 1$, the constant e appearing in front of the integral (19.63) represents the *charge per unit volume* of the brane, rather than the charge, see Problem 19.9. Notice that, as in the case of the particle, the expression (19.63) is actually not a *function*, but rather a *distribution* in $\mathcal{S}'(\mathbb{R}^D)$.

Reparameterization invariance. By construction, the expression (19.63) satisfies the requirements (1) and (2). We are thus left with the proofs of properties (3) and (4). To verify property (3) we perform in the integral (19.63) a generic reparameterization $\lambda^a \to \lambda'^a(\lambda)$, leading to the transformed current

$$j'^{\mu_1\cdots\mu_n}(x) = e \int U_{a_1}'^{\mu_1} \cdots U_{a_n}'^{\mu_n} \varepsilon^{a_1\cdots a_n} \delta^D(x - y'(\lambda'))\, d^n\lambda', \tag{19.65}$$

where $U_a'^\mu = \partial y'^\mu(\lambda')/\partial\lambda'^a$. Introducing the Jacobian matrix of the reparameterization $K_a{}^b = \partial\lambda^b/\partial\lambda'^a$, we derive the transformation laws[2]

$$y'^\mu(\lambda') = y^\mu(\lambda), \quad U_a'^\mu = \frac{\partial\lambda^b}{\partial\lambda'^a} \frac{\partial y^\mu(\lambda)}{\partial\lambda^b} = K_a{}^b U_b^\mu, \quad d^n\lambda' = \frac{d^n\lambda}{\det K}. \tag{19.66}$$

Inserting them in the transformed current (19.65) we find

$$j'^{\mu_1\cdots\mu_n}(x) = e \int \frac{1}{\det K} U_{b_1}^{\mu_1} \cdots U_{b_n}^{\mu_n} K_{a_1}{}^{b_1} \cdots K_{a_n}{}^{b_n} \varepsilon^{a_1\cdots a_n} \delta^D(x - y(\lambda))\, d^n\lambda. \tag{19.67}$$

Thanks to the *determinant identity*, holding for an arbitrary $n \times n$ matrix $K_a{}^b$,

$$K_{a_1}{}^{b_1} \cdots K_{a_n}{}^{b_n} \varepsilon^{a_1\cdots a_n} = (\det K)\, \varepsilon^{b_1\cdots b_n}, \tag{19.68}$$

---

[2] For simplicity we assume that the reparameterization $\lambda \to \lambda'(\lambda)$ preserves the *orientation* of the brane, in which case by definition we have $\det K > 0$. In fact, in general the measure would transform as $d^n\lambda' = d^n\lambda/|\det K|$.

Equation (19.67) then reduces to $j'^{\mu_1 \cdots \mu_n}(x) = j^{\mu_1 \cdots \mu_n}(x)$, as we wanted to show. From this proof we also conclude that the polynomial $\mathcal{P}(U)$ corresponding to the choice (19.63) is the unique polynomial of degree n ensuring the reparameterization invariance of the *ansatz* (19.61).

Conservation. To verify property (4) we need to evaluate the D-divergence of the current (19.63) in the distributional sense. For this purpose we proceed in complete analogy to Eqs. (2.103) and (2.105), applying the D-divergence of the current to a generic test function $\varphi(x) \in \mathcal{S}(\mathbb{R}^D)$. Since the current (19.63) when applied to a test function yields

$$j^{\mu_1 \cdots \mu_n}(\varphi) = e \int U_{a_1}^{\mu_1} \cdots U_{a_n}^{\mu_n} \varepsilon^{a_1 \cdots a_n} \varphi(y(\lambda)) \, d^n\lambda,$$

for its D-divergence applied to a test function we obtain

$$(\partial_{\mu_1} j^{\mu_1 \cdots \mu_n})(\varphi) = -j^{\mu_1 \cdots \mu_n}(\partial_{\mu_1}\varphi) = -e \int U_{a_1}^{\mu_1} \cdots U_{a_n}^{\mu_n} \varepsilon^{a_1 \cdots a_n} \partial_{\mu_1}\varphi(y(\lambda)) \, d^n\lambda$$

$$= -e \int U_{a_2}^{\mu_2} \cdots U_{a_n}^{\mu_n} \varepsilon^{a_1 \cdots a_n} \frac{\partial\varphi(y(\lambda))}{\partial\lambda^{a_1}} \, d^n\lambda,$$

where for the tangent vector $U_{a_1}^{\mu_1}$ we have used its definition (19.59). Performing an integration by parts we then find

$$\partial_{\mu_1} j^{\mu_1 \cdots \mu_n}(\varphi) = -e \int \frac{\partial}{\partial\lambda^{a_1}} \left[U_{a_2}^{\mu_2} \cdots U_{a_n}^{\mu_n} \varepsilon^{a_1 \cdots a_n} \varphi(y(\lambda)) \right] d^n\lambda$$

$$+ e \int \left(\frac{\partial U_{a_2}^{\mu_2}}{\partial\lambda^{a_1}} \cdots U_{a_n}^{\mu_n} + \cdots + U_{a_2}^{\mu_2} \cdots \frac{\partial U_{a_n}^{\mu_n}}{\partial\lambda^{a_1}} \right) \varepsilon^{a_1 \cdots a_n} \varphi(y(\lambda)) \, d^n\lambda.$$

The terms in the second row all vanish, since the factors

$$\frac{\partial U_a^\mu}{\partial\lambda^b} = \frac{\partial^2 y^\mu}{\partial\lambda^a \partial\lambda^b} \tag{19.69}$$

are symmetric in a and b, while the Levi-Civita tensor is antisymmetric. Moreover, as the integrand of the first row is an n-divergence, thanks to Gauss's theorem in n dimensions, also this term is zero. Indeed, on the one hand, the *spatial* world volume of the brane parameterized by λ is boundaryless, and on the other, along the non-compact coordinate λ^0 the test function $\varphi(y(\lambda)) = \varphi(y^0(\lambda), \mathbf{y}(\lambda))$ goes rapidly to zero for $\lambda^0 \to \pm\infty$. In fact, for $\lambda^0 \to \pm\infty$ we have $y^0(\lambda) \to \pm\infty$. Since the functionals $(\partial_{\mu_1} j^{\mu_1 \cdots \mu_n})(\varphi)$ vanish for all $\varphi \in \mathcal{S}(\mathbb{R}^D)$, we have thus proven that $\partial_{\mu_1} j^{\mu_1 \cdots \mu_n} = 0$ in $\mathcal{S}'(\mathbb{R}^D)$.

19.2.4   Maxwell's Equations

Once we know the current associated with a p-brane we can ask which electromagnetic field it generates, or more concretely, which are the *generalized Maxwell's equations* tieing the field to the current. As we shall see, in the formalism of differential forms the sought generalization emerges quit naturally. We being by generalizing the current 1-form of a particle (19.27). As the expression (19.63) is a completely antisymmetric tensor of rank n, it is natural to introduce the associated n-form

$$j_n = \frac{1}{n!}\, dx^{\mu_n} \wedge \cdots \wedge dx^{\mu_1}\, j_{\mu_1\cdots\mu_n}. \tag{19.70}$$

Performing the same steps as in (19.37) one easily verifies that in the language of differential forms the continuity equation (19.60) can be recast in the equivalent form

$$\partial_{\mu_1} j^{\mu_1\cdots\mu_n} = 0 \quad \Leftrightarrow \quad d * j_n = 0. \tag{19.71}$$

The Hodge dual $*j_n$ of the n-form j_n is thus a closed $(D-n)$-form. It is actually not necessary to prove the double implication (19.71) explicitly, in that it is implied by Lorentz invariance. In fact, thanks to the identity (19.13) we have the double implication

$$d * j_n = 0 \quad \Leftrightarrow \quad * d * j_n = 0.$$

On the other hand, $* d * j_n$ is an $(n-1)$-form, corresponding hence to a completely antisymmetric tensor of rank $n-1$, linear in $\partial_\nu j^{\mu_1\cdots\mu_n}$. Due to Lorentz invariance, this tensor is thus necessarily proportional to the unique non-vanishing contracted tensor of rank $n-1$ one can form, i.e. $\partial_{\mu_1} j^{\mu_1\cdots\mu_n}$. This implies the relation (19.71).

Generalized Maxwell tensor. Since $*j_n$ is a closed $(D-n)$-form, if we want to preserve the structure of Maxwell's equations for point-particles (19.28) and (19.29), the generalized electromagnetic field strength produced by a p-brane must be represented by a *completely antisymmetric* tensor field of rank $n+1 = p+2$

$$F_{\mu_1\cdots\mu_{n+1}}(x). \tag{19.72}$$

In fact, with such a tensor we can associate the $(n+1)$-form

$$F_{n+1} = \frac{1}{(n+1)!}\, dx^{\mu_{n+1}} \wedge \cdots \wedge dx^{\mu_1}\, F_{\mu_1\cdots\mu_{n+1}}, \tag{19.73}$$

and then we can postulate the generalized set of Maxwell's equations[3]

---

[3]The sign in front of the current $*j_n$ in Eq. (19.75) is purely conventional. Our choice has the advantage of leading, in tensor notation, to the simple equation (19.77).

$$dF_{n+1} = 0, \tag{19.74}$$

$$d * F_{n+1} = (-)^{D+n+1} * j_n. \tag{19.75}$$

Equation (19.74) represents a generalized Bianchi identity for F_{n+1}. The consistency of equation (19.75) is ensured by two properties. In the first place, both members are forms of degree $D - n$. In fact, $*F_{n+1}$ is a $(D - n - 1)$-form, so that $d * F_{n+1}$ is a $(D - n)$-form, as is $*j_n$. This counting determines, ultimately, the rank $n + 1$ of the generalized Maxwell tensor (19.72). In the second place, thanks to current conservation (19.71), Eq. (19.75) equals a closed form to a closed form. For the same reasons already explained in the electrodynamics of point-particles, see Sect. 2.3, for consistency also Eqs. (19.74) and (19.75) must be considered as partial differential equations in the space of distributions $\mathcal{S}'(\mathbb{R}^D)$. Finally, it is easy to translate equations (19.74) and (19.75) in the traditional tensor notation, see Problem 19.2,

$$\partial_{[\mu_1} F_{\mu_2 \cdots \mu_{n+2}]} = 0, \tag{19.76}$$

$$\partial_\mu F^{\mu \mu_1 \cdots \mu_n} = j^{\mu_1 \cdots \mu_n}. \tag{19.77}$$

Written in this way, these equations represent a *manifest* generalization of the Bianchi identity (2.40) and of the Maxwell equation (2.22) for particles, respectively, and they reduce to them for $n = 1$.

Electric and magnetic fields. Being a completely antisymmetric tensor of rank $n + 1$ in D dimensions, the electromagnetic field (19.72) possesses $\binom{D}{n+1}$ independent component fields. Its independent spatial component fields are the tensors

$$E^{i_1 \cdots i_n} = F^{i_1 \cdots i_n 0}, \tag{19.78}$$

$$B^{i_1 \cdots i_{n+1}} = F^{i_1 \cdots i_{n+1}}, \tag{19.79}$$

where the indices i_k assume the spatial values $1, \ldots, D - 1$. Equations (19.78) and (19.79) define the generalized electric and magnetic fields, respectively, generated by a p-brane. For $n > 1$ both fields are completely antisymmetric spatial tensors. The electric field is of rank n and has $\binom{D-1}{n}$ independent components, and the magnetic field is of rank $n + 1$ and has $\binom{D-1}{n+1}$ independent components. The binomial identity $\binom{D-1}{n} + \binom{D-1}{n+1} = \binom{D}{n+1}$ ensures the matching of the total number of components. Clearly, one could also represent the electric and magnetic fields as the $(D - 1)$-dimensional spatial Hodge duals of the fields (19.78) and (19.79). However, this would lower or raise their rank depending on the values of n and D.

Generalized n-form potential and gauge invariance. We address now the problem of the general solution of the generalized Bianchi identity (19.74). This equation states that F_{n+1} is a closed $(n + 1)$-form in $\mathcal{S}'(\mathbb{R}^D)$, and so the Poincaré lemma for distribution-valued differential forms ensures that it is also exact, see Sect. 19.1.1. Therefore, there exists an n-form potential

$$A_n = \frac{1}{n!} \, dx^{\mu_n} \wedge \cdots \wedge dx^{\mu_1} A_{\mu_1 \cdots \mu_n}, \tag{19.80}$$

such that

$$F_{n+1} = dA_n. \tag{19.81}$$

This means that a p-brane generates not a vector potential A_μ, but rather a completely antisymmetric *tensor potential* $A_{\mu_1 \cdots \mu_n}$ of rank $n = p + 1$. As in the case of the electrodynamics of point particles, this potential is, however, not unique. In fact, given two generic n-form potentials A'_n and A_n we have

$$F_{n+1} = dA_n = dA'_n \quad \Rightarrow \quad d\big(A'_n - A_n\big) = 0.$$

Therefore, there exists a generalized *gauge function*, i.e. an $(n-1)$-form

$$\Lambda_{n-1} = \frac{1}{(n-1)!} \, dx^{\mu_{n-1}} \wedge \cdots \wedge dx^{\mu_1} \Lambda_{\mu_1 \cdots \mu_{n-1}},$$

such that $A'_n - A_n = d\Lambda_{n-1}$, or

$$A'_n = A_n + d\Lambda_{n-1}. \tag{19.82}$$

The relation (19.82) tells us that the potential A_n is determined modulo a *gauge transformation*, whereas the electromagnetic field F_{n+1} is *gauge invariant*. The general solution of the Bianchi identity (19.74) is thus expressed, schematically, by the relations

$$dF_{n+1} = 0 \quad \Leftrightarrow \quad F_{n+1} = dA_n, \quad \text{with } A_n \approx A_n + d\Lambda_{n-1}, \tag{19.83}$$

which resemble the analogous relations for a vector potential (19.33). To rewrite the above relations between the potential and the field strength in tensor notation, we insert the n-form potential (19.80) in Eq. (19.81), and compare the resulting $(n+1)$-form dA_n with the right hand side of equation (19.73). In this way, we find that the electromagnetic field and the tensor potential are related by

$$F_{\mu_1 \cdots \mu_{n+1}} = (n+1) \, \partial_{[\mu_1} A_{\mu_2 \cdots \mu_{n+1}]}, \tag{19.84}$$

an equation that generalizes the analogous relation for the Maxwell tensor in electrodynamics (19.31). Finally, in tensor notation the gauge transformation (19.82) reads

$$A'_{\mu_1 \cdots \mu_n} = A_{\mu_1 \cdots \mu_n} + n \, \partial_{[\mu_1} \Lambda_{\mu_2 \cdots \mu_n]}, \tag{19.85}$$

and it is immediately seen that it leaves the electromagnetic field (19.84) invariant. To illustrate the above relations we write them out explicitly for the case $n = 2$, corresponding to the electromagnetic field generated by a *string*. In this case the potential is an antisymmetric tensor $A_{\mu\nu}$, and the Maxwell tensor (19.84) assumes the form

$$F_{\mu\nu\rho} = 3 \, \partial_{[\mu} A_{\nu\rho]} = \partial_\mu A_{\nu\rho} + \partial_\nu A_{\rho\mu} + \partial_\rho A_{\mu\nu}.$$

This tensor is invariant under the gauge transformation (19.85)

$$A'_{\mu\nu} = A_{\mu\nu} + \partial_\mu \Lambda_\nu - \partial_\nu \Lambda_\mu,$$

as expressed by the relations $F'_{\mu\nu\rho} = 3 \, \partial_{[\mu} A'_{\nu\rho]} = 3 \, \partial_{[\mu} A_{\nu\rho]} = F_{\mu\nu\rho}$.

19.3 Lorentz Equation and Variational Method

The problem of reparameterization invariance. To complete the dynamics of the *coupled* system, formed in the simplest case by a charged p-brane and the electromagnetic field, it remains to establish the equation that governs the dynamics of a brane, i.e. its *Lorentz equation*. The latter should constitute an appropriate generalization of the Lorentz equation for the particle (2.33). However, *naive* attempts to generalize the latter fail due to the difficulties related with the implementation of reparameterization invariance. To illustrate the problem we reconsider an isolated particle with world line $y^\mu(\lambda)$, obeying thus the *free* Lorentz equation

$$m \frac{d^2 y^\mu}{ds^2} = 0.$$

We recall that d/ds is the Lorentz- and reparameterization-invariant derivative (2.6)

$$\frac{d}{ds} = \left(\frac{dy^\nu}{d\lambda} \frac{dy_\nu}{d\lambda} \right)^{-1/2} \frac{d}{d\lambda}.$$

In the case of a p-brane the role of the single derivative $d/d\lambda$ is played by the n partial derivatives $\partial/\partial\lambda^a$. A possible generalization of the proper-time derivative d/ds, which is invariant under the rescaling $\lambda^a \to \lambda'^a = \lambda^a/k$, could then be

$$\frac{d}{ds} \to \frac{\partial}{\partial s^a} \equiv \left(\frac{\partial y^\nu}{\partial \lambda^b} \frac{\partial y_\nu}{\partial \lambda^b} \right)^{-1/2} \frac{\partial}{\partial \lambda^a},$$

in which case the resulting equation of motion of the free brane would be

$$m \frac{\partial}{\partial s^a} \frac{\partial}{\partial s^a} y^\mu = 0. \tag{19.86}$$

However, the partial derivatives $\partial/\partial s^a$ are not invariant under a generic reparameterization $\lambda^a \to \lambda'^a(\lambda)$, and neither is Eq. (19.86). The latter is, therefore, not a physically acceptable equation of motion. To overcome this impasse we resort, once more, to the *variational method*. In fact, the construction of a reparameterization-invariant action turns out to be notably simpler than the search for a reparameterization-invariant equation of motion.

19.3.1   Action for the Electromagnetic Field

As prototype for an action of a p-brane interacting with the electromagnetic field we consider the action of the electrodynamics of point-particles (4.11). In fact, the structure of the latter suggests for the action governing the propagation of the electromagnetic field, and its interaction with the p-brane, the functional of the tensor potential $A_{\mu_1 \cdots \mu_n}(x)$ and of the world volume coordinates $y^\mu(\lambda)$

$$I_1 + I_2 = \frac{(-)^n}{n!} \int \left(\frac{1}{2(n+1)} F^{\mu_1 \cdots \mu_{n+1}} F_{\mu_1 \cdots \mu_{n+1}} + A_{\mu_1 \cdots \mu_n} j^{\mu_1 \cdots \mu_n} \right) d^D x. \tag{19.87}$$

The relative coefficient between the two terms of the integrand has been chosen in such a way that $I_1 + I_2$ gives rise to Maxwell's equation (19.77), see below, whereas the overall coefficient $1/n!$ is conventional. Conversely, the global sign $(-)^n$ is required by the positivity of energy, as we shall see in Sect. 19.3.4. In the action I_1 it is understood that the Maxwell tensor is expressed in terms of $A_{\mu_1 \cdots \mu_n}$ according to (19.84), and the current $j^{\mu_1 \cdots \mu_n}$ is given by Eq. (19.63). The action (19.87) is manifestly invariant under Lorentz transformations and under reparameterizations and, in addition, it is also gauge invariant. In fact, I_1 is not affected by the transformation (19.85) as the Maxwell tensor (19.84) is gauge invariant, whereas the variation of I_2 is

$$\begin{aligned} I_2' - I_2 &= \frac{(-)^n}{n!} \int \left(A'_{\mu_1 \cdots \mu_n} - A_{\mu_1 \cdots \mu_n} \right) j^{\mu_1 \cdots \mu_n} d^D x \\ &= \frac{(-)^n}{(n-1)!} \int \partial_{\mu_1} \Lambda_{\mu_2 \cdots \mu_n} j^{\mu_1 \cdots \mu_n} d^D x \\ &= \frac{(-)^n}{(n-1)!} \int \left(\partial_{\mu_1} \left(\Lambda_{\mu_2 \cdots \mu_n} j^{\mu_1 \cdots \mu_n} \right) - \Lambda_{\mu_2 \cdots \mu_n} \partial_{\mu_1} j^{\mu_1 \cdots \mu_n} \right) d^D x. \end{aligned}$$

The first term of the last line, being a D-divergence, is irrelevant, and the second term vanishes thanks to current conservation. The action (19.87) is, thus, gauge invariant. It remains to verify that this action gives rise to the Maxwell equation (19.77). For this purpose we must consider a generic infinitesimal variation of the tensor potential $\delta A_{\mu_1 \cdots \mu_n}$, vanishing at the boundary of the integration domain, which in the space-time integral (19.87) is understood, i.e. at the times $t = t_1$ and $t = t_2$, and compute the corresponding variation of the action. For the variation of I_2 we obtain simply

$$\delta I_2 = \frac{(-)^n}{n!} \int \delta A_{\mu_1 \cdots \mu_n} j^{\mu_1 \cdots \mu_n} d^D x,$$

while for the variation of I_1, using the definition (19.84), we find

$$\delta I_1 = \frac{(-)^n}{(n+1)!} \int F^{\mu_1 \cdots \mu_{n+1}} \delta F_{\mu_1 \cdots \mu_{n+1}} d^D x$$

$$= \frac{(-)^n}{n!} \int F^{\mu_1 \cdots \mu_{n+1}} \partial_{\mu_1} \delta A_{\mu_2 \cdots \mu_{n+1}} d^D x$$

$$= \frac{(-)^n}{n!} \int \left(\partial_{\mu_1} \left(F^{\mu_1 \cdots \mu_{n+1}} \delta A_{\mu_2 \cdots \mu_{n+1}} \right) - \partial_{\mu_1} F^{\mu_1 \cdots \mu_{n+1}} \delta A_{\mu_2 \cdots \mu_{n+1}} \right) d^D x.$$

The first term, a D-divergence, gives a vanishing contribution to the integral. In fact, at spatial infinity vanishes the electromagnetic field $F^{\mu_1 \cdots \mu_{n+1}}$, and at the temporal boundaries t_1 and t_2 vanishes the variation $\delta A_{\mu_1 \cdots \mu_n}$. We obtain thus

$$\delta (I_1 + I_2) = \frac{(-)^n}{n!} \int \left(j^{\mu_2 \cdots \mu_{n+1}} - \partial_{\mu_1} F^{\mu_1 \cdots \mu_{n+1}} \right) \delta A_{\mu_2 \cdots \mu_{n+1}} d^D x.$$

Imposing that this integral vanishes for arbitrary variations $\delta A_{\mu_2 \cdots \mu_{n+1}}$ we retrieve equation (19.77). Of course, the same conclusion is reached if one writes out the Euler-Lagrange equations

$$\partial_\mu \frac{\partial \mathcal{L}}{\partial (\partial_\mu A_{\mu_1 \cdots \mu_n})} - \frac{\partial \mathcal{L}}{\partial A_{\mu_1 \cdots \mu_n}} = 0,$$

relative to the Lagrangian

$$\mathcal{L} = \frac{(-)^n}{n!} \left(\frac{1}{2(n+1)} F^{\mu_1 \cdots \mu_{n+1}} F_{\mu_1 \cdots \mu_{n+1}} + A_{\mu_1 \cdots \mu_n} j^{\mu_1 \cdots \mu_n} \right).$$

19.3.2 Action for a Free p-Brane

Induced metric. To complete the description of the electrodynamics of a p-brane within the variational method, it remains to find an appropriate generalization of the action of the free particle $-m \int ds$. From a geometric point of view, the integral $\int ds$ represents the length of the particle's world line in space-time. It is thus natural to assume that in the case of a p-brane this integral must be replaced by the $(p+1)$-dimensional world *volume* $\int dV$ traced by the brane during its time evolution. This choice is also supported by the fact that this volume – by definition – is reparameterization invariant. As we will see, in the definition of the infinitesimal volume element dV a crucial role will be played by the *induced metric* on the brane

$$g_{ab}(\lambda) = U_a^\mu(\lambda) U_b^\nu(\lambda)\, \eta_{\mu\nu}, \tag{19.88}$$

involving the tangent vectors (19.59). The relation (19.88) defines a *Lorentz-invariant* and *symmetric* $n \times n$ matrix. For sufficiently *smooth* branes this matrix is invertible, and we denote its inverse – likewise a Lorentz-invariant and symmetric matrix – by the symbol with upper indices g^{ab}. We thus have

$$g^{ab} g_{bc} = \delta_c^a. \tag{19.89}$$

Volume element in Euclidean space. Before presenting the expression of dV for the world volume of a p-brane propagating in a D-dimensional Minkowski space-time, we illustrate the construction in the case of an n-dimensional region $\mathcal{B}$ embedded in the *Euclidean* space $\mathbb{R}^D$. If $\mathcal{B}$ is sufficiently regular, in the language of *differential geometry* such a region represents a *submanifold* of $\mathbb{R}^D$ of dimension n, see e.g. Ref. [3]. If we parameterize $\mathcal{B}$ by the embedding coordinates $y^\mu(\lambda)$, where $\lambda = \{\lambda^1, \ldots, \lambda^n\}$ and $\mu = 1, \ldots, D$, in this case the induced metric is defined by

$$g_{ab}(\lambda) = U_a^\mu(\lambda) U_b^\nu(\lambda) \delta_{\mu\nu} = U_a^\mu(\lambda) U_b^\mu(\lambda), \tag{19.90}$$

where the Minkowski metric $\eta_{\mu\nu}$ of formula (19.88) has been replaced by the Kronecker symbol $\delta_{\mu\nu}$. The metric (19.90) is *positive* definite, see Problem 19.7. According to a well-known result of Euclidean geometry, the volume element dV of $\mathcal{B}$ is then expressed in terms of the determinant g of the induced metric by the formula

$$dV = \sqrt{g}\, d^n\lambda, \qquad g = \det g_{ab} > 0. \tag{19.91}$$

Consequently, the volume occupied by the region $\mathcal{B}$ is given by

$$V_\mathcal{B} = \int_\mathcal{B} \sqrt{g}\, d^n\lambda.$$

Surface element of a sphere. As an example, we determine the volume element, i.e. the surface element dS, of a two-sphere embedded in $\mathbb{R}^3$, parameterized by the embedding coordinates (19.43)–(19.45). In this case we have $D = 3$ and $n = 2$, and the two tangent vectors are

$$\mathbf{U}_1 = \frac{\partial \mathbf{y}}{\partial \lambda^1} = \left(\cos\lambda^1 \cos\lambda^2, \cos\lambda^1 \sin\lambda^2, -\sin\lambda^1\right),$$

$$\mathbf{U}_2 = \frac{\partial \mathbf{y}}{\partial \lambda^2} = \left(-\sin\lambda^1 \sin\lambda^2, \sin\lambda^1 \cos\lambda^2, 0\right).$$

The induced metric (19.90) then becomes

$$g_{ab} = \begin{pmatrix} \mathbf{U}_1 \cdot \mathbf{U}_1 & \mathbf{U}_1 \cdot \mathbf{U}_2 \\ \mathbf{U}_2 \cdot \mathbf{U}_1 & \mathbf{U}_2 \cdot \mathbf{U}_2 \end{pmatrix} = \begin{pmatrix} 1 & 0 \\ 0 & \sin^2 \lambda^1 \end{pmatrix}, \qquad \sqrt{g} = \sin \lambda^1. \qquad (19.92)$$

In basis of the identifications $\lambda^1 = \vartheta$ and $\lambda^2 = \varphi$, the volume element (19.91) then reproduces the familiar surface element of the unit sphere[4]

$$dS = \sqrt{g}\, d\lambda^1 d\lambda^2 = \sin \vartheta\, d\vartheta\, d\varphi,$$

with total area

$$S = \int dS = \int_0^{2\pi} d\varphi \int_0^{\pi} \sin \vartheta\, d\vartheta = 4\pi.$$

Using the alternative parameterization – or embedding coordinates – of the sphere (19.46)–(19.48), we obtain instead the induced metric

$$g'_{ab} = \frac{1}{1 - (\lambda'^1)^2 - (\lambda'^2)^2} \begin{pmatrix} 1 - (\lambda'^2)^2 & \lambda'^1 \lambda'^2 \\ \lambda'^1 \lambda'^2 & 1 - (\lambda'^1)^2 \end{pmatrix}, \quad \sqrt{g'} = \frac{1}{\sqrt{1 - (\lambda'^1)^2 - (\lambda'^2)^2}},$$

leading to the surface element

$$dS' = \sqrt{g'}\, d\lambda'^1 d\lambda'^2 = \frac{d\lambda'^1 d\lambda'^2}{\sqrt{1 - (\lambda'^1)^2 - (\lambda'^2)^2}}.$$

Using the transformation rules (19.49) one easily verifies the equality

$$dS' = dS,$$

which expresses the basic geometrical feature that the area of a surface does not depend on the way we parameterize it.

Reparameterization invariance of dV. Actually, it is not difficult to show that the Euclidean volume element (19.91) is invariant under an arbitrary reparameterization $\lambda \to \lambda'(\lambda)$. In fact, from the transformation laws (19.66) we find that the Euclidean metric (19.90) transforms according to

$$g'_{ab} = U'^{\mu}_a U'^{\nu}_b \delta_{\mu\nu} = K_a{}^c U^{\mu}_c K_b{}^d U^{\nu}_d \delta_{\mu\nu} = K_a{}^c K_b{}^d g_{cd}. \qquad (19.93)$$

Denoting the $n \times n$ matrix associated with g_{ab} by the symbol $\mathcal{G}$, in matrix notation the relation above reads

$$\mathcal{G}' = K \mathcal{G} K^T.$$

Taking the determinant of both members of this equation we obtain

---

[4]We recall that polar coordinates are *singular* at the North and South Poles, i.e. for $\vartheta = 0$ and $\vartheta = \pi$, and consequently at these points the induced metric is singular, too, and in particular its determinant g vanishes.

$$g' = \left(\det K\right) g \left(\det K^T\right) = \left(\det K\right)^2 g \quad \Rightarrow \quad \sqrt{g'} = |\det K|\sqrt{g}. \quad (19.94)$$

Since the measure transforms according to the rule $d^n \lambda' = d^n \lambda/|\det K|$, we have thus proven that the volume elements are equal

$$\sqrt{g'}\, d^n \lambda' = \sqrt{g}\, d^n \lambda.$$

Volume element in Minkowski space-time and causal motions. In a space-time endowed with the Minkowski metric (19.1) the volume element dV associated with the world volume $y^\mu(\lambda)$ of a p-brane is defined in terms of the tangent vectors (19.59) by the relations

$$dV = \sqrt{g}\, d^n \lambda, \qquad g = (-)^p \det g_{ab}, \qquad g_{ab} = U_a^\mu U_b^\nu \eta_{\mu\nu}. \quad (19.95)$$

As the induced metric g_{ab} is manifestly invariant under Lorentz transformations, so is the volume element dV. Its reparameterization invariance can be proven in the same way as in the Euclidean case. It is sufficient to replace in the transformation law (19.93) the Euclidean metric $\delta_{\mu\nu}$ with the Minkowski metric $\eta_{\mu\nu}$, and then to multiply both members of the first equation in (19.94) with the sign $(-)^p$, before extracting the square root. The additional sign $(-)^p$ in the definition of g in (19.95) is necessary to make this radicand *semi-positive* definite,

$$g = (-)^p \det g_{ab} \geq 0, \quad (19.96)$$

whenever the brane performs a *causal* motion. We call the motion of a brane causal, if the *physical velocity* of each point of the profile $y^i(t, \boldsymbol{\lambda})$ (19.51), namely the component of the *nominal* velocity $dy^i(t, \boldsymbol{\lambda})/dt$ *normal* to the brane, is smaller or equal than the speed of light. For a technical definition of the physical velocity, and for a proof of the connection between causality and the condition (19.96) in the particular case of a 1-brane, i.e. of a *string*, see Problem 19.6. Below we exemplify this important connection in the case of a *flat p*-brane, a configuration which, in a sense, generalizes that of a particle in uniform linear motion.

Flat branes, static branes, and causality. By definition, the world volume of a *flat p*-brane admits the *linear* parameterization

$$y^\mu(\lambda) = U_a^\mu \lambda^a, \quad (19.97)$$

meaning that the tangent vectors U_a^μ (19.59) are all constant, as is the induced metric g_{ab} in (19.95). The latter, being a symmetric matrix, can be diagonalized by means of an appropriate reparameterization, and, if it satisfies the causality constraint (19.96) as a strict inequality, being a $(p+1)$-dimensional matrix it admits at least one *positive* eigenvalue. Taking λ^0 as the parameter corresponding to this eigenvalue, we then have $g_{00} = U_0^\mu U_0^\nu \eta_{\mu\nu} > 0$, or, upon rescaling λ^0, $g_{00} = U_0^\mu U_0^\nu \eta_{\mu\nu} = 1$. This implies that there exists a Lorentz transformation which sends the tangent vector U_0^μ into $U_0^\mu = (1, 0, \ldots, 0)$. Finally, by shifting the parameter λ^0 into $\lambda^0 - \sum_{a=1}^p U_a^i \lambda^a$, we get furthermore $U_0^0 = (1, 0, \ldots, 0)$. In this way, the parameterization (19.97) takes the *static* form

$$y^0(\lambda) = \lambda^0, \qquad y^i(\lambda) = \sum_{a=1}^p U_a^i \lambda^a. \tag{19.98}$$

In fact, its spatial profile $y^i(\lambda)$ has become independent of λ^0, i.e. of time. In conclusion, via an appropriate choice of the Lorentz frame and of the parameterization, a flat brane satisfying the condition (19.96) can always be made static, and so it respects causality in a trivial way. Such a brane resembles thus a particle in uniform linear motion – with world line $y^\mu(\lambda) = u^\mu \lambda$ in place of (19.97) – traveling with a speed less than the speed of light, i.e. $u^2 > 0$. Notice also that, within the parameterization (19.98), the induced metric when restricted to the last p parameters λ^a is *negative* definite, i.e. *space-like*. In fact, from the relations (19.59), (19.95) and (19.98) we obtain $g_{ab} = -U_a^i U_b^i$, for $a, b = (1, \ldots, p)$. A canonical way to embed such a brane into space-time is to dispose it along the first p spatial coordinates. Accordingly, separating the space-time coordinates $\{x^\mu\}$ in the two groups $\{x^a\}$ and $\{x^I\}$, where $a = (0, \ldots, p)$ and $I = (p + 1, \ldots, D - 1)$, the parameterization (19.98) takes the simplified form

$$y^\mu(\lambda) \rightarrow \begin{cases} y^a(\lambda) = \lambda^a, \\ y^I(\lambda) = 0. \end{cases} \tag{19.99}$$

The $p + 1$ coordinates $\{x^a\}$, containing the time $x^0 = t$, are thus *parallel* to the brane, and the $D - p - 1$ coordinates $\{x^I\}$ are *orthogonal* to it. In particular, the p-dimensional region occupied by the brane is described by the $D - p - 1$ equations $x^I = 0$. For the simplified parameterization (19.99) the tangent vectors become

$$U_b^\mu = \frac{\partial y^\mu}{\partial \lambda^b} \rightarrow \begin{cases} U_b^a = \delta_b^a, \\ U_b^I = 0, \end{cases} \tag{19.100}$$

so that the induced metric reduces to the $(p + 1)$-dimensional Minkowski metric

$$g_{ab} = U_a^\mu U_b^\nu \eta_{\mu\nu} = \delta_a^c \delta_b^d \eta_{cd} = \eta_{ab}, \tag{19.101}$$

as expected. Its determinant is given by

$$\det g_{ab} = \det \eta_{ab} = (-)^p,$$

so that – in agreement with the condition (19.96) we started from – we obtain the positive radicand $g = 1$. Actually, as anticipated above, it can be shown that if a p-brane performs a generic causal motion, then we have the inequality $\det g_{ab} \geq 0$

if p is even, and $\det g_{ab} \leq 0$ if p is odd; the proof for the particular case $p = 1$ is given in Problem 19.6. Finally, for a particle, $p = 0$, the relations (19.95) reduce to

$$g = \det g_{ab} = \frac{\partial y^\mu}{d\lambda}\frac{\partial y^\nu}{d\lambda}\eta_{\mu\nu} = (1 - v^2)\left(\frac{dt}{d\lambda}\right)^2 \Rightarrow dV = \sqrt{g}\,d\lambda = \sqrt{1 - v^2}\,dt.$$

$$(19.102)$$

The volume element dV coincides, thus, with the proper time ds, and the condition $g \geq 0$ reduces to the usual causality constraint $v \leq 1$:

Action and equation of motion of a free brane. As action for a free p-brane we choose the invariant functional of the coordinates $y^\mu(\lambda)$

$$I_3[y] = -m\int dV = -m\int \sqrt{g}\,d^n\lambda. \qquad (19.103)$$

This functional generalizes the action $-m\int ds$ of a free particle, and reduces to it for $p = n - 1 = 0$. However, for $n > 1$ the parameter m in the action (19.103) represents the *mass per unit volume* of the brane, see Sect. 19.3.4. To determine the equation of motion deriving from the action $I_3[y]$ we must evaluate its response under arbitrary variations $\delta y^\mu(\lambda)$ of the coordinates, vanishing at the boundary of the integration volume. We begin by determining the variation of the determinant of a generic matrix M, under generic infinitesimal variations δM of its elements

$$\delta \det M = \det(M + \delta M) - \det M = \det\left(M\left(1 + M^{-1}\delta M\right)\right) - \det M$$
$$= \det M \det\left(1 + M^{-1}\delta M\right) - \det M \approx \det M\,\mathrm{tr}\left(M^{-1}\delta M\right).$$

Setting $M_{ab} = g_{ab}$, and recalling the definition of the induced metric (19.95) and of its inverse (19.89), we then obtain

$$\delta\sqrt{g} = \frac{\delta g}{2\sqrt{g}} = \frac{(-)^p}{2\sqrt{g}}\,\delta\det g_{ab} = \frac{(-)^p}{2\sqrt{g}}\det g_{ab}\left(g^{cd}\delta g_{cd}\right) = \frac{1}{2}\sqrt{g}\,g^{cd}\delta g_{cd}$$
$$= \frac{1}{2}\sqrt{g}\,g^{cd}\delta(U_c^\mu U_d^\nu \eta_{\mu\nu}) = \sqrt{g}\,g^{cd}U_c^\mu\delta U_{\mu d} = \sqrt{g}\,g^{cd}\frac{\partial y^\mu}{\partial\lambda^c}\frac{\partial\delta y_\mu}{\partial\lambda^d}.$$

The variation of the brane action (19.103) then becomes

$$\delta I_3[y] = -m\int \delta\sqrt{g}\,d^n\lambda = -m\int \sqrt{g}\,g^{ab}\frac{\partial y^\mu}{\partial\lambda^a}\frac{\partial\delta y_\mu}{\partial\lambda^b}\,d^n\lambda$$
$$= -m\int \left(\frac{\partial}{\partial\lambda^b}\left(\sqrt{g}\,g^{ab}\frac{\partial y^\mu}{\partial\lambda^a}\delta y_\mu\right) - \frac{\partial}{\partial\lambda^b}\left(\sqrt{g}\,g^{ab}\frac{\partial y^\mu}{\partial\lambda^a}\right)\delta y_\mu\right)d^n\lambda$$
$$= m\int \frac{\partial}{\partial\lambda^b}\left(\sqrt{g}\,g^{ab}\frac{\partial y^\mu}{\partial\lambda^a}\right)\delta y_\mu\,d^n\lambda. \qquad (19.104)$$

The first term of the second line vanishes because its integrand is an n-divergence, and at the boundary of the integration domain the variations δy^μ are zero.

Imposing that $\delta I_3[y]$ vanishes for otherwise arbitrary variations δy^μ, we then obtain as *equation of motion of a free brane*

$$m\frac{\partial}{\partial\lambda^b}\left(\sqrt{g}\,g^{ab}\frac{\partial y^\mu}{\partial\lambda^a}\right) = 0. \tag{19.105}$$

As this equation descends from a reparameterization-invariant action, it is automatically invariant under reparameterizations, in contrast to Eq. (19.86).

19.3.3 Lorentz Equation

In complete analogy with the action for charged particles (4.11), we write the total action for the electrodynamics of a charged p-brane as the sum of three terms

$$I[A,y] = \frac{(-)^n}{n!}\int\left(\frac{1}{2(n+1)}F^{\mu_1\cdots\mu_{n+1}}F_{\mu_1\cdots\mu_{n+1}} + A_{\mu_1\cdots\mu_n}j^{\mu_1\cdots\mu_n}\right)d^D x$$

$$- m\int\sqrt{g}\,d^n\lambda = I_1 + I_2 + I_3. \tag{19.106}$$

The functional $I[A,y]$ is invariant under Lorentz transformations, under reparameterizations, and under gauge transformations, exactly as the action (4.11) governing the electrodynamics of point-particles. As we have already seen, by requiring the action $I[A,y]$ to be stationary under arbitrary variations of the tensor potential $A_{\mu_1\cdots\mu_n}$ one obtains Maxwell's equation (19.77). On the other hand, by requiring $I[A,y]$ to be stationary under arbitrary variations of the coordinates y^μ one derives the equation of motion of the brane. The action I_1 is independent of y^μ, and the variation of I_3 has already been computed in (19.104). It remains therefore only to evaluate the variation of the functional I_2. For this purpose it is convenient to rewrite this functional in a different form, by inserting the definition of the current (19.63),

$$I_2 = \frac{(-)^n}{n!}\int A_{\mu_1\cdots\mu_n}(x)\,j^{\mu_1\cdots\mu_n}(x)\,d^D x$$

$$= (-)^n\frac{e}{n!}\int A_{\mu_1\cdots\mu_n}(x)\left(\int U^{\mu_1}_{a_1}\cdots U^{\mu_n}_{a_n}\varepsilon^{a_1\cdots a_n}\delta^D(x-y(\lambda))\,d^n\lambda\right)d^D x$$

$$= (-)^n\frac{e}{n!}\int A_{\mu_1\cdots\mu_n}(y(\lambda))\,U^{\mu_1}_{a_1}\cdots U^{\mu_n}_{a_n}\varepsilon^{a_1\cdots a_n}\,d^n\lambda$$

$$= (-)^n\frac{e}{n!}\int \mathcal{O}\,d^n\lambda, \tag{19.107}$$

where the last line defines the integrand $\mathcal{O}$. To determine δI_2 we must vary the coordinates y^μ appearing in $A_{\mu_1\cdots\mu_n}(y(\lambda))$, as well as the y^μ appearing in the tangent vectors $U^\mu_a = \partial y^\mu/\partial\lambda^a$. In this way we find the variation

$$\delta\mathcal{O} = \varepsilon^{a_1\cdots a_n}\left(\delta y^\alpha \partial_\alpha A_{\mu_1\cdots\mu_n}U_{a_1}^{\mu_1}\cdots U_{a_n}^{\mu_n} + n A_{\mu_1\cdots\mu_n}\frac{\partial\delta y^{\mu_1}}{\partial\lambda^{a_1}}U_{a_2}^{\mu_2}\cdots U_{a_n}^{\mu_n}\right)$$

$$= \varepsilon^{a_1\cdots a_n}\left(\delta y^\alpha \partial_\alpha A_{\mu_1\cdots\mu_n}U_{a_1}^{\mu_1}\cdots U_{a_n}^{\mu_n} - n\frac{\partial A_{\mu_1\cdots\mu_n}}{\partial\lambda^{a_1}}\delta y^{\mu_1}U_{a_2}^{\mu_2}\cdots U_{a_n}^{\mu_n}\right) + \frac{\partial\mathcal{H}^{a_1}}{\partial\lambda^{a_1}},$$

where we have set

$$\mathcal{H}^{a_1} = n\varepsilon^{a_1\cdots a_n}A_{\mu_1\cdots\mu_n}\delta y^{\mu_1}U_{a_2}^{\mu_2}\cdots U_{a_n}^{\mu_n}.$$

Above we have applied Leibniz's rule, and we have used that the second derivatives (19.69) are symmetric in a and b, while the Levi-Civita tensor is antisymmetric. Noting that

$$\frac{\partial A_{\mu_1\cdots\mu_n}}{\partial\lambda^{a_1}} = \frac{\partial y^\alpha}{\partial\lambda^{a_1}}\partial_\alpha A_{\mu_1\cdots\mu_n} = U_{a_1}^\alpha\,\partial_\alpha A_{\mu_1\cdots\mu_n},$$

rearranging the indices we then obtain

$$\delta\mathcal{O} = \varepsilon^{a_1\cdots a_n}(\partial_\alpha A_{\mu_1\cdots\mu_n} - n\partial_{\mu_1}A_{\alpha\mu_2\cdots\mu_n})U_{a_1}^{\mu_1}\cdots U_{a_n}^{\mu_n}\delta y^\alpha + \frac{\partial\mathcal{H}^a}{\partial\lambda^a}$$

$$= \varepsilon^{a_1\cdots a_n}(n+1)\,\partial_{[\alpha}A_{\mu_1\cdots\mu_n]}U_{a_1}^{\mu_1}\cdots U_{a_n}^{\mu_n}\delta y^\alpha + \frac{\partial\mathcal{H}^a}{\partial\lambda^a}$$

$$= \varepsilon^{a_1\cdots a_n}F_{\alpha\mu_1\cdots\mu_n}U_{a_1}^{\mu_1}\cdots U_{a_n}^{\mu_n}\delta y^\alpha + \frac{\partial\mathcal{H}^a}{\partial\lambda^a}.$$

When we insert this expression in the variation of the integral (19.107), the n-divergence $\partial\mathcal{H}^a/\partial\lambda^a$ gives no contribution because at the boundary the variations δy^μ vanish. Knowing the variation of the action I_3 (19.104), we then find that under a generic variation of the coordinates $y^\mu(\lambda)$ the total action (19.106) changes by

$$\delta I_2 + \delta I_3 = \int\left(m\frac{\partial}{\partial\lambda^b}\left(\sqrt{g}g^{ab}\frac{\partial y^\mu}{\partial\lambda^a}\right)\right.$$

$$\left. + (-)^n\frac{e}{n!}\,F^{\mu\mu_1\cdots\mu_n}U_{\mu_1 a_1}\cdots U_{\mu_n a_n}\varepsilon^{a_1\cdots a_n}\right)\delta y_\mu\,d^n\lambda.$$

Imposing that this expression vanishes for arbitrary variations δy^μ we obtain the *Lorentz equation for the p-brane*

$$m\frac{\partial}{\partial\lambda^b}\left(\sqrt{g}g^{ab}\frac{\partial y^\mu}{\partial\lambda^a}\right) = (-)^{n+1}\frac{e}{n!}\,F^{\mu\mu_1\cdots\mu_n}U_{\mu_1 a_1}\cdots U_{\mu_n a_n}\varepsilon^{a_1\cdots a_n} \tag{19.108}$$

$$= (-)^{n+1}e\,F^{\mu\mu_1\cdots\mu_n}U_{\mu_1 0}\cdots U_{\mu_n n-1}.$$

As the action (19.106) is gauge invariant, also the equations of motion must be so. It is for this reason that in Eq. (19.108) there appears the gauge-invariant Maxwell tensor $F_{n+1} = dA_n$, rather than the tensor potential A_n. In the case of a particle, $n = 1$, Eq. (19.108) reduces to the usual Lorentz equation, see formulas (19.102).

19.3.4 Energy-Momentum Tensor

As the dynamics of our system descends from the total action (19.106), which is Poincaré-invariant in a D-dimensional space-time, Noether's theorem yields a canonical conserved energy-momentum tensor. And, as in four dimensions, this tensor can always be symmetrized, see Sect. 3.4. Without repeating explicitly the procedure carried out for the electrodynamics of charged particles in Sect. 4.3, here we limit ourselves to present the resulting *symmetric* energy-momentum tensor $T^{\mu\nu}$. As in the case of charged particles, this tensor is composed of two terms, one representing the electromagnetic field, and one representing the brane,

$$T^{\mu\nu} = T_{\text{em}}^{\mu\nu} + T_{\text{b}}^{\mu\nu}. \tag{19.109}$$

The contribution of the electromagnetic field is given by

$$T_{\text{em}}^{\mu\nu} = \frac{(-)^n}{n!} \left(F^{\mu\alpha_1\cdots\alpha_n} F^\nu{}_{\alpha_1\cdots\alpha_n} - \frac{1}{2(n+1)} \eta^{\mu\nu} F^{\alpha_1\cdots\alpha_{n+1}} F_{\alpha_1\cdots\alpha_{n+1}} \right), \tag{19.110}$$

and the contribution of the brane reads

$$T_{\text{b}}^{\mu\nu}(x) = m \int \sqrt{g} g^{ab} U_a^\mu U_b^\nu \, \delta^D(x - y(\lambda)) \, d^n\lambda, \tag{19.111}$$

formulas which generalize the tensors (2.137) and (2.138) of the electrodynamics of point-particles. Notice that the tensor $T_{\text{em}}^{\mu\nu}$ is gauge invariant, and that the tensor $T_{\text{b}}^{\mu\nu}$ is invariant under reparameterizations. The total energy-momentum tensor (19.109) is conserved, $\partial_\mu T^{\mu\nu} = 0$, as long as the brane coordinates y^μ satisfy the Lorentz equation (19.108), and the field F_{n+1} satisfies Maxwell's equations (19.74) and (19.75), see Problem 19.4. The overall sign as well as the relative coefficient of the electromagnetic energy-momentum tensor (19.110) are fixed by the total action (19.106). In particular, thanks to the sign $(-)^n$ multiplying the term I_1 of the action (19.106), the energy density of the electromagnetic field is *positive* definite. We have, in fact, see the definitions (19.78) and (19.79) and Problem 19.5,

$$T_{\text{em}}^{00} = \frac{1}{2n!} \left(E^{i_1\cdots i_n} E^{i_1\cdots i_n} + \frac{1}{n+1} B^{i_1\cdots i_{n+1}} B^{i_1\cdots i_{n+1}} \right) \geq 0, \tag{19.112}$$

as both terms are positive. Formula (19.112) generalizes the formula for the energy density of the electromagnetic field produced by a particle (2.133).

To exemplify the meaning of the brane's energy-momentum tensor (19.111) we consider again the *flat p-brane* (19.97), written in the simplified static coordinates (19.99). In this case the induced metric g_{ab} is the flat one (19.101), and so its inverse g^{ab} appearing in Eq. (19.111) is equal to the Minkowski metric η^{ab}. Furthermore, we have $g = (-)^{n+1} \det g_{ab} = 1$. Since, in addition, the tangent vectors (19.100) are constant, Eq. (19.111) so reduces to

$$T_b^{\mu\nu}(x) = m\, \eta^{ab} U_a^\mu U_b^\nu \int \delta^D(x - y(\lambda))\, d^n\lambda. \qquad (19.113)$$

To carry out the integral of the δ-function we denote the n coordinates parallel to the brane collectively by $\{x^a\} = x^\parallel$, and the $D - n$ coordinates orthogonal to the brane by $\{x^I\} = x^\perp$. By inserting the world-volume coordinates (19.99) in the δ-function appearing in Eq. (19.113) we then find for the energy-momentum tensor of the brane the simple expression

$$T_b^{\mu\nu}(x) = m\, \eta^{ab} U_a^\mu U_b^\nu \int \delta^{D-n}(x^\perp)\, \delta^n(x^\parallel - \lambda)\, d^n\lambda = m\, \eta^{cd} U_c^\mu U_d^\nu\, \delta^{D-n}(x^\perp). \qquad (19.114)$$

The tensor $T_b^{\mu\nu}(x)$ is hence non-vanishing only for $x^I = 0$, i.e. in the space occupied by the brane, as expected.

Energy and mass. In view of the tangent vectors (19.100), the parallel, mixed, and orthogonal components of the energy-momentum tensor (19.114) are given by

$$T_b^{ab}(x) = m\, \eta^{ab} \delta^{D-n}(x^\perp), \qquad T_b^{aI}(x) = 0, \qquad T_b^{IJ}(x) = 0,$$

respectively. In particular, the time-space components of this tensor are zero, $T_b^{0i}(x) = 0$, implying that the momentum $P_V^i = \int_V T_b^{0i}(x)\, d^{D-1}x$ contained in an arbitrary volume V vanishes, as must happen for a static brane. Vice versa, the energy density is non-vanishing and has the simple expression

$$T_b^{00}(x) = m\, \delta^{D-n}(x^\perp). \qquad (19.115)$$

Integrating it over a spatial volume V which extends to infinity along all coordinates x^I orthogonal to the brane, and which corresponds to a hypercube of side L in the p spatial directions parallel to the brane, placed for instance in the region $0 < x^a < L$, for the energy contained in V we obtain

$$\varepsilon_V = \int_V T_b^{00}(x)\, d^{D-1}x = m \prod_{a=1}^p \left(\int_0^L dx^a \right) \int \delta^{D-n}(x^\perp)\, d^{D-n}x^\perp = mL^p.$$

Since for a *static* extended object the energy density is the same as the mass density, and since L^p is the volume of the portion of the brane enclosed in V, from the above formula for ε_V we derive that the parameter m has the meaning of *mass per unit volume* of the brane, as anticipated in Sect. 19.3.2.

Summary. We conclude this short introduction to the classical theory of charged extended objects by recapitulating the system of fundamental equations which govern the electrodynamics of a single charged p-brane, and which generalizes the corresponding system of equations for charged particles (2.20)–(2.22).

• Lorentz equation:

$$m\frac{\partial}{\partial\lambda^b}\left(\sqrt{g}\,g^{ab}\frac{\partial y^\mu}{\partial\lambda^a}\right) = (-)^{n+1}\frac{e}{n!}\,F^{\mu\nu_1\cdots\nu_n}U_{\nu_1 a_1}\cdots U_{\nu_n a_n}\varepsilon^{a_1\cdots a_n}.$$

$$(19.116)$$

• Bianchi identity:

$$\partial_{[\mu_1}F_{\mu_2\cdots\mu_{n+2}]} = 0.\qquad(19.117)$$

• Maxwell equation with source:

$$\partial_\mu F^{\mu\mu_1\cdots\mu_n} = j^{\mu_1\cdots\mu_n}.\qquad(19.118)$$

The current, the tangent vectors, and the induced metric are given by

$$j^{\mu_1\cdots\mu_n} = e\int U_{a_1}^{\mu_1}\cdots U_{a_n}^{\mu_n}\varepsilon^{a_1\cdots a_n}\,\delta^D(x - y(\lambda))\,d^n\lambda,\qquad U_a^\mu = \frac{\partial y^\mu}{\partial\lambda^a},$$

$g_{ab} = U_a^\mu U_b^\nu\eta_{\mu\nu}$, respectively. Furthermore, $g = (-)^{n+1}\det g_{ab}$ and $n = p+1$. The generalization of these equations to a system of p-branes is straightforward.

Unified theories and p-branes. As we observed in the introduction to this chapter, charged p-branes – although a direct evidence of their physical reality in our four-dimensional space-time is still missing – constitute elementary excitations of *superstring theories* and of *M-theory*, which are candidates for the unification of the fundamental interactions.[5] In particular, the elementary excitations of M-theory, living in an eleven-dimensional space-time, are 2-branes and 5-branes – actually electromagnetic *duals* of each other, see Sects. 20.1 and 20.2.1, and in particular the analysis following Eq. (20.23) – while in superstring theories living in ten-dimensional space-times *all* p-branes with $p = 0, 1, \ldots, 9$ appear as elementary excitations. An, *a priori* rather surprising, aspect is that both theories *predict* that for each charged p-brane there exists a tensor potential A_n of degree $n = p+1$, and that the corresponding Maxwell tensor $F_{n+1} = dA_n$ interacts with the brane precisely according to the set of fundamental equations (19.116)–(19.118), or, equivalently, according to the action (19.106).

We emphasize, however, that in this chapter we have derived the electrodynamics of charged p-branes, represented by the set of equations (19.116)–(19.118),

---

[5]The dimensions of the spaces where these theories live are determined by *internal consistency conditions*. Actually, in the case of superstring theories there is a plethora of possible consistent theories, but their maximum space-time dimension is $D = 10$. The structure of M-theory, which should unify all superstring theories in a space-time with one spatial dimension more, i.e. $D = 11$, has not yet been established in a definitive way.

independently, i.e. without invoking any requirement of unification, but rather by relying on very general principles: *relativistic invariance*, *gauge invariance*, *invariance under reparameterizations*, *current conservation*, and – last but not least – the *variational principle*. In fact, the latter ensures, in turn, via Noether's theorem in connection with Poincaré invariance, the conservation of energy, momentum, and D-dimensional angular momentum. M-theory and superstring theories – beyond their phenomenological relevance which is still under investigation – thus confirm the *universal validity* of these general principles, which goes far beyond the electrodynamics of charged particles in four dimensions.

19.4 Problems

19.1 Prove the graded Leibniz's rule for the product of differential forms (19.20).

19.2 Verify that the equations for differential forms (19.74) and (19.75) are equivalent to Maxwell's equations in tensor notation (19.76) and (19.77).
Hint: Perform first the Hodge dual of equation (19.75), rewriting it in the equivalent form $* d * F_{n+1} = (-)^{nD} j_n$, see formula (19.13), and then proceed as in (19.35).

19.3 Prove the identity for the square of the Hodge-duality operator (19.13), starting from the defining relations of the Hodge dual (19.10) and (19.11), and using the identities (1.37) and (19.14).

19.4 Verify that the total energy-momentum tensor (19.109) for the electrodynamics of a p-brane is conserved, whenever the brane coordinates y^μ and the Maxwell tensor $F^{\mu_1 \cdots \mu_{n+1}}$ satisfy the fundamental equations (19.116)–(19.118).
Hint: If the tensor $F^{\mu_1 \cdots \mu_{n+1}}$ satisfies Maxwell's equations (19.117) and (19.118), you can show that the electromagnetic energy-momentum tensor (19.110) satisfies the equation

$$\partial_\mu T_{\mathrm{em}}^{\mu\nu} = \frac{(-)^n}{n!} F^{\nu\mu_1 \cdots \mu_n} j_{\mu_1 \cdots \mu_n}$$

$$= \frac{(-)^n e}{n!} \varepsilon^{a_1 \cdots a_n} \int F^{\nu\mu_1 \cdots \mu_n} U_{\mu_1 a_1} \cdots U_{\mu_n a_n} \delta^D(x - y(\lambda)) \, d^n\lambda.$$

By an integration by parts you can show that the brane's energy-momentum tensor (19.111) satisfies the identity

$$\partial_\mu T_{\mathrm{b}}^{\mu\nu} = m \int \frac{\partial}{\partial\lambda^b} \left(\sqrt{g} \, g^{ab} \frac{\partial y^\nu}{\partial\lambda^a} \right) \delta^D(x - y(\lambda)) \, d^n\lambda.$$

19.5 Verify that the 00 component of the energy-momentum tensor (19.110) has the form (19.112).

19.6 *Causal motion of a string*. Consider a string moving in a D-dimensional space-time sweeping out a generic world sheet parameterized by $y^\mu(t, \sigma)$, meaning that we have chosen $\lambda^0 = t$, $\lambda^1 = \sigma$, so that $y^0(t, \sigma) = t$.

(a) Show that the induced metric (19.88) is given by

$$g_{ab} = \begin{pmatrix} 1 - v^2 & -\mathbf{v} \cdot \mathbf{w} \\ -\mathbf{v} \cdot \mathbf{w} & -w^2 \end{pmatrix}, \quad v^i = \frac{\partial y^i}{\partial t}, \quad w^i = \frac{\partial y^i}{\partial \sigma}, \quad i = 1, \ldots, D-1,$$

where $\mathbf{v}(t, \sigma)$ represents the *nominal* velocity of the point P of the string corresponding to the coordinate σ, at time t.

(b) Show that the radicand g defined in formulas (19.95) has the form

$$g = (-1)^p \det g_{ab} = w^2 - v^2 w^2 + (\mathbf{v} \cdot \mathbf{w})^2. \tag{19.119}$$

(c) Introducing the unit vector $\mathbf{n} = \mathbf{w}/w$, and noting that $\mathbf{w}(t, \sigma)$ is a vector tangent to the string at the point P at time t, show that the *physical* velocity of this point is given by

$$\mathbf{v}_{\text{ph}} = \mathbf{v} - (\mathbf{v} \cdot \mathbf{n})\,\mathbf{n}.$$

(d) Noting that the physical speed of the point P is given by $v_{\text{ph}} = \sqrt{v^2 - (\mathbf{v} \cdot \mathbf{n})^2}$, conclude that the string performs a *causal* motion, i.e. $v_{\text{ph}} \leq 1$, if and only if the radicand g in Eq. (19.119) is semi-positive definite, as stated after Eq. (19.95). A completely analogous proof shows that a generic p-brane performs a causal motion, i.e. $v_{\text{ph}} \leq 1$ for all points of the brane, if and only if the radicand g in (19.95) satisfies for all λ^a the inequality $g \geq 0$.

19.7 Show that the metric g_{ab} (19.90) is positive definite.
Hint: Analyze the matrix elements $W^a g_{ab} W^b$.

19.8 Show that the exterior derivative of the 1-form Φ_1 (19.24) is given by

$$d\Phi_1 = dx \wedge dy \left(\frac{\partial f}{\partial y} - \frac{\partial g}{\partial x} \right).$$

Verify that when Φ_1 is exact, it is also closed.

19.9 Consider a flat p-brane in static coordinates, parameterized as in Eqs. (19.99) and (19.100).

(a) Show that in this case the only non-vanishing components of the current (19.63) are those tangent to the brane, being given by (see the notation of Sect. 19.3.4)

$$j^{b_1 \cdots b_n}(t, \mathbf{x}) = e\, \delta^{b_1}_{a_1} \cdots \delta^{b_n}_{a_n}\, \varepsilon^{a_1 \cdots a_n} \int \delta^{D-n}(x^\perp)\, \delta^n(x^\| - \lambda)\, d^n\lambda$$

$$= e\, \varepsilon^{b_1 \cdots b_n} \delta^{D-n}(x^\perp).$$

(b) Recall that the charge density corresponds to the components $j^{0b_1\cdots b_p}(t,\mathbf{x})$. Integrating the latter over a p-cube of volume L^p along the parallel spatial directions, and over all space in the $D-n$ orthogonal spatial directions, show that the charge contained in this volume V is given by

$$Q^{b_1\cdots b_p} = \int_V j^{0b_1\cdots b_p}(t,\mathbf{x})\,d^{D-1}x = eL^p\,\varepsilon^{b_1\cdots b_p}. \qquad (19.120)$$

As the only independent charge component is $Q^{12\cdots p} = eL^p$, conclude that the constant e represents the *charge per unit volume* occupied by the p-brane.

References

1. M. Born, L. Infeld, Foundations of the new field theory. Proc. R. Soc. Lond. **144**, 425 (1934)
2. R. Bott, L.W. Tu, *Differential Forms in Algebraic Topology* (Springer, New York, 1982)
3. Y. Choquet-Bruhat, C. DeWitt-Morette, M. Dillard-Bleick, *Analysis, Manifolds and Physics* (North-Holland, Amsterdam, 1982)

Chapter 20
Magnetic Monopoles in Classical Electrodynamics

Though in the theoretical interpretation of the electromagnetic phenomena of nature the electric field $\mathbf{E}$ and the magnetic field $\mathbf{B}$ play *specular* roles in some respects, they exhibit substantial differences in others. A *mirror symmetry* between these fields is, for instance, evident in Maxwell's equations in empty space (20.1)–(20.4), which govern the propagation of the electromagnetic waves. These equations involve, in fact, the fields $\mathbf{E}$ and $\mathbf{B}$ on the same footing, except for a minus sign in Faraday's law of induction (20.3). Similarly, in the Poynting vector (2.147), which quantifies the electromagnetic energy flux, the interchange of $\mathbf{E}$ and $\mathbf{B}$ produces just a minus sign. Furthermore, these fields contribute in exactly the same way to the electromagnetic energy density (2.133). In maximum contrast to these mirror properties, in the Lorentz equation

$$\frac{d\mathbf{p}}{dt} = e\left(\mathbf{E} + \frac{\mathbf{v}}{c} \times \mathbf{B}\right)$$

the fields $\mathbf{E}$ and $\mathbf{B}$ play completely different roles. In particular, the effects of the magnetic field are suppressed by a relativistic factor v/c with respect to those of the electric field. However, the most significant distinction between these fields emerges in the presence of non-vanishing sources. In this case we have, in fact, the contrasting Gauss's laws

$$\mathbf{\nabla} \cdot \mathbf{E} = \rho, \qquad \mathbf{\nabla} \cdot \mathbf{B} = 0,$$

according to which a (static) charge distribution generates an electric field, but no magnetic one. In other words, conventional electrodynamics hosts electric charges, but no *magnetic* ones.

In this chapter we explore the possibility of introducing in classical electrodynamics elementary particles endowed with magnetic charge, so-called *magnetic monopoles*, and, more generally, elementary particles carrying both electric and magnetic charges, so-called *dyons*. At the same time, we want to maintain the

© Springer International Publishing AG, part of Springer Nature 2018
K. Lechner, *Classical Electrodynamics*, UNITEXT for Physics,
https://doi.org/10.1007/978-3-319-91809-9_20

same foundational basis of the theory in the absence of magnetic charges. *A priori* this purpose seems, however, to have little chance of success, as the conceptual framework of *classical electrodynamics* appears rather rigid. This theory, as formulated in Chap. 2, rests indeed on a series of basic cornerstones, in delicate balance between each other: relativistic invariance, gauge invariance, local conservation of four-momentum, of angular momentum, and of the electric charge, and, last but not least, the variational principle. Accordingly, any *ad hoc* modification of the fundamental equations of electrodynamics (2.20)–(2.22) runs the risk of clashing with one of these cornerstones. Although the hypothesis of the existence of magnetic monopoles has been considered already at the beginning of the last century, by H. Poincaré [1] and by J.J. Thomson [2, 3], in light of the rigidity of the theoretical paradigm we want to preserve, the main conclusion of this chapter, namely that *in the presence of magnetic monopoles classical electrodynamics remains a consistent theory*,[1] has to be regarded as a highly non-trivial result.

Charge quantization. Once assured that magnetic monopoles are compatible with the theoretical structure of *classical* electrodynamics, the hypothesis of the existence of this new type of particles becomes stimulating also from an experimental point of view. The – firmly established – experimental datum in question is the *quantization of the electric charge*, namely the observation that all electric charges present in nature are integer multiples of a fundamental charge, a phenomenon that still awaits for a theoretical explanation. However, as shown by P.A.M. Dirac in his seminal work [4] only a few years after the advent of quantum mechanics, if somewhere in the Universe there exists even only a single magnetic monopole, then the internal consistency of *quantum* electrodynamics inevitably predicts the quantization of the electric charge. We will provide a *semiclassical* derivation of *Dirac's quantization condition* in Sect. 20.4, and we will derive it in a *quantum mechanical* framework in Chap. 21. We begin the present chapter with a precise formulation of the "mirror symmetry" between the electric and magnetic fields invoked above – better known as *electromagnetic duality* – which comprehensibly is closely related with the possible existence of magnetic charges in nature.

20.1   Electromagnetic Duality

In empty space Maxwell's equations read

$$-\frac{\partial \mathbf{E}}{\partial t} + \boldsymbol{\nabla} \times \mathbf{B} = 0, \tag{20.1}$$

$$\boldsymbol{\nabla} \cdot \mathbf{E} = 0, \tag{20.2}$$

$$\frac{\partial \mathbf{B}}{\partial t} + \boldsymbol{\nabla} \times \mathbf{E} = 0, \tag{20.3}$$

---

[1]Of course, the inconsistencies arising from the self-interaction of charged particles survive, see Chap. 15.

$$\nabla \cdot \mathbf{B} = 0. \tag{20.4}$$

This system of equations remains unchanged if we operate the replacements

$$\mathbf{E} \to \mathbf{B}, \qquad \mathbf{B} \to -\mathbf{E}. \tag{20.5}$$

The transformations (20.5) represent thus a *discrete* symmetry of Maxwell's equations in empty space, called *electromagnetic duality*. Obviously, in the presence of electric charges this symmetry is violated, due to the presence of source terms in Eqs. (20.1) and (20.2), but not in Eqs. (20.3) and (20.4). Despite their three-dimensional appearance, the duality transformations (20.5) actually respect relativistic invariance. To show it, let us introduce the *electromagnetic dual* of the Maxwell tensor $F^{\mu\nu}$ as

$$\widetilde{F}^{\mu\nu} = \frac{1}{2}\, \varepsilon^{\mu\nu\rho\sigma} F_{\rho\sigma}, \tag{20.6}$$

which is likewise an antisymmetric covariant tensor. Applying the duality operation twice, and recalling the first identity in (1.41)

$$\varepsilon^{\alpha\beta\gamma\delta}\, \varepsilon_{\alpha\beta\mu\nu} = -4\, \delta^{\gamma}_{[\mu}\, \delta^{\delta}_{\nu]},$$

we find

$$\widetilde{\widetilde{F}}^{\mu\nu} = \frac{1}{2}\, \varepsilon^{\mu\nu\rho\sigma} \widetilde{F}_{\rho\sigma} = -F^{\mu\nu}, \tag{20.7}$$

meaning that the *square* of the electromagnetic duality corresponds to *minus* the identity map. In terms of the Maxwell tensor $F^{\mu\nu}$ and its dual $\widetilde{F}^{\mu\nu}$ the transformations (20.5) amount to the Lorentz-covariant replacements

$$F^{\mu\nu} \to \widetilde{F}^{\mu\nu}, \qquad \widetilde{F}^{\mu\nu} \to -F^{\mu\nu}. \tag{20.8}$$

Actually, the second replacement in (20.8) follows from the *basic* first replacement $F^{\mu\nu} \to \widetilde{F}^{\mu\nu}$, via the identity (20.7). To verify that the substitutions (20.8) are equivalent to the substitutions (20.5) it suffices to calculate the dual electric and magnetic fields entailed by the definition (20.6)

$$
\begin{aligned}
\widetilde{E}^i &= \widetilde{F}^{i0} = \frac{1}{2}\, \varepsilon^{i0jk} F_{jk} = -\frac{1}{2}\, \varepsilon^{ijk} F^{jk} = B^i, \\
\widetilde{B}^i &= -\frac{1}{2}\, \varepsilon^{ijk} \widetilde{F}^{jk} = -\frac{1}{2}\, \varepsilon^{ijk} \varepsilon^{jkl0} F_{l0} = -\frac{1}{2}\, \varepsilon^{ijk} \varepsilon^{ljk} E^l = -E^i.
\end{aligned}
\tag{20.9}
$$

Electromagnetic duality and Hodge duality. In the language of *differential forms* the electromagnetic duality admits a simple geometric interpretation. In fact, in Sect. 19.1.2 we have associated with the Maxwell tensor $F^{\mu\nu}$ the 2-form F (19.26), and, according to the definitions (19.10) and (19.11), we have introduced its Hodge dual 2-form $*F$

$$F = \frac{1}{2}\, dx^\nu \wedge dx^\mu F_{\mu\nu}, \qquad *F = \frac{1}{2}\, dx^\nu \wedge dx^\mu \widetilde{F}_{\mu\nu}. \qquad (20.10)$$

The substitutions (20.8) then simply amount to the replacements

$$F \to *F, \qquad *F \to -F.$$

In the language of differential forms, the *electromagnetic* duality is hence mapped into *Hodge* duality. In particular, within this formalism the minus sign in the second replacement in (20.8) is a consequence of the general identity for the square of the $*$ operator (19.13), which for $p = 2$ and $D = 4$ gives indeed

$$*^2 = -1. \qquad (20.11)$$

Finally, in the presence of electric sources Maxwell's equations (2.21) and (2.22) can be rewritten in terms of the tensors $F^{\mu\nu}$ and $\widetilde{F}^{\mu\nu}$ in the equivalent form

$$\partial_\mu F^{\mu\nu} = j_e^\nu, \qquad (20.12)$$

$$\partial_\mu \widetilde{F}^{\mu\nu} = 0. \qquad (20.13)$$

Henceforth, we denote the current j^μ with the symbol j_e^μ to signal that it represents the *electric* four-current. In light of equations (20.12) and (20.13) it is clear that, if we want Maxwell's equations to remain invariant under the duality transformations (20.8) also in the presence of electric sources, then we must turn on a *magnetic* four-current at the right hand side of the Bianchi identity (20.13).

20.2   Electrodynamics of Dyons

In this section we introduce the equations which govern the electrodynamics of a system of N particles carrying generic electric charges e_r and magnetic charges g_r. These equations generalize, thus, the fundamental equations (2.20)–(2.22) relative to a system of only electrically charged particles. If for a particle we have $e_r \neq 0$ and $g_r \neq 0$ it is called a *dyon*, if $e_r \neq 0$ and $g_r = 0$ it is called an (electric) *charge*, and if $e_r = 0$ and $g_r \neq 0$ it is called a (magnetic) *monopole*. In Sect. 20.2.1 we first adapt Maxwell's equations to a generic system of dyons, and in Sect. 20.2.3 we then derive the corresponding new Lorentz equations, by imposing that the total four-momentum remains locally conserved, despite the changes made in Maxwell's equations.

20.2.1   Generalized Maxwell Equations

Denoting the world lines of the particles, as usual, by $y_r^\mu(s_r)$ with $r = 1, \ldots, N$, we associate to our system of dyons the electric and magnetic four-currents

$$j_e^\mu = \sum_r e_r \int \delta^4(x - y_r)\, dy_r^\mu, \qquad j_m^\mu = \sum_r g_r \int \delta^4(x - y_r)\, dy_r^\mu. \quad (20.14)$$

In the same way as in Sect. 2.3.2 we have shown that the electric current j_e^μ is identically conserved, one proves that also the magnetic current j_m^μ is conserved. We thus have the local conservation laws

$$\partial_\mu j_e^\mu = 0, \qquad \partial_\mu j_m^\mu = 0, \quad (20.15)$$

irrespective of the values of the charges e_r and g_r. In particular, the total magnetic charge $G = \int j_m^0\, d^3x = \sum_r g_r$ is a constant of motion. Given the currents (20.14), we introduce the *generalized Maxwell equations*

$$\partial_\mu \widetilde{F}^{\mu\nu} = j_m^\nu, \quad (20.16)$$
$$\partial_\mu F^{\mu\nu} = j_e^\nu. \quad (20.17)$$

These equations are, in fact, consistent with the conservation laws (20.15), in that both $F^{\mu\nu}$ and $\widetilde{F}^{\mu\nu}$ are antisymmetric tensors. As in the case of the original electrodynamics, see Eqs. (2.39)–(2.41), it is straightforward to show that the generalized Bianchi "identity" (20.16) can be written in the three equivalent ways

$$-\frac{1}{2}\, \varepsilon^{\mu\nu\rho\sigma} \partial_\nu F_{\rho\sigma} = j_m^\mu, \quad (20.18)$$
$$\partial_\mu F_{\nu\rho} + \partial_\nu F_{\rho\mu} + \partial_\rho F_{\mu\nu} = \varepsilon_{\alpha\mu\nu\rho}\, j_m^\alpha, \quad (20.19)$$
$$\partial_{[\mu} F_{\nu\rho]} = \frac{1}{3}\, \varepsilon_{\alpha\mu\nu\rho}\, j_m^\alpha. \quad (20.20)$$

Failure of the variational principle. The introduction of a non-vanishing magnetic current at the right hand side of the original Bianchi identity (20.13) raises *a priori* a series of problems, potentially mining the internal consistency of the electrodynamics of dyons. Here we focus on the most problematic aspect of the modified Bianchi identity (20.16), which regards the variational principle. As the right hand side of this equation is a four-vector, the resulting new dynamics naturally preserves Poincaré invariance. Therefore, according to Noether's theorem, the conservation of the total four-momentum and angular momentum would be automatically guaranteed, whenever there exists a Poincaré-invariant *action* giving rise to the generalized Maxwell equations (20.16) and (20.17). However, we know that in order to write an action producing the dynamics of the electromagnetic field, we need a vector potential A^μ. If the magnetic current j_m^μ is zero, the Bianchi identity (20.13) itself is equivalent to the existence of a vector potential, but for a non-vanishing magnetic current this identity is violated, and so there exists no *natural* way to introduce a vector potential in the theory. This leads us to the unpleasant conclusion that in the presence of magnetic sources there exists no *canonical* action giving rise to the generalized fundamental equations of electrodynamics.[2] In fact, as a consequence, the validity of the main conservation laws is no longer guaranteed *a priori*. Despite this rather discouraging starting situation, in Sect. 20.2.3 we will provide a consistent generalization of the Lorentz equations (2.20) to the case of dyons, which preserves Poincaré invariance and maintains, in addition, *all* conservation laws of the electrodynamics without magnetic charges.

Differential forms. Some properties of the generalized Maxwell equations become more transparent if they are translated into the language of differential forms. For this purpose, we keep the definitions of the 2-forms F and $*F$ of formulas (20.10), and we introduce in addition the current 1-forms

$$j_e = dx^\mu j_{e\mu}, \qquad j_m = dx^\mu j_{m\mu}.$$

If we rewrite the modified Bianchi identity (20.16) as in Eq. (20.20), it is then easy to recognize that in this formalism equations (20.16) and (20.17) take the form (compare with Eqs. (19.28) and (19.29))

$$dF = - * j_m, \tag{20.21}$$

$$d * F = *j_e. \tag{20.22}$$

Similarly, the continuity equations (20.15) go over into the identities $d * j_e = 0 = d * j_m$. The latter guarantee, in turn, the *algebraic* consistency of equations (20.21) and (20.22), in the sense that these equations equate closed forms to closed forms.

Can electric and magnetic charges be identified? From the generalized Maxwell equations we read off that the electric and magnetic charges e_r and g_r carry the same units of measure, at the same footing of the tensors F and $*F$. A systematic identification of these two types of charges would then require to enforce one of the two constraints

$$* F = \pm F, \tag{20.23}$$

namely, to require F to be an *(anti-)self-dual* 2-form, see the definitions (19.15). In fact, in this case equations (20.21) and (20.22) would imply the identifications

---

[2]In the presence of magnetic charges, in order to write an action one must renounce to at least one of the fundamental properties commonly required for a relativistic action, for instance, *locality* or *manifest* Lorentz invariance [5]. Nevertheless, the equations of motion derived from such actions are local and Lorentz invariant. However, the absence of a local manifestly Lorentz-invariant action raises major problems concerning the *quantization* of the theory. In fact, the difficulties involved in the quantization process, deriving from the absence of a canonical action, have considerably delayed the construction of an internally consistent *relativistic quantum field theory of dyons*, which indeed has been completed only in 1979 [6].

$j_e^\mu = \pm j_m^\mu$. However, as shown in (19.15)–(19.17), in four space-time dimensions there are no (anti-)self-dual 2-forms, meaning that the constraints (20.23) admit the only solution $F = 0$. Obviously, the same conclusion is reached if one rewrites the constraints (20.23) in three-dimensional notation. Choosing, for instance, the plus sign, from formulas (20.9) one derives that the single equation $*F = F$ amounts to the equations

$$\mathbf{B} = \mathbf{E}, \qquad -\mathbf{E} = \mathbf{B},$$

which imply $\mathbf{E} = \mathbf{B} = 0$. We conclude, thus, that in four space-time dimensions there is no way to identify an electric charge, or current, with a magnetic one. In other words, there exist no *self-dual* dyons.

Branes, dual branes, and self-dual branes. For comparison, we briefly analyze the problem considered above in the more general case of *extended* charged objects, in a space-time of arbitrary dimension D. In a space-time of dimension D different from four, insisting for the moment on charged *particles*, the first obstacle we encounter is that the Hodge dual $*F$ of a 2-form F is no longer a 2-form, but rather a $(D-2)$-form. In fact, in this case, whereas Eq. (20.22) equates a $(D-1)$-form to a $(D-1)$-form, Eq. (20.21) equating a 3-form to a $(D-1)$-form would be meaningless. Actually, in D dimensions the electromagnetic dual of a charged particle is no longer a magnetic *particle*, but rather a magnetic $(D-4)$-*brane*. To see it, more in general, we compare the system of dual equations (20.21) and (20.22) with the system (19.74) and (19.75), describing an *electric p-brane* coupled to an electromagnetic field strength represented by a $(p+2)$-form F. We recall that for a p-brane, according to formulas (19.70) and (19.63), the electric current j_e in the generalized Maxwell equation (20.22) rises to a $(p+1)$-form. The exterior derivative dF is then a $(p+3)$-form, and so the magnetic current j_m of the generalized Bianchi identity (20.21) rises to a $(D-p-3)$-form. But, again according to formulas (19.70) and (19.63), such a current would represent a *magnetic $(D-p-4)$-brane*.

In conclusion, in a D-dimensional space-time the dual of an electric p-brane is a magnetic $(D-p-4)$-brane. As a consequence, the potential identification of a p-brane with its dual $(D-p-4)$-brane is tied to the condition $p = D-p-4$, which, in turn, constrains the space-time to be of the *even* dimension $D = 2(p+2)$. Indeed, in this case the dual $*F$ of the $(p+2)$-form F is again a $(p+2)$-form, so that now it makes again sense – at least *a priori* – to impose on F an (anti-)self-duality condition like (20.23). Equations (20.21) and (20.22) would then again imply the equality $j_e = \pm j_m$, which would indeed identify a p-brane as an *(anti-)self-dual* p-brane. However, as shown in (19.15)–(19.18), (anti-)self-dual $(p+2)$-forms F in a $2(p+2)$-dimensional space-time exist only for *odd* values of p. This implies that, eventually, (anti-)self-dual branes exist only in space-times with dimensions

$$D = 2, 6, 10, 14, \ldots \tag{20.24}$$

In particular, in a six-dimensional space-time there exist (anti-)self-dual *strings*, in a ten-dimensional space-time there exist (anti-)self-dual *3-branes*, and so on. Finally, by writing out the constraint (20.23) in terms of the generalized electric and magnetic fields (19.78) and (19.79) we obtain the relation

$$B^{i_1 \cdots i_{p+2}} = \pm \frac{1}{(p+1)!} \, \varepsilon^{i_1 \cdots i_{p+2} j_1 \cdots j_{p+1}} E^{j_1 \cdots j_{p+1}}.$$

This means that (anti-)self-dual p-branes are distinguished by the property of generating electric and magnetic fields which are linearly dependent from each other, in the sense that one is $\pm$ the *spatial* Hodge dual of the other.

Self-dual instantons. As suggested by the sequence of values (20.24), in a two-dimensional space-time, $D = 2$, there exist (anti-)self-dual (-1)-branes, also called *instantons*. Instantons, which also play an important role in superstring theory, see, for instance, Ref. [7], are elementary defects which, in a D-dimensional space-time, carry a 0-form current of the type $j_e(x) = e \, \delta^D(x - y)$. These defects are localized at the space-time point $y^\mu = (y^0, \mathbf{y})$, and so they live only at the *instant* y^0. (Anti-)self-dual instantons in two space-time dimensions create thus, according to Eqs. (20.21)–(20.23), an (anti-)self-dual field-strength 1-form $F = dx^\mu F_\mu$. It is straightforward to show that for a self-dual instanton ($*F = F$, $g = e$) localized at the space-time origin $y^\mu = (0, 0)$, these equations entail the unique causal solution $F^0(t, x) = -F^1(t, x) = e \, H(t) \, \delta(t + x)$. This solution represents an electromagnetic shock-wave field, see Sect. 6.3.2, created at time $t = 0$ at the point $x = 0$, which then propagates at the speed of light along the negative x axis. Conversely, an anti-self-dual instanton ($*F = -F$, $g = -e$) localized at $y^\mu = (0, 0)$ would create the shock-wave field $F^0(t, x) = F^1(t, x) = e \, H(t) \, \delta(t - x)$ propagating along the positive x axis.

Generalized Maxwell equations and electromagnetic duality. The generalized Maxwell equations (20.16) and (20.17) are now duality-invariant, if we associate with the transformations of the fields (20.8), equivalent to the replacements (20.5), the transformations of the currents

$$j_e^\mu \to j_m^\mu, \qquad j_m^\mu \to -j_e^\mu. \tag{20.25}$$

Correspondingly, the charges of the dyons transform according to

$$e_r \to g_r, \qquad g_r \to -e_r. \tag{20.26}$$

In the language of differential forms the above transformations amount to

$$F \to *F, \qquad *F \to -F, \qquad j_e \to j_m, \qquad j_m \to -j_e. \tag{20.27}$$

Ultimately, the form of the modified Bianchi identity (20.21) is dictated by duality invariance. This equation follows, in fact, from the Maxwell equation (20.22), if one operates in the latter the substitutions (20.27). In other words, duality exchanges the (modified) Bianchi identity with the (dynamical) Maxwell equation. Finally, in three-dimensional notation the generalized Maxwell equations (20.16) and (20.17) assume the (still manifestly duality-invariant) form

$$\nabla \cdot \mathbf{E} = j_e^0,$$

$$\nabla \times \mathbf{B} - \frac{\partial \mathbf{E}}{\partial t} = \mathbf{j}_e,$$

$$\nabla \cdot \mathbf{B} = j_m^0, \qquad (20.28)$$

$$-\nabla \times \mathbf{E} - \frac{\partial \mathbf{B}}{\partial t} = \mathbf{j}_m.$$

The magnetic field is now generated not only by moving electric charges, but also by *static magnetic monopoles*, whereas the electric field is now generated not only by static electric charges, but also by *moving magnetic monopoles*.

20.2.2 *SO(2)-Duality and Z_4-Duality*

The set of linear transformations that leave the generalized Maxwell equations (20.28) invariant is, actually, larger than the discrete map represented by the replacements (20.5) and (20.25). In fact, these equations are invariant under the family of transformations depending on the continuous parameter $\varphi \in [0, 2\pi]$

$$\mathbf{E}' = \cos\varphi\,\mathbf{E} + \sin\varphi\,\mathbf{B}, \qquad (20.29)$$

$$\mathbf{B}' = -\sin\varphi\,\mathbf{E} + \cos\varphi\,\mathbf{B}, \qquad (20.30)$$

$$j_e'^\mu = \cos\varphi\,j_e^\mu + \sin\varphi\,j_m^\mu, \qquad (20.31)$$

$$j_m'^\mu = -\sin\varphi\,j_e^\mu + \cos\varphi\,j_m^\mu, \qquad (20.32)$$

the proof being left as exercise. With this assertion we understand that, if the configuration $\{\mathbf{E}, \mathbf{B}, j_e^\mu, j_m^\mu\}$ satisfies the system of equations (20.28), then so does the configuration $\{\mathbf{E}', \mathbf{B}', j_e'^\mu, j_m'^\mu\}$. For $\varphi = \pi/2$ the transformations (20.29)–(20.32) reduce to the original discrete electromagnetic duality. We can give the transformations of the fields (20.29) and (20.30) a manifestly Lorentz-invariant form, by noting that they can be rewritten in the equivalent form

$$F'^{\mu\nu} = \cos\varphi\,F^{\mu\nu} + \sin\varphi\,\widetilde{F}^{\mu\nu}, \qquad (20.33)$$

$$\widetilde{F}'^{\mu\nu} = -\sin\varphi\,F^{\mu\nu} + \cos\varphi\,\widetilde{F}^{\mu\nu}. \qquad (20.34)$$

Notice that these transformation laws are compatible with each other, in that the replacement (20.34) is nothing else than the Hodge dual of (20.33). The invariance of the generalized Maxwell equations under the transformations (20.29)–(20.34) can be made *manifest*, by adopting a two-dimensional vector notation. In fact, by introducing the two-component vectors

$$\mathcal{F}^{\mu\nu} = \begin{pmatrix} F^{\mu\nu} \\ \widetilde{F}^{\mu\nu} \end{pmatrix}, \qquad J^\mu = \begin{pmatrix} j_e^\mu \\ j_m^\mu \end{pmatrix}, \qquad Q_r = \begin{pmatrix} e_r \\ g_r \end{pmatrix}, \qquad (20.35)$$

Equations (20.16) and (20.17) can be recast as the single equation for these doublets

$$\partial_\mu \mathcal{F}^{\mu\nu} = J^\nu. \tag{20.36}$$

Moreover, by introducing the matrix

$$\mathcal{R}(\varphi) = \begin{pmatrix} \cos\varphi & \sin\varphi \\ -\sin\varphi & \cos\varphi \end{pmatrix},$$

the transformations (20.31)–(20.34) can be recast in the compact form

$$\mathcal{F}'^{\mu\nu} = \mathcal{R}(\varphi)\,\mathcal{F}^{\mu\nu}, \qquad J'^\mu = \mathcal{R}(\varphi)\,J^\mu, \qquad Q'_r = \mathcal{R}(\varphi)\,Q_r, \tag{20.37}$$

so that Eq. (20.36) is now manifestly invariant. The matrix $\mathcal{R}(\varphi)$ corresponds to a two-dimensional rotation by an angle φ, and the set of these matrices forms the continuous group $SO(2)$. We have thus shown that in the presence of an arbitrary set of dyons Maxwell's equations are invariant under the *continuous* duality group $SO(2)$. As observed above, the original duality transformations (20.27) are represented by the $SO(2)$-matrix corresponding to the angle $\varphi = \pi/2$

$$\mathcal{R}\left(\frac{\pi}{2}\right) = \begin{pmatrix} 0 & 1 \\ -1 & 0 \end{pmatrix}. \tag{20.38}$$

As this matrix satisfies the identity $\mathcal{R}(\frac{\pi}{2})\mathcal{R}(\frac{\pi}{2}) = -\mathbf{1}$, it generates the discrete subgroup of $SO(2)$ formed by the four elements $\{\mathbf{1}, -\mathbf{1}, \mathcal{R}(\frac{\pi}{2}), -\mathcal{R}(\frac{\pi}{2})\}$, which is isomorphic to the group Z_4. The original electromagnetic duality corresponds, thus, to the *discrete* duality group Z_4.

In conclusion, the *classical* electrodynamics of dyons – represented by the generalized Maxwell equations (20.16) and (20.17) – is invariant under the duality group $SO(2)$, which generalizes the original invariance subgroup Z_4. In contrast, as we will explain at the end of the chapter, in *quantum* theory the dynamics of dyons can be formulated in two physically *inequivalent* ways, according to whether one implements the duality group $SO(2)$, or rather its subgroup Z_4, see Sect. 20.4.2.

20.2.3 Generalized Lorentz Equation and Conservation Laws

If we want to promote the generalized Maxwell equations (20.16) and (20.17) to new *fundamental* equations, then it is essential that, also in the presence of magnetic charges, there exists a conserved total energy-momentum tensor $T^{\mu\nu}$. As these equations admit no canonical action we can longer resort to Noether's theorem, and so, to find such a tensor, we proceed heuristically taking duality invariance as a guiding principle. First of all, the contribution of the particles $T_p^{\mu\nu}$ (2.138), depending neither on the charges, nor on the electromagnetic field, is trivially duality invariant. But also the electromagnetic energy-momentum tensor $T_{\text{em}}^{\mu\nu}$ (2.137) is invariant under the whole duality group $SO(2)$. The simplest way to see it is to perform in the component expressions (2.144)–(2.146) the replacements (20.29) and (20.30), and to verify that they remain unaffected. This suggests to maintain for both tensors $T_{\text{em}}^{\mu\nu}$ and $T_{\text{p}}^{\mu\nu}$ their original expressions

$$T_{\text{em}}^{\mu\nu} = F^{\mu\alpha} F_\alpha{}^\nu + \frac{1}{4}\, \eta^{\mu\nu} F^{\alpha\beta} F_{\alpha\beta}, \quad T_{\text{p}}^{\mu\nu} = \sum_r m_r \int u_r^\mu\, u_r^\nu\, \delta^4(x - y_r)\, ds_r,$$

and to set again $T^{\mu\nu} = T_{\text{em}}^{\mu\nu} + T_{\text{p}}^{\mu\nu}$. To find out under which conditions the energy-momentum tensor $T^{\mu\nu}$ remains conserved, we evaluate separately the four-divergence of its two addends. We begin with the electromagnetic contribution[3]

$$\begin{aligned}
\partial_\mu T_{\text{em}}^{\mu\nu} &= j_{\text{e}}^\alpha F_\alpha{}^\nu + F^{\mu\alpha} \partial_\mu F_\alpha{}^\nu + \frac{1}{2} F^{\alpha\beta} \partial^\nu F_{\alpha\beta} \\
&= -F^{\nu\alpha} j_{\text{e}\alpha} + \frac{1}{2} F_{\alpha\beta} \left(\partial^\alpha F^{\beta\nu} + \partial^\beta F^{\nu\alpha} + \partial^\nu F^{\alpha\beta} \right) \\
&= -F^{\nu\alpha} j_{\text{e}\alpha} - \frac{1}{2} F_{\alpha\beta}\, \varepsilon^{\alpha\beta\nu\mu} j_{\text{m}\mu} \\
&= -F^{\nu\alpha} j_{\text{e}\alpha} - \widetilde{F}^{\nu\alpha} j_{\text{m}\alpha} \\
&= -\sum_r \int \left(e_r F^{\nu\alpha} + g_r \widetilde{F}^{\nu\alpha} \right) u_{r\alpha}\, \delta^4(x - y_r)\, ds_r.
\end{aligned}$$

In the first line we replaced $\partial_\mu F^{\mu\alpha}$ with j_{e}^α, enforcing Maxwell's equation (20.17), in the third line we have used the modified Bianchi identity (20.19), and in the last line we have inserted the currents (20.14). As the four-divergence of the tensor $T_{\text{p}}^{\mu\nu}$ has the standard expression (2.141)

$$\partial_\mu T_{\text{p}}^{\mu\nu} = \sum_r \int \frac{dp_r^\nu}{ds_r}\, \delta^4(x - y_r)\, ds_r,$$

by summing these four-divergences we obtain

$$\partial_\mu T^{\mu\nu} = \sum_r \int \left(\frac{dp_r^\nu}{ds_r} - \left(e_r F^{\nu\alpha} + g_r \widetilde{F}^{\nu\alpha} \right) u_{r\alpha} \right) \delta^4(x - y_r)\, ds_r.$$

Therefore, if we want to keep the total energy-momentum tensor locally conserved, we are forced to replace the original Lorentz equations (2.20) with the *generalized Lorentz equations*

$$\frac{dp_r^\nu}{ds_r} = \left(e_r F^{\nu\alpha} + g_r \widetilde{F}^{\nu\alpha} \right) u_{r\alpha}. \tag{20.39}$$

---

[3]For simplicity, we neglect here the difficulties related to the singularities generated by the self-interaction of the dyons, which can, however, be resolved with the same method applied in Chap. 16, for details, see Ref. [8].

By construction, these new equations are invariant under $SO(2)$-duality. In fact, the tensors appearing at second member can be rewritten as the two-dimensional scalar products (see the definitions (20.35) and the transformations (20.37))

$$e_r F^{\nu\alpha} + g_r \widetilde{F}^{\nu\alpha} = Q_r \cdot \mathcal{F}^{\nu\alpha}.$$

Equations (20.16), (20.17) and (20.39) represent the *fundamental equations of the classical electrodynamics of dyons*, in replacement of the original system (2.20)–(2.22). Using the expressions (20.9) it is straightforward to rewrite the new Lorentz equations (20.39) in the three-dimensional notation. Reinserting the speed of light c we obtain

$$\frac{d\mathbf{p}_r}{dt} = e_r \left(\mathbf{E} + \frac{\mathbf{v}_r}{c} \times \mathbf{B} \right) + g_r \left(\mathbf{B} - \frac{\mathbf{v}_r}{c} \times \mathbf{E} \right), \qquad (20.40)$$

$$\frac{d\varepsilon_r}{dt} = \mathbf{v}_r \cdot \left(e_r \, \mathbf{E} + g_r \, \mathbf{B} \right). \qquad (20.41)$$

With respect to standard electrodynamics, a dyon of electric charge e_r and magnetic charge g_r is, thus, subject to the additional Lorentz force $g_r \left(\mathbf{B} - \mathbf{v}_r \times \mathbf{E}/c \right)$. From the fundamental equations in three-dimensional notation (20.28), (20.40) and (20.41) we infer another important consequence of duality invariance: the dynamics of a system of only monopoles ($e_r = 0$ for all r) is *physically equivalent* to the dynamics of a system of only charges ($g_r = 0$ for all r), i.e. to the original electrodynamics. This means that a Universe containing only charges would be indistinguishable from a Universe containing only monopoles. Actually, in such a world, which set of particles we call *charges* or *monopoles* is merely a matter of convention.

As the total energy-momentum tensor is conserved and symmetric, the standard angular-momentum density

$$M^{\mu\alpha\beta} = x^\alpha T^{\mu\beta} - x^\beta T^{\mu\alpha}$$

is still conserved, $\partial_\mu M^{\mu\alpha\beta} = 0$. Moreover, as the tensors $T^{\mu\nu}$ and $M^{\mu\alpha\beta}$ have the same form of the electrodynamics of only charges, the conserved quantities maintain the same expressions derived in Sect. 2.4. For future reference, we recall the expression of the total spatial angular momentum (2.161)

$$\mathbf{L} = \sum_r (\mathbf{y}_r \times \mathbf{p}_r) + \frac{1}{c} \int \mathbf{x} \times (\mathbf{E} \times \mathbf{B}) \, d^3x = \mathbf{L}_\mathrm{p} + \mathbf{L}_\mathrm{em}. \qquad (20.42)$$

Notice that for a *static* system of only charges, or of only monopoles, the angular momentum of the electromagnetic field $\mathbf{L}_\mathrm{em}$ is zero, because in the first case $\mathbf{B}$ vanishes, and in the second case $\mathbf{E}$ vanishes. In contrast, in Sect. 20.3 we shall see that for a static system composed of charges *and* monopoles $\mathbf{L}_\mathrm{em}$ will be different from zero, with far-reaching consequences.

Self-interaction and Lorentz-Dirac equation. As in standard electrodynamics, the second member of the Lorentz equations (20.39) involve the electromagnetic field created by the rth dyon itself, and so they are divergent. However, as in that case, these divergences can be absorbed by an (infinite) renormalization of the dyons' masses. The form of the resulting Lorentz-Dirac equations, namely

$$\frac{dp_r^\nu}{ds_r} = \frac{e_r^2 + g_r^2}{6\pi}\left(\frac{dw_r^\nu}{ds_r} + w_r^2 u_r^\nu\right) + \left(e_r F_r^{\nu\alpha} + g_r \widetilde{F}_r^{\nu\alpha}\right) u_{r\alpha}, \qquad (20.43)$$

where $F_r^{\mu\nu}$ is the total external field acting on the rth dyon, is essentially determined by the $SO(2)$-duality invariance. In fact, the intensity of the radiation reaction force in Eq. (20.43), proportional to $e_r^2 + g_r^2 = Q_r \cdot Q_r$, is $SO(2)$ invariant, see (20.35) and the $SO(2)$ transformations (20.37). Finally, also for the electrodynamics of dyons it can be shown that the Lorentz-Dirac equations (20.43) are *imposed* by the conservation of the *renormalized* total energy-momentum tensor of the system, see Ref. [8].

20.3 Two-Dyon System

From the analysis carried out so far it is clear that genuinely new physical phenomena arise only in particle systems including both charges and monopoles. The simplest system of this kind is composed of a charge and of a monopole or, if we want to keep track of duality invariance in a manifest manner, of two dyons. The main difference that we will find in the interaction between two dyons with respect to the interaction between two charges consists in the appearance of a mutual interaction force of *first* order in $1/c$, that we call the *dyon force*. In contrast, in the mutual interaction forces $\mathbf{F}_{12}$ and $\mathbf{F}_{21}$ between two charges (15.105) there are no terms of order $1/c$, in that the relativistic corrections start with terms of order $1/c^2$. The dyon force arises in the Lorentz equations (20.40) because – according to the generalized Maxwell equations (20.28) – (i) the magnetic field acquires *Coulomblike* terms of order zero in $1/c$ and, (ii) the electric field acquires *magnetic* terms of first order $1/c$. Both these terms are absent in the non-relativistic expansions of the fields (7.78) in standard electrodynamics. As this section is mainly devoted to a study of the effects of this new force of order $1/c$ on the dynamics of a two-dyon system, we will neglect all contributions of order $1/c^2$ and higher.

20.3.1 Relative Motion and Dyon Force

We begin by writing out the electric field $\mathcal{E}$ and the magnetic field $\mathcal{B}$ generated by a dyon with charges (e, g), up to terms of first order in $1/c$. These fields can be easily derived using i) the expansions (7.78), ii) the duality rules (20.5) and (20.26), and iii) the fact that the generalized Maxwell equations (20.28) are still linear. One obtains

$$\mathcal{E} = \frac{e\mathbf{R}}{4\pi R^3} - \frac{g}{4\pi}\frac{\mathbf{V}}{c} \times \frac{\mathbf{R}}{R^3},$$

$$\mathcal{B} = \frac{g\mathbf{R}}{4\pi R^3} + \frac{e}{4\pi}\frac{\mathbf{V}}{c} \times \frac{\mathbf{R}}{R^3}, \tag{20.44}$$

where $\mathbf{R} = \mathbf{x} - \mathbf{y}(t)$, $\mathbf{y}(t)$ is the trajectory of the dyon, and $\mathbf{V}(t) = \dot{\mathbf{y}}(t)$ its velocity. Consider now two dyons, one with charges (e_1, g_1) and mass m_1, and one with charges (e_2, g_2) and mass m_2, and denote their trajectories, velocities, and accelerations by $\mathbf{y}_i$, $\mathbf{v}_i$, and $\mathbf{a}_i$, respectively ($i = 1, 2$). From formulas (20.44), for the total electric and magnetic fields produced by the two dyons we then obtain

$$\mathbf{E} = \mathbf{E}_1 + \mathbf{E}_2 = \left(\frac{e_1\mathbf{r}_1}{4\pi r_1^3} - \frac{g_1}{4\pi}\frac{\mathbf{v}_1}{c} \times \frac{\mathbf{r}_1}{r_1^3}\right) + \left(\frac{e_2\mathbf{r}_2}{4\pi r_2^3} - \frac{g_2}{4\pi}\frac{\mathbf{v}_2}{c} \times \frac{\mathbf{r}_2}{r_2^3}\right),$$

$$\mathbf{B} = \mathbf{B}_1 + \mathbf{B}_2 = \left(\frac{g_1\mathbf{r}_1}{4\pi r_1^3} + \frac{e_1}{4\pi}\frac{\mathbf{v}_1}{c} \times \frac{\mathbf{r}_1}{r_1^3}\right) + \left(\frac{g_2\mathbf{r}_2}{4\pi r_2^3} + \frac{e_2}{4\pi}\frac{\mathbf{v}_2}{c} \times \frac{\mathbf{r}_2}{r_2^3}\right), \tag{20.45}$$

where we have set

$$\mathbf{r}_i = \mathbf{x} - \mathbf{y}_i.$$

We now use these fields to write out the Lorentz equations (20.40) for the two particles. To derive the equation for dyon 2, we must set in (20.40) $\mathbf{E} = \mathbf{E}_1$ and $\mathbf{B} = \mathbf{B}_1$, and we must evaluate these fields at the position $\mathbf{x} = \mathbf{y}_2$. Notice that we do not include the (divergent) contributions due to $\mathbf{E}_2$ and $\mathbf{B}_2$. In fact, the latter must actually be replaced with the *radiation reaction forces* of the Lorentz-Dirac equations (20.43). However, from Sect. 15.2.5 we know that these forces are of order $1/c^3$. In this way, introducing the relative coordinate

$$\mathbf{r} = \mathbf{y}_2 - \mathbf{y}_1,$$

and considering that at first order in in $1/c$ we have $\mathbf{p}_i = m_i\mathbf{v}_i$, the Lorentz equation (20.40) for dyon 2 becomes

$$m_2\mathbf{a}_2 = e_2\left(\mathbf{E}_1(\mathbf{y}_2) + \frac{\mathbf{v}_2}{c} \times \mathbf{B}_1(\mathbf{y}_2)\right) + g_2\left(\mathbf{B}_1(\mathbf{y}_2) - \frac{\mathbf{v}_2}{c} \times \mathbf{E}_1(\mathbf{y}_2)\right)$$

$$= e_2\left(\left(\frac{e_1\mathbf{r}}{4\pi r^3} - \frac{g_1}{4\pi}\frac{\mathbf{v}_1}{c} \times \frac{\mathbf{r}}{r^3}\right) + \frac{\mathbf{v}_2}{c} \times \frac{g_1\mathbf{r}}{4\pi r^3}\right)$$

$$+ g_2\left(\left(\frac{g_1\mathbf{r}}{4\pi r^3} + \frac{e_1}{4\pi}\frac{\mathbf{v}_1}{c} \times \frac{\mathbf{r}}{r^3}\right) - \frac{\mathbf{v}_2}{c} \times \frac{e_1\mathbf{r}}{4\pi r^3}\right), \tag{20.46}$$

where again we have neglected all terms of order $1/c^2$ or higher. In the same way one can derive the equation of motion for dyon 1. Introducing the relative velocity

$$\mathbf{v} = \frac{d\mathbf{r}}{dt} = \mathbf{v}_2 - \mathbf{v}_1,$$

the equations of motion of the dyons then turn out to be

$$m_1\mathbf{a}_1 = -\frac{e_1e_2 + g_1g_2}{4\pi}\frac{\mathbf{r}}{r^3} - \frac{e_2g_1 - e_1g_2}{4\pi}\frac{\mathbf{v}}{c}\times\frac{\mathbf{r}}{r^3} = \mathbf{F}_{12},$$

$$m_2\mathbf{a}_2 = \frac{e_1e_2 + g_1g_2}{4\pi}\frac{\mathbf{r}}{r^3} + \frac{e_2g_1 - e_1g_2}{4\pi}\frac{\mathbf{v}}{c}\times\frac{\mathbf{r}}{r^3} = \mathbf{F}_{21}, \tag{20.47}$$

where we have introduced the mutual interaction forces $\mathbf{F}_{12}$ and $\mathbf{F}_{21}$. As we see from their explicit expressions, at first order in $1/c$ a two-dyon system still satisfies Newton's principle of *action and reaction* $\mathbf{F}_{12} = -\mathbf{F}_{21}$. As a consequence, the center of mass of the system $\mathbf{r}_{CM} = (m_1\mathbf{y}_1 + m_2\mathbf{y}_2)/(m_1 + m_2)$ performs a uniform linear motion

$$\mathbf{a}_{CM} = \frac{m_1\mathbf{a}_1 + m_2\mathbf{a}_2}{m_1 + m_2} = 0. \tag{20.48}$$

Dyon force. Dividing the first equation in (20.47) by m_1 and the second by m_2, and by subtracting the resulting equations term by term, we obtain the *equation of motion of the relative dyon*

$$m\frac{d\mathbf{v}}{dt} = \frac{e_1e_2 + g_1g_2}{4\pi}\frac{\mathbf{r}}{r^3} + \mathbf{F}_{\mathrm{d}}, \tag{20.49}$$

where we have introduced the *dyon force*

$$\mathbf{F}_{\mathrm{d}} = \frac{e_2g_1 - e_1g_2}{4\pi}\frac{\mathbf{v}}{c}\times\frac{\mathbf{r}}{r^3}, \tag{20.50}$$

and $m = m_1 m_2/(m_1 + m_2)$ is the reduced mass of the particles. Equation (20.49) is the main result of this section, as it codifies the leading effects of the magnetic charges on the non-relativistic dynamics of the two-dyon system. In fact, with respect to a non-relativistic two-body system of two *charges*, the *Coulomb force* has changed its intensity – the product e_1e_2 has been replaced by the sum $e_1e_2 + g_1g_2$ – albeit remaining *central* and *spherically symmetric*. However, in addition, a new force of order $1/c$ appeared, the *dyon force* $\mathbf{F}_{\mathrm{d}}$ (20.50), with an intensity proportional to the difference $e_2g_1 - e_1g_2$. As this force is of the *magnetic* type $\mathbf{v}\times\mathbf{b}$, it conserves energy, see Sect. 20.3.2. However, since $\mathbf{F}_{\mathrm{d}}$ is not *central*, it *twists* the motion of the relative dyon in such a way that its trajectory is not confined to a *plane*, and that its angular momentum is not conserved, see Sect. 20.4. Finally, by construction, the coupling constants $e_1e_2 + g_1g_2$ and $e_2g_1 - e_1g_2$ appearing in the equation of motion (20.49) are both invariant under $SO(2)$-duality. In fact, in the two-dimensional vector notation of Sect. 20.2.2 the former equals the scalar product $Q_1 \cdot Q_2$ between the charge vectors $Q_1 = \binom{e_1}{g_1}$ and $Q_2 = \binom{e_2}{g_2}$, and the latter equals their cross product $Q_2 \times Q_1$, and both are invariant under two-dimensional rotations.

20.3.2   Conservation Laws

As shown in Sect. 20.2.3, the fundamental equations of the electrodynamics of dyons still guarantee the validity of the main conservation laws. In this section we explore the way these conservation laws are realized in practice in a two-dyon system, again up to terms of order $1/c$. For each of the conserved quantities we shall determine separately the contribution of the dyons and that of the electromagnetic field. Regarding the latter, we can then resort to the expressions of the fields (20.45), which are indeed correct up to terms of order $1/c$.

Energy. By taking the scalar product of equation (20.49) with the relative velocity $\mathbf{v}$ the dyon force drops out, and we remain with

$$\frac{d}{dt}\left(\frac{1}{2}\,mv^2\right) = \frac{e_1 e_2 + g_1 g_2}{4\pi}\,\frac{\mathbf{v}\cdot\mathbf{r}}{r^3} = -\frac{d}{dt}\left(\frac{e_1 e_2 + g_1 g_2}{4\pi r}\right).$$

Since at order $1/c$ the energy of the dyons is given by

$$\varepsilon_1 + \varepsilon_2 = m_1 c^2 + \frac{1}{2}\,m_1 v_1^2 + m_2 c^2 + \frac{1}{2}\,m_2 v_2^2$$
$$= m_1 c^2 + m_2 c^2 + \frac{1}{2}\,mv^2 + \frac{1}{2}\,(m_1 + m_2)v_{CM}^2,$$

and since the center-of-mass velocity $\mathbf{v}_{CM}$ is constant, we find the conservation law for the total energy

$$\frac{d}{dt}\left(\varepsilon_1 + \varepsilon_2 + \frac{e_1 e_2 + g_1 g_2}{4\pi r}\right) = 0.$$

This relation determines the energy of the electromagnetic field $\varepsilon_{\mathrm{em}}$ up to a constant. The simplest choice

$$\varepsilon_{\mathrm{em}} = \frac{e_1 e_2 + g_1 g_2}{4\pi r} \tag{20.51}$$

would imply that $\varepsilon_{\mathrm{em}}$ carries no corrections of order $1/c$. To test this guess we start from the general expression of the total electromagnetic energy

$$\varepsilon_{\mathrm{em}} = \frac{1}{2}\int \left(E^2 + B^2\right) d^3x. \tag{20.52}$$

First we evaluate the contribution of the electric field. From the first formula in (20.45) we obtain for its square the expression

$$E^2 = \left(\frac{e_1 \mathbf{r}_1}{4\pi r_1^3} + \frac{e_2 \mathbf{r}_2}{4\pi r_2^3}\right)^2 + \frac{e_1 g_2 \mathbf{v}_2 - e_2 g_1 \mathbf{v}_1}{(4\pi)^2\, c}\cdot\frac{\mathbf{r}_1 \times \mathbf{r}_2}{r_1^3\, r_2^3} + O\!\left(\frac{1}{c^2}\right). \tag{20.53}$$

If the guess (20.51) is correct, then when inserting the expression (20.53) in the integral (20.52) the terms of order $1/c$ must cancel. To verify this it suffices to prove the vanishing of the integral

$$\int \frac{\mathbf{r}_1 \times \mathbf{r}_2}{r_1^3 r_2^3} d^3x = \mathbf{r} \times \int \frac{\mathbf{x}}{\left|\mathbf{x}+\frac{\mathbf{r}}{2}\right|^3 \left|\mathbf{x}-\frac{\mathbf{r}}{2}\right|^3} d^3x = 0, \qquad (20.54)$$

where we have performed the change of variable $\mathbf{x} \to \mathbf{x} + (\mathbf{y}_1 + \mathbf{y}_2)/2$. The second integral in (20.54) is zero, because its integrand is odd under $\mathbf{x} \to -\mathbf{x}$. Therefore, only the first term of the expression (20.53), corresponding just to the energy density of two non-relativistic *charges*, contributes to the total energy. A standard argument of classical electrodynamics then yields (see Problem 2.8)

$$\frac{1}{2} \int E^2 d^3x = \frac{1}{2} \int \left(\frac{e_1 \mathbf{r}_1}{4\pi r_1^3} + \frac{e_2 \mathbf{r}_2}{4\pi r_2^3}\right)^2 d^3x + O\left(\frac{1}{c^2}\right) \to \frac{e_1 e_2}{4\pi r} + O\left(\frac{1}{c^2}\right).$$

We recall that this *argument* involves, in particular, the subtraction of the infinite energy coming from the self-interaction of the two charges, see the general method of Chap. 16. Since the same calculation yields for the magnetic energy

$$\frac{1}{2} \int B^2 d^3x = \frac{g_1 g_2}{4\pi r} + O\left(\frac{1}{c^2}\right),$$

equation (20.52) reduces to the total duality-invariant electromagnetic energy (20.51), as expected.

Momentum. Apart from terms of order $1/c^2$, the total momentum of the dyons is given by $\mathbf{p} = m_1 \mathbf{v}_1 + m_2 \mathbf{v}_2$ and, thanks to the equation of motion for the center of mass (20.48), it is conserved. It follows that also the total momentum of the electromagnetic field

$$\mathbf{P}_{\text{em}} = \frac{1}{c} \int \mathbf{E} \times \mathbf{B} \, d^3x \qquad (20.55)$$

must be conserved, modulo terms of order $1/c^2$. Actually, it turns out that the momentum $\mathbf{P}_{\text{em}}$ is of order $1/c^2$, and so it does not contribute to the momentum balance at order $1/c$. To show it we compute from formulas (20.45) the cross product

$$\mathbf{E} \times \mathbf{B} = \frac{e_1 g_2 - e_2 g_1}{(4\pi)^2} \frac{\mathbf{r}_1 \times \mathbf{r}_2}{r_1^3 r_2^3} + O\left(\frac{1}{c}\right). \qquad (20.56)$$

Thanks to the identity (20.54), the integral over all space of the leading term of this cross product vanishes, and so the momentum $\mathbf{P}_{\text{em}}$ (20.55) is indeed a quantity of order $1/c^2$.

Angular momentum. Modulo terms of order $1/c^2$, the total angular momentum of the dyons can be written as

$$\mathbf{L}_p = \mathbf{y}_1 \times m_1\mathbf{v}_1 + \mathbf{y}_2 \times m_2\mathbf{v}_2 = \mathbf{r} \times m\mathbf{v} + \mathbf{r}_{CM} \times (m_1 + m_2)\,\mathbf{v}_{CM}.$$

Since the angular momentum of the center of mass is conserved, using the equation of motion of the relative dyon (20.49) we find for the time derivative of $\mathbf{L}_p$

$$\frac{d\mathbf{L}_p}{dt} = \mathbf{r} \times m\frac{d\mathbf{v}}{dt} = \frac{e_2 g_1 - e_1 g_2}{4\pi c}\left(\frac{\mathbf{v}}{r} - \frac{(\mathbf{v}\cdot\mathbf{r})\,\mathbf{r}}{r^3}\right). \tag{20.57}$$

Thus, the dyon force prevents the conservation of the dyons' angular momentum. As a consequence, also the angular momentum of the electromagnetic field $\mathbf{L}_{em}$ cannot be conserved and, in particular, it must be a non-vanishing quantity of order $1/c$. To evaluate it explicitly we substitute the cross product (20.56) in the general expression of $\mathbf{L}_{em}$ (20.42), keeping only the terms of order $1/c$,

$$\begin{aligned}
\mathbf{L}_{em} &= \frac{1}{c}\int \mathbf{x} \times (\mathbf{E}\times\mathbf{B})\,d^3x = \frac{e_1 g_2 - e_2 g_1}{(4\pi)^2 c}\int \mathbf{x} \times \left(\frac{\mathbf{r}_1\times\mathbf{r}_2}{r_1^3 r_2^3}\right)d^3x \\
&= \frac{e_1 g_2 - e_2 g_1}{(4\pi)^2 c}\int \mathbf{r}_1 \times \left(\frac{\mathbf{r}_1\times\mathbf{r}_2}{r_1^3 r_2^3}\right)d^3x \\
&= \frac{e_1 g_2 - e_2 g_1}{(4\pi)^2 c}\int (\mathbf{x}+\mathbf{r}) \times \left(\frac{(\mathbf{x}+\mathbf{r})\times\mathbf{x}}{|\mathbf{x}+\mathbf{r}|^3\,|\mathbf{x}|^3}\right)d^3x.
\end{aligned} \tag{20.58}$$

The second line differs from the first by a term proportional to the vanishing integral (20.54), since the position vectors $\mathbf{y}_i$ are independent of the integration variable $\mathbf{x}$ (recall the definitions $\mathbf{r}_i = \mathbf{x} - \mathbf{y}_i$ and $\mathbf{r} = \mathbf{y}_2 - \mathbf{y}_1$). To obtain the third line we have performed the change of variable $\mathbf{x} \to \mathbf{x} + \mathbf{y}_2$. To evaluate the integral (20.58) we recast its integrand in the form

$$\begin{aligned}
(\mathbf{x}+\mathbf{r}) \times \left(\frac{(\mathbf{x}+\mathbf{r})\times\mathbf{x}}{|\mathbf{x}+\mathbf{r}|^3\,|\mathbf{x}|^3}\right) &= \frac{[(\mathbf{x}+\mathbf{r})\cdot\mathbf{x}]\,(\mathbf{x}+\mathbf{r})}{|\mathbf{x}+\mathbf{r}|^3\,|\mathbf{x}|^3} - \frac{\mathbf{x}}{|\mathbf{x}+\mathbf{r}|\,|\mathbf{x}|^3} \\
&= -\frac{x^i}{|\mathbf{x}|^3}\,\partial_i\left(\frac{\mathbf{x}+\mathbf{r}}{|\mathbf{x}+\mathbf{r}|}\right).
\end{aligned}$$

Introducing then in the integral (20.58) for the variable $\mathbf{x}$ polar coordinates, with $|\mathbf{x}| = R$ and $d^3x = R^2 dR d\Omega$, the above integrand becomes

$$(\mathbf{x}+\mathbf{r}) \times \left(\frac{(\mathbf{x}+\mathbf{r})\times\mathbf{x}}{|\mathbf{x}+\mathbf{r}|^3\,|\mathbf{x}|^3}\right) = -\frac{1}{R^2}\frac{\partial}{\partial R}\left(\frac{\mathbf{x}+\mathbf{r}}{|\mathbf{x}+\mathbf{r}|}\right).$$

The integral (20.58) reduces so to the simple expression of order $1/c$ ($\mathbf{n} = \mathbf{x}/R$)

$$\mathbf{L}_{\text{em}} = -\frac{e_1 g_2 - e_2 g_1}{(4\pi)^2 c} \int \left(\int_0^\infty \frac{\partial}{\partial R} \left(\frac{\mathbf{x} + \mathbf{r}}{|\mathbf{x} + \mathbf{r}|} \right) dR \right) d\Omega$$

$$= -\frac{e_1 g_2 - e_2 g_1}{(4\pi)^2 c} \int \left(\mathbf{n} - \frac{\mathbf{r}}{r} \right) d\Omega$$

$$= \frac{e_1 g_2 - e_2 g_1}{4\pi c} \frac{\mathbf{r}}{r}, \tag{20.59}$$

where we have used that the invariant integral $\int \mathbf{n} \, d\Omega$ is zero. The expression of $\mathbf{L}_{\text{em}}$ (20.59) is invariant under $SO(2)$-duality and, furthermore, unaffected by an exchange of the dyons ($1 \leftrightarrow 2$, $\mathbf{r} \leftrightarrow -\mathbf{r}$). As expected, the electromagnetic angular momentum is not conserved

$$\frac{d\mathbf{L}_{\text{em}}}{dt} = \frac{e_1 g_2 - e_2 g_1}{4\pi c} \frac{d}{dt} \left(\frac{\mathbf{r}}{r} \right) = \frac{e_1 g_2 - e_2 g_1}{4\pi c} \left(\frac{\mathbf{v}}{r} - \frac{(\mathbf{v} \cdot \mathbf{r}) \mathbf{r}}{r^3} \right).$$

However, in light of equation (20.57) the total angular momentum

$$\mathbf{L} = \mathbf{L}_{\text{p}} + \mathbf{L}_{\text{em}} = \mathbf{y}_1 \times m_1 \mathbf{v}_1 + \mathbf{y}_2 \times m_2 \mathbf{v}_2 + \frac{e_1 g_2 - e_2 g_1}{4\pi c} \frac{\mathbf{r}}{r} \tag{20.60}$$

is, actually, a constant of motion.

20.4 Dirac's Quantization Condition: A Semiclassical Argument

As we have seen so far, the classical electrodynamics of a system of dyons based on the fundamental equations (20.16), (20.17) and (20.39) is perfectly consistent, for arbitrary values of the dyon charges (e_r, g_r). In what follows, we present a semi-classical argument suggesting that the consistency of the *quantum* dynamics of such a system requires, instead, these charges to be suitably tied to each other, by *Dirac's quantization condition*. In Chap. 21 we will then show that this condition is, actually, *necessary and sufficient* for the existence of a consistent *quantum mechanics* of a *non-relativistic* system of dyons.

20.4.1 Asymptotic Scattering of Two Dyons

The argument is based on the two-dyon system analyzed in Sect. 20.3, in particular on the peculiar properties of its angular momentum. The fact that the angular momentum of the dyons $\mathbf{L}_{\text{p}}$ is not conserved has, in fact, two important consequences. In the first place, the relative motion is no longer a *planar* motion, as happens, instead, for two particles interacting via a central force. In the second place, in an *asymptotic scattering* experiment, in which the minimum distance between the dyons, and hence their impact parameter b, tend to *infinity*, their angular momentum changes, nevertheless, by a *finite* amount, subtracting it from the electromagnetic field. In fact, the variation of the dyons' angular momentum $\Delta\mathbf{L}_p$ between the initial and final states can be deduced from Eq. (20.60), by enforcing the conservation of the total angular momentum

$$\Delta\mathbf{L}_p = -\Delta\mathbf{L}_{em} = -\frac{e_1 g_2 - e_2 g_1}{4\pi c}\left(\frac{\mathbf{r}}{r}\bigg|_f - \frac{\mathbf{r}}{r}\bigg|_i\right). \qquad (20.61)$$

As the particles always move at a large distance from each other, they perform practically uniform linear motions, because at large relative distances the mutual interaction force decreases as $1/r^2$, see the equations of motion (20.47). Therefore, also the relative dyon performs a nearly uniform linear motion, that we can take along the z direction. Correspondingly, the asymptotic radial directions become

$$\frac{\mathbf{r}}{r}\bigg|_i = -\mathbf{u}_z, \qquad \frac{\mathbf{r}}{r}\bigg|_f = \mathbf{u}_z.$$

Formula (20.61) then yields for the change of the dyons' angular momentum during the asymptotic scattering the simple result

$$\Delta\mathbf{L}_p = \frac{e_2 g_1 - e_1 g_2}{2\pi c}\mathbf{u}_z. \qquad (20.62)$$

Explicit evaluation of $\Delta\mathbf{L}_p$. To better understand the mechanism that produces in the limit for $b \to \infty$ a non-vanishing variation of the angular momentum – whereas in the same limit the initial and final relative velocities remain the same – we evaluate $\Delta\mathbf{L}_p$ by a direct calculation. First we determine the variation of the relative velocity during the scattering process $\Delta\mathbf{v}$, as a function of the impact parameter b. Obviously, in the limit for $b \to \infty$ the variation $\Delta\mathbf{v}$ must vanish. It is, therefore, sufficient to determine $\Delta\mathbf{v}$ by means of a perturbative calculation around an *unperturbed* linear trajectory, taking as expansion parameter $1/b$. Let us suppose that the unperturbed trajectory $\mathbf{r}(t)$ lies in the xz plane *of incidence*, and that it is parallel to the z axis. Then its kinematics is represented by the equations

$$\mathbf{r}(t) = b\,\mathbf{u}_x + vt\,\mathbf{u}_z, \qquad r^2(t) = b^2 + v^2 t^2, \qquad \mathbf{v}(t) = v\,\mathbf{u}_z, \qquad (20.63)$$

where $v\mathbf{u}_z$ is the initial velocity of the exact trajectory. At first order in $1/b$, the change of velocity $\Delta\mathbf{v}$ during the scattering can then be obtained by integrating the equation of motion of the relative dyon (20.49) along this trajectory

$$\Delta \mathbf{v} = \frac{e_1 e_2 + g_1 g_2}{4\pi m} \int_{-\infty}^{\infty} \frac{b\,\mathbf{u}_x + vt\,\mathbf{u}_z}{(b^2 + v^2 t^2)^{3/2}}\, dt$$

$$+ \frac{e_2 g_1 - e_1 g_2}{4\pi mc} \int_{-\infty}^{\infty} \frac{v\,\mathbf{u}_z \times (b\,\mathbf{u}_x + vt\,\mathbf{u}_z)}{(b^2 + v^2 t^2)^{3/2}}\, dt$$

$$= \frac{b}{4\pi m}\left((e_1 e_2 + g_1 g_2)\,\mathbf{u}_x + \frac{v}{c}\,(e_2 g_1 - e_1 g_2)\,\mathbf{u}_y \right) \int_{-\infty}^{\infty} \frac{dt}{(b^2 + v^2 t^2)^{3/2}}$$

$$= \frac{1}{2\pi mbv}\left((e_1 e_2 + g_1 g_2)\,\mathbf{u}_x + \frac{v}{c}\,(e_2 g_1 - e_1 g_2)\,\mathbf{u}_y \right) = \Delta v_x\,\mathbf{u}_x + \Delta v_y\,\mathbf{u}_y.$$

$$(20.64)$$

We have used the integral

$$\int_{-\infty}^{\infty} \frac{dt}{(b^2 + v^2 t^2)^{3/2}} = \frac{2}{vb^2}.$$

The terms proportional to vt drop out from the integral by symmetric integration. In the final state, the dyon's velocity acquires hence two components $\Delta \mathbf{v}$ orthogonal to the initial velocity $v\mathbf{u}_z$. The component Δv_x of the velocity (20.64) lies in the xz plane of the incident trajectory (20.63), and represents the usual hyperbolic deflection of a charged particle in a Coulomb field, characterized by the *scattering angle*

$$\varphi \approx \frac{\Delta v_x}{v} = \frac{e_1 e_2 + g_1 g_2}{2\pi mbv^2}. \qquad (20.65)$$

For comparison, recall that a non-relativistic unitary charge in the Coulomb potential $V(r) = \alpha/r$ is subject to the exact scattering angle

$$\varphi = 2\arctan\left(\frac{\alpha}{mbv^2} \right).$$

For $\alpha = (e_1 e_2 + g_1 g_2)/4\pi$, in the limit of large b this formula reduces indeed to the angle (20.65). On the other hand, the y component of the final velocity (20.64)

$$\Delta v_y = \frac{e_2 g_1 - e_1 g_2}{2\pi mbc}, \qquad (20.66)$$

caused by the dyon force, is *orthogonal* to the xz plane. In order to evaluate $\Delta \mathbf{L}_p$ in the limit for $b \to \infty$, it is convenient to integrate the identity

$$\frac{d\mathbf{L}_p}{dt} = \mathbf{r} \times m\,\frac{d\mathbf{v}}{dt}$$

between $t = -\infty$ and $t = \infty$. Substituting for $\mathbf{r}$ the unperturbed trajectory (20.63), and for the total change of velocity $\Delta \mathbf{v}$ the result (20.64), in this way we obtain

$$\Delta \mathbf{L}_p = \int_{-\infty}^{\infty} \mathbf{r} \times m \, d\mathbf{v} = m \int_{-\infty}^{\infty} (b \, \mathbf{u}_x + vt \, \mathbf{u}_z) \times d\mathbf{v} = mb \, \mathbf{u}_x \times \int_{-\infty}^{\infty} d\mathbf{v}$$

$$= mb \, \mathbf{u}_x \times \Delta \mathbf{v} = mb \, \Delta v_y \, \mathbf{u}_z = \frac{e_2 g_1 - e_1 g_2}{2\pi c} \mathbf{u}_z, \tag{20.67}$$

in agreement with the result (20.62). In the integral above we have neglected the term proportional to vt since – using for $d\mathbf{v}$ the equation of the relative dyon (20.49), and substituting at its right hand side the unperturbed trajectory (20.63) – it happens that, by symmetric integration, this term drops out. In the line (20.67) we see that the change of velocity Δv_x – caused by the Coulomb force – does not contribute to $\Delta \mathbf{L}_p$. Conversely, the change Δv_y – caused by the dyon force, and forcing the relative dyon to abandon the xz plane of incidence – produces a non-vanishing variation $\Delta \mathbf{L}_p$, parallel to the direction of incidence $\mathbf{u}_z$. In fact, in the second term of line (20.67) the *moment arm* b compensates the decay factor $1/b$ present in the change of velocity Δv_y (20.66).

20.4.2   Charge Quantization and Physical Implications

Let us now interpret the result of the classical scattering experiment between two dyons of the previous section in a *quantum mechanical* framework. Since we have enforced the limit for $b \to \infty$, it is consistent to regard the dyons both in the initial and in the final state as particles in uniform linear motion in the same z direction. According to the quantum mechanical commutation relations between the total momentum operator $\mathbf{p}$ and the total angular momentum operator $\mathbf{L}_p$ of the dyons

$$[L_p^i, p^j] = i\hbar \, \varepsilon^{ijk} p^k,$$

the z components of the momentum and of the angular momentum are thus *compatible observables*

$$[L_p^z, p^z] = 0.$$

This implies that we can measure the observable L_p^z both in the initial and final states, without disturbing the motion of the dyons along the z direction. According to the *measurement paradigm* of quantum mechanics, the values that we obtain for L_p^z in the two states then belong necessarily to the spectrum of the corresponding operator, namely they are integer – or half-integer, if the dyons have spin – multiples of $\hbar$. Therefore, their difference is quantized according to the rule

$$\Delta L_p^z = n\hbar,$$

where n is a negative or position *integer*. If we identify these values with those given by the classical formula (20.62), then we derive *Dirac's quantization condition*[4]

$$e_2 g_1 - e_1 g_2 = 2\pi n\hbar c, \qquad n = 0, \pm 1, \pm 2, \ldots \qquad (20.68)$$

We are thus led to conclude that – at the semiclassical and non-relativistic level – a necessary condition for a consistent quantum *coexistence* of charges and monopoles and, more generally, of dyons, is that any pair of such particles satisfies Dirac's quantization condition. However, only rather recently has it been proven that the validity of Dirac's condition (20.68), actually – as we will soon explain – of its slightly more restrictive variant (20.70), is also *sufficient* for the self-consistency of a *relativistic quantum field theory* of dyons, see Ref. [6]. Despite these encouraging theoretical results, the experimental research of magnetic monopolies, which is however still in progress, has given negative results so far. Nevertheless, below we list some interesting important consequences that would derive from the existence of magnetic monopoles in nature.

Quantization of the electric charge. Let us suppose that in our Universe there exists at least a dyon, or a monopole, with magnetic charge g_0. In this case, in basis of Dirac's condition (20.68), the electric charge e_r of any *known* elementary particle – having vanishing magnetic charge g_r – should satisfy the relation $e_r g_0 = 2\pi n_r \hbar c$, that can be recast as

$$e_r = n_r e_0, \qquad e_0 = \frac{2\pi \hbar c}{g_0}, \qquad n_r \in \mathbb{Z}.$$

We would thus conclude that all electric charges present in nature are necessarily integer multiples of a fundamental charge e_0. This feature represents, indeed, a basic physical phenomenon, known as the *quantization of the electric charge*, which is confirmed by experiments with extremely high precision. The relative difference between the absolute values of the charges of the proton and of the electron is, in fact, measured to be less than 10^{-21}, see Ref. [11]. In this way, the existence of magnetic monopoles would yield an elegant theoretical explanation of a basic experimental datum, which at the moment still appears as a pure *coincidence*. By interchanging in the above argument the roles of charges and monopoles, one concludes that also the magnetic charges g_r of the monopoles – if they exist – are necessarily quantized in integer multiples of a fundamental charge g_0. Finally, the hypothesis of the existence of magnetic monopoles remains compatible with observations, even if we regard the *unconfined* quarks of the primordial quark-gluon plasma as *elementary* particles, with electric charges $\pm e/3$ and $\pm 2e/3$, where $-e$ is the charge of the

---

[4]In his 1931 paper P.A.M. Dirac considers a two-particle system formed by a charge and by a monopole, deriving his original quantization condition $eg = 2\pi n\hbar c$ [4]. The generalization of the latter to the condition (20.68) for two dyons has been established by J. Schwinger in 1968 [9]. The discovery that the integer n must, eventually, be *even*, has been made by J. Schwinger already in 1966 [10], by relying on arguments of relativistic quantum field theory, see the discussion around Eq. (20.70).

electron. In this case it suffices, in fact, to choose as fundamental charge the value $e_0 = e/3$.

Strong-weak coupling duality. Consider a system formed by a *charge* of electric charge e and by a *monopole* of magnetic charge g. Dirac's quantization condition

$$eg = 2\pi n\hbar c \qquad (20.69)$$

then establishes a relation between the *electric* and *magnetic* fine structure constants

$$\alpha_e = \frac{e^2}{4\pi\hbar c}, \qquad \alpha_m = \frac{g^2}{4\pi\hbar c},$$

which in quantum field theory represent a measure for the dimensionless strength of the interaction. In fact, the condition (20.69) implies the relation.

$$\alpha_e \alpha_m = \frac{n^2}{4}.$$

If we assume that n is of order unity, then if $\alpha_e \ll 1$, we have $\alpha_m \gg 1$, and vice versa. This means that in a theory in which the electric charges are weakly coupled, the magnetic charges are strongly coupled, and vice versa. Since duality interchanges electric charges with magnetic ones, it is not surprising that a symmetry of this kind turns out to be very useful for analyzing a theory in a regime in which the coupling constant is large – in which case the perturbative method would fail – in terms of a weakly coupled *dual theory*, which can instead be analyzed by means of perturbative methods. The electromagnetic duality belongs, in fact, to the so-called class of *strong-weak coupling dualities*, which represent a wider family of duality symmetries, see, for instance, Refs. [12, 13].

Magnetic monopoles in grand unified theories (GUT). The study of magnetic monopoles, which have been introduced by us *ad hoc* in the framework of classical electrodynamics, is also motivated by the fact that – as shown by G. 't Hooft [14] and A.M. Polyakov [15] – in all *grand unified theories* of the fundamental gauge-interactions, as, for instance, the theory based on the simple group $SU(5)$, the existence of magnetic monopoles is an unavoidable *prediction*. Actually, the fact that until now there is no experimental evidence for the existence of these particles does not contradict these theories. Indeed, the masses M_m of the magnetic monopoles possibly existing in the Universe could be well above the energies that we are able to produce in the accelerators available today. In particular, in grand unified theories, typical monopole masses are expected to be of the order of the unification scale, say 10^{16} GeV, divided by the $SU(5)$ fine structure constant $g_{GUT}^2/4\pi\hbar c \sim 0.1$, namely $M_m \sim 10^{17}$ GeV.

Dyons in relativistic quantum field theory. Dirac's quantization condition (20.68) has been derived as a *necessary* condition for the existence of dyons, via a *non-relativistic* and *semiclassical* argument. However, as happens for all elementary particles, the full quantum dynamics of dyons can be realized only in the framework of *relativistic quantum field theory*. Actually, as anticipated by J. Schwinger in 1975 [16], for dyons there exist two physically *inequivalent* theories of this type, characterized by the distinct duality-symmetry groups $SO(2)$ and Z_4. Below we briefly explain their basic differences.

SO(2)-theory. As shown in [6], for a given system of dyons with charges (e_r, g_r) one can formulate a consistent relativistic quantum field theory, if the charges satisfy *Schwinger's quantization conditions* [9]

$$(e_r g_s - e_s g_r) = 4\pi n_{rs} \hbar c, \quad \text{for all } r \text{ and } s, \quad (20.70)$$

where the n_{rs} are positive or negative integers. Notice that these conditions differ by a factor of 2 from Dirac's original condition (20.68). The corresponding theory is invariant under the continuous duality group $SO(2)$, as are Schwinger's conditions. The *necessity* of the conditions (20.70) for the consistency of the related quantum field theory of dyons has been established by J. Schwinger [9, 10], who called this version of the theory *symmetrical formulation* [16]. A particular aspect of this theory is that no quantization condition is required between the charges e_r and g_r of the same dyon.

Z_4-*theory.* In Ref. [6] it has, furthermore, been shown that it is possible to formulate a consistent relativistic quantum field theory of dyons, also if their charges satisfy the *Dirac-Schwinger quantization conditions*

$$e_r g_s = 2\pi n_{rs} \hbar c, \quad \text{for all } r \text{ and } s, \quad (20.71)$$

where the n_{rs} are again positive or negative integers. Notice the factor 2π in these conditions, in place of the factor 4π of Schwinger's conditions (20.70). The corresponding theory is invariant only under the discrete duality group Z_4. Actually, although under the action of the generator (20.38) of Z_4 – entailing the replacements of the charges (20.26) – the product $e_r g_s$ appearing in the conditions (20.71) is not invariant, the transformed product $-g_r e_s$ is still an integer multiple of $2\pi \hbar c$, and it can be seen that this property suffices to guarantee the invariance of the quantum field theory under the Z_4-duality. In his last foundational paper on magnetic monopoles [16], J. Schwinger called this version of the theory of dyons *unsymmetrical formulation*. Both Schwinger's conditions (20.70) and the Dirac-Schwinger conditions (20.71) imply Dirac's original conditions (20.68), whereas the opposite is not true. In fact, the appearance of the more restrictive conditions (20.70) and (20.71) in the framework of relativistic quantum field theories is a genuine *relativistic* effect. Finally, the physical *inequivalence* between the $SO(2)$ and Z_4 theories can also be inferred from the fact that none of the consistency conditions (20.70) and (20.71) implies the other. On the other hand, if a set of dyons satisfies both conditions, for instance, if $e_r g_s = 4\pi n_{rs} \hbar c$ for all r and s, then the two theories are actually identical [5].

Spin-statistics transmutation. A basic difference between the conditions (20.70) and (20.71) is that the latter imposes also a constraint between the charges e_r and g_r of the *same* dyon

$$e_r g_r = 2\pi n_{rr} \hbar c, \quad \text{for all } r. \tag{20.72}$$

This relation signals the occurrence of a non-trivial *self-interaction* of the rth dyon in the Z_4-theory, which is absent in the $SO(2)$-theory. An interesting consequence of this self-interaction is a peculiar phenomenon known as *spin-statistics transmutation*. To explain its meaning, suppose that we switch off the charges e_r and g_r of all dyons, transforming them in neutral particles. Of course, this process implies also the switch-off of the electromagnetic field they produce. The resulting *bare* dyons are thus *free* particles, experiencing no interaction at all. The *spin-statistics theorem* then states that each bare dyon is a *boson* (having integer spin, satisfying quantum-*commutation* relations, and entailing symmetric wave functions), or a *fermion* (having half-integer spin, satisfying quantum-*anticommutation* relations, and entailing antisymmetric wave functions), see, for instance, Ref. [17]. If we now *dress* the particles, turning on the charges, it happens that in the Z_4-theory the spin S_r of the rth dressed dyon is tied to the spin S_r^0 of the corresponding bare dyon by the relation (see Ref. [18])

$$S_r = S_r^0 + \frac{e_r g_r}{4\pi c} = S_r^0 + \frac{n_{rr}}{2}\hbar, \quad \text{mod } \hbar\mathbb{Z}, \tag{20.73}$$

where we have used the quantization conditions (20.72). Notice that, thanks to the fact that the numbers n_{rr} are integers, as the spins S_r^0 are integer or half-integer multiples of $\hbar$, so are the spins S_r of the dressed particles. Heuristically, the spin-shift formula (20.73) can be understood in terms of the classical equation (20.59) for the angular momentum of the electromagnetic field produced by two dyons. In fact, consider a dyon pair formed by a *charge* ($e_1 = e_r$, $g_1 = 0$) and a monopole ($e_2 = 0$, $g_2 = g_r$). According to Eq. (20.59), the electromagnetic angular momentum of the dyon pair is then given by $\mathbf{L}_{\text{em}} = (e_r g_r / 4\pi c)\mathbf{u}_r$, an expression that is independent of the distance r between the dyons. If we now let r tend to zero along a fixed direction $\mathbf{u}_r$, we obtain a single *composite* dyon with charges (e_r, g_r), endowed with an *intrinsic* angular momentum of magnitude $e_r g_r / 4\pi c$, as in Eq. (20.73). Notice that, according to the equation of motion of the relative dyon (20.49), a charge and a monopole exert no mutual Coulomb repulsion, and if they approach each other along a straight line, also the dyon force (20.50) vanishes. Coming back to the spin-shift formula (20.73), we derive that if the integer n_{rr} is *even* we have

$$S_r = S_r^0, \quad \text{mod } \hbar\mathbb{Z}, \tag{20.74}$$

whereas if n_{rr} is *odd* we have

$$S_r = S_r^0 + \frac{\hbar}{2}, \quad \text{mod } \hbar\mathbb{Z}. \tag{20.75}$$

Correspondingly, it can be shown that if the rth bare dyon obeys canonical commutation (anticommutation) relations, and n_{rr} is *even*, then also the rth dressed dyon obeys canonical commutation (anticommutation) relations. Vice versa, if the rth bare dyon obeys canonical commutation (anticommutation) relations, and n_{rr} is *odd*, then the rth dressed dyon obeys, instead, canonical anticommutation (commutation) relations [18]. This ensures that the spin shifts (20.74) and (20.75) occur in *observance of the spin-statistics theorem*.

In conclusion, in the Z_4-theory, due to a subtle electromagnetic self-interaction, a dyon with charges (e, g) undergoes a *spin-statistics transmutation* – becoming a boson, if the bare dyon was a fermion, and vice versa – if $eg/2\pi\hbar c$ is an odd integer. Conversely, if $eg/2\pi\hbar c$ is an even integer, no spin-statistics transmutation occurs. In contrast, no spin-statistics transmutation ever occurs in the $SO(2)$-theory.

In the physics of elementary particles, a spin-statistics transmutation represents a peculiar phenomenon in that it changes *the rules of the game*. In fact, in classical field theory any particle, charged or neutral, is described by a field with a well-defined tensorial character. Examples are scalar fields, vector fields, two-index tensor fields, and spinor fields. On the other hand, in relativistic quantum field theory, the spin and the statistics of the particle associated with a field are uniquely determined by the tensorial character of the field: the scalar, vector and symmetric two-index tensor fields describe *bosons*, with spin 0, $\hbar$, and $2\hbar$, respectively, while spinor fields describe *fermions*, with spin $\hbar/2$ or $3\hbar/2$. According to the analysis above it happens, instead, that if a (bare) dyon with an odd value of $eg/2\pi\hbar c$ is described, for instance, by a scalar field, then – as a result of the spin-statistics transmutation – the resulting (dressed) dyon is no longer a boson, but rather a fermion.

We conclude this short overview with the observation that in one and two spatial dimensions the spin-statistics transmutation is a rather common phenomenon, see, for instance, Ref. [19]. In these dimensions, in particular, the spin of a particle is not quantized in integer or half-integer multiples of $\hbar$ and can take any real value. In contrast, in three spatial dimensions there are only two known ways in which a spin-statistic transmutation can take place: the one related to dyons illustrated above, and the topology-induced transmutation characterizing the *skyrmions* [20], see also Ref. [21].

References

1. H. Poincaré, Remarques sur une expérience de M. Birkeland. Compt. Rendus **123**, 530 (1896)
2. J.J. Thomson, *Electricity and Matter* (Charles Scribner's Sons, New York, 1904)
3. J.J. Thomson, *Elements of the Mathematical Theory of Electricity and Magnetism*, 3rd edn. (Cambridge University, London, 1904)
4. P.A.M. Dirac, Quantized singularities in the electromagnetic field. Proc. R. Soc. Lond. A **133**, 60 (1931)
5. K. Lechner, P.A. Marchetti, Duality invariant quantum field theories of charges and monopoles. Nucl. Phys. B **569**, 529 (2000)
6. R.A. Brandt, F. Neri, D. Zwanziger, Lorentz invariance from classical particle paths in quantum field theory of electric and magnetic charge. Phys. Rev. D **19**, 1153 (1979)

7. M.B. Green, M. Gutperle, Effects of D-instantons. Nucl. Phys. B **498**, 195 (1997)
8. K. Lechner, Radiation reaction and four-momentum conservation for point-like dyons. J. Phys. A **39**, 11647 (2006)
9. J. Schwinger, Sources and magnetic charge. Phys. Rev. **173**, 1536 (1968)
10. J. Schwinger, Magnetic charge and quantum field theory. Phys. Rev. **144**, 1087 (1966)
11. G. Bressi, G. Carugno, F. Della Valle, G. Galeazzi, G. Ruoso, G. Sartori, Testing the neutrality of matter by acoustic means in a spherical resonator. Phys. Rev. A **83**, 052101 (2011)
12. C. Montonen, D. Olive, Magnetic monopoles as gauge particles? Phys. Lett. B **72**, 117 (1977)
13. N. Seiberg, Electric-magnetic duality in supersymmetric non-abelian gauge theories. Nucl. Phys. B **435**, 129 (1995)
14. G. 't Hooft, Magnetic monopoles in unified gauge theories. Nucl. Phys. B **79**, 276 (1974)
15. A.M. Polyakov, Particle spectrum in the quantum field theory. JETP Lett. **20**, 194 (1974)
16. J. Schwinger, Magnetic charge and the charge quantization condition. Phys. Rev. D **12**, 3015 (1975)
17. R.F. Streater, A.S. Wightman, *PCT, Spin and Statistics, and All That* (Princeton University Press, Princeton, 2000)
18. K. Lechner, P.A. Marchetti, Spin-statistics transmutation in relativistic quantum field theory of dyons. JHEP **0012**, 028 (2000)
19. P.A. Marchetti, Spin-statistics transmutation in quantum field theory. Found. Phys. **40**, 746 (2010)
20. T. Skyrme, A unified field theory of mesons and baryons. Nucl. Phys. **31**, 556 (1962)
21. I. Zahed, G.E. Brown, The Skyrme model. Phys. Rept. **142**, 1 (1986)

Chapter 21
Magnetic Monopoles in Quantum Mechanics

In this chapter we examine the dynamics of a two-dyon system in the framework of non-relativistic *quantum mechanics*. In quantum mechanics, the description of a physical system requires the introduction of a *Hilbert space* $\mathcal{H}$, whose elements ψ represent the *states* of the system, and of a self-adjoint *Hamiltonian operator H* which determines the time evolution of the states. The operator H is obtained from the classical Hamiltonian via the recipe of *canonical quantization*, and the classical Hamiltonian descends, in turn, from the classical Lagrangian L of the system, via a *Legendre transformation*. Finally, the Lagrangian is uniquely determined by the classical equations of motion, modulo total derivatives. A prototypical example of this procedure is provided by a non-relativistic charged particle in the presence of an external electromagnetic field $F^{\mu\nu} = (\mathbf{E}, \mathbf{B})$, whose equation of motion is thus the Lorentz equation

$$m\mathbf{a} = e\left(\mathbf{E} + \frac{\mathbf{v}}{c} \times \mathbf{B}\right). \tag{21.1}$$

If the tensor $F^{\mu\nu}$ satisfies the Bianchi identity $\partial_{[\mu}F_{\nu\rho]} = 0$, we can introduce a vector potential $A^{\mu} = (A^0, \mathbf{A})$ such that $F_{\mu\nu} = \partial_\mu A_\nu - \partial_\nu A_\mu$, and in this case we know that there exists a Lagrangian giving rise to the equation of motion (21.1), see formula (21.4) below. With this Lagrangian as starting point, the above process of canonical quantization then proceeds without meeting any obstacle, see Sect. 21.1.

Let us now analyze what happens for the non-relativistic two-dyon system considered in Sect. 20.3. After the decoupling of the center-of-mass motion, the classical dynamics of this system is entirely encoded by the equation of motion of the relative dyon (20.49)

$$m\mathbf{a} = \frac{e_1 e_2 + g_1 g_2}{4\pi} \frac{\mathbf{r}}{r^3} + \frac{e_2 g_1 - e_1 g_2}{4\pi} \frac{\mathbf{v}}{c} \times \frac{\mathbf{r}}{r^3}. \tag{21.2}$$

This equation formally has exactly the same structure as the Lorentz equation (21.1), where now the external fields $\mathbf{E}$ and $\mathbf{B}$ are both static and proportional to $\mathbf{x}/|\mathbf{x}|^3$.

© Springer International Publishing AG, part of Springer Nature 2018
K. Lechner, *Classical Electrodynamics*, UNITEXT for Physics,
https://doi.org/10.1007/978-3-319-91809-9_21

However, in this case the canonical quantization has to cope with the – apparently insurmountable – obstacle that the magnetic field is no longer a *solenoidal* vector field, $\nabla \cdot \mathbf{B} \neq 0$. In fact, we know that in the presence of magnetic charges the Maxwell tensor $F^{\mu\nu}$ violates the Bianchi identity, see Eq. (20.20). Therefore, in this case we do no longer know how to introduce a vector potential $\mathbf{A}$, and consequently the equation of motion of the relative dyon (21.2) can no longer be derived from a Lagrangian, see the general discussion after Eq. (20.20). To overcome this difficulty we need to make two mathematical digressions: the first regards the construction of a *generalized* Hilbert space, Sect. 21.2, and the second is about the possibility of introducing a vector potential for non-solenoidal magnetic fields, the famous *Dirac potential*. As we present the quantum mechanics of dyons within two different approaches – within the *canonical quantization* in Sect. 21.5, and within Feynman's *path-integral quantization* in Sect. 21.6 – we introduce this potential in two different mathematical frameworks: as a *regular* vector field defined in a restricted domain of $\mathbb{R}^3$, Sect. 21.3, and as a *distribution* of $\mathcal{S}'(\mathbb{R}^3)$, Sect. 21.4. Since, furthermore, a fundamental role in the construction of a consistent quantum dynamics for dyons is played by *gauge invariance*, we begin this chapter by explaining how this local symmetry is implemented, in general, in quantum mechanics.

Most textbooks consider in place of a two-dyon system a *charge e* moving in the presence of a static *monopole g*. This configuration can be retrieved as particular case from our two-dyon system by setting $e_1 = 0$, $g_1 = g$, $e_2 = e$, $g_2 = 0$, and by considering the limit for $m_1 \to \infty$. Under this limit the reduced mass $m = m_1 m_2/(m_1 + m_2)$ tends indeed to the mass m_2 of the charge, and the position of the monopole $\mathbf{y}_1$ tends to the center-of-mass position $\mathbf{r}_{CM}$, see Sect. 20.3.1.

21.1   Gauge Invariance in Quantum Mechanics

Consider a non-relativistic charged particle in the presence of an electromagnetic field $F^{\mu\nu}$, satisfying the equation of motion (21.1). If $F^{\mu\nu}$ satisfies the Bianchi identity $\partial_{[\mu} F_{\nu\rho]} = 0$, there exists a vector potential A^μ such that the fields can be expressed as

$$\mathbf{E} = -\nabla A^0 - \frac{1}{c}\frac{\partial \mathbf{A}}{\partial t}, \qquad \mathbf{B} = \nabla \times \mathbf{A}. \tag{21.3}$$

From Sect. 4.2 we then know that Eq. (21.1) follows from the Lagrangian (see Problem 4.2)

$$L = \frac{1}{2} m v^2 - eA^0 + \frac{e}{c}\mathbf{v} \cdot \mathbf{A}. \tag{21.4}$$

The Legendre transformation of L then leads to the known classical minimal-coupling Hamiltonian H:

$$\mathbf{P} = \frac{\partial L}{\partial \mathbf{v}} = m\mathbf{v} + \frac{e}{c}\mathbf{A} \quad \to \quad H = \mathbf{P} \cdot \mathbf{v} - L = \frac{1}{2m}\left(\mathbf{P} - \frac{e}{c}\mathbf{A}\right)^2 + eA^0.$$

According to the canonical-quantization paradigm, the Hamiltonian *operator*, which we denote by the same symbol, is then obtained by operating in the above expression of H the substitution $\mathbf{P} \rightarrow -i\hbar\nabla$. The Hamiltonian operator is thus given by

$$H = -\frac{\hbar^2}{2m}\left(\nabla - \frac{ie}{\hbar c}\mathbf{A}\right)^2 + eA^0. \tag{21.5}$$

Similarly, in the presence of an electromagnetic field satisfying the Bianchi identity, the momentum of the particle $\mathbf{p} = m\mathbf{v} = \mathbf{P} - e\mathbf{A}/c$, and its angular momentum $\mathbf{L} = \mathbf{x} \times \mathbf{p}$, are represented by the operators

$$\mathbf{p} = \frac{\hbar}{i}\nabla - \frac{e}{c}\mathbf{A}, \qquad \mathbf{L} = \mathbf{x} \times \left(\frac{\hbar}{i}\nabla - \frac{e}{c}\mathbf{A}\right). \tag{21.6}$$

Finally, the time evolution of the wavefunction $\psi(t, \mathbf{x})$, belonging to the Hilbert space $\mathcal{H} = L^2(\mathbb{R}^3)$, is governed by the Schrödinger equation

$$i\hbar\frac{\partial\psi}{\partial t} = H\psi. \tag{21.7}$$

21.1.1   Gauge Transformations and Physical Symmetries

The quantum mechanical observables H, $\mathbf{p}$ and $\mathbf{L}$ introduced above depend explicitly on the vector potential A^μ. However, this potential is defined only modulo the gauge transformations[1]

$$A'^0 = A^0 - \frac{1}{c}\frac{\partial\Lambda}{\partial t}, \qquad \mathbf{A}' = \mathbf{A} + \nabla\Lambda, \tag{21.8}$$

where the gauge parameter $\Lambda(t, \mathbf{x})$ in general is a function of space and time. This would imply that different vector potentials give rise to different observables, and, consequently, measurable quantities, like the expectation values of the energy and of the momentum, appear to depend on the chosen gauge fixing. We address this problem for a *static* field $F^{\mu\nu}(\mathbf{x})$, in order that the resulting physical system – a charged particle moving in the presence of a static external electromagnetic field – turns out to be *conservative*, i.e. its energy is a constant of motion. In this case we can choose a time-independent four-potential $A^\mu(\mathbf{x})$, so that the resulting Hamiltonian (21.5) is time-independent as well. The allowed gauge transformations (21.8) are then only those with a parameter $\Lambda(\mathbf{x})$ not depending on time, i.e.

$$A'^0 = A^0, \qquad \mathbf{A}' = \mathbf{A} + \nabla\Lambda. \tag{21.9}$$

---

[1] With respect to the notation adopted in previous chapters we have flipped the sign of Λ.

Notice that even under these restricted gauge transformations the observables H, $\mathbf{p}$ and $\mathbf{L}$ of formulas (21.5) and (21.6) are not invariant.

Physical symmetries. According to *Wigner's theorem*, in quantum mechanics the correspondence between physical states and vectors of the Hilbert space, on one hand, and the correspondence between observable quantities and self-adjoint operators, on the other hand, are not uniquely determined. In fact, different correspondences are physically equivalent, if they are related by a *physical symmetry*, i.e. by a unitary[2] operator U

$$U^\dagger U = U U^\dagger = 1.$$

More precisely, a correspondence in which the states and the observables are represented by the vectors ψ and by the operators O, respectively, and another correspondence in which they are represented by the vectors ψ' and by the operators O', respectively, give rise to the same measurable quantities if they are connected by the relations

$$\psi' = U\psi, \qquad O' = U O U^\dagger,$$

for a certain unitary operator U. It follows, in particular, that all matrix elements of the observables in the two correspondences are the same

$$(\psi'_1, O'\psi'_2) = (\psi_1, O\psi_2).$$

The problem of the gauge-dependence of the Hamiltonian (21.5), and of the operators in (21.6), can actually be reduced to a physical symmetry. In fact, the gauge-transformed Hamiltonian

$$H' = -\frac{\hbar^2}{2m}\left(\boldsymbol{\nabla} - \frac{ie}{\hbar c}\mathbf{A}'\right)^2 + eA^0, \tag{21.10}$$

where $\mathbf{A}'$ is the potential in (21.9), is related to the original Hamiltonian (21.5) through a unitary operator. The corresponding unitary transformation consists simply in the multiplication of the wavefunction $\psi(\mathbf{x})$ by the $\mathbf{x}$-dependent phase $e^{ie\Lambda(\mathbf{x})/\hbar c}$:

$$\psi'(\mathbf{x}) = (U\psi)(\mathbf{x}) = e^{ie\Lambda(\mathbf{x})/\hbar c}\psi(\mathbf{x}).$$

In fact, using the operator identities

$$U\,\boldsymbol{\nabla}\,U^\dagger = \boldsymbol{\nabla} - \frac{ie}{\hbar c}\boldsymbol{\nabla}\Lambda, \qquad U\left(\boldsymbol{\nabla} - \frac{ie}{\hbar c}\mathbf{A}\right)U^\dagger = \boldsymbol{\nabla} - \frac{ie}{\hbar c}\mathbf{A}', \tag{21.11}$$

---

[2]In the present context, *antiunitary* operators do not play any role.

it is immediately seen that the above Hamiltonians are related by the transformation

$$H' = UHU^\dagger. \tag{21.12}$$

Similarly, formulas (21.6) and (21.11) imply the relations

$$\mathbf{p}' = U\,\mathbf{p}U^\dagger, \qquad \mathbf{L}' = U\,\mathbf{L}U^\dagger.$$

We, therefore, conclude that in quantum mechanics a gauge transformation corresponds to a *physical symmetry*. All measurable quantities are hence gauge invariant.

21.2   Generalized Hilbert Space

The natural Hilbert space hosting the quantum dynamics of a non-relativistic spinless particle is $L^2(\mathbb{R}^3)$, i.e. the set of all square-integrable complex functions $\psi(\mathbf{x})$ on $\mathbb{R}^3$:

$$\int |\psi(\mathbf{x})|^2 \, d^3x < \infty.$$

However, as we will see in Sect. 21.5, for a charged particle moving in the presence of the electromagnetic field generated by a magnetic monopole, or, more generally, by a dyon, it will no longer be possible to introduce a *single* wavefunction $\psi(\mathbf{x})$ covering the whole space $\mathbb{R}^3$. This will force us to resort to a slightly more involved realization of the Hilbert space. The construction of the Hilbert space, needed for our purposes, starts from the choice of two open subsets V_1 and V_2 of $\mathbb{R}^3$, with the properties

$$V_1 \cup V_2 = \mathbb{R}^3, \qquad V_1 \cap V_2 = V_0 \neq \emptyset. \tag{21.13}$$

The non-empty intersection V_0 of these sets is thus again an open set. We then introduce a vector space $\mathcal{H}$, whose elements are defined as the pairs of complex functions $\psi = \{\psi_1, \psi_2\}$ such that

$$\psi_1 \in L^2(V_1), \quad \psi_2 \in L^2(V_2). \tag{21.14}$$

The function ψ_1 (ψ_2) is thus square-integrable on its domain of definition V_1 (V_2). As we will see in Sect. 21.5, in the case of the relative dyon satisfying the equation of motion (21.2), the need for *two* wavefunctions, each one being defined only in a subset of $\mathbb{R}^3$, derives from the fact that neither of the two can be *continued* to the whole space $\mathbb{R}^3$. The pairs of wavefunctions introduced above form naturally a linear vector space. In fact, the sum is defined as

$$\{\psi_1, \psi_2\} + \{\phi_1, \phi_2\} = \{\psi_1 + \phi_1, \psi_2 + \phi_2\}, \tag{21.15}$$

and the multiplication by a constant $k \in \mathbb{C}$ acts on a pair as

$$k\{\psi_1, \psi_2\} = \{k\psi_1, k\psi_2\}. \tag{21.16}$$

21.2.1 Transition Function, Scalar Product, and Operators

To promote the vector space $\mathcal{H}$ to a (pre)-Hilbert space, we must endow it with a *scalar product*. This is actually a non-trivial task, since in the case at hand the standard definition $(\psi, \phi) = \int_{\mathbb{R}^3} \psi^* \phi \, d^3x$ conflicts with the fact that in the subset V_0 of $\mathbb{R}^3$ we have introduced two different wavefunctions, and so this integral would not yield a uniquely determined number. This mathematical ambiguity would imply the physical contradiction that the probability density to find the particle in a position $\mathbf{x} \in V_0$ would have two different determinations, according to whether one considers $|\psi_1(\mathbf{x})|^2$ or $|\psi_2(\mathbf{x})|^2$. Therefore, if we want to keep the probabilistic interpretation of the wavefunction, we must require the equality

$$|\psi_1(\mathbf{x})| = |\psi_2(\mathbf{x})|, \quad \text{for all } \mathbf{x} \in V_0. \tag{21.17}$$

This condition is, in turn, equivalent to the existence of a *real* function $D(\mathbf{x})$ defined in V_0, such that

$$\psi_2(\mathbf{x}) = e^{iD(\mathbf{x})} \psi_1(\mathbf{x}), \quad \text{for all } \mathbf{x} \in V_0. \tag{21.18}$$

Notice that, as $D(\mathbf{x})$ is a function of the position $\mathbf{x}$, the relation (21.18) does not express the elementary fact that wavefunctions are defined modulo (constant) phases. If we now consider another element $\phi = \{\phi_1, \phi_2\}$ of $\mathcal{H}$, for the same reason we must have $\phi_2(\mathbf{x}) = e^{iE(\mathbf{x})} \phi_1(\mathbf{x})$, for all $\mathbf{x} \in V_0$, where *a priori* E could be a real function different from D. However, the operations of vector addition (21.15) and of scalar multiplication (21.16) must preserve the uniqueness property of the probability density (21.17). This implies that for all $\mathbf{x} \in V_0$ and for all complex constants a and b we must have

$$|a\psi_1 + b\phi_1| = |a\psi_2 + b\phi_2| = |ae^{iD}\psi_1 + be^{iE}\phi_1| = |a\psi_1 + be^{i(E-D)}\phi_1|,$$

a condition that holds only if

$$E(\mathbf{x}) - D(\mathbf{x}) = 2n\pi, \quad \text{with } n \in \mathbb{Z}.$$

This means that the exponential $e^{iD(\mathbf{x})}$, called *transition function*, that relates the wavefunctions of a pair according to Eq. (21.18), must be the same for all elements $\{\psi_1, \psi_2\}$ of $\mathcal{H}$. It characterizes, thus, the Hilbert space we are going to construct.

Scalar product. Thanks to the relation (21.18), we can now define an unambiguous scalar product between the elements $\psi = \{\psi_1, \psi_2\}$ and $\phi = \{\phi_1, \phi_2\}$ of $\mathcal{H}$ as

$$(\psi, \phi) = \int_{V_1} \psi_1^* \phi_1 \, d^3x + \int_{V_2 \setminus V_0} \psi_2^* \phi_2 \, d^3x = \int_{V_1 \setminus V_0} \psi_1^* \phi_1 \, d^3x + \int_{V_2} \psi_2^* \phi_2 \, d^3x.$$

(21.19)

With this scalar product, the space $\mathcal{H}$, i.e. the set of all pairs $\{\psi_1, \psi_2\}$ satisfying the conditions (21.14) and (21.18), is now a pre-Hilbert space, characterized by the transition function $e^{iD(\mathbf{x})}$. Finally, it is easy to see that, endowed with the *norm* induced by the scalar product (21.19), the space $\mathcal{H}$ is also *complete*, so that it is, actually, a *Hilbert space*. To distinguish it from $L^2(\mathbb{R}^3)$, henceforth, we will refer to the Hilbert space $\mathcal{H}$ as *generalized* Hilbert space.

Operators. In the Hilbert space $L^2(\mathbb{R}^3)$ a generic operator O can be built from the canonical operators $\mathbf{x}$ and $\mathbf{p} = -i\hbar \nabla$. Correspondingly, in what follows we restrict our attention to operators of the form $O(\mathbf{x}, \nabla)$, involving the partial derivatives with respect to $\mathbf{x}$ and the multiplication by functions of $\mathbf{x}$, as, for instance, the Hamiltonian (21.5). Operators of this type are *naturally* defined on a space of sufficiently regular functions. We can thus introduce an operator O in the generalized Hilbert space $\mathcal{H}$, by starting from two operators O_1 and O_2 defined in $L^2(V_1)$ and $L^2(V_2)$, respectively,[3]

$$\psi_i \in L^2(V_i) \quad \rightarrow \quad O_i \psi_i \in L^2(V_i), \quad i = 1, 2.$$

The resulting operator O in $\mathcal{H}$ is then defined by

$$\psi = \{\psi_1, \psi_2\} \in \mathcal{H} \quad \rightarrow \quad O\psi = \{O_1 \psi_1, O_2 \psi_2\}. \quad (21.20)$$

However, this definition is well posed only if the pair $\{O_1 \psi_1, O_2 \psi_2\}$ belongs again to $\mathcal{H}$, namely if it satisfies the relation (21.18)

$$O_2 \psi_2 = e^{iD} O_1 \psi_1, \quad \text{for all } \mathbf{x} \in V_0.$$

Since for $\mathbf{x} \in V_0$ we have $\psi_2 = e^{iD} \psi_1$, we thus infer that the operators O_1 and O_2 must be related by

$$O_2 = e^{iD} O_1 e^{-iD}, \quad \text{for all } \mathbf{x} \in V_0. \quad (21.21)$$

This relation is well defined because the operators O_1 and O_2 are constructed from the basic operators $\mathbf{x}$ and ∇, and so they admit natural restrictions from the spaces V_1 and V_2, respectively, to V_0. In conclusion, the operators O_1 and O_2 cannot be chosen arbitrarily in that in V_0 they must be related to each other by the unitary operator $U = e^{iD}$, the same operator that relates the wavefunctions according to Eq. (21.18). Thanks to these relations, all matrix elements $(\phi, O\psi)$, defined according to the scalar product (21.19), are unambiguously determined.

---

[3] We shall return to the question of the *domain* of these operators below, when we analyze the regularity properties of the transition function $e^{iD(\mathbf{x})}$.

Domains of operators and continuity of the transition function. An important restriction for the allowed transition functions $e^{iD(\mathbf{x})}$ – regarding their regularity properties – comes from the particular kind of observables we need to introduce on $\mathcal{H}$. In the conventional case of the quantum mechanics of a particle, an observable is represented by a *self-adjoint* operator in the Hilbert space $L^2(\mathbb{R}^3)$. However, all fundamental observables, as the position, the momentum, the angular momentum, and the Hamiltonian itself, are represented by *unbounded* self-adjoint operators, and as such their *domain* is not the whole space $L^2(\mathbb{R}^3)$, but rater one of its *dense* subspaces. These observables involve in particular the derivative operator ∇, and correspondingly their domain of self-adjointness is formed by sufficiently regular functions, which in general are *continuous*[4]. As an example, consider in the one-particle Hilbert space $L^2(\mathbb{R}^3)$ the free Hamiltonian $H_0 = -\nabla^2$, whose domain of self-adjointness is known to be

$$\mathcal{D}(H_0) = \{\psi(\mathbf{x}) \in L^2(\mathbb{R}^3)/\, |\mathbf{k}|^2\widehat{\psi}(\mathbf{k}) \in L^2(\mathbb{R}^3)\}, \qquad (21.22)$$

where $\widehat{\psi}(\mathbf{k})$ is the Fourier transform of $\psi(\mathbf{x})$. Since also $\widehat{\psi}(\mathbf{k})$ belongs to $L^2(\mathbb{R}^3)$, we can rewrite the Fourier transform of a wavefunction $\psi(\mathbf{x})$ belonging to $\mathcal{D}(H_0)$ as the product of two functions belonging to $L^2(\mathbb{R}^3)$

$$\widehat{\psi}(\mathbf{k}) = \left(\widehat{\psi}(\mathbf{k}) + |\mathbf{k}|^2\widehat{\psi}(\mathbf{k})\right) \cdot \frac{1}{1 + |\mathbf{k}|^2}.$$

As the scalar product of two functions of $L^2(\mathbb{R}^3)$ is always a finite number, we thus infer that $\widehat{\psi}(\mathbf{k}) \in L^1(\mathbb{R}^3)$. From the Riemann-Lebesgue lemma, Theorem I of Sect. 5.2, it then follows that $\psi(\mathbf{x})$ is a continuous function. We have thus proven that every wavefunction $\psi(\mathbf{x})$ of the domain of self-adjointness of the free Hamiltonian (21.22) is *continuous*. This property can be extended to more general, interacting, Hamiltonians, by means of the *Kato-Rellich theorem*, see, for instance, the textbook [1].

Returning to our generalized Hilbert space $\mathcal{H}$ constructed above, let us consider a generic element $\{\psi_1, \psi_2\}$ belonging to the domain of self-adjointness of one of the fundamental observables, and let us hence assume that both functions $\psi_1(\mathbf{x})$ and $\psi_2(\mathbf{x})$ are continuous in V_1 and V_2, respectively. Then, the relation (21.18) implies the important restriction that *the transition function $e^{iD(\mathbf{x})}$ must be continuous* in its domain of definition V_0. Otherwise, it would not be possible to introduce the fundamental observables as self-adjoint operators on $\mathcal{H}$. Notice that this does not necessarily imply that the function $D(\mathbf{x})$ itself is continuous. In fact, it can have discontinuities of integer multiples of 2π. The continuity of the transition function will play a fundamental role in the derivation of Dirac's quantization condition within the approach of canonical quantization, see Sect. 21.5.1.

---

[4]Frequently, the *eigenfunctions* and *eigenfunctionals* of the fundamental observables of quantum mechanics are even more regular than the generic functions of their domain of self-adjointness, typically they are of class C^∞.

21.3   A Vector Potential for the Magnetic Monopole

A magnetic monopole of charge g, sitting at the origin, generates the magnetic field

$$\mathbf{B} = \frac{g\mathbf{x}}{4\pi r^3},\tag{21.23}$$

satisfying the equation

$$\mathbf{\nabla} \cdot \mathbf{B} = g\delta^3(\mathbf{x}).\tag{21.24}$$

As the second member of this equation is different from zero, there exists no vector potential $\mathbf{A}$ such that $\mathbf{B} = \mathbf{\nabla} \times \mathbf{A}$ in the whole space $\mathbb{R}^3$. In this section we address the problem of the existence of vector potentials for the magnetic field (21.23) in *restricted* domains of $\mathbb{R}^3$. As anticipated in the introduction to this chapter, the related construction, leading to the so-called *Dirac potentials*, will allow us to implement the quantum mechanics of dyons in the canonical-quantization approach, see Sect. 21.5.

21.3.1   Dirac Potentials and Dirac Strings

Poincaré lemma and Dirac string. To address the problem of a vector potential for the magnetic field (21.23) it is convenient to first translate the static equation (21.24) in the language of differential forms in $\mathbb{R}^3$, and then to resort to the *Poincaré lemma*, see Sect. 19.1. We recall that a differential p-form Φ is called closed if $d\Phi = 0$, and that it is called exact if there exists a $(p-1)$-form Ψ such that $\Phi = d\Psi$. Since $d^2 = 0$, every exact form is closed. The Poincaré lemma ensures, in turn, that a form which is closed in a *contractible* domain $\mathcal{C}$, is exact in $\mathcal{C}$. Qualitatively, a set is called contractible if it can be deformed continuously to one of its points.

We introduce as basic differential forms the 2-form

$$B = \frac{1}{2}\,dx^j \wedge dx^i \varepsilon^{ijk} B^k = \frac{g}{8\pi}\,dx^j \wedge dx^i \varepsilon^{ijk}\frac{x^k}{r^3},\tag{21.25}$$

representing the magnetic field B^k (21.23), and the 3-form

$$J = \frac{1}{6}\,dx^k \wedge dx^j \wedge dx^i \varepsilon^{ijk}\delta^3(\mathbf{x}),\tag{21.26}$$

representing the unit charge density. The scalar equation (21.24) then goes over to the equation between 3-forms, see Problem 21.1,

$$dB = gJ.\tag{21.27}$$

This equation can also be derived directly from the generalized Bianchi identity
(20.21), by setting $\mathbf{E} = 0$, $\mathbf{j}_m = 0$, and $j_m^0 = g\delta^3(\mathbf{x})$. In fact, in this case we obtain
$F = \frac{1}{2} dx^\mu \wedge dx^\nu F_{\nu\mu} = \frac{1}{2} dx^j \wedge dx^i F_{ij} = -B$, and $*j_m = gJ$. As the 3-form J
(21.26) vanishes in whole space except the origin O, Eq. (21.27) implies that the 2-
form B is closed in the domain $\mathbb{R}^3 \setminus \{O\}$. We are thus tempted to apply the Poincaré
lemma in this domain, searching for a 1-form

$$A = dx^i A^i$$

such that in $\mathbb{R}^3 \setminus \{O\}$ we have the relation $B = dA$, which is equivalent to $\mathbf{B} = \nabla \times \mathbf{A}$, see Problem 21.1. However, this strategy fails because the domain $\mathbb{R}^3 \setminus \{O\}$ is *not* contractible,[5] as instead required by the Poincaré lemma. To proceed
along this path we must, therefore, restrict the domain $\mathbb{R}^3 \setminus \{O\}$ in such a way that
it becomes contractible. A *minimal* way to operate such a restriction is to introduce
a curve γ – called *Dirac string* – which has one of its endpoints at the origin, where
the monopole sits, and whose free end extends to infinity. The *restricted domain*
defined as $V_\gamma = \mathbb{R}^3 \setminus \gamma$ is then contractible, and so the Poincaré lemma ensures that
in V_γ the 2-form B is *exact*. This means that in V_γ there exists a 1-form A_γ such
that

$$B = dA_\gamma, \quad \text{i.e.} \quad \mathbf{B} = \nabla \times \mathbf{A}_\gamma, \quad \text{for all } \mathbf{x} \in V_\gamma. \tag{21.28}$$

Of course, the 1-form A_γ is determined only modulo closed 1-forms. But, since
V_γ is contractible, this implies that it is determined modulo exact 1-forms: $A_\gamma \approx A_\gamma + dh(\mathbf{x})$, where $h(\mathbf{x})$ is an arbitrary scalar function defined on V_γ.

Dirac potential. There exists, however, a *canonical* vector potential A_γ – called
Dirac potential – which can be written in terms of a line integral along the Dirac
string γ. For this purpose, we parameterize γ as $\mathbf{y}(\lambda)$, where the parameter λ ranges
over the half-line $\lambda \geq 0$ and $\mathbf{y}(0) = O$. Dirac's vector potential is then defined by
the line integral

$$A_\gamma^i(\mathbf{x}) = \frac{g}{4\pi} \varepsilon^{ijk} \int_\gamma \frac{x^j - y^j}{|\mathbf{x} - \mathbf{y}|^3} \, dy^k. \tag{21.29}$$

To verify that this potential satisfies indeed equation (21.28) we calculate its curl for
a generic position $\mathbf{x} \in V_\gamma$:

$$(\nabla \times \mathbf{A}_\gamma)^l = \varepsilon^{lmn} \partial_m A_\gamma^n = \frac{g}{4\pi} \varepsilon^{lmn} \partial_m \varepsilon^{nij} \int_\gamma \frac{x^i - y^i}{|\mathbf{x} - \mathbf{y}|^3} \, dy^j$$

$$= \frac{g}{4\pi} \left(\delta^{li}\delta^{mj} - \delta^{lj}\delta^{mi}\right) \int_\gamma \partial_m \frac{x^i - y^i}{|\mathbf{x} - \mathbf{y}|^3} \, dy^j$$

$$= \frac{g}{4\pi} \int_\gamma \partial_j \frac{x^l - y^l}{|\mathbf{x} - \mathbf{y}|^3} \, dy^j - \frac{g}{4\pi} \int_\gamma \partial_i \frac{x^i - y^i}{|\mathbf{x} - \mathbf{y}|^3} \, dy^l \tag{21.30}$$

---

[5]For instance, a sphere enclosing the origin, which is a subset of $\mathbb{R}^3 \setminus \{O\}$, cannot be deformed
continuously to a point without crossing the origin.

$$= -\frac{g}{4\pi} \int_\gamma d\left(\frac{x^l - y^l}{|\mathbf{x} - \mathbf{y}|^3}\right) = \frac{gx^l}{4\pi r^3} = B^l. \tag{21.31}$$

We have used the identity $\nabla \cdot (\mathbf{x} - \mathbf{y}/|\mathbf{x} - \mathbf{y}|^3) = 0$, holding for $\mathbf{x} \notin \gamma$, so that the integrand of the second term in the line (21.30) is identically zero. Due to the denominator $|\mathbf{x} - \mathbf{y}|^3$, the Dirac potential (21.29) is *singular* for all $\mathbf{x} \in \gamma$, i.e. along the Dirac string, while it is regular for all $\mathbf{x} \in V_\gamma$, i.e. in the complement of the Dirac string. To make these properties more explicit, we evaluate the line integral (21.29) explicitly for the case of a *straight* Dirac string, directed, say, along the half-line $\mathbf{y}(\lambda) = \lambda\mathbf{n}, |\mathbf{n}| = 1$ (see Problem 21.2)

$$A_\gamma^i = \frac{g\varepsilon^{ijk}}{4\pi} \int_0^\infty \frac{x^j - n^j\lambda}{|\mathbf{x} - \mathbf{n}\lambda|^3} n^k d\lambda = \frac{g\varepsilon^{ijk}x^j n^k}{4\pi} \int_0^\infty \frac{d\lambda}{|\mathbf{x} - \mathbf{n}\lambda|^3} = \frac{g\varepsilon^{ijk}x^j n^k}{4\pi r(r - \mathbf{n} \cdot \mathbf{x})}. \tag{21.32}$$

For a linear Dirac string the Dirac potential takes thus the simple form

$$\mathbf{A}_\gamma(\mathbf{x}) = \frac{g}{4\pi} \frac{\mathbf{x} \times \mathbf{n}}{r(r - \mathbf{n} \cdot \mathbf{x})}. \tag{21.33}$$

This expression is indeed regular for all $\mathbf{x} \in \mathbb{R}^3$, except the points of the Dirac string $\mathbf{x} = \lambda\mathbf{n}$. Actually, the vector potential originally introduced by Dirac in [2] corresponds to the choice $\mathbf{n} = (0, 0, -1)$, which amounts to direct the Dirac string along the negative z axis.

In conclusion, the magnetic field created by a static monopole admits *locally*, i.e. in a restricted domain V_γ, a vector potential, which has as *minimal* singularity locus a curve γ connecting the monopole to infinity. On the other hand, this curve cannot be of any empirical relevance – it cannot be *observable* – and so one is naturally faced with the problem of what happens when one chooses another curve.

21.3.2 Moving the Dirac String

Let us choose two different Dirac strings γ_1 and γ_2, intersecting only at the origin O, parameterized by $\mathbf{y}_1(\lambda)$ and $\mathbf{y}_2(\lambda)$, respectively. We can then introduce the associated Dirac vector potentials A_1 and A_2, according to formula (21.29), which are defined in the domains $V_1 = \mathbb{R}^3 \setminus \gamma_1$ and $V_2 = \mathbb{R}^3 \setminus \gamma_2$, respectively. These potentials have as common domain of definition the region $V_0 = V_1 \cap V_2 = \mathbb{R}^3 \setminus \Gamma$, where $\Gamma = \gamma_1 \cup \gamma_2$ is a curve extending from "minus" infinity to "plus" infinity, passing through the position of the monopole. We can parameterize it as

$$\Gamma = \gamma_1 \cup \gamma_2 \quad \leftrightarrow \quad \mathbf{y}(\lambda) = \begin{cases} \mathbf{y}_1(-\lambda), & \text{for } -\infty < \lambda \le 0, \\ \mathbf{y}_2(\lambda), & \text{for } 0 \le \lambda < \infty. \end{cases} \tag{21.34}$$

Since both potentials A_1 and A_2 satisfy equation (21.28) in the domains V_1 and V_2, respectively, in the intersection V_0 they fulfill the identities

$$dA_1 = B = dA_2 \quad \Rightarrow \quad d(A_2 - A_1) = 0.$$

Hence, in V_0 the 1-form $A_2 - A_1$ is closed. However, the domain V_0, namely $\mathbb{R}^3$ without the infinite curve Γ, is *not* contractible.[6] Therefore, we cannot infer that the 1-form $A_2 - A_1$ is exact in V_0. Once more, we must restrict the domain V_0 in such a way that it becomes contractible. A minimal way to do this is to introduce an infinitely extended surface Σ whose boundary is the curve Γ,

$$\partial \Sigma = \Gamma. \tag{21.35}$$

Obviously, there are infinitely many ways to choose such a surface. We can parameterize the chosen surface Σ via a vector function of two variables λ and u,

$$\mathbf{y}(\lambda, u), \quad -\infty < \lambda < \infty, \quad 0 \leq u < \infty. \tag{21.36}$$

To fulfill the constraint (21.35) this function must obey the boundary condition

$$\mathbf{y}(\lambda, 0) = \mathbf{y}(\lambda), \tag{21.37}$$

where $\mathbf{y}(\lambda)$ is the curve (21.34) parameterizing Γ. From a topological point of view, Σ corresponds to an infinitely extended half-plane, and consequently the *restricted domain* $\mathbb{R}^3 \setminus \Sigma$ is contractible. The Poincaré lemma then ensures that in $\mathbb{R}^3 \setminus \Sigma$ the 1-form $A_2 - A_1$ is exact. Therefore, there exists a scalar function $\Lambda(\mathbf{x})$ – a *gauge function* – such that

$$A_2 - A_1 = d\Lambda, \quad \text{i.e. } \mathbf{A}_2 - \mathbf{A}_1 = \boldsymbol{\nabla}\Lambda, \quad \text{for all } \mathbf{x} \in \mathbb{R}^3 \setminus \Sigma. \tag{21.38}$$

In the restricted domain $\mathbb{R}^3 \setminus \Sigma$ the Dirac potentials $\mathbf{A}_1$ and $\mathbf{A}_2$ differ, hence, by a *gauge transformation*, so that commonly one says that a *change of the Dirac string is equivalent to a gauge transformation*. This statement, however, must be considered with some caution in that, whereas the potentials $\mathbf{A}_1$ and $\mathbf{A}_2$ are regular in the region V_0, which includes the surface Σ, the gauge function Λ is *discontinuous* on Σ, as we will see below. In other words, the gauge transformation that moves the Dirac string from one position to another is *singular*. In the next section we derive an integral representation for $\Lambda(\mathbf{x})$ that will allow us to analyze its singularities in more detail.

---

[6]For instance, a closed loop circling the curve Γ – a subset of V_0 – cannot be deformed continuously to a point without crossing Γ.

21.3.3   The Gauge Function

Using for the Dirac potentials A_1 and A_2 the representations (21.29), we find for their difference the line-integral representation

$$A_2^i - A_1^i = \frac{g}{4\pi} \int_\Gamma Q^{ik} dy^k, \qquad Q^{ik} = \varepsilon^{ijk} \frac{x^j - y^j}{|\mathbf{x} - \mathbf{y}|^3}, \qquad (21.39)$$

where Γ is the curve (21.34), and we understand that $\mathbf{y} = \mathbf{y}(\lambda)$. In the following, we want to apply to the integral (21.39) Stokes' theorem, allowing to transform a line integral around a closed curve in a surface integral. For this purpose, we first rewrite the integral (21.39) as the limit for $L \to \infty$ of an integral along a closed curve α_L, that is made up of two curves, $\alpha_L = \Gamma_L \cup \beta_L$. The curve Γ_L is the finite segment of the curve Γ bounded by $-L < \lambda < L$. For β_L we choose a curve of the form of a "semicircle" of "radius" L, lying on the surface Σ, whose endpoints coincide with the endpoints of Γ_L. By construction, the closed curve α_L lies on the surface Σ, and hence there exists a surface $\Sigma_L \subset \Sigma$ whose boundary is the curve α_L, $\partial \Sigma_L = \alpha_L$. In the limit for $L \to \infty$ the surface Σ_L tends to Σ and the curve β_L moves to infinity. Consequently, in the limit for $L \to \infty$ the line integral

$$\int_{\beta_L} Q^{ik} dy^k$$

tends to zero, since in the limit for $|\mathbf{y}| \to \infty$ the function Q^{ik} in formula (21.39) vanishes as $1/y^2$. Therefore, by applying Stokes' theorem, the integral (21.39) can be rewritten as

$$\begin{aligned}
A_2^i - A_1^i &= \frac{g}{4\pi} \lim_{L \to \infty} \int_{\Gamma_L} Q^{ik} dy^k = \frac{g}{4\pi} \lim_{L \to \infty} \left(\int_{\Gamma_L} Q^{ik} dy^k + \int_{\beta_L} Q^{ik} dy^k \right) \\
&= \frac{g}{4\pi} \lim_{L \to \infty} \int_{\alpha_L = \partial \Sigma_L} Q^{ik} dy^k = \frac{g}{4\pi} \lim_{L \to \infty} \int_{\Sigma_L} \varepsilon^{kmn} \frac{\partial Q^{in}}{\partial y^m} d\Sigma^k \\
&= \frac{g}{4\pi} \int_\Sigma \varepsilon^{kmn} \frac{\partial Q^{in}}{\partial y^m} d\Sigma^k, \qquad (21.40)
\end{aligned}$$

where $d\Sigma^k$ is the area element of the surface Σ (21.36)

$$d\Sigma^k = \varepsilon^{krs} \frac{\partial y^r}{\partial \lambda} \frac{\partial y^s}{\partial u} d\lambda \, du. \qquad (21.41)$$

Since the tensor Q^{ik} (21.39) depends only on the difference $\mathbf{x} - \mathbf{y}$, the derivative $\partial/\partial y^m$ in the integral (21.40) can be replaced by the derivative $-\partial/\partial x^m = -\partial_m$, which can be taken out of the integral. Inserting in Eq. (21.40) the expression for Q^{ik} (21.39) and using the identity $\varepsilon^{kmn}\varepsilon^{ijn} = \delta^{ki}\delta^{mj} - \delta^{kj}\delta^{mi}$, we then obtain

$$A_2^i - A_1^i = -\frac{g}{4\pi}\,\partial_m \int_\Sigma \varepsilon^{kmn} Q^{in} d\Sigma^k$$

$$= \frac{g}{4\pi}\,\partial_i \int_\Sigma \frac{x^k - y^k}{|\mathbf{x} - \mathbf{y}|^3}\, d\Sigma^k - \frac{g}{4\pi}\,\partial_k \int_\Sigma \frac{x^k - y^k}{|\mathbf{x} - \mathbf{y}|^3}\, d\Sigma^i, \tag{21.42}$$

where now we understand the identification $\mathbf{y} = \mathbf{y}(\lambda, u)$. Finally, since for $\mathbf{x} \notin \Sigma$ we have $\nabla \cdot (\mathbf{x} - \mathbf{y}/|\mathbf{x} - \mathbf{y}|^3) = 0$, the last term of equation (21.42) vanishes. Comparing this equation with the relation (21.38), we derive that a possible gauge function is given by the surface integral over Σ

$$\Lambda(\mathbf{x}) = \frac{g}{4\pi} \int_\Sigma \frac{x^k - y^k}{|\mathbf{x} - \mathbf{y}|^3}\, d\Sigma^k = \frac{g}{4\pi}\, \varepsilon^{ijk} \int_\Sigma \frac{\partial y^i}{\partial \lambda} \frac{\partial y^j}{\partial u} \frac{x^k - y^k}{|\mathbf{x} - \mathbf{y}|^3}\, d\lambda\, du. \tag{21.43}$$

This function is, in fact, regular for all $\mathbf{x} \in \mathbb{R}^3 \setminus \Sigma$, and it is singular for $\mathbf{x} \in \Sigma$. However, it is not difficult to realize that if $\mathbf{x}$ tends to a point P of Σ, the integral (21.43) yields a *finite* value for $\Lambda(P)$. On the other hand, as we show below, this value depends also on the side of Σ from which P is approached. In other words, the singularity of the function $\Lambda(\mathbf{x})$ consists in a *finite discontinuity* when $\mathbf{x}$ crosses the surface Σ.

Discontinuity of the gauge function. To determine the discontinuity of Λ across Σ in a generic point $\mathbf{x} \in \Sigma$, it is convenient to introduce an arbitrary closed loop G passing through $\mathbf{x}$ and intersecting Σ only at the point $\mathbf{x}$. In this way, the loop G circles the curve Γ just once. The discontinuity of Λ across Σ at the point $\mathbf{x}$ can then be computed through the integral around a *quasi-closed* curve G

$$\Delta\Lambda(\mathbf{x}) = \int_G d\Lambda = \int_G \nabla\Lambda \cdot d\mathbf{x} = \int_G (\mathbf{A}_2 - \mathbf{A}_1) \cdot d\mathbf{x}, \tag{21.44}$$

where we understand that the "endpoints" of G are the two points $\mathbf{x} \pm \varepsilon$, lying at opposite sides of Σ. It is understood that in formula (21.44), at the end, one must take the limit for $\varepsilon \to 0$. Consequently, since the potentials $\mathbf{A}_1$ and $\mathbf{A}_2$ are well defined in the domain $\mathbb{R}^3 \setminus \Gamma$, the integrals in the last term of equation (21.44) can be considered as integrals along the whole loop G, and as such they can be evaluated by Stokes' theorem. However, although both potential $\mathbf{A}_1$ and $\mathbf{A}_2$ are well defined along the loop G, since G goes around the combined Dirac string $\Gamma = \gamma_1 \cup \gamma_2$, there exists no surface S with boundary G such that on S both $\mathbf{A}_1$ and $\mathbf{A}_2$ are well defined. In fact, S necessarily intersects Γ, and consequently it intersects at least one of the Dirac strings γ_1 and γ_2. There exist, however, two distinct surfaces S_1 and S_2, with $\partial S_1 = G = \partial S_2$, such that $S_1 \subset V_1 = \mathbb{R}^3 \setminus \gamma_1$ and $S_2 \subset V_2 = \mathbb{R}^3 \setminus \gamma_2$. Correspondingly, A_1 is regular on S_1 and A_2 is regular on S_2. Moreover, the combined surface $S_1 \cup S_2$ is closed and encloses the monopole, sitting at the origin. Applying Stokes' theorem to these two surfaces, conveniently oriented, from formula (21.44) we then obtain

$$\Delta\Lambda(\mathbf{x}) = \int_G \mathbf{A}_2 \cdot d\mathbf{x} - \int_G \mathbf{A}_1 \cdot d\mathbf{x} = \int_{S_2} \boldsymbol{\nabla} \times \mathbf{A}_2 \cdot d\boldsymbol{\Sigma} - \int_{S_1} \boldsymbol{\nabla} \times \mathbf{A}_1 \cdot d\boldsymbol{\Sigma}$$

$$= \int_{S_2} \mathbf{B} \cdot d\boldsymbol{\Sigma} - \int_{S_1} \mathbf{B} \cdot d\boldsymbol{\Sigma} = \int_{S_1 \cup S_2} \mathbf{B} \cdot d\boldsymbol{\Sigma} = g,$$

where in the last integral we have inserted the magnetic field of the monopole (21.23). We therefore conclude that the discontinuity of the gauge function $\Lambda(\mathbf{x})$ across the surface Σ, conveniently oriented, is independent of the transition point $\mathbf{x} \in \Sigma$ and equals the charge of the monopole

$$\Delta\Lambda(\mathbf{x}) = g. \tag{21.45}$$

This result will be essential for the derivations of Dirac's quantization condition in Sects. 21.5.1 and 21.6.2.

Geometric interpretation of the gauge function. The integral representation of the gauge function (21.43) admits a simple geometric interpretation. Consider a generic infinitesimal surface element $d\boldsymbol{\Sigma}$ placed at a point $\mathbf{y}$, and choose a generic point $\mathbf{x} \neq \mathbf{y}$. The unit vector of the line joining the two points, and their distance, are then given by

$$\mathbf{u} = \frac{\mathbf{x} - \mathbf{y}}{|\mathbf{x} - \mathbf{y}|}, \qquad R = |\mathbf{x} - \mathbf{y}|.$$

To an *observer* in $\mathbf{x}$ the surface $d\boldsymbol{\Sigma}$ then shows the area $\mathbf{u} \cdot d\boldsymbol{\Sigma}$, and, since the observer is at a distance R, he thus sees this surface under the solid angle

$$d\Omega(\mathbf{x}) = \frac{\mathbf{u} \cdot d\boldsymbol{\Sigma}}{R^2} = \frac{\mathbf{x} - \mathbf{y}}{|\mathbf{x} - \mathbf{y}|^3} \cdot d\boldsymbol{\Sigma}.$$

A generic finite surface Σ is then seen by the observer in $\mathbf{x}$ under the solid angle

$$\Omega_\Sigma(\mathbf{x}) = \int_\Sigma \frac{\mathbf{x} - \mathbf{y}}{|\mathbf{x} - \mathbf{y}|^3} \cdot d\boldsymbol{\Sigma}.$$

We thus recognize that the integral (21.43) can be recast in the form

$$\Lambda(\mathbf{x}) = \frac{g}{4\pi} \, \Omega_\Sigma(\mathbf{x}). \tag{21.46}$$

Since the surface Σ of the integral (21.43) extends to infinity, in light of the geometric meaning of the solid angle $\Omega_\Sigma(\mathbf{x})$ it is then straightforward to read off from formula (21.46) the discontinuity of the gauge function (21.45).

21.3.4   An Example

We exemplify the general procedure described above in the case of two *straight*
Dirac strings γ_1 and γ_2, parameterized by $\mathbf{y}_1(\lambda) = \lambda \mathbf{n}_1$ and $\mathbf{y}_2(\lambda) = \lambda \mathbf{n}_2$, respectively. The corresponding Dirac potentials $\mathbf{A}_1$ and $\mathbf{A}_2$ are then given by formula (21.33). Furthermore, for $\mathbf{A}_1$ we choose as Dirac string the negative z axis, $\mathbf{n}_1 = (0, 0, -1)$, and for $\mathbf{A}_2$ the positive z axis, $\mathbf{n}_2 = (0, 0, 1)$. Formula (21.33) then yields the expressions

$$\mathbf{A}_1 = \frac{g(-y, x, 0)}{4\pi r(r + z)}, \quad \mathbf{A}_2 = \frac{g(y, -x, 0)}{4\pi r(r - z)}, \quad \mathbf{A}_2 - \mathbf{A}_1 = \frac{g(y, -x, 0)}{2\pi(x^2 + y^2)}. \quad (21.47)$$

In this case the combined Dirac string $\Gamma = \gamma_1 \cup \gamma_2$ coincides thus with the z axis, parameterized by $\mathbf{y}(\lambda) = (0, 0, \lambda)$, $\lambda \in \mathbb{R}$. To determine the gauge function $\Lambda(\mathbf{x})$ we first must choose a surface Σ whose boundary is Γ, i.e. the z axis. We choose for Σ the xz half-plane $x > 0$, that we parameterize as

$$\mathbf{y}(\lambda, u) = (u, 0, \lambda), \qquad -\infty < \lambda < \infty, \qquad 0 \le u < \infty.$$

To compute the gauge function (21.43) we need the vectors

$$\frac{\partial \mathbf{y}(\lambda, u)}{\partial \lambda} = (0, 0, 1), \quad \frac{\partial \mathbf{y}(\lambda, u)}{\partial u} = (1, 0, 0), \quad \mathbf{x} - \mathbf{y}(\lambda, u) = (x - u, y, z - \lambda).$$

Performing in the integral (21.43) the shift of variables $\lambda \to \lambda + z$ and $u \to u + x$, we find

$$\Lambda(\mathbf{x}) = \frac{gy}{4\pi} \int_{-x}^{\infty} du \int_{-\infty}^{\infty} \frac{d\lambda}{(u^2 + \lambda^2 + y^2)^{3/2}} = \frac{gy}{2\pi} \int_{-x}^{\infty} \frac{du}{u^2 + y^2}.$$

To evaluate the last integral we rescale u by the positive number $|y|$:

$$\int_{-x}^{\infty} \frac{du}{u^2 + y^2} = \frac{1}{|y|} \left(\frac{\pi}{2} + \arctan\left(\frac{x}{|y|} \right) \right).$$

In this way we obtain the gauge function

$$\Lambda(\mathbf{x}) = \frac{g}{2\pi} \left(\frac{\pi}{2} \varepsilon(y) + \arctan\left(\frac{x}{y} \right) \right), \quad (21.48)$$

where $\varepsilon(\cdot)$ is the *sign function*. The function (21.48) is continuous on the xz half-plane $x < 0$, in that for $x < 0$ it entails the limits $\lim_{y \to 0\pm} \Lambda(\mathbf{x}) = 0$. Conversely, on the xz half-plane $x > 0$, i.e. on the surface Σ, $\Lambda(\mathbf{x})$ is discontinuous. On this half-plane it entails, in fact, the different limits

$$\lim_{y \to 0^{\pm}} \Lambda(\mathbf{x}) = \pm \frac{g}{2}.$$

The gauge function (21.48) has thus the expected discontinuity across Σ (21.45). These properties become more transparent if one rewrites the gauge function (21.48) in the form

$$\Lambda(\mathbf{x}) = \frac{g}{2\pi}\,\varphi, \qquad (21.49)$$

where φ is the polar angle of the xy plane, with the conventions that $\varphi = 0$ corresponds to the negative x axis, and that the angle varies in the interval $\varphi \in (-\pi, \pi)$. Finally, it is easy to verify that for $\mathbf{x} \notin \Sigma$ the expressions (21.47) and (21.48) satisfy the relation $\nabla \Lambda = \mathbf{A}_2 - \mathbf{A}_1$.

Summary. We end the section with a collection of the main results, that we will need later on in Sects. 21.5 and 21.6.

- B is a closed 2-form in the complement of the monopole's position

$$dB = 0, \quad \text{for all } \mathbf{x} \in \mathbb{R}^3 \setminus \{O\}.$$

- B is an exact 2-form in the complement of the Dirac string γ

$$B = dA_\gamma, \quad \text{for all } \mathbf{x} \subset V_\gamma = \mathbb{R}^3 \setminus \gamma.$$

- A change of the Dirac string $\gamma_1 \to \gamma_2$ corresponds to a gauge transformation

$$A_2 - A_1 = d\Lambda, \quad \text{for all } \mathbf{x} \in \mathbb{R}^3 \setminus \Sigma, \quad \text{where } \partial\Sigma = \Gamma = \gamma_1 \cup \gamma_2.$$

- The gauge function Λ has a finite constant discontinuity when it crosses the surface Σ, which equals the charge of the monopole

$$\Delta\Lambda = g.$$

21.4   Dirac Potential in the Space of Distributions

Irrespective of the presence of magnetic monopoles, Maxwell's equations are well posed only in the space of distributions. In fact, the classical analysis of Sect. 21.3 can be rephrased in this complementary mathematical framework, where the fields and the potentials are reinterpreted as *differential forms in the space of distributions*, see Sect. 19.1.1. In this alternative perspective, which will turn out to be particularly suitable for the path-integral quantization of dyons in Sect. 21.6, the analysis is simplified by the fact that the Poincaré lemma holds in a *strong* sense: it guarantees, in fact, that *every distribution-valued closed differential form is exact*. An additional advantage of the distributional approach is that it allows to *quantify* the singularities of the Dirac potential. For these reasons, and additional ones of more geometric nature that will emerge during the analysis, in this section we readdress the problem of the existence of a vector potential for the magnetic monopole in the context of distribution theory. Since our monopole is *static*, the appropriate space of distributions is $\mathcal{S}'(\mathbb{R}^3)$.

21.4.1   Distributional Exterior Derivative of the Dirac Potential

We start again from the identity between 3-forms (21.27)

$$dB = gJ, \tag{21.50}$$

which, actually, holds in the sense of distributions. In fact, the differential forms B (21.25) and J (21.26) are distribution-valued, and Eq. (21.50) is equivalent to the *classical* identity (2.116). Once more, we introduce a Dirac string γ and consider the associated Dirac-potential 1-form $A_\gamma = dx^i A^i_\gamma$, see Eq. (21.29),

$$A_\gamma = \frac{g}{4\pi} \, dx^i \varepsilon^{ijk} \int_\gamma \frac{x^j - y^j}{|\mathbf{x} - \mathbf{y}|^3} \, dy^k. \tag{21.51}$$

It is not difficult to realize that also the integral (21.51) – despite its singularities for $\mathbf{x} \in \gamma$ – defines a *distribution* in $\mathcal{S}'(\mathbb{R}^3)$, see Problem 21.3. This implies that also its distributional exterior derivative dA_γ is well defined, so that we can introduce a further distribution-valued 2-form C_γ, by setting

$$gC_\gamma = B - dA_\gamma. \tag{21.52}$$

This 2-form entails the properties:

- $dC_\gamma = J$;
- the support of C_γ is γ.

The first property follows from Eqs. (21.50) and (21.52), using that $d^2 = 0$. The second property is based on the fact that in the domain $\mathbb{R}^3 \setminus \gamma$ the 1-form A_γ is smooth, so that in this domain its exterior derivative can be computed in the ordinary sense. In Eq. (21.31) we have, in fact, shown that in $\mathbb{R}^3 \setminus \gamma$ the 2-form dA_γ equals precisely B. Therefore, the support of $B - dA_\gamma$ is necessarily (a subset of) the curve γ. Below we will, in fact, show that C_γ is proportional to a δ-function supported on γ.

Explicit form of C_γ. In order to determine C_γ we must evaluate the exterior derivative dA_γ, i.e. the curl $\nabla \times \mathbf{A}_\gamma$, in the sense of distributions. Actually, it is not necessary to perform this computation *ex novo*, since we have already done it in Sect. 21.3.1. In fact, it is sufficient to use in the intermediate result (21.30) the distributional identity (2.116)

$$\partial_i \left(\frac{x^i - y^i}{|\mathbf{x} - \mathbf{y}|^3} \right) = 4\pi \delta^3(\mathbf{x} - \mathbf{y}), \tag{21.53}$$

implying that now also the second integral of (21.30) gives a non-vanishing contribution. In this way, in place of equation (21.31) we now obtain the distributional relation

$$\nabla \times \mathbf{A}_\gamma = \mathbf{B} - g\mathbf{C}_\gamma, \quad \text{where } \mathbf{C}_\gamma = \int_\gamma \delta^3(\mathbf{x} - \mathbf{y}) \, d\mathbf{y}. \tag{21.54}$$

In the language of differential forms it reads

$$B = dA_\gamma + gC_\gamma, \quad \text{where } C_\gamma = \frac{1}{2} dx^j \wedge dx^i \, \varepsilon^{ijk} \int_\gamma \delta^3(\mathbf{x} - \mathbf{y}) \, dy^k. \tag{21.55}$$

The support of C_γ is thus the Dirac string γ, as anticipated above. Equation (21.55) admits the following mathematical interpretation. Its first member B is a 2-form which is regular in whole space, excluded the origin. On the other hand, the 1-form A_γ is regular in whole space, excluded the curve γ, and consequently its exterior derivative dA_γ contains a singular contribution, proportional to a δ-function supported on γ. The 2-form gC_γ precisely cancels this contribution, in such a way that the sum $dA_\gamma + gC_\gamma$ is regular on γ, as is B. In a sense, the 2-form $-gC_\gamma$ *quantifies* the singularities of the Dirac potential A_γ.

The magnetic flux. Let us check that the decomposition of the magnetic field (21.54), i.e. $\mathbf{B} = \nabla \times \mathbf{A}_\gamma + g\mathbf{C}_\gamma$, is consistent with the fact that the flux of $\mathbf{B}$ across an arbitrary closed surface S (not containing the monopole) is equal to (zero) g. The flux of the term $\nabla \times \mathbf{A}_\gamma$, being the distributional curl of a vector field, is zero across *any* closed surface. We remain, therefore, with the flux of the term $g\mathbf{C}_\gamma$, see (21.54),

$$\Phi_\gamma = g \int_S \mathbf{C}_\gamma \cdot d\mathbf{\Sigma} = g \int_{S \times \gamma} \delta^3(\mathbf{x} - \mathbf{y}) \, d\mathbf{y} \cdot d\mathbf{\Sigma} = g \int_{S \times \gamma} \delta^3(\mathbf{x} - \mathbf{y}) \, dV, \tag{21.56}$$

where dV is an *oriented* three-dimensional volume element. In the above integrand $\mathbf{x}$ is a point on S and $\mathbf{y}$ is a point on γ. The integral (21.56) can thus be different from zero only if γ intersects S. Every time γ intersects S *leaving* the surface, the integral receives a contribution $+g$, while every time γ intersects S *entering* the surface, it receives a contribution $-g$, see Problem 21.6. If S does not contain the monopole, the curve γ enters the surface and comes out from it the same number of times, and so $\Phi_\gamma = 0$. Conversely, if S contains the monopole, γ comes out from the surface $2n + 1$ times and enters it $2n$ times, and so the total flux is $\Phi_\gamma = g$. We have thus verified the expected equality

$$\Phi_\gamma = \int_S g\mathbf{C}_\gamma \cdot d\mathbf{\Sigma} = \int_S \mathbf{B} \cdot d\mathbf{\Sigma} \tag{21.57}$$

for all closed surfaces S. In a sense, it is as if the singular field $\mathbf{C}_\gamma$ would concentrate all magnetic field lines along the Dirac string, while the *true* magnetic field $\mathbf{B}$ spreads them isotropically in all directions, albeit producing the same flux.

21.4.2 Poincaré Duality

By construction, the differential forms J and C_γ obey the identity

$$J = dC_\gamma. \tag{21.58}$$

This relation can be given a geometric interpretation by observing that the operator d associates to the 2-form C_γ, which is a δ-function with support γ, the 3-form J, which is a δ-function with support the origin O. On the other hand, the origin is nothing else than the boundary of γ

$$O = \partial \gamma. \tag{21.59}$$

We thus see that d associates to a form which is a δ-function supported on a subset of $\mathbb{R}^3$, a form which is a δ-function supported on the boundary of this subset. In particular, the relations (21.58) and (21.59) follow from each other via the replacements

$$d \leftrightarrow \partial, \qquad J \leftrightarrow O, \qquad C_\gamma \leftrightarrow \gamma. \tag{21.60}$$

Moreover, in the same way as the exterior derivative d is a *nilpotent* operator in the space of p-forms, $d^2 = 0$, the boundary operator ∂ is a nilpotent operator acting in the space of n-dimensional submanifolds of $\mathbb{R}^3$, $\partial^2 = 0$. In fact, the boundary of a submanifold is boundaryless.

General case. The above correspondence between the operators d and ∂ has general validity and constitutes, in fact, a fundamental trait of an important map of *differential topology*, called *Poincaré duality*. This map associates to every p-dimensional submanifold M_p of $\mathbb{R}^D$ a *dual* $(D - p)$-form C_{D-p}, proportional to a δ-function with support M_p. More precisely, the map is such that the equality between integrals of differential forms[7]

$$\int_{\mathbb{R}^D} C_{D-p} \wedge \Phi_p = \int_{M_p} \Phi_p \tag{21.61}$$

holds for all smooth *test* p-forms Φ_p. The interpretation of this equality is essentially that the δ-function C_{D-p} restricts the integral of a generic p-form Φ_p from the total space $\mathbb{R}^D$ to its subspace M_p. In the same way, Poincaré duality asso-

---

[7] For the definition of the integral of a p-form over a p-dimensional manifold we refer the reader to the textbook [3].

ciates to the boundary $M_{p-1} = \partial M_p$, which is a $(p-1)$-dimensional submanifold of $\mathbb{R}^D$, a $(D-p+1)$-form J_{D-p+1}, which is proportional to a δ-function with support M_{p-1}. As anticipated in the above example (21.58)–(21.60), these differential forms are related by the exterior derivative according to the duality rule[8]

$$ M_{p-1} = \partial M_p \quad \leftrightarrow \quad J_{D-p+1} = dC_{D-p}. \tag{21.62} $$

It can be seen that, with the (standard) normalization chosen for C_{D-p} in the definition (21.61), in three spatial dimensions the Poincaré duals of a generic point O and of a generic curve γ are, indeed, given by formulas (21.26) and (21.55), respectively. Another important feature of Poincaré duality arises from its very definition (21.61), if one allows also the test p-form Φ_p to become the Poincaré dual C_p of a $(D-p)$-dimensional submanifold M_{D-p}. In this case, one can in fact prove that the real number defined by the integrals[9]

$$ \#(M_p, M_{D-p}) = \int_{\mathbb{R}^D} C_{D-p} \wedge C_p = \int_{M_p} C_p = \int_{M_{D-p}} C_{D-p} \tag{21.63} $$

is an *integer*, counting the algebraic number of intersections between the manifolds M_p and M_{D-p}. It is not difficult to recognize in the formula for the magnetic flux of the monopole (21.56) a particular example of the equalities (21.63). In fact, if we denote the Poincaré dual of the closed surface S by C_S, a 1-form, this flux can be recast in the form

$$ \Phi_\gamma = g \int_S C_\gamma = g \int_{\mathbb{R}^3} C_S \wedge C_\gamma = g \, \#(\gamma, S). $$

Since the number of intersections $\#(\gamma, S)$ between the Dirac string γ and the surface S is equal to one or zero according to whether or not S encloses the monopole, the above relation is thus the same as Eq. (21.57). Comprehensibly, being a one-to-one map between differential forms and submanifolds, Poincaré duality plays a fundamental role in the topological classification of differentiable manifolds. For more details on this subject we refer the reader to the seminal book [4] by G. de Rham.

---

[8]In *(co)homology theory*, Poincaré duality regards boundaryless submanifolds, $\partial M_p = \emptyset$, and, correspondingly, closed dual forms, $dC_{D-p} = 0$. In addition, this theory considers *smooth* representatives for the forms C_{D-p}, differing from the latter by exact $(D-p)$-forms. The variant of Poincaré duality presented by us in (21.61), where the dual forms C_{D-p} are *singular*, i.e. distribution-valued, has been developed by G. de Rham, see Ref. [4].

[9]If D is *even*, the last integral in (21.63) carries the additional overall sign $(-)^{(D+1)p}$, coming from the graded commutation relation between differential forms (19.8).

21.4.3   Moving the Dirac String

Let us now analyze how the Dirac potential A_γ of formula (21.51) responds to a change of the Dirac string. Given two generic Dirac strings γ_1 and γ_2, we introduce the corresponding Dirac potentials A_1 and A_2 as in Eq. (21.51), and their Poincaré dual 2-forms C_1 and C_2 as in Eq. (21.55). We then have the equalities

$$dA_1 + gC_1 = B = dA_2 + gC_2,$$

which imply the relation

$$g(C_2 - C_1) = d(A_1 - A_2). \tag{21.64}$$

The 2-form $C_2 - C_1$, being the exterior derivative of $(A_1 - A_2)/g$, is hence exact. On the other hand, the 1-form whose exterior derivative is $C_2 - C_1$ is determined only modulo exact 1-forms. Therefore, as $C_2 - C_1$ is essentially a δ-function on a curve, we may ask whether there exists a simpler *representative* than $(A_1 - A_2)/g$. The answer is affirmative, thanks to Poincaré duality. In fact, the exact 2-form $C_2 - C_1$ is the Poincaré dual of the boundaryless curve $\Gamma = \gamma_1 \cup \gamma_2$, which, in turn, is the boundary of a surface Σ, see Sect. 21.3.2. The Poincaré dual of Σ is a 1-form, which we denote by $M = dx^i M^i(\mathbf{x})$. According to the correspondence (21.62), the relation $\partial \Sigma = \Gamma$ then implies the relation between 2-forms

$$C_2 - C_1 = dM. \tag{21.65}$$

Comparing equations (21.64) and (21.65) we then derive the further equation

$$d(A_2 - A_1 + gM) = 0, \tag{21.66}$$

which implies the existence of a scalar function $\widetilde{\Lambda}(\mathbf{x})$ such that, eventually, the difference between two Dirac potentials can be expressed as

$$A_2 - A_1 = d\widetilde{\Lambda} - gM. \tag{21.67}$$

The forms M and $\widetilde{\Lambda}$. Once a surface Σ with boundary Γ has been chosen, the 1-form M is uniquely determined. Equation (21.67) then determines the function $\widetilde{\Lambda}$ uniquely, modulo a constant. To determine M and $\widetilde{\Lambda}$ explicitly it is convenient to resort to the calculations of Sect. 21.3.3, in particular to the intermediate equation (21.42). In the space of distributions, however, due to the identity (21.53), now also the last term of this equation gives a non-vanishing contribution. In fact, in place of $\mathbf{A}_2 - \mathbf{A}_1 = \nabla \Lambda$ we now obtain the equality

$$\mathbf{A}_2 - \mathbf{A}_1 = \nabla \Lambda - g \int_\Sigma \delta^3(\mathbf{x} - \mathbf{y})\, d\Sigma, \tag{21.68}$$

where Λ is the (distribution-valued) gauge function (21.43). The comparison between equations (21.67) and (21.68) then yields the explicit expression

$$M = dx^i \int_\Sigma \delta^3(\mathbf{x} - \mathbf{y})\, d\Sigma^i = dx^i \varepsilon^{ijk} \int_\Sigma \delta^3(\mathbf{x} - \mathbf{y}) \frac{\partial y^j}{\partial \lambda} \frac{\partial y^k}{\partial u}\, d\lambda\, du, \quad (21.69)$$

where we have inserted the measure (21.41), and, furthermore, the identification $\widetilde{\Lambda}(\mathbf{x}) = \Lambda(\mathbf{x})$. The expression of the 1-form M (21.69) represents, actually, the Poincaré dual of a generic surface Σ imbedded in a three-dimensional space. Finally, Eq. (21.67) admits an interpretation similar to the one given above for Eq. (21.55). The 1-form $A_2 - A_1$ is regular in the complement of the curve Γ, including the surface Σ, whereas the function Λ is discontinuous on Σ. Correspondingly, the exterior derivative $d\Lambda$ contains a δ-function with support Σ, which is canceled by the 1-form gM. This implies that the 1-form $d\Lambda - gM$ is *regular* on Σ, as is the 1-form $A_2 - A_1$.

An example. We exemplify the differential forms introduced above, by choosing as Dirac strings γ_1 and γ_2 two half-lines directed along the negative and positive z axes, respectively, as in Sect. 21.3.4. For the surface Σ we take again the xz half-plane $x > 0$. We then obtain the explicit expressions, see Problem 21.4,

$$\begin{aligned}
J &= dz \wedge dy \wedge dx\, \delta(x)\, \delta(y)\, \delta(z), \\
C_2 &= dy \wedge dx\, H(z)\, \delta(x)\, \delta(y), \\
C_1 &= -dy \wedge dx\, H(-z)\, \delta(x)\, \delta(y), \qquad (21.70) \\
C_2 - C_1 &= dy \wedge dx\, \delta(x)\, \delta(y), \\
M &= dy\, H(x)\, \delta(y),
\end{aligned}$$

where H is the Heaviside function.

Summary. We end the section by collecting the main results that we will use later on in Sect. 21.6.

- The 2-form B associated with the magnetic field of the monopole satisfies the modified Bianchi identity

$$dB = gJ,$$

where J is the 3-form associated with a unit charge density.
- The magnetic field can be written in terms of a Dirac potential A_γ relative to a Dirac string γ as

$$B = dA_\gamma + gC_\gamma,$$

where the 2-form C_γ is the Poincaré dual of the curve γ.
- Under a change of Dirac string $\gamma_1 \to \gamma_2$ the Dirac potential changes by

$$A_2 - A_1 = d\Lambda - gM, \qquad (21.71)$$

where the 1-form M is the Poincaré dual of a surface Σ such that $\partial\Sigma = \gamma_1 \cup \gamma_2$.

- The above differential forms satisfy the Poincaré-duality relations

$$dC_1 = J = dC_2, \qquad C_2 - C_1 = dM. \tag{21.72}$$

21.5 Canonical Quantization of Dyons

Let us return to the two-dyon system analyzed in Sect. 20.3 at the classical level. As the center of mass of the system performs a uniform linear motion, it is sufficient to consider the dynamics of the relative dyon, obeying the equation of motion (21.2). We can recast the latter in the form of a non-relativistic Lorentz equation,

$$m\mathbf{a} = \mathbf{E} + \frac{\mathbf{v}}{c} \times \mathbf{B}, \qquad \mathbf{E} = \frac{q\mathbf{x}}{4\pi r^3}, \qquad \mathbf{B} = \frac{g\mathbf{x}}{4\pi r^3}, \tag{21.73}$$

where the *effective charges* are given by

$$q = e_1 e_2 + g_1 g_2, \qquad g = e_2 g_1 - e_1 g_2. \tag{21.74}$$

Equation (21.73) can be interpreted as an equation of motion of a particle having *unitary* electric charge and vanishing magnetic charge, moving in the presence of the electromagnetic field generated by a static dyon, sitting at the origin, having electric charge q and magnetic charge g. The electric field $\mathbf{E}$ admits the scalar potential

$$A^0 = \frac{q}{4\pi r}, \qquad \mathbf{E} = -\boldsymbol{\nabla} A^0. \tag{21.75}$$

Moreover, from Sect. 21.3 we know that, if we exclude from $\mathbb{R}^3$ a Dirac string γ, then the magnetic field $\mathbf{B}$ admits the vector potential $\mathbf{A}_\gamma$ (21.29), where now $g = e_2 g_1 - e_1 g_2$. We thus have $\mathbf{B} = \boldsymbol{\nabla} \times \mathbf{A}_\gamma$, for $\mathbf{x} \in V_\gamma = \mathbb{R}^3 \setminus \gamma$. Since in the restricted domain V_γ the electromagnetic field admits the four-potential $(A^0, \mathbf{A}_\gamma)$, in this domain the equation of motion of the relative dyon (21.73) can be derived from the Lagrangian (21.4) for $e = 1$

$$L_\gamma = \frac{1}{2} m v^2 - A_0 + \frac{1}{c} \mathbf{v} \cdot \mathbf{A}_\gamma. \tag{21.76}$$

Correspondingly, in the domain V_γ the quantum dynamics of the relative dyon is described by the Hamiltonian (21.5)

$$H_\gamma = -\frac{\hbar^2}{2m} \left(\boldsymbol{\nabla} - \frac{i}{\hbar c} \mathbf{A}_\gamma \right)^2 + A^0. \tag{21.77}$$

Generalized Hilbert space. As the vector potential $\mathbf{A}_\gamma$ (21.29) is defined only in the complement of the Dirac string γ, the Hamiltonian (21.77) can, however, not be traded for an operator in the whole Hilbert space $L^2(\mathbb{R}^3)$. This problem can be solved by resorting to a *generalized Hilbert space*, see Sect. 21.2, as originally proposed by T.T. Wu and C.N. Yang [5].[10] According to the procedure of Sect. 21.2, first of all we must introduce two domains V_1 and V_2 with the properties (21.13). For this purpose we introduce two non-intersecting Dirac strings γ_1 and γ_2, and we define these domains as in Sect. 21.3.2

$$V_1 = \mathbb{R}^3 \setminus \gamma_1, \qquad V_2 = \mathbb{R}^3 \setminus \gamma_2.$$

In this way, the conditions (21.13) hold, indeed, in virtue of the relations

$$V_1 \cup V_2 = \mathbb{R}^3, \qquad V_0 = V_1 \cap V_2 = \mathbb{R}^3 \setminus \Gamma, \qquad \Gamma = \gamma_1 \cup \gamma_2.$$

The generalized Hilbert space $\mathcal{H}$ is then formed by the pairs of wavefunctions $\psi = \{\psi_1, \psi_2\}$, where $\psi_1 \in L^2(V_1)$ and $\psi_2 \in L^2(V_2)$, and an operator $O = \{O_1, O_2\}$ acts on these pairs as in (21.20). The wavefunctions and the operators defined in V_1 and V_2 must be connected in V_0 by the relations (21.18) and (21.21), i.e.

$$\psi_2 = e^{iD}\psi_1, \qquad O_2 = e^{iD}O_1 e^{-iD}, \qquad \text{for } \mathbf{x} \in V_0, \tag{21.78}$$

for a, still to be determined, transition function e^{iD}.

21.5.1 Transition Function and Dirac's Quantization Condition

To determine the transition function e^{iD} we introduce two Hamiltonians H_1 and H_2 of the form (21.77), defined in terms of the Dirac potentials $\mathbf{A}_1$ and $\mathbf{A}_2$ associated with the Dirac strings γ_1 and γ_2, which are well defined in the domains V_1 and V_2, respectively,[11]

$$H_1 = -\frac{\hbar^2}{2m}\left(\boldsymbol{\nabla} - \frac{i}{\hbar c}\mathbf{A}_1\right)^2 + A^0, \qquad H_2 = -\frac{\hbar^2}{2m}\left(\boldsymbol{\nabla} - \frac{i}{\hbar c}\mathbf{A}_2\right)^2 + A^0. \tag{21.79}$$

---

[10] Actually, it is possible to construct in the Hilbert space $L^2(\mathbb{R}^3)$ a well-defined self-adjoint Hamiltonian, based on the formal expression H_γ (21.77) relative to a *fixed* Dirac string γ, see Ref. [6]. However, in this case the derivation of the quantization condition (21.82) as a *necessary* condition for a consistent quantum theory of dyons becomes an extremely delicate task.

[11] As the potentials $\mathbf{A}_{1,2}$ are singular on the Dirac strings $\gamma_{1,2}$, if we want the operators $H_{1,2}$ to be well behaved, actually, we must slightly restrict the domains $V_{1,2}$ by excluding from them small tubular neighborhoods around the strings $\gamma_{1,2}$. The resulting deformed domains $V_{1,2}^*$ still conform with our construction of the generalized Hilbert space of Sect. 21.2.

According to the compatibility condition for operators (21.78), we must find a transition function such that in V_0 these Hamiltonians are related by

$$H_2 = e^{iD} H_1 e^{-iD}. \tag{21.80}$$

For this purpose, we take advantage from the fact that in a region $\mathbb{R}^3 \setminus \Sigma$, where Σ is a surface whose boundary is the combined Dirac string Γ, there exists a gauge transformation connecting the two Dirac potentials (see Eqs. (21.38) and (21.43))

$$\mathbf{A}_2 = \mathbf{A}_1 + \nabla \Lambda. \tag{21.81}$$

In this case, from Sect. 21.1.1 on the implementation of gauge invariance in quantum mechanics, we know that the associated Hamiltonians are related by (see Eqs. (21.9)–(21.12))

$$H_2 = e^{i\Lambda/\hbar c} H_1 e^{-i\Lambda/\hbar c}.$$

The Hamiltonians (21.79) satisfy thus indeed the relation (21.80), if we choose the transition function

$$e^{iD(\mathbf{x})} = e^{i\Lambda(\mathbf{x})/\hbar c}.$$

However, the consistency of our generalized Hilbert-space construction of Sect. 21.2 requires the transition function e^{iD} to be *continuous* in $V_0 = \mathbb{R}^3 \setminus \Gamma$. On the other hand, in Sect. 21.3.3 we have seen that whereas the gauge function $\Lambda(\mathbf{x})$ (21.43) is continuous for $\mathbf{x} \in \mathbb{R}^3 \setminus \Sigma$, on Σ it entails the discontinuity $\Delta\Lambda(\mathbf{x}) = g$, see Eq. (21.45). For the discontinuity of the transition function across Σ we then find

$$\Delta\big(e^{iD}\big) = e^{i(\Lambda+g)/\hbar c} - e^{i\Lambda/\hbar c} = e^{i\Lambda/\hbar c}\big(e^{ig/\hbar c} - 1\big).$$

Therefore, since we want e^{iD} to be continuous in the whole intersection domain V_0, we must require $g/\hbar c$ to be an integer multiple of 2π. In conclusion, a necessary condition for the consistency of the quantum mechanics of a non-relativistic system of dyons is that the charges of any dyon pair satisfy the condition

$$\frac{g}{\hbar c} = \frac{e_1 g_2 - e_2 g_1}{\hbar c} \in 2\pi\mathbb{Z}. \tag{21.82}$$

We so retrieve Dirac's quantization condition (20.68), previously derived by means of a semiclassical argument. The derivation we gave above is frequently interpreted by saying that Dirac's quantization condition is necessary to make the Dirac string *invisible*. With this one means that the transition function e^{iD} – which moves the Dirac string from one position to another, making it unobservable – is regular, and hence physically admissible, only if the dyon charges satisfy the condition (21.82). We stress that what we have, actually, shown is that this condition is *necessary* for a consistent quantum dynamics of non-relativistic dyons, whereas its *sufficiency* would require the explicit construction of a domain of (essential) self-adjointness for the Hamiltonian $H = \{H_1, H_2\}$, in the generalized Hilbert space $\mathcal{H}$.

Finally, the relation (21.78) between the operators O_1 and O_2 must hold in V_0 for all observables $O = \{O_1, O_2\}$ defined on $\mathcal{H}$. For observables that do not involve derivatives, i.e. observables that result in the multiplication of the wavefunction $\{\psi_1(\mathbf{x}), \psi_2(\mathbf{x})\}$ by a real function $W(\mathbf{x}) = \{W_1(\mathbf{x}), W_2(\mathbf{x})\}$, as for instance the scalar potential $A^0(\mathbf{x})$, this condition is trivially satisfied, in that it leads to the simple identification $W_2(\mathbf{x}) = e^{iD(\mathbf{x})} W_1(\mathbf{x}) e^{-iD} = W_1(\mathbf{x})$. Conversely, observables that involve derivatives are represented in V_1 and V_2 by different operators. Examples are the *momentum* and the orbital *angular momentum*, which in the case at hand are represented by the operators (21.6)

$$\mathbf{p}_n = \frac{\hbar}{i} \boldsymbol{\nabla} - \frac{1}{c} \mathbf{A}_n, \qquad \mathbf{L}_n = \mathbf{x} \times \left(\frac{\hbar}{i} \boldsymbol{\nabla} - \frac{1}{c} \mathbf{A}_n \right), \qquad n = 1, 2.$$

In fact, due to Eq. (21.81), in V_0 we have the non-trivial relations

$$\mathbf{p}_2 = e^{iD(\mathbf{x})} \mathbf{p}_1 e^{-iD(\mathbf{x})} \neq \mathbf{p}_1, \qquad \mathbf{L}_2 = e^{iD(\mathbf{x})} \mathbf{L}_1 e^{-iD(\mathbf{x})} \neq \mathbf{L}_1.$$

21.6 Dyons in Feynman's Path-Integral Quantization

The paradigm of *canonical quantization* is a universal instrument for the formulation of the quantum dynamics of a generic physical system. An alternative method, physically equivalent to the canonical quantization, is represented by Feynman's path-integral quantization. One of the most significant advantages of this approach is that it does not explicitly involve a *quantum Hamiltonian* as an operator on a Hilbert space, in that it relies directly on the *classical Lagrangian* of the system. In fact, its basic outcome is the *time-evolution operator*, represented by the Schrödinger kernel. As this operator is unitary, and hence bounded, its domain is the total Hilbert space $L^2(\mathbb{R}^3)$, and so no issue ever arises about the regularity properties of the wavefunctions. In this section, we illustrate how non-relativistic dyons can be quantized in this alternative framework, without dwelling on the conceptual aspects of the approach. For a comprehensive presentation of the path-integral quantization we refer the reader to the standard textbook [7] by R.P. Feynman and A.R. Hibbs.

21.6.1 Schrödinger Kernel

In quantum mechanics the dynamics of a particle is governed by the Schrödinger equation

$$i\hbar \frac{\partial \psi}{\partial t} = H\psi.$$

Being a first-order differential equation in time, it admits a unique solution $\psi(t, \mathbf{x})$ once the initial wavefunction $\psi(0, \mathbf{x})$ is known. As the Schrödinger equation is linear, for a time-independent Hamiltonian $H(\mathbf{x}, \mathbf{p})$, with $\mathbf{p} = -i\hbar\nabla$, we can try to solve it via the integral representation

$$\psi(t, \mathbf{x}) = \int K(t, \mathbf{z}, \mathbf{x})\, \psi(0, \mathbf{z})\, d^3 z, \tag{21.83}$$

where $K(t, \mathbf{z}, \mathbf{x})$ is a complex function of the time and of the two spatial coordinates $\mathbf{x}$ and $\mathbf{z}$, called *Schrödinger kernel*. Substituting this representation in the Schrödinger equation we find that the kernel must satisfy the conditions

$$i\hbar\frac{\partial K(t, \mathbf{z}, \mathbf{x})}{\partial t} = HK(t, \mathbf{z}, \mathbf{x}), \qquad K(0, \mathbf{z}, \mathbf{x}) = \delta^3(\mathbf{z} - \mathbf{x}), \tag{21.84}$$

where it is understood that H acts on the coordinate $\mathbf{x}$ of $K(t, \mathbf{z}, \mathbf{x})$. If the kernel satisfies the conditions (21.84), for uniqueness reasons formula (21.83) yields an integral representation of the known solution of the Schrödinger equation

$$\psi(t) = e^{-itH/\hbar}\psi(0). \tag{21.85}$$

As recalled at the beginning of the chapter, the Hamiltonian $H(\mathbf{x}, \mathbf{p})$ is the Legendre transform of the Lagrangian $L(\mathbf{x}, \mathbf{v})$ generating the classical equation of motion of the particle. Without entering into details, we present Feynman's formula, which expresses the solution of the system (21.84) in terms of a path integral:

$$K(T, \mathbf{z}, \mathbf{x}) = \int_{\mathbf{z}}^{\mathbf{x}} \{\mathcal{D}\mathbf{r}(t)\}\, e^{\frac{i}{\hbar}\int_0^T L(\mathbf{r}(t), \mathbf{v}(t))\, dt}. \tag{21.86}$$

In this expression the symbol $\{\mathcal{D}\mathbf{r}(t)\}$ denotes a *functional measure* on the space of all *paths* $\mathbf{r}(t)$ satisfying the boundary conditions

$$\mathbf{r}(0) = \mathbf{z}, \quad \mathbf{r}(T) = \mathbf{x}. \tag{21.87}$$

The most relevant aspect of the expression (21.86), for our purposes, is that it its integrand contains in the exponent the classical action

$$I = \int_0^T L(\mathbf{r}(t), \mathbf{v}(t))\, dt, \qquad \mathbf{v}(t) = \frac{d\mathbf{r}(t)}{dt},$$

involving the Lagrangian in place of the Hamiltonian. If we want to describe, as particular case, the quantum dynamics of the relative dyon, we must choose a Dirac string γ, construct the associated Dirac potential $\mathbf{A}_\gamma$ of formula (21.29), and finally insert in the path integral (21.86) the Lagrangian (21.76)

$$L(\mathbf{r}, \mathbf{v}) = \frac{1}{2} m v^2 - A_0(\mathbf{r}) + \frac{1}{c} \mathbf{v} \cdot \mathbf{A}_\gamma(\mathbf{r}), \tag{21.88}$$

to obtain a Schrödinger kernel $K(T, \mathbf{z}, \mathbf{x})$.[12] As observed above, in the perspective of Feynman's approach the wavefunctions are generic elements of $L^2(\mathbb{R}^3)$, and the Schrödinger kernel governs their time evolution according to formula (21.83). However, since $\mathbf{A}_\gamma$ depends on the chosen Dirac string, *a priori* also the Schrödinger kernel will do so. But this would imply that observers using different Dirac strings, in general would realize different time evolutions. In other words, the Dirac string would become *visible*. However, if we were able to show that a change of the Dirac string results in a *physical symmetry*, see Sect. 21.1.1, then the Dirac string would become again invisible. In the section to follow we show that this is indeed the case – provided that the dyon charges satisfy Dirac's quantization condition.

21.6.2 Unobservability of the Dirac String

An observer adopting a different Dirac string γ' introduces via formula (21.29) the Dirac potential $\mathbf{A}_{\gamma'}$, and the corresponding Lagrangian

$$L'(\mathbf{r}, \mathbf{v}) = \frac{1}{2} m v^2 - A_0(\mathbf{r}) + \frac{1}{c} \mathbf{v} \cdot \mathbf{A}_{\gamma'}(\mathbf{r}), \tag{21.89}$$

and then he constructs the Schrödinger kernel

$$K'(T, \mathbf{z}, \mathbf{x}) = \int_{\mathbf{z}}^{\mathbf{x}} \{\mathcal{D}\mathbf{r}(t)\} \, e^{\frac{i}{\hbar} \int_0^T L'(\mathbf{r}(t), \mathbf{v}(t)) \, dt}. \tag{21.90}$$

He will introduce a wavefunction $\psi'(t, \mathbf{x})$ and write the solution (21.83) in the form

$$\psi'(t, \mathbf{x}) = \int K'(t, \mathbf{z}, \mathbf{x}) \, \psi'(0, \mathbf{z}) \, d^3 z. \tag{21.91}$$

To find the relation between the kernels K and K' we introduce a surface Σ such that $\partial \Sigma = \gamma \cup \gamma'$, and then we recall from Sect. 21.3.2 that – in the domain $\mathbb{R}^3 \setminus \Sigma$ – the potentials $\mathbf{A}_{\gamma'}$ and $\mathbf{A}_\gamma$ differ by the gauge transformation

$$\mathbf{A}_{\gamma'} = \mathbf{A}_\gamma + \nabla \Lambda.$$

---

[12]If the Lagrangian is given by (21.88), the exponent in Feynman's formula (21.86) contains the integral of the Dirac potential (21.29) along a generic path $\mathbf{r}(t)$, i.e. $\int \mathbf{A}_\gamma \cdot d\mathbf{r}$. Since $\mathbf{A}_\gamma$ is singular on the curve γ, this integral diverges if the path $\mathbf{r}(t)$ intersects γ. However, it can be seen that the set of paths $\mathbf{r}(t)$ which intersect a given curve has vanishing measure with respect to the path-integral measure $\{\mathcal{D}\mathbf{r}(t)\}$, and so these divergences are irrelevant.

Therefore, whenever a path $\mathbf{r}(t)$ does *not* intersect Σ, the Lagrangians (21.88) and (21.89) differ by

$$L' - L = \frac{1}{c} \mathbf{v} \cdot (\mathbf{A}_{\gamma'} - \mathbf{A}_{\gamma}) = \frac{1}{c} \mathbf{v} \cdot \boldsymbol{\nabla} \Lambda = \frac{1}{c} \frac{d\Lambda(\mathbf{r}(t))}{dt}. \tag{21.92}$$

We now perform the analysis separately for paths which have (possibly multiple) intersections with Σ, and for paths which do not intersect this surface.

Paths which do not intersect Σ. From Eq. (21.92), using the boundary conditions (21.87), we find that for paths $\mathbf{r}(t)$ which have no intersections with Σ the actions appearing in the exponents of the Schrödinger kernels (21.86) and (21.90) differ by

$$\int_0^T L' \, dt - \int_0^T L \, dt = \frac{1}{c} \left(\Lambda(\mathbf{r}(T)) - \Lambda(\mathbf{r}(0)) \right) = \frac{1}{c} \left(\Lambda(\mathbf{x}) - \Lambda(\mathbf{z}) \right). \tag{21.93}$$

As this difference is independent of the path $\mathbf{r}(t)$, we can take the corresponding exponentials out of the path integral. Restricted to such paths, the kernel (21.90) would thus be related to the kernel (21.86) by

$$K'(T, \mathbf{z}, \mathbf{x}) = e^{i(\Lambda(\mathbf{x}) - \Lambda(\mathbf{z}))/\hbar c} K(T, \mathbf{z}, \mathbf{x}). \tag{21.94}$$

The solution (21.91) could then be recast in the form

$$e^{-i\Lambda(\mathbf{x})/\hbar c} \psi'(t, \mathbf{x}) = \int K(t, \mathbf{z}, \mathbf{x}) \, e^{-i\Lambda(\mathbf{z})/\hbar c} \psi'(0, \mathbf{z}) \, d^3 z.$$

This equation, actually, coincides with the time-evolution formula of the original observer (21.83), if the wavefunctions are related by the *unitary* transformation

$$\psi'(\mathbf{x}) = e^{i\Lambda(\mathbf{x})/\hbar c} \psi(\mathbf{x}) = (U\psi)(\mathbf{x}). \tag{21.95}$$

We thus see that, if the domain of the path integral would contain only paths which do not intersect Σ, then a change of the Dirac string would indeed amount to a *physical symmetry*, see Sect. 21.1.1. Notice that – as all unitary operators – U is a well-defined operator in the whole Hilbert space $L^2(\mathbb{R}^3)$, irrespective of the regularity properties of the real function $\Lambda(\mathbf{x})$.

Paths which intersect Σ. Consider now a path $\mathbf{r}(t)$ which intersects the surface Σ at the points $\{\mathbf{r}_i\}$. In this case, we can split up the path into arcs which start at a point close to $\mathbf{r}_i$ (or at the initial point $\mathbf{z}$) and which end at a point close to $\mathbf{r}_{i+1}$ (or at the endpoint $\mathbf{x}$), without intersecting Σ. For the integral of $L' - L$ along each arc $(\mathbf{r}_i, \mathbf{r}_{i+1})$ we can then apply formula (21.92). However, due to the discontinuity of the gauge function Λ (21.45), the difference of the actions (21.93) now acquires an additional contribution $\pm g/c$ for every intersection point $\mathbf{r}_i$, depending on the direction of the intersection. Therefore, we now obtain

$$\int_0^T L' \, dt - \int_0^T L \, dt = \frac{1}{c} \left(\Lambda(\mathbf{x}) - \Lambda(\mathbf{z}) - Ng \right), \qquad (21.96)$$

where N is the *algebraic* number of times the path $\mathbf{r}(t)$ intersects the surface Σ. Using this equation in the kernel (21.90) we then find the relation

$$K'(T, \mathbf{z}, \mathbf{x}) = e^{i(\Lambda(\mathbf{x}) - \Lambda(\mathbf{z}))/\hbar c} \int_{\mathbf{z}}^{\mathbf{x}} \{\mathcal{D}\mathbf{r}(t)\} \, e^{-iNg/\hbar c} \, e^{\frac{i}{\hbar} \int_0^T L(\mathbf{r}(t), \mathbf{v}(t)) \, dt}.$$

$$(21.97)$$

This formula has, actually, general validity, because for paths which do not intersect Σ the integer N vanishes. Notice that the phase $e^{-iNg/\hbar c}$ is a function of $\mathbf{r}(t)$, and hence it cannot be taken out of the path integral. If we again want the kernel K' to be related to K by the transformation law (21.94) – which is the condition ensuring that a change of the Dirac string amounts to a physical symmetry – we must impose the equality $e^{-iNg/\hbar c} = 1$ for all integers N. Once more, we so obtain Dirac's quantization condition

$$\frac{g}{\hbar c} = \frac{e_2 g_1 - e_1 g_2}{\hbar c} \in 2\pi \mathbb{Z}.$$

Dirac's quantization condition from distribution theory. To illustrate the power of the theory of differential forms in the space of distributions, we rederive equation (21.96) relying on this alternative, conceptually simpler, setting. From the Lagrangians (21.88) and (21.89) we obtain the general formula

$$\int_0^T L' \, dt - \int_0^T L \, dt = \frac{1}{c} \int_0^T \mathbf{v} \cdot (\mathbf{A}_{\gamma'} - \mathbf{A}_{\gamma}) \, dt = \frac{1}{c} \int_{[\mathbf{z}, \mathbf{x}]} (A_{\gamma'} - A_{\gamma}),$$

$$(21.98)$$

where the last expression represents the line integral of the 1-form $A_{\gamma'} - A_{\gamma}$ along the path $\mathbf{r}(t)$. Using the distributional relation (21.71), with $\gamma_1 = \gamma$ and $\gamma_2 = \gamma'$, we can rewrite the difference (21.98) in the form

$$\int_0^T L' \, dt - \int_0^T L \, dt = \frac{1}{c} \int_{[\mathbf{z}, \mathbf{x}]} (d\Lambda - gM), \qquad (21.99)$$

where the 1-form M is the Poincaré dual of Σ, see formula (21.69). We now use an important result of the theory of distributions, namely that the *definite* integrals of distributions obey the *fundamental theorem of calculus*.[13] Therefore, the first integral at the second member of equation (21.99) becomes

---

[13] The validity of this theorem requires the distributional integrand not to be *singular* on the *boundary* of the domain of integration. Consequently, the integral (21.100) holds as long as the endpoints $\mathbf{z}$ and $\mathbf{x}$ do not belong to Σ.

$$\int_{[\mathbf{z},\mathbf{x}]} d\Lambda = \Lambda(\mathbf{x}) - \Lambda(\mathbf{z}). \tag{21.100}$$

The value of the second integral in (21.99) follows, instead, from Poincaré duality. In fact, as M is the Poincaré dual of Σ, the equalities (21.63) imply that

$$\int_{[\mathbf{z},\mathbf{x}]} M = N, \tag{21.101}$$

where, as above, N is the algebraic number of times the path $\mathbf{r}(t)$ intersects the surface Σ. Equation (21.99) reduces thus again to

$$\int_0^T L' \, dt - \int_0^T L \, dt = \frac{1}{c} \left(\Lambda(\mathbf{x}) - \Lambda(\mathbf{z}) - Ng \right),$$

in agreement with the previous result (21.96). Notice, however, the different mechanisms producing the term $-Ng/c$ in the two derivations. Finally, the result (21.101) can also be checked via a direct calculation, by inserting for the 2-form M its explicit expression (21.69),

$$\int_{[\mathbf{z},\mathbf{x}]} M = \int_{[\mathbf{z},\mathbf{x}] \times \Sigma} \delta^3 \big(\mathbf{r}(t) - \mathbf{y}(\lambda, u) \big) \, d\mathbf{r} \cdot d\mathbf{\Sigma}.$$

For the evaluation of this integral, see Problem 21.6.

21.7   Problems

21.1 Verify that in the language of differential forms Eq. (21.24) amounts to Eq. (21.27), and that the relation $\mathbf{B} = \nabla \times \mathbf{A}$ is equivalent to $B = dA$.

21.2 Complete the proof of equation (21.32).
Hint: Use the definite integral

$$\int_x^\infty \frac{dt}{(1+t^2)^{3/2}} = 1 - \frac{x}{\sqrt{1+x^2}}.$$

21.3 Show that the Dirac potential (21.33), relative to the straight Dirac string $\mathbf{y}(\lambda) = \lambda \mathbf{n}$, defines a distribution in $\mathcal{S}'(\mathbb{R}^3)$.
Hint: Use rotation invariance to transform the potential (21.33) in the potential $\mathbf{A}_1$ given in (21.47), having as Dirac string the negative z axis. To isolate the singularities occurring for $z < 0$ at $x = y = 0$, multiply both the denominator and the numerator of $\mathbf{A}_1$ by $r - z$. Recall that, in two dimensions, the function $1/\sqrt{x^2 + y^2}$ is integrable around the origin.

21.4 Derive the expressions of the differential forms C_1, C_2 and M given in formulas (21.70), by evaluating explicitly the integral representations (21.55) and (21.69) for Dirac strings directed along the positive and negative z axes, respectively. Verify that these forms satisfy the identities (21.72).

21.5 Show that the distributional exterior derivative of the gauge function $\Lambda(\mathbf{x})$ (21.48) is equal to

$$d\Lambda(\mathbf{x}) = \frac{g}{2\pi} \frac{y\,dx - x\,dy}{x^2 + y^2} + g\,dy\,H(x)\,\delta(y). \tag{21.102}$$

What is the relation between this equality and the general relation (21.71)?
Hint: The function $\Lambda(\mathbf{x})$ can also be rewritten as in Eq. (21.49).

21.6 Consider in a three-dimensional space a generic smooth surface S, parameterized by $\mathbf{x}(u, v)$, and a generic smooth curve α, parameterized by $\mathbf{y}(\lambda)$. Denote the surface element of S by $d\mathbf{\Sigma}$. Show that the real number $\#(S, \alpha)$ defined by the three-dimensional integral

$$\#(S, \alpha) = \int_{S \times \alpha} \delta^3\big(\mathbf{x}(u, v) - \mathbf{y}(\lambda)\big)\, d\mathbf{y}(\lambda) \cdot d\mathbf{\Sigma} \tag{21.103}$$

is an *integer*, counting the intersections, with sign, between S and α, whenever the integral is well defined.

Solution: The integral (21.103) receives non-vanishing contributions only from the points P_i where S and α intersect. It can thus be written as a sum of integrals over small volumes around each P_i. Choosing one of these points, we can use translation and rotation invariance to let P_i coincide with the origin $(0, 0, 0)$, and to let the tangent plane to S at P_i become the xy plane. In this way, near P_i the surface S and the curve α entail the linearized parameterizations

$$\mathbf{x}(u, v) = (u, v, 0), \qquad \mathbf{y}(\lambda) = \mathbf{V}\lambda,$$

where $\mathbf{V}$ is the tangent vector to α at the origin. Near P_i the surface element of S has then the simple expression $d\mathbf{\Sigma} = (0, 0, 1)\, du\, dv$. The contribution of the intersection point P_i to the integral (21.103) takes thus the simple form

$$\#_i(S, \alpha) = \int \delta^3\big(u - V_x\lambda, v - V_y\lambda, -V_z\lambda\big) V_z\, d\lambda\, du\, dv = \frac{V_z}{|V_z|}, \tag{21.104}$$

which equals, in fact, ± 1 according to the orientation of the intersection. Notice that when the intersection number between S and α is ill defined, e.g. if α is a curve lying entirely on the surface S, then $V_z = 0$ and hence the integral (21.103) is ill defined as well.

References

1. M. Reed, B. Simon, *Methods of Modern Mathematical Physics - II Fourier Analysis, Self-adjointness* (Academic Press, New York, 1975)
2. P.A.M. Dirac, Quantized singularities in the electromagnetic field. Proc. R. Soc. Lond. A **133**, 60 (1931)
3. Y. Choquet-Bruhat, C. DeWitt-Morette, M. Dillard-Bleick, *Analysis, Manifolds and Physics* (North-Holland, Amsterdam, 1982)
4. G. de Rham, *Differentiable Manifolds: Forms, Currents, Harmonic Forms* (Springer, Berlin, 1984)
5. T.T. Wu, C.N. Yang, Concept of nonintegrable phase factors and global formulation of gauge fields. Phys. Rev. D **12**, 3845 (1975)
6. C.A. Hurst, Charge quantization and nonintegrable Lie algebras. Ann. Phys. **50**, 51 (1968)
7. R.P. Feynman, A.R. Hibbs, *Quantum Mechanics and Path Integrals* (McGraw-Hill, New York, 1965)

Index

© Springer International Publishing AG, part of Springer Nature 2018
K. Lechner, *Classical Electrodynamics*, UNITEXT for Physics,
https://doi.org/10.1007/978-3-319-91809-9

Printed in the United States
By Bookmasters